T0073562

DURHAM WEATHER AND CLIMATE SINCE 1841

Endorsements

Durham University has a venerable history of observational climate science. When Gordon Manley, perhaps the greatest British climatologist of the twentieth century, arrived in Durham in 1928 to establish the Department of Geography, he resolved to place the Durham Observatory weather records on the same basis as those of the Radcliffe Observatory in Oxford, which have long been recognized as a valuable resource. This book updates and extends Manley's pioneering work.

Karen O'Brien, *Vice-Chancellor, Durham University*

This lovingly crafted history will be the envy of all long-term weather stations around the world. Stephen and Tim have respectfully interpreted the painstaking efforts of those who came before them, delivering an engaging and useful volume which transports you to the University grounds throughout the seasons and the decades. As the Earth continues to warm, these kinds of careful histories will only become more important.

Linden Ashcroft, *The University of Melbourne, Australia*

Durham Weather and Climate since 1841 undertakes a comprehensive rescue and analysis of this hugely valuable long-term meteorological station record, including an in-depth reconstruction of the station history. The resulting meticulous data analysis provides key new insights into long-term UK climate changes that are essential to understanding our rapidly changing climate.

Peter Thorne, *ICARUS Climate Research Centre, Maynooth University, Ireland*

This definitive book beautifully discusses the variations in the weather and climate in Durham over nearly two centuries, including all the highs and lows. The long-term view provided by these detailed records clearly highlights the warming of our climate and the fingerprint of human influence on our weather, even at this local scale.

Ed Hawkins MBE, *University of Reading, UK*

Durham has long been known for its eminence in meteorology and climatology. In this beautifully illustrated volume, Stephen Burt and Tim Burt place Durham's long record of observations in their complete historical and social context. They describe the struggles and accomplishments of the observers, both the famous and those who quietly carried out their daily duties. Burt and Burt take these centuries' worth of observations and turn them into analytical descriptions of Durham's climate, month by month and season by season, linking climatic events with citizens' daily lives. Packed with statistics, meteorological and climatological analysis, and historical commentary, this book is for anyone interested in long-term climate change, observational records, historical climatology, weather analysis, and the history of meteorology.

Victoria Slonosky, *McGill University, Montreal, Canada*

Climate science relies on long, carefully re-evaluated meteorological records. It is this long-term view that allows changes in weather and climate to be assessed and put into perspective. In *Durham Weather and Climate since 1841*, Stephen Burt and Tim Burt, two widely-known experts in the field, present another long record. The book describes the history of weather and climate in northern England and the role of weather in daily lives. It tells the story of meteorological measurements in Durham, which, at the same time, is a story about astronomy, the University, and about the life-long dedication of individuals such as Gordon Manley—and the authors of this book.

Stefan Brönnimann, *University of Bern, Switzerland*

Durham Weather and Climate since 1841

Stephen Burt

Department of Meteorology, University of Reading

and

Tim Burt

Department of Geography, Durham University

OXFORD
UNIVERSITY PRESS

Great Clarendon Street, Oxford, OX2 6DP,
United Kingdom

Oxford University Press is a department of the University of Oxford.
It furthers the University's objective of excellence in research, scholarship,
and education by publishing worldwide. Oxford is a registered trade mark of
Oxford University Press in the UK and in certain other countries

Impression: 1

Published in the United States of America by Oxford University Press
198 Madison Avenue, New York, NY 10016, United States of America

British Library Cataloguing in Publication Data
Data available

Library of Congress Control Number: 2021952585

ISBN 978–0–19–887051–7

DOI: 10.1093/oso/9780198870517.001.0001

Printed and bound by
CPI Group (UK) Ltd, Croydon, CR0 4YY

For our children: Fiona and Jennifer; Emma and Tom

Forewords

Professor Stuart Corbridge FRGS
Vice-Chancellor and Warden of Durham University, 2015–21

Stephen Burt and Tim Burt provide us in this volume with a wonderfully concise account of Durham's weather and climate over the last 180 years. They show us how data on temperature, precipitation, and atmospheric pressure have been collected over this period, by whom, and according to which protocols. They also highlight the contributions made by an extraordinary collection of individuals who were centrally involved in the making of one of the world's most successful Geography Departments. Gordon Manley is rightly at the centre of this narrative, and not only for the work he oversaw at Durham University from 1928–39, when he helped establish the Department. Professor Manley stayed engaged for another forty years. Joan Kenworthy is also praised for her contributions, as are Ray Harris, Audrey Warner—the last resident of the splendid Durham Observatory, Nicholas Cox, and most recently Matthew Eglise, who has extended the Durham weather and climate record back to 1784.

Stephen Burt and Tim Burt have produced a book that is learned, accessible, and beautifully illustrated. They link the climate record of north-east England to the history of meteorology and its instrumentation and to the history of England's third-oldest University. I warmly congratulate them on their achievement. The data reported here are of global significance. I hope their book finds the wide audience it deserves.

Professor Penny Endersby FREng
Chief Executive Officer, Met Office, Exeter

In 2019 I made the long journey from Devon to Durham to drop off my son and his worldly goods for the start of his undergraduate degree. When this came up in conversation with the Met Office's Chair, Rob Woodward, and my personal coach, Val Wark, it transpired that they were both Durham geography graduates, and had in common the shared experience of the cold trudge uphill on winter mornings to take their turn recording the measurements from the weather station there. It was evidently a formative experience: numb fingers and damply flapping notebooks loomed large in the memory.

They were, of course, part of a proud tradition of meteorological endeavour that stretches back to the Enlightenment in the United Kingdom. We are fortunate to have some of the longest-running weather and climate records in the world, but there are only a handful that extend back over 150 years, and Durham University Observatory is part of that handful. These long time series are especially precious: we still use them in the Met Office to validate other techniques for extrapolating climate records further into the past, such as tree-ring measurements.

The very foundation of the Met Office by Admiral Fitzroy in 1854 was driven by the realization that the ability to measure aspects of the current weather, in combination with the existence of the electric telegraph—which allowed those measurements to be rapidly collated—would enable a prediction of imminent weather. He named this new technique a forecast, and was passionate about the potential to save lives, particularly at sea. We have been forecasting, and saving lives, ever since.

Our archive at the Met Office contains almost 6 km of shelving, much of it taken up with physical weather records. Digitizing this volume of data is a Herculean task, but one which we embarked on some time ago. Making weather records available digitally is more important than ever before: some of our overseas development work is dedicated to recovering scarce and even more fragile records from across the world, in nations where the climate record is far sparser.

It is a pleasure to see this testament to the dedication to the records from Durham Observatory come into being. Long may the tradition of intellectual curiosity, rationality, and expertise that created it continue.

Preface

We published *Oxford Weather and Climate since 1767* in 2019. Beautifully produced by Oxford University Press, we were delighted by how well it was received, which encouraged us (and OUP) to compile a sister volume for the Durham University Observatory record. Durham's weather records are not quite as long as those for Oxford's Radcliffe Observatory, but only six locations in Western Europe—including Oxford—have longer records maintained at the same site throughout. With climate change a matter of major societal importance to present and future generations, long- and well-maintained climatological series such as Durham's become more valuable with every passing year.

The Department of Geography at Durham University was founded in 1928 by Gordon Manley, who was to become the best-known British climatologist of his generation. Manley immediately recognized the importance of the early Durham Observatory records, ensured their preservation and continuation, and thereby effectively laid the foundations for this book. Gordon Manley remains very familiar in Durham's Department of Geography: coffee and tea are drunk in the Manley Room twice daily, the Department's social centre (where of course his photograph takes pride of place). When he arrived in Durham, Manley was a resident don in Durham's Hatfield College and married the daughter of the then-Master of the College. It is an amazingly small world: Tim Burt was Master of Hatfield College for 21 years!

Durham's long tradition of manual weather observations ended in 1999; since then, an automatic weather station has continued the lengthy and important record. Automatic weather stations have their advantages (hourly observations available over a telecoms link, day and night, 365 days per year), but also their disadvantages (occasional technical failures, their inability to report thunderstorms or measure snow depths, for example), but at least most of the Durham Observatory record continues into the twenty-first century as we celebrate 180 years of operation. Through the wonders of modern communications, one of us continues to 'run' the weather station remotely (from Devon): files arrive daily from the Met Office, so we know very quickly if something is amiss. However, physical distance from Durham means that we have been especially reliant on support from staff in the Department of Geography at Durham University. David Hodgson not only deals with computer files and the website, but also maintains another smaller AWS on the roof of the West Building, a few hundred metres from the Observatory site and a helpful back-up in event of the occasional data interruption. Before David took on these tasks, Steven Allan maintained the website. Michele Allan has taken many photographs for us and has a wonderful archive of Durham weather photos; she has also mined the departmental photograph collection to make sure nothing relevant was missed. We are very grateful to Michele and the department for making these photographs available. As with our Oxford book, Chris Orton has again prepared many maps and diagrams for us with his usual skill and good humour, especially when modifications were requested. Thanks also to three Chief Technicians who helped with any equipment issues whenever they could over the years: Derek Hudspeth, Frank Davies, and Eleanor Ross.

We are also very grateful to staff from Durham University Library, who have been especially helpful in retrieving material from the special collections (the Palace Green Library has an extensive and catalogued collection of documents from the earliest days of the Durham Observatory). Owing to lockdown restrictions during the coronavirus pandemic, we were unable to make several visits to Durham to examine these in person as originally planned, but fortunately we were able to review many of these early documents online once they had been scanned. For this, and many other library requests answered promptly and professionally, grateful thanks are due in particular to Michael Stansfield, Richard Higgins, and Gemma Lewis.

We could not have compiled this history without access to Met Office sources, both digital and paper-based. We gratefully acknowledge help and support from Catherine Ross and Mark Beswick at the Met Office Library and Archives, and from Mike Kendon and Mark McCarthy in the National Climate Information Centre, at the Met Office HQ in Exeter. From the Met Office Observations team, Adam Barber, Karl Shepherdson, and Chris Mueller provided invaluable information and support on all aspects of the post-1999 site and equipment. (By happy coincidence, Chris is a graduate of the Geography Department at Durham University).

There are also many others we need to thank in Durham and the surrounding area. First and foremost, Joan Kenworthy, the former Principal of St Mary's College, who knew and worked with Gordon Manley; she was kind enough to share much archival material, including her notes on the Manley archive in Cambridge. Joan also provided a digital record of pre-1850 temperatures, transcribed from a ledger unknown to Gordon Manley in his lifetime, although he suspected it must exist somewhere. Joan, together with Nicholas Cox, supervised Andrew Joyce, who digitized all the records from 1850 to 1997, funded by a grant from the Leverhulme Trust. These datasets have been invaluable to us. Previously little known outside Durham, they form the basis of the digital archive which we publish online alongside this book (see https://durhamweather.webspace.durham.ac.uk). We have also been able to make excellent use of Matthew Eglise's 2003 PhD thesis in order to extend the Durham temperature record back to the 1780s. Outside Durham, we must give special thanks to Ken Cook from Copley, who has always been extremely helpful and supportive whenever comparative data have been needed to check data or infill gaps. Debbie Smith has always been very helpful too, and her book about the James Losh diaries in early nineteenth-century Newcastle has been an important source for us. Within Durham City, Michael Rider has also kindly provided data from his weather station in Newton Hall whenever requested. We thank John Moreels (curator of the Ward Philipson Collection) and Michael Richardson (Gilesgate Archive) who kindly allowed us to use numerous weather-related photographs. Dan Hudachek, Head of Collections at Beamish Museum, kindly supplied the photo of snowbound Rowley Station in 1910. Reanalysis data and plots are courtesy of Gil Compo and team at the Cooperative Institute for Research in Environmental Sciences, University of Colorado, Boulder, USA. And once again, we are also very grateful to Ed Hawkins, University of Reading, for provision of Durham's climate stripe.

Of course, none of what you now hold in your hands would have been possible without the help and professionalism of our team at Oxford University Press. We particularly thank Fran McMahon and Giulia Lipparini, our Commissioning Editors at OUP, and Sonke Adlung, our Editor.

Finally, as we wrote in the preface to our Oxford book, in case you are wondering, we are not related! We dedicated 'Oxford' to our long-suffering wives, Elizabeth and Helen. This book is dedicated to our children.

Stephen Burt and Tim Burt
Stratfield Mortimer, Berkshire and Sampford Peverell, Devon
February 2022

Contents

Abbreviations and Glossary

Air frost   An air frost is recorded when the minimum temperature in a Stevenson screen at 1.25 m above ground level falls below 0 °C during the 24 hours ending 0900 GMT on the specified date. For periods when air temperatures were recorded only to the nearest degree Fahrenheit, a minimum temperature of 32 °F has also been included as an air frost.

AWS   Automatic Weather Station

CET   The Central England Temperature series; a monthly temperature series, originally prepared by Gordon Manley [1, 2] and currently updated by the Met Office Hadley Centre [3]. The series extends back to 1659 and is available online at https://www.metoffice.gov.uk/hadobs/hadcet/data/download.html

Cumecs   Cubic metres per second (m^3 s^{-1}), usually with reference to gauged river flows

Fog   By meteorological convention, 'fog' is defined as 'visibility less than 1000 m'. Thick fog is defined as 'visibility less than 200 m'.

GMT   Greenwich Mean Time; used as the time standard of all meteorological records since 1885. For consistency, GMT is used throughout although today the term UTC is used in preference: in all practical senses UTC is identical to GMT.

Ground frost   A ground frost is recorded when the minimum temperature recorded above short grass (or, since July 2015 at Durham Observatory, over an artificial grass surface) falls below 0 °C during the 24 hours ending 0900 GMT on the specified date. For periods when grass minimum temperatures were recorded only to the nearest degree Fahrenheit, a grass minimum temperature of 32 °F has also been included as a ground frost.

Maximum temperature   By convention, the maximum temperature is the highest temperature recorded in a Stevenson screen at 1.25 m above ground level during the 24 hours ending 0900 GMT on the specified date. As it is most likely to have occurred during the

	afternoon of the previous day, the reading made at 0900 GMT is 'thrown back' to the preceding day's date.
MIDAS	Met Office Integrated Data Archive System. Climatological datasets held at the Centre for Environmental Data Analysis (CEDA) https://data.ceda.ac.uk/badc/ukmo-midas
Minimum temperature	By convention, the minimum temperature is the lowest temperature recorded in a Stevenson screen at 1.25 m above ground level during the 24 hours ending 0900 GMT on the specified date. At Durham, until 1960, the minimum temperature was measured over the period 2100–2100 GMT. Grass minimum temperatures refer to the minimum temperature over a level grass surface (or, since July 2015 at Durham Observatory, over an artificial grass surface).
MSL	Mean sea level
MSL pressure	The reading of a barometer, reduced to Mean Sea Level (MSL). As barometric pressure decreases with height, the reading of a barometer at any altitude above MSL needs to be corrected or 'reduced' to MSL in order to allow observations from different stations to be compared. At Durham Observatory, the average correction to MSL of the observed barometer reading is 13 hPa, although this varies with both atmospheric pressure itself and with the temperature of the barometer and the external air. See Appendix 5.
NGR	National Grid Reference
Rainfall day	By convention, the rainfall day is the 24 hours commencing at 0900 GMT on the specified date.
Rain day	A 'rain day' is one in which 0.2 mm or more of precipitation is recorded. Precipitation in this context includes rain, drizzle, snow, sleet, hail, and occasionally fog or dew.
Snow cover	A 'morning with snow cover' is counted when more than half the ground representative of the observing site is covered with snow, regardless of its depth, at the time of the 0900 GMT observation.
Visibility	Visibility is the greatest distance at which it is just possible to see and identify with the unaided eye a prominent dark object against the sky at the horizon. Observations are made by reference to a set of objects at known distances, the 'visibility objects'. See also *Fog.*
Wet day	A 'wet day' is one in which 1.0 mm or more of precipitation is recorded.

Notes

Averages	Throughout the text, the period of averages used relates to the current World Meteorological Organization standard averaging period of the 30 years 1991–2020, except where specifically stated otherwise. Appendix 10 lists monthly and annual averages for this period for the various climatological parameters observed at Durham University Observatory.
Figure and Table numbering	Figures and Tables are numbered by chapter, and consecutively within each chapter. Thus, Table 24.4 can be found in Chapter 24.
Ranking	In the monthly, annual and seasonal tables, the most extreme (highest and lowest) months/years/seasons are ranked in order. For precipitation and sunshine, where monthly totals are rarely duplicated exactly, this causes few problems with rank order, but for temperature records this is not the case, because the mean temperatures of extreme months may differ by less than 0.1 degrees Celsius. This is approximately equivalent to the level of accuracy and repeatability of current measurements, but for older measurements (pre-1900 and particularly pre-1860, as explained in Appendix 2) uncertainty in measurements may be as high as 0.3 degrees Celsius. In the monthly/annual/seasonal tables, therefore, months are shown as equally ranked where the mean temperatures are the same when rounded to 0.1 degrees Celsius. In the text, where sometimes it is important to distinguish records on the basis of the second decimal place (such as in annual mean temperatures), rankings may be stated based upon means to two decimal places, and accordingly the stated rank order may differ slightly from that given in the relevant table/s.

References   References are shown within the text by square brackets thus:
[56]— details of the reference can be found by referring to this
number in the References section starting on page 568.

Temperatures   Each of the monthly and seasonal chapters includes a table listing
the extremes of temperature. Temperatures recorded prior to
the introduction of a Stevenson screen in November 1899 are
shown in italics; these are regression estimates from Glaisher stand
readings from January 1860 to October 1899, as read from the
North Shed October 1851 to December 1859 and as read from
the North Wall exposure prior to October 1851. Temperatures
were recorded in degrees Fahrenheit until October 1971, read to
a precision of 0.1 °F during the period 1843 to 1918 and 1 °F
for the remainder of the record. All values in degrees Fahrenheit
have been converted to degrees Celsius, but those for the 1 °F
precision period are subject to approximately 0.2 °C uncertainty
as a result. Since November 1971 temperatures have been recorded
to a precision of 0.1 °C. See Appendix 2 and 3 for more details.
Appendix 3 provides more details on thermometer instrumentation
and exposures at Durham.

Units   Units are SI standard throughout, except where required for
historical continuity (such as references to the 'five-inch rain
gauge') or where they remain in current meteorological usage,
such as wind speeds in knots. Pressure observations are given in
hectopascals (hPa), a unit which is numerically identical to the
millibar. Exceptions are stated and conversion factors to SI units
given.

Part 1

Durham's weather and climate

1

Durham—its regional, economic, and physical setting

Durham City: a brief history

The historic city of Durham lies on the banks of the River Wear in north-east England, 29 km south of Newcastle-upon-Tyne and 21 km south-west of Sunderland (Figure 1.1). The River Wear flows through Durham; an incised meander encloses the historic city centre on three sides to form the high Durham peninsula, its steep slopes constituting a natural line of defence. The city can clearly be traced back to AD 995, when a group of monks from Lindisfarne chose the strategic peninsula as a place to settle with the body of St Cuthbert. Their stone-built church was later replaced by the famous Norman cathedral.

Durham Cathedral was built between the late eleventh and early twelfth century to house the bodies of St Cuthbert (AD 634–687) and the Venerable Bede (AD 672/3–735). It attests to the importance of the early Benedictine monastic community and is the largest and finest example of Norman architecture in England. The innovative audacity of its vaulting foreshadowed Gothic architecture. The nave vault of Durham Cathedral is its most significant architectural element because it marks a turning point in the history of architecture. The pointed arch as a structural element was used successfully for the first time here in this building; the use of stone 'ribs' forming pointed arches to support the stone ceiling of the nave was an important achievement [4].

The cathedral lies within the precinct of Durham Castle, whose construction began in the late eleventh century by order of William the Conqueror. Durham Castle was of strategic importance both to defend the troublesome border with Scotland and to control local English rebellions, which were common in the years immediately following the Norman Conquest. The castle was the stronghold and residence of the Prince Bishops of Durham, who were given virtual autonomy in return for protecting the northern boundaries of England, and thus held both religious and secular power. Given how far Durham was from London, the Bishop of Durham was always accorded special status in the defence of England against the Scots: from 1075, the Bishop of Durham became a Prince Bishop, with the right to raise an army, mint his own coins, and levy taxes. As long

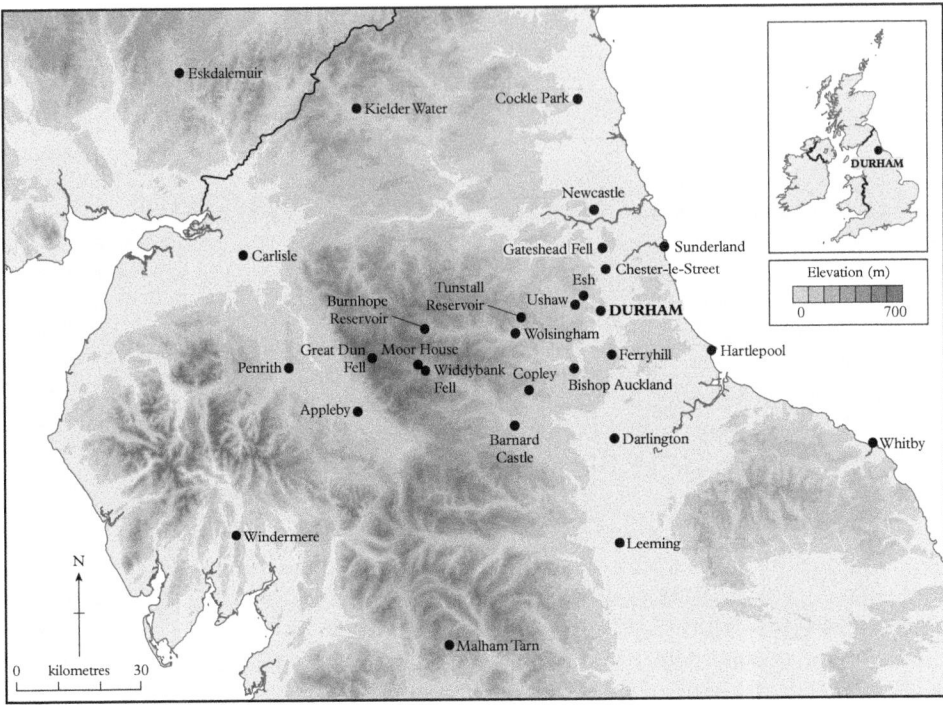

Figure 1.1 *Durham's regional location. Important locations referred to later in the text are included on the map (drawn by Chris Orton)*

as he remained loyal to the king of England, he could govern as a virtually autonomous ruler, reaping the revenue from his territory, but also remaining mindful of his role of protecting England's northern frontier. The city played an important part in the defence of the north, and Durham Castle is the only Norman castle keep never to have suffered a breach [5].

The cathedral and castle were designated a UNESCO World Heritage Site in 1986. In addition to recognizing the exceptional architecture heritage, the UNESCO designation also emphasizes the visual drama of the cathedral and castle on the peninsula and the associations of the site with notions of romantic beauty.

Durham City has experienced mixed fortunes in more recent times. Durham Priory was a Benedictine priory associated with the cathedral, but the priory was dissolved in 1540, and the cathedral taken over by the Church of England. Durham suffered greatly during the English Civil War (1642–51) and Commonwealth (1649–60). This was not due to direct assault by Cromwell or his allies, but to the abolition of the Church of England and the closure of religious institutions pertaining to it. The city has always relied upon the cathedral as an economic force. The industrial revolution saw the city develop at the heart of the coalfields, County Durham's main industry until the 1970s. Practically every village around the city had a coal mine and, although these have since disappeared as part of the regional decline in heavy industry, the traditions, heritage, and community spirit are still evident.

Durham University was founded in 1832 thanks to the benevolence of Bishop van Mildert. Durham Castle became the first college in 1837. Bishop Hatfield's Hall (later Hatfield College) was added in 1846 specifically for the sons of poorer families, and the Principal, David Melville, inaugurated a system new to English university life of advanced fees to cover accommodation and communal dining; this is now the accepted model throughout the university system in the UK and indeed around the world. Early in the twentieth century coal became depleted, with a particularly important seam worked out in 1927, and in the following depression, Durham and its surrounding villages were among those places that suffered the most severe hardship. Yet the university continued to expand throughout the century, especially to the south of the city, with a new campus on an old colliery site (Lower Mountjoy) and many new colleges still further south, starting with St Mary's College moving to new premises in 1952 and a completely new college, Grey College, close by in 1959 (Figure 1.2).

Most recently, Durham has seen the usual suburban expansion, especially at Newton Hall to the north-west, spreading out to meet the nearest villages. While County Durham is not the most affluent area in England, the city centre continues to thrive given a large and expanding student population. Thankfully, urban expansion has not yet infringed significantly on the Durham Observatory site, which remains relatively well-exposed.

The population of Durham City was 48,069 in 2011, with surrounding villages adding significantly to that number.

The topography of Durham City and the Wear Valley

The River Wear rises in the Pennine Hills and flows generally eastwards through County Durham to meet the North Sea at Sunderland. The source of the Wear is further east than some of the Pennine rivers, with the headwaters of the Tees and South Tyne further west, capturing areas that might otherwise have fed into the Wear. The Wear rises within the Alston Block, a simply tilted massif of Carboniferous limestone dipping gently east to pass beneath the strata of the Durham coalfield. Beneath the limestone is the Weardale granite, an igneous intrusion of Devonian age (362 million years) [6], which has led to important lead and zinc mining in the past [7], now long gone. The outcrop of the coal measures is almost 50 km wide in County Durham. Durham coal was used for coking and gas production, but the demise of 'town' gas provision in favour of North Sea natural gas meant that coal extraction in County Durham soon ceased [8]. The path of the Wear is blocked further east by an escarpment of Permian rocks. The Wear may once have flowed directly to the coast, but now appears to be a 'subsequent' river, flowing north along the Magnesian limestone escarpment from Bishop Auckland to Chester-le-Street before cutting across to the sea. This complex history, made more complicated still by several phases of considerable glacial disruption, may account for the apparently odd passage of the Wear through Durham City, having cut an incised meander south around the peninsula when a route to the north of the peninsula appears the more simple and direct course (Figures 1.3 and 1.4). It is on the top of the Wear valley meander, 800 m

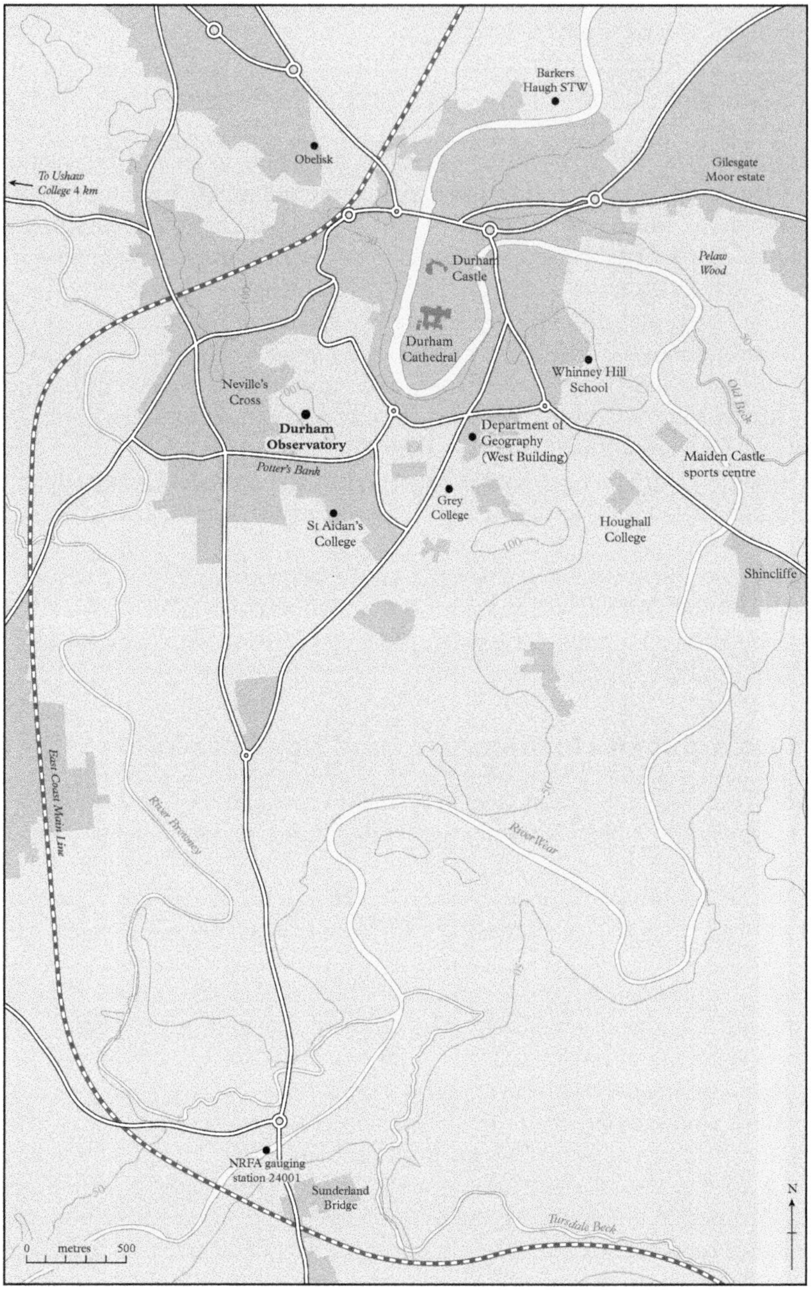

Figure 1.2 *The location of Durham Observatory within the Durham City area (drawn by Chris Orton)*

Figure 1.3 *Aerial view of Durham city in July 1947. The photograph shows well the meander of the River Wear around the peninsula, where castle, cathedral, and Palace Green are easily identified. Quite a lot has changed since then, including the construction of the Leases Road bridge, Grey College (to the east of St Mary's College, which was then under construction), and development of the Lower Mountjoy campus. The Observatory is in the top right-hand corner of the photograph but hidden by trees (arrowed); Observatory House is visible. (Courtesy of Department of Geography, University of Cambridge, Cambridge air photo BG88, 11 July 1947)*

south-west from the cathedral across the steep valley, that the Durham Observatory was built. Due north of the Observatory across an undulating plateau is an obelisk, erected by William Lloyd Wharton in 1850, which marks the line running due north from the Observatory used for checking the alignment of the main telescope [9].

To the west of the plateau on which the Observatory stands, the ground falls sharply to the Browney valley which drains the area around Lanchester. The Browney joins the

Figure 1.4 *Aerial view of Durham city from the east. In the foreground is Durham Racecourse (University cricket ground). The cathedral and castle are clearly seen on the peninsula with the incised meander of the River Wear cutting around to the south. The Observatory dome is just visible at the top of the photograph, about a quarter of the way in from the left-hand side (arrowed). (Tim Burt)*

Wear at Sunderland Bridge, 8 km upstream of Durham City. There has been a gauging station at Sunderland Bridge since 1957 (National Rivers Flow Archive, NRFA, gauge 24001) with an upstream catchment area of 658 km². Downstream of Sunderland Bridge is the former agricultural college at Houghall and the adjacent Durham University sports ground at Maiden Castle. The famous British climatologist Gordon Manley argued that Houghall was one of the coldest places in England, owing to cold-air drainage down the Wear valley in favourable conditions. After several episodes of serious flooding in Durham in recent years (see Chapter 22), there is now an additional gauging station in Durham City itself, opposite the Royal County hotel (Environment Agency station 8288).

Durham's weather in its regional context

It is clear from the description of its geological and topographical situation that Durham lies to the east of the Pennine Hills, closer in distance and indeed climatic conditions to the North Sea coast than to the Pennine moorlands. The Pennine Hills, although relatively modest in altitude, are sufficiently high to create a harsh local environment of low temperatures, high rainfall, and frequent cloud cover [10]. Durham lies in the rain shadow of the Pennines with a very different precipitation regime compared to the hills to the west.

As its name suggests, north-east England's position underlies its prevailing coolness. The region's northern location leaves it more frequently exposed to active depressions and less subject to the advantages gained from the extensions of the Azores anticyclone [10]. The coldest waters around Britain's shores are those of the western North Sea between Aberdeen and the Humber [11]. While the North Sea exercises a moderating influence along the coast, it is especially important in keeping summer conditions among the coolest in England [10]. Although sea breezes and coastal fog (locally known as a 'fret') only rarely penetrate inland as far as Durham, the Durham temperature record nevertheless very much reflects its north-eastern situation.

At various points, we refer to other weather stations in the north-east region to which Durham may usefully be compared: Malham Tarn, Moor House, and Cockle Park. Table 1.1 compares monthly mean air temperatures at Malham Tarn, Durham Observatory, and Cockle Park, just to the north of Morpeth and much closer to the coast than Durham (Figure 1.1).

Malham Tarn

A simple comparison between Durham (altitude 102 m above sea level) and Malham Tarn (54.096°N, 2.167°W, altitude 395 m ASL, 85 km south-west of Durham, Figure 1.1) illustrates the contrast immediately (Tables 1.1 and 1.2). Climatological records have been made at the Malham Tarn Field Studies Centre since 1950 (Figure 1.5), although a monthly rainfall record exists from 1870 [12].

Figure 1.5 *The climatological station at Malham Tarn Field Studies Centre in North Yorkshire (Stephen Burt)*

Table 1.1 *Monthly mean air temperatures (°C) at Durham, Malham Tarn, Moor House, and Cockle Park for the standard averaging period 1991–2020. See Figure 1.1 for locations*

	Jan	Feb	Mar	Apr	May	Jun	Jul	Aug	Sep	Oct	Nov	Dec	Year
Durham	4.1	4.6	6.2	8.3	11.0	13.6	15.8	15.6	13.3	10.0	6.6	4.2	9.5
Malham Tarn	2.5	2.6	4.1	6.4	9.3	11.9	13.7	13.5	11.4	8.3	5.2	2.9	7.6
Moor House	1.0	0.9	2.2	4.2	7.3	10.1	12.1	11.7	9.7	6.6	3.6	1.5	5.9
Cockle Park	4.0	4.5	5.8	7.6	10.2	12.9	15.0	15.0	12.9	9.8	6.6	4.3	9.1

Table 1.2 *Monthly mean precipitation (mm) at Durham and Malham Tarn, period 1991–2020*

	Jan	Feb	Mar	Apr	May	Jun	Jul	Aug	Sep	Oct	Nov	Dec	Year
Durham	53	45	41	51	45	61	62	66	57	64	73	61	680
Malham Tarn (MT)	164	138	116	92	91	100	116	133	129	154	166	182	1581
MT as % Durham	307	306	280	180	204	164	186	201	227	242	225	297	232

Malham is 1.9 degrees Celsius cooler on average and receives almost two and a half times as much precipitation as Durham [12]. Burt and Ferranti [13] studied precipitation along a transect of sites from Armagh, across the Lake District and Pennine Hills, to Durham. They noted how the nature of the Durham precipitation record was very often quite different from the gauges to the west, including those on the Cumbrian coast and those in the lee of the Lake District fells. For example, the winter:summer rainfall ratio at Malham is 1.35 whereas at Durham it is 1.02. In terms of the weather types generating precipitation, Durham has more rainfall from cyclonic conditions than stations further west, in both winter and summer, and less rainfall from westerly and southwesterly air flow; in summer a greater percentage of rainfall is recorded at Durham during anticyclonic synoptic types than further west [13]. Thus, Durham weather is more typical of lowland, eastern England—in many ways more like Lincolnshire or East Anglia than the neighbouring western hills, if a little cooler, given its more northerly location.

Moor House

Gordon Manley established a simple weather station at Moor House (54.6895°N, 2.3755°W, 556 m ASL) in the early 1930s [11, 14–16]. For decades Moor House was an important site for research into upland ecosystems and their management, for example, the link between heather burning and 'crop' productivity. Important research was also undertaken on the hydrology of blanket peat and the impact of drainage on runoff and erosion [17]. The research station was eventually abandoned in the 1980s and the old house knocked down, having become unsafe. However, research there was reinvigorated by the UK Environmental Change Network (ECN), which installed an

Figure 1.6 *Moor House automatic weather station (Tim Burt)*

automatic weather station in 1991 (Figure 1.6). Like other ECN sites, there is now regular monitoring of a number of hydrological and biogeochemical parameters such as dissolved organic carbon. Rain gauges in the Moor House area such as the one at Widdybank Fell have allowed a long composite record to be constructed for Moor House [18]. Being at a considerably greater altitude than Malham Tarn (556 m AMSL vs 395 m), Table 1.1 shows that Moor House is significantly cooler, the more so in summer.

Cockle Park

In the other direction from the Pennines, near the coast, the climatological record at Cockle Park near Morpeth (55.2129°N, 1.6861°W, 95 m ASL, Figure 1.7) dates back to 1897, and accordingly it was one of the first of the World Meteorological Organization's 'Centennial' sites in the UK (https://tinyurl.com/y47nn77f). This long record provides an invaluable comparison with Durham Observatory, which is further inland, yet still a relatively lowland site: both share an 'eastern, lowland' situation (see Chapter 3). Table 1.1 shows that Cockle Park, much nearer the coast than Durham, is cooler on average, more so in spring and summer. Wheeler [10] notes that not only that the North

Figure 1.7 *The climatological station at Morpeth, Cockle Park in Northumberland in June 2014 (Karl Shepherdson, Met Office: Crown Copyright)*

Sea exercises a direct control on coastal temperatures, but also that easterly winds and sea breezes can extend their influence several kilometres inland.

Ushaw College and Houghall College

Two other stations much closer to Durham deserve special mention. Over the decades, the records from Ushaw College (54.7867°N, 1.6594°W, 181 m AMSL), some 5 km north-west of the Observatory, have been used as a cross-check, for example, by Gordon Manley when 'reducing' Durham temperature data in the 1920 and early 1930s [19]. Another climatological record, even nearer, was maintained at Houghall College (54.7648°N, 1.5601°W, 37 m AMSL), on the Wear floodplain, just 1.5 km south-south-east from the Observatory, from 1925 to 1977. Sadly, neither of these stations are still active; as Wheeler [10] notes, the loss of Houghall is particularly to be lamented as the site of one of Britain's most notorious frost hollows (Figure 1.8). Wheeler [10] asserts that no discussion of frost hollows in north-east England would be complete without reference to Houghall, one of the most pronounced and well-known examples of its type in the country. Houghall lies on the floodplain of the River Wear, just to the east of Durham, and just across the road from Durham University's Maiden Castle sports

Figure 1.8 *The frostiest sports ground in the country? Durham University's Maiden Castle sports ground sits next to Houghall College and shares its susceptibility to very low minimum temperatures and a high frequency of ground frosts. The steep, wooded valley-side slope behind Houghall College is visible in the distance (Tim Burt)*

ground. It is here, at an altitude of just 36 m in an otherwise sheltered location, and not on the inhospitable Northern Pennine summit at Dun Fell [15], that the north-east region's lowest temperatures have been recorded [20]. On 5 January 1941 Houghall noted a screen minimum of −21.1 °C, the lowest air temperature yet recorded in County Durham [10], a value subsequently equalled at this site on 4 March 1947. On that date in January 1941, the minimum at the Durham Observatory site was −13.3 °C. During another spell of very cold anticyclonic weather in 1929, Houghall fell to −18.3 °C on 16 February when −11.7 °C was recorded at Durham Observatory. Two days earlier, on 14 February 1929, the maximum temperature reached only −8.1 °C at Houghall compared to −1.7 °C at the Observatory. In contrast, Cross Fell's absolute minimum temperature is −13.5 °C [10]. Houghall's extreme minimum regime is not apparent in its long-term means, which differ little from those at the Observatory. Manley [20] advises readers that the katabatic drainage of cold air down the Wear valley is not a vigorous breeze; nevertheless, the aggregate effect of a gentle breeze down the valley and shelter from mixing with warmer air above the inversion afforded by the topography can clearly produce extraordinary variations in nocturnal temperatures in and around Durham.

Manley [20, pp. 165–167] gives an interesting comparison between the three 'Durham' sites: Durham Observatory, Ushaw, and Houghall, barely 10 km apart:

Table 1.3 *Comparison of temperatures over the period 1925–40 at Durham Observatory (102 m AMSL), Ushaw (181 m AMSL), and Houghall (36 m)*

		Jan	Feb	Mar	Apr	May	Jun	Jul	Aug	Sep	Oct	Nov	Dec
Tmax	Ushaw	5.8	5.9	8.5	10.4	13.6	17.2	19.2	19.2	16.2	11.8	8.2	5.8
	Durham	6.2	6.5	9.1	10.9	13.9	17.6	19.9	19.6	16.6	12.6	8.8	6.2
	Houghall	6.7	7.1	9.6	11.4	14.4	18.2	20.3	20.1	17.2	13.2	9.3	6.6
Tmin	Ushaw	0.3	0.5	1.8	3.2	5.7	8.4	11.0	10.6	8.6	5.4	2.9	1.1
	Durham	0.4	0.5	1.5	2.8	5.2	8.1	10.6	10.3	8.1	5.0	2.7	0.9
	Houghall	−0.5	0.0	0.9	2.7	5.0	8.0	10.5	10.0	7.8	4.4	2.0	0.1
Range	Ushaw	5.4	5.4	6.7	7.3	7.9	8.8	8.2	8.6	7.6	6.3	5.3	4.7
	Durham	5.8	6.0	7.6	8.1	8.8	9.5	9.3	9.3	8.5	7.6	6.1	5.3
	Houghall	7.2	7.1	8.7	8.7	9.4	10.2	9.8	10.1	9.4	8.7	7.3	6.6
Absmin	Ushaw	−4.9	−4.6	−3.2	−1.6	0.2	3.6	6.9	6.5	3.5	−0.4	−1.8	−3.8
	Durham	−6.2	−5.8	−4.4	−3.0	−1.6	2.5	5.3	4.6	2.0	−1.6	−3.4	−4.7
	Houghall	−8.4	−7.4	−6.5	−4.6	−3.1	0.7	4.0	2.9	−0.2	−3.0	−5.2	−6.8
D–H	Tmax	−0.5	−0.6	−0.6	−0.5	−0.4	−0.6	−0.4	−0.5	−0.6	−0.6	−0.5	−0.4
	Tmin	0.9	0.5	0.6	0.2	0.2	0.1	0.1	0.3	0.3	0.6	0.7	0.8
	Absmin	2.2	1.6	2.1	1.6	1.6	1.8	1.3	1.7	2.2	1.4	1.8	2.1

Extended and recalculated from Manley [20] p. 166.
Tmax: mean monthly maximum temperature. Tmin: mean monthly minimum temperature. Range: Tmax minus Tmin. Absmin: mean monthly absolute minimum temperature. D–H: Durham Observatory minus Houghall. All data in degrees Celsius.

Table 1.3 lists various aspects of their maximum and minimum temperatures. Houghall consistently has the highest mean maxima and the lowest mean minima; thus, it sees the largest diurnal ranges in temperature. The valley microclimate therefore favours added daytime warmth (probably through reduced mixing in the boundary layer as a result of topographical shelter, in addition to altitudinal effects) as well as night-time cold. Mean minima were always lower at Houghall compared to Durham Observatory, and by a bigger margin in winter, although September is an interesting departure from this seasonal pattern. Manley [20] further supplemented mean data for these sites with some examples of the observed differences on cold, clear nights with a deep snow cover (Table 1.4). At the time that Manley wrote, the screen minimum of −21.1 °C observed at Houghall on 4 March 1947 was a national record for March, equalling minimum temperatures observed the same month at Peebles and Braemar [20]. Chapter 22 includes a comparison of minimum temperatures for the very cold 1962/63 winter at Durham

Table 1.4 *Absolute minimum temperatures (°C) at the three Durham sites on several notably cold winter nights*

	4 Mar 1928	16 Feb 1929	21 Jan 1940	5 Jan 1941	26 Feb 1941	8 Feb 1942	24 Jan 1945	23 Feb 1947	4 Mar 1947
Ushaw	−5.6	−11.1	−11.1	−7.2	−7.2	−7.8	−8.3	−7.8	−10.0
Durham D	−7.8	−11.7	−16.1	−13.3	−10.0	−10.0	−12.8	−10.0	−15.0
Houghall H	−13.3	−18.3	−20.0	−21.1	−16.1	−13.9	−16.7	−16.7	−21.1
H—D	5.6	6.7	3.9	7.8	6.1	3.9	3.9	6.7	6.1

Recalculated from [20], p. 167.
H–D is Houghall minus Durham Observatory.

Observatory, Houghall, and Moor House, as well as for other very cold nights at this site, in both winter and summer.

The catchment of the River Wear

Given that the river provides such an important landscape feature through Durham City and the continued prevalence of flooding (Figure 1.9), some coverage of the hydrology of the River Wear and its catchment area is appropriate. As noted, there has been a gauging station at Sunderland Bridge since October 1957 (NRFA gauge 24001 located at NGR NZ 265 377) with an upstream catchment area of 658 km^2 [21]. The catchment rises in the east Pennines and has a relatively narrow shape aligned west–east, lying between the Tees and the Tyne. Its main tributary is the River Gaunless (78 km^2), which flows east-north-east from near Copley to its confluence with the Wear just downstream of Bishop Auckland. NRFA describes the Wear as predominantly an upland catchment on mainly Carboniferous formations (limestone, millstone grit and coal measures). Half the catchment is overlain by superficial deposits (peat in the headwaters, boulder clay in the lower reaches). Land use is half grass pasture, with moorland headwaters, some arable farming, and urban development mainly in the lower reaches from Bishop Auckland towards Durham City. Altitudes in the gauged catchment range from 745 m to 40 m at Sunderland Bridge. To date, the maximum gauged flow at Sunderland Bridge has been 410 cubic metres per second or 'cumecs' on 18 July 2009. The runoff regime is significantly influenced by three large reservoirs, Burnhope, Waskerley, and Tunstall, with a combined surface area of 45 km^2. Sewage effluent forms a significant portion of low flows; in drought years, minimum flows are supported by transfer from Kielder Water in Northumberland.

Given its eastern, lowland location, Durham is not the ideal location at which to relate precipitation and river discharge; upland locations further west are likely to provide a much better correlation. Long rainfall records are available for Burnhope and Tunstall reservoirs. Burnhope Reservoir (54.744°N, 2.244°W) lies at an elevation of 398 m near

Figure 1.9 *The River Wear has a notably flashy response, like many Pennine rivers. These two views at the Hatfield College boathouse, show typical low-flow conditions (upper, 17 November 2006) and a moderate flood response (lower, 16 April 2005, peak flow 163 cumecs, 99th highest peak 1957–2019) (Tim Burt)*

the head of the Wear catchment, 1 km south-west of the village of Wearhead; construction began in 1931 and was completed in 1937. A continuous daily precipitation record is available from January 1931 to date. Tunstall Reservoir (54.765°N, 1.900°W) lies further east at an elevation of 220 m, 3.5 km north of the village of Wolsingham. A near-continuous daily precipitation record exists from 1 January 1896. Together, these two records provide a helpful guide to high and low flows on the Wear (see also comments on drought in Chapter 21).

Comparison with the Durham Observatory precipitation record for the period 1981–2010 is provided in Table 1.5. Burnhope is clearly much wetter than Tunstall, despite having much the same number of rain days; mean rainfall per rain day is 50 per cent greater at Burnhope. Durham is much the drier location of the three, except in summer when it is more similar to Tunstall in term of totals and rainfall intensity; Durham's lowland location may give it some advantage in terms of convectional rainfall. It is notable that the maximum recorded daily total at all three locations occurred on 11 September 1976 (87.8 mm at Durham, 113.3 mm at Tunstall, and 103.7 mm at Burnhope); this event is discussed in more detail subsequently.

Table 1.5 *Comparison of monthly, seasonal and annual precipitation data (mm) at three locations in the River Wear catchment for the period 1981–2010. 'Wettest day' is greatest daily fall within the 1981–2010 period.*

Durham Observatory (102 m)

	Mean total (mm)	Rain days	Mean rain per rain day (mm)	Days per decade ≥10 mm	Days per decade ≥25 mm	Wettest day (mm)
January	53.8	18	2.9	29	0	24.5
February	42.8	15	2.8	20	0	23.0
March	44.6	16	2.8	27	1	31.7
April	52.7	15	3.6	44	4	43.0
May	44.2	14	3.2	20	5	41.0
June	55.5	15	3.7	42	4	54.0
July	54.9	13	4.2	43	7	44.4
August	60.7	14	4.4	50	7	69.1
September	54.7	14	3.8	40	6	45.8
October	61.4	17	3.5	45	2	38.1
November	69.2	18	3.9	53	8	43.2
December	59.4	18	3.3	40	2	28.3
Winter	156.0	53	3.0	87	2	24.5
Spring	141.5	44	3.2	91	10	43.0
Summer	171.1	42	4.1	135	18	69.1
Autumn	185.3	53	3.8	138	16	45.8
Annual	653.8	187	3.5	453	46	69.1

Continued

Table 1.5 *Continued*

Tunstall Reservoir (220 m)

	Mean total (mm)	Rain days	Mean rain per rain day (mm)	Days per decade ≥10 mm	Days per decade ≥25 mm	Wettest day (mm)
January	87.2	23	3.8	68	1	41.2
February	69.4	18	3.8	59	5	29.2
March	69.6	21	3.3	54	4	33.7
April	70.9	18	4.0	57	10	46.0
May	56.8	16	3.5	36	3	56.4
June	63.2	17	3.8	52	4	63.2
July	65.5	16	4.2	57	9	67.2
August	67.7	17	4.0	51	10	59.0
September	71.4	17	4.1	60	12	57.0
October	85.3	21	4.1	80	7	51.0
November	96.1	22	4.3	77	16	53.8
December	91.4	22	4.2	79	12	44.5
Winter	247.3	62	4.0	205	18	44.5
Spring	197.4	55	3.6	147	17	56.4
Summer	196.4	49	4.0	160	23	67.2
Autumn	249.7	60	4.2	212	35	57.0
Annual	889.7	226	3.9	724	93	67.2

Continued

Table 1.5 *Continued*

Burnhope Reservoir (398 m)

	Mean total (mm)	Rain days	Mean rain per rain day (mm)	Days per decade ≥10 mm	Days per decade ≥25 mm	Wettest day (mm)
January	154.0	22	6.9	159	31	68.5
February	121.6	18	6.7	113	27	55.5
March	115.1	20	5.7	107	13	45.9
April	86.1	18	4.8	72	11	49.9
May	73.5	15	4.8	54	5	57.5
June	76.1	16	4.9	72	4	47.7
July	78.0	15	5.1	63	7	83.0
August	85.5	16	5.3	68	12	54.0
September	100.9	16	6.2	95	17	56.2
October	136.5	21	6.6	141	19	55.6
November	152.8	22	7.0	146	30	57.5
December	156.7	22	7.2	153	33	59.1
Winter	434.8	63	6.9	427	90	68.5
Spring	274.8	53	5.2	233	29	57.5
Summer	239.7	47	5.1	203	23	83.0
Autumn	390.1	59	6.7	382	66	57.5
Annual	1336.9	222	6.0	1243	209	83.0

2

Meteorological observations in Durham

The arrival of astronomy in Durham

Astronomy at the newly-founded University of Durham began with the arrival of the Reverend Temple Chevallier from Cambridge in 1835 (Figure 2.1). Described by Joan Kenworthy [22] as 'a remarkable Victorian polymath', Chevallier was a scholar of considerable ability and remarkable scope. His influence on the university was profound—he was Professor of Mathematics from his arrival and added 'Astronomy' to his professorial title from 1841. Other duties included Reader in Hebrew, Curator of the Library, and Registrar, not forgetting his parochial duties as Vicar of Esh, just outside Durham [23].

Initially, Chevallier's teaching of astronomy was theoretical—and treated purely as a branch of mathematics—but in 1838 the possibility arose of observational work as the University was given the opportunity to purchase an excellent set of astronomical instruments [23]. Following positive advice from the then-Astronomer Royal at Greenwich Observatory, George Biddell Airy (1801–92, a native of Northumberland with an interest in the new northern University, and Astronomer Royal 1835–81; later Sir George), an appeal was launched for funds, both for purchase of the instruments and to build an observatory on a site leased from the Dean and Chapter at very favourable terms. The architect for the new Observatory building (Figure 2.2) was Anthony Salvin, an architect with a national reputation who had grown up in County Durham and had attended Durham School; Salvin was already busy with restoration of the Durham Castle keep for student accommodation in the recently established University College. Nevertheless, funds remained tight and the first report of the Curators of the Observatory on 23 June 1840 lamented that further large expenses would be required before the Observatory could be brought into actual operation.

Inevitably, Chevallier could only devote a fraction of his time to astronomy, but he was fortunate in having an Observer, a paid employee who could work full-time on astronomical observations. A list of Durham Observers from 1840 is given in

Figure 2.1 *A three-quarter length portrait oil on canvas painting of The Reverend Temple Chevallier (1794–1873) by the artist Charles West Cope. (Reproduced with permission from Durham University Library and Heritage Collections)*

Appendix 1. Of the astronomers who occupied that post over the years to 1939, the late Sir Arnold Wolfendale, himself once Astronomer Royal, identified Richard C. Carrington (Observer October 1849 to March 1852) as the most distinguished [23]. Carrington was the son of a wealthy brewer and Fellow of the Royal Astronomical Society; he continued and extended the work on sunspots started by Chevallier and his observations after he left Durham, which led to his discovery of the differential rotation of the Sun—namely that the solar surface rotates more rapidly at the solar equator than elsewhere. Regrettably, Carrington was so disillusioned with the University's lack of support for the observatory that he ceased employment as Observer there after fewer than three years in post. There is a copy of his resignation letter in the Durham University Library archives ([24], item Fac. 5/19, dated 10 February 1852) describing in detail his reasons for leaving Durham—it is six pages long!—and stating his intentions to either find a post with good equipment or to set up a private observatory. He moved to Redhill, Surrey, where he did found his own observatory and built a distinguished career in his own right, most notable for his observation of the first recorded solar flare on 1 September 1859 [25, 26]; this solar storm is still known today as the 'Carrington Event'. Wolfendale [23] commented that there appears to have been little work of note following Carrington's departure; no doubt, Wolfendale had astronomy in mind, but the same might be said of meteorology, at least until Plummer's work on the records in the early 1870s [27].

Figure 2.2 *Lizars' 1840 engraving of the Durham Observatory building (Reproduced by permission of Durham University Library and Collections)*

The start of meteorological observations

As at other astronomical observatories, meteorological observations were necessary from the outset because air temperature and atmospheric pressure data were needed for the calculation of atmospheric refraction and the accurate determination of the position of celestial objects. The earliest known meteorological records from the Observatory (comprising barometric pressure, air temperature, and rainfall) are from June 1841, when an incomplete series of daily observations, in weekly summary form, were published in the *Durham Advertiser* [28]. The earliest surviving collected manuscript observation register dates from 23 July 1843, preserved in a bound volume within Durham University Library entitled *Observatory Durham Meteorological Observations—Commencing July 23, 1843, ending Dec 31, 1847*. The records consist of temperatures at 9 a.m. and 9 p.m. local time, together with the day's maximum and minimum temperatures, all in degrees Fahrenheit; Figure 2.3 shows the first page of the register. During the period when

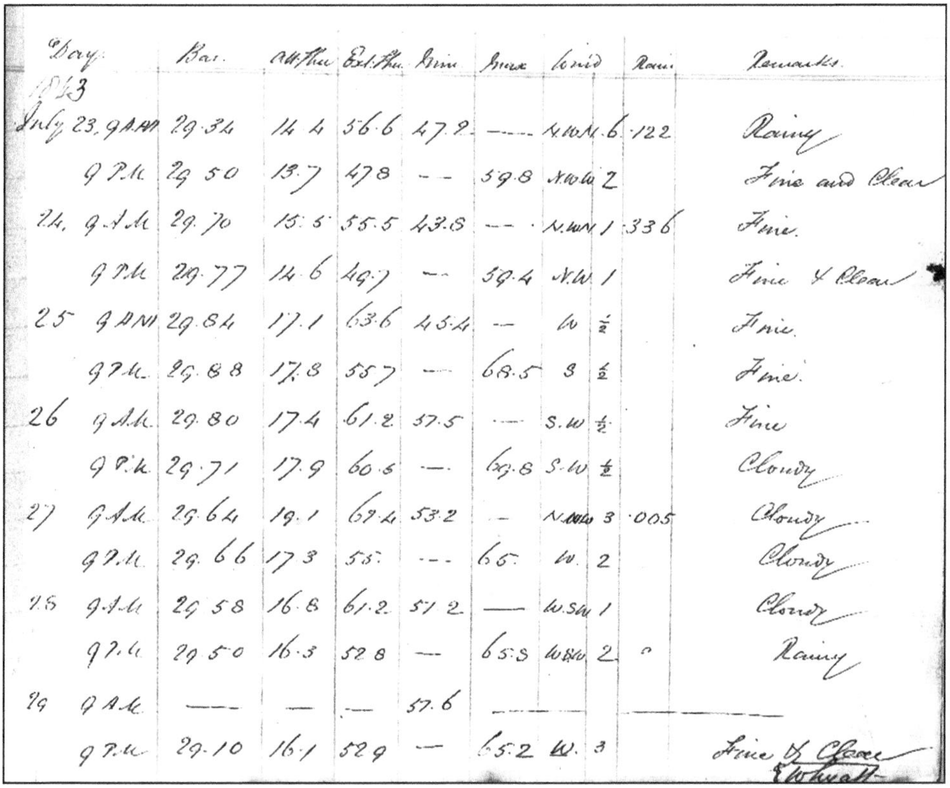

Day 1843	Bar.	Att. Ther.	Ext. Ther.	Min	Max	Wind		Rain	Remarks.
July 23. 9 A.M.	29.34	14.4	56.6	47.9	——	N.W.N.	6	·122	Rainy
9 P.M.	29.50	13.7	47.8	—	59.8	N.W.W.	2		Fine and Clear
24. 9 A.M.	29.70	15.5	55.5	43.8	—	N.W.N.	1	·336	Fine.
9 P.M.	29.77	14.6	49.7	—	59.4	N.W.	1		Fine & Clear
25 9 A.M.	29.84	17.1	63.6	45.4	—	W	½		Fine.
9 P.M.	29.88	17.8	55.7	—	68.5	S	½		Fine.
26 9 A.M.	29.80	17.4	61.2	37.5	—	S.W	½		Fine
9 P.M.	29.71	17.9	60.6	—	69.8	S.W	½		Cloudy
27 9 A.M.	29.64	19.1	69.4	53.2	—	N.W.W.	3	·005	Cloudy
9 P.M.	29.66	17.3	55.5	—	65	W.	2		Cloudy
28 9 A.M.	29.58	16.8	61.2	51.2	—	W.S.W	1		Cloudy
9 P.M.	29.50	16.3	52.8	—	65.3	W.S.W.	2	0	Rainy
29 9 A.M.	—	—	—	57.6	—			—	
9 P.M.	29.10	16.1	52.9	—	65.2	W.	3		Fine & Clear

Figure 2.3 *The first page in the Durham Observatory meteorological logbook, starting with the 9 a.m. observation on 23 July 1843. (Reproduced by permission of Durham University Library and Collections: the first two volumes, for 1843–47 and 1848–50, have been scanned and are available online [29])*

Gordon Manley analysed the early Durham records, the existence of this early register was unknown (it was thought lost), for it was only re-discovered in the university library in 1982 [28]. It has recently been scanned by Durham University Library and is available online, together with the second volume, which covers the period January 1848 to April 1850 [29].

There were frequent changes of Observer in the early years, due partly to the low salary offered and partly to the mandatory obligation that the Observer should be a bachelor. The position tended to attract young men early in their careers or, after the restriction on marriage was lifted in 1866, those with private means [30]. As noted, Richard Carrington was particularly distinguished, despite his limited time in post as Observer; Gordon Manley described him as ‘a particularly keen and careful observer’ [19]. For a long time, Carrington’s manuscript volume of careful meteorological observations for 1 January 1850 to 31 March 1852, in his distinctive careful handwriting (Figure 2.4), was thought to be the first surviving logbook, although Manley was aware that observations had been made from

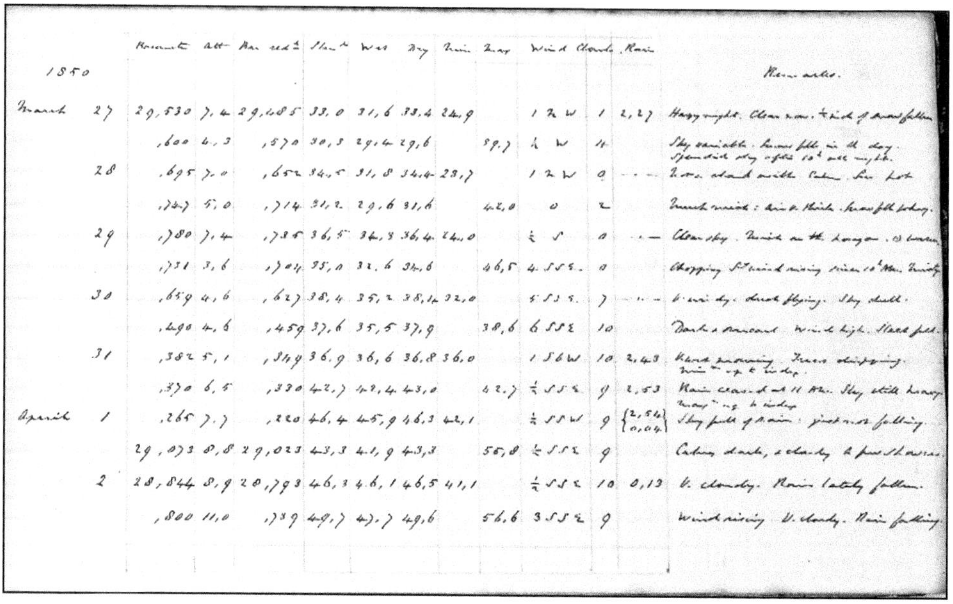

Figure 2.4 *A page from the Durham Observatory register for 27 March to 2 April 1850, in Richard Carrington's neat hand. (Reproduced by permission of Durham University Library and Collections)*

at least 1847 because summaries had been included in James Glaisher's meteorological tables in the Registrar General's *Quarterly Returns of Births, Marriages and Deaths*. When earlier volumes were discovered in the basement of the University Library after Manley's death, it was found that Carrington had copied the first four months of 1850 from an earlier ledger to his beautifully kept volume, adding corrections for instrumental errors. Carrington's 1850 volume includes a useful preface [19].

Manley also noted that daily observations were made, mostly at 0900 and 2100 local time, of temperature, barometric pressure, humidity, rainfall, wind, and cloud amount, with notes on the day's weather; careful application was made of the instrumental corrections, the instruments being by well-known makers [19]. Manley commented that since 1850 this routine had in general been followed well. Air temperature measurements appear to have been from instruments exposed on the north wall of the observatory until October 1851, after which the thermometers were exposed in a ventilated penthouse or 'North Shed' until December 1859. It is unfortunate, however, that most of the air temperature measurements were made in a Glaisher stand from January 1860 to October 1899, for it was not until November 1899 that a Stevenson screen was installed. Appendix 3 provides more details on the temperature metadata.

Other instruments added from time to time included a Robinson anemometer in 1866, black-bulb and grass minimum thermometers in 1874, a Campbell–Stokes sunshine recorder in May 1880, and a Dines anemograph, rain recorder, and a thermograph (the latter to act as a check on the temperature observations) in 1936–37. Given gaps

in the earlier record, Manley [19] followed the convention that the Durham record was unbroken since 1850, although Manley himself was well aware of significant gaps, particularly in 1854–55. This convention was established by Carrington who had himself rejected observations made before 1850 [9]. While there remain rumours of a 'drunken observer', the gaps in July–August 1854 and October 1854 to mid-February 1855 probably have more to do with the legacy of Carrington's departure and an ongoing lack of careful oversight, given Temple Chevallier's inability to devote enough time to the Observatory. It seems as if standards slipped for a while, and the daily routine of observations suffered as a result. Even as late as 1865–67, rainfall totals were sometimes accumulated, i.e. not read every day; it appears that the Observer M. R. Dolman was not properly trained in this regard (see Appendix 4 for rainfall metadata). It was probably not until John Plummer's time as Observer (from November 1867) that the meteorological record began to be regarded as important in its own right, and increasing attention was paid to the quality and continuation of the twice-daily observations. Plummer's major contribution was to attempt to determine corrections to render the temperature record 'throughout the whole of the twenty-three years (i.e. since 1850) strictly comparable' [27], as described in Appendix 4.

The resources required in keeping up the high-quality meteorological record, with two observations per day, were not inconsiderable in terms of staff time for making and writing-up the observations, submitting monthly returns to the Meteorological Office, maintaining and occasionally re-siting or replacing instruments, answering local enquiries, and so on. There were times when the continuation of the meteorological record was far from certain, particularly during the difficult years of the First World War when the main meteorological observer, Frederick Carpenter, was serving in the Army: he was later badly wounded on the Western Front and invalided back to England (see Appendix 1). Letters in the Durham University Library archive from January 1917 ([24], records 861–862) seek advice as to whether the meteorological observations should be suspended for the time being, 'as they are a great burden in war time'. This was agreed, only for a rapid respite following a letter from Sir Napier Shaw, then Director of the Meteorological Office, assuring the university that 'the records are of real and continuing value'.

Figure 2.5 shows a map of the observatory grounds, indicating the historical position of the 'North Shed' on the rear parapet (see also Appendix 2), and of the Glaisher stand and the location of past and present instruments on the front lawn. Figure 2.6 is a recent photograph of the current site. Appendix 2 provides more details on the Observatory site and common aspects to the observations, such as site description, observation hours, and the like. Subsequent appendices contain more specific metadata by element (including instruments in use by period, exposure information where known, details of dataset quality control procedures, and any adjustments made to the original records)—for temperature in Appendix 3, precipitation in Appendix 4, barometric pressure in Appendix 5, sunshine records in Appendix 6, and wind records in Appendix 7.

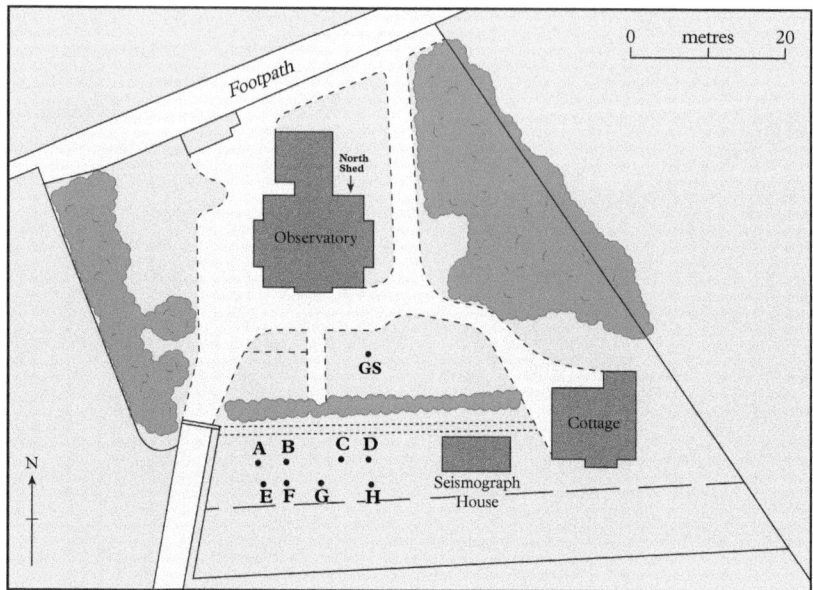

Figure 2.5 *Map of the Durham Observatory grounds showing the historical location of the Glaisher stand, Stevenson screens, and other meteorological equipment. GS: Glaisher stand; A and B: Stevenson screens; C: rain gauge; D: Dines tilting siphon recording rain gauge; E and F: earth thermometers; G: grass minimum thermometer; H: solar radiation thermometer; and North Shed. Today, the Metspec Stevenson screen is located close to position C, with grass minimum, 30 cm and 100 cm earth temperature sensors nearby, and the tipping-bucket rain gauge at position H (Figure 2.6). The Campbell–Stokes sunshine recorder was located on the front parapet of the Observatory building. Redrawn by Chris Orton, Durham University, from Baxter (1956) with additional information from Manley [19](1941).*

Gordon Manley and the Durham temperature record

Gordon Valentine Manley (3 January 1902–29 January 1980; Figure 2.7) was the best-known climatologist of his generation. His volume in the New Naturalist series *Climate and the British Scene* [20] remains one of his greatest contributions to British climatology: it not only appealed to the target popular audiences of the series, but is still regarded by many contemporary climate scientists as a classic text [31–34]. Nationally, his greatest legacy and contribution in terms of data analysis was the Central England Temperature (CET) series [2].

Gordon Manley was born in Douglas, Isle of Man, and brought up in Blackburn, Lancashire, where he attended Queen Elizabeth's Grammar School. After obtaining degrees in engineering and geography at the Victoria University of Manchester and

Figure 2.6 *The Durham Observatory site, photographed on 27 July 2021. (Michele Allan)*

Gonville and Caius College, Cambridge respectively, Manley joined the Meteorological Office in 1925, but resigned the following year to join the Cambridge Expedition to East Greenland in the summer of 1926. Later that same year he began a lengthy career in academia when he became an assistant lecturer in geography at Birmingham University. His enthusiasm for his subject, his joy of learning, and his wit made him an excellent teacher. In 1928 he was appointed Lecturer in Geography at the University of Durham to lead the new geography degree programme. He subsequently became a Senior Lecturer, founding Head of Department, and Curator of the University's Observatory [31].

    Given the lack of attention paid to the Durham temperature record since Plummer in the 1870s, Manley's arrival in Durham seems little short of miraculous. Manley himself noted:

> Numerous investigators in recent years have commented on the value of long and comparable series of meteorological observations and in this country the published tables of the Oxford record from 1815 have proved to be especially welcome. During my sojourn as Lecturer-in-Charge of the School of Geography at Durham I resolved to try to place the records from the University Observatory, of which I was a curator, on a similar basis. In the light of other recent work the temperature record has been considered first [19].

The complexity of problems met in standardizing the temperature series make his account a classic illustration of the fact that it is seldom enough simply to accumulate instrumental measurements of climatic conditions—'differences in exposure, in hours

Figure 2.7 *Gordon Manley.*
(Geography Department, Durham University)

of observation, in instrumental corrections and even in the observers' predilections must all be accounted for.' Manley hoped that the task had proved worthwhile and that the long Durham series of temperature records would prove to be a useful addition to the climatological history of north-east England; at the same time his 1941 paper celebrated the centenary of the foundation of the Observatory. In 1943 Manley was awarded the Buchan Prize (jointly with T. E. W. Schumann) by the Royal Meteorological Society for this paper on the Durham meteorological record. The various methods by which Manley derived his 'adopted means' are described subsequently (Appendix 3). His 'adopted means' of temperature for 1847–1940 were taken up to 1950 by Baxter in 1956 [35], and the series was updated to 1979 by Manley himself in his last, posthumous, publication in 1980 [36].

While at Durham, Manley also established his own weather stations in the Pennines, first at Moor House, on the exposed moorlands of Upper Teesdale, 'the coldest part of England' [11] (see Figure 1.1 for locations), and later near the exposed summit of Great Dun Fell at 847 m, where conditions are often particularly bleak [15]. He was a pioneer of such upland weather studies, and his run of personally collected data helped to establish Great Dun Fell as one of England's best-known (perhaps 'notorious') mountain weather stations [37]. Manley's principal aim there was to discover under what circumstances the Helm Wind (England's only named wind) blew and to provide some measurements of the strength of the wind [38, 39]. Moor House became an important site for investigations of upland ecology and, apart from one major gap (1980–91), climate data continue to be collected there to this day (see Chapter 1).

Manley held particular affection for the Pennines throughout his life. This was an area in which he regularly walked alone, with his students, or with his wife Audrey Fairfax Robinson after their marriage in June 1930 (see Chapter 22). Manley maintained detailed commentaries on the walks he had undertaken and there remain four notebooks, now held in the Gordon Manley papers in the Department of Manuscripts and University Archives, Cambridge University, recording these walks and associated field and weather observations from his teenage years into the 1930s [31, 38].

Gordon Manley left Durham in 1939 to become a Demonstrator in Geography at Cambridge University. From 1942 to 1945 he was a Flight Lieutenant in Cambridge University Air Squadron, but he continued his research and teaching of students from Cambridge and Bedford College, London (the latter institution having been evacuated to Cambridge). From 1948 to 1964, Manley was Professor of Geography at Bedford College (then an all-female college) in the University of London. In 1964, at the age of 62, he took on the challenge of founding the Department of Environmental Studies at the newly established Lancaster University. In 1967, aged 65, he retired to Cambridge and remained active until shortly before his death in January 1980. His work on the extension of the Durham temperature series was undertaken in retirement, when he visited Durham frequently, and suggested privately, when made an honorary Doctor of Science in 1979, that the extended series would be his return gift to the University.

In working on the material such as the Durham temperature record, Gordon Manley acquired an unmatched knowledge of early meteorological instruments and the problems of their use, of the history of their exposure and the character of the pioneer observers. His monumental CET work, the series of monthly mean temperatures representative of a typical lowland site in central England for every year from 1698, was first published in 1953 [1] and updated and extended back to 1659 in 1974 [2]. It remains the longest meteorological instrument record for anywhere in the world, maintained today by the Hadley Centre of the Met Office in Exeter [3, 40]. Manley worked on it for thirty years, developing and refining it from numerous shorter records maintained at various sites during overlapping periods. His method of reducing the data to a single homogenized record was first devised at Durham (see Appendix 3). It is clear that Manley established new standards of scholarship and reliability in this field [41]. We make use of Manley's CET record at various points later in the book.

Meteorological observations at Durham Observatory post-Gordon Manley

Members of the Geography Department at Durham University continued to oversee the daily meteorological observations after Manley's departure in 1939, although details are scant for earlier in that period. The day-to-day work was undertaken by a series of resident observers (see Appendix 1). Joan Kenworthy (*pers. comm.*) has provided some details of the more recent period. She comments that climatology in the Department was given a boost by Ray Harris (now Emeritus Professor of University

College, London), who taught climatology in the Durham department from 1976. Joan herself joined the University the following year to take up the position of Principal of St Mary's College, also teaching in the Geography Department. Ray Harris established a local meteorological society in Durham; this was later adopted as a regional branch of the Royal Meteorological Society (RMetS) and, in consequence, Joan Kenworthy was elected to the Council of the Royal Meteorological Society 1985–87.

As might be expected, technicians in the Geography Department have taken a close interest in the Observatory weather station over the years. In particular, Derek Hudspeth, Departmental Superintendent (Honorary MSc, 2008; MBE 2009), was closely involved in maintenance of the instruments, liaison with Met Office, annual inspections, and making observations if needed. Academic members of staff such as Ray Harris, Michael Tooley, and Helen Goldie prepared returns for the Met Office and produced monthly and annual reports; these used to be printed but today are distributed via the Department website **durhamweather.webspace.durham.ac.uk**. The Observatory building, once entirely managed by the Department of Geography, was taken over by the Department of Physics, except for two rooms that continued to be used for the meteorological work and for pollution observations. Tim Burt took over running the weather station in October 2001 following Helen Goldie's retirement.

The Geography Department's continuing involvement in climatology was also evident in conferences held at St Mary's College [42, 43]. Joan Kenworthy and Nicholas Cox obtained an award from the Leverhulme Trust in 1995 for the digitization and analysis of the Durham records, which they completed with Andrew Joyce as Research Assistant [44]. Much use is made in this book of the digital record thus started, suitably updated to the end of 1999 when manual observations ceased, and kept updated from the automatic weather station records since (see below, and Appendices). Kenworthy and Cox also supervised Matthew Eglise's PhD research [45], which resulted in a Durham temperature series extended back to 1784, as outlined later in this chapter.

The end of manual observations at Durham Observatory

Mrs Audrey Warner (Figure 2.8), Durham Observer 1969–99, was the last observer to live at the Observatory (in the observatory cottage); daily manual observations largely ceased with her retirement in September 1999, although most observations were maintained by Geography staff through to the end of the year. Audrey Warner received a special presentation from the Met Office and an Honorary M.A. from Durham University in recognition of her 30-year service. Investment in a full-time member of staff with a right to residence was, alas, not deemed good value for money by the University, notwithstanding the length of the meteorological record from the 1840s, and she was not replaced.

Figure 2.8 *Mrs Audrey Warner, the last Durham Observer to live at the Observatory, photographed during the late 1990s. (Courtesy of the Department of Geography, Durham University)*

Automatic weather station records: October 1999 to date

Following the withdrawal of manual observations, an automatic weather station (AWS) was installed, commencing records on 1 October 1999. At first this was owned and operated by Durham University's Department of Geography; the Met Office took over ownership and responsibility for maintenance in 2006. The Met Office would not, however, assume responsibility for the Campbell–Stokes sunshine recorder on the roof, or the Dines anemograph exposed on a mast high above the Observatory dome; when the old mast failed early in 2011, it was not replaced. More recently, a separate AWS on the roof of the Geography Department (West Building, Lower Mountjoy campus) has been installed, a little over 800 m east-south-east of the Observatory, providing helpful

backup data and some measure of continuity, albeit with lower-quality instruments and at a much more built-up and imperfectly exposed site. The AWS at the Observatory itself has proved largely reliable, connectivity via telephone lines being a greater issue than the AWS itself. Local observers, especially Ken Cook at Copley, the Environment Agency (Barkers Haugh sewage treatment works), and the station at Cockle Park (Newcastle University), have all kindly provided data to infill various minor gaps as detailed in Appendices 2–4.

In 1956, when Baxter wrote up the history of Durham Observatory in *Weather* [35], he stated that the building was:

> ... awaiting demolition as the ground floor is so damp as to be nearly useless and policy is no longer to carry out astronomical work. In any case the dome would require much money spent on it before it could be used. The present plan is for a two-storey building with a basement to the west of the site, which will be used only for meteorological and seismological work.

Fortunately, this early Victorian building, significant to both astronomy and meteorology, was finally saved from demolition, and indeed was refurbished in 1961. A suitable plaque to commemorate its history and its observers was unveiled in March 1962 (Figure 2.9). At the time of writing (summer 2021) Ustinov College, the largest college of Durham University, uses part of the Observatory building as student activity/common

Figure 2.9 *Building plaque installed at the Observatory in March 1962. There is a complete list of the Observers in Appendix 1 (Stephen Burt)*

room space. There is an intention to install external air quality monitoring instrumentation in the grounds, while trees to the north are being lopped to improve instrument exposure.

Extending the 'Durham' temperature record back to 1784

While at Durham, Gordon Manley resolved to place the record of the Durham Observatory on a similar basis to that of the Radcliffe Observatory at Oxford, for which published tables of the temperature and precipitation records from 1815 were then available. His resulting account of a series of 'adopted means' of temperature for the period 1847–1940 already referred to [19] established the Durham record as the second-longest for any university in Britain and, in the year of its publication, happened to celebrate the centenary of the foundation of the Observatory. He remained unaware that observation registers prior to 1850 existed—these were only re-discovered after Manley's death.

In the later years of his retirement, Manley began the challenging (and somewhat frustrating) work of constructing a broader temperature series for north-east England, using such records as were known at the time, in an attempt to extend a 'Durham' series back to 1794, using similar methods to those for his Central England series [1, 2] and for Lancashire [46]. His accounts stress the vital importance of keeping sources for each series separate, in order to create independent series as far as possible given limited or missing records or gaps in records. Kenworthy [28] describes in some detail how Manley went about this task, in particular his identification of other records from the north-east region that could be used to extend the series—and some that were definitely not suitable. Most important to Manley were the daily meteorological readings of James Losh of Jesmond Dene in Newcastle-upon-Tyne, about 25 km due north of Durham Observatory. Losh's records cover a period of almost 32 years (1802–33) and consist of three fixed-hour observations each day. Manley believed that Losh may well have sited his thermometer on the north wall of the house, possibly outside his study window, but later suggested the east wall. Losh's house (The Grove) and its park were well up on the gentle slope that declines towards Jesmond Dene, to the east of Newcastle's city centre, at about 52 m above MSL.[*] In a letter to Joan Kenworthy dated 27 September 1978 Manley mentions that the difference in average temperature between Jesmond Dene and Durham Observatory (at 102 m above MSL) should be close to 0.6 °F (0.33 degrees C), very close to what would be expected for the 50 m height difference between the two sites using a standard environmental lapse rate of 0.65 degrees Celsius per 100 m [47]. Using the Losh records to form the backbone of the series for 1802–32, Manley used data from stations further afield to arrive at a tentative series of

---

[*] Kenworthy (op. cit) gives the altitude of The Grove as 46 m but close examination of maps from the 1860s onwards suggests that the house stood just upslope of the 50 m contour. Matthew Eglise's PhD thesis also uses 46 m (p. 106); he states that the house was situated in a slight valley, with slightly higher ground to either side.

mean monthly temperatures for north-east England (see Kenworthy [28], Table 2.4 that volume). Clearly, Manley felt that it would be quite some time before it was completed and written up, and stated explicitly, '... this isn't yet for publication, merely retention' in a letter to Joan Kenworthy dated 5 November 1979. Kenworthy [28] describes in some detail the stations and methods adopted by Manley in attempting to infill the gap between 1833 and 1843, as well as the stations used for the earlier period in addition to Jesmond. While Manley hoped very much to get back to the well-known cold winter of 1794–95, sadly he died before the work was completed.

Recently, Smith [48] has published a detailed account of the James Losh diaries, including analysis of both the temperature data and transcription of the weather diary, adding great detail to our knowledge of weather and climate in north-east England in the early nineteenth century.

The Eglise Durham series 1784–1849

Manley was working in haste in ill-health by late 1979. Close examination of his working papers suggested that some arithmetical errors had been made, and over 20 years after his death the decision was made to replicate his work digitally [9]. This task was undertaken by Matthew Eglise, a PhD student supervised by Joan Kenworthy and Nicholas Cox. After careful examination of Manley's working papers, letters and the original archived sources of temperature data for the north of England used by Manley, Eglise was able to produce an improved or corrected 'Manley-based' series back to 1801 [45]. Moreover, by examining contemporary relationships between temperatures at Durham and a series of stations in the north of England, together with Manley's original data and archival sources that had become available since Manley's time, Eglise was able to create a new and statistically justified extended series for 'Durham' back to 1784—admittedly with a few gaps, particularly between October 1791 and December 1794, for which no suitable regional records could be traced. No doubt, Manley would have approved of Eglise's skilful use of statistical analysis and would have envied the benefits of modern computers and powerful spreadsheets; in return, Eglise thanked Manley for inspiration in his thesis.

By careful examination and analysis of Manley's surviving papers, Eglise was able to ascertain how Manley arrived at his various temperature 'reductions' to form the series by bridging between sometimes disparate records scattered in space and time. His thesis carefully documents and analyses Manley's methods, correcting the occasional error in the derivation of the monthly series, and where appropriate, refining and enhancing Manley's methods using modern statistical techniques to improve reliability and accuracy. In addition, Eglise also had the benefit of the two 'rediscovered' Durham Observatory ledgers from 1843 not available to Manley.

The resulting Eglise Durham series, from January 1784 to December 1849, can be briefly compared with other temperature series for the British Isles. Figure 2.10 plots the extended 'Eglise Durham' annual mean air temperature series alongside CET for the period 1784 to 1849: as expected, there is a high correlation between the two series. Indeed, one concern is the perhaps too close agreement between Durham and CET in the period when the Losh record from Jesmond forms the backbone of the Durham

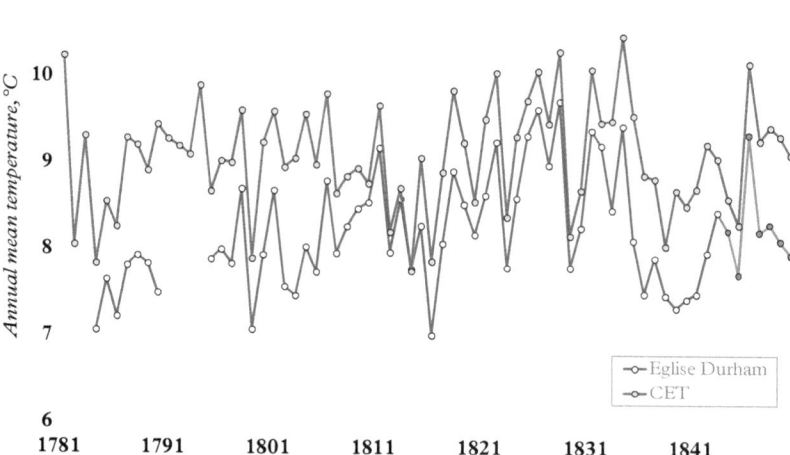

Figure 2.10 *Annual mean temperatures (°C) for the 'Eglise Durham' series and Central England for the period 1784 to 1843 (Durham Observatory instrumental values 1844–1850 in green). Central England Temperature series from https://www.metoffice. gov.uk/hadobs/hadcet/cetml1659on.dat*

extended series, 1802–33. Adjusted monthly means from Smith's more recent work (see Appendix C in [48]) suggest that temperature estimates based on the Losh data alone (suitably adjusted for altitude) are about 1 degree Celsius lower on average than the Eglise series. There is very little surviving information regarding the exposure of Losh's thermometer, and thus how it might compare with a modern Stevenson screen record. As a result, the Eglise extension suggests slightly warmer conditions in the 1810s compared to the Losh estimate.

Perhaps most notable are the extremes: the very cold years of 1784, 1799, and 1816, and the very warm years of 1828 and 1846. At the start of the extended series, it appears that north-east England temperatures emerged from the latter stages of the Little Ice Age more slowly than further south. Both regions experienced the cool periods of the 1810s and the late 1830s/early 1840s and warmer years in the 1820s and early 1830s. Either way, the series quantifies numerous very cold years prior to 1850, of which 1816 is perhaps the most notable (see also Chapters 21 and 22).

The impact of urban growth on the Durham Observatory temperature record

As noted in Chapter 1, suburban expansion around Durham City has not infringed significantly on the Durham Observatory site, which remains relatively well exposed, if becoming more sheltered by trees than for much of the 180-year instrumental record.

There have been small housing estates built to the west and south-west of the site, along the eastern side of the A167 by-pass, but these do not closely approach the observatory site. To the south, new houses on the other side of Potters Bank Road are at minimum 200 m distant (Figure 1.2). The newest local development, Ustinov College's new site at Sheraton Park to the north-west, on the old Neville's Cross College site, is also more than 200 m away (Figure 2.11). This is quite unlike the Radcliffe Observatory site in Oxford where new hospital and university buildings have come very close to the weather station, and the expansion of north Oxford from the nineteenth century onwards has completely changed a site that was on the edge of the built-up area when the observatory was built in 1772 [49]. As a result, the long Oxford temperature record shows a small urban heat island signal of about 0.2 degrees Celsius.

We have compared the Durham temperature record with the CET record, a series first created by Gordon Manley during his time in Durham: Manley took care to keep the two series entirely separate, so we can regard the values as independent of each other. Tables 2.1 to 2.3 show mean air temperatures for CET and Durham for five 30-year periods including the most recent standard averaging period 1981–2010 and the new 1991–2020 averages. Negative values for the differences in individual periods indicate that Durham is colder than CET, which is no surprise given Durham's north-eastern location. The difference between the two series has gradually narrowed over 150 years: for the period 1861–90 the difference in mean temperature was 1.00 degrees Celsius but this narrowed to 0.79 degrees in the 1991–2020 averaging period. The change is barely

Figure 2.11 *Durham Observatory (centre of frame) and its surroundings (© Google Earth, May 2020)*

Table 2.1 *Mean air temperature,* °C

Period	CET mean °C	Durham mean °C	Difference, degC
1861–90	9.14	8.14	−1.00
1891–1920	9.25	8.20	−1.05
1921–50	9.58	8.64	−0.94
1951–80	9.47	8.53	−0.94
1981–2010	10.00	9.16	−0.84
1991–2020	10.25	9.46	−0.79
1991–2020 minus 1861–90	**+1.11**	**+1.32**	**+0.21**

Table 2.2 *Mean maximum air temperature,* °C

Period	CET mean °C	Durham mean °C	Difference, degC
1861–90	*Not available*		
1891–1920	12.84	11.93	−0.91
1921–50	13.17	12.46	−0.71
1951–80	13.06	12.18	−0.88
1981–2010	13.70	12.88	−0.82
1991–2020	14.00	13.19	−0.81
1991–2020 minus 1891–1920	**+1.16**	**+1.26**	**+0.10**

Table 2.3 *Mean minimum air temperature,* °C

Period	CET mean °C	Durham mean °C	Difference, degC
1861–90	*Not available*		
1891–1920	5.66	4.46	−1.20
1921–50	5.98	4.83	−1.15
1951–80	5.87	4.87	−1.00
1981–2010	6.30	5.42	−0.88
1991–2020	6.48	5.71	−0.77
1991–2020 minus 1891–1920	**+0.84**	**+1.25**	**+0.41**

Table 2.4 *Monthly mean temperature differences between CET and Durham, 1991–2020: units ° C*

	J	F	M	A	M	J	J	A	S	O	N	D
Mean temperature												
CET	4.66	4.89	6.74	8.95	11.90	14.69	16.79	16.51	14.17	10.92	7.41	4.97
Durham	4.08	4.56	6.16	8.26	11.01	13.61	15.78	15.56	13.27	10.01	6.64	4.25
Difference	−0.58	−0.32	−0.57	−0.69	−0.89	−1.08	−1.01	−0.95	−0.90	−0.90	−0.78	−0.73
Mean daily maximum temperature												
CET	7.36	7.91	10.39	13.32	16.47	19.24	21.40	20.92	18.27	14.31	10.35	7.70
Durham	6.87	7.75	9.87	12.48	15.49	17.96	20.23	19.84	17.32	13.51	9.66	7.07
Difference	−0.49	−0.16	−0.52	−0.84	−0.97	−1.28	−1.17	−1.08	−0.95	−0.80	−0.70	−0.63
Mean daily minimum temperature												
CET	1.97	1.84	3.08	4.57	7.31	10.13	12.16	12.09	10.09	7.51	4.48	2.24
Durham	1.29	1.38	2.46	4.04	6.52	9.26	11.32	11.26	9.21	6.52	3.62	1.43
Difference	−0.68	−0.46	−0.62	−0.54	−0.79	−0.88	−0.83	−0.83	−0.88	−0.99	−0.87	−0.81

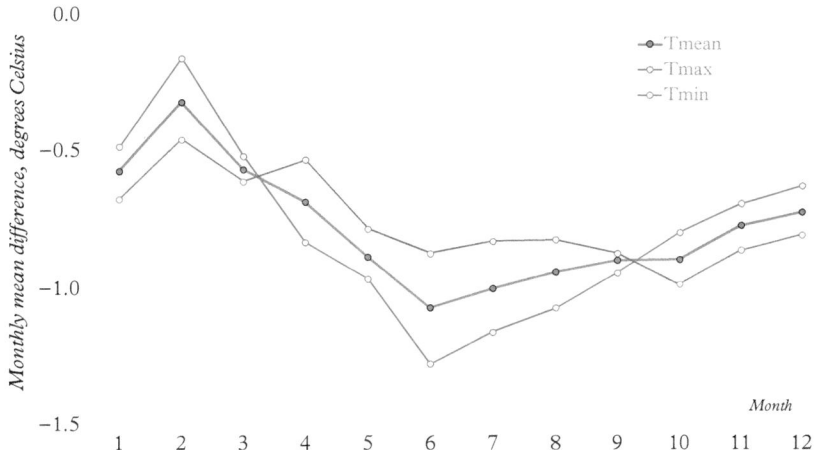

Figure 2.12 *Monthly mean temperature differences between CET and Durham, 1991–2020: units ° C*

detectable in the mean maximum temperature between the averages for 1991–2020 compared with those for exactly 100 years previously (Table 2.2), but there is a greater relative increase of 0.41 degrees Celsius in the mean minimum temperature (Table 2.3). Comparing the monthly averages for 1991–2020 between CET and Durham (Table 2.4 and Figure 2.12), we can see that the mean differences for both maximum and minimum temperatures are greatest (most negative) in summer, as expected from the differential warming rates of land surfaces (Central England) versus that of sea (the North Sea coast) during the warmer months.

Table 2.4 shows the difference (in degrees Celsius) between CET and Durham air temperatures over successive 30-year periods since 1861. There is a suggestion of a small urban heat island effect as suburban development has gradually encroached closer to the Observatory, of a similar magnitude to that seen in Oxford [49]. Minor rounding errors may result in one-digit differences in the second place of decimals. Note that CET mean monthly maxima and minima are available only from 1878 and thus there are no entries for the 1861–90 averages for those elements. Source of CET data: https://www.metoffice.gov.uk/hadobs/hadcet/data/download.html

The Durham record has therefore become relatively warmer, by about 0.2 degrees Celsius in 150 years, and this is most likely partly due to a small heat island effect on the Durham record from the growing city (see Figure 1.2). Other reasons—such as changes in atmospheric circulation which affect Durham in a slightly different way to Central England—may also have played a role. However, these are very small differences and do not, in our opinion, invalidate the value and importance of the long Durham temperature record.

Part 2

Durham weather through the year

3

The annual cycle

This chapter provides a brief introduction to the analysis and summaries of individual months and seasons which follow. Figures 3.1, 3.2, and 3.3 show respectively the average temperature (daily mean maximum, daily mean, and daily mean minimum) and daily mean precipitation for every day of the year (with the exception of 29 February) over the standard averaging period, 1991–2020. For daily average sunshine duration (hours), the averaging period is the 30 years 1970–99 as daily sunshine observations ceased at the end of 1999.

While there is of course considerable variation year to year, these serve to illustrate the general progression of the main weather elements throughout the annual cycle. There are also some brief comments about wind speed and direction using data from a shorter period (0900 GMT observations, 1961–97). Appendix 9 shows the monthly and annual averages of temperature, precipitation, and sunshine for Durham over the 1991–2020 period. For comparison, Appendix 8 provides the 1981–2010 averages in the same format.

Temperature

At first sight, the temperature curves appear remarkably smooth, as might be expected for 30-year averages. There is nevertheless some evidence of persistent spells of weather: *singularities*, as Hubert Lamb called them [50]. A singularity may be defined as the tendency of some weather characteristic to recur about a specific date in the year [51]. Over the most recent 30-year averaging period, the coldest day of the year on average has been 20 December, with a mean air temperature of 3.0 °C; five of the ten lowest mean minimum temperatures of the year also occur between 20 and 30 December, most of the remaining five being scattered between mid-January and mid-February (see also Table 23.1). The lowest mean daily minimum temperature, −0.1 °C, occurs on 29 December: this is also the only night of the year which has a mean minimum temperature below freezing in Durham. By day, seven of the ten coldest days (on average) occur in the fortnight commencing 20 December, with 20 December seeing the lowest mean daily maximum, at 5.5 °C. By mean temperature, January is the coldest month (mean temperature 4.1 °C), with December just one tenth of a degree less cold. Mean minimum temperatures differ only by a tenth of a degree between December and February, and there is often a cold snap in mid-February, but by February the increasing power of the Sun is beginning to lift daytime temperatures above those of midwinter days.

After the cold of February, a period of significant warming is characteristic of early March. The temperature plots are noticeably asymmetric, a fairly steady warming in spring set against more rapid cooling in autumn. The warmest days and nights, again on average, are at the end of July and the beginning of August; 5 August has the highest mean air temperature (16.8 °C) and the highest mean maximum temperature (21.3 °C), while 1 August has the highest mean minimum temperature (12.4 °C). The warmest month, by a small margin is July, with a mean temperature of 15.8 °C, just 0.2 degrees warmer than August. Air frosts are unlikely between mid-May and mid-October, although ground frosts have occurred in every month.

The warming over the past 150 years is clearly evident from a comparison of daily averages for 1871–1900 with the most recent 30-year period (Figure 3.1, thinner lines).

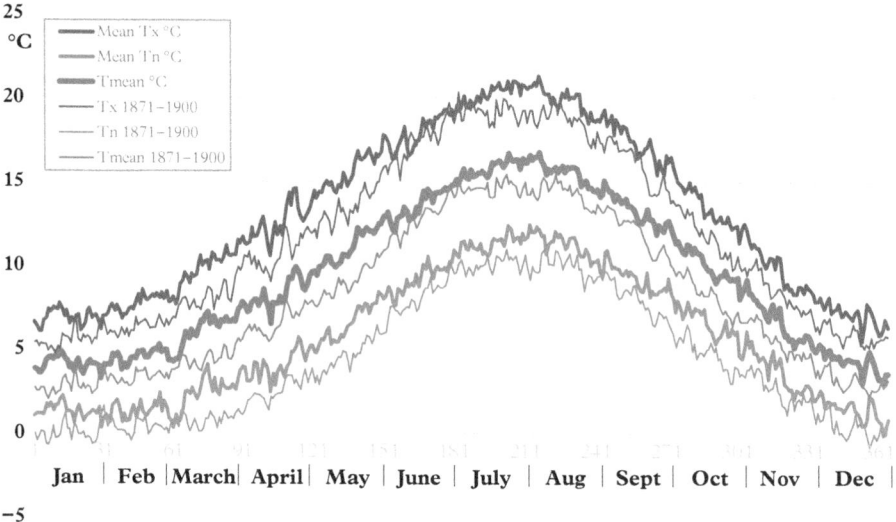

Figure 3.1 *Daily mean temperatures (mean daily maximum and minimum, and mean temperature, which is the average of the two) for every day of the year in Durham, for the current 30-year averaging period 1991–2020 (dark solid lines) and for the 30-year period 1871–1900 (thinner lines). Mean daily maximum temperature Tx in red, mean daily minimum temperature Tn in blue, and mean daily temperature Tmean in green. Yearday numbers in pale grey; 29 February data omitted. Units °C*

Precipitation

The pattern of daily precipitation throughout the year at Durham is much more irregular than for temperature, and to allow for this, Figure 3.2 shows *weekly means* of daily precipitation averages. The driest period of the year tends to be mid-March, with weekly means of daily precipitation under 1 mm per day 15–23 March; other dry periods are evident in early May and in late August. During the 1991–2020 period, March became the driest month of the year, with an average fall of 41 mm; during 1981–2010 February held the title of driest month, with 43 mm. Thereafter, there is an uneven trend upwards to reach a maximum in November, with a monthly mean of 75 mm; August (66 mm) and October (64 mm) are next-wettest (Appendix 9). The wettest week (on average) occurs in mid-November when daily rainfall amounts approach 3 mm per day. It should be borne in mind, however, that individual very wet days during the 30-year period considered will skew the averages for specific dates, and that while these wettest and driest periods generally hold true over decades, the actual dates and amounts will vary considerably from one averaging period to another. In addition, the frequency and intensity of rainfall also varies throughout the year, so while 2 mm may fall in a short shower on a summer afternoon, the same amount may take 12 hours to fall in midwinter.

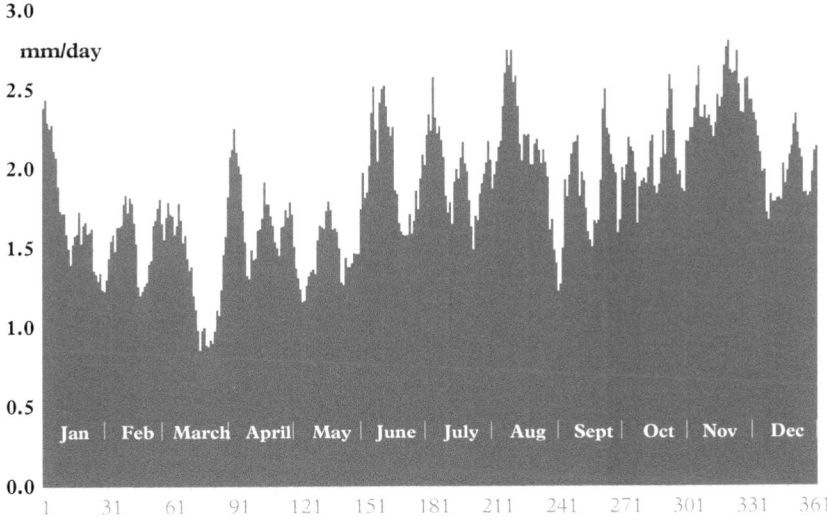

Figure 3.2 *Weekly means of daily mean precipitation amounts centred upon every day of the year in Durham; period 1991–2020. Yearday numbers in pale grey; 29 February data omitted. Units mm*

Sunshine

Not surprisingly, the plot of weekly means of average daily sunshine during the year shows a clearer seasonal variation (Figure 3.3), with least sunshine in midwinter but a broader maximum from early May to late August—daily mean sunshine averaging 5 to 6 hours per day during the latter period. (Note that, because daily sunshine records ceased at the end of 1999, these averages refer to the 30-year period 1970–99 only, and not to the same span of years as the temperature and precipitation plots). Over the period 1970–99, the sunniest day of the year, on average, was 29 June, with an average 7.18 hours of sunshine, while the sunniest week, on average, was that centred on 16 May, with a daily mean of 6.45 hours. At the other end of the year, the least sunny day was 17 December, with just 0.83 hours (50 minutes) daily, and the dullest week was that centred on 14 and 15 December, averaging just 1.24 hours (74 minutes) sunshine. By month—and using the current monthly averages for 1991–2020 (Appendix 9)—the sunniest month of the year is May, with 187 hours of bright sunshine (a daily average of 6.03 hours), and December the dullest with just 57 hours (1.84 hours per day).

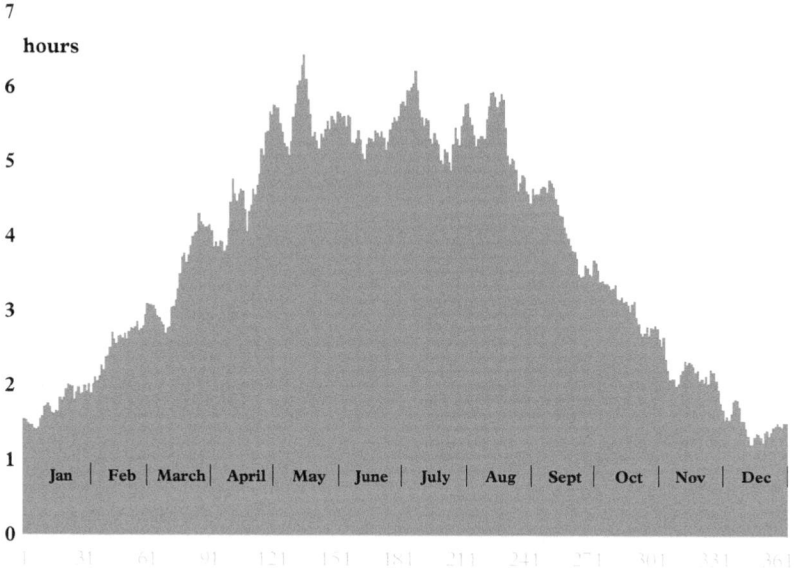

Figure 3.3 *Weekly means of daily sunshine duration centred upon every day of the year in Durham; period 1970–99. Yearday numbers in pale grey; 29 February data omitted. Units hours*

Wind speed and direction

Wind velocity is a vector involving both direction and magnitude. It is desirable therefore, whenever possible, to deal with both components, direction and speed. In summaries, this can be done using a two-way frequency table of speed against direction, or, diagrammatically, by constructing a wind rose [52]. At Durham Observatory, wind speed and direction data were first recorded with the installation of the Robinson anemometer in 1866 (Appendix 7), although methodology, instruments, and units of measurement have changed frequently over the decades and there is a dearth of accompanying metadata. As a result, unfortunately it has not been possible to produce a single, homogenous series.

Table 3.1 shows the percentage frequency of winds of different speeds and directions, and of calms, at the 0900 GMT observation for the year over the period 1961–97 inclusive.

Table 3.1 *Percentage frequency of winds of different speeds and directions, and of calms, at Durham Observatory at 0900 GMT for the year over the period 1961–97. Wind speed classes are in Beaufort Force (Bft) from 0 (Calm) to 7 or more, with the wind speed ranges (knots) shown for each class. Wind directions are in degrees clockwise from north (090 is east, 180 is south, etc.) and by 30-degree sectors centred on that shown*

Bft scale	Bft 0	Bft 1	Bft 2	Bft 3	Bft 4	Bft 5	Bft 6	Bft 7 or more	
Degrees True	*knots* < 1	1–2	3–6	7–10	11–16	17–21	22–27	28 or more	Total
	19.0								19.0
360		0.9	1.8	1.3	0.8	0.3	0.01	0.01	5.1
030		1.0	2.1	1.1	0.8	0.1	0.04	0.01	5.1
060		1.1	1.9	0.9	0.7	0.1	0.01	0	4.7
090		0.7	1.1	0.6	0.5	0.1	0.01	0	3.0
120		0.5	0.6	0.3	0.2	0.04	0.01	0	1.6
150		0.8	1.0	0.4	0.3	0.1	0	0.01	2.6
180		2.1	3.7	2.1	1.5	0.3	0.1	0.03	9.8
210		2.8	5.0	2.9	2.4	0.6	0.1	0.1	13.8
240		1.5	2.8	2.4	2.5	1.0	0.2	0.2	10.6
270		0.9	1.9	1.9	2.4	0.9	0.3	0.2	8.4
300		1.0	2.6	2.0	2.4	0.6	0.3	0.1	8.9
330		1.0	2.5	1.7	1.6	0.4	0.1	0.1	7.4
Total	19.0	14.3	27.0	17.4	16.0	4.3	1.2	0.7	100.0

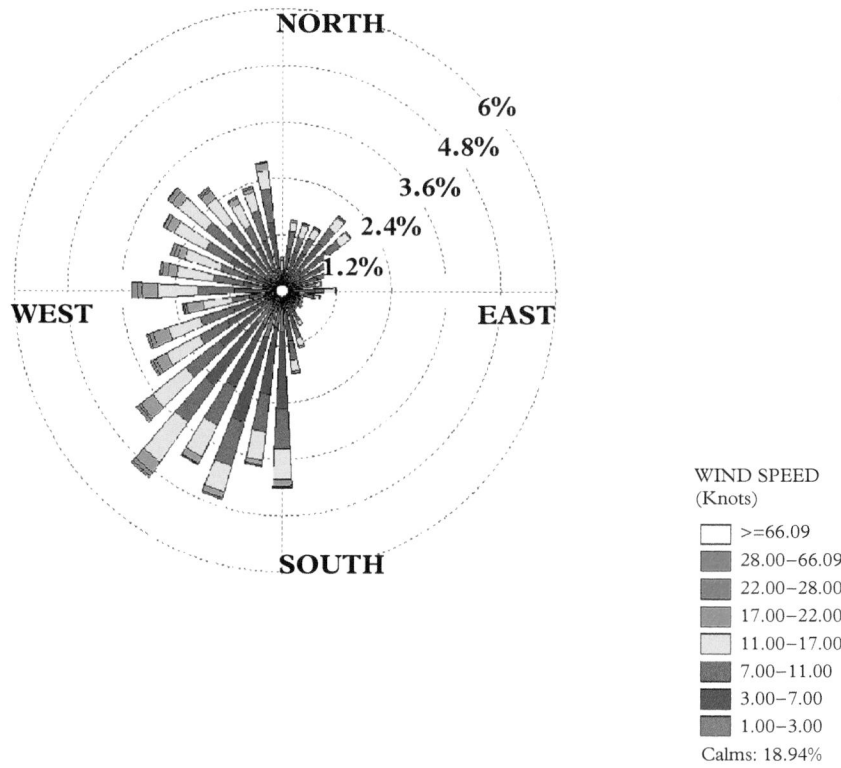

Figure 3.4 *Wind rose for Durham 0900 GMT wind speed and direction observations, all months 1961–97 (see also Table 3.1)*

While these data obviously do not refer to the current averaging period, there is no reason to believe that, other than increased shelter at the site from surrounding tree growth, the general pattern of wind speed and direction at Durham has varied significantly in more recent years. The table emphasizes the dominance of wind direction from between south and north through west, with the majority of observations in the range from 3 to 6 knots (Beaufort Force 2, or 1.5 to 3 m s^{-1}). Figure 3.4 presents the same data in the form of a traditional wind rose. Table 3.2 shows monthly mean wind speeds at 0900 GMT, again over the period 1961–97, although it should be noted that increasing shelter around the Observatory site has resulted in a progressive and significant reduction in wind speed over this period.

Table 3.2 *Mean monthly wind speeds at Durham Observatory at 0900 GMT; period 1961–97*

Monthly means	Jan	Feb	Mar	Apr	May	June	July	Aug	Sept	Oct	Nov	Dec	Annual
knots	7.8	7.2	8.2	7.7	6.5	6.3	6.0	6.1	6.5	6.3	7.2	7.5	6.9
m s^{-1}	4.0	3.7	4.2	4.0	3.3	3.2	3.1	3.1	3.3	3.2	3.7	3.9	3.5

The main climatic elements in relation to atmospheric circulation

Figure 3.5 shows monthly scores of Lamb Weather Types (LWT) for the standard averaging period, 1991–2020; the methodology follows Jones et al. [53]. Cyclonic weather is at a minimum in February, coinciding with a greater likelihood of colder and drier conditions. A minimum of westerlies from April to June helps explain dry spells at this time of year and why May is the sunniest month. Frequent wet weather from autumn until the end of the year is clearly related to a reduction in anticyclonic weather with a greater frequency of cyclonic weather and westerly airflow.

Table 3.3 shows seasonal correlations between mean air temperature, total precipitation, and sunshine at Durham and seasonal and annual values of the Lamb Weather Type indices anticyclonic (A), cyclonic (C), and westerly (W), and the North Atlantic Oscillation index (NAO). As before, the LWT scoring methodology follows Jones et al. [53]. Rainfall at Durham is strongly correlated with the incidence of cyclonic weather systems and thus negatively correlated with the incidence of anticyclonic weather. In contrast, mean air temperature is inversely correlated with cyclonic weather, suggesting that cyclonic weather can include both cold air masses of continental origin and cold polar maritime air in winter.

Apart from sunshine, the only correlation with westerly airstreams is winter temperature, but this is strongly positive, implying mild winters with a strong westerly airflow, and vice versa. There is a similar correlation in winter with the NAO index,

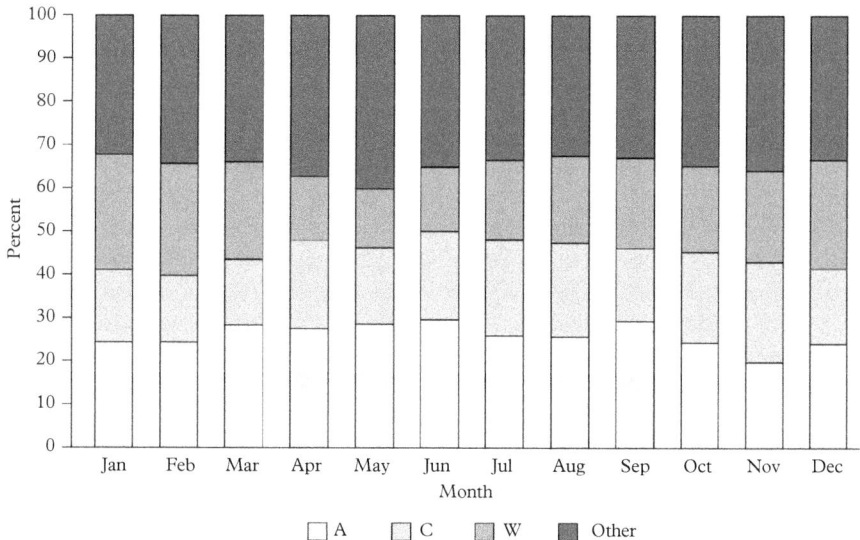

Figure 3.5 *The main Lamb Weather Type scores by month over the British Isles: A- Anticyclonic, C-Cyclonic, W-Westerly, plus others combined in one category, expressed as a percentage for each month over the period 1991–2020*

Table 3.3 *Seasonal correlations between mean air temperature, total precipitation and hours of bright sunshine at Durham, and seasonal/annual values of the Lamb Weather Type indices cyclonic (C) and westerly (W), and the North Atlantic Oscillation index (NAO).*

		C	W	NAO
Mean air temperature	Winter	−0.40	**0.69**	**0.61**
	Spring	−0.39	..	..
	Summer	−0.38	..	..
	Autumn	..	..	..
	Annual	..	..	..
Total precipitation	Winter	**0.65**	..	..
	Spring	**0.75**	..	..
	Summer	**0.68**	..	−0.41
	Autumn	**0.81**	..	..
	Annual	**0.76**	..	..
Sunshine duration	Winter	..	0.36	0.43
	Spring	−0.37	..	0.36
	Summer	*−0.55*	..	..
	Autumn	−0.39	*0.56*	..
	Annual	..	..	..

Cells marked [..] indicate no significant correlation. Ordinary font indicates significance at $p = 0.05$; *italic font* indicates significance at $p = 0.01$; **bold font** indicates significance at $p = 0.001$.

again reflecting the strength of influence of Atlantic air masses. There is an inverse correlation between NAO and summer rainfall, a relationship observed for many sites in lowland, southern, and eastern England [54]. This negative correlation in summer is not unexpected: summer brings a weaker NAO, and the Azores anticyclone extends to cover more of the North Atlantic. It is notable that the inverse relationship between summer rainfall totals and the NAO index holds over a much longer timescale than the standard period 1991–2020: for a sample length of 160 years, the correlation is significant at 1% probability ($p = 0.01$). For this same long series, winter rainfall at Durham is negatively correlated with NAO, $p = 0.01$ [54]. These correlations are again an indication of Durham's 'eastern, lowland' location, rather different to the weather patterns observed in the Pennine uplands to the west. Sunshine is negatively correlated with cyclonic conditions but positively correlated with westerly airflow in autumn and winter. The correlation of sunshine hours with the NAO index suggests that a strong westerly airflow may offer the best chance of sunny periods in between the showers.

4

January

_____

January is on average the coldest month of the year in Durham, although just one-tenth of a degree colder than December. In the 30 years used for the current standard averaging period, 1991–2020, January was the coldest month of the winter 13 times, December ten times, February six times, and March once (in 2013). January is also the second-windiest month of the year, and wind chill can often be significant. The mean temperature for the month over 1991–2020 was 4.1 °C (average daily maximum temperature 6.9 °C, average daily minimum 1.3 °C). The 'annual cycle' plots in Chapter 3 show that the beginning and the end of the month tend to be colder, and there is often a milder period mid-month. The monthly mean precipitation for January, averaged over the same period 1991–2020, is 53 mm, approximately midway between the normal for December and February. A 'rain day' is one in which 0.2 mm or more is recorded during the 24 hours commencing at 0900 GMT on the day of measurement, and by this measure rain falls on 19 days in an average January in Durham. Although rainfall is frequent and can be persistent, heavy falls are rare in January: more modest totals are much more common, although a daily fall of 10 mm or more can be expected on at least one day in most Januarys. Sunshine duration in January creeps up a little from the December minimum, averaging 62 hours or two hours of bright sunshine daily, although this is slightly misleading as typically 12 days will remain sunless in January.

Temperature

Since temperature records commenced at the Durham Observatory in July 1843, temperatures in January have ranged from −16.9 °C on 17 January 1881 to 16.1 °C just seven years later on 9 January 1888, the latter one of the very rare days when Durham briefly became the warmest place in Europe. Both are 'Stevenson screen equivalents' from temperatures originally measured in a Glaisher stand (see Appendix 3 for details); since Stevenson screen records commenced in November 1899, the recorded extremes have been −16.1 °C on 21 January 1940 (when the temperature fell to −20 °C at Houghall), equalled on 11 January 1982, and 15.6 °C on 25 January 1918. In the earliest records, we should also mention the −15.7 °C recorded unscreened on the Observatory's north wall on 29 January 1848 (with a grass minimum temperature of −20.0 °C).

Table 4.1 *Highest and lowest maximum and minimum temperatures at Durham Observatory in January, 1844–2022*

Rank	Mildest days	Coldest days	Mildest nights	Coldest nights
1	*16.1 °C,* *9 Jan 1888*	*−6.2 °C,* *1 Jan 1871*	11.4 °C, 3 Jan 1932	*−16.9 °C,* *17 Jan 1881*
2	15.6 °C, 25 Jan 1918	*−4.8 °C,* *25 Jan 1881*	10.6 °C, 25 Jan 2016	−16.1 °C, 21 Jan 1940, 11 Jan 1982
3	15.0 °C, 30 Jan 1944	−4.5 °C, 7 Jan 1982	*10.3 °C, 1 Jan 1851,* 19 Jan 1932	*−15.8 °C,* *26 Jan 1881*
4	14.9 °C, 6 Jan 1916	−4.4 °C, 5 Jan 1941	10.1 °C, 30 Jan 1898	*−15.7 °C,* *29 Jan 1848*
5	14.4 °C, 4 Jan 1921, 28 Jan 1921 and 26 Jan 2003	*−4.3 °C,* *2 Jan 1854*	10.0 °C, 19 Jan 1932, 15 Jan 1947	−14.6 °C, 8 Jan 1982
Last 50 years to 2022				
	14.4 °C, 26 Jan 2003	−4.5 °C, 7 Jan 1982	10.6 °C, 25 Jan 2016	−16.1 °C, 11 Jan 1982

Temperatures recorded prior to the introduction of a Stevenson screen in Nov 1899 are shown in italics; these are regression estimates from Glaisher stand readings from Jan 1860 to Oct 1899, as read from the 'North Shed' Oct 1851 to Dec 1859 and as read from the north wall exposure prior to Oct 1851. Temperatures were recorded in degrees Fahrenheit until Oct 1971, read to a precision of 0.1 °F during the period 1843 to 1918 and 1 °F for the remainder of the record. All values in degrees Fahrenheit have been converted to degrees Celsius, but those for the 1 °F precision period are subject to approximately 0.2 °C uncertainty as a result. Since November 1971 temperatures have been recorded to a precision of 0.1 °C. See Notes on page xvii and Appendix 3 for more details.

Table 4.1 shows the mildest days and nights, and the coldest days and nights, in January at Durham since 1844.

Very cold nights were more common in the nineteenth century. In the first 50 years of record, allowing for the variations in thermometer exposure prior to 1900, the temperature fell below −10 °C on 31 occasions in January. In the following 123 years to 2022, such cold nights occurred only 25 times in January, most recently in 1982 when there were six such observations.

January is the month most likely to record an 'ice day', one in which the temperature fails to reach 0 °C. Since 1900, and to January 2022, there have been 85 such days in January, but in the 50 years preceding 1900 there were 81—further evidence that very cold weather was more likely in the nineteenth century (although the thermometer exposures in use in Victorian times also made this more likely). On New Year's Day in 1871, the maximum temperature reached only −6.2 °C, a very low value that has been surpassed only once since, in December 1878 (Table 24.4), and not remotely approached since. Only four of January's 85 ice days since 1900 have occurred in the twenty-first century, most recently (at the time of writing) on 26 January 2017, when a maximum temperature of −0.3 °C was recorded.

The Met Office's *Monthly Weather Report* describes January 1940 as 'exceptionally cold', and this is certainly no exaggeration. In Durham this was the coldest month since January 1881: seven days failed to reach zero and there were 28 air frosts. The record cold night on 21 January 1940 was associated with a ridge of high pressure moving east over the UK introducing an arctic airstream from the north-west. Equally severe winter weather occurred during the first half of January 1982, when nine consecutive days 6–14 January remained below zero all day, while between 7–15 January, six of nine nights fell below −10 °C. The synoptic situation on 7 January 1982, Durham's coldest day for over a century when the maximum temperature was only −4.5 °C, is shown in Figure 4.1. Later in this spell, the minimum temperature on the morning of 11 January fell to −16.1 °C, equalling the January 1940 record as the coldest night of the twentieth century. The very cold weather was associated a ridge of high pressure resulting in clear

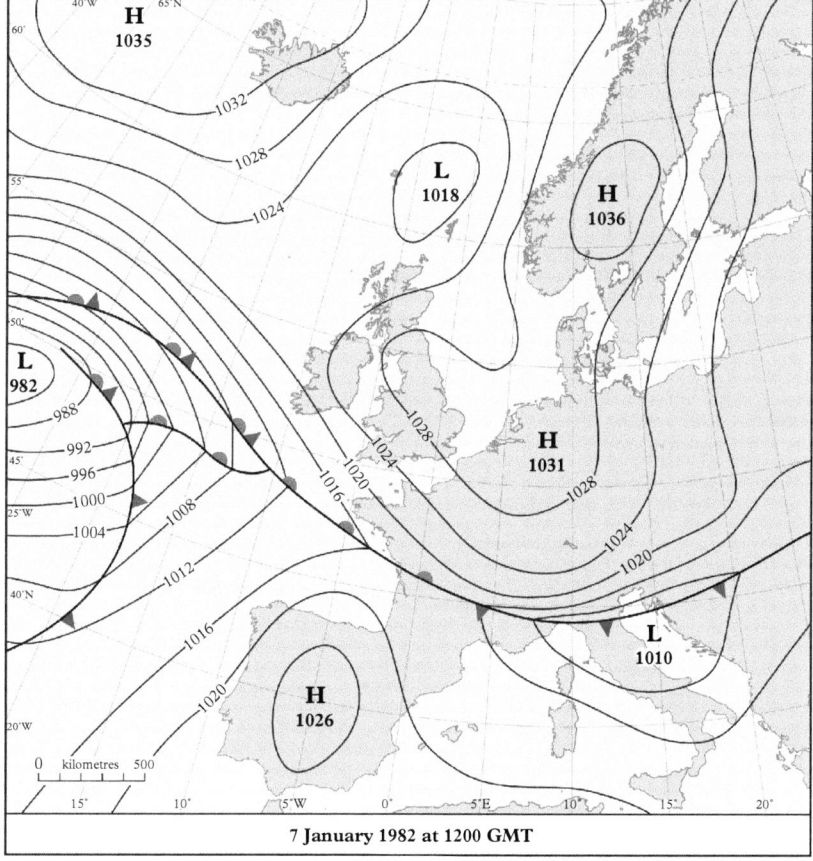

Figure 4.1 *The synoptic chart for 1200* GMT *7 January 1982, Durham's coldest day of the twentieth century, when the day's maximum temperature reached only −4.5° C. (Drawn by Chris Orton, Durham University)*

skies, light winds, and intense cooling over a fresh snow cover. The morning observation on 11 January 1982 noted only 1 okta of cloud cover, the wind 1 knot from the west-south-west, and 5 cm of lying snow.

Mild nights and mild days occur regularly in January. Three January days have attained 15 °C, although the most recent such occasion was back in 1944: every month of the year has recorded considerably colder days than these examples of extreme mid-winter mildness in England's north-east, notably 7.2 °C in June (1953) and 9.4 °C in August (1919). To date, January 1916 remains the mildest January on record in Durham (Table 4.2). More recently, the winter of 2019/20 became the second-mildest on record, with again a very mild January being an important contributor.

The mildest January night on record occurred on 3 January 1932, when the minimum temperature fell no lower than 11.4 °C (Table 4.1), the only January night to remain above 11 °C. Perhaps the most remarkable statistic within Table 4.1 is the minimum of 10.0 °C on 15 January 1947, during a mid-month spell when four consecutive days reached 10 °C; this, in a famously cold and snowy winter before bitter easterly winds set in, for by 30 January the temperature had fallen below −8 °C.

At the time of writing, both the coldest and mildest January on Durham's records are more than a century old. January 1881 remains the coldest on record at Durham: the mean air temperature was −1.7 °C, 5.8 degrees below the current January average. Figure 4.2 shows maximum and minimum temperatures for the month; note that these

Table **4.2** *January mean temperatures at Durham Observatory, 1844–2022*

January mean temperature 4.1°C (average 1991–2020)

Mildest months			Coldest months		
Mean temperature, °C	Departure from 1991–2020 normal degC	Year	Mean temperature, °C	Departure from 1991–2020 normal degC	Year
6.7	+2.6	1916	−1.7	−5.8	*1881*
6.3	+2.2	1989, 2007	−1.5	−5.6	1940
6.2	+2.1	*1898, 2020*	−0.5	−4.6	*1879, 1895, 1941, 1963, 1979*
5.7	+1.6	1923, 1983	−0.4	−4.5	*1871*
5.6	+1.5	1975	*0.1*	−4.0	*1850*
Last 50 years to 2022					
6.3	+2.2	1989	−0.5	−4.6	1979

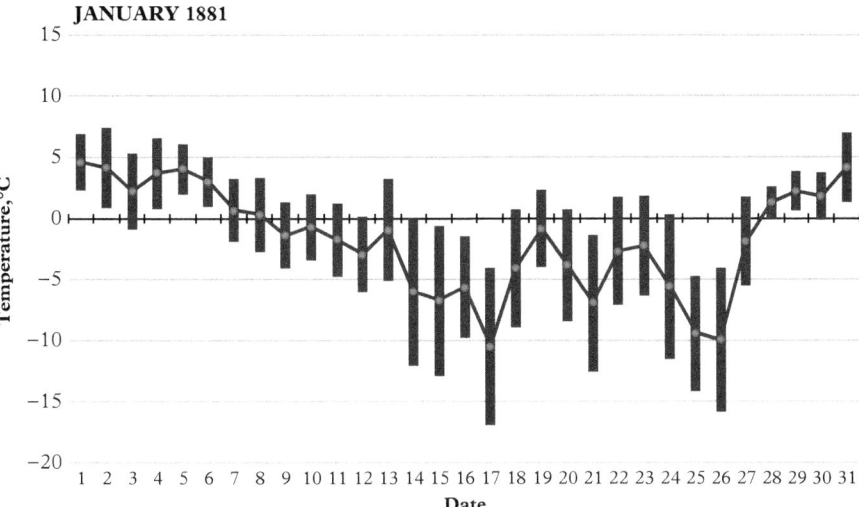

Figure 4.2 *Daily maximum and minimum temperatures during January 1881—Durham's coldest January*

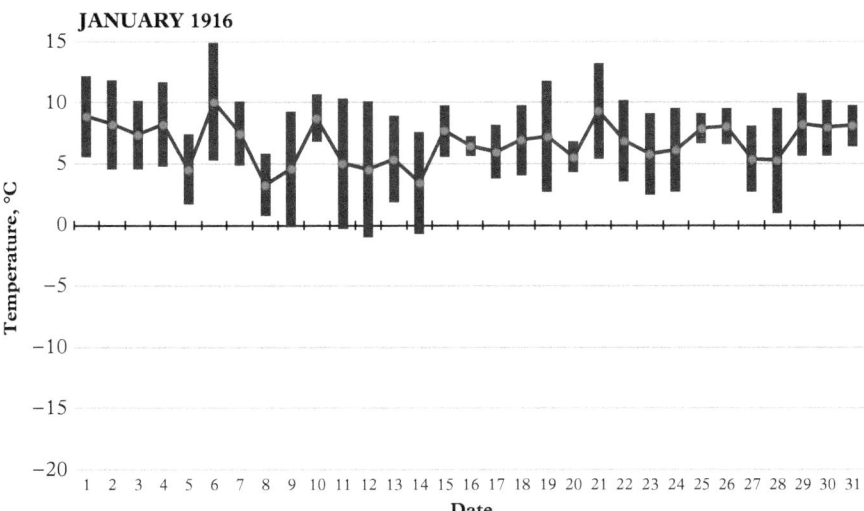

Figure 4.3 *Daily maximum and minimum temperatures during January 1916—Durham's mildest January*

are 'Stevenson screen equivalent temperatures' derived by regression from thermometers housed in a Glaisher stand and are thus subject to some uncertainty. Nevertheless, it is clear that this was a severely cold month, with seven nights in a fortnight falling below −10 °C, and as noted earlier, the minimum of −16.9 °C on 17 January remains the lowest

January temperature yet recorded at the Durham Observatory, and indeed second only to one night in February 1895 as Durham's coldest-ever night. The 'grass' minimum temperature (probably above a snow surface) that morning was −19.4 °C. The daily mean temperature for 17 January 1881 (the mean of the daily maximum and minimum temperatures) was −10.5 °C, equalling 25 December 1860 as Durham's coldest-ever day (Table 24.6). Almost as cold was 26 January that year, with a mean daily temperature of −10.0 °C. Literally, at the other end of the scale, the exceptionally mild January 1916 (Figure 4.3) recorded only two air frosts and a remarkable 17 days with maxima of 10 °C or greater, including 14.9 °C on 6 January that year.

Cold winter weather has become less frequent in recent decades, but notable in this context was January 2021's mean temperature of 2.3 °C, 1.8 degrees below normal and the coldest January since 2010.

Precipitation

January is a relatively dry month, at least in terms of total precipitation, averaging 53 mm during the standard averaging period 1991–2020: only the months February to May are drier. Monthly totals have ranged from just 7.2 mm in 1989 (13 per cent of normal) to 188 mm in 1948 (352 per cent of normal—see Table 4.3).

Only eight Januarys have received more than 100 mm since 1868, with 1948 being the wettest by a large margin (until 2021), almost 70 mm in excess of the then second-wettest, January 1939. January 2021, with 137 mm, then took over this position, and contributed to Durham's wettest winter on record; in contrast, January 2022 recorded just 16 mm. Very wet Januarys are more frequent than very dry; within the last 150 years,

Table 4.3 *January precipitation at the Durham Observatory 1868–2022*

January mean precipitation 53 mm (average 1991–2020)

Wettest months			Driest months			Wettest days	
Total mm	Per cent normal	Year	Total mm	Per cent normal	Year	Date	Total mm
188.0	352	1948	7.2	13	1989	32.8	12 Jan 1921
137.2	257	2021	9.1	17	1905	29.7	2 Jan 1961
121.8	228	1939	9.9	19	1876	29.1	4 Jan 1948
116.6	218	2016	10.8	20	1964	28.6	4 Jan 2016
116.0	217	1972	12.0	22	2019	25.0	9 Jan 1892
Last 50 years to 2022							
137.2	257	2021	7.2	13	1989	28.6	4 Jan 2016

only three Januarys have received less than 10 mm total precipitation; January 1989 was the driest, and very unusual in being both very mild and exceptionally dry (Tables 4.2 and 4.3).

Rain can be expected to fall on 19 days in an average January, the highest frequency of any month (with November and December), but heavy falls of rain are decidedly uncommon—falls of 25 mm or more within a day have occurred only five times since 1868.

Durham's wettest January day was 12 January 1921, when 32.8 mm fell—the only January day to exceed 30 mm (Table 4.3) and at the start of what turned out to be an exceptionally dry year. *British Rainfall* noted that

> a primary depression passed across central Ireland and the north of England during the 11th and 12th and was accompanied by widespread though not remarkably heavy rainfall. Over a well-defined area, lying entirely in Co. Durham, rainfall totals on the 12th slightly exceeded 50 mm. Serious flooding occurred in various parts of the north of England.

The highest total listed in the County was for Biddick Hall just west of South Shields, where 56.4 mm was recorded.

Sunshine

On average, January is the second dullest month, with 62 hours of bright sunshine (2.0 hours per day), a slight but welcome increase on December's average of 57 hours, resulting from a combination of a gradual increase in day length towards the end of the month, and fractionally reduced cloud cover. Januarys are becoming sunnier, partly due to changes in atmospheric circulation and partly the result of improvements in air quality: the January average sunshine for the most recent 30-year period average (1991–2020) was almost one-third higher than the average for the period 1881–1910, namely 47 hours. Sunless days occur frequently in January; on average, 12 days will remain sunless. This figure has hardly changed since sunshine records began in 1880.

Table 4.4 shows the extremes of January sunshine since 1881. January 1996 was remarkable for its lack of sunshine; only 7.6 hours was recorded at the Durham Observatory, just 3 per cent of the possible duration, the dullest month of any in Durham's long record. There were 24 days with no bright sunshine at all, more than in any previous month—surpassing the previous record of 23 sunless days in December 1903. The Observatory's monthly report remarked that 'January 1996 distinguished itself by its spectacular dullness!'. In January 1913, 22 days remained sunless, with 21 in the Januarys of 1917, 1941, and 1955; in contrast, only four sunless days occurred in January 1882, and five in 1908. Spells of a week or more without any sunshine are not uncommon in January. Durham's longest spell without sunshine was one of 17 days (equalled in December 1968) which commenced on 24 January 1940, a particularly raw and cheerless period: the mean daily maximum temperature for these 17 days was only 1.2 °C. The 4.1 hours of sunshine on 10 February 1940, the first day to record

Table 4.4 *January sunshine duration at Durham Observatory. Observed monthly extremes 1881–2022, sunniest days 1882–1999 only*

January mean sunshine duration 62 hours, 2.0 hours per day (average 1991–2020)
Possible day length: 246 hours. Mean sunshine duration as percentage of possible: 25.1

Sunniest months			Dullest months			Sunniest days 1882–1999	
Duration, hours	Per cent of possible	Year	Duration, hours	Per cent of possible	Year	Duration, hours	Date
112	45.5	2022	7.6	3.1	1996	7.6	31 Jan 1987 and 28 Jan 1999
94.0	38.3	1959	12.5	5.1	1942	7.5	27 Jan 1962 and 25 Jan 1986
86	35.0	2000	15.5	6.3	1913	7.4	29 Jan 1997
85	34.6	2015	19.5	7.9	1917	7.3	26 Jan 1986
84.7	34.5	1991	21.4	8.7	1929	7.2	26 Jan 1902, 24 Jan 1949 and 16 Jan 1998
Last 50 years to 2022							
112	45.5	2022	7.6	3.1	1996	7.6	31 Jan 1987 and 28 Jan 1999

Daily sunshine records ceased at the Observatory on 31 December 1999: monthly sunshine totals from January 2000 onwards are estimated (to the nearest hour) by regression from the Met Office East and North-east England sunshine series. See Appendix 6 for record metadata and missing months.

any sunshine since 23 January, must have been particularly welcome, even though the temperature struggled to reach 0 °C.

At this time of year, six hours of sunshine in a day represents almost unbroken sunshine, and only towards the end of the month is seven hours in a day possible as the days begin to lengthen. Only four January days (to 1999) have managed 7.5 hours or more sunshine, all within the final week (Table 4.4). January 2022 established a new sunshine record with an estimated total of 112 hours - far exceeding the previous sunniest of 94 hours in 1959. Four of the five sunniest Januarys have occurred since 1991, although this period does also include the very dull January in 1996. Other than 1996, the least sunny Januarys all date from 1942 or earlier. Indeed, of the 40 least-sunny Januarys, only 2016 comes from the present century and even then, ranks only 38th. Of the 40 sunniest Januarys, only seven occurred before 1950. It seems clear, therefore, that improvements in air quality have had a marked effect on hours of bright sunshine in Durham, especially so in winter; this is also evident in the reduction of fogs in recent years, as detailed below.

Wind and weather

Snow cover

Information on snow cover is rather limited—the digital record does not begin until 1960, and manual observations ended in late 1999, although we can ascertain snow cover duration from some months before 1960 by reference to the Durham tabulations in the *Monthly Weather Report*. By convention, 'snow cover' is counted when more than half the ground representative of the observing site is covered with snow, regardless of its depth, while the usual statistical measure is 'mornings with snow cover', referring to a snow cover at the time of the 0900 GMT observation.

Over the period 1960 to 1999, there were an average of 5.5 mornings with snow cover at the Durham Observatory; for the period from 1971–99, this decreased slightly to 5.0. By far the snowiest winter on the data available was the winter of 1962/63 (see also Chapter 22), which saw snow on the ground for 70 consecutive mornings from 27 December 1962 to 6 March 1963, inclusive. The snow lay 8 cm deep on 27 December 1962 and by 17–19 January 1963 had increased in depth to 23 cm. January 1979 recorded 25 mornings with a snow cover, the greatest depth being 18 cm on New Year's Day, while January 1985 saw a snow cover on 20 mornings,

Figure 4.4 *An accumulation of snow in the garden of Kingsgate House, Bow Lane, Durham, on 9 January 2010; heavy snowfall in January has become much less common in the present century (Elizabeth Burt)*

Figure 4.5 *Deeper snow in the Wear Valley at Copley: Ken Cook's Lead Mill station (218 m AMSL) on 6 January 2010 (Ken Cook)*

13 cm deep on 17th. In earlier years, January 1940 recorded a snow cover on 16 mornings, but we have no information on maximum depths attained. In general, recent winters have been milder and the frequency of snow cover has become much less common in the present century, with occasional exceptions such as January 2010 (Figures 4.4 and 4.5).

Fog

We are similarly limited to the period 1962–97 for information on the frequency of fog, which is determined from the visibility at the 0900 GMT observation—'visibility' being determined by the maximum distance at which a defined set of visibility objects are visible. In the meteorological sense, 'fog' is defined as a visibility less than 1000 metres, and 'thick fog' below 200 metres (see Appendix 8).

Over the period 1962–97, fog at 0900 GMT occurred on three mornings in a typical January, one of which was 'thick fog'. In January 1971, fog was noted on ten mornings, and five of these were thick fog. The frequency of fog has been declining steadily since the early 1960s, following the introduction of the Clean Air Acts of the 1950s and the gradual phasing out of coal fires, albeit that this probably happened rather later in County Durham than in most areas of the country.

Barometric pressure (1844–1960)

It is in midwinter that we see the greatest range in barometric pressure, and January has seen a range in pressure in excess of 100 hPa in the period 1844–1960. The highest pressure on record for the month, and for the year, occurred on 23 January 1907, when at 0900 GMT MSL pressure stood at 1052.5 hPa. This value appears slightly high (at 0800 GMT North Shields reported 1050.8 hPa): the true maximum probably lay between 1051 and 1051.5 hPa (see also Chapter 22). At the time of writing, this value has not since been exceeded in north-east England. The only other occasion of 1050 hPa on Durham's records occurred on 31 January 1902, when the barometer stood at 1050.8 hPa at 0900 GMT and 1050.7 hPa at the 2100 observation. That evening, the barometer rose to 1053.6 hPa in Aberdeen, a value which remains the highest on record within the UK [55]. Also notable was 1049.7 hPa at 0900 GMT on 26 January 1932.

At the other end of the scale, the MSL pressure at the 2200 GMT observation on 26 January 1884 stood at only 950.7 hPa, January's lowest on record, although it may have been lower earlier in the evening. The exceptionally deep depression responsible reached its minimum over Perthshire at about 2145 GMT, when the MSL barometer fell to 925.6 hPa at Ochtertyre [56]. On 1 January 1949, Durham's barometer fell to 956.0 hPa at 0900 GMT and 953.7 hPa at 2100 GMT, the latter reading the second-lowest on the available record for January. Of course, observations made just twice daily are unlikely to capture the absolute highest or lowest barometric pressures. (The extreme values quoted here, and in the other monthly and annual chapters, must be regarded as the extremes on the observational record, rather than the true absolute extremes, which could occur at any time of the day or night.)

The mean MSL pressure for the month stood at 1027.4 hPa in January 1880, the highest on record for the month over the period of available record; in January 1948, the mean MSL pressure was only 994.5 hPa, the lowest on record for any month. It is no coincidence, of course, that January 1948 remains the wettest January on record (Table 4.3). Also notable are January 1936 (mean MSLP 997.5 hPa) and January 1872 (998.8 hPa).

Wind speed and direction

Figure 4.6 shows the distribution of wind speed and direction at the 0900 GMT observation in January over the period 1961–97. January is the second-windiest month of the year, with a mean wind speed at 0900 GMT of 7.8 knots, although there is little variation in mean wind speeds between November and April. Winds from between south and south-west are dominant, while winds from between north and south-east are almost absent at this time of year. Some of the highest gusts on Durham's record have occurred in January: notable are 65 knots on 13 January 1948, and 66 knots on 31 January 1953—the night of the devastating North Sea floods, the worst natural

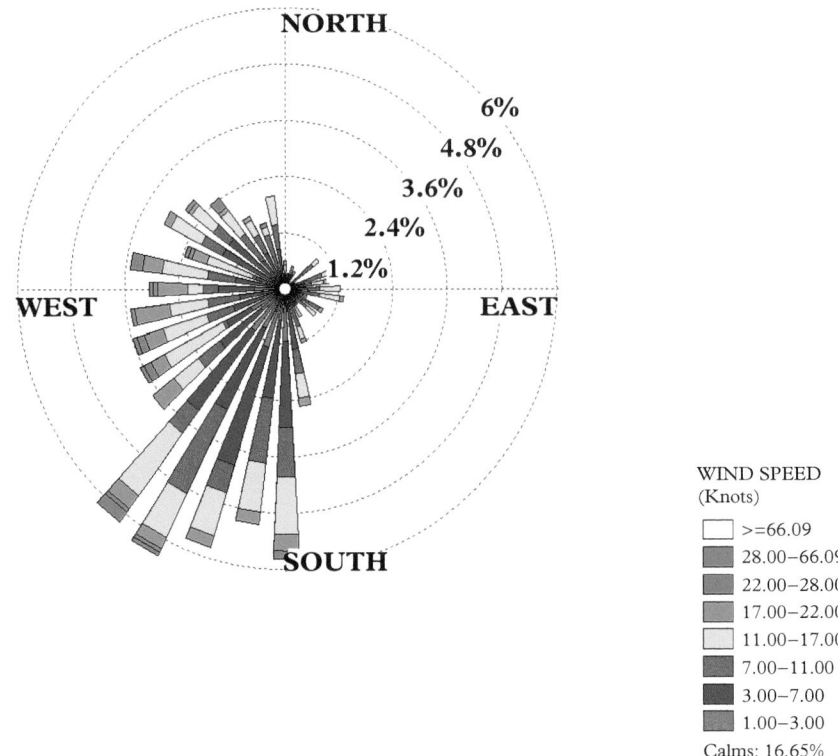

Figure 4.6 *Wind rose for 0900 GMT observations in January, period 1961–97*

disaster to befall Britain during the twentieth century, in which 307 people lost their lives [57]. More recently, the passage of storm *Malik* on the morning of 29 January 2022 resulted in a gust of 56 knots being recorded at the AWS on the roof of the Geography Department on the University's Lower Mountjoy campus, about 800 m east of the Observatory.

Durham temperature, precipitation, and sunshine in graphs—January

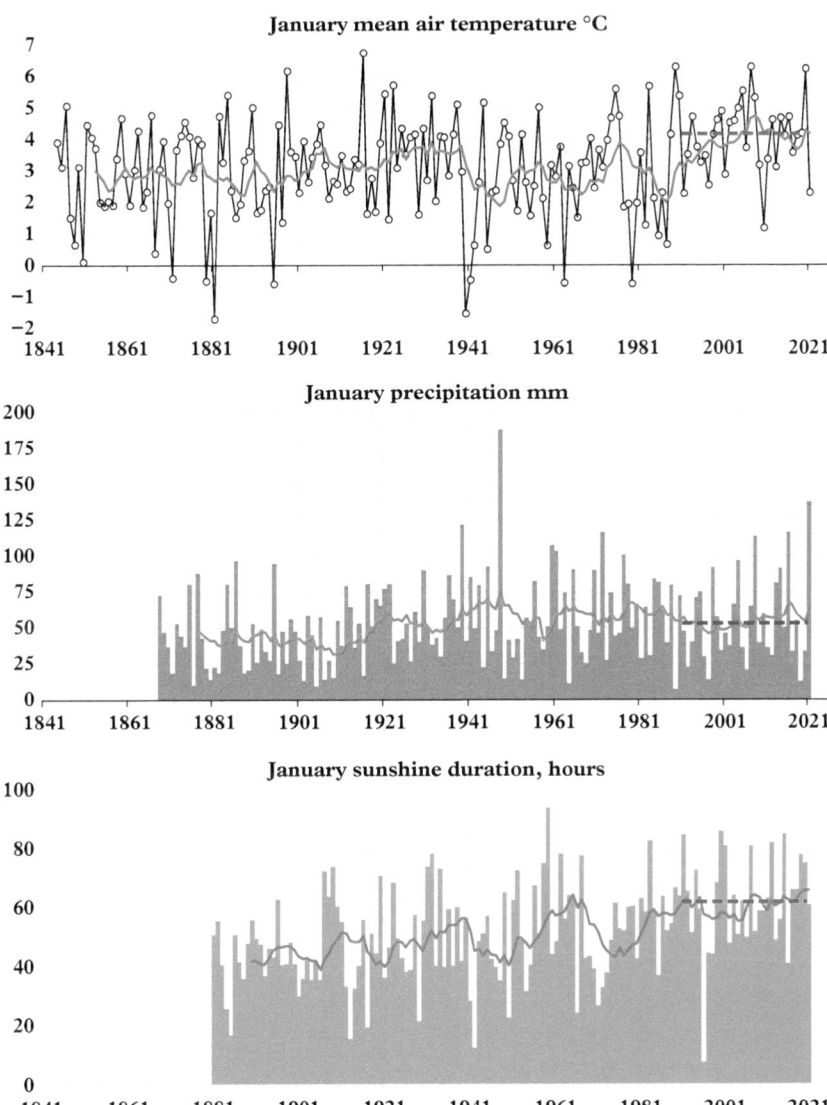

Figure 4.7 *Monthly values of (from top) January's mean temperature (° C, since 1844), total precipitation (mm, since 1868), and sunshine duration (hours, since 1881) at Durham Observatory. The 1991–2020 averages are indicated by the thick blue dashed line, while the ten-year running mean ending at the year shown is indicated by the red line*

5

February

It is in February that mean temperatures begin their seasonal rise: on average, February in Durham is just slightly less cold than January, albeit only by half a degree Celsius, and of course the picture varies from year to year. But February can still be cold, and there is often a cold snap mid-month when snow is likely. There are, on average, ten air frosts and 15 ground frosts in February, both only slightly fewer than in January despite the shorter month. February used to be the driest month of the year in Durham, but a number of wet months in recent years have shifted that title to March, and February is now only slightly drier than January when considering the mean rainfall per day. A typical February will see measurable rainfall on 16 days, three fewer than in December or January in line with the shorter month. It is in February that increasing day length and lighter evenings start to become obvious after the gloomy midwinter months. The monthly mean duration of bright sunshine is 85 hours, a shade more than three hours per day, although on average nine days in the month will remain sunless.

Temperature

Although January is, on average, the coldest month of the year in Durham, February holds the record for both the coldest night and the coldest month on Durham's records. On the morning of Friday 8 February 1895, the minimum temperature in the Glaisher stand then in use (see Appendix 3 for details) was −18.8 °C, for which the 'Stevenson screen equivalent' can be taken as −18.0 °C, with a small margin of uncertainty.* On this, Durham's coldest-ever morning, the temperature at 0900 GMT was still an extremely chilly −16.0 °C, with 6/10 cloud cover (probably little more than high cloud) and calm conditions. Figure 5.1 shows the synoptic situation at 0800 that morning; a bitter easterly airflow around an intense anticyclone over northern Scandinavia had been in place for

---

* The original manuscript entry in the Durham observations register clearly reads −1.8 °F (i.e. −18.8 °C). The Met Office MIDAS database shows this record as 2 °F (or −16.7 °C), that is, the minus sign has been omitted, and the value rounded to the nearest degree Fahrenheit. The February 1895 *Monthly Weather Report* entry for Durham shows the minimum temperature as −2 [°F] on the 8th. Thus, we believe the omission of the minus sign to be a keying error and based on the evidence of the original record and *MWR* entry we have reinstated the original value.

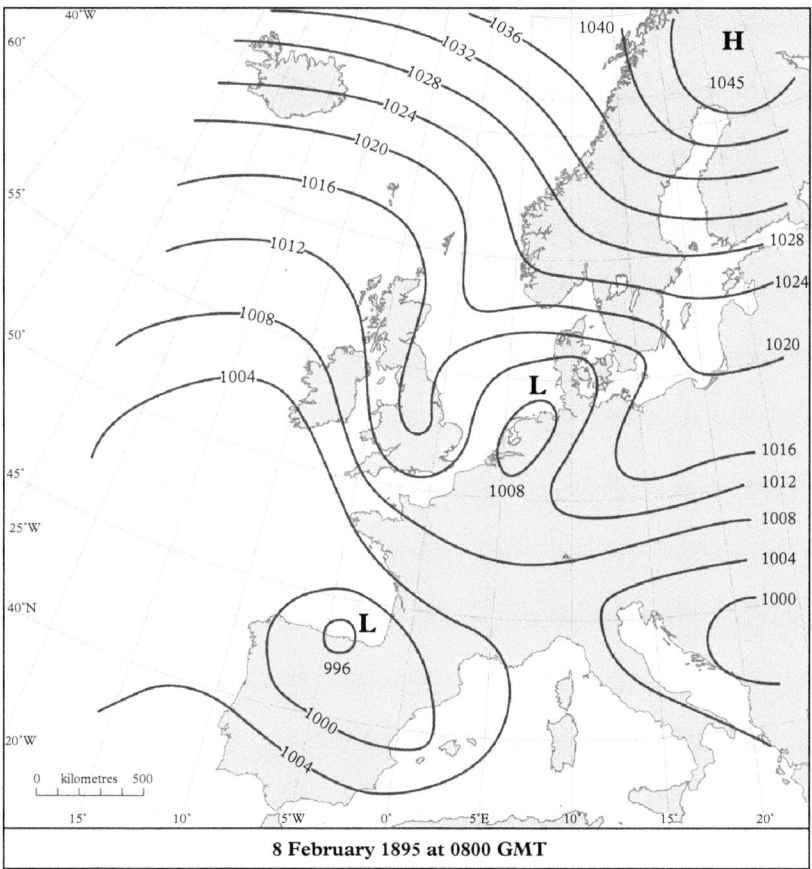

Figure 5.1 *The synoptic situation at 8 a.m. on 8 February 1895, Durham's coldest-ever morning, when the overnight temperature fell to −18° C. (Based upon the Met Office* Daily Weather Report *and the 20th Century Reanalysis version 3 (20CRv3) plot [58]; drawn by Chris Orton, Durham University)*

several days and continued to feed very cold air from the European continent, warmed only slightly by its passage over the North Sea. Around north-east England that morning, the temperature fell to −20 °C at Morpeth and −16 °C at York, but the coastal strip was slightly less cold, with 'only' −14 °C recorded at Shields. At Durham, the temperature fell below −12 °C for five consecutive nights 7–11 February 1895; these five nights contain three of the five coldest nights yet recorded in Durham (Tables 5.1 and 16.1). All February's coldest nights come from the nineteenth century as listed in Table 5.1. Since 1895, the coldest February night has been 1 February 1972, when −13.1 °C was recorded: since the turn of the twenty-first century, the coldest February night has been −9.4 °C on 12 February 2021.

Two very cold Februarys stand out in the pre-1900 period—those of 1855 and 1895. The former remains by some margin the coldest month yet recorded in Durham, with

Table 5.1 *Highest and lowest maximum and minimum temperatures at Durham Observatory in February, 1844–2021*

Rank	Mildest days	Coldest days	Mildest nights	Coldest nights
1	17.4 °C, 28 Feb 2012	*−4.3 °C, 16 Feb 1855*	11.1 °C, 5 Feb 2004	*−18.0 °C, 8 Feb 1895*
2	16.8 °C, 7 Feb 1993, 26 Feb 2019	−3.5 °C, 10 Feb 1986	10.7 °C, 8 Feb 1903	*−16.5 °C, 10 Feb 1895*
3	16.7 °C, 13 Feb 1998	−3.4 °C, 12, 13 and 15 Feb 1929	10.6 °C, 26 Feb 1945	*−14.7 °C, 19 Feb 1892*
4	16.6 °C, 14 Feb 1998	−3.3 °C, 2 Feb 1956	*10.2 °C, 6 Feb 1854*	*−14.6 °C, 16 Feb 1855*
5	16.5 °C, 27 Feb 2019	*−3.2 °C, 6 and 7 Feb 1895*	10.1 °C, 24 Feb 1990, 12 Feb 1998, and 4 Feb 2004	*−13.8 °C, 11 Feb 1895*
Last 50 years to 2021				
	17.4 °C, 28 Feb 2012	−3.5 °C, 10 Feb 1986	11.1 °C, 5 Feb 2004	−13.1 °C, 1 Feb 1972

Temperatures were recorded in different thermometer screens during the period of record; see Notes on page xvii and Appendix 3 for more details

a mean temperature about −2.2 °C (Table 5.2: the figure is a partial estimate, as observations are missing for the first half of the month: see also details in Chapter 22). The lowest recorded temperature from the available observations that month was −14.6 °C in the 'North Shed' (see Appendix 3) on the 16th. The maximum temperature that day of −4.3 °C remains the coldest February day on Durham's record, although not recorded under modern conditions: the Stevenson screen equivalent would probably be about a degree less cold. The other very cold February in Victorian times was 1895, with a mean temperature in Durham of −1.1 °C, which included the very cold spell in the second week detailed above. Further information about this very cold month is given in Chapter 22; in Durham the River Wear was frozen, allowing skating and curling on the ice (Figure 5.2). Notable also is the very cold spell in mid-February 1929, when the temperature remained below 0 °C for five consecutive days, three of which remained below −3 °C (Table 5.1), in heavily overcast conditions with a bitter south-easterly wind (Figure 5.3). The Februarys of 1947 and 1963 were of a similar degree of cold to 1895; in more recent years, the mean temperature of −0.6 °C in February 1986 was the coldest of only four months within the last 50 years (at the time of writing) to have recorded a mean temperature below 0 °C. February 1986 included the coldest February day in over 130 years, namely −3.5 °C on the 10th, a cold, foggy day with 4 cm of snow on the ground. In the tally of Durham's 12 coldest months of all time, there are five Februarys, six Januarys, and one December (Table 24.8), with February 1986 being the most recent entry at the time of writing.

Table 5.2 *February mean temperatures at Durham Observatory, 1844–2021*

February mean temperature 4.6°C (average 1991–2020)

Mildest months			Coldest months		
Mean temperature, °C	Departure from 1991–2020 normal degC	Year	Mean temperature, °C	Departure from 1991–2020 normal degC	Year
7.6	+3.0	1998	*(−2.2)*	*(−6.8)*	*1855*
6.5	+1.9	1945	−1.2	−5.8	1963
6.4	+1.8	2019	−1.1	−5.7	1947
6.3	+1.7	1990	*−1.0*	*−5.6*	*1895*
6.1	*+1.5*	*1846*	−0.6	−5.2	1986
Last 50 years to 2021					
7.6	+3.0	1998	−0.6	−5.2	1986

Figure 5.2 *Skating on the frozen River Wear at Prebends Bridge, February 1895 (courtesy of Gilesgate Archive)*

Figure 5.3 *Durham schoolboys skating on the ice near Framwellgate Bridge during the cold spell in mid-February 1929 (courtesy of Gilesgate Archive)*

With sea temperature off the north-east coast close to its lowest in February, very mild nights are few and far between, and only a few have remained above 10 °C, although these have become more frequent in recent decades. The mildest February night was on 5 February 2004, minimum temperature 11.1 °C, surpassing a record set just over 100 years earlier in 1903 (Table 5.1). Notable also is the minimum of 10.2 °C as far back as 6 February 1854, although this was recorded in the 'North Shed' exposure and over the period 2100–2100 local time, rather than today's standard 0900–0900 GMT period.

From the middle of February onwards, sunny days with mild southerly winds in anti-cyclonic conditions can produce remarkably warm days. Six February days have reached 16 °C, warmer than an average day in May: warmest of all was 17.4 °C recorded in a mild anticyclonic south-westerly airflow on 28 February 2012 (Table 5.1), but espe-cially remarkable for occurring so early in the month was 16.8 °C on 7 February 1993. More recently, sunny anticyclonic conditions saw the temperature reach 16.5 °C on 27 February 2019.

February 1998 stands out as, by some margin, Durham's mildest February (Table 5.2) with a mean temperature of 7.6 °C, 3 degrees above the 1991–2020 average, surpassing the previous record holder (February 1945) by over a degree. The month was so mild that it ranked as only the sixth coldest month of the year, ahead of January, March, April, November, and December; winds were predominantly light and from the south-west. Unusually for mild winter months, it was also very dry—the fourth driest on record (Table 5.3). Mild Februarys are more common today than in the early part of the record, but notable amongst those are those of 1846 (mean temperature 6.1 °C, 1.5 degrees above the current February average) and 1869 (mean 5.9 °C, and the third

Table 5.3 *February precipitation at Durham Observatory 1868–2021*

February mean precipitation 45.3 mm (average 1991–2020)

Wettest months			Driest months			Wettest days	
Total fall, mm	Per cent of normal	Year	Total fall, mm	Per cent of normal	Year	Daily fall, mm	Date
152.5	337	1941	2.1	5	1891	50.0	19 Feb 1941
103.6	229	2001	3.7	8	1921	27.7	26 Feb 1900
98.6	218	1966	5.7	13	1985	27.0	29 Feb 1936
94.1	208	1958	6.1	13	1998	25.0	18 Feb 1941
93.8	207	1900	8.1	18	1946	23.9	22 Feb 1967
Last 50 years to 2021							
87.2	192	2020	10.2	23	2012	23.0	25 Feb 2002

of three consecutive very mild Februarys). In a time of generally very cold winters, such mild months must have felt very unusual indeed.

Precipitation

February is usually one of the drier months of the year: during the 1981–2010 averaging period it was the driest month of the year, although in the current (1991–2020) averages that title goes to March, with 41 mm against February's 45 mm. The driest February on Durham's records was in 1891, an exceptionally anticyclonic month in which 13 per cent of all rainfall sites in England recorded no rainfall whatever (*British Rainfall 1891*); in County Durham, more than one-tenth of the rain gauges remained dry throughout the month, and the average fall across the county was just 2 mm, matching the total at the Durham Observatory (Table 5.3) where only three days recorded 0.2 mm or more, the wettest just 0.5 mm. February 1891 remains Durham's third-driest month on record (Table 25.3); another February, 1921, lies in seventh place. Oddly enough, Durham's driest February within the last 50 years was in 2012, with 10.2 mm, a year which went on to become the wettest in Durham's history (see Chapter 22).

By far the wettest February on record at Durham was 1941, with well over three times the usual amount (Table 5.3). *British Rainfall* remarks that February 1941 was rather cold, and remarkable for frequent snowfalls. The snowfall of 18-20 February that year was notably heavy in north-east England and south-east Scotland, with Durham recording 107 cm of snow—the highest figure quoted in the *British Rainfall* national report and quite possibly Durham's greatest measured snow depth; 74 cm fell at Newcastle-upon-Tyne (see also Chapter 22 for more on this event). The precipitation map for February 1941 shows remarkably high percentages along the east coast from Morayshire to Hull. Compared to the 1881–1915 average in use at the time (which was a comparatively dry period), Durham's monthly fall was 451 per cent of normal, just pipped by Middlesbrough's 455 per cent.

Only one other February since 1868 has recorded in excess of 100 mm precipitation—2001, with 104 mm. Despite being the wettest February in 60 years, and rain falling on 19 of the month's 28 days, February 2001 was also sunny, 15 per cent above average. The reason behind this apparent contradiction lies in the fact that the beginning and end of the month were wet (64 mm fell in the first seven days, including 19 mm on 4 February), followed by a mostly dry and often very sunny two weeks thereafter.

February 1941's record precipitation total can largely be attributed to the heavy snowfall event on the 18th/19th, when in two days 75 mm of rainfall equivalent was recorded (see also Chapter 22, Chronology for more details on this notable snowstorm). Durham recorded 50 mm on 19 February (a retrospective estimate for 50–60 cm of snow, as the day's maximum temperature was just 0.6 °C), while 75 mm fell at Yarm, 70 mm at Ushaw College, and 62 mm at Cockle Park. The high total on 26 February 1900 may well also have fallen as snow; *British Rainfall* mentions that snow yielded 32 mm rainfall equivalent at Auckland Castle, while Durham recorded a cold day (maximum 5 °C) with a brisk north-easterly wind. On early and less reliable records, 63.5 mm precipitation was recorded on 21 February 1853; this was almost certainly mostly snow, as temperatures remained at or below freezing for much of the 24 hours (minimum −1.7 °C, maximum 0.0 °C). More details on snow cover (Figure 5.4) and snow depths appear in the section on Wind and weather.

Figure 5.4 *Angus Macfarlane-Grieve coaching in the snow, 8 February 1923. Macfarlane-Grieve was a well-known figure in Durham with a particular interest in rowing. He was jointly Master of University and Hatfield Colleges during the Second World War. (Reproduced by permission of Durham University Library and Collections and the MacFarlane-Grieve family)*

Sunshine

Not surprisingly, given increasing day length, February is on average the sunniest winter month, with 85 hours of bright sunshine, just over three hours per day, and the sunniest days all come, again not surprisingly, towards the end of the month. Many of the sunniest days listed in Table 5.4 are from the 1990s, shortly before the daily sunshine record ceased when the automatic weather station was installed in late 1999, the sunniest February day of all being 10.2 hours on 26 February 1995 (96 per cent of the possible daylight for that date).

The end of February 1946 was remarkably sunny; the seven days commencing 23 February of that year all managed five hours or more daily, the week amounting to 53.6 hours (almost two-thirds of the current monthly average) giving a daily average of 7.7 hours. There were also four consecutive February days with six hours or more in 1899, 1937, 1996 and 1999, after which the daily record ceased. However, it is two recent Februarys that top the sunshine table: until 2008, February 1907 was Durham's sunniest with almost 116 hours of sunshine, but this was exceeded in 2008 with (an areal-based estimate) of 118 hours, beaten again by some margin in February 2019 with 128 hours. At the other end of the scale, February 1940 was the dullest on record, with little over an hour of sunshine per day. Apart from 1972, the other dull Februarys listed in Table 5.4 are all from earlier in the 20th century, quite possibly influenced by poor air quality as well as meteorological conditions. Notable among these is February 1929

Table 5.4 *February sunshine duration at Durham Observatory. Observed monthly extremes 1881 to 2021, sunniest days 1882–1999 only*

February mean sunshine duration 85 hours, 3.0 hours per day (average 1991–2020)
Possible day length: 272 hours (282 h in leap years). Mean sunshine duration as percentage of possible: 31.1

Sunniest months			Dullest months			Sunniest days 1882–1999	
Duration, hours	Per cent of possible	Year	Duration, hours	Per cent of possible	Year	Duration, hours	Date
128	46.7	2019	30.6	11.2	1940	10.2	26 Feb 1995
118	43.0	2008	32.4	11.8	1929	9.5	27 Feb 1996
115.8	42.3	1907	32.6	11.9	1923	9.4	27 Feb 1946
115.4	42.1	1943	33.4	12.2	1972	9.3	27 Feb 1943 and 28 Feb 1996
114.1	41.6	1970	34.4	12.6	1930	9.2	22 Feb 1999
Last 50 years to 2021							
128	46.7	2019	35.6	13.0	1980	10.2	26 Feb 1995

Daily sunshine records ceased at the Observatory on 31 December 1999: monthly sunshine totals from January 2000 onwards are estimated (to the nearest hour) by regression from the Met Office East and North-east England sunshine series. See Appendix 6 for record metadata and missing months.

with just 32 hours of bright sunshine, a particularly cold and cheerless month whose mean temperature of −0.4 °C ranked just outside the five coldest Februarys listed in Table 5.2.

On average, around nine days during February will remain sunless, and as with January this figure has changed little in the last 140 years. No February has ever recorded sunshine on every day in the month, although up to 1999 three (1907, 1989 and 1999) saw only one sunless day. This is in contrast to several Februarys that have had no sunshine on half or more of the days in the month—including 19 sunless days in February 1940 (including a 17-day sunless spell ending on 9 February) and 18 in the Februarys of 1947 (another bitterly cold, cheerless month) and 1980.

Wind and weather

Snow cover (1960–1999)

Over the period 1960 to 1999, there were an average of 4.3 February mornings with snow cover at the Durham Observatory; for the period from 1971–99, this decreased almost by half, to 2.8, mostly owing to the removal of 1963 from the calculation. February 1963 was the only month of that name with snow cover on every day; the winter of 1962/63 saw snow on the ground for 70 consecutive mornings from 27 December 1962 to 6 March 1963, inclusive (see also Chapter 22). In February 1963 snow lay 40 cm deep or more on 14 mornings, reaching 43 cm deep 15–18 February; this is the greatest depth of snow on the 1960–97 Durham record. Between 1960 and 1997, nine Februarys recorded at least one day with 15 cm or more of level snow—all 28 days in 1963, five days in February 1966, one in 1968, nine in 1969 (23 cm on 21st), two in 1970, three in 1978 (24 cm on 14th), one in 1983 (18 cm on 11th), one in 1986 (15 cm on 25th), and seven in 1991 (30 cm or more for three days 10–12 February, maximum depth 37 cm on 11th—Table 25.9). February 1969 and 1978 both saw snow cover on 17 mornings, and 13 in 1979; in the very cold February of 1986, snow lay on 23 mornings (Figure 5.5).

Fog (1962–97)

Over the period 1962–97, fog at 0900 GMT occurred on three mornings in a typical February; a morning of 'thick fog' could be expected about one year in two. In February 1980, fog was noted on eight mornings. As noted in January, the frequency of fog has been declining steadily since the early 1960s, following the introduction of the Clean Air Acts of the 1950s and the gradual phasing out of coal fires.

Barometric pressure (1844–1960)

The highest MSL barometric pressure on Durham's February records was 1048.2 hPa, which occurred at 0900 GMT 1 February 1902, as an intense anticyclone began to decline (the barometer had reached 1050.7 hPa at the previous evening's 2100 observation). Also notable is 1047.4 hPa at 2100 GMT on 15 February 1934, and 1047.3 hPa at 0900 GMT on 7 February 1960.

Figure 5.5 *Hazardous conditions underfoot on Durham's Kingsgate Bridge, 3 February 2003. (Tim Burt)*

The lowest barometric pressures on record for February are 956.4 hPa at 2100 GMT on 4 February 1951 (960.6 hPa at 0900 GMT the following morning), followed by 957.0 hPa at 2100 GMT on 19 February 1900. At 0900 GMT on 22 February 1914, the MSL pressure was 963.2 hPa.

February 1932 saw the highest mean MSL pressure for any month on Durham's long record; the average of the 0900 and 2100 GMT observations was an astounding 1034.6 hPa. Figure 5.6 shows the monthly mean pressure for the month, taken from the *Monthly Weather Report*. At Malin Head in the north of Ireland, the monthly mean MSL pressure reached 1035.4 hPa, the highest on record for any month in the British Isles. Perhaps surprisingly, Durham's precipitation total for the month was as much as 26 mm. Other particularly anticyclonic Februarys in Durham included 1891 (mean 1030.3 hPa, and the driest February on record—Table 5.3) and 1959 (mean 1031.6 hPa).

In contrast, the monthly mean for the Februarys of 1910 and 1937 was only 998.8 hPa, the lowest on record for the month, although neither month was particularly wet.

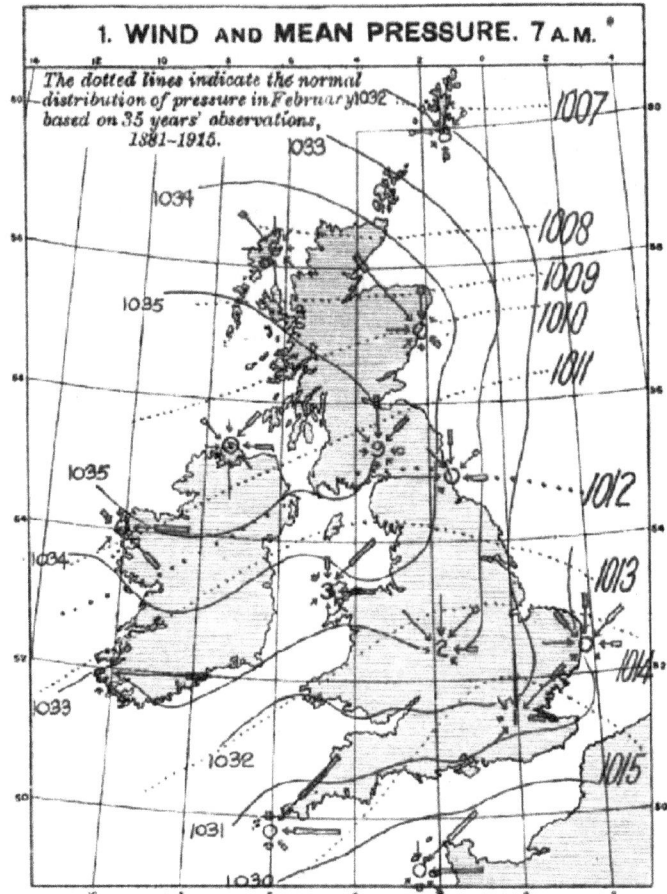

Figure 5.6 *The monthly mean MSL pressure at 0700 GMT for February 1932, the highest monthly mean on Durham's long record.* (Monthly Weather Report, *courtesy Met Office)*

Wind speed and direction

Figure 5.7 shows the distribution of wind speed and direction at the 0900 GMT observation in February over the period 1961–97. February tends to be less windy, and with a lower risk of stronger winds, than its neighbouring months of January and March; the mean wind speed at 0900 GMT is 7.2 knots. A gust of 70 knots from the west-south-west occurred at 1225 GMT on 3 February 1970, and later that month an hourly mean

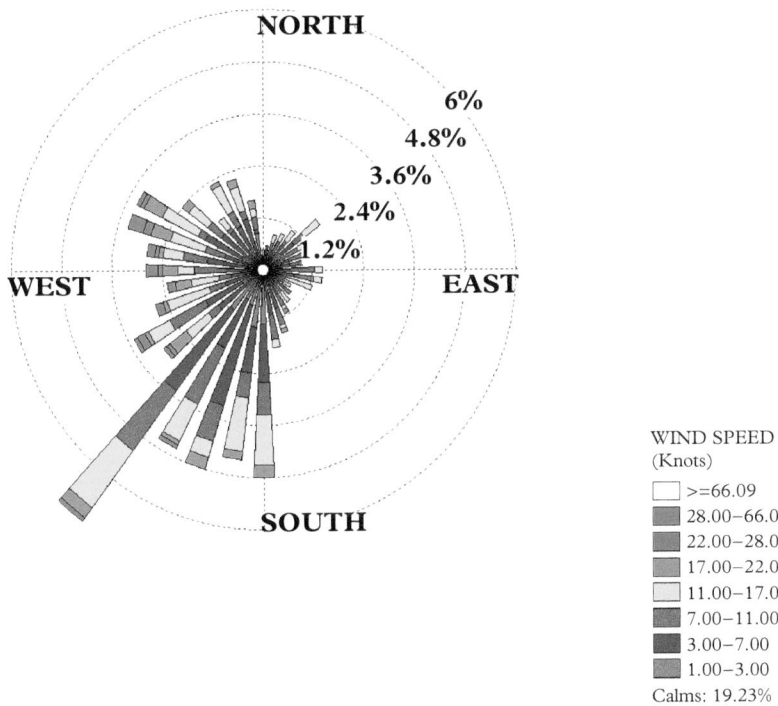

NORTH

6%

4.8%

3.6%

2.4%

1.2%

WEST

EAST

WIND SPEED
(Knots)

☐ >=66.09
■ 28.00–66.09
■ 22.00–28.00
■ 17.00–22.00
☐ 11.00–17.00
■ 7.00–11.00
■ 3.00–7.00
■ 1.00–3.00
Calms: 19.23%

SOUTH

Figure 5.7 *Wind rose for 0900* GMT *observations in February, period 1961–97*

of west-south-west 37 knots was reported on the 19th. Winds from between south and south-west remain dominant in February, although there is a slightly greater frequency of winds from between north and south-east than in January.

Figure 5.8, 'Durham Zephyrs', is a cartoon from the University archives. The caption states that it was 'Designed, Etched and Published by Edward Bradley Feby 20th 1849'. We were surprised to find that this etching clearly related to a specific event, for the Observatory registers for February 1849 show the following (Table 5.5 temperatures and pressures have been converted to modern units, MSL pressure derived from observations at station level as described in Appendix 5): Beaufort Force 9, a 'strong gale', was noted at the 0900 observations on both 18 and 19 February, the day before the etching. Force 9 corresponds to a mean wind speed around 45 knots or 23 m s^{-1}.

It is also interesting to note the topmost figure in the frame, with the mortar board, labelled 'Professor Airy'. Was this just a pun on the wind, or was it a reference to the then Astronomer Royal, George Airy?

Figure 5.8 *'Durham Zephyrs', by Cuthbert Bede (Edward Bradley), February 1849 (reproduced with permission from Durham University Library and Heritage Collections)*

Table 5.5 *Meteorological observations made at Durham Observatory, February 1849, at the time of Bradley's 'Durham Zephyrs'*

Date February		MSL pressure	Dry bulb	Min temp	Max temp	Wind dir'n and speed	Cloud cover	
1849	Time	hPa	°C	°C	°C	Bft Force	tenths	Weather notes
17	0900	1034.4	4.8			SSE 1	0	Fair
	2100	1026.8	5.1	2.1	9.4	WSW 5	1	
18	0900	1021.1	9.2			**W 9**	1	Fine
	2100	1022.4	8.9	3.9	11.9	WSW 7	9	
19	0900	1005.6	8.7			**WSW 9**	7	Fine, mostly cloudy
	2100	1009.6	5.0	8.2	10.0	W 7	0	Slight aurora
20	0900	1007.6	4.7			SW 3	10	Generally overcast
	2100	1003.4	3.1	2.5	5.8	WNW 3	0	Quite clear—aurora

Durham temperature, precipitation, and sunshine in graphs—February

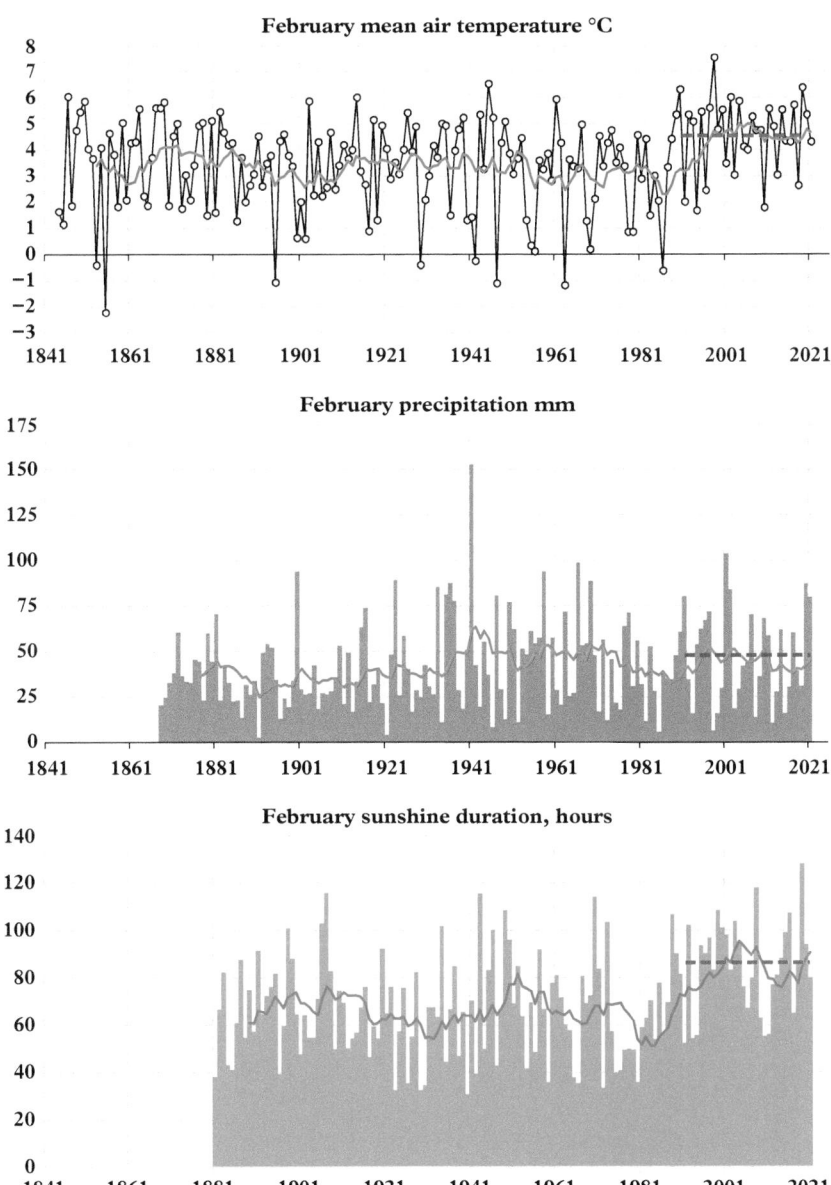

Figure 5.9 *Monthly values of (from top) February's mean temperature (°C, since 1844), total precipitation (mm, since 1868), and sunshine duration (hours, since 1881) at Durham Observatory. The 1991–2020 averages are indicated by the thick blue dashed line, while the ten-year running mean ending at the year shown is indicated by the red line*

6

March

___

In most years, March marks the transition from winter to spring. The old saying 'March comes in like a lion and goes out like a lamb' often has more than a grain of truth in it, as March in Durham has the highest range in temperature of any month in the year. Despite year-to-year variations, on average there is marked warming throughout the month: during the current 30-year averaging period of 1991–2020, mean daily maximum temperatures rise from below 8 °C at the beginning of the month to 11.5 °C in the closing days. The weather can often feel very changeable in March, and snowfalls and warm days can occur within in the same month—sometimes within the same week! Over the most recent 30-year averaging period, March has become the driest month of the year in Durham, taking the title held by February in the 1981–2010 period. Compared to February, there is almost an extra hour of bright sunshine each day.

Temperature

In Durham's long record, March has experienced the widest range in temperature of any month (Figure 6.1). Particularly at the beginning of the month, cold northerly or easterly winds can bring heavy snowfalls and temperatures almost as low as the depths of winter, whereas longer days, stronger sunshine, and milder southerly winds later on in the month can produce remarkably warm conditions the equal of many a midsummer afternoon. Very occasionally, both extremes occur within the same month, as happened in March 1929 and in 1965.

Table 6.1 shows the highest and lowest temperatures recorded in March since 1844. The coldest March night came towards the end of the very cold and snowy winter of 1947, a difficult period so soon after the end of the Second World War when austerity still prevailed. There were 48 consecutive air frosts from 28 January to 16 March that year, including the coldest March night of −15.0 °C on the 4th and one of the lowest grass (or probably snow surface) minimum temperatures on the Durham record at −19.4 °C. Later in the day, with the benefit of almost four hours of sunshine, the temperature rose to 3.3 °C, which must have seemed positively mild after conditions over the preceding six weeks or so. There was a similarly cold start to March in 1963 at the tail end of that

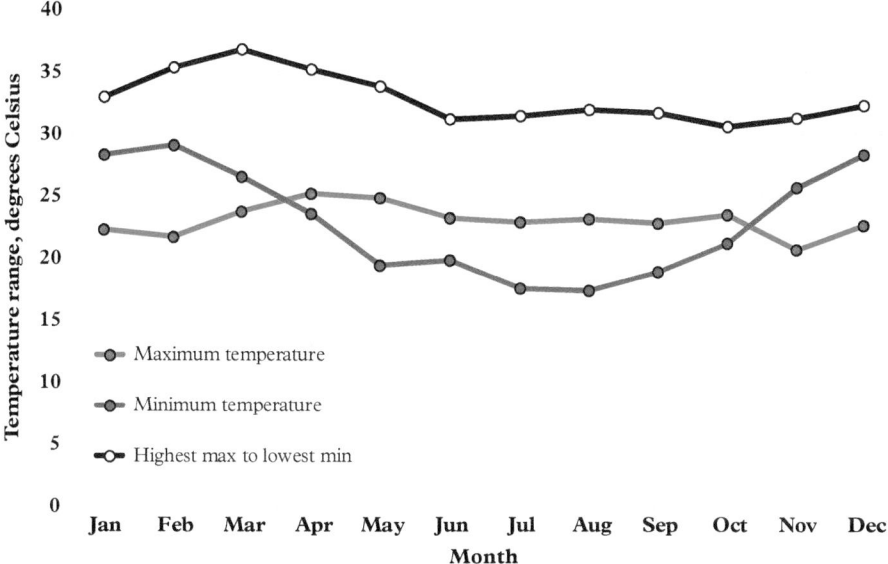

Figure 6.1 *The range in maximum (red line) and minimum (blue line) temperatures by month in Durham over the entire period of record. The black line shows the absolute range, from the highest maximum to the lowest minimum, which is greatest of all in March, at a remarkable 36.8 degrees Celsius*

long, snowy winter (see Chapter 22), with the temperature falling to −11.7 °C on the 2nd.

Ice days—those when the maximum temperature remains below 0 °C—are rare but not unknown in March, although they have become rarer still in recent decades. Most of the coldest March days listed in Table 6.1 occurred in the nineteenth century, but notable are maxima of −1.7 °C as late as 6 March 1942, −0.3 °C on 17 March 1979, and most recently −0.1 °C on 1 March 2018.

To date, ten March days have reached or surpassed July's average maximum temperature of 20.2 °C (Table 6.1). The earliest day in the year to reach 20 °C was 8 March 1929, when the afternoon temperature reached 20.4 °C (Table 24.1). The minimum temperature that morning was −2.5 °C, and the daily temperature range—an extraordinary 22.9 degrees Celsius—remains the second-highest on Durham's long record (Table 24.9), aided by dry ground conditions and almost ten hours of sunshine. Coming only three weeks after severe cold in mid-February of that year, it must have seemed particularly out of season. The warmest day in March to date was 21.8 °C on 28 March 2012, occurring exactly a month to the day after February's record warmth that year. Such warmth in late March, with the trees still largely without leaves and the strength of the Sun greater than in late September and increasing daily, can feel quite surreal.

Table 6.1 *Highest and lowest maximum and minimum temperatures at Durham Observatory in March, 1844–2021*

Rank	Mildest days	Coldest days	Mildest nights	Coldest nights
1	21.8 °C, 28 Mar 2012	−1.9 °C, 18 and 19 Mar 1846, and 8 Mar 1858	11.5 °C, 5 Mar 1859	−15.0 °C, 4 Mar 1947
2	21.7 °C, 24 Mar 1953 and 29 Mar 1965	−1.7 °C, 6 Mar 1942	11.1 °C, 8 Mar 1981	−12.2 °C, 3 Mar 1965
3	21.6 °C, 27 Mar 1929, 28 Mar 1929, and 27 Mar 2012	−0.8 °C, 18 Mar 1853	10.8 °C, 31 Mar 2017	−11.7 °C, 2 Mar 1963
4	20.8 °C, 31 Mar 2021	−0.7 °C, 4 Mar 1845 and 9 Mar 1931	10.6 °C, 30 Mar 1938	−11.6 °C, 9 Mar 1917
5	20.4 °C, 8 Mar 1929, 29 Mar 1929, and 26 Mar 2012	−0.6 °C, 9 Mar 1958	10.5 °C, 16 Mar 2004 and 27 Mar 2006	−11.3 °C, 4 Mar 1889
Last 50 years to 2021				
	21.8 °C, 28 Mar 2012	−0.3 °C, 17 Mar 1979	11.1 °C, 8 Mar 1981	−10.8 °C, 3 Mar 2001

Temperatures were recorded in different thermometer screens during the period of record; see Notes on page xvii and Appendix 3 for more details.

In 2021, a summer-like 20.8 °C was reached on 31 March—followed by a return to winter with a maximum of 7.8 °C on 1 April.

As well as the greatest range in temperature of any month, one March—1965—produced the greatest range in temperature of any month on Durham's records. The month began particularly cold, with heavy snowfalls: on 3 March Durham recorded a minimum temperature of −12.2 °C, the lowest of the 'winter'. Milder conditions returned mid-month, followed by another cold week with frosts—and then sudden warmth towards the end of the month. On 28 March the temperature reached 15.6 °C, but the following day, aided by 11.3 hours of sunshine, 21.7 °C was recorded—giving a range in temperature for the month of a remarkable 33.9 degrees Celsius. March 1965 is thus the only entry to appear on both sides of Table 6.1. In a particularly cruel 'tease', the summer of 1965 was very cool, dull, and wet, and only three days that year subsequently surpassed the temperature reached on 29 March—the warmest day in July managing just 21.1 °C, cooler than March's warmest day. (Similarly, 21.4 °C on 24 March 1953 was another seasonal reversal when on 2 June that year, the Coronation Day of Queen Elizabeth II, the maximum temperature reached just 7.2 °C, the first of four cold, sunless days.)

Table 6.2 shows that Durham's warmest March remains that of 1938, although several recent Marches have closely approached 1938's figure. March 1938 was also the third driest on record (Table 6.3). March 1883 remains the coldest with a mean temperature of 1.7 °C (Stevenson screen equivalent), fractionally colder than March 1947. In recent years, March 2013 was only 0.6 degrees Celsius less cold than March 1947.

It is not unusual for March to be colder than February (Figure 6.2). This has happened 39 times in 177 years, or on average about once every five years, most recently in 2013. The largest reversal came in 1867 when February had a mean air temperature of 5.6 °C but March only 2.2 °C. February 1867 was a mild month with only three air frosts, which fell on the last three days of the month; there were nine days with maxima above 10 °C. March was dismal by comparison: 16 air frosts and no day above 10 °C until the 24th—the maximum on the 13th was a miserable 0.5 °C. Whenever February has been warmer than March, a relatively mild February heralding spring only resulted in those hopes being dashed in March. It may just be happening a little less often now: there were 16 years when this occurred between 1844 and 1900, and 21 times in the

Table 6.2 *March mean temperatures at Durham Observatory, 1844–2021*

March mean temperature 6.2 °C (average 1991–2020)

Mildest months			Coldest months		
Mean temperature, °C	Departure from 1991–2020 normal degC	Year	Mean temperature, °C	Departure from 1991–2020 normal degC	Year
9.3	+3.1	1938	*1.7*	−4.5	*1883*
8.7	+2.5	2012	1.8	−4.4	1947
8.1	+1.9	1961, 1990	2.0	−4.2	*1853*, 1916, 1962
7.9	+1.7	1997, 2017	2.2	−4.0	1917, 1969
7.8	+1.6	1957	*2.3*	−3.9	*1855, 1867*
Last 50 years to 2021					
8.7	+2.5	2012	2.4	−3.8	2013

Figure 6.2 *St Chad's College crew carrying their boat in the snow, March 1923. (Reproduced by permission of Durham University Library and Collections and the MacFarlane-Grieve family)*

twentieth century, but only in 2006 and 2013 this century to date—statistically, it is probably too soon to tell.

Precipitation

In the current set of 30-year averages (1991–2020), March has overtaken February as the driest month of the year, with an average of just 41 mm. Measurable rain can be expected to fall on 15 days, one day less than February but one day more than April.

Table 6.3 sets out the extremes of March precipitation at Durham since 1868. Monthly totals have exceeded 100 mm only four times in that span, and by far the wettest of all—and the only month to have received four times the normal rainfall based upon current averages—was March 1979, with 166 mm. This was a cold, wet, and cheerless month, with snow lying on ten mornings (maximum depth of 31 cm on the 18th and 19th): two days in this month account for the wettest and third-wettest March days on record, notably 33.9 mm on the 17th (almost all as snow, with a maximum temperature of −0.3 °C) and 50.6 mm on 28th (most of which probably fell as cold rain or wet sleet, the temperature varying between 4.1 °C and 0.4 °C). The second-wettest

Table 6.3 *March precipitation at Durham Observatory 1868–2021*

March mean precipitation 41.3 mm (average 1991–2020)

Wettest months			Driest months			Wettest days	
Total fall, mm	Per cent of normal	Year	Total fall, mm	Per cent of normal	Year	Daily fall, mm	Date
166.0	402	1979	1.3	3	1953	50.6	28 Mar 1979
113.5	275	1981	4.2	10	1929	35.8	29 Mar 1888
108.1	262	1909	7.6	18	1938	33.9	17 Mar 1979
103.4	250	1888	9.1	22	1973	31.7	21 Mar 1981
97.9	237	1947	9.6	23	1966	28.1	25 Mar 1877
Last 50 years to 2021							
166.0	402	1979	9.1	22	1973	50.6	28 Mar 1979

March day, 35.8 mm on Maundy Thursday, 29 March 1888, was probably similar, with temperatures ranging only between 0 and 5 °C, and a brisk cyclonic north to north-easterly wind as a particularly deep depression, below 968 hPa, passed not far to the south. In March 1916 rain fell on 27 days, although the month does not figure in even the top 30 wettest Marches.

In contrast, dry Marches are not uncommon: Chaucer mentions the 'drought (droghte) of March' at the start of his General Prologue to the *Canterbury Tales*. About one March in five will receive only half the normal precipitation, or about 20 mm, and with warming days and strengthening sunshine, and sometimes boisterous March winds, this meagre amount can quickly be lost by evaporation and soils can dry out rapidly. The extremely anticyclonic March 1953 remains the driest month of any on the post-1868 record in Durham (Tables 6.3 and 25.3), with just 1.3 mm in total, all of which fell on 30 March, ending a long dry spell of 41 days comprising just 0.3 mm rainfall. March 1929 was also dry, and it is no coincidence that both months recorded high temperatures (Tables 6.1 and 6.3) and persistent high barometric pressure (see Barometric pressure section).

Sunshine

The duration of bright sunshine in March averages 120 hours, equivalent to 3.9 hours per day. As with precipitation, there is a huge range in monthly sunshine totals (Table 6.4), with a factor of six between the sunniest and dullest Marches on record. Dealing with the dullest first, all but one have occurred since 1960, with March 1996 the leader with a mere 29 hours. After such a poor January for sunshine, March 1996 must have been a great disappointment at a time when people look forward to spring sunshine after the gloom of winter: as well as being very dull indeed, it was milder than February but colder than January. Only in March 1964 and 1996 have sunshine totals failed to attain even one-tenth of the possible duration.

Table 6.4 *March sunshine duration at Durham Observatory. Observed monthly extremes 1881 to 2021, sunniest days 1882–1999 only*

March mean sunshine duration 120 hours, 3.9 hours per day (average 1991–2020)
Possible day length: 368 hours. Mean sunshine duration as percentage of possible: 32.6

Sunniest months			Dullest months			Sunniest days 1882–1999	
Duration, hours	Per cent of possible	Year	Duration, hours	Per cent of possible	Year	Duration, hours	Date
191.4	52.0	1894	29.3	8.0	1996	11.8	29 Mar 1907 and 31 Mar 1997
187.1	50.8	1907	33.1	9.0	1964	11.6	30 Mar 1997
180	49.0	2012	48.2	13.1	1928	11.5	30 Mar 1892
172	46.9	2003	51.5	14.0	1984	11.4	21 Mar 1968
166.3	45.2	1893	53.5	14.5	1960	11.3	31 Mar 1883, 28 Mar 1910, and 29 Mar 1965
Last 50 years to 2021							
180	49.0	2012	29.3	8.0	1996	11.8	31 Mar 1997

Daily sunshine records ceased at the Observatory on 31 December 1999: monthly sunshine totals from January 2000 onwards are estimated (to the nearest hour) by regression from the Met Office East and North-east England sunshine series. See Appendix 6 for record metadata and missing months.

In terms of greatest March sunshine, 1894 remains at the top of the list, with 191 hours, 52 per cent of possible; this followed another very sunny March in 1893. More recently, 2012 managed an estimated 180 hours to take third place in Table 6.4. As expected, the sunniest days on record all occur towards the end of the month: 11.8 hours was recorded on 29 March 1907 and equalled on 31 March 90 years later. In contrast, March 1969 and March 1996 both saw runs of 11 consecutive days without sunshine, a record for any spring month. The sunniest spells in March since 1882 have included 18–26 March 1893 (nine days, 80.8 hours of sunshine, an average of 9.0 hours per day), 22–30 March 1894 (also nine days, 80.2 hours, average 8.9 hours daily) and 22–28 March 1933 (seven days, 66.3 hours, average 9.5 hours daily).

Wind and weather

Snow cover (1960–1999)

Over the period 1960 to 1999, there were an average of 2.1 mornings in March with snow cover (Figure 6.3) at the Durham Observatory; for the period from 1971–99, this decreased to 1.6 mornings. The winter of 1962/63 saw snow on the ground for 70 consecutive mornings from 27 December 1962 to 6 March 1963, inclusive (see also Chapter 22). In March 1979, snow lay for ten days, at the end of this cold, wet, protracted

Figure 6.3 *Snowy view from the A66. Compared to Durham, snow cover is much more frequent and long-lasting in the Pennine hills (Tim Burt)*

winter, and eight mornings in March 1965 and 1969. The greatest depth of snow in March during this period was on 1 March 1963, when it lay 38 cm deep; the snow was still 20 cm deep on 6 March but had reduced to below 50 per cent cover by the following morning. On 1 March 1965, the snow depth was 33 cm, and 31 cm on 18 and 19 March 1979. County Durham also suffered heavy snowfalls in early March 2018.

Fog (1962–97)

Over the period 1962–97, fog at 0900 GMT occurred on two mornings in a typical March, with thick fog occurring about one year in two. In March 1964, and again in 1972, fog was noted on seven mornings. As previously commented upon, the frequency of fog has been declining steadily since the early 1960s because of the Clean Air Acts of the 1950s and the gradual phasing out of coal fires, and several recent Marches have passed without any morning fogs.

Barometric pressure (1844–1960)

The highest MSL barometric pressure on Durham's March records occurred in 1852, when the barometer stood at 1044.9 hPa at 0900 on 6 March; this value was equalled at the 0900 GMT observation on 9 March 1935. Almost as high was the 1044.8 hPa at 1000 on 3 March 1867, equalled at 2100 GMT on 9 March 1953. The lowest barometric pressures on record for March occurred back in 1876–963.5 hPa at the 2200 observation on 9 March, followed by 962.3 hPa at 1000 the following morning.

In March 1953, the mean MSL pressure for the month was 1029.9 hPa; not surprisingly, this was Durham's driest-ever March (Table 6.3). Some way behind was March 1929, with a monthly mean of 1027.4 hPa: fittingly, Durham's second-driest March to date. In contrast, the mean pressure for March 1876 stood at only 997.6 hPa, and 999.9 hPa in March 1914. Neither month was very wet—March 1876 was only slightly wetter than normal, and March 1914 was actually quite dry.

Wind speed and direction (1961–97)

March is the windiest month of the year in Durham; the mean wind speed at 0900 GMT is 8.2 knots. March 1967 was the windiest month on Durham's records during this period, with a mean wind speed at 0900 GMT of 18.2 knots; wind speeds exceeded gale force on two mornings, the 2nd (mean speed 35 kn) and the 17th (40 kn). The latter date saw a mean wind of 300° (west-north-westerly) 41 knots ended at 1100 GMT, including one gust in that hour of 72 knots. The month was exceptionally windy across the whole country, as the monthly mean isobaric map makes clear (Figure 6.4); compare this with February 1932 in February's chapter (Figure 5.5). Figure 6.5 shows the distribution of wind speed and direction at the 0900 GMT observation in March over the period 1961–97.

Winds between south-west and north-west remain dominant in March, and are sometimes strong to gale force, but there is a somewhat greater frequency of winds from between north and south-east than in January or February.

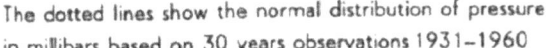

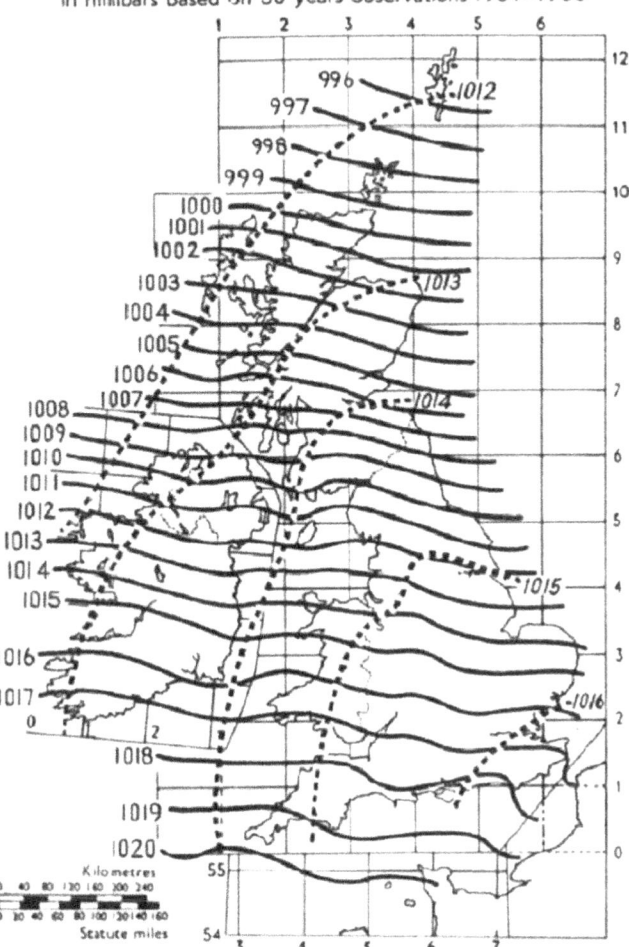

Figure 6.4 *The monthly mean* MSL *pressure at 0900 GMT for March 1967, one of the windiest months on record across the United Kingdom (from the* Monthly Weather Report, *courtesy Met Office)*

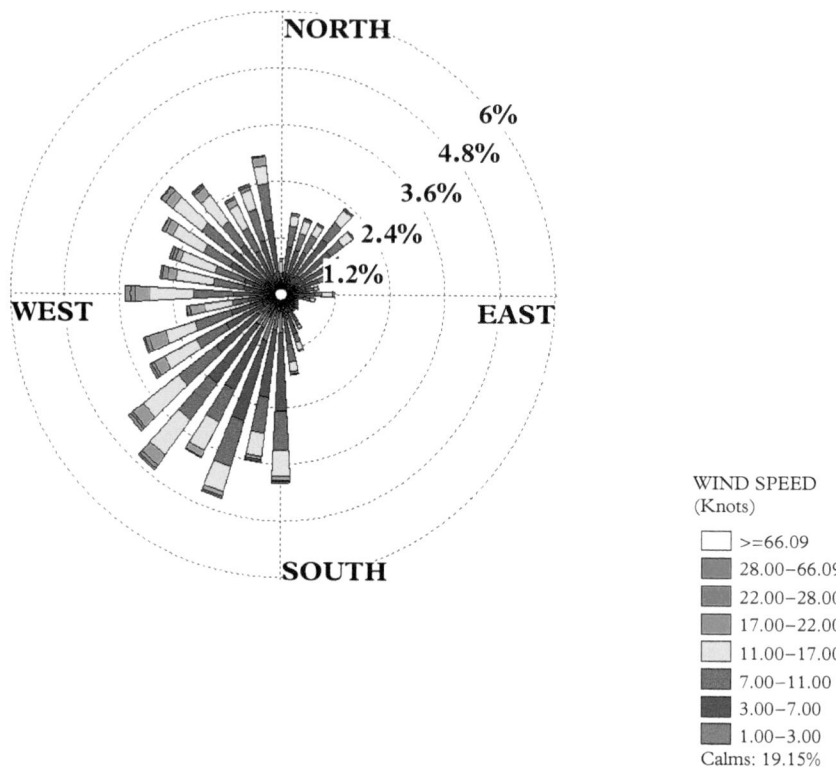

Figure 6.5 *Wind rose for 0900* GMT *observations in March, period 1961–97*

Durham temperature, precipitation, and sunshine in graphs—March

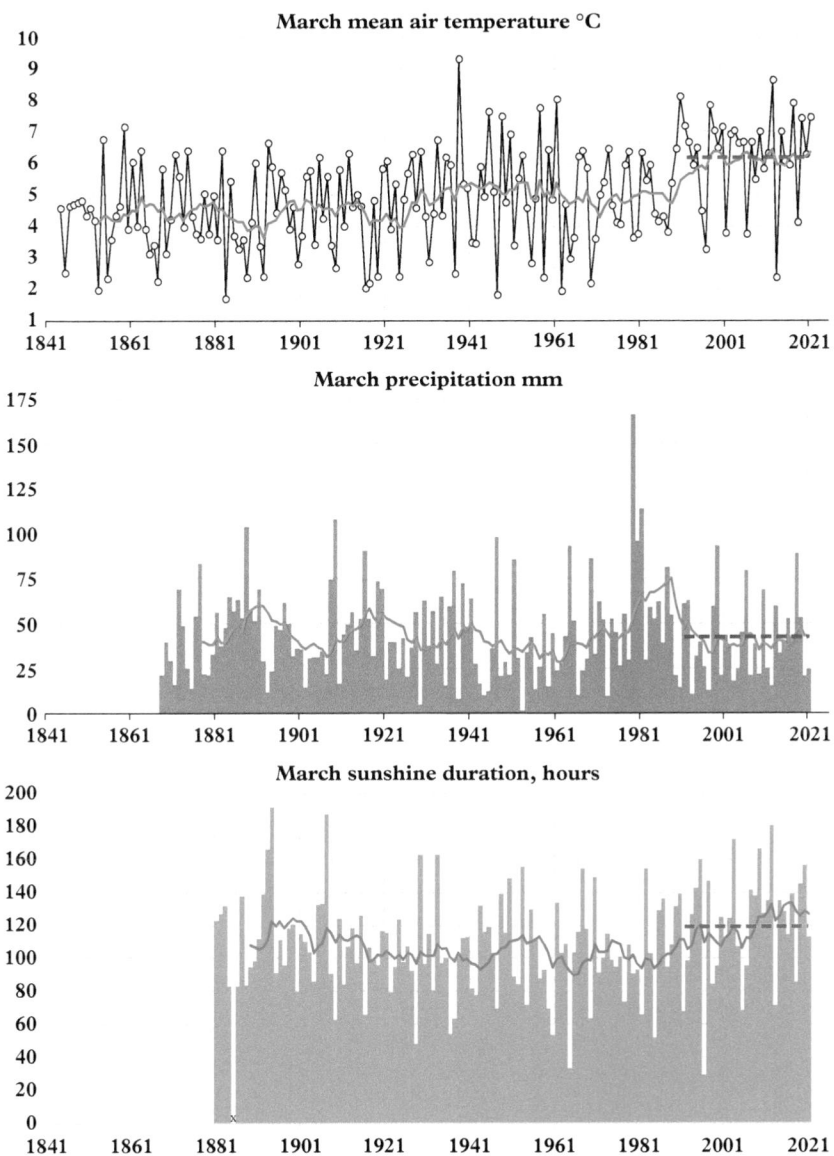

Figure 6.6 *Monthly values of (from top) March's mean temperature (°C, since 1844), total precipitation (mm, since 1868), and sunshine duration (hours, since 1881, missing data 1885) at Durham Observatory. The 1991–2020 averages are indicated by the thick blue dashed line, while the ten-year running mean ending at the year shown is indicated by the red line*

7

April

The middle month of spring, we expect April to be warmer than March if cooler than May, but the rate of warming can be disappointing and on average April is closer to the former than the latter; Figure 3.1 confirms the slow rate of temperature increase through the central part of spring with acceleration not normally until May. A normal April is wetter than February and March, almost as wet as January; other than a drier May, all other months to the end of the year are wetter. With longer day length, average sunshine increases by more than an hour per day compared to March. Snowfall is not uncommon, and frosts remain a problem: there are on average three air frosts and ten ground frosts during the month. In the standard averaging period 1991–2020, only one April (2007) escaped without recording an air frost—gardeners beware!

Temperature

Since 1844, air temperatures in April have varied from −11.1 °C on 2 April 1917 to 24.1 °C on 16 April 2003 (Table 7.1). Clear days can produce welcome sunshine and warmth, but clear nights can bring frost, particularly if the ground is dry. Air frosts can be expected at any time during the month (three in an average April, but 16 in 1917, 15 in 1906 and 1922, and 12 in 2021), and about one night in three will experience a ground frost—an ongoing problem for gardeners as the growing season progresses.

April 1917 was Durham's coldest April on record, and clearly deserves some comment. The month began very cold over the whole country, with deep snowfalls in some places; in Durham the temperature rose to just −1.1 °C on the first day of the month, still the coldest April day on record, and the only April day on Durham's long record to remain below freezing throughout (Figure 7.1). The following night was clear and calm, and over a snow cover the screen minimum fell to an extraordinary −11.1 °C, a value rarely attained even on the coldest winter nights; the grass (or snow) minimum fell to −17.8 °C, the fifth-lowest such value on the entire record, all winter nights included. That same morning −15.4 °C was recorded at Eskdalemuir Observatory in the Scottish uplands, a value which remains the lowest April temperature on record in the UK. Durham's maximum temperature of −1.1 °C on 1 April was followed by one of only 0.7 °C on the 2nd, the two coldest April days on record (Table 7.1). Snow lay at Durham

Table 7.1 *Highest and lowest maximum and minimum temperatures at Durham Observatory in April, 1844–2021*

Rank	Warmest days	Coldest days	Mildest nights	Coldest nights
1	24.1 °C, 16 Apr 2003	−1.1 °C, 1 Apr 1917	12.5 °C, 28 Apr 1994	−11.1 °C, 2 Apr 1917
2	23.9 °C, 15 Apr 1949	0.7 °C, 2 Apr 1917	12.4 °C, 25 Apr 2007	−8.0 °C, 24 Apr 1908
3	23.7 °C, 19 Apr 2018	*1.1 °C, 17 Apr 1849*	12.2 °C, 24 Apr 2007	−7.2 °C, 6 Apr 1917
4	23.3 °C, 16 Apr 1949 and 16 Apr 2007	1.7 °C, 7 Apr 1905 and 5 Apr 1911	11.9 °C, 23 Apr 2007	*−6.5 °C, 16 Apr 1892*
5	23.1 °C, 21 Apr 1900	2.2 °C, 3 Apr 1958	11.4 °C, 19 Apr 2018	*−5.9 °C, 4 Apr 1881*
Last 50 years to 2021				
	24.1 °C, 16 Apr 2003	3.6 °C, 7 Apr 1986	12.5 °C, 28 Apr 1994	−4.4 °C, 14 Apr 1999

Temperatures were recorded in different thermometer screens during the period of record; see Notes on page xvii and Appendix 3 for more details

on 11 mornings in April 1917: in Northumberland, hundreds of lambs and sheep died in the cold and snow. The *Monthly Weather Report* headline for the month read 'Abnormal and persistent cold. Frequent snow, then drought'. No day was warmer than 10 °C until the 19th, although all but one of the remaining days passed this threshold. There were air frosts on 16 of the first 18 days, the same for ground frosts; in the latter part of the month there were no air frosts and just two ground frosts.

The minimum of −8.0 °C on 24 April 1908 is remarkable for so late in the spring, the latest date in the year on which the minimum temperature has fallen below −5 °C (Table 24.1). More recently, the minimum of −3.2 °C on 10 April 2021 made this the coldest April night for more than 20 years. In contrast, three consecutive nights 23–25 April 2007 remained at or close to 12 °C.

The warmest Aprils at Durham (Table 7.2) have all occurred since 1949, with April 2007 exceeding 1949's record warmth only to be surpassed itself just four years later by the exceptionally warm, dry and sunny April in 2011. The mean daily maximum temperature that month, 16.5 °C, was 4.0 degrees above the April average, and more than a degree above the May normal. Only on the 13th was the maximum temperature below the normal for the month; four days had maximum temperatures above 20 °C, and twenty reached 15 °C. The mildest night, 11.1 °C on 6 April 2011, was

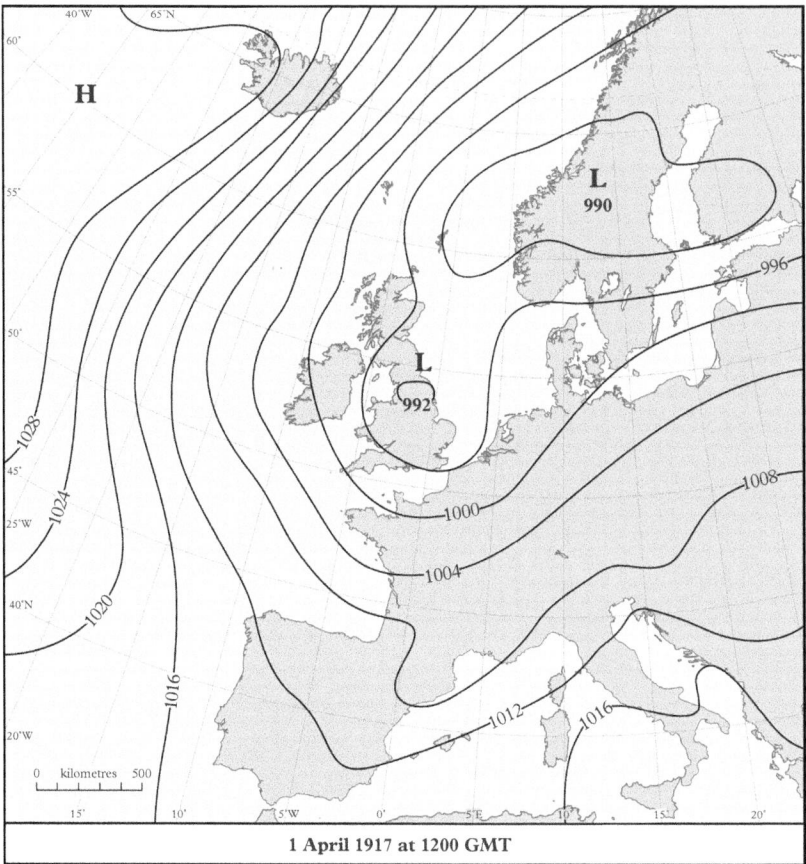

Figure 7.1 *The synoptic situation at noon* GMT *on 1 April 1917, the coldest April day on Durham's record: the temperature remained below freezing all day and fell to −11°C the following morning. (Based upon the 20th Century Reanalysis version 3 (20CRv3) plot [58] and the Met Office* Daily Weather Report; *drawn by Chris Orton, Durham University)*

a remarkably warm night for April, ranking equal sixth warmest and just outside the leaders in Table 7.1. April 2011 was also a very dry month, and one of the sunniest.

April 2007 was the warmest on record to that date and would still be regarded as exceptional if not having been beaten by 2011. It remains only the fifth April since 1948 to have no air frosts. Only three April days have been warmer than the highest temperature that month, namely 23.3 °C on the 16th, while there were six minima above 10 °C and 18 maxima above 15 °C. Very high temperatures in April 2007 and 2011 reflect a notable warming in April since the early 1970s (Figure 7.2); ten of the top twenty warmest Aprils have occurred since and including 2002, whereas the most recent of the 'bottom 20' coolest Aprils was back in 1986. April 2021 showed that warming is not unbroken, however: its mean temperature of 6.4 °C made this the coldest April since

Table 7.2 *April mean temperatures at Durham Observatory, 1844–2021*

April mean temperature 8.3 °C (average 1991–2020)

Warmest months			Coldest months		
Mean temperature, °C	Departure from 1991–2020 normal degC	Year	Mean temperature, °C	Departure from 1991–2020 normal degC	Year
11.2	+3.9	2011	4.4	−3.9	1917
10.7	+3.4	2007	4.5	−3.8	1922
9.8	+1.5	1949	4.9	−3.4	1903
9.6	+1.3	2014	5.0	−3.3	1879, 1891, 1908
9.5	+1.2	2004	5.1	−3.2	1986
Last 50 years to 2021					
11.2	+3.9	2011	5.1	−3.2	1986

1989. The month's 12 air frosts—more than some recent entire winter seasons—was the highest for the month since 1922, and its mean minimum temperature of just 0.7 °C was the lowest since 1956. Clear skies and polar air made for a cold month, but also a sunny one—the second-sunniest April on record (Table 7.4).

Precipitation

With an average monthly fall of just under 40 mm during the 1941–70 average period, April became the second-driest month of the year (behind March), but several wet Aprils in recent decades have bumped up the month's normal fall to 51 mm for the current 1991–2020 average, an increase of 28 per cent since 1970. Monthly totals are very variable, however, and recent years have included three very wet months (April 1998, 2000 and 2012 are the three wettest on record) and three very dry (1997, 2011 and 2020—ranking fourth, sixth, and third driest, respectively)—Table 7.3. Three Aprils feature in the ten driest months on Durham's records (1938, 1912, and 2020, Table 25.3), more than any other month.

Large falls of rain are rare in April. The fall of 43 mm on 1 April 1992 was a result of many hours of cold rain turning to sleet and snow, from near-stationary fronts to the north of a depression located over southern England; the day's maximum temperature was just 4.8 °C, and snow lay 3 cm deep next morning. Similar conditions may have prevailed during the falls of 40 mm on 15 April 2005 and 34 mm on 21 April 1872, the second- and third-wettest April days, for the maximum temperature on both days was

Table 7.3 *April precipitation at Durham Observatory 1868–2021*

April mean precipitation 51.2 mm (average 1991–2020)

Wettest months			Driest months			Wettest days	
Total fall, mm	Per cent of normal	Year	Total fall, mm	Per cent of normal	Year	Daily fall, mm	Date
151.0	295	1998	2.2	4	1938	43.0	1 Apr 1992
149.8	293	2000	2.4	5	1912	39.8	15 Apr 2005
134.4	263	2012	3.8	7	2020	33.8	21 Apr 1872
109.5	214	1986	5.8	11	1997	32.2	23 Apr 1971
106.5	208	1877	6.9	13	1957	31.7	21 Apr 1935
Last 50 years to 2021							
151.0	295	1998	3.8	7	2020	43.0	1 Apr 1992

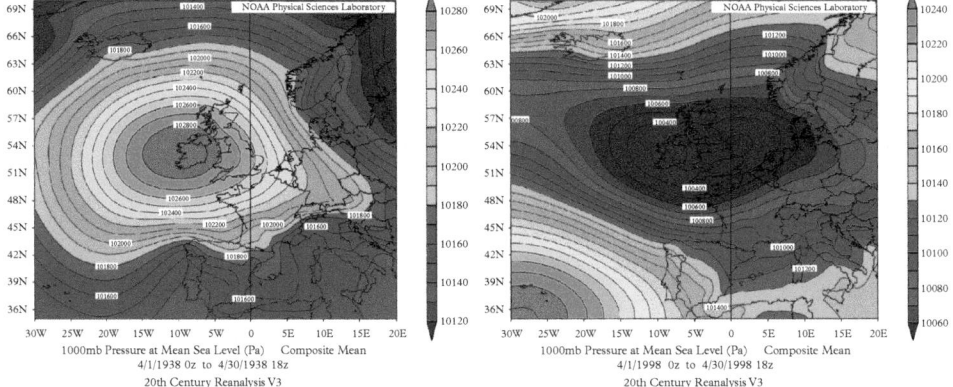

Figure 7.2 *Mean* MSL *pressure distribution during the Aprils of 1938 (left, very dry) and 1998 (right, very wet). Isobars are in Pascals (divide by 100 for hPa = millibars), at 100 Pa intervals (from the 20th Century Reanalysis version 3 (20CRv3) plot [58])*

only 6 °C. The very wet days of 21 April 1935 and 23 April 1971 also resulted from persistent cyclonic rainfall to the north of a depression.

April 1998 was very wet, dull and cool—cooler than February that year, and with a particularly cold and dreary Easter period: temperatures remained below 10 °C from the 8th to 17th, while Easter Sunday (the 12th) started with frost and reached only 8 °C by day. That month, rain fell on 25 days, an April record, and daily totals of 10 mm or more were recorded on six days, including 25.2 mm on the 2nd. Just two years later, April 2000 was almost as wet once again, just over a millimetre short of the 1998 record; this was the start of a very wet 12 months, for by the end of

March 2001 Durham's rainfall accumulation amounted to 996 mm—almost half as much again as the usual amount in a year. After a very dry start to 2012, with just 56 mm in the first three months, April 2012 was also extremely wet, with 134 mm rainfall recorded by month-end. The rest of the year continued in much the same vein, and 2012 ended up the wettest year in Durham's record, a total of 1018 mm, the only time 1000 mm has been exceeded in the calendar year (see also Chapter 22, Chronology).

Figure 7.2 compares the mean distribution of barometric pressure during the Aprils of 1938 (very dry) and 1998 (very wet). The Aprils of 1938 and 1912 were the fourth and fifth driest months on the post-1868 Durham rainfall record (Table 7.3); the former marked a widespread and severe spring drought across England and Wales, and included a 26-day rainless spell—at the time Durham's longest (Table 25.1). April 2020, our most recent dry April (and third-driest on record with just 3.8 mm in all) deserves particular attention. The month was dominated by anticyclonic conditions, and only 1.6 mm had fallen by the 28th; the monthly total of 3.8 mm made this the driest month in Durham since March 1953. It also just missed inclusion in the 'Top 5' warmest Aprils by a tenth of a degree, while sunshine was also well above average. Many will remember it as the first of the coronavirus lockdowns in 2020; further details are provided in Chapter 22.

Sunshine

April's mean monthly sunshine in the period 1991 to 2020 was 154 hours, equivalent to just over five hours per day. As expected, the sunniest days all come towards the end of the month (Table 7.4) as day length continues to increase, although a normal April will see five days without any recorded sunshine. April 1937 was the dullest April on record: that month saw 13 days without any sunshine, while only three days managed ten hours or more. Two Aprils from the 1960s feature in the least sunny Aprils. The dullest April this century to date was the very wet April 2012, with only 104 hours sunshine, although ranking only nineteenth dullest overall.

Three of the five sunniest Aprils are from the present century, 2021 being the sunniest of these with 229 hours in total and a daily average of 7.6 hours; this is closely followed by April 2015 with 219 hours, and April 2020 with 218 hours (and incidentally followed by a near-record sunny May—see Chapter 22). April 1914 remains the sunniest April on record, with 238 hours: every day had some sunshine, although only 18 minutes on the 24th; nine days had over ten hours. With 56.2 per cent of possible sunshine, April 1914 also remains the second-sunniest of any month on record when ranked by the percentage of possible sunshine—only June 1940's 57.6 per cent has surpassed this level of brightness (Table 26.2).

April is normally sunnier than March, given a longer day length and reduced cloud cover, but March has been sunnier than April in 34 years since 1882, most recently in 2014. The greatest 'retrospective step' came in 2012, when March's sparkling sunshine total of 180 hours was followed by a very dull 104 hours in April.

Table 7.4 *April sunshine duration at Durham Observatory. Observed monthly extremes 1881–2021, sunniest days 1882–1999 only*

April mean sunshine duration 154 hours, 5.1 hours per day (average 1991–2020)
Possible day length: 424 hours. Mean sunshine duration as percentage of possible: 36.3

Sunniest months			Dullest months			Sunniest days 1882–1999	
Duration, hours	Per cent of possible	Year	Duration, hours	Per cent of possible	Year	Duration, hours	Date
238.1	56.2	1914	71.3	16.8	1937	14.0	30 April 1921
229	54.0	2021	78.5	18.5	1889	13.7	29 and 30 April 1990
219	51.7	2015	78.7	18.6	1930	13.6	29 April 1921
218	51.5	2020	81.0	19.1	1966	13.5	30 April 1893 and 24 April 1921
212.7	50.2	1990	81.6	19.3	1961	13.3	25 April 1941, 16 April 1942 and 25 April 1942
Last 50 years to 2021							
229	54.0	2021	91.7	21.6	1979	13.7	29 and 30 April 1990

Daily sunshine records ceased at the Observatory on 31 December 1999: monthly sunshine totals from January 2000 onwards are estimated (to the nearest hour) by regression from the Met Office East and North-east England sunshine series. See Appendix 6 for record metadata and missing months.

Wind and weather

Snow cover (1960–1999)

Snow cover is unusual but far from unknown in April, occurring in about one year in three, usually at the beginning of the month. The Aprils of 1966 and 1973 both recorded three mornings with snow lying; the deepest April snowfall during this period was a remarkable 25 cm on 5 April 1975, while 15 cm lay on the morning of 2 April 1966 and 10 cm on 7 April 1977. The latest date with a morning snow cover during this period was 25 April 1981, when the snow lay 7 cm deep. This was a very heavy late snowfall in many parts of England and Wales [59], and up to 30 cm fell over the moors in Northumberland. On the very early records, the 9 a.m. observation on 1 April 1847 notes 'Snow 7 or 8 inches deep [17–20 cm]'.

Fog (1962–97)

Morning fog is increasingly unlikely by April, the average being just one morning in a typical month, although there were seven morning fogs in the Aprils of 1964 and 1974, probably 'sea fret' blown in on a north-easterly wind.

Barometric pressure (1844–1960)

The highest MSL barometric pressure on Durham's April records occurred in 1906; at 2100 GMT on 8 April of that year, the MSL pressure reached 1042.5 hPa. This value was equalled at the 2100 GMT observation on 11 April 1938, while not far behind was 1042.0 hPa noted at 2100 GMT on 25 April 1948. This latter statistic is particularly noteworthy, for April's lowest observed pressure also occurred that month—970.2 hPa at 0900 on the first day of the month, giving a range in pressure that month of 72 hPa—greater than that seen in many an entire calendar year. Other notably low barometric pressure occurred on 23 April 1947 (975.2 hPa at MSL at the 2100 GMT observation), and 975.3 hPa at 2100 GMT on 14 April 1919.

The most anticyclonic April was 1938, when the mean for the month stood at 1026.9 hPa—not surprisingly, this was also the driest April on record (Table 7.3). The most cyclonic Aprils were those of 1920 (mean at MSL 1005.1 hPa) and 1932 (1006.0 hPa)—the former was fairly wet, but the latter month was rather dry, with less than two-thirds normal rainfall.

Wind speed and direction (1961–97)

Figure 7.3 shows the distribution of wind speed and direction at the 0900 GMT observation in April during the period 1961–97. April is the third-windiest month of the year in Durham, behind only March and January; the mean wind speed at 0900 GMT is 7.7 knots. April 1973 recorded the greatest mean wind speed on Durham's records during this period, namely 14.2 knots; there was a gust of 55 knots on 6 April that year. On earlier records, there was a gust of 62 knots on 8 April 1950.

April's wind rose differs considerably from that of previous months: winds between south and west are less frequent, while winds from between north and east are much more frequent, and often quite strong.

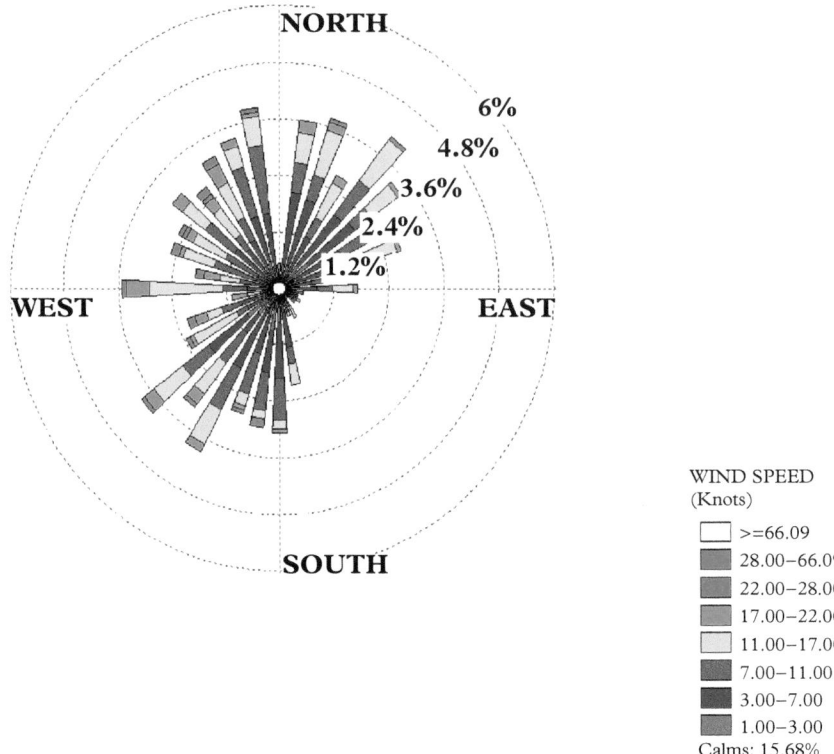

Figure 7.3 *Wind rose for 0900 GMT observations in April, period 1961–97*

Durham temperature, precipitation, and sunshine in graphs—April

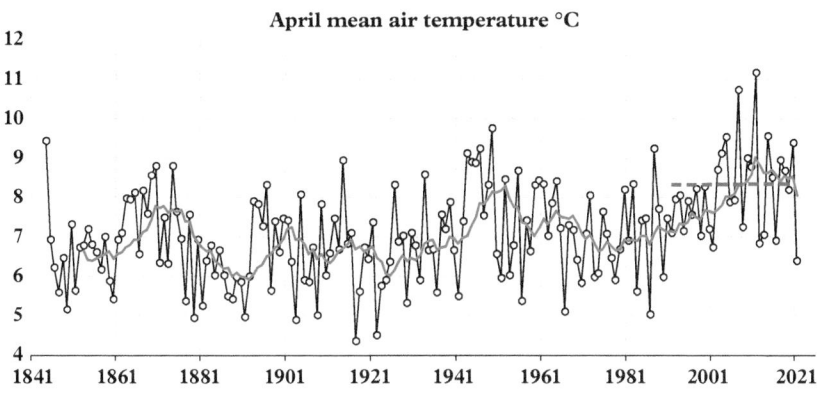

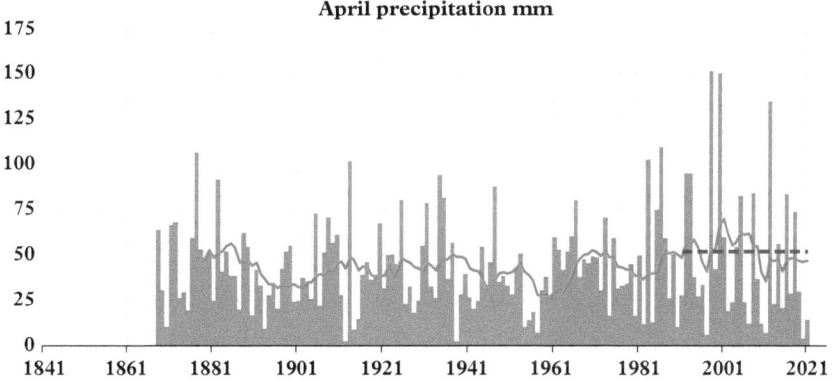

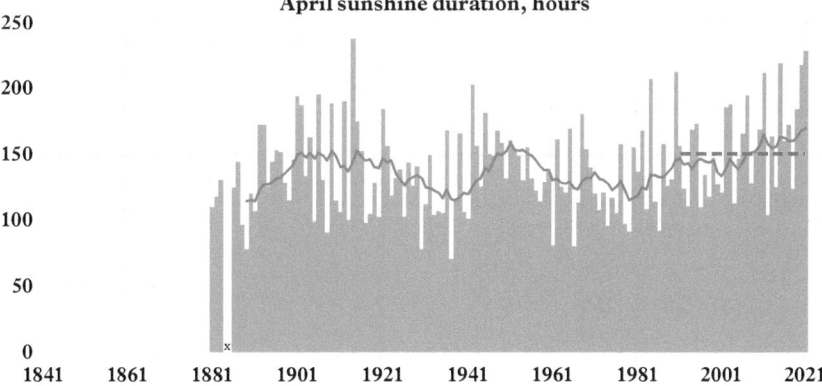

Figure 7.4 *Monthly values of (from top) April's mean temperature (°C, since 1844), total precipitation (mm, since 1868), and sunshine duration (hours, since 1881, missing data 1884 and 1885) at Durham Observatory. The 1991–2020 averages are indicated by the thick blue dashed line, while the ten-year running mean ending at the year shown is indicated by the red line*

8

May

May is the final month of spring—but in a good year it can seem like the start of summer! Like other spring months, it is a time of transition: days can be cool, nights frosty and snowfalls cannot be entirely ruled out. There is a steady increase in temperatures through the month, as daytime temperatures rise from just below 14 °C on average at the start of the month to above 17 °C by the close. May is, on average, the sunniest month of the year in Durham and days in the middle of the month rival days in June and July as the sunniest of the year; on average, just four days in the month will remain sunless. May is also the second-driest month of the year, after March.

Temperature

Like its spring predecessors, May can be a capricious month, and the records for the month feature both sharp frosts and summer heatwaves. The coldest May morning was back in 1891, when as late as the 18th the minimum temperature was −4.8 °C (Table 8.1); the grass minimum that morning fell to −9.7 °C—a very sharp spring frost indeed. Within the last 50 years, no air minimum below −2.5 °C has been recorded in May, although at least one air frost can be expected in most Mays, and typically 3–4 ground frosts, particularly in dry, clear weather. Since 2000, only the Mays of 2003 and 2009 have escaped a ground frost altogether, whereas May 1996 recorded 15, with 12 the following year. Conditions were a little worse in the nineteenth century: May 1879 recorded six air frosts, while in more recent years the Mays of 1979 and 1997 both notched up four: in May 1980, there were 17 ground frosts.

The earliest date in the year on which 25 °C—a good threshold for 'summer heat'—has been attained came in 2018, when this was exceeded on 7 May (Table 24.1). The average highest temperature for May in Durham during the period 1991–2020 was 22.7 °C, and only about one in five Mays will reach 25 °C. In May 1894, no day reached 15 °C—not surprisingly, this was one of Durham's coldest Mays (Table 8.2). In contrast, the temperature reached 29.0 °C on 23 May 2001, surpassing by over a degree a May record that had stood since 1922. It is not unknown for May to produce the highest temperature of the year—this has happened eight times since 1844, most recently in 2012. The top four warmest May days in Table 8.1 were all the warmest days of that

Table 8.1 *Highest and lowest maximum and minimum temperatures at Durham Observatory in May, 1844–2021*

Rank	Warmest days	Coldest days	Mildest nights	Coldest nights
1	29.0 °C, 23 May 2001	*4.0 °C, 8 May 1883*	14.6 °C, 19 May 2004	*−4.8 °C, 18 May 1891*
2	27.9 °C, 31 May 1922	*4.2 °C, 4 May 1855*	14.3 °C, *13 May 1848*	−4.2 °C, 1 May 1927 and 1 May 1929
3	27.2 °C, 29 May 1947 and 23 May 2010	*4.4 °C, 16 May 1873*	14.1 °C, 30 May 2003	−3.9 °C, 3 May 1967
4	26.7 °C, 29 May 1944	*4.5 °C, 18 May 1873*	14.0 °C, *16 May 1845,* 14 May 2004 and 26 May 2017	−3.5 °C, 2 May 1914
5	26.1 °C, 25 May 2017 and 7 May 2018	5.0 °C, 4 May 1947	13.9 °C, 28 May 2017	−3.3 °C, *6 May 1868* and 3 May 1915
Last 50 years to 2021				
	29.0 °C, 23 May 2001	6.0 °C, 1 May 1979	14.6 °C, 19 May 2004	−2.5 °C, 12 May 2010

Temperatures were recorded in different thermometer screens during the period of record; see notes on page xvii and Appendix 3 for more details

year—even in 1947, which saw a very warm, dry, and sunny August. Note also that May 1947 appears in both 'warmest' and 'coldest' days in Table 8.1, revealing a particularly marked improvement during the month.

Of the five coldest Mays in Durham's records, the most recent is 1902 (Table 8.2); all the others date from the nineteenth century. In May 1902, maximum temperatures remained at or below 10 °C for ten consecutive days up to the 16th. The coldest May within the last five decades was 1975, with a mean temperature just under 8.0 °C, and the coldest at Durham since 1923. The warmest day in the month (the 19th) reached only 20.0 °C. Contrary to expectations at the time, it was followed by many dry and warm months, the summer of 1975 turning out to be the warmest in Durham since 1947 (equalling 1959), followed of course by the even hotter summer of 1976. It was not all sunshine and warmth, however—May 1976 failed to reach even May 1975's highest temperature! These two Mays, in an otherwise very warm period, remind us that this is a transitional spring month and fine weather certainly cannot be taken for granted.

An exception to the cooler Mays of the nineteenth century was 1848, a particularly warm month that ranks second in Table 8.2. We only have north wall temperatures to go on; however, the average temperature (the mean of all available daily maxima and

Table 8.2 *May mean temperatures at Durham Observatory, 1844–2021*

May mean temperature 11.0 °C (average 1991–2020)

Warmest months			Coldest months		
Mean temperature, °C	Departure from 1991–2020 normal degC	Year	Mean temperature, °C	Departure from 1991–2020 normal degC	Year
13.4	+2.4	2004	7.3	−3.7	*1869, 1877, 1902*
13.1	*+2.1*	*1848*	*7.4*	*−3.6*	*1856, 1885, 1894*
12.7	+1.7	2017	7.6	−3.4	*1855*
12.4	+1.4	1952	7.7	−3.3	*1879, 1899, 1923*
12.3	+1.3	2018	8.0	−3.0	1975
Last 50 years to 2021					
13.4	+2.4	2004	8.0	−3.0	1975

minima) was 13.1 °C, very close to Manley's estimate for the month of 12.9 °C [19], so the records seem reasonably reliable. This remained Durham's warmest May for over 150 years, until surpassed by the warmth of May 2004. Notably, two very recent Mays, 2017 and 2018, feature in the 'Top 5' warmest in Table 8.2, while May 2020 (mean 12.1 °C) just failed to make an entry in the table, ranking sixth. Like the month before, anticyclonic conditions dominated the weather and wind direction determined whether individual days were warm or cool (see Chapter 22).

Precipitation

May is the second-driest month of the year in Durham, after March, with a mean monthly fall of 44.5 mm. As April has become wetter, so recent Mays have become somewhat drier, the May average for the period 1911–40 standing at 53 mm. Monthly totals below 15 mm or above 100 mm occur only once in 20 years or less. Table 8.3 shows the wettest and driest Mays on record.

Durham's wettest May came in 1924, when 70.2 mm fell on the last day of the month—Durham's wettest May day, and third-wettest day on the post-1868 record; only one day in 1901 had been wetter until then. *British Rainfall 1924* comments that the thunderstorms of 31 May and 1 June produced some of the most remarkable rainfall totals of that year. A shallow low-pressure area, a secondary to a depression in the Atlantic, drifted north from the English Channel to become centred over Lincolnshire by the morning of 1 June, deepening as it did so, with winds gradually backing from

Table 8.3 *May precipitation at Durham Observatory 1868–2021*

May mean precipitation 44.5 mm (average 1991–2020)

Wettest months			Driest months			Wettest days	
Total fall, mm	Per cent of normal	Year	Total fall, mm	Per cent of normal	Year	Daily fall, mm	Date
154.0	346	1924	7.7	17	1989	70.2	31 May 1924
120.0	270	1954	8.1	18	1905	41.0	13 May 1993
104.2	234	1932	13.3	30	1888, 1959	38.9	19 May 1906
102.4	230	1983	13.5	30	1922	35.2	28 May 1954, 28 May 1998
102.0	229	1969	13.6	31	1896	30.4	12 May 1954
Last 50 years to 2021							
102.4	230	1983	7.7	17	1989	41.0	13 May 1993

southerly to easterly at Durham. The largest totals, mostly on 31 May, were in the West Midlands and north-east England: Durham received 13.5 mm, 70.2 mm and 14.3 mm on three successive days 30–31 May and 1 June. The highest total in the Durham area was 103.6 mm at Sunderland's Southwick cemetery, also on 31 May. Figure 8.1, from *British Rainfall 1924*, shows the combined rainfall totals over the two days 31 May and 1 June, when Durham's total amounted to 84.5 mm. Angus MacFarlane-Grieve was busy with his camera the next day, recording dramatic flooding through Durham City (Figures 8.2, 8.3, and 8.4). This was well before river flow was measured on the River Wear, of course, but comparison with more recent floods in November 2009, in November's chapter, suggests an equally extreme flood; the flood peak on 18 November 2009 (277 cumecs, Figure 14.1) ranks seventeenth-highest on the streamflow record (October 1957 to September 2019).

In other notably wet days in May, the year 1954 includes two of the five entries in Table 8.3, 16 days apart, contributing to Durham's second-wettest May that year.

The most recent very wet May was in 2013 with a total rainfall amounting to 100.8 mm. This came at the end of a very wet period commencing in April 2012, which included the record calendar year total for 2012 (1018 mm), while the 12-month running total remained above 1000 mm at the end of May 2013—almost half as much again as a normal year, that is, 18 months' rainfall in 12—for the sixth consecutive month—a remarkable sequence. Heavy falls on the 14th (18.8 mm) and especially on the 17th/18th (two-day total of 47.6 mm) resulted in extensive flooding in the Durham area. On the morning of 18 May, 22.4 mm fell in two hours commencing 0700 GMT.

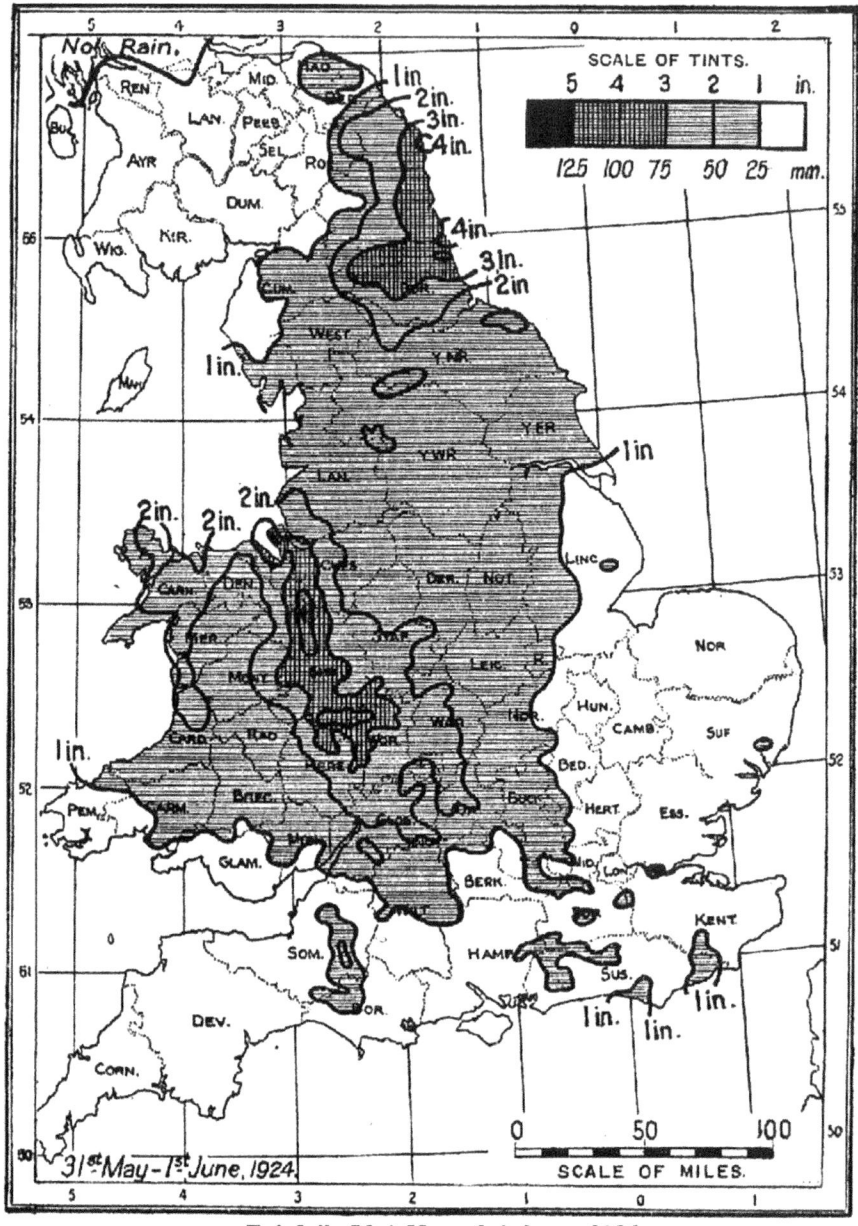

Rainfall, 31st May—1st June, 1924.

Figure 8.1 *Rainfall totals for the 48 hours commencing 0900* GMT *on 31 May 1924, the first day being Durham's wettest-ever May Day when 70 mm fell. The isohyets are in inches: the shaded scale gives approximate millimetre equivalents. From* British Rainfall 1924, *p. 51 (courtesy Met Office)*

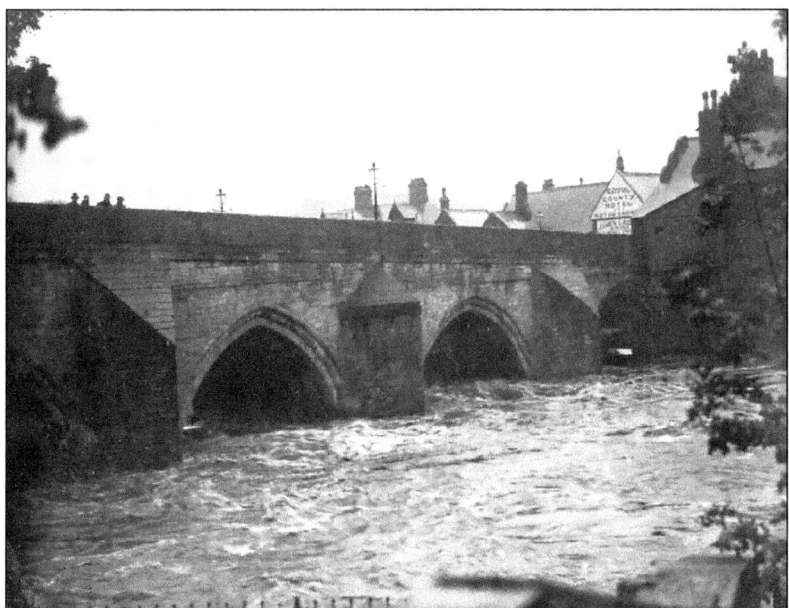

Figure 8.2   *Durham's Elvet Bridge on 1 June 1924, following the near-record rainfall the day previously. (Reproduced by permission of Durham University Library and Collections and the MacFarlane-Grieve family)*

Figure 8.3   *Spectators observing the River Wear in flood below Durham's Framwellgate Bridge (now known as Framwellgate Waterside) on 1 June 1924. Crook Hall is visible in the distance. (Reproduced by permission of Durham University Library and Collections and the MacFarlane-Grieve family)*

Figure 8.4 *Looking towards Durham's Framwellgate Bridge during the 1 June 1924 flood. (Reproduced by permission of Durham University Library and Collections and the MacFarlane-Grieve family)*

Durham's driest May was in 1989, when just 7.7 mm fell in a month which was also one of the sunniest on record—just short of an entry in the 'Top 5' (Table 8.4).

Sunshine

May is, on average, the sunniest month of the year in Durham, with 187 hours of bright sunshine in the standard averaging period 1991–2020, equivalent to just over six hours per day—37 per cent of the possible total. Typically, only four days will remain sunless. In the period 1970–1999, the week centred on 16 May was the sunniest period in the year, averaging 6.45 hours daily, just ahead of the week centred on 8 July (average 6.23 hours per day). Four Mays feature within the ten sunniest months in Durham since 1880—namely 1881, 1971, 2018, and 2020 (Table 26.1).

The sunniest Mays can expect many days of long-lasting sunshine (Figure 8.5): May 1921, for example, recorded ten days with ten hours or more of bright sunshine and just two sunless days, while May 1971 went one better with 12 days with ten hours or more of bright sunshine and, like 1921, just two sunless days.

Not surprisingly, sunny months tend to be warm and dry, while dull months tend to be cool and wet. In the latter camp stands May 1932, with entries in both Tables 8.3 and 8.4. The remarkably sunny May of 2020, with 256 hours of bright sunshine, or 8.26 hours daily average, ranks equal second with May 1881, fewer than three hours behind the sunniest May on record. Add this to a sunny March (156 hours, 30 per cent greater than average) and the third-sunniest April on record (218 hours, 41 per cent above

Table 8.4 *May sunshine duration at Durham Observatory. Observed monthly extremes 1881 to 2021, sunniest days 1882–1999 only*

May mean sunshine duration 187 hours, 6.0 hours per day (average 1991–2020).
Possible day length: 500 hours. Mean sunshine duration as percentage of possible: 37.4

Sunniest months			Dullest months			Sunniest days 1882–1999	
Duration, hours	Per cent of possible	Year	Duration, hours	Per cent of possible	Year	Duration, hours	Date
258.8	51.8	1971	86.6	17.3	1932	15.4	27 May 1883
256	51.2	1881	97.5	19.5	1887	15.3	29 May 1952
256	51.1	2020	98.9	19.8	1886	15.0	21 May 1921, 25 May 1929, and 18 May 1940
248	49.6	2018	100.9	20.2	1933	14.9	21 May 1888 and 25 May 1921
247.6	49.5	1921	103.4	20.7	1969	14.8	*Numerous occasions*
Last 50 years to 2021							
258.8	51.8	1971	104.4	20.9	1983	14.7	26 May 1989

Daily sunshine records ceased at the Observatory on 31 December 1999: monthly sunshine totals from January 2000 onwards are estimated (to the nearest hour) by regression from the Met Office East and North-east England sunshine series. See Appendix 6 for record metadata and missing months.

normal), and it is no surprise that spring 2020 was Durham's sunniest-ever spring—sunnier, indeed, than most summers (see Spring, Chapter 18).

Wind and weather

Snow cover (1960–1999)

Snowfall is not uncommon in May, although no morning snow cover has been recorded later than 25 April during this period of record.

Fog (1962–97)

Morning fog is similarly unlikely in May, the average being less than one such occurrence during the month, although three mornings with fog were noted in the Mays of 1968, 1969, 1973, and 1992; some of these were certainly 'sea fret' blown in on a north-easterly wind rather than slow-to-clear radiation fog following a clear night.

Figure 8.5 *Cricket on the Racecourse, 10 May 2003 (Tim Burt)*

Barometric pressure (1844–1960)

The highest MSL barometric pressure on Durham's May records occurred on 9–10 May 1881. At the 2200 observation on 9 May of that year, the MSL barometer stood at 1041.6 hPa, rising a little further to 1042.3 hPa by the 1000 observation next morning, 10 May 1881. May's lowest observed pressure stands at 972.0 hPa at 2100 GMT on 8 May 1943, having fallen from 977.6 hPa at 0900 GMT that morning. At 1000 on 28 May 1877, the barometer stood at 978.6 hPa. The May 1943 event is particularly notable, for only eight days later the pressure had risen to 1040.3 hPa at 0900 GMT on the 16th, a range of 68 hPa in little more than week—unusual in a winter month, but unprecedented before or since in a spring or summer month.

The most anticyclonic May was in 1896, when the mean for the month stood at 1025.8 hPa—in what remains as the fifth-driest May on record (Table 8.3). May 1844, with a mean of 1023.2 hPa, and May 1876, with a mean of 1022.4 hPa, were also very anticyclonic. The most cyclonic Mays were those of 1925 (mean at MSL 1007.7 hPa), a dull, cool, and very thundery month, and 1878 (1007.9 hPa), which was rather wet.

Wind speed and direction (1961–97)

Figure 8.6 shows the distribution of wind speed and direction at the 0900 GMT observation in May during the period 1961–97. Mean wind speeds begin their summer decline in May, although some of the decline may be due to trees in leaf sheltering the wind instruments in the latter years of the record. The mean wind speed at 0900 GMT is 6.5 knots. May 1974 was the windiest May during this period with a mean speed of 11.8 knots.

May sees the annual maximum frequency of winds between north and east, particularly north-easterlies, which outnumber south-westerly winds. It is the only month in the year when this is the case, and the winds are frequently quite strong.

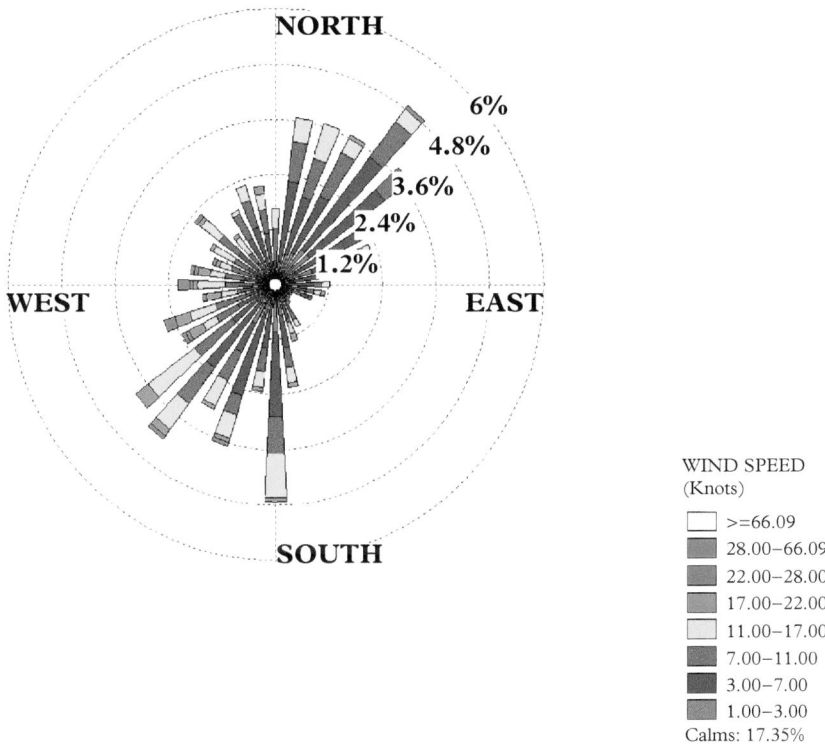

Figure 8.6 *Wind rose for 0900 GMT observations in May, period 1961–97*

Durham temperature, precipitation, and sunshine in graphs—May

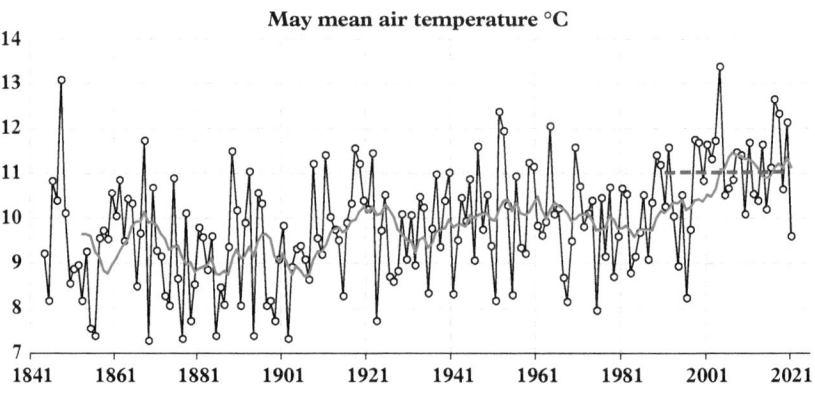

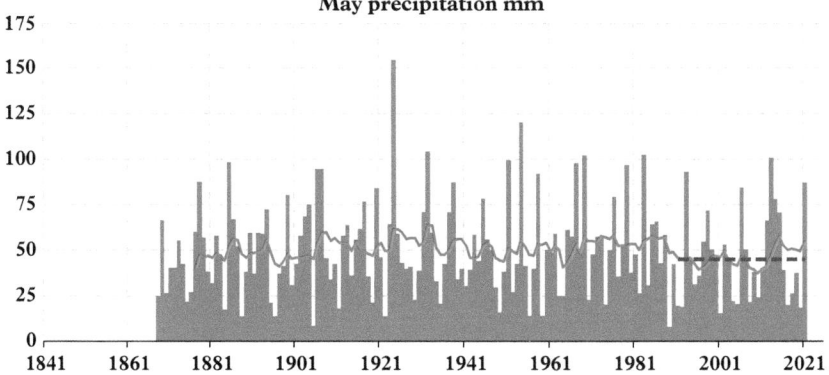

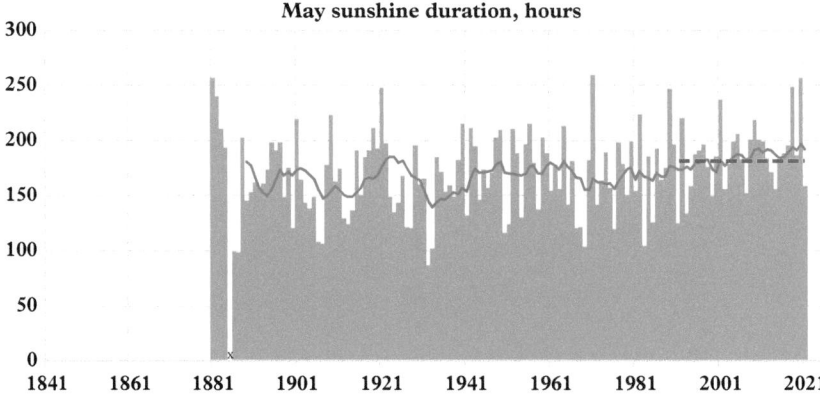

Figure 8.7 *Monthly values of (from top) May's mean temperature (°C, since 1844), total precipitation (mm, since 1868), and sunshine duration (hours, since 1881, missing data 1885) at Durham Observatory. The 1991–2020 averages are indicated by the thick blue dashed line, while the ten-year running mean ending at the year shown is indicated by the red line*

9

June

The meteorological summer consists of the three months June, July, and August. As the first of these, June in Durham is, on average, almost 3 degrees Celsius warmer than May, but is also significantly wetter and somewhat less sunny. Average temperatures continue to rise steadily as the month progresses, the mean daily maximum temperature climbing from 16 °C in the first week to above 19 °C by month-end. Nights can be chilly at the start of the month and ground frosts (and even a rare air frost) are not unknown. 'Flaming June' is occasionally the sunniest month of the summer (two or three times per decade), but much less frequently the hottest month of the summer (only four times in the last 100 years or so). Rain can be expected on one day in two, while the increasing warmth brings with it a greater risk of thunderstorms. These can produce heavy short-period falls of rain, and a fall of 25 mm or more in a day (often in just a few hours) can be expected about once in five years.

Temperature

The beginning of June can often produce remarkably cool nights, and even air frosts are not unknown at this time of year: five are on record, the most recent being on 8 June 1991 when the screen minimum fell to −0.6 °C (Table 9.1). In 1974, there was an air frost as late as 28 June (−0.5 °C), the latest air frost on Durham's records (Table 24.1). Ground frosts used to be more frequent in June—nine in 1921 and seven in 1974, for example—but have become much less frequent since the turn of the century (although the change to the automatic weather station in late 1999, and the introduction of an artificial turf surface for the sensor in July 2015, have both made ground frosts less likely). June can also produce unseasonably cold days, and heading the list here is 2 June 1953, Queen Elizabeth II's Coronation Day. Chilly northerly winds blew around a depression in the North Sea (Figure 9.1) and were particularly raw on north-east coasts [60]. The day was sunless, very cold, and very wet in Durham: the temperature reached only 7.2 °C by day, about the same as an average January day, and 13 mm of rain fell. Contemporary news reports comment on the dreadful weather on that occasion, which must have put more than a dampener on street parties and other festivities planned for the event. Only slightly less cold was the same date 65 years earlier, 2 June 1888, when the maximum temper-

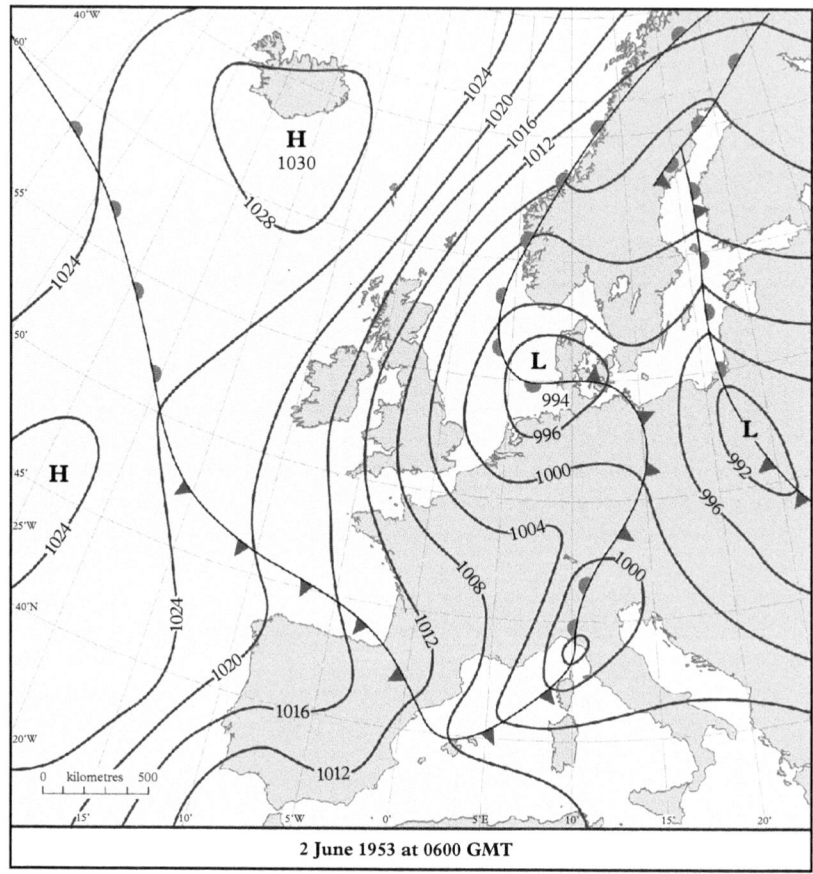

2 June 1953 at 0600 GMT

Figure 9.1 *Synoptic situation at 0600* GMT *on Tuesday 2 June 1953, Queen Elizabeth II's Coronation Day, and the coldest June day on Durham's records, with a maximum temperature of only 7.2° C. (Based upon the Met Office* Daily Weather Report; *drawn by Chris Orton, Durham University)*

ature was 7.4 °C, another sunless day with a chilly easterly wind and 7 mm of rainfall. Notable also was 13 June 1916, which reached only 8.2 °C. (Attentive readers may recall that both 1888 and 1916 saw extremely mild days in January—in 1888 16.1 °C on the 9th, and in 1916 14.9 °C on the 6th—making for a cruel seasonal reversal of midwinter in midsummer.) Maxima of 10 °C or lower are not uncommon in June, occurring almost 80 times in the record to 2020, although less frequently in recent decades. Four Junes (1862, 1882, 1972, and 1985) have been and gone without reaching 20 °C even once.

In June 1972 the warmest day attained just 18.7 °C, although a 'normal' June can be expected to reach 25 °C on at least one day. June air temperatures have reached 30 °C on three occasions—although at the time of writing, the most recent of these was over 80 years ago, on 8 June 1940 (Table 9.1). In June 1939, 30 °C was attained on the

Table 9.1 *Highest and lowest maximum and minimum temperatures at Durham Observatory in June, 1844–2021*

Rank	Hottest days	Coldest days	Warmest nights	Coldest nights
1	30.4 °C, 10 June 1925	7.2 °C, 2 June 1953	17.8 °C, 5 June 1980	−0.8 °C, 15 June 1927
2	30.0 °C, 6 June 1939 and 8 June 1940	*7.4 °C, 2 June 1888*	*17.6 °C, 27 June 1856*	−0.7 °C, 19 June 1915
3	29.4 °C, 22 June 1941	7.8 °C, *1 June 1886* and 3 June 1953	*17.5 °C, 27 June 1878*	−0.6 °C, 8 June 1920 and 8 June 1991
4	29.3 °C, 29 June 1976	8.0 °C, 1 June 1902	17.2 °C, 30 June 1941	−0.5 °C, 28 June 1974
5	29.2 °C, 20 June 1989	8.2 °C, 13 June 1916	16.8 °C, 19 June 2005	0.3 °C, 3 June 1924
Last 50 years to 2021				
	29.3 °C, 29 June 1976	8.3 °C, 5 June 2009	17.8 °C, 5 June 1980	−0.6 °C, 8 June 1991

Temperatures were recorded in different thermometer screens during the period of record; see Notes on page xvii and Appendix 3 for more details.

6th, the earliest such date on record (Table 24.1). In the nineteenth century, 28 °C was reached three times (in the north wall unscreened exposure) in the hot June of 1846, and again on 27 June 1878 (Stevenson screen equivalent). These extremes were not surpassed until the very dry and sunny June of 1925, when 30.4 °C was reached on the 10th—a value that still stands as the hottest June day on Durham's records, although 6 June 1939 and 8 June 1940 were almost as hot. More recently, 29 °C was attained in June 1976, and again in 1989 (Figure 9.2).

Warm nights—minimum temperatures of 15 °C or greater—also begin to make their appearance in June. The minimum temperature of 16.1 °C on 2 June 2003 was the earliest date in any summer on which the minimum temperature remained above 15 °C all night (Table 24.1). The warmest June night on record was 5 June 1980, when a minimum temperature of 17.8 °C was recorded, following the previous afternoon's maximum temperature of 26 °C.

It is certainly counter-intuitive that, after almost 180 years of records and three decades of pronounced global warming, that Durham's hottest June remains that of 1846, only three years into the record (Table 9.2). The surprise is enhanced by the discovery that the second-hottest June was barely a decade later, in 1858. Indeed, the warmth of June 1846 was not surpassed on Durham's records until August 1975! Your authors certainly have some doubts regarding the exposure of the thermometers in those early years (an unscreened north wall exposure in 1846, the 'North Shed' in 1858—see

Figure 9.2 *Sunbathing on Palace Green, 5 June 2016. At its best, June can have some very warm and pleasant days (Tim Burt)*

Table 9.2 *June mean temperatures at Durham Observatory, 1844–2021*

June mean temperature 13.6 °C (average 1991–2020)

Hottest months			Coolest months		
Mean temperature, °C	Departure from 1991–2020 normal degC	Year	Mean temperature, °C	Departure from 1991–2020 normal degC	Year
17.5	+3.9	*1846*	10.7	−2.9	*1888*, 1916
15.8	+2.2	*1858*	10.9	−2.7	1909
15.7	+2.1	1940	11.0	−2.6	1927
15.4	+1.8	1976	11.1	−2.5	*1871*, 1928
15.3	+1.7	2003	11.2	−2.4	1971
Last 50 years to 2021					
15.4	+1.8	1976	11.2	−2.4	1971

Appendix 3 for details), but there is plenty of evidence from other sources to back up the claim of both months. In Manley's long Central England Temperature (CET) series—which begins in 1659—June 1846 remains the hottest June to date, with a CET mean of 18.2 °C, while 1858 ranks third-hottest, behind 1976.

Since that hot June in 1858, the warmest June has been 1940, when the mean maximum temperature for the first 15 days of the month was above 20 °C, with 25 °C surpassed every day from 3–10 June. Altogether, there were 23 days with maxima above 20 °C, a June record, followed by 1970 (19 such days) and 2010 (18). Of course, June 1940 was a key period in the Second World War, seeing the end of the Dunkirk evacuation, the fall of Paris, and the Allied evacuation of Norway. It must have been a difficult time in Britain to have such glorious weather at a time of national crisis.

Of the five coolest Junes on record, none is more recent than 1928 (Table 9.2); 1888 and 1916 tie as the coldest. Of the two, June 1888 had the lower mean maximum temperature (14.3 °C, compared to 14.6 °C in 1916)—and, of course, the mean maximum temperature is often more relevant to public perception than the mean minimum during the summer months. But there were many poor Junes between 1971 and 1991, which lay only just outside the 'Top 5' coolest, including 1971 (mean maximum 14.8 °C), 1972 (15.4 °C), 1987 (15.1 °C), and 1991 (15.7 °C).

Precipitation

The headline summary for the national *Monthly Weather Report* for June 1980 was 'Mostly unsettled, cool and thundery', and indeed it was in Durham, where this became the wettest June on record and the fifth-wettest month on record (Table 25.3). It was so wet that the total rainfall during the month (191 mm, more than three times normal) was almost twice the fall of the previous wettest (and very cool) June 1928, which received 117 mm. Rain was recorded on 22 days in June 1980: eight days recorded totals in excess of 10 mm, including 33 mm on the 11th, 28 mm on the 25th, and a further 24 mm on the last day of the month. Many of these wet days resulted from thunderstorms: we do not have records of the number of days in that month where thunder was heard at the Durham Observatory, but elsewhere in the county, Low Etherley reported eight days 'thunder heard', as did Hartburn Grange in Cleveland.

Since 1980, there have been several other very wet Junes—1997, 2007, and 2012 all featuring in the 'Top 5' in Table 9.3.

June 2007 was a notable month for flooding further south in the UK, in Yorkshire and Humberside in particular. Although Durham's rainfall total for that month was well below the record levels a little further south, it still ranked as the third-wettest June on record at the time (although since surpassed in 2012). However, June 2007 did not produce any exceptional short-period falls of rain, in Durham or further up the River Wear catchment, and the resultant floods were relatively modest as a result. Six days recorded daily totals over 10 mm; five days in the middle of the month (11–15 June) accumulated 64 mm in total, but fortunately individual falls were fairly well distributed and major flooding was averted. Nevertheless, the month saw the fourth-highest mean

Table 9.3 *June precipitation at Durham Observatory 1868–2021*

June mean precipitation 61.2 mm (average 1991–2020)

Wettest months			Driest months			Wettest days	
Total fall, mm	Per cent of normal	Year	Total fall, mm	Per cent of normal	Year	Daily fall, mm	Date
191.4	313	1980	1.7	3	1925	54.0	3 June 2000
174.4	285	1997	4.1	7	1891	48.0	13 June 1928
136.6	223	2012	5.3	9	1960	44.1	9 June 1914
120.8	197	2007	6.4	10	1887	38.2	3 June 1936
117.4	192	1928	8.1	13	1918	33.8	11 June 2020
Last 50 years to 2021							
191.4	313	1980	8.6	14	1996	54.0	3 June 2000

daily flow of the River Wear at Sunderland Bridge (16.2 cumecs) since records began in October 1957; June 2012, an even wetter month than 2007, currently holds the record (20.7 cumecs).

Durham's wettest June day came on 3 June 2000, when 54.0 mm fell during the 0900–0900 GMT rain day, including 46.8 mm in the 12 hours commencing 1700 GMT (Figures 9.3 and 9.4). This caused extensive flooding next day, most notably at the Maiden Castle sports centre, where university examinations were disrupted (see Chapter 22). At 1300 GMT on 4 June, the level of the River Wear peaked at 3.48 m, the

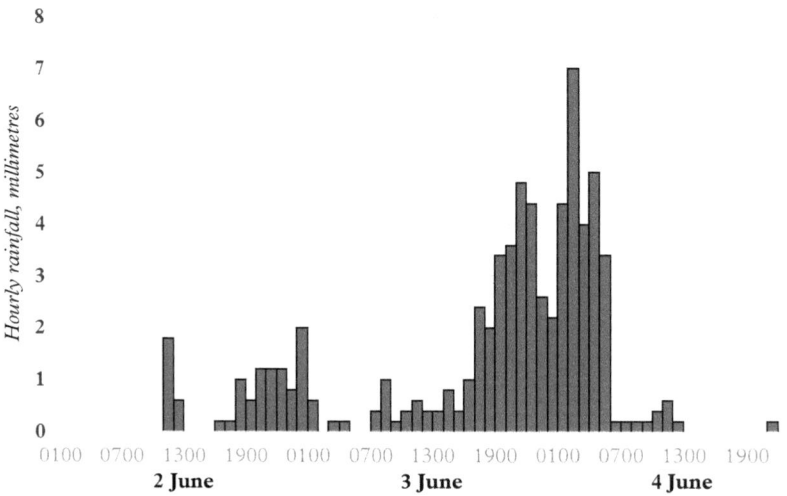

Figure 9.3 *Hourly rainfall totals (mm, in hour ending at time shown) from the Durham Observatory AWS, 2–4 June 2000; time is in* GMT

Figure 9.4 *Durham's Fulling Mill weir almost completely drowned out during the flood of 4 June 2000 (Tim Burt)*

second-highest on record. More recently, 33.8 mm fell on 11 June 2020: rain persisted for 18 hours, with three hours having more than 4 mm in the hour. This rain was welcome, for the total for April and May combined amounted to only 21.8 mm, but the flooding was not!

The driest June on record was 1925, when only 1.7 mm was recorded; rain fell on six days that month, but none amounted to even 0.5 mm. This was Durham's driest month on record until March 1953 (1.3 mm), and it remains the second-driest to date (Table 25.3, Figure 9.5). On earlier and less reliable records, rain fell on 27 days in June 1852.

Sunshine

During the 30-year period 1991–2020, June averaged 166 hours of bright sunshine, equivalent to 5.5 hours per day, or 32 per cent of daylight hours. June is a little less sunny than May and July; it is also less sunny than August but has almost the same daily average given one day fewer in the month. June was the sunniest summer month 60 times between 1882 and 2020. On average, three June days will remain sunless.

With just under 300 hours of sunshine, a fraction under ten hours per day and 57.6 per cent of possible daylight hours, June 1940 was not only Durham's sunniest June

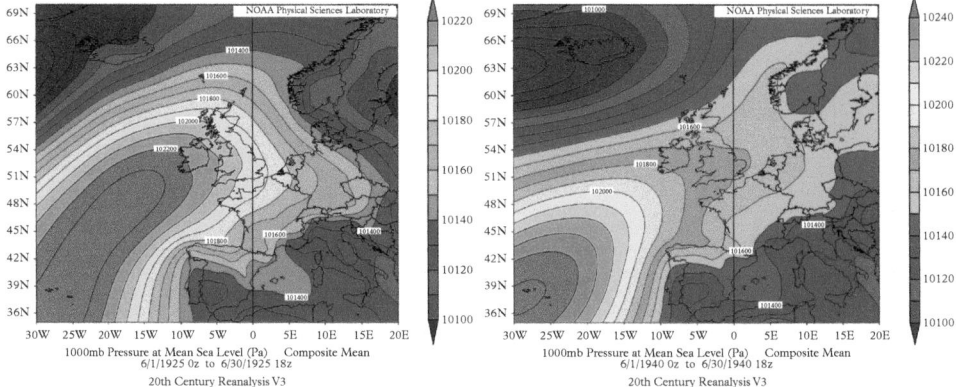

Figure 9.5 *Mean* MSL *pressure distribution during the Junes of 1925 (left, driest on record) and 1940 (right, sunniest month on record). Isobars are in Pascals (divide by 100 for hPa = millibars), at 100 Pa intervals. (From the 20th Century Reanalysis version 3 (20CRv3) plot [58])*

on record, it was also *the* sunniest month on record; no month before or since has come within 20 hours (Table 26.1, Figure 9.5). As noted earlier, this was also the warmest June on record, although its monthly rainfall of 33 mm was unremarkable. At the opposite extreme, June 1987 became Durham's dullest, with only 89 hours of sunshine, a value exceeded in several winter months—February 2019 with 128 hours, and January 2022 with 112 hours, for example. Five days missed out on any sunshine, and the month was also very wet. Three Junes have recorded ten days without sunshine, namely those of 1953, 1958, and 1982; the first of these includes 2 June 1953, Coronation Day.

Until 1987, June 1912 was the dullest June on the record, with only 92 hours of sunshine; six days remained sunless, while the sunniest day managed only 9.3 hours, the lowest June maximum pre-2000. This was the start of a notoriously dull summer, attributed by Gordon Manley to the eruption of the Mount Katmai volcano in Alaska in April of that year; summer 1912 remains by some margin the dullest summer in Durham since sunshine records commenced in 1880 (Chapter 19, Table 19.4). Exactly 100 years later, June 2012 was only slightly less dull, with a mere 102 hours of bright sunshine, barely half that recorded in March that year—180 hours.

The sunniest single day on Durham's records up to 1999 was on 5 June 1940, equalled on 15 June 1957, with a near maximum-possible 15.8 hours of sunshine (Table 9.4). No day has been sunnier than 14.7 hours (on 26 May 1989) since July 1963 (and none have reached or surpassed 15.0 hours in any month since 1960), which almost certainly represents a gradual increase in obstructions due to tree growth limiting early morning and late evening midsummer sunshine.

Table 9.4 *June sunshine duration at Durham Observatory. Observed monthly extremes 1880–2021, sunniest days 1882–1999 only*

June mean sunshine duration 166 hours, **5.5 hours per day** (average 1991–2020). Possible day length: 516 hours. Mean sunshine duration as percentage of possible: 32.1

Sunniest months			Dullest months			Sunniest days 1882–1999	
Duration, hours	**Per cent of possible**	**Year**	**Duration, hours**	**Per cent of possible**	**Year**	**Duration, hours**	**Date**
297.0	57.6	1940	89.4	17.3	1987	15.8	5 June 1940 and 15 June 1957
253.5	49.1	1957	91.7	17.8	1912	15.7	14 June 1959
250.9	48.6	1949	97.5	18.9	1916	15.6	26 June 1937 and 5 June 1962
245.7	47.6	1970	102	19.7	2012	15.5	29 June 1931
237.6	46.0	1960	103.4	20.0	1997	15.4	6 June 1940 and 30 June 1940
Last 50 years to 2021							
224.0	43.4	1989	89.4	17.3	1987	14.6	19 June 1989 and 27 June 1995

Daily sunshine records ceased at the Observatory on 31 December 1999: monthly sunshine totals from January 2000 onwards are estimated (to the nearest hour) by regression from the Met Office East and North-east England sunshine series. See Appendix 6 for record metadata and missing months.

Wind and weather

Fog (1962–97)

Morning fog is uncommon in June, averaging about one morning in two years, although a few years have recorded two fogs at 0900 GMT.

Barometric pressure (1844–1960)

The highest observed MSL barometric pressures in June at the Durham Observatory occurred in 1959, when at 2100 GMT on 13 June the MSL pressure reached 1043.1 hPa. The barometer remained above 1040 hPa over 13–14 June 1959—the only instance of 1040 hPa during the summer months. The next-highest for the month was 1039.1 hPa at 2200 on 27 June 1867, which was closely approached on 15 June 1874 when at 1000 the pressure was 1038.8 hPa.

June's lowest observed pressure was 982.4 hPa, the reading at 2100 GMT on 28 June 1938—an unusually deep depression for midsummer, and a sharp contrast to the prolonged anticyclonic conditions that prevailed throughout a very dry spring, when the mean for April was the highest on record. Values almost as low were noted at 0900 GMT on 18 June 1933 (984.6 hPa) and at 2200 on 12 June 1862 (985.1 hPa).

The most anticyclonic June was in 1865, when the mean for the month reached 1024.1 hPa; not far behind was June 1887, with a mean of 1023.7 hPa, and June 1921, 1022.5 hPa—all three were in the 'very dry' category (1887 and 1925 feature in Table 9.3), although we are less sure of a reliable rainfall total for June 1865 than for later years. The most cyclonic Junes were those of 1852 (a very wet but also extremely warm month; mean at MSL 1004.9 hPa), and 1860 (1006.4 hPa), a disappointingly cool and wet month.

Wind speed and direction (1961–97)

Figure 9.7 shows the distribution of wind speed and direction at the 0900 GMT observation in June during the period 1961–97. June is the windiest of the three summer months, with a mean wind speed at 0900 GMT of 6.3 knots, but there is normally little to choose between June, July, and August. June 1973 was the windiest month during this period, mean speed 11.3 knots. Strong winds are not unknown in June—in an earlier period, the deep depression in June 1938 referred to earlier registered a mean hourly wind speed at the Durham Observatory of west-south-westerly 25 knots on the 27th, with a peak gust of 53 knots at 1255 GMT. June also sees 'the return of the westerlies', with significantly greater frequencies of winds between south-west and north-west when compared to May (compare with Figure 8.6) and a reduction in north-easterly winds.

Figure 9.6 *A sunny day for Durham Regatta, 12 June 2011. (Tim Burt)*

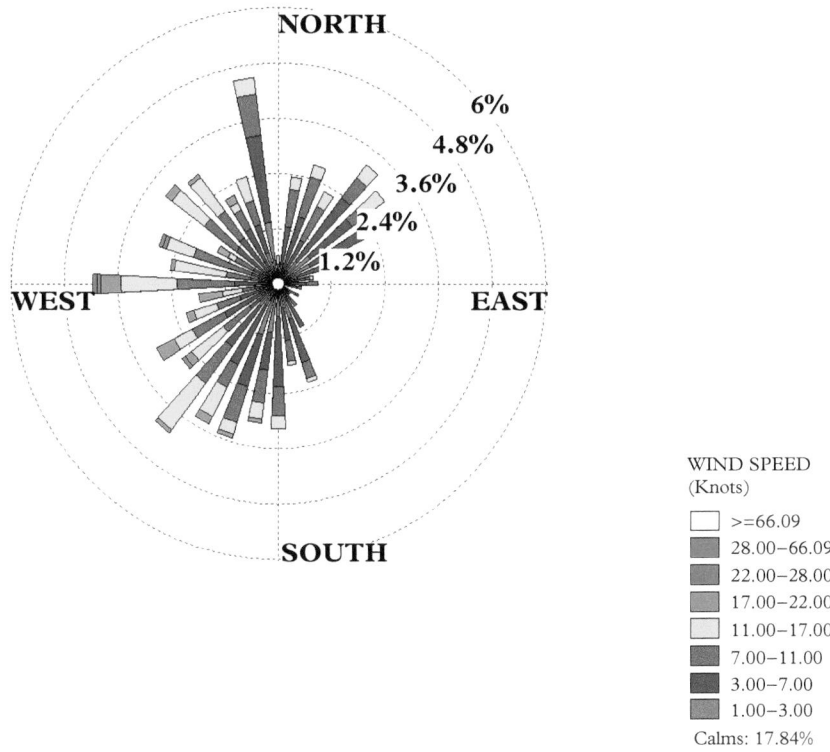

NORTH

6%
4.8%
3.6%
2.4%
1.2%

WEST   EAST

SOUTH

WIND SPEED
(Knots)

☐	>=66.09
■	28.00–66.09
■	22.00–28.00
☐	17.00–22.00
☐	11.00–17.00
■	7.00–11.00
■	3.00–7.00
☐	1.00–3.00

Calms: 17.84%

Figure 9.7 *Wind rose for 0900* GMT *observations in June, period 1961–97*

Durham temperature, precipitation, and sunshine in graphs—June

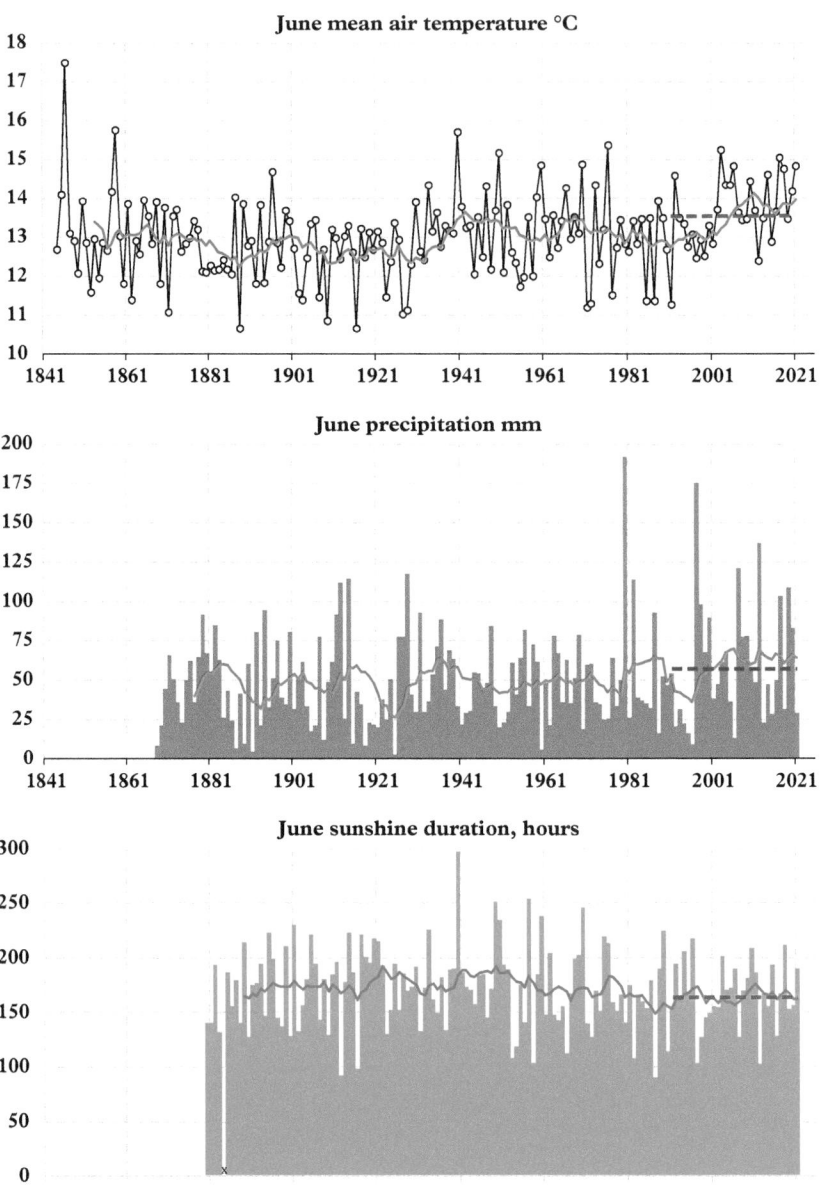

Figure 9.8 *Monthly values of (° from top) June's mean temperature (°C, since 1844), total precipitation (mm, since 1868), and sunshine duration (° hours, since 1880, missing data 1884) at Durham Observatory. The 1991–2020 averages are indicated by the thick blue dashed line, while the ten-year running mean ending at the year shown is indicated by the red line*

10

July

July is, on average (and by just two-tenths of a degree Celsius), the warmest month of the year in Durham; August lies close behind. The mean air temperature during the standard averaging period 1991–2020 was 15.8 °C (mean daily maximum 20.2 °C, mean daily minimum 11.3 °C), while the mean daily maximum temperature increases from a little above 19 °C at the beginning of the month to 21 °C during the fourth week, the warmest week of the year. Within the 100 summers 1921–2020 July was the warmest month of the year 54 times, in other words just a little more than one year in two. Measurable rainfall can be expected on one day in two, about the same as in June (despite July being one day longer), and the mean monthly rainfall is 62 mm. Thunderstorms are not uncommon and can deposit up to a month's average rainfall in just a few hours. July is also the second sunniest month of the year, after May, with 175 hours of bright sunshine on average, equivalent to 5.7 hours daily; typically, just two days are sunless.

Temperature

Normally at least one day in Durham in July will reach or exceed 26 °C, but 30 °C is much less frequent—between 1843 and 2021 only 17 days attained this temperature, eight of these in July (three in June, five in August, and one in September). Of these eight, half have occurred since and including 2006, while 2019 and 2020 became the first two consecutive years to attain 30 °C in Durham, both in July.

Perhaps it is most appropriate to start with 25 July 2019, the day when a maximum temperature of 32.9 °C was measured, the highest temperature yet recorded at the Durham Observatory (Figure 10.1 and Table 24.4). This was on the same day upon which a new United Kingdom maximum temperature was attained, namely 38.7 °C at the Cambridge Botanic Garden (Figure 10.2). The new record just surpassed Durham's previous record of 32.5 °C set on 3 August 1990. The highest July temperature previously observed at Durham—a record which had stood for over 140 years—was on 16 July 1876 when the maximum temperature in the Glaisher stand reached 33.6 °C, from which a Stevenson screen equivalent maximum temperature is estimated at 31.3 °C (based also upon the maximum that same day of 30.1 °C in the 'North Shed'). The previous record for the highest daily *mean* temperature for any July day was

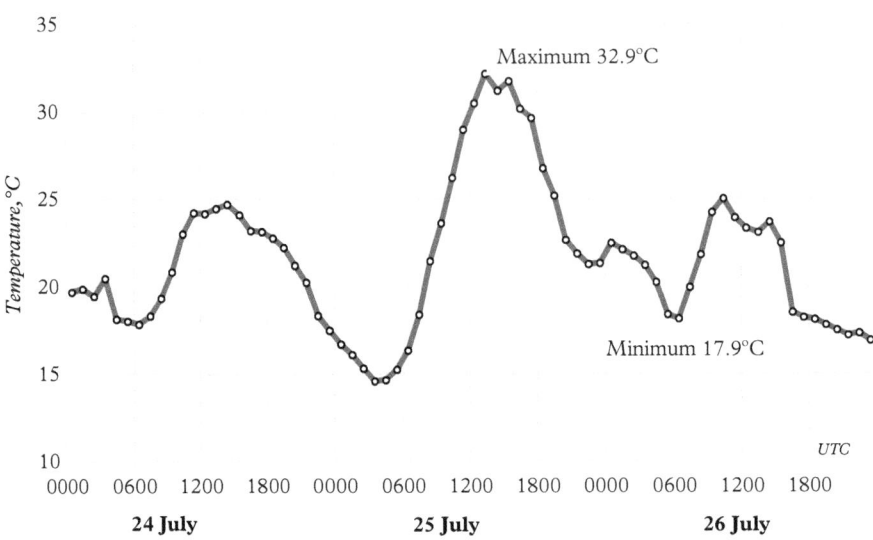

Temperature, °C

35

30

25

20

15

10

Maximum 32.9°C

Minimum 17.9°C

UTC

| 0000 | 0600 | 1200 | 1800 | 0000 | 0600 | 1200 | 1800 | 0000 | 0600 | 1200 | 1800 |

24 July   **25 July**   **26 July**

Figure 10.1 *Hourly temperatures at the Durham Observatory, 24–26 July 2019, from the Observatory AWS record; 25 July became Durham's hottest day on record, reaching 32.9° C. Times are in* GMT (UTC)

also surpassed on 25 July 2019 (Figure 10.1): the average of the daily maximum and minimum temperatures was 23.6°C, surpassing the previous July record of 23.2 °C, set on 19 July 1901 (Table 24.6). Further details of 2019's record-breaking temperatures are included in Chapter 22.

July 2020 was a rather mixed month, and the great heat of the last day of the month came as something of a surprise; it was not until 12 July when the daily maximum temperature exceeded even 20 °C and indeed, up until the final day of the month, the highest temperature had been 23.2 °C on 16 July. It was all the more surprising, therefore, when the temperature surged to 30.8 °C on the 31st—the fourth-hottest July day on record.

Table 10.1 shows Durham's hottest days and nights during the month of July. Durham's warmest-ever night was on 20 July 2016, when the minimum temperature was 18.9 °C, following a maximum temperature of 29.2 °C the previous afternoon. Notable for its era was the minimum temperature of 18.1 °C on 12 July 1859, a value which remains the second-warmest night on Durham's records (Table 24.5), although there is more uncertainty about the reading considering its exposure, which would be non-standard by today's recommendations.

No air frost has ever been recorded at Durham in July; the coldest night on record was 1 July 1929, when the minimum temperature in the Stevenson screen fell to 1.4 °C. There was a ground frost that night, grass minimum temperature −1.7 °C, one of only 22 ground frosts in July since such records began in 1874. In July 1918 there were five ground frosts, and four in both the Julys of 1922 and 1939, although there has not been a ground frost at Durham in July since 1940. Notable also is the inclusion of July 1921 in

Table 10.1 *Highest and lowest maximum and minimum temperatures at Durham Observatory in July, 1843–2021*

Rank	Hottest days	Coldest days	Warmest nights	Coldest nights
1	32.9 °C, 25 July 2019	10.0 °C, *21 July 1883* and 4 July 1920	18.9 °C, 20 July 2016	1.4 °C, 1 July 1929
2	31.3 °C, 16 July 1876	*10.1 °C, 5 July 1888*	18.1 °C, 12 July 1859	1.7 °C, 11 July 1917 and 5 July 1921
3	31.0 °C, 1 July 2015	*10.4 °C, 6 July 1888*	17.9 °C, 26 July 2019	2.2 °C, 10 July 1911 and 2 July 1917
4	30.8 °C, 31 July 2020	*10.5 °C, 16 July 1888*	17.7 °C, 24 July 2019	2.4 °C, 2 July 1909
5	30.6 °C, 10 July 1921 and 31 July 1943	*10.7 °C, 11 July 1888*	17.6 °C, 24 July 1900	2.5 °C, 20 July 1898
Last 50 years to 2021				
	32.9 °C, 25 July 2019	12.3 °C, 1 July 1980	18.9 °C, 20 July 2016	3.1 °C, 24 July 1990

Temperatures were recorded in different thermometer screens during the period of record; see notes on page xvii and Appendix 3 for more details.

both the 'coldest nights' (a minimum temperature of 1.7 °C on the 5th, grass minimum −2.8 °C) and the 'hottest days' (maximum temperature of 30.6 °C on the 10th). This seeming contradiction can be explained by clear skies (14 hours of sunshine on the 10th) and dry ground conditions—dry soil both warms and cools more quickly than when moist; 1921 remains one of the driest years on Durham's records.

In contrast, the coldest summer days are usually wet; on 21 July 1883 and again on 4 July 1920, the maximum temperature was just 10.0 °C, during unseasonably cold and wet spells. Three days in July 1888 remained below 11 °C, contributing to that July becoming the coldest on record (Table 10.2), and the summer of 1888 was one of Durham's poorest (Chapter 19).

July 1888 must have come as something of a shock after the previous July: the mean daily maximum temperature in July 1887 was 21.6 °C, a July record at the time, but the average for July 1888 was only 15.6 °C—thereby establishing another July record, but at the low extreme, and one which has yet to be beaten (Figure 10.3). Within the last 50 years, July 1978 remains the coolest; the highest temperature that month was just 23.2 °C, although even this was an improvement on July 1974, whose warmest day (the 7th) reached only 20.5°C. The hottest day in July 1965 was 21.1 °C on 22nd, thus failing to attain the 21.7 °C reached on 29 *March* that year.

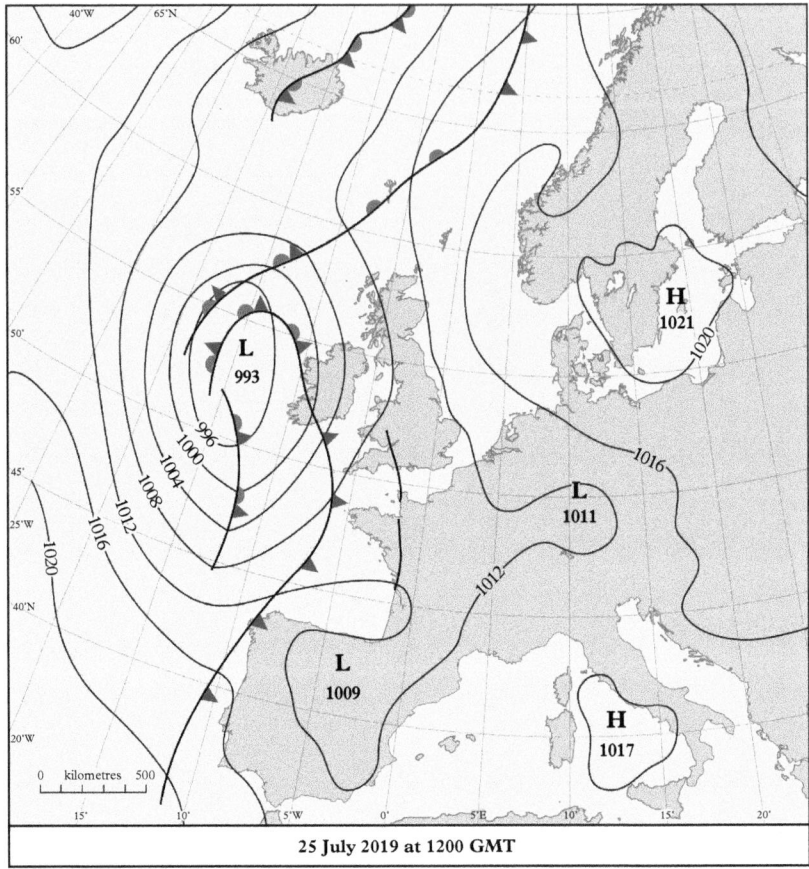

Figure 10.2 *Synoptic chart for 1200* GMT *on 25 July 2019, Durham's (and the United Kingdom's) hottest day to date, the temperature reaching twin record levels, 32.9 °C in Durham and 38.7 °C in Cambridge. (Based upon the Met Office* Daily Weather Summary; *drawn by Chris Orton, Durham University)*

July 1852 was the second-hottest month of the nineteenth century in Durham, with a mean temperature of 17.4 °C; only June 1846 was hotter. July 1852 retained the July record until 1983, but July 1983 now stands fourth behind 2006, 2013, and 2018; July 2019 tied with 1901 as the seventh-warmest overall, although the mean minimum temperature in 2019 (12.9 °C) remains the equal second-highest on record for any month, behind 13.2 °C in July 1852.

All the warmest Julys in Table 10.2 have occurred since 1983, with the exception of 1852; temperatures at that time were measured in the 'North Shed' penthouse and so should be reasonably reliable (Appendices 2 and 3). At the time of writing,

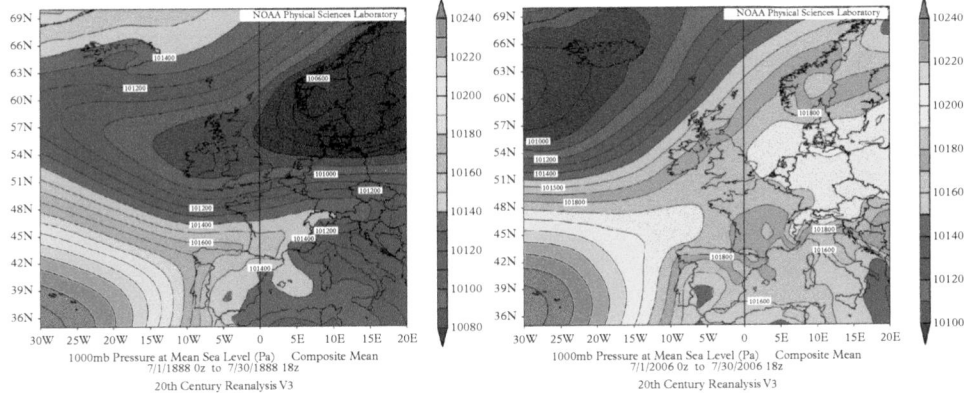

Figure 10.3 *A contrast in summer conditions: mean* MSL *pressure distribution during the very cool and wet July of 1888 (left) and July 2006 (right) the warmest month of any on Durham's records, and one of the sunniest and driest too. Isobars are in Pascals (divide by 100 for hPa = millibars), at 100 Pa intervals. (From the 20th Century Reanalysis version 3 (20CRv3) plot [58])*

July 2006 remains Durham's hottest-ever month, as measured by both mean temperature (18.3 °C) and mean daily maximum temperature (24.2 °C) (Table 24.7 and Figure 10.3), as well as the sunniest (Table 10.4). Only four days that month were cooler than July's long-term mean daily maximum temperature (20.2 °C), while 17 days exceeded 25 °C, including a run of nine consecutive days at the end of the month. The hottest day was the 18th, at 30.5 °C, just missing out on a place in Table 10.1. Two other recent Julys appear in Table 10.2, namely 2013 and 2018. After a rather cool start to July 2013, it ended up as Durham's second-hottest month on record, although mainly due to a succession of warm nights as the highest maximum was 'only' 28.4 °C that month. July 2018 was also persistently warm rather than hot: eight days reached 25 °C, although the highest maximum was just 27.6 °C. That month's high mean temperature was also boosted by particularly warm nights, four of which remained above 16°C. The Julys of 2006, 2013, 2018, and 1983 also take the top four places in the hottest months on Durham's instrumental record (Table 24.7); July 1852 appears in seventh place, and July 2019 in ninth. July 2021 just misses out on a place in the table, with a mean temperature of 17.2 °C, owing mainly to high night-time temperatures rather than exceptional heatwave conditions. The mean minimum temperature of 12.8 °C was only a tenth of a degree behind the highest July mean in the last 160 years: only one night fell below 10 °C during the entire month.

Precipitation

July's average monthly rainfall in Durham was 62 mm over the 30 years ending 2020. Since 1868, July's rainfall has varied from just 3.6 mm in 1878 to just under 210 mm ten years later (Table 10.3).

Table 10.2 *July mean temperatures at Durham Observatory, 1844–2021*

July mean temperature 15.8 °C (average 1991–2020)

Hottest months			Coolest months		
Mean temperature, °C	Departure from 1991–2020 normal degC	Year	Mean temperature, °C	Departure from 1991–2020 normal degC	Year
18.3	+2.5	2006	12.1	−3.7	1888
18.0	+2.2	2013	*12.2*	*−3.6*	*1862*
17.9	+2.1	2018	12.5	−3.3	1965
17.7	+1.9	1983	12.7	−3.1	*1892, 1922*
17.4	*+1.6*	*1852*	12.8	−3.0	1907
Last 50 years to 2021					
18.3	+2.5	2006	13.8	−2.0	1978

Table 10.3 *July precipitation at Durham Observatory 1868–2021*

July mean precipitation 62.0 mm (average 1991–2020)

Wettest months			Driest months			Wettest days	
Total fall, mm	Per cent of normal	Year	Total fall, mm	Per cent of normal	Year	Daily fall, mm	Date
209.7	338	1888	3.6	6	1878	61.4	8 July 1893
183.9	297	1930	7.7	12	1935	57.1	25 July 1888
168.8	272	2009	9.4	15	1977	44.7	22 July 1930
134.0	216	2008	10.0	16	2006	44.4	16 July 2009
133.4	215	1940	10.1	16	1989	43.9	14 July 1961
Last 50 years to 2021							
168.8	272	2009	9.4	15	1977	44.4	16 July 2009

July 1878 remains the driest July on record at Durham, with just 6 per cent of normal rainfall, falling on only six days. July 1977 was the driest July in the last five decades, a welcome dry month in the middle of a wet period that included the 1976 autumn and the winter of 1976/77, both very wet seasons at the end of the long drought of 1975/76. As might be expected, July 2006, the sunniest and warmest July on record as noted earlier, was a very dry month. July 1989, another fine July and the third-sunniest on record to that date, also appears in Table 10.3 among the driest Julys on record.

July 1888 was not only the *coldest* July on record (Table 10.2), but also it was the *wettest* (Table 10.3)—and, for good measure, the wettest month of any recorded in Durham in the past 150 years (Table 25.3, Figure 10.3). Rainfall that year was very varied: March, July, and November were all very wet (above 100 mm), while January, May, and October were very dry (below 20 mm). The rainfall statistics for July 1888 make dire reading: rain fell on 20 days, six of which amounted to 10 mm or more, including 26.4 mm on the 4th, 27.3 mm on the 17th, 57.1 mm on the 25th, and 38.0 mm on the 26th. The two-day total for 25–26 July 1888, 95.1 mm, remains the third-largest two-day fall on record at Durham, surpassed only in August 1986 (96.8 mm) and September 1976 (120.1 mm) (Table 25.2). *British Rainfall* noted heavy falls in a 'compact area' in County Durham on 25 July 1888: the largest total was at Seaham Hall (77.0 mm), between Durham and the east coast.

July's wettest day was 8 July 1893, when 61.4 mm fell; *British Rainfall* noted five locally heavy falls of rain in the north-east—in Sunderland, Gateshead, and three sites in Newcastle, where 62 mm fell at the Clifton Road gauge. In July 1960, and again in July 2015, rain fell on 26 days. Neither of these months have particularly high totals and come nowhere close enough to be included in Table 10.3; the weather must have been very disappointing, nevertheless.

Severe flooding resulted in Durham from a fall of 44.4 mm on 16 July 2009 (11.6 mm falling in the hour commencing 1400 GMT), followed by a further 39.8 mm the following day, including 15.8 mm in two hours from 1400 GMT. The combined two-day total of 84.2 mm was the wettest such spell in July since 1888. Large summer floods are rare on the River Wear, but the highest flood discharge ever recorded on the river level gauge at Sunderland Bridge (where records began in late 1957) followed this event: at 0730 GMT on 18 July the water level peaked at 3.71 m. The University's Maiden Castle sports centre was inundated for only the third time since being built in the early 1960s. Just across the road from the sports centre, there was severe erosion on Houghall College land when the river burst its banks, eroding a spectacular 'canyon' incised several metres deep into the floodplain sediments (Figure 10.4).

Sunshine

July lies behind May as, on average, the second-sunniest month of the year in Durham, with an average of 175 hours of bright sunshine in the standard averaging period 1991–2020, equivalent to just under 5.7 hours per day.

The five sunniest Julys are all from notable British summers. July 1911 was a memorable month and an even more memorable summer—Durham's hottest since 1857. July 1955, the sunniest July on record to that date (Table 10.4), began wet, but rain fell on only three days, with a 27-day absolute drought starting on 5 July lasting beyond the end of the month. Thirteen days that month recorded ten hours of sunshine, the highest for any July. Both July and August 1955 were warm and dry, a fine summer

Figure 10.4 *The 'Durham Canyon' resulting from flooding due to heavy rainfall on 16–17 July 2009. This photograph was taken by Jeff Warburton on 19 July 2009: note the pipework left suspended above the channel floor. See also Chapter 22.*

sandwiched between two dreadful summers in 1954 and 1956—blamed in contemporary media on atomic bomb tests. July 1989 was another hot, dry, sunny summer, with a warm, dry, and sunny July. Summer 1976 saw the culmination of the long drought (see Chapter 22), while July contributed to a brilliant summer, being warm and dry as well as the fifth sunniest. July 2006 was a truly exceptional month—hot, dry, and sunny, making an appearance in Tables 10.2, 10.3, *and* 10.4. July's daily record for sunshine duration is held by 13 July 1929, with 15.5 hours, in what was otherwise an average month for sunshine.

Durham's dullest July came in 1912, the 74 hours total representing just 14 per cent of the possible duration. To date, 17 Januarys and 58 Februarys have recorded more sunshine than July 1912: it must have been an especially disappointing month given the 'flaming summer' of the previous year. July's meagre hours of sunshine contributed to a dismal, wet summer and indeed a very dull year (August 1912's sunshine total was even lower than July's), for 1912 remains both the dullest summer and dullest calendar year on Durham's records. Gordon Manley attributed this poor state of affairs to the volcanic eruption of Mount Katmai in Alaska the previous month [19].

Table 10.4 *July sunshine duration at Durham Observatory. Observed monthly extremes 1880 to 2021, sunniest days 1882–1999 only*

July mean sunshine duration 175 hours, 5.7 hours per day (average 1991–2020). Possible day length: 517 hours. Mean sunshine duration as percentage of possible: 33.9

Sunniest months			Dullest months			Sunniest days 1882–1999	
Duration, hours	Per cent of possible	Year	Duration, hours	Per cent of possible	Year	Duration, hours	Date
273	52.9	2006	73.9	14.3	1912	15.5	13 July 1929
254.5	49.3	1955	75.3	14.6	1944	15.4	7 July 1905
252.0	48.8	1911	81.9	15.9	1968	15.3	11 July 1911 and 3 July 1933
237.2	46.0	1989	86.9	16.8	1965	15.2	20 July 1906 and 10 July 1927
236.7	45.9	1976	93.3	18.1	1917	15.1	8 July 1901 and 1 July 1917
Last 50 years to 2021							
273	52.9	2006	122	23.6	2000	14.3	Numerous

Daily sunshine records ceased at the Observatory on 31 December 1999: monthly sunshine totals from January 2000 onwards are estimated (to the nearest hour) by regression from the Met Office East and North-east England sunshine series. See Appendix 6 for record metadata and missing months.

Wind and weather

Fog (1962–97)

Fog is uncommon in July, with fog at 0900 GMT being recorded in only seven years between 1962 and 1967, although twice in 1980 and 1988. The average is for morning fog to be recorded less than once in three years.

Barometric pressure (1844–1960)

The highest MSL barometric pressures recorded at the Durham Observatory in July all occurred in 1911, a fine, warm, dry, and sunny month when the barometer rose to 1037.7 hPa at 0900 GMT on the 10th. The next-highest was 1035.9 hPa at 2100 GMT on 4 July 1952, followed by 1035.8 hPa at 1000 on 24 July 1868—the latter also during a fine, dry, and hot spell.

July's lowest observed pressure was 977.2 hPa at 0900 GMT on 6 July 1922, another unusually deep depression for midsummer, and a sharp contrast to the prolonged anti-cyclonic conditions that prevailed throughout the previous summer. Values not quite as low were recorded at 2100 GMT on 3 July 1924 (982.3 hPa) and at 0900 on 20 July 1848 (983.0 hPa).

The most anticyclonic Julys were in 1885 (which was otherwise unremarkable) and the sunny and mostly dry July of 1955 (Tables 10.3 and 10.4), tying at a mean for the month of 1021.7 hPa, just ahead of July 1911 with a mean of 1020.9 hPa. The most cyclonic Julys were those of 1861 (mean at MSL 1005.1 hPa), 1936 (1006.3 hPa), and 1931 (1006.9 hPa). All three were cool and wet; 1931 and 1936 were also very dull, as almost certainly was 1861, although this predates the start of sunshine records.

Wind speed and direction (1961–97)

Figure 10.5 shows the distribution of wind speed and direction at the 0900 GMT observation in July during the period 1961–97. Mean wind speeds are at their lowest in July (mean at 0900 GMT 6.0 knots), but there is normally little to choose between June, July, and August. July 1970 was remarkably breezy, the mean speed of 12.8 knots being more than twice the long-term mean. July also sees a return closer to the mean annual pattern

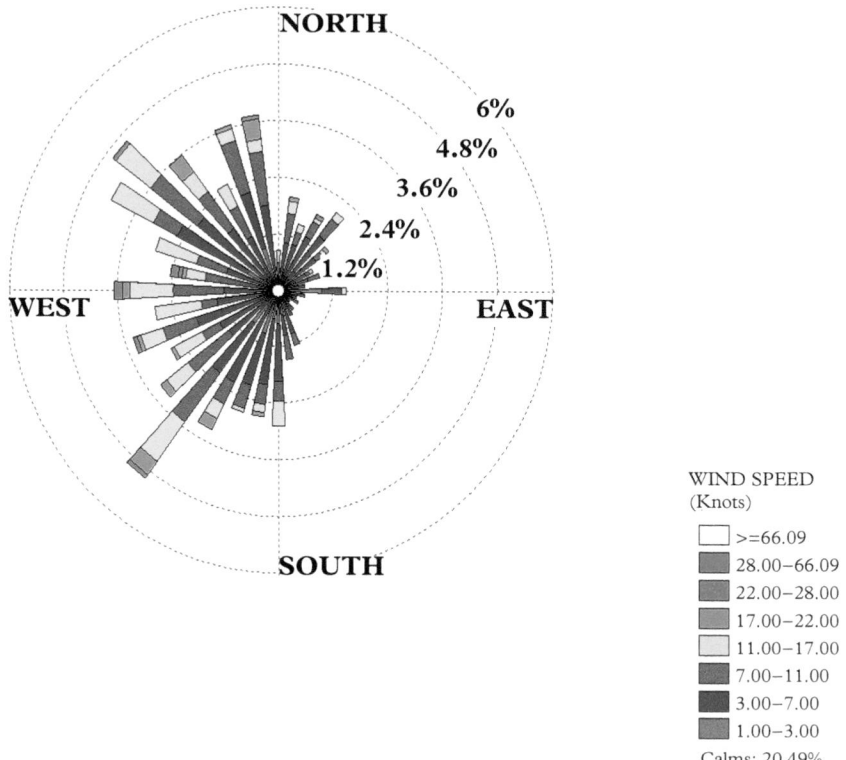

Figure 10.5 *Wind rose for 0900 GMT observations in July, period 1961–97*

of the strongest and most frequent winds blowing from between south-west and north-west, markedly different from May, for example (compare with Figure 8.6), which sees a much greater incidence of north-easterly winds. Strong winds do occur occasionally in July; the heavy rainfall on the afternoon of 16 July 2009 described earlier (Table 10.3, Figure 10.2) was accompanied by a strong and very gusty west-north-westerly wind, the highest gust 29 knots at 1450 GMT.

Durham temperature, precipitation, and sunshine in graphs—July

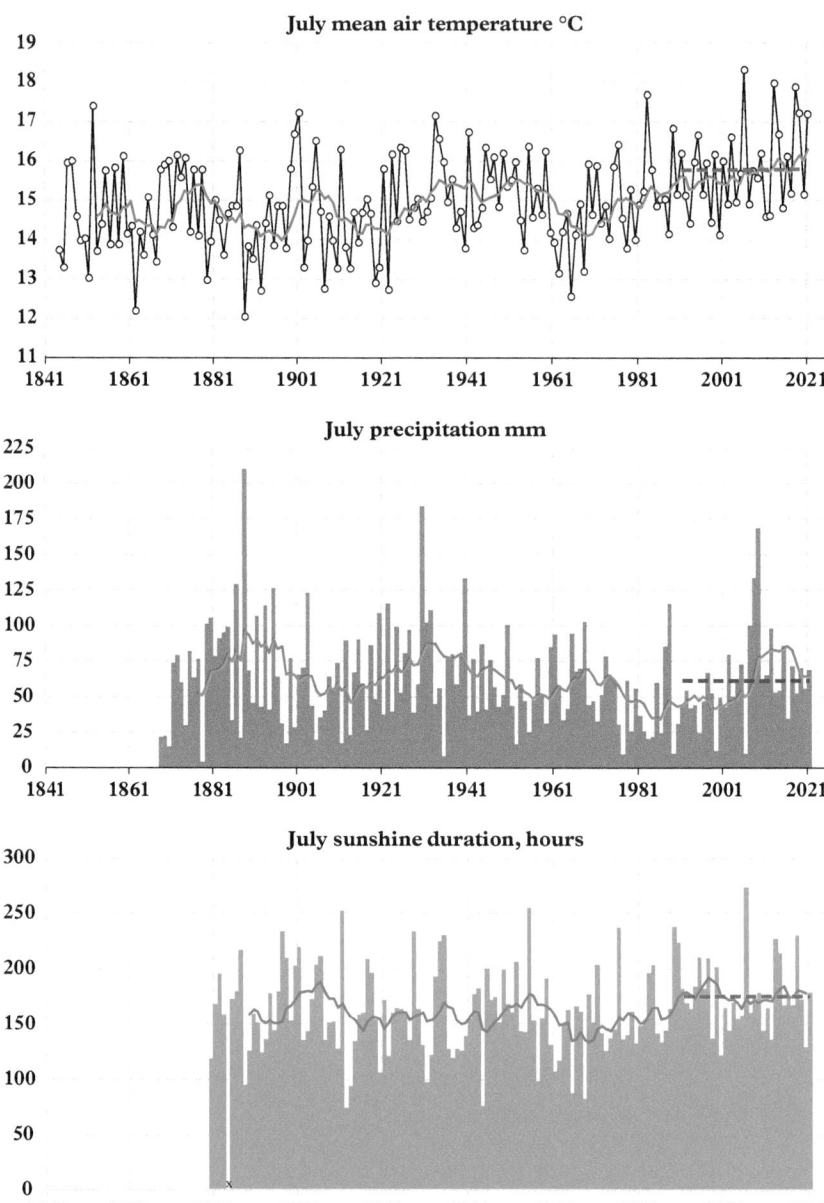

Figure 10.6 *Monthly values of (from top) July's mean temperature (°C, since 1844), total precipitation (mm, since 1868), and sunshine duration (hours, since 1880, missing data 1884) at Durham Observatory. The 1991–2020 averages are indicated by the thick blue dashed line, while the ten-year running mean ending at the year shown is indicated by the red line*

11

August

In the minds of many, at least in England and Wales, August is the traditional high summer month—long school holidays, visits to the beach, light and warm evenings. However, the slow decline into autumn has begun and perhaps the Scots are sensible to return to school mid-month rather than waiting for September to arrive. On average, August in Durham is just two-tenths of a degree Celsius cooler than July, although it is the hottest month of the year in about one year in three. Temperatures normally begin to decline, slowly at first (mean daily maximum temperatures falling from around 21 °C in the first week, to below 19 °C in the closing days), and by the end of the month the evenings are drawing in and often starting to feel cool. Mid-August is often one of the sunniest periods of the year, but as day length begins to decline, so does the average sunshine, with August slightly less sunny than July. There are of course wide variations from year to year, but on average August is the wettest of the three summer months, and the second-wettest month of the year, with occasional heavy or thundery falls of rain featuring in the annals of the Observatory.

Temperature

August is, on average, slightly cooler than July, with a mean temperature of 15.6 °C: the mean daily maximum and minimum temperatures over the 1991–2020 period were 19.8 °C and 11.3 °C, respectively.

Several notable heatwaves have occurred in August, although only five August days have ever attained 30 °C—including the only instance of two consecutive days at this level on Durham's records (Table 11.1, and Figures 11.1 and 11.2). At the start of August 1990, advection of very warm continental air around an anticyclone centred near Denmark resulted a short but intense hot spell [61]. In Durham, the temperature reached 31.9 °C on 2 August and 32.5 °C the next day—at the time the two hottest days ever recorded in Durham (surpassing the previous all-time record set back in July 1876). Both were (slightly) surpassed in July 2019, but the mean daily temperature on 3 August 1990, 24.4 °C, remains Durham's highest on record (Table 24.6). Prior to 1990, the hottest August day was 9 August in the hot summer of 1911, when the maximum temperature reached 30.1 °C; this was also exceeded in 1995, with 30.3 °C as

Table 11.1 *Highest and lowest maximum and minimum temperatures at Durham Observatory in August 1843–2021*

Rank	Hottest days	Coldest days	Warmest nights	Coldest nights
1	32.5 °C, 3 Aug 1990	9.4 °C, 28 Aug 1919	17.9 °C, 14 Aug 2001 and 6 Aug 2006	*0.5 °C, 22 Aug 1864*
2	31.9 °C, 2 Aug 1990	*10.5 °C, 16 Aug 1888*	17.8 °C, 26 Aug 1959	*0.7 °C, 27 Aug 1864*
3	30.3 °C, 21 Aug 1995	10.7 °C, 22 Aug 1904	17.7 °C, 3 Aug 2018	*0.8 °C, 23 Aug 1864*
4	30.1 °C, 9 Aug 1911	10.9 °C, 8 Aug 1900	17.5 °C, 3 Aug 1933	*1.1 °C, 24 Aug 1847*
5	30.0 °C, 12 Aug 1953	*11.0 °C, 6 Aug 1860*	17.1 °C, 20 Aug 2009	*1.4 °C, 22 Aug 1850*
Last 50 years to 2021				
	32.5 °C, 3 Aug 1990	11.5 °C, 26 Aug 1986	17.9 °C, 14 Aug 2001 and 6 Aug 2006	1.9 °C, 31 Aug 2012

Temperatures were recorded in different thermometer screens during the period of record; see Notes on page xvii and Appendix 3 for more details.

late as 21 August that year, the latest date in the year on which 30 °C has been reached (Table 24.1).

Durham's position near the cooler North Sea coast does provide welcome relief from the very warm nights in heatwave conditions typical of central and southern England, where overnight temperatures occasionally remain above 20 °C; in Durham, only a handful of August nights have remained above 17 °C (Table 11.1), notably 17.9 °C on 14 August 2001, equalled on 6 August 2006.

Cool, wet August days are far from unknown, however, although to date only one has failed to reach 10 °C—on 28 August 1919 the maximum temperature was just 9.4 °C, colder than an average day in March or November, on a sunless day with 22 mm of rainfall and a chilly north-west to northerly wind. Summer 1888 makes yet another appearance, with the second-coldest August day on record on the 16th of that year, at just 10.4 °C: each of the three summer months of 1888 holds a place in the 'Top 5' coldest days in June, July, and August. Within the last 50 years, the coldest August day has been 26 August 1986—when, in the second of two very wet days associated with the remains of ex-Hurricane *Charley* (see Precipitation, and Chapter 22), the maximum temperature reached only 11.5 °C in the coldest August day for 65 years.

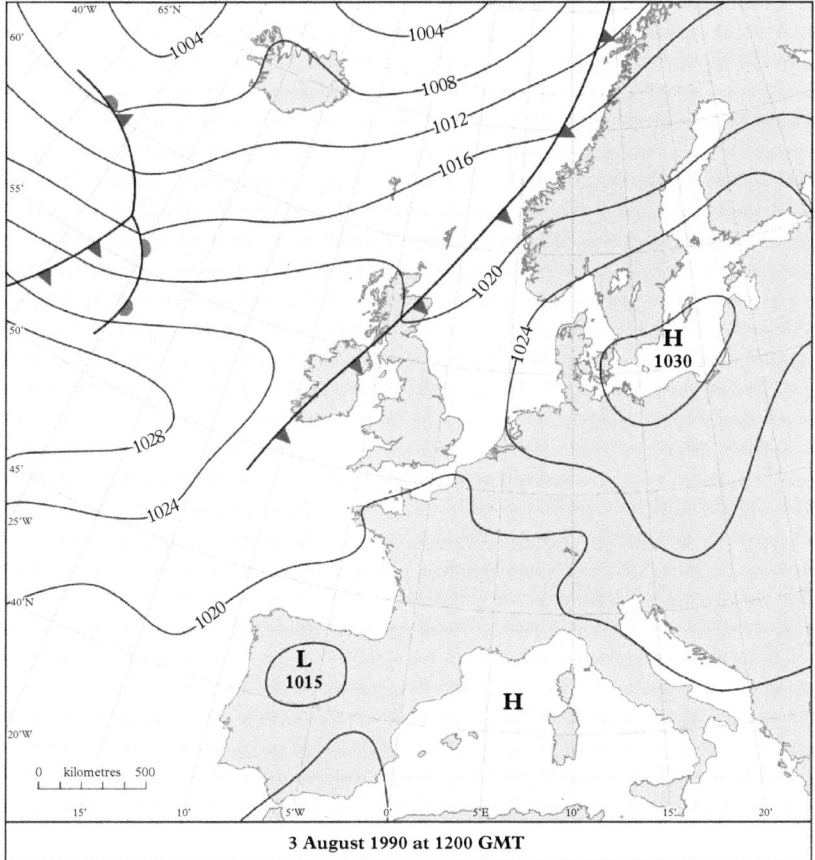

3 August 1990 at 1200 GMT

Figure 11.1 *The synoptic situation at noon* GMT *on 3 August 1990—an exceptionally hot day over almost the whole of the United Kingdom, the temperature reaching a then-new national record 37.1 °C at Cheltenham. In Durham, 30 °C was reached on both the 2nd and 3rd—the only occasion when 30 °C has been attained on two consecutive days in the city (Based upon the Met Office* Daily Weather Report*)*

No air frost has ever been recorded in August in Durham, although three nights in August 1864 came close to doing so, the coldest night being the 22nd, when the Stevenson screen equivalent minimum was 0.5 °C (0.3 °C on the Glaisher stand). Since 1900, the lowest August minimum in a Stevenson screen has been 1.7 °C on 23 August 1938. Ground frosts are rare but not unknown; at the time of writing, none have occurred in August since 1992, but in the Augusts of 1885 and 1920 there were four. On 30 August 1885, the grass minimum was −2.7 °C.

The five warmest Augusts on record are all from memorable hot, sunny, dry summers, with 1933 the earliest of those on Table 11.2. August 1947 must have been

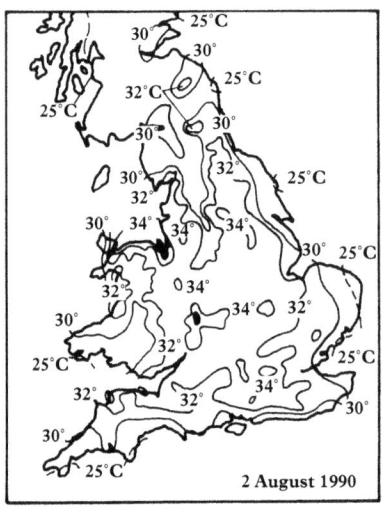 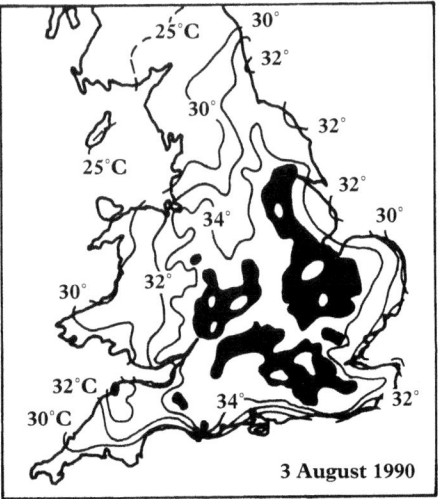

Figure 11.2 *Maximum temperatures on 2 and 3 August 1990 over England and Wales. The shaded area represents 35 °C, and the white area within the 35 °C area 36 °C. From reference [61] (Stephen Burt)*

Table 11.2 *August mean temperatures at Durham Observatory, 1843–2021*

August mean temperature 15.6 °C (average 1991–2020)

Hottest months			Coolest months		
Mean temperature, °C	Departure from 1991–2020 normal degC	Year	Mean temperature, °C	Departure from 1991–2020 normal degC	Year
17.6	+2.0	1975	11.7	−3.9	1912
17.2	+1.6	1947, 1995	*12.2*	−*3.4*	*1848*
17.1	+1.5	1997	12.3	−3.3	*1885, 1956, 1986*
17.0	+1.4	1990	12.4	−3.2	1902
16.8	+1.2	1933	12.6	−3.0	1922
Last 50 years to 2021					
17.6	+2.0	1975	12.3	−3.3	1986

very welcome after a very cold, late winter and ongoing post-war austerity. The mean maximum temperature for the month was 23.0 °C, the highest on record to that date, although the warmth was consistent rather than spectacular throughout the month, as the warmest day (the 16th, equalled on the 28th) attained only 26.7 °C, while under clear summer skies minimum temperatures fell below 10 °C on nine nights. The warmest August of all, 1975, was in the first of two summers bracketing the 1975/76 drought. That month again featured steady warmth rather than extremely hot days, for although no day reached 29 °C, the mean maximum temperature was 22.8 °C, at the time second only to August 1947. Despite the absence of extreme heat, the seven-day spell 4–10 August 1975 remains the warmest week on Durham's records, with a mean daily temperature of 20.8 °C (Table 24.11). August 1995 just pipped the mean maximum temperatures reached in 1947 and 1975, at 23.1 °C.

August 1912 remains the coldest August on Durham's records (Table 11.2), its mean air temperature of 11.7 °C more like late May than the height of summer, and after the exceptional summer of 1911, this must have been a very disappointing August indeed. As previously noted, Gordon Manley attributed the poor weather of 1912 to the eruption of Mount Katmai in Alaska earlier in the year. The warmest day reached only 19.1 °C; the mean daily maximum temperature of 15.3 °C was down over five degrees on the previous August and remains the lowest on record for the month. The most recent of the coolest Augusts was 1986, a wet and stormy month which saw the lowest mean maximum temperature (15.9 °C) since 1912; only one day managed to surpass 20 °C. It was clearly not an August to encourage the usual outdoor activities.

Precipitation

Durham's August rainfall averages 66 mm, the second-wettest month of the year—only November is wetter.

Four of the five driest Augusts come from unforgettably fine summers; two of the driest also appear in Table 11.2 as among the warmest Augusts on record. The driest of all was 1959, a prolonged fine summer, with only 6.4 mm rainfall during August, falling on just four days. In August 1947, rain was recorded on only two days; a 32-day rainless spell began on 5 August, one that remains the longest dry spell on Durham's records (Table 25.1) and one of only two to extend beyond four weeks. In August 1976 rain fell on only three days, although one of these was the 28th when 16 mm fell—the first day in almost six weeks to receive as much as 2 mm of rain in that summer of blazing heat. August 1972, the fourth driest on record, is the exception—a cool month with near-normal sunshine.

The wettest Augusts have all seen well over twice normal rainfall, with 1956 topping the list (Table 11.3) with 176 mm, almost three times normal, closely followed by August 1986's 169 mm. Similarly, it is also no surprise to see both months feature in Table 11.2 as among the coolest Augusts on record. An 'average' August will see 16 days with measurable rainfall; the Augusts of 1860, 1912, 1956, and 2008 all had 23 days with rain,

Table 11.3 *August precipitation at Durham Observatory 1868–2021*

August mean precipitation **66.2 mm** (average 1991–2020)

Wettest months			Driest months			Wettest days	
Total fall, mm	Per cent of normal	Year	Total fall, mm	Per cent of normal	Year	Daily fall, mm	Date
175.7	272	1956	6.4	10	1959	69.1	25 Aug 1986
169.1	262	1986	7.7	12	1947	52.2	9 Aug 2004
158.3	245	1927	14.2	22	1995	50.3	13 Aug 1971
157.0	243	1917	14.6	23	1972	48.2	11 Aug 1948
156.8	243	2004	15.6	24	1955	46.4	13 Aug 1953
Last 50 years to 2021							
169.1	262	1986	14.2	22	1995	69.1	25 Aug 1986

and 1896, 1917, and 2011 had 24. In August 1986, the re-invigorated former hurricane *Charley* crossed the United Kingdom on 25th/26th, bringing widespread rain and disrupting traffic and sporting events, particularly on 25th, the late summer Bank Holiday Monday. Over the UK, the area of rainfall exceeding 25 mm and 50 mm surpassed previous records [62], and this remained the wettest day in the UK until 3 October 2020 [63]. In the north-east, the highest fall was 104.6 mm at Bar Gap Farm in the upper Tees catchment within County Durham. At Durham Observatory, 69.1 mm fell on 25th (the wettest August day on record) and a further 27.7 mm fell the following day. The two-day total of 96.8 mm remains the second-highest such fall on record at Durham, surpassed only in September 1976 (Table 25.2). Chapter 22 provides more details about this event, including a map of the rainfall over the United Kingdom.

August 2004 also deserves mention, the fifth wettest August on record—including the second-highest daily rainfall total for the month, 52.2 mm on 9th (Table 11.3, and 88.4 mm in four days (9th–12th) resulting in flooding in Durham (Figure 11.3). There was a second heavy fall later that month of 25 mm on 19 August. Unusually, August 2004 was also warm as well as very wet, although sunshine was well below normal. Another wet August day deserves mention—on 11 August 1948, 48.2 mm fell in Durham, on a day when destructive flooding occurred in the Tweed valley in southern Scotland; 158 mm fell at one gauge in Kelso, resulting in the highest known flood on the Tweed, even higher than the flood of 1831.

Sunshine

The average daily sunshine for August is 15 fewer minutes than in July and 7 fewer minutes than in June, although of course there are wide variations from year to year.

Figure 11.3 *High river level at the Durham racecourse, 20 August 2004. This flood ranks fourteenth-highest on record, with a maximum discharge of 288 cumecs (Tim Burt)*

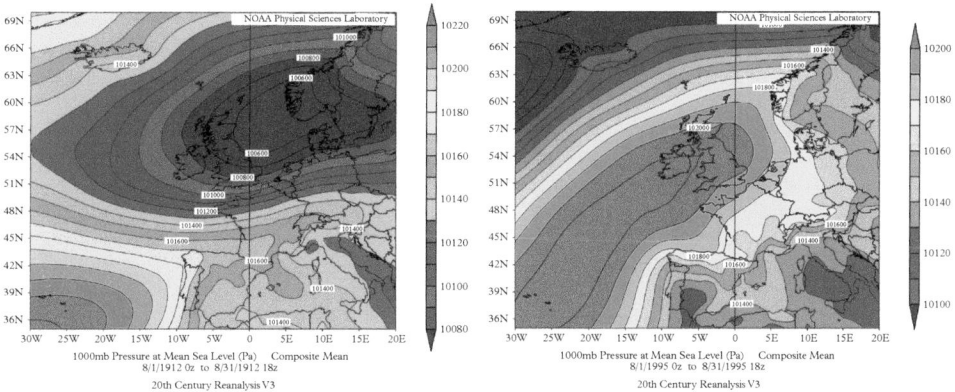

Figure 11.4 *The dullest and sunniest Augusts on Durham's records—mean MSL pressure distribution during August 1912 (left) and August 1995 (right). Isobars are in Pascals (divide by 100 for hPa = millibars), at 100 Pa intervals. (From the 20th Century Reanalysis version 3 (20CRv3) plot [58])*

Over the most recent 30-year averaging period 1991–2020, August averages 168 hours, equivalent to 5.4 hours per day.

Not surprisingly, the sunniest Augusts all coincide with memorably hot summers, with 1947 the earliest in the list (Table 11.4). Both hot summers in the mid-1970s feature, and more recently 1990 and 1995. The latter became the sunniest August of them all, with but a single sunless day (on the very last day of the month) and more than ten hours of bright sunshine on 11 days. The duration of bright sunshine, 249.3 hours, rated 54.1 per cent of the possible daylight hours. Very few months surpass 50 per cent, and by this measure August 1995 ranks as the third-sunniest month of any on Durham's records, behind only June 1940 (57.6 per cent) and April 1914 (56.2 per cent); see also Figure 11.4 and Table 26.2.

At the other extreme comes August 1912 with 57 hours in all, just 12 per cent of possible, and less sunshine than a normal January. Eight days remained sunless, and the sunniest day recorded just 6.1 hours (on the 8th)—less than many a sunny midwinter's day and by far the least sunny 'sunniest day' for any August in the period 1882–1999. As stated previously, August 1912 was also the coolest August on record (Table 11.2,

Table 11.4 *August sunshine duration at Durham Observatory. Observed monthly extremes 1880 to 2021, sunniest days 1882–1999 only*

August mean sunshine duration 168 hours, 5.4 hours per day (average 1991–2020). Possible day length: 461 hours. Mean sunshine duration as percentage of possible: 36.4

Sunniest months			Dullest months			Sunniest days 1882–1999	
Duration, hours	Per cent of possible	Year	Duration, hours	Per cent of possible	Year	Duration, hours	Date
249.3	54.1	1995	56.8	12.3	1912	14.3	14 Aug 1930
234.5	50.9	1947	95.6	20.7	1929	14.0	3 Aug 1882 and 5 Aug 1916
223.7	48.5	1975	97.1	21.1	1896	13.9	15 Aug 1960
220.5	47.8	1976	98.5	21.4	1891	13.8	16 Aug 1930
214.2	46.5	1990	101.5	22.0	1954	13.7	1 Aug 1899, 16 Aug 1925, and 15 Aug 1995
Last 50 years to 2021							
249.3	54.1	1995	105.2	22.8	1978	13.7	15 Aug 1995

Daily sunshine records ceased at the Observatory on 31 December 1999: monthly sunshine totals from January 2000 onwards are estimated (to the nearest hour) by regression from the Met Office East and North-east England sunshine series. See Appendix 6 for record metadata and missing months.

Figure 11.4) as well as the dullest, in what surely must have been a great disappointment after the glorious summer of 1911. More recently, August 2021 was cloudy but dry—the sunshine duration of 122 hours (73 per cent of normal), while more than twice 1912's record low figure, was also the second-lowest for the month in the previous 30 years.

Wind and weather

Fog (1962–97)

Fog is uncommon in August, and incidence has declined since the 1960s. Between 1962 and 1970, fog at 0900 GMT occurred 21 times in August, an average of 2.3 days per month; but between 1971 and 1997, 27 years, only nine fogs were recorded during the month, an average of about once every three years.

Barometric pressure (1843–1960)

The highest MSL barometric pressures recorded at the Durham Observatory in August occurred in 1874, when on the 20th the 2200 observation recorded 1035.0 hPa, with little change the following day. Other high values in August include 1034.4 hPa at 2100 GMT on 28 August 1920, and again at 0900 GMT the following day; and 1034.3 hPa at 2100 on 31 August 1854.

Deep depressions are extremely rare over the United Kingdom in August, but a most unseasonable storm in late August 1917 established low barometric pressure records for the month, which remain in place for most of England and Wales over a century later. At Durham, the MSL barometer reading of 968.5 hPa at 0900 GMT that day (and 976.1 hPa at 2100 GMT that evening) was by far the lowest on record for August, before or since. The next lowest values are more than 10 hPa higher—981.7 hPa at 2100 GMT on 9 August 1957, and 982.3 hPa at 0900 GMT on 15 August 1903.

The most anticyclonic Augusts were in 1947 (mean at MSL 1022.2 hPa), 1869 (1020.9 hPa), and 1955 (1020.8 hPa—following the record high 1021.7 hPa in July of that year). August 1947 was the warmest, driest, and sunniest August on record to that date (Tables 11.2, 11.3, and 11.4), and August 1955 the second driest to that date (Table 11.3), although sunshine was close to normal. The most cyclonic August was that of 1860, when the mean at MSL was a remarkably low 1003.4 hPa; the month was very cool and wet. Not far behind are August 1917 (mean 1004.1 hPa) and August 1912 (1005.3 hPa). August 1917 remains one of the wettest Augusts on record (Table 11.3), while August 1912 remains the coldest (Table 11.3) and dullest (Table 11.4).

Wind speed and direction (1961–97)

Figure 11.5 shows the distribution of wind speed and direction at the 0900 GMT observation in August during the period 1961–97. Mean wind speeds in August differ little from June or July (the mean wind speed at 0900 GMT is 6.1 knots), but there is a higher

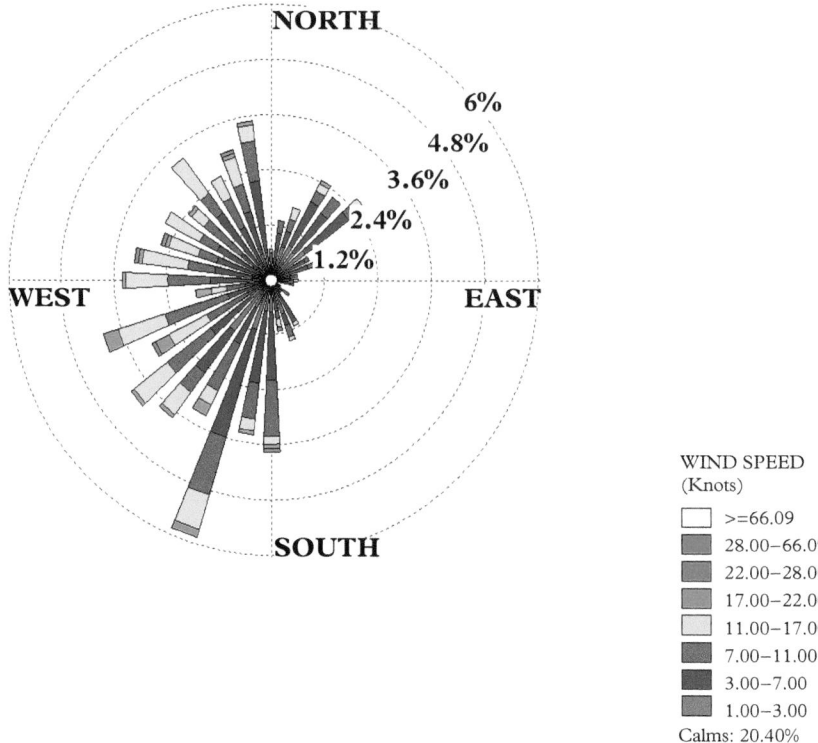

Figure 11.5 *Wind rose for 0900* GMT *observations in August, period 1961–97*

likelihood of stronger winds after mid-August. August 1962 was the windiest during this period of record, with an average speed of 11.2 knots at the morning observation. August sees further backing in wind directions compared to June and July, with more frequent winds from between south and west, although north-easterlies also increase a little this month.

Durham temperature, precipitation, and sunshine in graphs—August

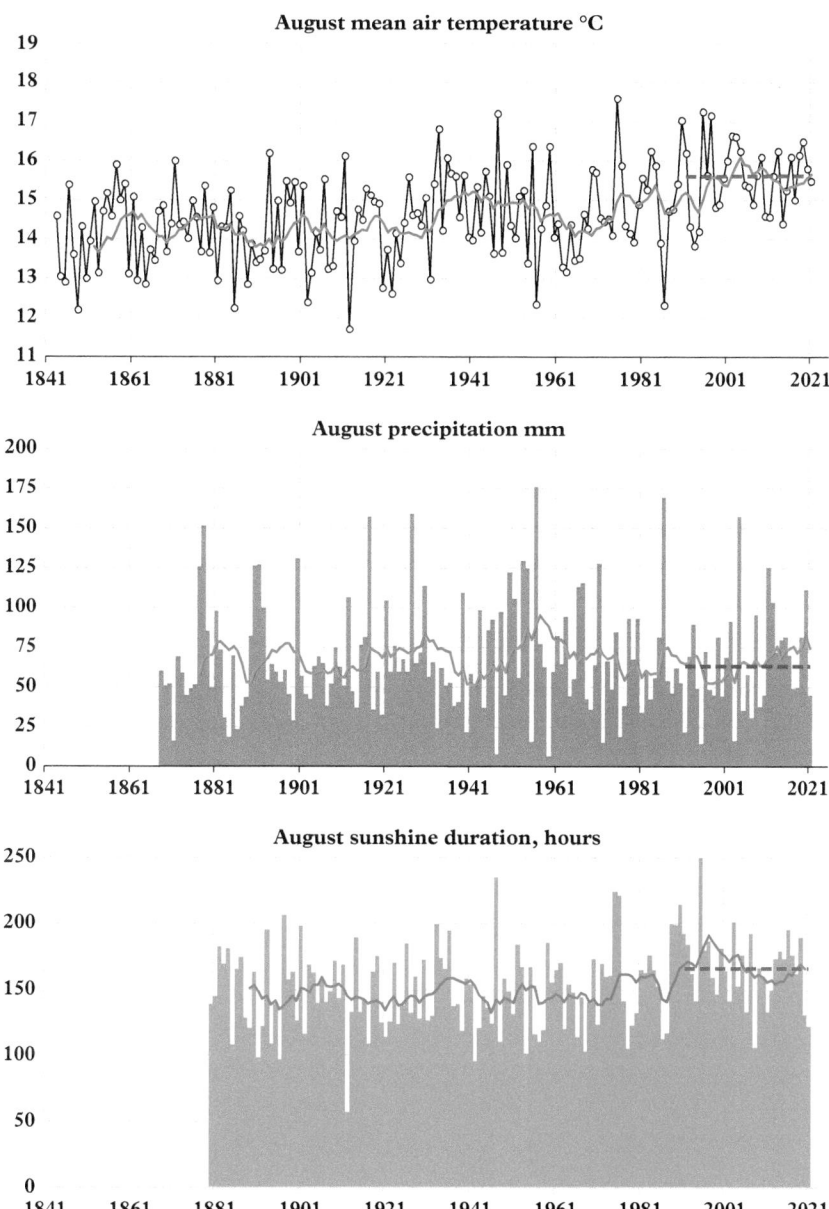

August mean air temperature °C

August precipitation mm

August sunshine duration, hours

Figure 11.6 *Monthly values of (from top) August's mean temperature (°C, since 1843), total precipitation (mm, since 1868), and sunshine duration (hours, since 1880) at Durham Observatory. The 1991–2020 averages are indicated by the thick blue dashed line, while the ten-year running mean ending at the year shown is indicated by the red line*

12

September

If we may transfer a term from North America to Europe, then September can be very much the month of the Indian summer. The Met Office's *Meteorological Glossary*, first published in 1916, defines an Indian summer as 'a warm, calm spell of weather occurring in autumn, especially in October and November', but September is more likely to have fine, dry, warm days, especially earlier in the month. Such spells are reminiscent of high summer and indeed may match the warmth of July and August. September is thus a month of transition, from summer to autumn, and warm, calm periods may alternate with stormy, wet weather as the North Atlantic circulation begins its winter spin-up. September's mean air temperature is 13.3 °C, over two degrees lower than July and August, although only a fraction of a degree cooler than June. Autumn sees a much faster decline in mean temperature than the more gradual rise through spring and early summer (Figure 3.1). Only in 1890 has September been the warmest month of the year. September is, on average, the driest month of the second half of the year, although in contrast the month also includes Durham's wettest-ever day, in a dramatic ending to the long drought of 1975/76.

Temperature

At least one September day normally reaches 23 °C; only three September days have ever reached 28 °C, two of which were the opening days of the month in 1906 [87] (Table 12.1). More recently, a short hot spell in September 2021 saw temperatures reach their highest level for the month since 1906, namely 27.2 °C on 7th and 28.0 °C on 8th. Five years previously, 27.3 °C was reached on 13 September 2016, by five days the latest date in the year on which 27 °C has been reached (Table 24.1), although the following day turned out to be 11 degrees colder. September has recorded the hottest day of the year eight times since 1856, most recently in 2004. In 2021, 8 September was within half a degree of the warmest day of that year. In the poor summers of 1954 and 1956, warmth in September must have been welcome, even though the warmest day was just 23.3 °C in both years.

Table 12.1 *Highest and lowest maximum and minimum temperatures at Durham Observatory in September, 1843–2021*

Rank	Hottest days	Coldest days	Warmest nights	Coldest nights
1	30.0 °C, 1 Sept 1906	*7.2 °C, 27 Sept 1852*	17.2 °C, 12 Sept 1945, 5 Sept 1949 and 7 Sept 2016	−1.7 °C, 27 Sept 1954
2	29.3 °C, 2 Sept 1906	*7.6 °C, 24 Sept 1889*	*16.4 °C, 3 Sept 1898*	*−1.2 °C, 27 Sept 1847*
3	28.0 °C, 8 Sept 2021	*7.7 °C, 23 Sept 1893*	16.3 °C, 2 Sept 1999	−1.1 °C, 4 Sept 1907
4	*27.5 °C, 4 Sept 1898*	7.8 °C, *24 and 25 Sept 1872 and 25 Sept 1942*	16.2 °C, 8 Sept 2016	−0.8 °C, 29 Sept 1932
5	27.3 °C, 13 Sept 2016	*8.2 °C, 30 Sept 1852*	16.1 °C, 5 Sept 1951	*−0.7 °C, 28 Sept 1887*
Last 50 years to 2021				
	28.0 °C, 8 Sept 2021	9.4 °C, 11 Sept 1976	17.2 °C, 7 Sept 2016	−0.5 °C, 12 Sept 1991

Temperatures were recorded in different thermometer screens during the period of record; see Notes on page xvii and Appendix 3 for more details.

With the warmth of summer still lingering in the ground, warm nights are not unknown in September; there have been several above 16 °C, and three above 17 °C (Table 12.1), most recently again in 2016.

A fine September with sunny days and clear nights can result in early frosts, particularly if the ground is dry—most often ground frosts (at least one can be expected before the end of the month in most years), but sometimes air frosts too. Durham's earliest air frost came on 4 September 1907, when the temperature fell to −1.1 °C; notably, 1907 was one of the driest Septembers. At the time of writing, no air frost has been recorded in September since 1991, although the frost on 12 September that year (−0.5 °C) bookended the shortest frost-free season on record—just 95 days since the last air frost of the previous winter season, namely −0.6 °C on 8 June. Frosts are more likely as the evenings draw in rapidly following the equinox, and September's coldest night was 27 September 1954, with an air minimum of −1.7 °C (Table 12.1). Late September ground frosts can be very sharp, particularly where the air is clear and the ground dry; notable here are grass minima of −6.1 °C on 29 September 1899, and −6.7 °C on 28 September 1919. In September 1986, 13 ground frosts were recorded, 11 on consecutive nights (9th–19th), although the lowest air minimum was +1.0 °C (11th).

Chilly north or north-easterlies off the North Sea, never very warm at the best of times, can bring the chill of winter days as soon as this first month of autumn—where these also include persistent rain, the daily maximum temperature can be depressingly low. Back in 1976, only a few days after the end of the long, hot summer when rain had fallen on only ten days in the previous 11 weeks, 11 September produced two records in Durham—the coldest September day since 1942, the maximum temperature reaching only 9.4 °C (Table 12.1), and the wettest day on record, when 87.8 mm of rain fell.

Until 2006, 1949 was the warmest September on Durham's record, with a mean daily temperature of 15.2 °C (Table 12.2). This was surpassed by 0.7 degrees in 2006, and again in 2016 and 2021. September 2006's mean maximum temperature was 20.0 °C, only a fraction of a degree below July's average (20.2 °C), although even this was equalled in the warm, dry September of 1865, when the mean maximum (Stevenson screen equivalent) reached 20.2 °C.

All but one of the 'Top 5' coldest Septembers date from more than a century ago, showing how September's mean temperatures have risen since then. September 1912 remains the coldest on record, following a chilly, dull, and wet summer—no Indian summer that year—with the mean maximum temperature more than three degrees lower than the previous September, which followed a notably warm and sunny summer. September 1912 also follows August of that year as the coldest in Durham's long record.

Table 12.2 *September mean temperatures at Durham Observatory, 1843–2021*

September mean temperature 13.3 °C (average 1991–2020)

Warmest months			Coldest months		
Mean temperature, °C	Departure from 1991–2020 normal degC	Year	Mean temperature, °C	Departure from 1991–2020 normal degC	Year
15.9	+2.6	2006	10.2	−3.1	1912
15.5	+2.2	2021	10.3	−3.0	1918
15.3	+2.0	2016	*10.4*	*−3.0*	*1847, 1894*
15.2	+1.9	1949	10.5	−2.9	1952
14.9	*+1.6*	*1865*	10.6	−2.8	*1877*, 1922
Last 50 years to 2021					
15.5	+2.2	2021	11.2	−2.1	1974

There was an even greater contrast between daytime temperatures in the Septembers of 1917 and 1918.

Precipitation

Durham's driest September on the post-1868 record was in 1907, with just 6.3 mm rainfall, the fourteenth-driest month of any on Durham's records. Only three Septembers have recorded fewer than 10 mm of rainfall; the driest in recent years has been 1997, with 11.3 mm (Table 12.3).

The hot, dry, and sunny summer of 1976, with its famous drought, came to a dramatic end in Durham in September 1976. At Durham, the drought was broken by a rain fall of 16 mm on 28 August—as much rain as had fallen in the previous six weeks put together; but much more was to come. As a depression crossed the country during the second week of September, a modest 3.7 mm fell at Durham on 8 September and a further 7.0 mm the following day. A second depression (Figure 12.1) then gave two days of prolonged rainfall in the north-east, amounting at Durham to 32.3 mm on 10 September and 87.8 mm on 11th, the largest one-day and two-day totals on record, and as noted earlier, 11 September also became the coldest September day in 54 years (a map of the day's rainfall appears in Chapter 22). The combined fall contributed 120 mm of the month's total 193 mm, almost three and a half times the monthly average, and resulted in September 1976 becoming by far the wettest September in Durham's record—indeed, the fourth-wettest of any month on the post-1868 records.

Table 12.3 *September precipitation at Durham Observatory 1868–2021*

September mean precipitation 56.8 mm (average 1991–2020)

Wettest months			Driest months			Wettest days	
Total fall, mm	Per cent of normal	Year	Total fall, mm	Per cent of normal	Year	Daily fall, mm	Date
193.2	340	1976	6.3	11	1907	87.8	11 Sept 1976
135.1	238	1935	8.1	14	1898	57.4	24 Sept 2012
131.4	231	1944	9.5	17	1910	55.8	21 Sept 1935
128.3	226	1880	10.3	18	1870	55.6	3 Sept 1948
117.6	207	1993	11.3	20	1997	45.8	21 Sept 1992
Last 50 years to 2021							
193.2	340	1976	11.3	20	1997	87.8	11 Sept 1976

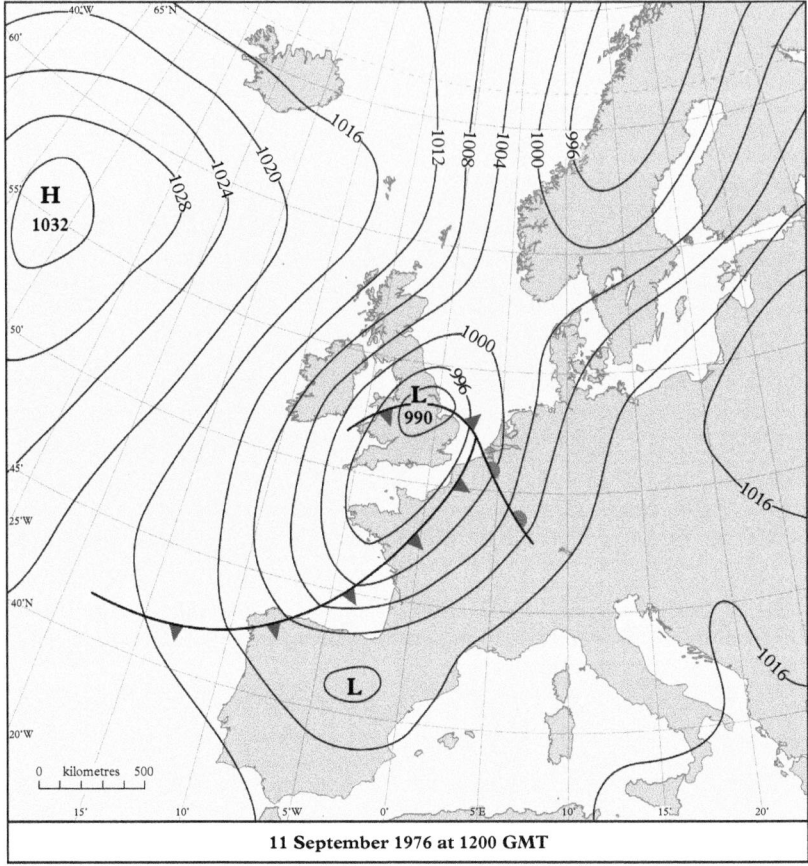

11 September 1976 at 1200 GMT

Figure 12.1 *The synoptic situation at 1200* GMT *on 11 September 1976, Durham's wettest-ever day, when 87.8 mm fell at Durham Observatory. (Based upon the Met Office* Daily Weather Report; *drawn by Chris Orton, Durham University)*

Elsewhere in County Durham, 113 mm fell at Tunstall Reservoir and 104 mm at Burnhope reservoir, both in the upper Wear catchment, on 11 September. Remarkably, this exceptionally wet spell starting in September 1976 remains Durham's wettest period for all timescales from 1–75 days (Table 25.2). Adding insult to injury, September 1976 was also the dullest September on record: more details are provided in the 'Sunshine' section.

For a month that is often benign, it is perhaps surprising that four of the ten wettest days on Durham's records have occurred in September (Table 25.5). At the time of writing, the most recent of these was the fall of 57.4 mm on 24 September 2012,

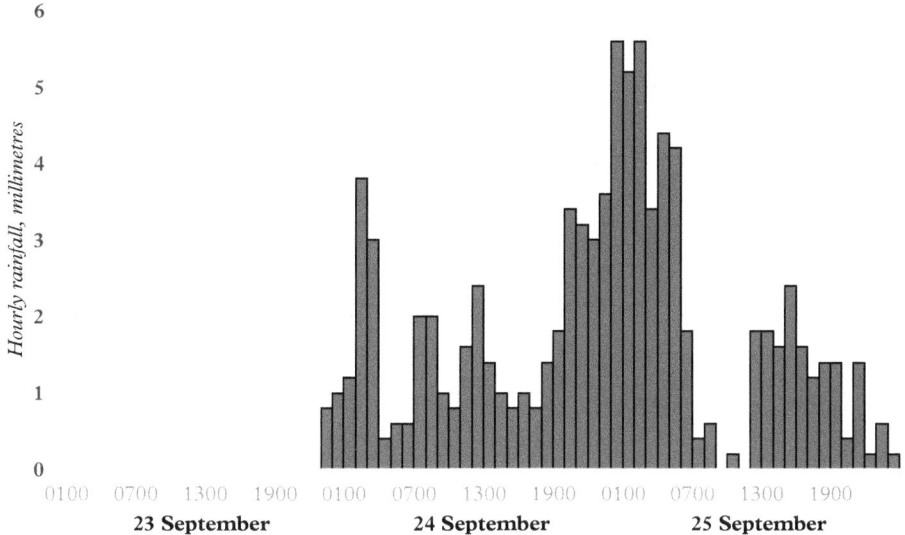

Figure 12.2 *Hourly rainfall totals 23–25 September 2012 recorded at Durham Observatory; during this three-day spell, just over 90 mm of rainfall was recorded, including the fourth-wettest September day on Durham's records*

part of a three-day fall 23–25 September of just above 90 mm. Of this, 41.6 mm fell in ten hours commencing 1900 GMT on 24 September (Figure 12.2). The resulting runoff caused extensive flooding in Durham city centre, culminating in a peak flow at Sunderland Bridge of 3.098 m (320 cumecs) at 1315 GMT on 25 September—the eighth-highest flood since river gauging began there in late 1957 (Figure 12.3). Among the buildings flooded were the University's Elvet Riverside car park—a 1960s construction, always vulnerable at times of high flow—and offices along Framwellgate Waterside (since demolished as part of a regeneration project unrelated to flood risk, we think—but let us hope that the necessary precautions will have been taken for the new buildings!).

Sunshine

In the standard averaging period 1991–2020, September averaged a total of 133 hours of bright sunshine, equivalent to 4.4 hours per day, 35 per cent of the theoretical possible.

Durham's sunniest September was in 1895. With only one sunless day and 21 days recording five hours or more of bright sunshine, this must have been a very welcome start to autumn. The cool September of 1986 became second-sunniest, although it is

Figure 12.3 *The flood of 25 September 2012 in Durham City. (Top left) Framwellgate Waterside looking south from the footbridge; (bottom left) Framwellgate Waterside looking north towards the newly-opened Radisson Hotel, which was marooned during the flood peak. The line of iron fencing is the same as shown in Figures 8.3 and 8.4. (right) Entrance to the University's Elvet Waterside car park; the River Wear is seen in the background. Hatfield College boathouse is on the wooded slope opposite, to the left. (All photographs courtesy of Michele Allan)*

noteworthy that no September has exceeded 160 hours since the turn of the millennium, despite the warmth of 2006 and 2016 (Table 12.4).

September 1976, the wettest September on record at Durham, also became the dullest. Its total sunshine of just over 59 hours is a mere 15.6 per cent of possible—less than half the average receipt, and about the same as a normal December. From 11 September 1976, Durham's wettest day, onwards there were 12 days with no sunshine at all by month-end; there were 13 in all, a record for September, surpassing the previous record holder, 1922, which had 12 sunless days. Until 1976, September 1922 was Durham's dullest September on record, following on from a particularly chilly and dull summer, and remains second-dullest to date.

As the days draw shorter, 12 hours of bright sunshine is possible only during the first few days of the month (Table 12.4). September's sunniest day on the record to 1999 was 2 September 1994, with 12.7 hours of bright sunshine.

Table 12.4 *September sunshine duration at Durham Observatory. Observed monthly extremes 1880 to 2021, sunniest days 1882–1999 only*

September mean sunshine duration 133 hours, 4.4 hours per day (average 1991–2020).

Possible day length: 381 hours. Mean sunshine duration as percentage of possible: 34.7

Sunniest months			Dullest months			Sunniest days 1882–1999	
Duration, hours	*Per cent of possible*	*Year*	*Duration, hours*	*Per cent of possible*	*Year*	*Duration, hours*	*Date*
203.9	53.5	1895	59.4	15.6	1976	12.7	2 Sept 1994
192.7	50.6	1986	64.6	17.0	1922	12.6	1 Sept 1951
183.2	48.1	1963	66	17.3	1881	12.1	1 Sept 1906, 3 Sept 1941 and 7 Sept 1991
176.0	46.2	1940	70.1	18.4	1931	12.0	2 Sept 1906
171.2	44.9	1997	72.4	19.0	1965	11.9	2 Sept 1987
Last 50 years to 2021							
192.7	50.6	1986	59.4	15.6	1976	12.7	2 Sept 1994

Daily sunshine records ceased at the Observatory on 31 December 1999: monthly sunshine totals from January 2000 onwards are estimated (to the nearest hour) by regression from the Met Office East and North-east England sunshine series. See Appendix 6 for record metadata and missing months.

Wind and weather

Fog (1962–97)

With the longer nights, the chances of fog begin to increase in September, and typically one morning in the month, most often towards the end of the month, is likely to see fog lingering at 0900 GMT. As with other autumn and winter months, the frequency of fog declined quite sharply during the 1980s and 1990s. Four or more mornings with fog were noted in 1966, 1971, 1973, and 1976 (the last four days of the month, no doubt owing to the relentless rainfall), but thereafter this frequency did not recur until 1992.

Barometric pressure (1843–1960)

The highest MSL barometric pressure recorded at the Durham Observatory during September was 1039.3 hPa at 0900 on 16 September 1851, a slight increase from 1038.9 hPa at 2100 on the previous evening. Only a fraction behind was the reading of 1038.8 hPa recorded at 2100 GMT on 21 September 1934, followed by 1038.5 hPa which has been reached three times—at 1400 on 24 September 1855, and at 0900 GMT on both 24 and 26 September 1931.

Three surprisingly deep depressions dominate the low-pressure extremes for September, starting with 971.9 hPa at 0900 GMT on 17 September 1935, followed by 973.6 hPa at 2100 GMT on 21 September 1953, and 973.9 hPa at 2100 GMT on 22 September 1896. It is noteworthy that August's minimum pressure (968.5 hPa in 1917) is lower than September's.

September 1865, a very dry month in the north-east, was the most anticyclonic on the record, with mean MSL pressure 1024.7 hPa, just ahead of 1024.6 hPa in September 1910, and 1023.9 hPa in September 1894. In contrast stand September 1918, with a monthly mean of just 1003.6 hPa, September 1866 with 1004.5 hPa, and September 1869 with 1005.0 hPa.

Wind speed and direction (1961–97)

Figure 12.4 shows the distribution of wind speed and direction at the 0900 GMT observation in September during the period 1961–97. When compared to the plot for August, there is a significant increase in both the frequency and strength of winds from between south-west and north-west, compared to the summer months. Mean wind speeds pick up a little from August, with a mean wind speed at 0900 GMT of 6.5 knots; September 1970 averaged 10.1 knots.

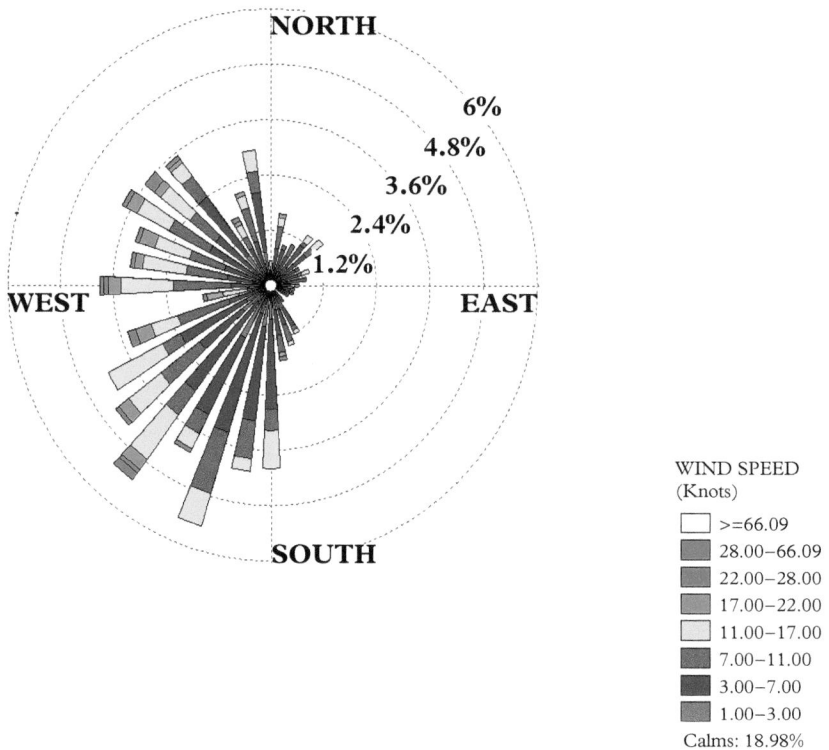

WIND SPEED (Knots)

	>=66.09
	28.00–66.09
	22.00–28.00
	17.00–22.00
	11.00–17.00
	7.00–11.00
	3.00–7.00
	1.00–3.00

Calms: 18.98%

Figure 12.4 *Wind rose for 0900 GMT observations in September, period 1961–97*

Durham temperature, precipitation, and sunshine in graphs—September

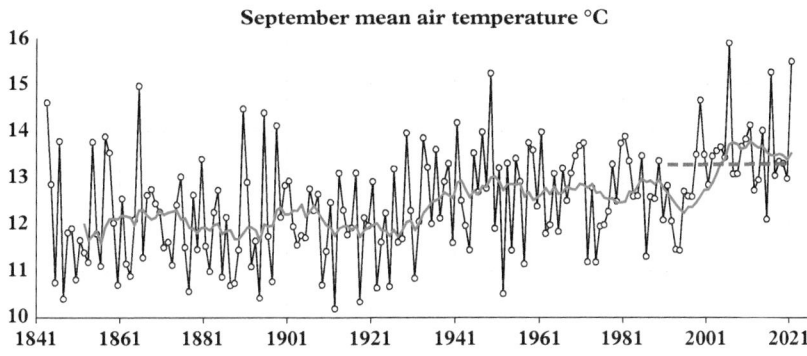

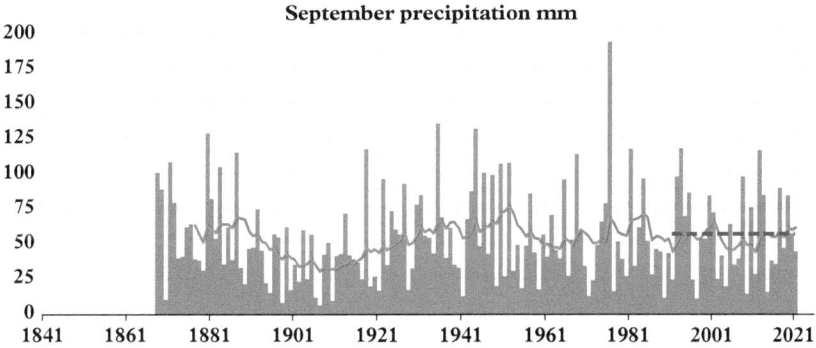

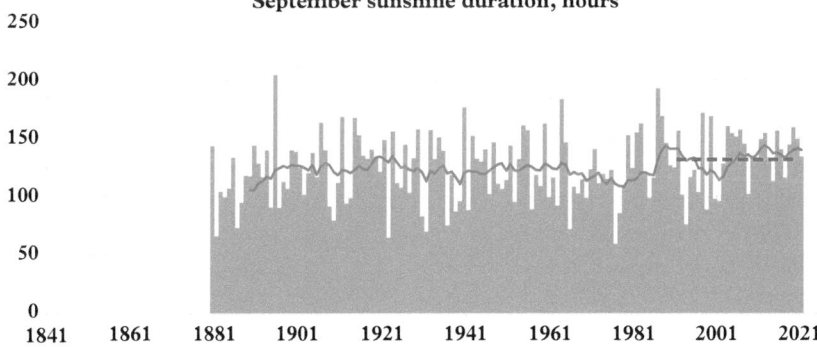

Figure 12.5 *Monthly values of (from top) September's mean temperature (°C, since 1843), total precipitation (mm, since 1868), and sunshine duration (hours, since 1880) at Durham Observatory. The 1991–2020 averages are indicated by the thick blue dashed line, while the ten-year running mean ending at the year shown is indicated by the red line*

13

October

October is another transitional month, with shorter, cloudier days, more frequent rainfall, and lower temperatures—the mean temperature for the month down over three degrees Celsius from September, while the mean daily maximum temperature can be expected to fall from 16 °C at the beginning of the month to below 12 °C by the closing days. A few years will see warm summer-like days in the first week or so (including, remarkably, one year in which Durham's warmest day of the year occurred in October), but the first air frosts of the autumn can normally be expected before the month is out. Surprisingly then, October has the lowest absolute range of temperature for any month on Durham's records, at just over 30 degrees Celsius. Of all months, October has shown the largest increase in mean air temperature since 1900, equivalent to 1.4 degrees per century. October is a wetter month than September, and with more frequent rainfall, too, and this together with sharply-reduced sunshine in most years (over an hour less sunshine each day, on average) brings home the fact that summer is over, and winter lies just around the corner.

Temperature

Not surprisingly, considering the rapid decline in solar strength and day length, October's warmest days are all early in the month. The warmest on Durham's records was 1 October 2011, when the maximum temperature reached 25.3 °C, the only occasion to date on which 25 °C has been attained in October (Table 13.1) and the second-warmest day of that year (one day in June reached 26.7 °C). This was a remarkably intense late hot spell across England and Wales, the first day of the month setting new UK maximum temperature records for the month in England (29.9 °C at Gravesend in Kent) and Wales (28.2 °C at Hawarden Airport in Flintshire) [64]. The second-warmest October night followed two days later, showing that night-time temperatures can remain high for some time in October as the ground loses the summer heat more slowly. This factor together with increased cloud cover partly accounts for October having the lowest range in temperature for any month, just 30.6 degrees Celsius since 1843.

The beginning of October also saw high temperatures in 1908, 1926, 1969, and 1985 (Table 13.1). The minimum temperature on 1 October 1985, 15.9 °C, remains the

Table 13.1 *Highest and lowest maximum and minimum temperatures at Durham Observatory in October, 1843–2021*

Rank	Warmest days	Coldest days	Warmest nights	Coldest nights
1	25.3 °C, 1 Oct 2011	*1.8 °C, 21 Oct 1859*	15.9 °C, 1 Oct 1985	−5.3 °C, 25 Oct 1926
2	24.9 °C, 4 Oct 1908	2.8 °C, 31 Oct 1934	15.2 °C, 3 Oct 2011	−5.2 °C, 31 Oct 1909
3	24.1 °C, 1 Oct 1985	3.2 °C, *18 Oct 1848* and 29 Oct 1932	15.1 °C, 14 Oct 2017	*−5 °C, 31 Oct 1881*
4	23.9 °C, 10 Oct 1969	3.3 °C, 28 Oct 1919	15.0 °C, 1 Oct 1953	−4.7 °C, 17 Oct 1916
5	23.8 °C, 2 Oct 1908 and 4 Oct 1926	3.8 °C, 27 Oct 1926	14.9 °C, *3 Oct 1859* and 2 Oct 2011	*−4.3 °C, 22 Oct 1880*
Last 50 years to 2021				
	25.3 °C, 1 Oct 2011	4.0 °C, 30 Oct 1974	15.9 °C, 1 Oct 1985	−3.8 °C, 29 Oct 1997

Temperatures were recorded in different thermometer screens during the period of record; see notes on page xvii and Appendix 3 for more details.

highest on October's records, while later that day the maximum temperature reached 24.1 °C. Remarkably, this was the equal-hottest day of that year (with 25 July), the only time on Durham's records when the hottest day of the year has fallen in October. A ridge of high pressure covered England and Wales while Scotland and Northern Ireland were affected by the passage of a frontal system associated with a depression over Iceland. The latest date in the year on which 20 °C has been reached in Durham was 16 October, in 1951, when the maximum temperature was 20.6 °C (Table 24.1). The same value was recorded on 15 October in 1945. In 2017, the minimum temperature of 15.1 °C on 14 October became the latest date in the year in which the temperature has remained above 15 °C throughout the 24 hours (Table 24.1).

Towards the end of the month, daytime temperatures occasionally struggle to reach 5 °C. The coldest October day on record came in 1859, when the temperature on the 21st reached only 1.8 °C in the 'North Shed' exposure. Cold northerly winds blew across the British Isles, with widespread snow or hail showers and night frosts. This period is a particularly notable one in British meteorological history, for it involved the preparation of the earliest series of daily weather maps across the Atlantic, using simultaneous observations made in Europe and America. Early on the morning of 25 October 1859, a severe north-easterly gale wrecked the 2719 tonne full-rigged luxury clipper *Royal*

Charter on the north coast of Anglesey, with the loss of over 400 lives, and a fortune in gold from the Australian goldfields [65, 66]. The same storm sank 135 other ships that night, with over 800 souls lost. The *Royal Charter* storm, coming early in the history of the Meteorological Department of the Board of Trade established by Admiral Robert Fitzroy in 1854 (the body that became the Meteorological Office in 1867), was a seminal moment in British meteorological history. It led directly to Fitzroy's introduction of embryonic storm warning services in 1861, which, although heavily criticized for their frequent inaccuracies in the early days, nonetheless saved the lives of thousands of seafarers in the following decades [67, 68].

By far the warmest October yet recorded was 2001, half a degree warmer than its nearest rival, 1969 (Table 13.2). The month's warmest day was the 12th, with a maximum temperature of 19.7 °C, and remarkably the coldest day was no lower than 11.7 °C, on the last day of the month. Three of the five warmest Octobers in Table 13.2 have occurred this century; 1969 and 1921 complete the listing. The Observatory weather station annual report for 1969 describes October as 'easily the most outstanding month of the year'. A persistent southerly airflow, related to a European high-pressure system, resulted in a very warm and very dry month (Figure 13.1).

The first air frosts are likely towards the end of October, although these are much less frequent in recent decades than they were in the early years of the record. In the 21 years to 2021, 13 Octobers remained frost-free, whereas the Octobers of 1895 and 1926 recorded nine air frosts each, and seven in 1880, 1887, 1931, and 1992. Temperatures

Table **13.2** *October mean temperatures at Durham Observatory, 1843–2021*

October mean temperature 10.0 °C (average 1991–2020)

Warmest months			Coldest months		
Mean temperature, °C	Departure from 1991–2020 normal degC	Year	Mean temperature, °C	Departure from 1991–2020 normal degC	Year
12.7	+2.7	2001	*6.1*	*−3.9*	*1896*
12.2	+2.2	1969	*6.3*	*−3.7*	*1892*
12.1	+2.1	2006, 2017	*6.4*	*−3.6*	*1880, 1895*
12.0	+2.0	1921	6.7	−3.3	*1885, 1905*
11.8	+1.8	1995	6.8	−3.2	*1887, 1992*
Last 50 years to 2021					
12.7	+2.7	2001	6.8	−3.2	1992

Figure 13.1 *Even in mid-October there can be some pleasantly mild and sunny days: Durham's Saddler Street on 13 October 2012 (Tim Burt)*

below −5 °C, a killing frost for many half-hardy garden plants, occurred in the last week of the month in 1881, 1909, and 1926, the lowest of all being −5.3 °C on 25 October 1926 (Table 13.1).

Since 1900 October has shown a larger increase in mean air temperature than any other month, so it comes as no surprise to find that most of the coldest Octobers were in the nineteenth century. Coldest of all was 1896, with a mean air temperature almost four degrees below the current average, the tally including six air frosts and 11 ground frosts. The month opened fairly mild (a maximum temperature of 16.3 °C on the 2nd, and a minimum the following night of 10.7 °C), but there were some raw days towards the end of the month with a chilly maximum on the 29th of only 4.2 °C following an overnight minimum of −2.3 °C. In more recent years, October 1992 just makes the 'Top 5' coldest in Table 13.2, in what was the coldest October for almost 90 years.

October 1926 features prominently in Table 13.1. For the month of October 1926 as a whole, there was a remarkable range in temperature of 29.1 degrees Celsius, from 23.8 °C on 4th to −5.3 °C just three weeks later on 25th, Durham's coldest October night on record (at Braemar the temperature fell to −10.6 °C on the same night), and the earliest 'minus 5' in any year to date (Table 24.1). The *Monthly Weather Report* noted that the end of the month was 'unusually cold with record low screen minima in many places'. Durham's daytime temperatures never made it above 10 °C in the last two weeks of the month and reached only 3.8 °C on the 27th.

Precipitation

October is wetter than September, both in terms of rainfall amount (monthly average 64 mm, an increase of about 10 per cent on September) and frequency (18 days with rain, compared to 15 in September).

October 1969 was the only October to receive less than 10 mm of rainfall in Durham, just 15 per cent of normal; it was the driest month of that year by a wide margin as the weather was dominated by southerly airflow around a high-pressure system over mainland Europe. Elsewhere in England, less than 10 per cent of normal rainfall was recorded in places, including no measurable rain at one station near Bedford. The *Monthly Weather Report* noted at the time that this was the driest October on record at both Oxford and Durham, known for their very long records. October 1972, Durham's second-driest with just 12 mm that month (falling on only eight days), was drier than normal across the whole country but especially so for the eastern regions—East Anglia in particular. There were only four days with measurable rainfall in October 1897, and five in 1951 and 1969, whereas October 1896 recorded rain on 29 days (the highest for any month on Durham's records); the Octobers of 1882 and 1903 both received rainfall on 28 days.

October 1903 remains the United Kingdom's wettest-ever month, and Durham's wettest October and second-wettest month; 202 mm fell, more than three times normal (one of only two months to surpass 200 mm, the other being July 1888—Table 25.3). The *Monthly Weather Report* noted that the weather remained 'in an exceptionally unsettled state' throughout the month. This included a prolonged cyclonic rainfall event on 8 October when 63.7 mm fell at Durham Observatory, Durham's wettest-ever

Table 13.3 *October precipitation at Durham Observatory 1868–2021*

October mean precipitation 63.7 mm (average 1991–2020)

Wettest months			Driest months			Wettest days	
Total fall, mm	Per cent of normal	Year	Total fall, mm	Per cent of normal	Year	Daily fall, mm	Date
201.8	323	1903	9.6	15	1969	63.7	8 Oct 1903
172.4	276	1960	11.9	19	1972	50.1	26 Oct 1900
148.8	238	1896	13.0	21	2007	45.8	14 Oct 1892
142.6	228	1939, 1976	13.2	21	1962	39.1	16 Oct 1967
142.0	224	1907	14.0	22	1947	38.4	7 Oct 1913
Last 50 years to 2021							
142.6	228	1976	11.9	19	1972	38.1	1 Oct 1981

Figure 13.2 *Daily rainfall totals for 8 October 1903 in north-east England; from* British Rainfall 1903, *p. [143]. The scale is in inches: 1 in = 25 mm, 2 in = 51 mm, and 3 in = 76 mm. The daily fall in Durham was 63.7 mm*

October day. Figure 13.2, from *British Rainfall 1903*, shows the distribution of rainfall on that date. Supporting Durham's large daily total, 93.5 mm fell at Newcastle's Town Moor site, 93.0 mm at the Newcastle Literary and Philosophical Society, and 87.6 mm at Hurworth Burn near Hartlepool. Considerable flooding resulted throughout the north-east, and flood-related damage was reported in Newcastle.

October 1960 was another very wet month over most parts of the country, the *Monthly Weather Report* describing the month as 'dull and exceptionally wet with extensive flooding in the south and east'. At Horncastle in Lincolnshire more than five times the normal October rainfall was recorded; Durham was not quite so wet, with some 30 mm below 1903's record fall, but 1960 was also the dullest October on record. Flood discharge was not exceptionally high on the River Wear but there were three peaks in quick succession on the 20th, 23rd, and 27th. The average daily flow in October 1960 remains the highest for any October since records began in October 1957. As noted in Chapter 12, September 1976 was the wettest September on record and October 1976 remains the fourth-wettest—two consecutive exceptionally wet months accumulating 336 mm rainfall, a fraction under half the normal yearly total, all of which fell in the 54 days ending 31 October. Autumn 1976 became the wettest autumn on record, following hard on the heels of the third-driest summer and more than a year of severe drought conditions.

Sunshine

October receives 99 hours of bright sunshine in an average year, equivalent to 3.2 hours daily, a reduction of more than an hour per day compared with September. Amounts received have ranged from 41 hours in 1960 to over 150 hours in 1893 (Table 13.4), so the sunniest October has seen nearly four times as much sunshine as the dullest.

Durham's sunniest October on record was in 1893, with just over 150 hours, or 46 per cent of possible sunshine. Sunshine was abundant especially in the northern and eastern parts of England; at Durham there was sunshine every day with a maximum of 10 hours on the 1st, one of the very few October days to record ten hours of bright sunshine. October 1981 became the sunniest October in almost 90 years, falling short of 1893's total by less than five hours. Three of the first five days were sunless, but there was plenty after that, with 13 days having more than six hours of bright sunshine with a maximum on the 10th of 9.9 hours. The following year, October 1982 presented a stark contrast, with barely 50 hours of bright sunshine, the third-dullest on record. Seven of the first nine days remained sunless, 13 in all during the month, a total only exceeded

Table 13.4 *October sunshine duration at Durham Observatory. Observed monthly extremes 1880 to 2021, sunniest days 1882–1999 only*

October mean sunshine duration 99 hours, 3.2 hours per day (average 1991–2020). Possible day length: 325 hours. Mean sunshine duration as percentage of possible: 30.3

Sunniest months			Dullest months			Sunniest days 1882–1999	
Duration, hours	Per cent of possible	Year	Duration, hours	Per cent of possible	Year	Duration, hours	Date
150.6	46.3	1893	40.9	12.6	1960	10.2	6 Oct 1917 and 2 Oct 1980
146.1	45.0	1981	42.1	13.0	1889	10.1	9 Oct 1930
138.5	42.6	1986	51.6	15.9	1982	10.0	1 Oct 1893, 5 Oct 1910 and 3 Oct 1927
134.9	41.5	1931	53.1	16.3	1894	9.9	Numerous
131.3	40.4	1913	54.8	16.9	1886	9.8	4 Oct 1918 and 9 Oct 1934
Last 50 years to 2021							
146.1	45.0	1981	51.6	15.9	1982	10.2	2 Oct 1980

Daily sunshine records ceased at the Observatory on 31 December 1999: monthly sunshine totals from January 2000 onwards are estimated (to the nearest hour) by regression from the Met Office East and North-east England sunshine series. See Appendix 6 for record metadata and missing months.

in October 1960 (16 days). Note that no October has been duller since 1982, perhaps another indication of October's changing climate. October 1986 was another very sunny October in Durham, despite three days without any sunshine.

Wind and weather

Snow cover (1960–99)

Snow falls occasionally in October, but during the period 1960–99, no October morning recorded a snow cover.

Fog (1962–96)

The frequency of fog increases rapidly during autumn, and over the period 1962–97 October was by a small margin the foggiest month of the year, averaging 3.4 mornings with fog. In the settled, dry, and warm October of 1969, fog was noted on 11 mornings, and nine in both 1975 and 1977.

Barometric pressure (1844–1960)

The earliest date in the winter half-year when the MSL barometric pressure surpassed 1040 hPa in Durham came in October 1945, when at the 0900 GMT observation on the second day of the month the pressure was 1040.1 hPa. The highest values on record for October during this period were all just above 1040 hPa: 1040.8 hPa at 2100 GMT on 31 October 1956, 1040.6 hPa at 1000 on 5 October 1882, and 1040.5 hPa at 2100 GMT on 30 October 1891.

The first of the deep cyclonic systems from the Atlantic can normally be expected during October. The lowest barometric pressures in October between 1843 and 1960 occurred on 27 October 1959, when the pressure at 0900 GMT stood at 963.1 hPa, and at 1000 on 14 October 1881, 963.2 hPa; the latter event is discussed in more detail below. Almost as low were the values of 964.1 hPa at 0900 GMT on 23 October 1925 and 968.0 hPa at 2200 on 23 October 1870.

The most anticyclonic Octobers were in 1919 and 1947, with a mean MSL pressure of 1022.2 hPa; the latter was the fifth-driest October on record (Table 13.3). The mean for October 1856, 1021.9 hPa, was almost as high. The most cyclonic Octobers were those of 1903 (mean at MSL 1000.0 hPa, 13 hPa below the 1931–60 normal), 1846 (1001.3 hPa) and 1932 (1001.5 hPa). October 1903 was Durham's wettest on record (Table 13.3); the mean pressure distribution across the north-east Atlantic is shown in Figure 13.3.

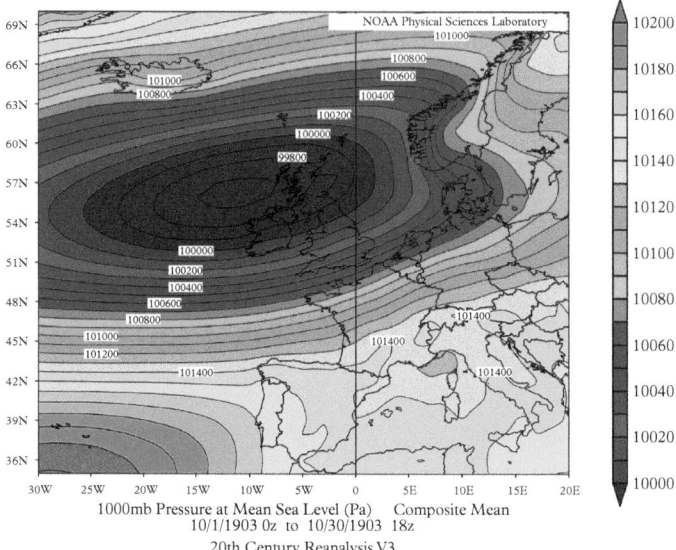

1000mb Pressure at Mean Sea Level (Pa)   Composite Mean
10/1/1903 0z to 10/30/1903 18z
20th Century Reanalysis V3

Figure 13.3 *Mean* MSL *pressure distribution during October 1903; this was one of the most cyclonic months in British weather history and remains the wettest month yet recorded in the United Kingdom. Isobars are in Pascals (divide by 100 for hPa = millibars), at 100 Pa intervals. (From the 20th Century Reanalysis version 3 (20CRv3) plot [58])*

Wind speed and direction (1961–97)

Figure 13.4 shows the distribution of wind speed and direction at the 0900 GMT observation in October during the period 1961–97. Mean wind speeds in October are little different from September (mean wind speed at 0900 GMT of 6.3 knots), although there have been numerous significant storms with very strong winds. October's wind pattern is dominated by winds from between south and north-west, with winds from other directions uncommon.

One of the worst October storms along the north-east coast is still remembered here, generations later: the 'Eyemouth storm' of 14–15 October 1881. A depression formed south of Nova Scotia on 10 October and crossed the Atlantic, deepening rapidly to below 960 hPa as it neared the British Isles. There was a widespread severe gale, at its most damaging along the east coasts of England and Scotland. As described by Hubert Lamb [69]:

> At Eyemouth in Berwickshire the morning of 14 October had dawned fine, with clear sky and a calm sea. The fishing fleet of 41 boats, mostly big deep-sea vessels, sailed out under a steady breeze ... In the middle of the day the wind fell light, and then the storm struck suddenly. Nineteen of the boats were lost, and 129 of the little town's manhood drowned. The barometer had been very low, and warnings had been issued. At Arbroath and at Berwick the fleets stayed in port.

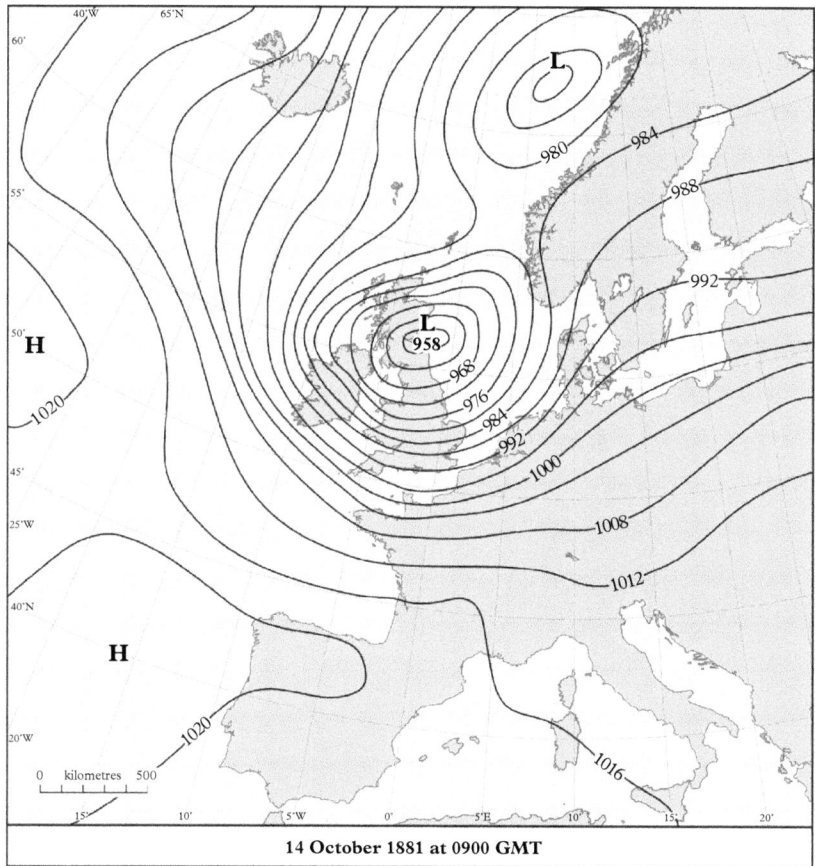

14 October 1881 at 0900 GMT

Figure 13.4 *The synoptic situation at 0900* GMT *on Friday 14 October 1881, the 'Eyemouth storm'. Based on Symons [71]. The innermost isobar is 960 hPa. (Drawn by Chris Orton, Durham University)*

In all, 189 east coast fishermen lost their lives in the storm, and there are several memorials in the villages along this coast. The *Daily Weather Report* chart for 8 a.m. on Friday 14 October showed the depression centred over Berwickshire, with central pressure about 961 hPa; a subsequent analysis of the storm by Symons [70] suggested the centre was a little further east, and somewhat more intense, at 958 hPa at 0900 GMT (Figure 13.4). On the Berwickshire coast, winds would have been light near the centre of the system at daybreak before a violent northerly gale set in as the depression moved rapidly eastward. The observations made at Durham, somewhat to the south of the depression centre, are shown in Table 13.5.

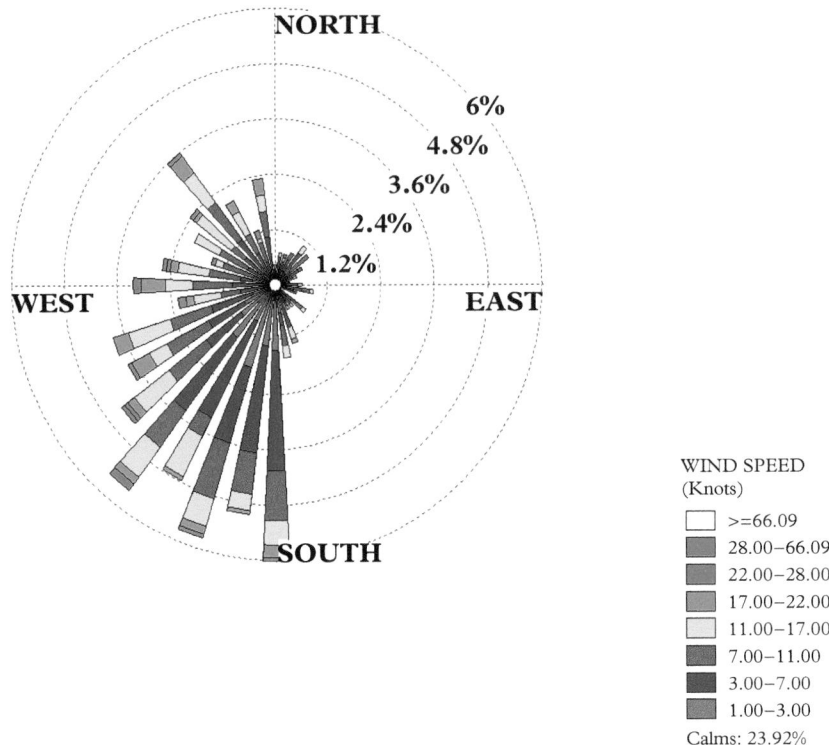

WIND SPEED
(Knots)

☐	>=66.09
■	28.00–66.09
■	22.00–28.00
■	17.00–22.00
☐	11.00–17.00
■	7.00–11.00
■	3.00–7.00
■	1.00–3.00

Calms: 23.92%

Figure 13.5 *Wind rose for 0900* GMT *observations in October, period 1961–97*

Table 13.5 *Meteorological observations at Durham Observatory 13–15 October 1881, the 'Eyemouth storm'*

Date	Time	Air temperature, °C	Barometer at MSL, hPa	Cloud cover, tenths	Wind direction	Wind speed, knots
13 October	1000	8.9	1004.2	3	W	26
	2200	4.5	997.0	10	SSE	13
14 October	1000	9.8	963.2	10	SW	15
	2200	2.9	1000.6	0	NW	19
15 October	1000	4.8	1009.5	7	NW	17

Durham temperature, precipitation, and sunshine in graphs—October

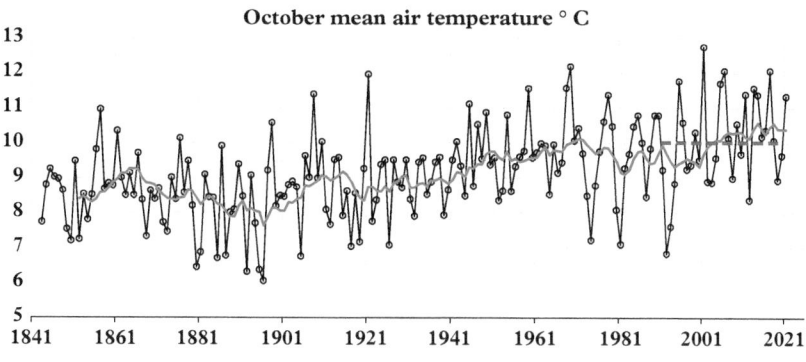

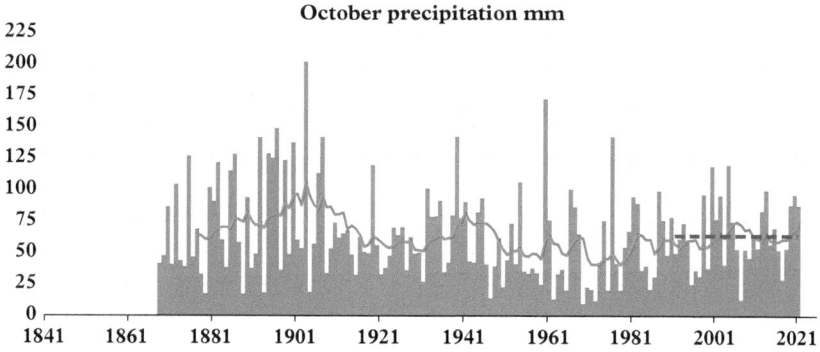

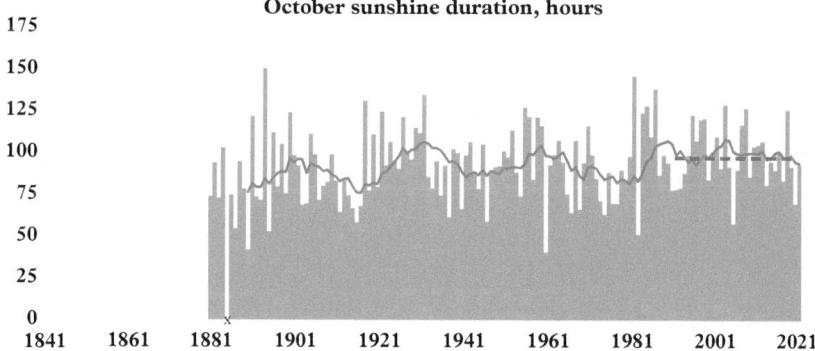

Figure 13.6 *Monthly values of (from top) October's mean temperature (°C, since 1843), total precipitation (mm, since 1868), and sunshine duration (hours, since 1880, missing data 1884) at Durham Observatory. The 1991–2020 averages are indicated by the thick blue dashed line, while the ten-year running mean ending at the year shown is indicated by the red line*

14

November

By the time we reach November, the final month of autumn, the days are getting rapidly shorter. Some Novembers see unseasonably mild or even warm days, but November can also be wintry, and the first snow of the winter is quite likely this month. November's mean air temperature falls almost four degrees when compared with October's, emphasizing the sharp decline in temperature at this time of year, and several frosty mornings can be expected in a normal month. November is also the wettest month of the year in Durham, the only month with average rainfall in excess of 70 mm. Reliable cloud cover and decreasing day length means that November has an hour less sunshine per day than October, although it is still sunnier than December or January. Morning fog can be expected on around three days in a typical November, although (as with other months) the frequency of fog has declined sharply within the last 50 years or so, at least partly as a result of reducing air pollution.

Temperature

During the standard averaging period 1991–2020, Durham's mean air temperature in November was 6.6 °C. Since 1843, monthly mean air temperatures in November have ranged from 2.4 °C in 1919 to 8.7 °C in 2011.

The first week of November occasionally sees remarkably high temperatures for the time of year (Table 14.1). Top of this list is 19.3 °C on 2 November 1927, supported by very similar temperatures recorded at Tynemouth and Chopwellwood on the same day, in a brisk, mild, west-south-westerly flow. With 4.4 hours of sunshine at Durham, this truly was a very late 'Indian summer'. A very similar synoptic situation prevailed 19 years later, when on 4 November 1946 Durham reached 18.9 °C—that same day, Prestatyn in North Wales recorded 21.7 °C, a value which remained the UK's highest November temperature until it was itself surpassed in 2015. More recently, afternoon sunshine in a mild south-westerly flow on 5 November 2020 lifted the temperature to 18.5 °C, Durham's warmest November day in 74 years. As is the case in the other autumn months, preceding warmth can ensure ground temperatures remain high, and resulting very mild nights are not unknown, even late in the month; the minimum temperature of 13.7 °C on 11 November 2015 was milder than many summer nights.

Table 14.1 *Highest and lowest maximum and minimum temperatures at Durham Observatory in November, 1843–2021*

Rank	Mildest days	Coldest days	Mildest nights	Coldest nights
1	19.3 °C, 2 Nov 1927	*−1.4 °C, 23 Nov 1853*	13.7 °C, 11 Nov 2015	−12.0 °C, 24 Nov 1993
2	18.9 °C, 4 Nov 1946	−1.1 °C, 12 Nov 1919 and 28 Nov 1969	13.3 °C, 13 Nov 1938 and 21 Nov 1947	−9.8 °C, *21 Nov 1880*
3	18.5 °C, 5 Nov 2020	−0.8 °C, 29 Nov 1912	13.1 °C, 2 Nov 1927	−9.3 °C, 25 Nov 1993
4	18.4 °C, 2 Nov 1844	−0.6 °C, *29 Nov 1856* and 29 Nov 1952	12.5 °C, 5 Nov 1978, 19 Nov 2003 and 2 Nov 2007	−8.9 °C, 16 Nov 1919
5	18.1 °C, 1 Nov 2015	−0.5 °C, 27 Nov 1874	12.4 °C, *2 Nov 1850*, 3 Nov 1996 and 10 Nov 2015	−8.7 °C, 28 Nov 2010
Last 50 years to 2021				
	18.5 °C, 5 Nov 2020	0.2 °C, 27 Nov 2010	13.7 °C, 11 Nov 2015	−12.0 °C, 24 Nov 1993

Temperatures were recorded in different thermometer screens during the period of record; see Notes on page xvii and Appendix 3 for more details

November 1994 was a remarkably mild month, surpassing November 1938 as Durham's mildest since records began in 1843, but this record lasted less than 20 years until surpassed by November 2011. Since then, 2015 has also surpassed both 1938 and 1994 (Table 14.2). November 1994 was also five degrees warmer than the previous year's November and recorded only two air frosts, compared with 12 the previous November.

November can also experience the bitter cold of winter, with a number of sub-zero maximum temperatures on record; the earliest such sub-zero date was 12 November 1919, when the maximum temperature was just −1.1 °C following an overnight minimum of −6.7 °C (Tables 14.1 and 24.1). The only colder November day was back in 1853, when on 23 November the maximum temperature (in the 'North Shed' exposure) was −1.4 °C. At the time of writing the most recent sub-zero November day was over 50 years ago in 1969, although 27 November 2010 attained only + 0.2 °C. November 2010 saw the greatest range in temperature for any November since 1880, when a maximum temperature of 16.1 °C on the 4th was followed by a minimum temperature of −8.7 °C on the 28th, giving a monthly range in temperature of 24.8 degrees Celsius.

Table **14.2** *November mean temperatures at Durham Observatory, 1843–2021*

November mean temperature **6.6** °C (average 1991–2020)

Mildest months			Coldest months		
Mean temperature, °C	Departure from 1991–2020 normal degC	Year	Mean temperature, °C	Departure from 1991–2020 normal degC	Year
8.7	+2.1	2011	2.4	-4.2	1919
8.5	+1.9	2015	2.5	-4.1	1910
8.3	+1.7	1994	2.7	-3.9	1915
8.0	+1.4	1938	2.9	-3.7	1862, 1925
7.9	+1.3	2021	3.0	-3.6	1923
Last 50 years to 2021					
8.7	+2.1	2011	3.3	-3.3	1993

November 1880 surpassed this, but only slightly, with the highest temperature of 15.2 °C on the 13th followed barely a week later by a minimum of −9.8 °C (until 1993, the lowest on record in November), for a monthly temperature range of 25.0 degrees Celsius.

November 1993's mean temperature of 3.3 °C was 3.3 degrees below normal, the coldest November since 1925 and the coldest month of 1993, a position more often held by January. A protracted cold spell in the second half featured two weeks when the temperature failed to reach 5 °C by day. This cold spell included nine consecutive air frosts, including November's coldest night on the 24th, when the minimum temperature was −12.0 °C, an extremely sharp frost for so early in the winter (Table 14.1) and the earliest 'minus 10' in any winter on record (Table 24.1). There was persistent freezing fog on both the 24th and 25th, the maximum temperature reaching only 0.6 °C on the 23rd and 0.4 °C on the 24th. Snow dominated the last third of the month, falling over several days to a total depth of 8 cm and lying for several more due to persistent low temperatures. November 1993 was also a very dull month—the seventh-dullest November on record—with just 34.5 hours of sunshine and 15 sunless days. It was the dullest month of the year, a position more normally held by December.

The five coldest Novembers all date from 1925 or earlier. Durham's coldest November remains 1919 (Figure 14.1), almost a degree colder still than 1993 and towards the end of a very disappointing year, one which remains the eighth-equal coldest year on Durham's records; all colder years were in the nineteenth century. A few mild days towards the end of the month (maximum 13.3 °C on the 23rd) failed to make up for some very cold days through the middle of the month including Durham's earliest sub-zero maximum temperature on the 12th as previously mentioned. Fifteen days failed

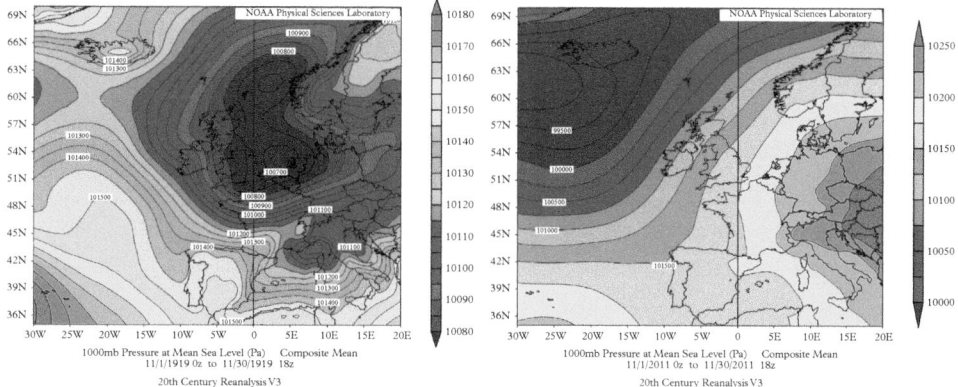

Figure 14.1 *The mean temperature in Durham differed by more than six degrees Celsius between these two Novembers—1919 (left) and 2011 (right). The maps show the mean distribution of barometric pressure; isobars are in Pascals (divide by 100 for hPa = millibars), at 100 Pa intervals. (From the 20th Century Reanalysis version 3 (20CRv3) plot [58])*

to get above 5 °C by day, while by night there were 13 air frosts and 23 ground frosts. The minimum of −8.9 °C on the 16th was, at that time, the second-coldest November night on record (Table 14.1). The *Monthly Weather Report* summarized November 1919 simply as 'exceedingly cold'.

Precipitation

November is, on average, Durham's wettest month of the year, with 73 mm precipitation; this is so for the current standard averaging period 1991–2020 and has been so since the 1971–2000 period, but in previous 30-year periods, August or October held that title. Rain can be expected on 19 days, which is about the same as December and January, despite those months having one extra day in the month.

November is rarely settled for very long, and only four Novembers since 1868 have recorded less than one-quarter of the monthly normal fall, with 1958 (12.2 mm or 16 per cent of average) being the driest (Table 14.3); that November less than 25 per cent of the expected rainfall was recorded along the east coast from Humberside to the Scottish border. In Durham, rain fell on only 11 days. The second half of the month was dominated by an anticyclone over Europe with a ridge extending westwards over the British Isles; weather was often cold and foggy, with the fog persisting all day in some places. More recently, November 2004 recorded only one-quarter of the month's normal rainfall, a dry and very mild month, although not very sunny. Rain fell on 18 days, although the wettest day (15th) received only 3.6 mm.

Table 14.3 *November precipitation at Durham Observatory 1868–2021*

November mean precipitation 73.4 mm (average 1991–2020)

Wettest months			Driest months			Wettest days	
Total fall, mm	Per cent of normal	Year	Total fall, mm	Per cent of normal	Year	Daily fall, mm	Date
186.1	244	1965	12.2	16	1958	72.8	12 Nov 1901
153.3	209	2010	16.3	21	1956	51.7	13 Nov 1875
148.8	195	1878	16.8	22	1942	48.0	26 Nov 2012
148.4	194	2000	18.2	24	2004	43.2	15 Nov 1995
146.6	192	2009	18.8	26	1932	40.2	5 Nov 1951
Last 50 years to 2021							
153.3	209	2010	18.2	24	2004	48.0	26 Nov 2012

Although November is, on average, Durham's wettest month of the year, there is only one November in Durham's 20 all-time wettest months—that of 1965, when 186 mm fell. This was also a wintry month, with a good deal of early snowfall: snow lay on 13 mornings, the earliest on 14 November. On 30 November, snow lay 41 cm deep at the morning observation—remarkable not only for arriving so early in the winter but also because this depth of snow was exceeded only once, in the depths of winter 1963, in recent records (1961–99; Table 25.8). The latter part of the month was particularly cold, with none of the last ten days reaching even 5 °C; there were air frosts every night from the 22nd, although persistent cloud cover prevented any very low minima. An airstream of arctic air with high pressure over Greenland and low pressure over Scandinavia brought the wintry weather to the British Isles. November 2010—the November with the highest range in temperature since 1880—was the wettest since 1965 and second only to it in total precipitation; rain or snow fell on 28 days, particularly snow during the last three days when daytime temperatures did not exceed 3 °C.

Three of the five wettest Novembers have occurred since and including 2000. November 2000 was almost the wettest month of a very wet year, beaten only marginally by April; 88.6 mm fell in three days 5th–7th, the 18th highest three-day fall on record. There was a major flood peak on the River Wear, the peak discharge of 363 cubic metres per second (cumecs) at 2200 GMT on 6 November amounting to the fourth-highest since records commenced in late 1957. November 2009, another wet month (147 mm), also included a major flood peak on the River Wear of 277 cumecs at 1445 GMT on 18th following rainfall of 23.4 mm on 17th (Figure 14.2).

Figure 14.2 *The River Wear in flood, 18 November 2009 (Tim Burt). Note that the riverside walk is impassable in such conditions. At a peak discharge of 277 cumecs, this flood ranks the seventeenth-highest on record. It appears to be a very similar flood level to that of 1 June 1924 (see Figure 8.2)*

Table 14.3 shows that November's wettest day was 12 November 1901, when 72.8 mm was recorded; 13 mm also fell on the previous day, and the two-day total of 85.8 mm remains the sixth-highest since 1868. A deep depression moved through causing heavy rain and severe gales; at 0900 GMT on 13 November the wind was recorded as north-easterly 35 knots, so measuring the rainfall that morning must have been a particularly unpleasant task. The heavy falls over these two days were significant enough to warrant a frontispiece map in *British Rainfall 1901*, noting a two-day fall of 121 mm at Wolsingham (107 mm on the 12th) and 110 mm at Wearhead. Many places in West and North Yorkshire received similar totals to Durham, and there was severe flooding at Todmorden in West Yorkshire.

Widespread heavy rainfall on 13 November 1875 led to a very uniform distribution of rainfall between Devon and North Wales, Lancashire, Yorkshire, Westmorland, Durham, and Northumberland, with falls typically between 38 and 55 mm; Durham recorded 51.7 mm, still the second-wettest November day on record. In November 2012, 88.6 mm fell in three days 24th–26th, including 48.0 mm on the final day. Once again, a significant flood resulted on the River Wear, with the seventh-highest gauged daily flow on record and the largest daily flow in 2012, Durham's wettest year (Figure 14.3).

Figure 14.3 *Durham's Framwellgate Waterside flooded, for the second time in 2012, on 27 November. At a peak discharge of 303 cumecs, this flood ranks ninth-highest on a record from late 1957 (Tim Burt)*

Sunshine

Monthly mean sunshine duration in November is 68 hours, equivalent to 2¼ hours per day, an hour per day less than October. The least sunny Novembers, 1897 and 1900, averaged less than one hour of sunshine per day, with 15 sunless days in the former and 17 in the latter. November 1947 remains by some margin the sunniest on record, with almost half the possible sunshine; there were 14 days with five or more hours of sunshine and only five sunless days: a bright end to the fourth-sunniest autumn on record—more pleasant weather to enjoy following the hot summer after the very harsh winter at the start of the year. Autumn 1922 experienced an unusual reversal of sunshine fortunes—a very dull September (just 64.6 hours, the dullest on record until 1976) was followed by a very sunny November (96.9 hours, half as much again as September), the sunniest on record until 1947.

All November's sunniest days come from early in the month; each must have enjoyed more or less unbroken sunshine as the amounts are very close to the maximum possible at that time of year (Table 14.4).

Table 14.4 *November sunshine duration at Durham Observatory. Observed monthly extremes 1880 to 2021, sunniest days 1882–1999 only*

November mean sunshine duration 68 hours, 2.3 hours per day (average 1991–2020). Possible day length: 254 hours. Mean sunshine duration as percentage of possible: 26.8

Sunniest months			Dullest months			Sunniest days 1882–1999	
Duration, hours	Per cent of possible	Year	Duration, hours	Per cent of possible	Year	Duration, hours	Date
120.0	47.2	1947	29.3	11.5	1897	8.4	2 Nov 1934
107	42.1	2005	29.3	11.5	1900	8.3	4 Nov 1922
101	39.7	2006	30.1	11.9	1968	8.1	4 Nov 1930, 2 Nov 1985, and 10 Nov 1985
96.9	38.1	1922	30.8	12.1	1885	8.0	7 Nov 1921 and 7 Nov 1941
96.3	37.9	1930	33.0	13.0	1997	7.9	Numerous
Last 50 years to 2021							
107	42.1	2005	33.0	13.0	1997	8.1	2 Nov 1985 and 10 Nov 1985

Daily sunshine records ceased at the Observatory on 31 December 1999: monthly sunshine totals from January 2000 onwards are estimated (to the nearest hour) by regression from the Met Office East and North-east England sunshine series. See Appendix 6 for record metadata and missing months.

Wind and weather

Snow cover (1960–98)

Snowfall is not uncommon in November, although lying snow is more so. The earliest date between 1960 and 1998 with a snow cover at 0900 GMT was 14 November in 1965, when there was a 5 cm cover. The monthly average for November during this period was 1.4 mornings with a snow cover, although the statistics are biased by a few snowy months—November 1965 with 13 snowy mornings, 1971 with six, and 1993 with eight. November 1965 included a spell of 14 consecutive mornings with 5 cm or more of snow on the ground starting on 25 November, 30 cm deep or greater 27 November to 2 December, including a remarkable 41 cm at 0900 GMT on 30 November. The only other occasions in November during this period with 10 cm or more of lying snow were 28 November 1969 (10 cm) and 23 November 1971 (11 cm).

Fog (1962–96)

The frequency of fog is similar between October and February, at about three mornings per month. Fog was much more frequent in the 1960s, averaging five mornings in

November, compared to the 1990s when the average was just two, with none in 1992 and 1995.

Barometric pressure (1844–1960)

The highest MSL barometric pressure recorded at the Durham Observatory in November occurred in 1922; at 0900 GMT on 15 November of that year the MSL pressure reached 1042.9 hPa. Similar values occurred at 1000 on 11 November 1857 (1042.5 hPa) and in 1899 (1041.5 hPa at 0900 GMT on 17 November). November's lowest observed pressure was 964.2 hPa at 1000 on 27 November 1881; 964.6 hPa was recorded at 2200 on 11 November 1877, and 965.9 hPa at 2100 GMT on 20 November 1926.

The most anticyclonic November was in 1867, with a monthly mean at MSL of 1026.2 hPa, some way ahead of the next most anticyclonic, which were 1879 (mean 1024.3 hPa) and 1942 (mean 1023.9 hPa). The most cyclonic were those of 1877 (mean at MSL 999.4 hPa), 1960 (1000.2 hPa), and 1951 (1000.3 hPa).

Wind speed and direction (1961–97)

Figure 14.4 shows the distribution of wind speed and direction at the 0900 GMT observation in November during the period 1961–97. Mean wind speeds are a little higher than in October (mean wind speed at 0900 GMT of 7.2 knots), but there is normally little to choose between any of the months in the winter half-year, and there is considerable variation from year to year. The Novembers of 1972 and 1973 were both very windy, the mean speeds at 0900 GMT being 13.0 and 13.7 knots, respectively, twice the period normal. As in October, winds are largely confined to between south and north-west. On 1 November 1970, the wind at 0900 GMT was recorded as mean 50 knots from the west-south-west. Two days later, at 1325 GMT on 3 November 1970, a gust of 80 knots was recorded, at that time the highest November gust on the Observatory's anemograph record, which commenced in 1937. For the hour ended 1400 GMT the mean wind speed was westerly, 47 knots. During the same storm, the mountain station at Great Dun Fell (857 m AMSL) recorded a gust of 102 knots. In a much earlier storm—the gale of 28 November 1892—surviving records indicate a mean speed of 75 mph (65 knots) was reached ([24], item 426), but the calibration factor of the Robinson anemometer in use at the time was subsequently found to be too high and applying the appropriate adjustment (see Appendix 7) reduces this to a more realistic 55 mph (48 knots).

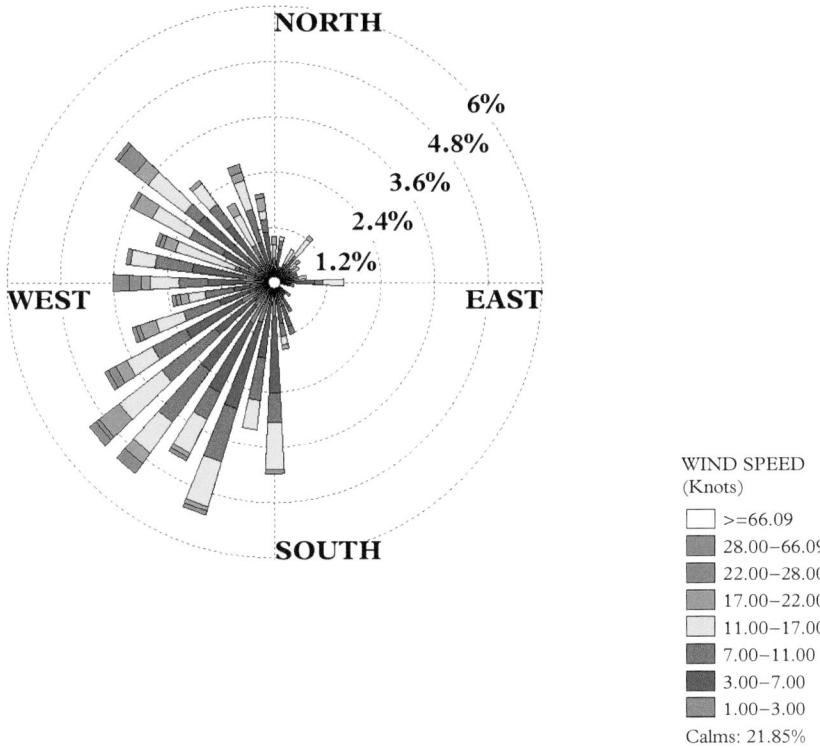

Figure 14.4 *Wind rose for 0900 GMT observations in November, period 1961–97*

Durham temperature, precipitation, and sunshine in graphs—November

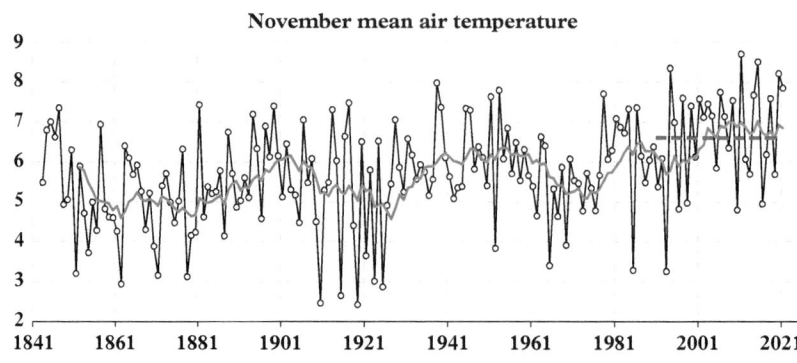

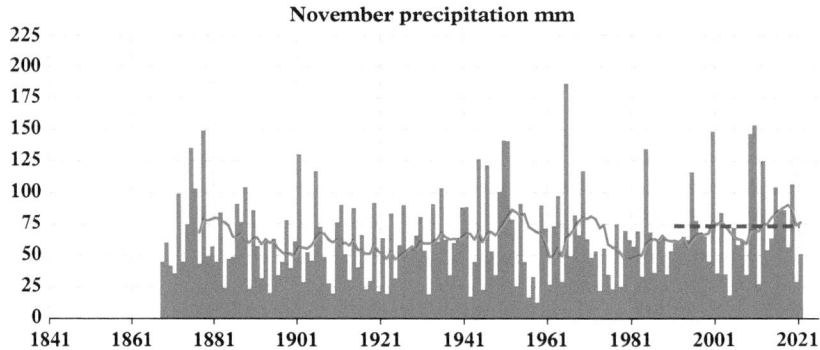

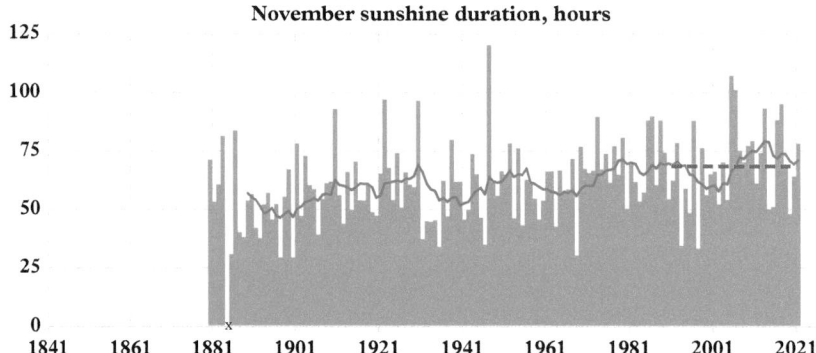

Figure 14.5 *Monthly values of (from top) November's mean temperature (°C, since 1843), total precipitation (mm, since 1868), and sunshine duration (hours, since 1880, missing data 1884) at Durham Observatory. The 1991–2020 averages are indicated by the thick blue dashed line, while the ten-year running mean ending at the year shown is indicated by the red line*

15

December

And so we come to December—the shortest and the least sunny days of the year, and the second-coldest month: January is but a tenth of a degree colder on average. The week leading up to Christmas has, on average, the least sunshine of any week in the year, and a normal December will see 13 days without any sunshine whatsoever. December can see some very cold days and nights, particularly towards the end of the month: in the most recent 30-year averaging period, the coldest day of the year was 20 December with a mean temperature of 3.0 °C. An average December is—perhaps surprisingly—almost 20 per cent drier than November, but with the same frequency of days with measurable rain or snow (19). Unlike the months August to November, December has *not* shown significant warming since 1900; indeed, it has the lowest rate of warming of any month, just 0.6 degrees per century. This is partly due to the five coldest December nights since Victorian times occurring in just three cold Decembers since and including 1981.

Temperature

December's mean temperature in Durham over the standard averaging period 1991–2020 was 4.2 °C, with monthly means in that period ranging from below zero to almost 8 °C.

Mild days in December are not uncommon (at least one day in the month can be expected to reach 12 or 13 °C), but very mild days—above 15 °C—are much rarer, with only five on record, and all of those since 1987 (Table 15.1). December 2015 was an exceptionally mild month, almost four degrees warmer than normal in Durham (Table 15.2), and by the huge margin of 1.6 degrees the mildest December (9.7 °C) in the Central England Temperature (CET) series dating from 1659 [72]. Two days that month surpassed all previous December warmth records in Durham, namely 15.8 °C on the 18th and 15.9 °C on the 19th. Notably, both values are slightly warmer than an average day in *May*! The month's weather was dominated by mild, tropical, maritime airstreams from the south-west, which also meant very cloudy conditions such that night-time temperatures also remained unusually high. Both mean daily maximum (10.9 °C) and minimum (5.0 °C) remain the highest for the month on record, as of

Table 15.1 *Highest and lowest maximum and minimum temperatures at Durham Observatory in December, 1843–2021*

Rank	Mildest days	Coldest days	Mildest nights	Coldest nights
1	15.9 °C, 19 Dec 2015	−6.6 °C, *13 Dec 1878*	11.9 °C, 18 Dec 1932	−16.4 °C, *25 Dec 1860*
2	15.8 °C, 18 Dec 2015	−5.2 °C, *31 Dec 1870*	11.3 °C, 17 Dec 1904 and 17 Dec 2015	−15.4 °C, *30 Dec 1874, and 14 Dec 1878*
3	15.1 °C, 20 Dec 1987 and 22 Dec 1991	−4.7 °C, *25 Dec 1860*	11.2 °C, *24 Dec 1843,* 5 Dec 1898 and 19 Dec 2015	−14.6 °C, *26 Dec 1860 and 4 Dec 1879*
4	15.0 °C, 10 Dec 1994	−4.2 °C, *19 Dec 1859*	11.0 °C, *10 Dec 1851* and 12 Dec 1994	−14.0 °C, *31 Dec 1870 and 3 Dec 1879*
5	14.8 °C, *10 Dec 1863* and 7 Dec 2016	−4.0 °C, *12 Dec 1882*	10.8 °C, 29 Dec 1987	−13.5 °C, *13 Dec 1878*
Last 50 years to 2021				
	15.9 °C, 19 Dec 2015	−3.2 °C, 31 Dec 1978	11.3 °C, 17 Dec 2015	−12.5 °C, 18 Dec 1981

Temperatures were recorded in different thermometer screens during the period of record; see notes on page xvii and Appendix 3 for more details

course is the monthly mean temperature (7.9 °C)—only slightly below a normal April. Only two minor air frosts were recorded, against the December norm of ten. Three consecutive nights (17th–19th) remained above 10 °C in December 2015, and two of these feature in Table 15.1, although 18 December 1932 remains Durham's mildest December night on record, with a minimum temperature of 11.9 °C during an extremely mild spell mid-month. The grass minimum that morning was 10.6 °C, one of only four December grass minima to remain above 10 °C.

Our first December in the Durham record, 1843, remains one of the mildest (Table 15.2), although of course we should be mindful of the differing standards of thermometer exposure at that time (see Appendix 3). There was only one air frost, the minimum for the month of −1.4 °C on the 2nd, while twelve days saw maximum temperatures in excess of 10 °C, reaching 13.3 °C on the 7th. On the CET series, December 1843 ranks as the eighth-equal mildest. It is perhaps ironic, therefore, that this should be the month when two iconic images of a Victorian Christmas first appeared: the publication of Charles Dickens's *Christmas Carol* on the 17th, and the world's first Christmas cards. The latter were commissioned by Sir Henry Cole in London from the artist

Table 15.2 *December mean temperatures at Durham Observatory, 1843–2021*

December mean temperature 4.2 °C (average 1991–2020)

Mildest months			Coldest months		
Mean temperature, °C	Departure from 1991–2020 normal degC	Year	Mean temperature, °C	Departure from 1991–2020 normal degC	Year
7.9	+3.7	2015	−0.7	−4.9	*1878*
7.3	+3.1	*1843*, 1988	−0.3	−4.5	*1874, 1981, 2010*
7.0	*+2.8*	*1857*	*0.6*	*−3.6*	*1846*
6.8	+2.6	1934	*0.7*	*−3.5*	*1879*
6.6	+2.4	1974	*1.0*	*−3.2*	*1859, 1890, 1950*
Last 50 years to 2021					
7.9	+3.7	2015	−0.3	−4.5	2010

John Calcott Horsley. While we lack any documentary evidence for Durham, we can be sure 1843 did not see a white Christmas.

At the opposite extreme to 1843's mild conditions, December has also seen some of the very lowest temperatures, by day and by night, yet recorded at Durham. Many of these were recorded in the nineteenth century when there are uncertainties with regard to the instruments and exposures in use at that time (see Appendix 3). In December 1859 temperatures remained below freezing for six consecutive days 14th–19th (Table 24.12), a record that stood until January 1982, but in more recent times very low temperatures have been recorded in the Decembers of 1981, 1995, and 2010.

December accounts for three of the five coldest days on record, all of them well over a century ago, all of which are Stevenson screen-equivalent temperatures from measurements originally made in a Glaisher stand (Table 24.6). The maximum temperature of −6.6 °C on 13 December 1878 remains without precedent on Durham's records and followed a minimum temperature that morning of −13.5 °C. A pool of extremely cold arctic air had become cut off over north-west Europe in the preceding days, resulting in the formation a deep depression over the Baltic Sea. Temperatures in Sweden fell below −20 °C. With an anticyclone over Greenland, north-easterly winds advected the cold air towards and across the British Isles and temperatures fell sharply, with widespread snowfalls along eastern and north-eastern coasts. On 13 December,

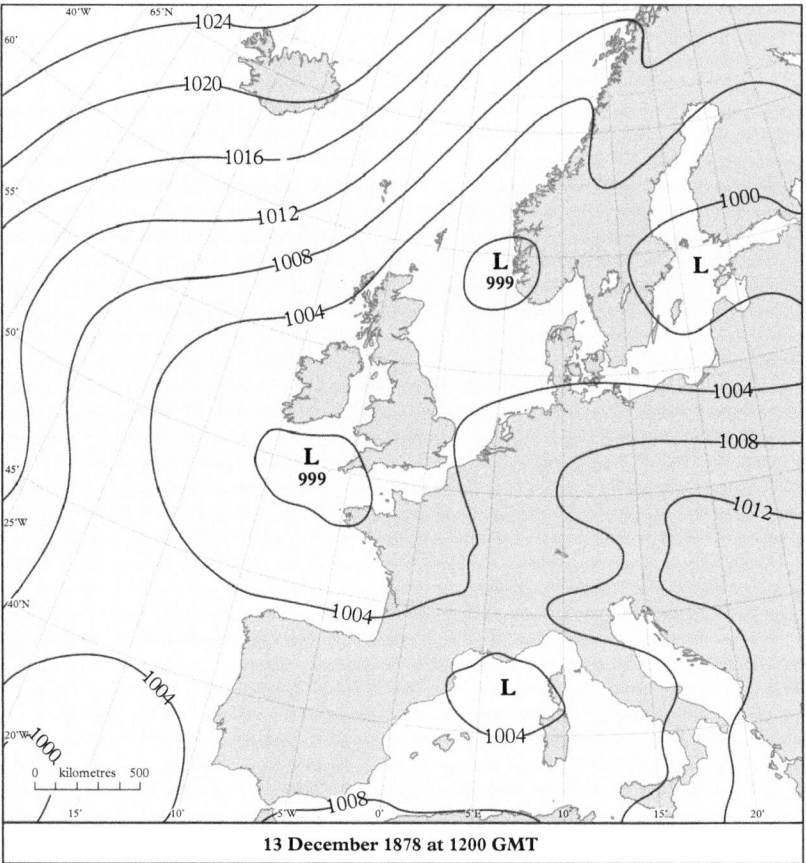

13 December 1878 at 1200 GMT

Figure 15.1 *The synoptic situation at 1200* GMT *on Friday 13 December 1878, Durham's coldest-ever day, when the maximum temperature reached only* −6.6°C, *between minima of* −13.5 °C *on 13th and* −15.4 °C *on 14th. (Based upon the 20th Century Reanalysis version 3 (20CRv3) plot [58] and the Met Office* Daily Weather Report; *drawn by Chris Orton, Durham University)*

the British Isles lay within a complex low-pressure area extending from the Baltic Sea south-westwards to the Azores, in a slack gradient between two shallow depression centres (Figure 15.1). Both the 1000 and 2200 Durham observations noted clear skies, although the low maximum temperature may have been partly as a result of freezing fog in light winds. At 1000 the temperature was −12.8 °C, and −13.2 °C at 2200, with a very light south-westerly breeze on both occasions. There was almost certainly a substantial snow cover, but the surviving records do not include that information. By the following morning, the temperature had fallen to −15.4 °C; three of these entries are included in Table 15.1. December 1878 remains the coldest on Durham's records (Table 15.2).

Christmas 1860 saw some of the coldest weather in modern instrumental records. At Oxford, Christmas Eve morning saw the temperature fall to −17.8 °C, a value that remains the lowest on more than 200 years' record there [49], while in Durham Christmas morning dawned sunny, still, and very cold. At 1000 the temperature was −11.0 °C, following the night's minimum temperature of −16.4 °C—the lowest December temperature yet recorded in Durham (Table 15.1). Since then, only two days —17 January 1881 and 8 February 1895—have recorded lower minima. Temperatures on Christmas Day 1860 rose to just −4.7 °C, the third-coldest December day on record, before falling back to −11.9 °C at 2200 and thence to a minimum the following morning of −14.6 °C. The events of this period are covered in more detail in Chapter 22, including a synoptic chart for Christmas morning.

Very severe conditions were also recorded in December 1879, the final act in a very cold year, when on 4 December the minimum temperature at the Durham Observatory fell to −14.6 °C (Stevenson screen equivalent). It was during this very early and very severe cold spell that the temperature fell to −26.7 °C at Kelso in the Scottish borders [73]. Seven consecutive nights fell below −7 °C in Durham, including four consecutive nights below −10 °C; only February 1895 has seen a longer run of such low temperatures. The month gradually became milder, however, and the temperature reached 11.6 °C on the 28th in near-gale-force west-south-westerly winds: this was the day on which the Tay Rail Bridge collapsed, derailing a train and costing 75 lives, as winds reached Force 11 in Dundee on the flanks of a rapidly-moving depression [74]. In Durham at 2200 that evening, skies were clear, air temperature 7.9 °C, and the wind west-south-westerly 25 knots. Further details about 1879 are included in Chapter 22.

After 1879, there were no instances of a December minimum temperature of −11 °C or lower for over 100 years. December 1981 began fairly mild, but between the 4th and the 8th waving cold fronts gradually introduced very cold arctic air southwards across the entire country, and widespread snowfalls resulted. In Durham, snow lay on the ground for 19 consecutive days from the 12th, reaching a maximum depth of 17 cm on 14 December. Short-lived anticyclones brought clearer skies at times, and in Durham overnight temperatures fell below −10 °C on 17th (−11.4 °C) and 18th (−12.5 °C), while remaining below zero by day; in all, there were 21 air frosts that month. At that time, only the Decembers of 1874 and 1878 had been colder in Durham.

December 1995 saw another severe cold spell towards the end of the month, with heavy snowfalls particularly in Scotland. Durham recorded −10.4 °C on 28 December; shortly afterwards, Altnaharra in north-west Scotland equalled the UK's lowest-ever minimum temperature when −27.2 °C was recorded there on 30 December [75].

December 2010 is the coldest December of the modern era, slightly colder even than 1981 and equal second coldest on Durham's records (Table 15.2). Since the second half of the twentieth century and at the time of writing, only six months have recorded a mean temperature below 0 °C in Durham (January and February 1963, January 1979, December 1981, February 1986, and December 2010). December 2010 opened in the midst of an exceptionally cold spell which lasted until the 9th. Strong north-easterly winds brought outbreaks of heavy snow to eastern Scotland, north-east England, the north Midlands, and parts of the Home Counties, with falls amounting to 25–30 cm in some areas; locally, the snow depth reached 40 cm at Copley (Co. Durham) on 2 December, while the Durham Observatory AWS recorded minima of −10.3 °C on the 3rd and −9.9 °C on the 4th. (Unfortunately, snow depths have not been recorded at the Observatory since automation in October 1999). Following the passage of a vigorous cold front on the 17th, airflow became dominated by polar air advected south around an anticyclone over Greenland. A second −10 °C followed on the 21st (−10.4 °C) and there were 21 air frosts in all.

Finally, it is interesting to comment on the remarkable differences between the Decembers of 2010 and 2015, just five years apart [72]. The former was the coldest since 1878, with a mean temperature of −0.3 °C; the latter became by far the mildest on Durham's records with a mean temperature of 7.9 °C, eight degrees above December 2010.

Figure 15.2 *The Durham Cow sculpture on the Racecourse on a frosty morning, 14 December 2012 (Tim Burt)*

Precipitation

December is, on average, almost 20 per cent drier than November—although conversely there are three Decembers, but only one November, within Durham's ten wettest-ever months (Table 25.3). Normal rainfall for the month is 61 mm, falling on 19 days. Heavy falls of rain are rare in December; since 1868 only four December days have exceeded 30 mm, and none 40 mm (Table 15.3). Wettest of all was 16 December 1872, when 37.4 mm fell. The weather in 1872 was dominated by cyclonic weather with locations in the south and east unusually wet compared to the uplands, which were simply wetter than normal [76]. In Durham, rain fell on 19 rain days in December 1872 with two other significant falls in addition to the wettest day: 22.2 mm fell on the 8th and 16.3 mm on the 17th, the monthly total of 125 mm ranking wettest December to that date, but fifth-wettest today.

In late December 1978, the fall of 34.3 mm on the 27th was a long-continued cyclonic rainfall associated with the stalling of an arctic front over northern districts [77, 78]. Rain fell every day in Durham during the last 12 days of the month, amounting to 133 mm, including significant falls on Christmas Eve (27.3 mm) and on 27, 28, and 29 December (respectively 34.3 mm, 23.1 mm, and 14.4 mm). The sixteenth-highest flood peak, 285 cumecs, occurred on the River Wear on 27 December. Very cold air behind the arctic front penetrated southwards and turned the rain to snow, introducing a widespread snow cover for New Year 1979 and ushering in Durham's coldest winter since 1962/63. The month's total precipitation of 195.5 mm made this the wettest December on record, the third-wettest month of any name since 1868 (Table 25.3) and, at the time of writing,

Table 15.3 *December precipitation at Durham Observatory 1868–2021*

December mean precipitation 61.2 mm (average 1991–2020)

Wettest months			Driest months			Wettest days	
Total fall, mm	Per cent of normal	Year	Total fall, mm	Per cent of normal	Year	Daily fall, mm	Date
195.5	329	1978	9.3	16	1873	37.4	16 Dec 1872
164.9	277	1876	13.1	22	1905	34.3	27 Dec 1978
161.4	271	1937	14.3	23	1885	33.6	25 Dec 2015
133.7	225	1915	14.5	24	1941	30.6	8 Dec 1954
124.8	204	1872	15.0	25	1879	28.3	8 Dec 1983
Last 50 years to 2021							
195.5	329	1978	19.7	33	1971	34.3	27 Dec 1978

Figure 15.3 *Flooding along the banks of the River Wear through central Durham, December 1981. This was taken further south than Figure 8.3 and shows the relatively new Millburngate Bridge, constructed in 1967. (Department of Geography, Durham University)*

the wettest month within the last 50 years. Just for good measure, it was also the dullest December on record (Table 15.4).

Of the driest Decembers, 1873 was the only December to receive less than 10 mm precipitation. Within the last 50 years, only the Decembers of 1971 and 2004 have been drier than 20 mm, and then only just (Table 15.3).

Sunshine

Durham's average sunshine duration during the period 1991–2020 was 57 hours, equivalent to 1.8 hours per day. December is the least sunny month of the year, not surprising given that this is the month of the winter solstice, and six of the eleven dullest months have been Decembers—the other five being Januarys (Table 26.1). Thirteen days in an average December, almost one day in two, remain sunless, ranging from 23 in 1882 and 1903 to one in 2010 and 2017, so it is no surprise that even the sunniest December (2014, 84 hours) managed only 37 per cent of possible sunshine. In the least sunny December, the notoriously wet 1978, only 9.7 hours of bright sunshine was recorded, just 4.3 per cent of possible and Durham's second-dullest month on record (Table 26.1). The Decembers of 1890 and 1903 were little different (Table 15.4).

Table 15.4 *December sunshine duration at Durham Observatory. Observed monthly extremes 1880 to 2021, sunniest days 1882–1999 only*

December mean sunshine duration 57 hours, 1.8 hours per day (average 1991–2020). Possible day length: 227 hours. Mean sunshine duration as percentage of possible: 23.8

Sunniest months			Dullest months			Sunniest days 1882–1999	
Duration, hours	Per cent of possible	Year	Duration, hours	Per cent of possible	Year	Duration, hours	Date
84	37.0	2014	9.7	4.3	1978	7.2	16 Dec 1962
78.8	34.7	1886	10.4	4.6	1903	7.0	1 Dec 1997
77	34.1	1999	10.8	4.8	1890	6.7	2 Dec 1942
72.8	32.1	1967	16.5	7.3	1927	6.6	5 Dec 1909
71.2	31.4	1946	17.8	7.8	1934	6.5	Numerous
Last 50 years to 2021							
84	37.0	2014	9.7	4.3	1978	7.0	1 Dec 1997

Daily sunshine records ceased at the Observatory on 31 December 1999: monthly sunshine totals from January 2000 onwards are estimated (to the nearest hour) by regression from the Met Office East and North-east England sunshine series. See Appendix 6 for record metadata and missing months.

December 1903 is a curiosity: a very dull month but also a dry one (26 mm); 1903 remains one of the wettest years on record in the UK, although (as at 2021) only Durham's fifteenth-wettest year, but December was its driest month. The *Monthly Weather Report* notes that the month was characterized by a great prevalence of dull, overcast skies. Despite the fact that cyclonic depressions were fairly numerous, rainfall was slight across the country. Bright sunshine was below average nearly everywhere; with only 1 per cent of possible duration at Newcastle, perhaps Durham was lucky to receive as much as 4.6 per cent of possible.

Wind and weather

Snow cover (1960–98)

On average, December had between two and three mornings with snow cover in recent decades, although this is skewed by a few snowy months—December 2010 saw 19 mornings with snow lying, while there were nine such in the Decembers of 1963, 1965, and 1976. In the final decade of this recorded period, half of the Decembers saw no lying snow at all.

The greatest snow depths in December between 1960 and 1998 were 36 cm and 41 cm on the 30th and 31st in 1961, and 36 and 38 cm on the 1st and 2nd in 1965. On 29 December 1968 snow lay 15 cm deep at 0900 GMT, and 14 cm deep at 0900 GMT on 31 December 1978. We have very few records of snow depths from before 1960, but one exception is an entry for 25–26 December 1869 made on the Met Office rainfall Ten Year Sheet, which noted that the average depth of snow on that occasion was about 25 cm, drifted in places to 90 cm.

Fog (1962–97)

Fog is fairly common throughout each of the months in the winter half-year, averaging two or three mornings with fog in December. The frequency of morning fog has reduced significantly since the 1960s and 1970s, when five or six foggy mornings were not unusual; but between 1979 and 1996, only December 1984 (seven fogs at 0900 GMT) recorded more than four.

Barometric pressure (1844–1960)

The highest MSL barometric pressures recorded at the Durham Observatory in December all occurred in 1926, when a particularly intense anticyclone centred over Scotland reached an intensity unseen in any month since, the highest land record being 1051.9 hPa at Wick at 1300 GMT on Christmas Eve that year [55]. At 2100 GMT on 23 December, Durham reached 1047.4 hPa, and just 0.1 hPa lower at 0900 GMT on 24 December. Aside from 1926, the December highest stand at 1045.6 hPa, recorded at 2100 GMT on 15 December 1946, and 1045.3 hPa at 2100 GMT on 12 December 1905.

December's lowest observed pressure is also Durham's lowest for any month, namely 936.2 hPa recorded at 2100 GMT on 8 December 1886, a further fall of 20 hPa from an already near-record lowest 956.3 hPa at 0900 GMT that morning. On this date, an exceptionally deep depression crossed the north of Ireland and southern Scotland; at Belfast the barometer fell to 927.2 hPa at 1330h as the depression centre passed nearby [56] (Figure 22.12). At Durham, the wind was east-south-easterly 29 knots at the 0900 GMT observation but had veered to southerly and decreased to 16 knots at 2100, suggesting that the centre of the depression lay not far to the north-west. By 0900 GMT next morning, the barometer had recovered to 953.7 hPa, while the wind had veered further to west-north-west, blowing at a brisk 18 knots. Aside from this event, the lowest December barometer reading on Durham's records occurred on 3 December 1909, when at 0900 GMT the MSL reading was 954.7 hPa.

The most anticyclonic Decembers were those of 1926, when the monthly mean was 1024.8 hPa, followed closely by 1951 (mean 1024.5 hPa) and the very cold 1879 (mean 1024.4 hPa). The most cyclonic Decembers came in 1868 (mean for the month 997.0 hPa), just ahead of the Decembers of 1876 and 1959, tying at 997.1 hPa. December 1876 remains the second-wettest December on record (Table 15.3), but those of 1868 and 1959, while wetter than normal, were not unduly so.

Wind speed and direction (1961–97)

Figure 15.4 shows the distribution of wind speed and direction at the 0900 GMT observation in December during the period 1961–97. Mean wind speeds are at their highest in the winter months: December's mean wind speed at 0900 GMT of 7.5 knots lies exactly midway between the averages for November and January. The mild December of 1974 was also windy, the mean wind speed at 0900 GMT of 12.9 knots was more than twice the long-term mean. December's wind rose shows the typical winter pattern in Durham—the strongest and most frequent winds blow from between south-west and north-west.

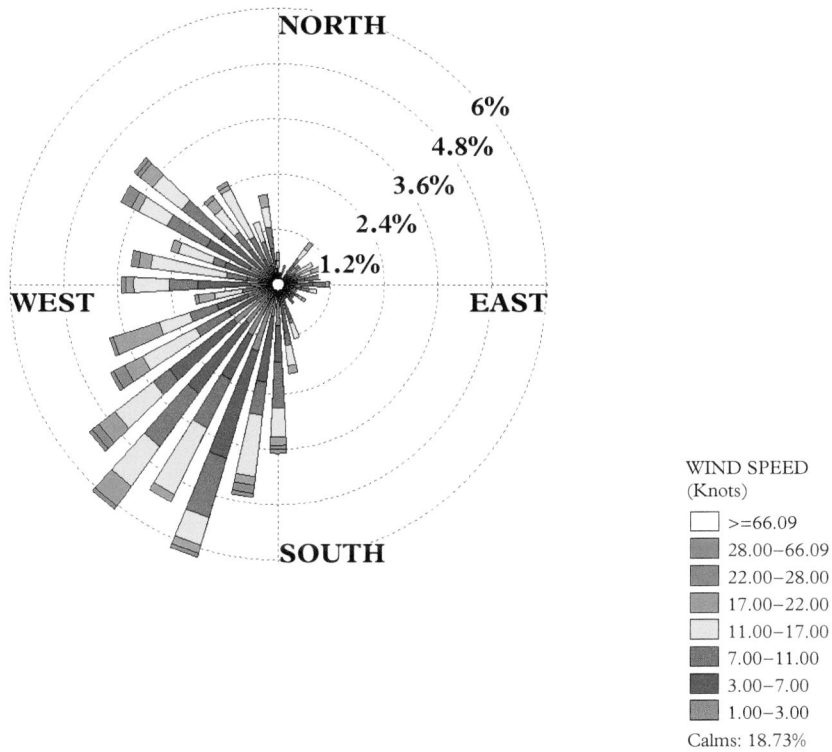

Figure 15.4 *Wind rose for 0900 GMT observations in December, period 1961–97*

Durham temperature, precipitation, and sunshine in graphs—December

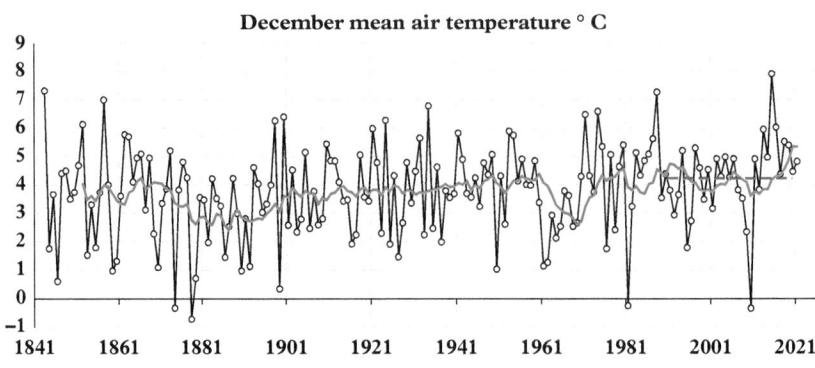

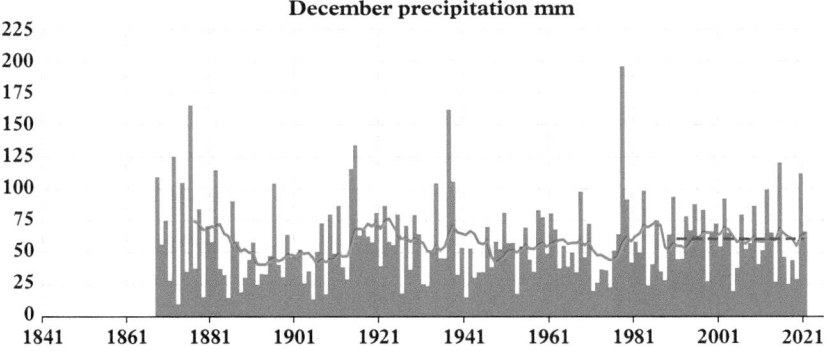

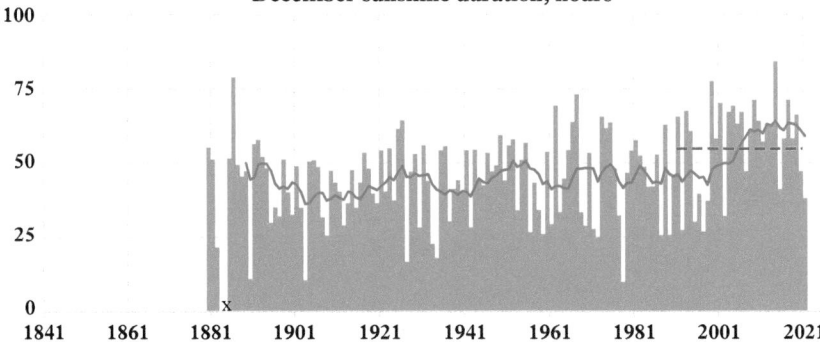

Figure 15.5 *Monthly values of (from top) December's mean temperature (°C, since 1843), total precipitation (mm, since 1868), and sunshine duration (hours, since 1880, missing data 1884 and 1885) at Durham Observatory. The 1991–2020 averages are indicated by the thick blue dashed line, while the ten-year running mean ending at the year shown is indicated by the red line*

16

The calendar year

The Pennine hills, although relatively modest in altitude, are sufficiently high to create an inhospitable climate of low temperatures, high rainfall, and frequent cloud cover. Durham lies in the rain shadow of the Pennines with a very different precipitation regime compared to the hills to the west. Indeed, Durham's climatic regime in general is closer to that of the North Sea coast than to the Pennine moorlands, but its more northerly location leaves it more exposed to active depressions (particularly those passing to the south, which may bring many hours of heavy cyclonic precipitation) and less subject to the advantages of either the Azores anticyclone or proximity to the European mainland, compared to places further south on the eastern side of England. Moreover, its proximity to the cold North Sea gives added emphasis to its prevailing coolness [10]. However, Durham is sufficiently far inland and to the west of the Magnesian Limestone hills to avoid the worst of the coastal fogs. Its wind regime is dominated by winds from between south and north-west, although northerly and easterly winds become more common in the spring and early summer. Compared to Oxford, for example [49], Durham is on average 1.6 degrees cooler and receives about 9 per cent less sunshine. Average precipitation amounts are very similar, although Durham records about 22 more rain days per year than Oxford.

Temperature

Since continuous (more or less) daily records began at Durham Observatory in July 1843—there are a few gaps, mainly before 1860—air temperatures have ranged between −18.0 °C (on 8 February 1895) and 32.9 °C (on 25 July 2019), a range in almost 180 years of a little over 50 degrees Celsius (Table 16.1). The record includes some regression-based assumptions to utilize the pre-1900 Glaisher stand thermometer records (January 1860 to October 1899), as set out in Appendix 3, but since the Stevenson screen record commenced in November 1899 the lowest temperature recorded has been −16.1 °C, on 21 January 1940 and equalled on 11 January 1982. The highest and lowest daily maximum and minimum air temperatures are listed in Table 16.1; an expanded list is given in Chapter 29 and more details of these events can be found in the monthly sections.

Table 16.1 *Highest and lowest maximum and minimum temperatures at Durham Observatory in any month, July 1843 to December 2021*

Rank	Hottest Days	Coldest days	Warmest nights	Coldest nights
1	32.9 °C, 25 Jul 2019	−6.6 °C, *13 Dec 1878*	18.9 °C, 20 Jul 2016	*−18.0 °C, 8 Feb 1895*
2	32.5 °C, 3 Aug 1990	*−6.2 °C, 1 Jan 1871*	*18.1 °C, 12 Jul 1859*	*−16.9 °C, 17 Jan 1881*
3	31.9 °C, 2 Aug 1990	*−5.1 °C, 31 Dec 1870*	17.9 °C, 14 Aug 2001, 6 Aug 2006, and 26 Jul 2019	*−16.5 °C, 10 Feb 1895*
4	*31.3 °C, 16 Jul 1876*	*−4.7 °C, 25 Jan 1881*	17.8 °C, 26 Aug 1959 and 5 Jun 1980	*−16.4 °C, 25 Dec 1860*
5	31.0 °C, 1 Jul 2015	*−4.6 °C, 25 Dec 1860*	17.7 °C, 3 Aug 2018 and 24 Jul 2019	−16.1 °C, 21 Jan 1940 and 11 Jan 1982
Last 50 years to 2021				
	32.9 °C, 25 Jul 2019	−4.5 °C, 7 Jan 1982	18.9 °C, 20 Jul 2016	−16.1 °C, 11 Jan 1982

Temperatures recorded prior to the introduction of a Stevenson screen in Nov 1899 are shown in italics; these are regression estimates from Glaisher stand readings from Jan 1860 to Oct 1899, as read from the 'North Shed' Oct 1851 to Dec 1859 and as read from the north wall exposure prior to Oct 1851. Temperatures were recorded in degrees Fahrenheit until Oct 1971, read to a precision of 0.1 °F during the period 1843 to 1918 and 1 °F for the remainder of the record. All values in degrees Fahrenheit have been converted to degrees Celsius, but those for the 1 °F precision period are subject to approximately 0.2 °C uncertainty as a result. Since November 1971 temperatures have been recorded to a precision of 0.1 °C. See notes on page xvii and Appendix 3 for more details

About one year in five, the lowest temperature recorded in the calendar year occurs outside of the three winter months (December, January, and February)—most frequently in March (29 occasions since 1844), and less frequently in November (ten occasions). Remarkably, the coldest night of the calendar year has occurred twice in April and once in October. In April 1934, the minimum temperature of −5.6 °C on the 7th was the lowest of the year, and similarly in April 1990, when −3.5 °C was recorded on the 4th, although this equalled that recorded on 2 March. In October 1926, the minimum temperature of −5.3 °C on the 25th was the equal coldest night of the year, with 1 November.

The *Monthly Weather Report* noted that 'January 1940 will long be remembered for intense cold'. This first winter of the Second World War was Durham's coldest month of any in the twentieth century, with a mean temperature of −1.5 °C; only February 1855 (mean temperature about −2.2 °C) and January 1881 (mean −1.9 °C) have been colder. Table 24.8 lists the ten coldest months on Durham's records; the most recent of these, at the time of writing, was February 1986, with a mean temperature of −0.6 °C. Figure 16.1 shows the lowest recorded temperature (°C) in every year since 1844, together with a ten-year unweighted running mean.

The highest temperature of the year is not always recorded during the three months of the meteorological summer (June, July, or August): between 1844 and 2021, and excluding 1854, 1855, and 1863 for which some months are missing, the hottest day of the year has occurred eight times each in both May and September—an average of about once in 20 years for each month. Only once in over 200 years has it occurred in October (1985). The earliest date on which the hottest day of the year has occurred is 13 May (in 1965, maximum temperature 25.0 °C), and the latest 1 October (in 1985, maximum temperature 24.1 °C, equalling 25 July that year). Figure 16.2 shows the

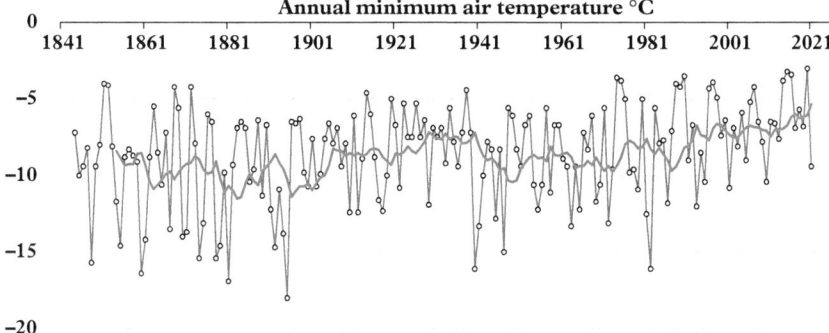

Figure 16.1 *Lowest daily minimum temperature (°C) in Durham by calendar year, 1844 to 2021, together with a ten-year unweighted running mean plotted at year ending*

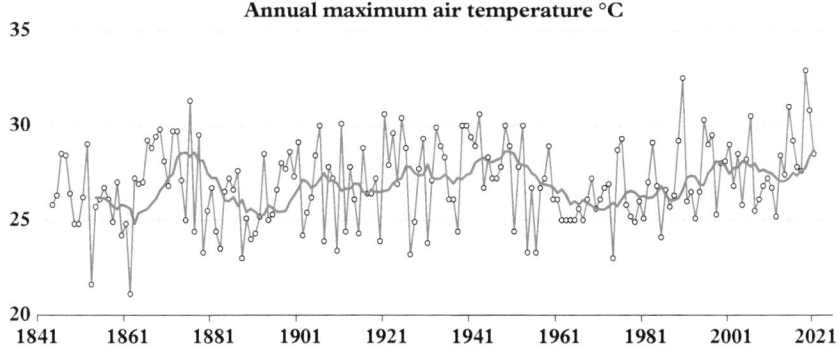

Figure 16.2 *Highest daily maximum temperature (°C) in Durham by calendar year, 1844 to 2021, together with a ten-year unweighted running mean plotted at year ending*

highest recorded temperature (°C) in every year between 1844 and 2021, together with a ten-year unweighted running mean.

The largest daily temperature range recorded was 23.7 degrees Celsius on 5 August 1916. The minimum temperature that morning was 5.1 °C; on a day with a light north-westerly wind and 14 hours unbroken sunshine, the afternoon maximum reached 28.8 °C. Daily ranges of temperature greater than 20 degrees are not uncommon at Durham, occurring on average roughly once every three years, and only January, February, October, and November have never recorded at least one such event. Within the last 50 years, the greatest daily range has been 20.6 degrees Celsius on 7 September 1991, a day with 12.1 hours of sunshine: the minimum temperature was 3.5 °C and the maximum 24.1 °C. Other high daily ranges in temperature are listed in Table 24.9.

Warm and cold years

Between 1844 and 2020, annual mean temperatures in Durham varied between 6.76 °C in 1879 and 10.21 °C in 2014—a range of 3.45 degrees Celsius. The warmest and coldest years are shown in Table 16.2. It is noteworthy that all six of the warmest years have occurred since 2004, whereas the most recent year in the 'Top 5 coldest' is 1888. This is also brought out graphically in Durham's 'climate stripe' (Figure 16.3), where every year's annual mean temperature is represented in colour by a thin vertical stripe, whose colour represents gradation from cold (blue) to warm (red) years. In a simple barcode-like graphic, this shows the progression from cold years to the series of particularly warm years within the last three decades or so.

Table 16.2 *Annual mean temperatures at Durham Observatory, 1844–2021.*

Annual mean temperature 9.4 °C (average 1991–2020)

Warmest years			Coldest years		
Mean temperature, °C	Departure from 1991-2020 normal degC	Year	Mean temperature, °C	Departure from 1991-2020 normal degC	Year
10.21	+0.8	2014	6.76	−2.6	1879
10.13	+0.7	2004	7.10	−2.3	1892
10.10	+0.7	2017	7.22	−2.2	1853
10.03	+0.6	2011, 2020	7.33	−2.1	1855
9.97	+0.6	2006	7.34	−2.1	1888
Last 50 years to 2021					
10.21	+0.8	2014	8.02	−1.4	1979

In order to rank years accurately, mean temperatures are shown to 0.01 degrees Celsius, although it should be noted that measurement errors are greater than this, particularly in the early years of the record.

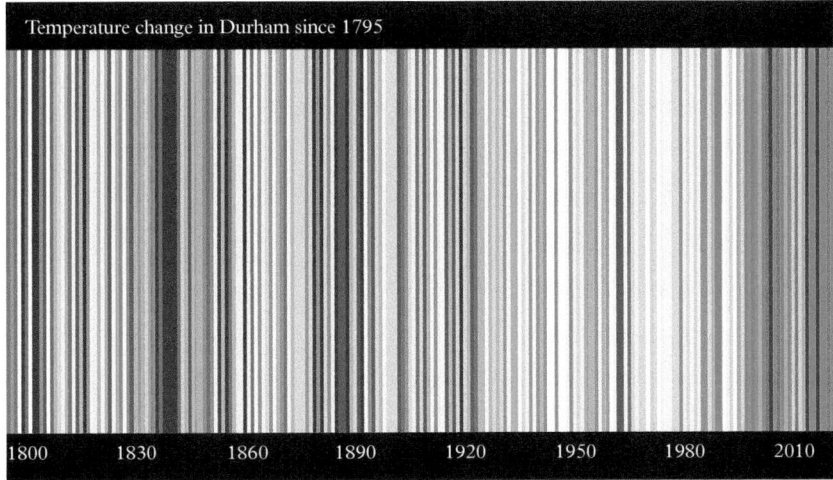

Figure 16.3 *Durham's 'climate stripe', showing the annual mean temperature for every year from 1795 to 2021 by thin coloured vertical stripes, graduated from blue for cold years to red for warm (courtesy of Professor Ed Hawkins, University of Reading)*

Durham's coldest calendar year was 1879. Both the 1878/79 winter and the spring of 1879 remain the second-coldest on record. Conditions remained very cool throughout the summer—only 13 days reached 20 °C, the warmest days being 11 and 12 August, at 23.3 °C—but autumn continued the cold theme. Just to round things off, December 1879 remains the fourth-coldest December on record (Table 15.2).

In 2014, Durham's warmest year to date, only August experienced below-normal monthly mean temperatures, while in 2004, only July and October were slightly cooler than normal.

Heatwaves and hot nights

A 'heatwave' is defined by the World Meteorological Organization [79] as:

> A period of marked unusual hot weather over a region persisting for at least three consecutive days during the warm period of the year based on local climatological conditions, with thermal conditions recorded above given thresholds.

For north-east England including Durham, the Met Office defines the 'heatwave' threshold as a maximum temperature reaching 25 °C for three consecutive days or more. Table 16.3 and Figure 16.4 show the frequency of all days reaching 25 °C or more by month and year since 1844 (excluding 1854–55 and 1863, for which data are incomplete). Durham averages five days per year when the maximum temperature reaches 25 °C, most likely in July, although these are of course not necessarily consecutive.

Between 1860 and 2021, 73 heatwaves (by the Met Office regional definition) have been recorded in Durham, an average of once every two or three years. Only four months have recorded ten or more days reaching 25 °C; July 2006 recorded 17 such days. Until August 1975 the record was held by June 1846, although we must be cautious about accepting thermometer exposures very early in the record as they are not directly comparable with modern measurements (Appendix 3). In annual rankings of 'heatwave days',

Table 16.3 *Annual frequency of individual 'heatwave days' (maximum temperature 25.0 °C or greater) at the Durham Observatory, 1844–2021 (see also Figure 16.4).*

Annual average 1991–2020 5.5 days

Rank	Highest monthly frequency	Month and year	Highest annual frequency	Year	Lowest annual frequency	Year
1	17	Jul 2006	21	1995	0	Numerous: none since 1985
2	16	Aug 1975	20	2006		
3	13	Aug 1995	19	1975		
4	*11*	*Jun 1846*	18	1911		
5	9	Jul 1901 Jul 1911	17	1959		
6	8	Jun 1940 Jul 1949 Jul 1989 Jul 2018	15	*1868,* 1933, 1949, 1976		

Data for periods prior to the introduction of the Stevenson screen in November 1899 are italicized. See Appendix 3 for detail of thermometer exposures and record adjustments.

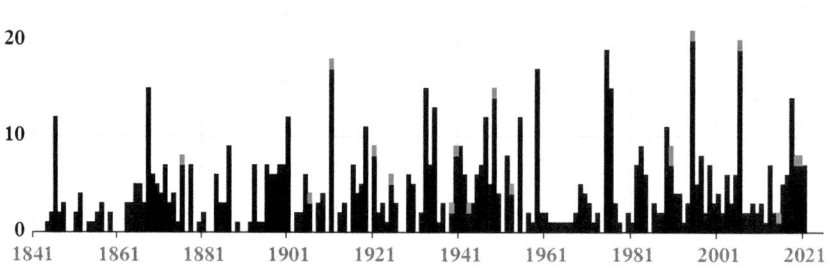

Figure 16.4 *Annual frequency of 'hot days' at Durham Observatory, 1844 to 2021, by calendar year. The black columns show the number of days with a maximum temperature 25.0°C or greater, the red columns 30°C or greater*

1995 comes out top, with 21 days, closely followed by 2006 with 20 days, and then 1975 with 19 days (Table 16.3).

The longest unbroken spell reaching 25 °C or more was the 13 consecutive days 2–14 August 1975, followed by nine consecutive days 25 June to 3 July 1976 and 21–29 July 2006 (Table 24.10). There were spells of eight consecutive days at or above 25 °C 3–10 June 1940 and 15–22 August 1995. Prior to June 1940, the longest heat-waves in Durham lasted six consecutive days, namely 2–7 June 1846 and 1–6 August 1868. In 2006, all but one day of the 14 consecutive days 16–29 July reached 25 °C or more (on 20 July the maximum was 'only' 24.7 °C).

Since 1844, 30 years have failed to reach 25 °C on any day in the year, and the general warming trend is apparent from a reduction in the numbers of years failing to reach this threshold. During the second half of the nineteenth century, 13 years in 50 failed to reach 25 °C; in the first half of the twentieth century, ten years; and in the second half of the twentieth century, only six years. At the time of writing, every year since 1985 has reached 25 °C. The lowest annual maximum temperature on the series was in 1862, when the warmest day attained just 21.1 °C (on 17 May). Since 1970, the lowest annual maximum temperature has been 23.0 °C in 1974 (on 20 June).

Sometimes the most uncomfortable aspect of prolonged hot weather is not so much daytime temperatures, but a succession of warm, close nights, which are most likely in August. Table 16.4 shows the incidence of warm nights (for Durham, this is taken as a minimum air temperature of 15 °C or more). The average annual frequency over 1991–2020 was five such nights, and the longest such spell on Durham's record was the five consecutive nights commencing 3 August 1933. Several months have recorded six or seven such nights; the greatest annual frequency was in 2016 and 2018, both with 13 nights. July and August 2018 were the first two consecutive months with six or more

Table 16.4 *Highest monthly and highest and lowest calendar year totals of 'warm nights' (15.0 °C or more) at Durham Observatory, 1844–2021.*

Annual average 1991–2020 5 nights

Rank	Highest monthly frequency	Month and year	Highest annual frequency	Year	Lowest annual frequency	Year
1	7	*Aug 1857* Aug 1933 Aug 2004 Aug 2018 Jul 2019	13	2016 2018	0	Numerous
2	6	*Jul 1859* Aug 1997 Aug 2002 Aug 2003 Jul 2018	12	*1857* 2003		
3	15	Jul 2018	11	2004		

Data for periods prior to the introduction of the Stevenson screen in November 1899 are italicized. See Appendix 3 for detail of thermometer exposures and record adjustments.

'warm nights' recorded. There are numerous years when no such threshold was reached, particularly in the early years of the record, but since 1993 every year has had at least one night where the temperature has remained at or above 15.0°C throughout.

Frosts and prolonged cold spells

Over the 1991–2020 period, Durham Observatory averaged 46 air frosts and 92 ground frosts annually. The longest spells of consecutive days with air frost on the record[*] since 1856 are shown in Table 16.5 (there are gaps in the record for some earlier years), while Table 16.6 shows the frostiest months since 1856/57.

A similar length of spell to these was probably attained in January/February 1855, the latter month being the coldest yet recorded on Durham's records, but unfortunately the records are incomplete for the first half of February.

Two spells are notable for only very short interruptions to unbroken runs of below-freezing minimum temperatures:

- The 31 days 21 January–20 February 1895 recorded 27 air frosts with a mean daily minimum temperature of −5.5 °C. During this period, five nights fell below −10 °C, the coldest being −18.0 °C on 8 February; the mildest night was just +0.8 °C, on 3 February.

- The 57 days 26 December 1939–20 February 1940 recorded 48 air frosts, two below −10 °C, including −16.1 °C on 21 January.

Table **16.5** *Longest spells of consecutive air frosts at Durham Observatory, 1856–2021, with the coldest night during these spells. Dates are inclusive.*

Length of spell	First and last dates with minimum ≤ 0 °C	Number of nights ≤ −10 °C	Coldest night of the spell
50 days	26 Jan to 16 Mar 1947	2	4 Mar 1947, −15.0 °C
36 days	22 Dec 1962 to 26 Jan 1963	4	23 Jan 1963, −13.3 °C
35 days	30 Jan to 5 Mar 1963	2	2 Mar 1963, −11.7 °C
31 days	9 Jan to 8 Feb 1942	1	8 Feb 1942, −10.0 °C
30 days	19 Jan to 18 Feb 1917	1	6 Feb 1917, −10.6 °C

[*] An air frost is normally defined as a screen minimum below 0 °C, i.e. −0.1 °C or lower. However, for this spell analysis the definition of an air frost has been taken as 0.0 °C or lower, in order to accommodate readings during the period when minimum temperatures were recorded in whole degrees Fahrenheit, where 32 °F has been converted to 0.0 °C. This will produce slightly higher results when compared to the standard definition, but is regarded as preferable to reckoning an air frost as 31 °F or lower. The same threshold applies to maximum temperatures for consideration of 'ice days' later in this chapter.

Table 16.6 *Frostiest calendar months at the Durham Observatory: air frosts period of record 1856 to 2021 (pre-Stevenson screen records prior to November 1899 are italicized), ground frosts Mar 1874– Jun 2021. Grass minimum records are inhomogeneous post-July 2015 (see Appendix 2) and may be missing during snowfalls.*

Frostiest months

Air frosts	Month and year	Ground frosts	Month and year
28	*Jan 1879*, Jan 1940, Feb 1963	30	Jan 1879, Dec 1886, Jan 1980, Jan 1985
27	Mar 1924	29	Jan 1891, Feb 1892, Jan 1895, Jan 1907, Jan 1940, Jan 1963, Mar 1975, Jan 1978, Jan 1979
26	*Jan 1850*, Feb 1942, Feb 1947, Jan 1959, Jan 1963	28	Jan 1881, Mar 1883, Jan 1886, Mar 1887, Dec 1887, Jan 1892, Jan 1893, Feb 1895, Dec 1965, Dec 1976, Jan 1987
25	*Dec 1878, Feb 1895, Mar 1883*	27	Various
24	*Jan 1871, Dec 1874*		

- The 65-day spell 9 January–14 March 1942 included only two (non-consecutive) nights above freezing, the highest minimum being 2.2 °C on 12 February. Only one night in this spell fell to −10 °C, namely 8 February.
- The 74-day spell 22 December 1962–5 March 1963 included only three consecutive nights above freezing, 27–29 January, the highest minimum being 2.2 °C on 27 January. This spell included six nights below −10 °C, the lowest minimum being −13.3 °C on 23 January.

Since the winter of 1962/63, the longest spell of consecutive air frosts has been 29 days, from 22 January to 19 February 1991. The coldest night in that spell was −8.5 °C, on 14 February.

Within the last 50 years, the frostiest months have been January 1979 and February 1986, both with 23 air frosts, and December 2010 with 21. Figure 16.5 shows the number of air frosts per calendar year since 1844; there are gaps in record in some years up to 1855 and in 1863. The frostiest year was 1879, hardly a surprise in that the calendar year 1879 was the coldest on record (Table 16.2). The ten-year period ending 1888 averaged 75 air frosts annually, almost twice as many as the decade 2011–20 with an average of just 38. Not surprisingly, a number of the years in the former period appear in the 'most frosty' list, although the winters of 1881/82 and 1883/84 were both mild and unsettled with relatively few air frosts.

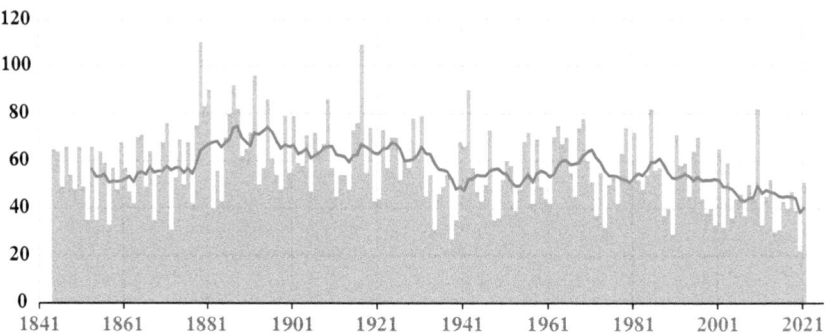

Figure 16.5 *Number of air frosts (minimum temperature below 0°C) in Durham, calendar years 1844 to 2021, with ten-year running mean in red*

Table 16.7 *Greatest number of ice days (maximum temperature at or below 0.0 °C) for months and winters at Durham Observatory, 1844–2021 excluding 1854 and 1855.*

Rank	Most ice days In a winter season	Year	In a month	Month
1	15	*1857*, 1941, 1947	12	Feb 1947
2	13	1963	11	Jan 1940
3	12	1982	*10*	*Jan 1867*
4	*11*	*1871*	9	Feb 1929, Jan 1982
5	10	*1867*, 1929	8	*Feb 1853*

Records made prior to the introduction of the Stevenson screen in November 1899 are italicized. Winters dated according to January.

An 'ice day' is one in which the maximum temperature fails to reach 0 °C. During 1991–2020 Durham averaged around seven ice days per decade, the distribution being very skewed towards cold winters—more than half of the years within the averaging period recorded no days below 0 °C. Table 16.7 shows the greatest number of ice days in a month and by winter during the period 1844–2021 excluding 1854 and 1855. Records are incomplete for parts of 1854 and 1855, but for the 14-day record that survives from the very cold month of February 1855, seven remained below 0 °C. The longest consecutive runs of ice days on Durham's records were six days, 14–19 December 1859, and nine days, 5–14 January 1982.

Monthly and annual ranges in temperature

Only two months since 1856 have recorded a monthly range in temperature of 30 degrees Celsius or greater, these being March 1965 (minimum temperature −12.2 °C on the 3rd, maximum 21.7 °C on the 29th, monthly range 33.9 degrees) and March 1929 (minimum −9.2 °C on the 1st, maximum 21.6 °C on the 27th and 28th, range 30.8 degrees)—see also Chapter 6. Almost one-third of all *calendar years* have recorded a smaller temperature range than March 1965. The greatest monthly range within the 50 years to 2021 was 29.7 degrees, in May 2010.

The least range of temperature in any month since 1856 was just 11.4 degrees, in January 1980 (−3.3 °C on the 3rd to 8.1 °C on the 30th), while the highest and lowest temperatures in March 1964, January 2001, and January 2014 differed by only 11.6 degrees.

During the period 1991–2020, the average *annual* temperature range was 34.6 degrees Celsius. The greatest and least annual ranges since 1856 are shown in Table 16.8. In 1974, the annual range in temperature amounted to only 26.6 degrees, compared with 46.1 degrees in 1940.

Table 16.8 *The greatest and least annual temperature ranges (calendar years) at the Durham Observatory, 1856–2021*

Mean annual temperature range 34.6 degrees Celsius (average 1991–2020).
Average highest maximum 27.9 °C, average lowest minimum −6.7 °C

Greatest annual ranges				Least annual ranges			
Temperature range, degC	Annual maximum temp °C	Annual minimum temp °C	Year	Temperature range, °C	Annual maximum temp °C	Annual minimum temp °C	Year
46.1	30.0	−16.1	1940	26.6	23.0	−3.6	1974
44.9	*29.5*	*−15.4*	*1878*	28.8	24.4	−4.4	1938
43.6	26.7	−16.9	*1881*	28.9	23.9	−5.0	1920
43.3	*25.3*	*−18.0*	*1895*	*29.4*	*23.0*	*−6.4*	*1888*
43.1	27.0	−16.1	1982	29.7	25.5	−4.2	2007

Data for periods prior to the introduction of the Stevenson screen in November 1899 are shown in italics—see Appendix 3 for detail of thermometer exposures and record adjustments.

Reversal in month-to-month temperature trends

As the days grow longer, the expectation is that from March to June the monthly mean temperature should rise each month. Conversely, into autumn we expect September to be cooler than August, October cooler than September, and so on. As the monthly chapters reveal, this is not always the case:

- March has been colder than February 39 times in almost 180 years, averaging out to roughly one year in five since 1844, most recently in 2013. The mean temperature in March 1867 was 3.4 degrees Celsius colder than February that year, and March 1883 3.0 degrees colder than February 1883. March colder than January occurs with a similar frequency, in about one year in six, most recently in 2018. March 1916 was 4.7 degrees colder than January that year, in what must have been a dramatic seasonal reverse.

- It is not particularly uncommon for April to be colder than March, this occurring about once every ten years or so; at the time of writing, the most recent such reversal was in 2021, when April was 1.1 degrees colder than March. In 1938, the dry April of that year was 2.1 degrees colder than March. Remarkably, there have been three Aprils colder than the preceding February—in 1849, 1903, and in 1998.

- June 1848 holds the unenviable record of being the only June on the entire record to have been colder than the previous May. June 1916 was only 4.0 degrees warmer than January of that year—normally June is almost 10 degrees warmer than January.

- September is, occasionally, warmer than August. This has happened seven times since 1843, or roughly once in 30 years on average, most recently in 2021. In 1865 and 1890 the difference was greater than a degree.

- No October has ever been warmer than the preceding September, although in 2001 October was only a tenth of a degree cooler than September.

- November has been slightly milder than the previous October twice, but not for more than a century at the time of writing; this happened in 1881 (when a very mild November followed a very cold October) and most recently in 1917.

It is not particularly uncommon for December to be milder than November—this has occurred 25 times, or on average once every seven years or so. In 1910, December was 3.0 degrees milder than November, and 2.8 degrees in 1862. This reversal happened most recently in 2016. A December milder than October has yet to occur, although in 1844 and 1974 the difference in mean temperature was very small (0.4 and 0.6 degrees, respectively). The remarkable December in 2015, the mildest on record by a large margin, was warmer than the subsequent January, February, March, and April.

Precipitation

The average annual precipitation at the Durham Observatory is 680 mm. The wettest and driest years, and the five wettest days, are set out in Table 16.9. Information on droughts and wet spells over various durations from days to years can be found in Chapter 25, while the wettest and driest months on record are given in Table 25.3. A longer table of the wettest days on record is included as Table 25.5.

Until 2012, the wettest calendar year on Durham's records was 1872, with an annual total of 895 mm,[*] falling on 216 rain days—the Durham record for 140 years. Only May

---

[*] An annual total of 915 mm appeared in Burt, T.P., Jones, P.D., and Howden, N.J.K. 2014. An analysis of rainfall across the British Isles in the 1870s. *International Journal of Climatology*, 35(10), 2934–2947. DOI: 10.1002/joc.4184. A transcription error for 21 April 1872 accounts for the difference, and the total given here is correct.

Table 16.9 *Annual precipitation extremes at Durham Observatory 1868–2021*

Annual mean precipitation 680.2 mm (average 1991–2020)

Wettest years			Driest years			Wettest days	
Total fall, mm	Per cent of normal	Year	Total fall, mm	Per cent of normal	Year	Daily fall, mm	Date
1018.0	150	2012	415.9	61	1989	87.8	11 Sept 1976
895.3	132	1872	439.8	65	1959	72.8	12 Nov 1901
887.3	131	2000	457.7	67	1870	70.2	31 May 1924
885.7	130	1930	470.8	69	1902	69.1	25 Aug 1986
850.2	125	1882	472.0	70	1949	63.7	8 Oct 1903
Last 50 years to 2021							
1018.0	150	2012	415.9	61	1989	87.8	11 Sept 1976

and June had below-average rainfall, and there was a particularly wet autumn. The 1870s were an exceptionally wet decade over most parts of the British Isles, Durham included, as noted elsewhere [76, 80]; further details regarding 1872 are given in Chapter 22.

There were several notably wet months in 2000, particularly April and November, although the first three months were also relatively dry; autumn's total was surpassed only by that of 1976, and the year as a whole became Durham's wettest since 1872 (Table 16.9).

Durham's wettest calendar year was 2012, despite another dry start to the year (only 56 mm had fallen by 31 March). Thereafter, five months in the remaining nine, including April, received in excess of 100 mm. The calendar year total was 1018 mm, half as much again as normal (Table 16.9), the first and so far only time on record the annual fall has surpassed 900 mm, and at the time a near-record number of rain days (215). Further details are provided in Chapter 22.

Durham's driest years were in 1870, 1959, and 1989, the latter being the driest on record with only 416 mm, just 61 per cent of the 1991–2020 average and less than half of the amount recorded in each of the five wettest years (Table 16.9). In 1989, 0.2 mm or more of precipitation fell on 149 days, but in 1959 only 135 (Table 16.11).

The wettest and driest 12 consecutive months in Durham's record (i.e. any 12 consecutive months, not necessarily January to December) are set out in Table 16.10. Note that the very wet last nine months of 2012 extended through to March 2013, making those twelve months the wettest 12-month period on Durham's record post-1868, and by a considerable margin. The second wettest of any consecutive 12-month period ended in August 1877, the second very wet period of the 1870s alongside calendar year 1872. Both prolonged wet spells and large rainfall totals were associated with a high

Table 16.10 *Consecutive 12-month precipitation extremes at Durham Observatory 1868–2021.*

Wettest 12 months			Driest 12 months		
Total fall, mm	Per cent of normal	Period ending	Total fall, mm	Per cent of normal	Period ending
1129.9	166	Mar 2013	362.8	53	Jul 1905
997.6	147	Aug 1877	367.8	54	Oct 1959
995.8	146	Mar 2001	375.9	55	Mar 1973
933.9	137	Jun 1969	380.1	56	Nov 1989
928.3	136	Jun 1931	389.5	57	Sep 1949

Only the wettest or driest 12-month period from each spell is included. The percentage of normal refers to the annual average of 680 mm 1991–2020.

frequency of cyclonic weather; at the same time, the frequency of anticyclonic weather and of westerly winds tended to be very low [76].

The driest 12-month period ended in July 1905, the culmination in the north-east of the so-called 'long drought' which persisted across large areas of the British Isles through the 1890s and 1900s. Low river flow and groundwater levels were a concern for many years during this period. The 1900s was the driest decade yet recorded in Durham (see Chapter 21), despite a very wet year in 1903, including the second-wettest month on record that October.

The next driest 12-month period includes most of 1959, a notably warm, dry year with a prolonged, fine, sunny summer (Table 16.9). Third on the list comes the period ending March 1973. While the severe drought of 1975/76 got much more attention, the early part of the 1970s was also a notably dry time, as this result confirms. Not surprisingly, the driest 60-month period at Durham ends in August 1976 (total 2571 mm, averaging just 514 mm/year), combining the two very dry periods in the first half of the 1970s. The second driest 60-month period ends in February 1909 (2661 mm, 532 mm/year).

Rain days and wet days

The average number of rain days (precipitation 0.2 mm or more) over 1991–2020 was 196 per annum. Figure 16.6 shows the annual frequency of rain days since 1868, together with a ten-year running mean. The highest and lowest monthly and annual frequency of rain days is given in Table 16.11. The early (pre-1868) rainfall records show very low annual totals in some years, particularly in 1866 and 1867, which may indicate multi-day accumulations rather than daily readings; given uncertainty about these data, we start the long-term precipitation analysis from 1868 (See Appendix 4).

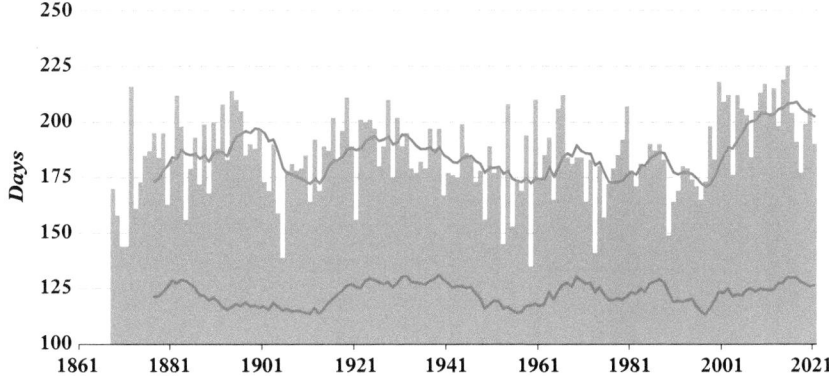

Figure 16.6 *Annual frequency of rain days at Durham since 1868, together with ten-year running means of rain days (red line) and wet days (1.0 mm or more, green line), plotted at year ending. See also text, and Appendix 4 metadata.*

Table 16.11 *Greatest and least monthly and annual precipitation frequency ('rain days', those with 0.2 mm or more) at the Durham Observatory, 1868–2021*

Greatest monthly		Least monthly		Greatest annual		Least annual	
Rain days	Month and year	Rain days	Month and year	Rain days	Year	Rain days	Year
29	Oct 1896	1	Mar 1953	225	2015	135	1959
28	Oct 1882 Oct 1903 Dec 2009 Nov 2010	2	Aug 1947	219	2014	139	1905
27	Dec 1868 Nov 1882 Mar 1916 Nov 1919 May 1967 Oct 1976 Dec 1978 Nov 2009 Nov 2014 Dec 2020	3	Feb 1891 Apr 1912 Jul 1955 Feb 1959 Aug 1976	218	2000	141	1973
26	Numerous	4	Numerous	217 216 215	2010 1872 2012	144	1870, 1871

In 2012, the wettest calendar year on record (Table 16.9), rain was recorded on 215 days; rain fell on 225 days in 2015, although the total fall was only 7 per cent wetter than normal (725 mm). The longest unbroken spells of consecutive rain days yet recorded were the 31 days 22 February–26 March 1916, 29 days 5 October–2 November 1896, and 28 days 24 February–23 March 1937.

The lowest frequency of rain days in a month since August 1976 has been four days, in April 2011.

Wettest days

Durham's wettest day on record was 11 September 1976; details of this and the other wettest days are provided in the monthly chapters and in Chapter 22. Table 25.5 lists the ten highest 24-hour precipitation totals (0900–0900 GMT) recorded at Durham Observatory since 1868, albeit with some reservations about the accuracy of some early measurements—see Appendix 4. Figure 16.7 shows a time series of the annual highest 24-hour precipitation totals since 1868, together with a ten-year unweighted running mean, and a 140-year trendline. The latter shows a slight increase, about 7 per cent, over the whole series, a value which is close to the expected theoretical value arising from the increase in atmospheric water vapour consequent in thermodynamic terms upon the observed 1.2 degrees rise in air temperatures over the same period (average 1851–80 8.21 °C, 1991–2020 9.44 °C).

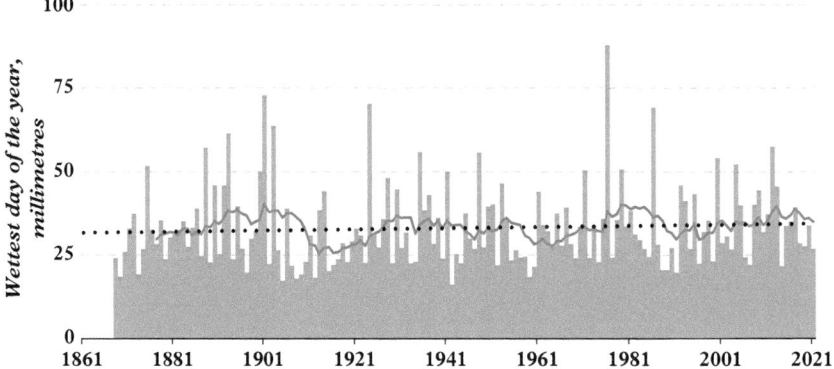

Figure 16.7 *The annual maximum series (highest daily precipitation total each year, mm) at Durham Observatory, 1868 to 2021, together with a ten-year unweighted running mean plotted at year ending (red line) and the 150-year trendline (dotted black line).*

Sunshine

A Campbell-Stokes sunshine recorder (Figure 16.8) was installed at the Durham Observatory in spring 1880, and although we have no daily sunshine data until 1 January 1882, monthly totals from June 1880 were given in an 1885 paper by Scott [81]. The Durham Observatory record therefore ranks among the earliest in the country. After some problems lasting from December 1883 to May 1885, during which period there are numerous gaps (see Appendix 1, note 4), the record is unbroken to the end of 1999 when the unfortunate decision was made to terminate the record. Since and including the year 2000, monthly sunshine durations for Durham have been estimated from the Met Office East and North-East England sunshine series (see Appendix 6 for details).

The average annual duration of 'bright sunshine' in Durham during 1991–2020 was 1473 hours. The sunniest and dullest calendar years since Durham's sunshine records commenced in June 1880 are shown in Table 16.12. During this period of almost 140 years, monthly sunshine amounts have varied from 297.0 hours in June 1940 to just 7.6 hours in January 1996 (Table 26.1), while annual amounts have varied by almost a factor of two, from only 982 hours in 1912 to about 1703 hours in 2003. Following the 'perfect summer' of 1911 [82], the summer of 1912 was wet, cool, and gloomy with less than half normal sunshine at Durham. The rainiest periods in June and August coincided with the hay and grain harvests, with calamitous impact on agriculture, and crops rotting in the fields. For 1912 as a whole, the only month with above-average sunshine was April; August saw barely one-third of average sunshine hours, and less than one-third of April's total. Gordon Manley attributed the low sunshine in 1912 to the volcanic eruption of Mount Katmai in Alaska earlier that summer [19]; see also reference [83] for details of the effects of the eruption. In 2003, on the other hand, February, March, and December were very sunny and April, June, and August through October also well above average; only January, May, and July were a little duller than normal.

Prior to 2000 (i.e. on the measured record, rather than a regression estimate from an areal value), 1989 was Durham's sunniest year, with 1646 hours (about 11 per cent more than usual), a little less than 37 per cent of possible sunshine. To make the list of sunny years, a sunny summer is essential, and at the time of writing only the summer of 1976 has been sunnier than 1989. However, 1976 does not feature in Table 16.12 because the very wet autumn that year turned out to be the dullest autumn on record, and as a result the annual total of 1382 hours was actually 10 per cent below normal. In 1995, another year with a hot, dry summer, the annual sunshine total ranks third, while two recent years, 2018 and 2019, also feature in the 'Top 5' sunny calendar years.

The months of May to August have similar average sunshine totals, and in about one year in four the sunniest month of the year occurs outside the three 'summer' months of June, July, and August; in 1907, March was, remarkably, the sunniest month of the year

Figure   16.8 *The   Campbell-Stokes   sunshine recorder in use at Durham Observatory; the observer is Mrs Audrey Warner. Although this is not the original instrument, serial number MO28, installed in 1880, remained in use until May 1981 (Department of Geography, Durham University)*

with 187 hours, about the average for May. April has been the sunniest month of the year ten times to 2021 (including 2021), and featuring including one year (1912) when April had more than three times August's sunshine; September has been the sunniest five times.

In about one year in eight, the dullest month occurs outside the three winter months (December, January, and February); in 2005 October was the dullest month of the year, while in 1936 and 1964 that title was held by March.

Figure 16.10 shows that there has been a slight increase in annual sunshine duration over the period of record, and particularly since the early 1980s, although it must be borne in mind that sunshine figures from 2000 onwards are regression estimates from a regional value and are thus not strictly comparable with the previous instrumental record. Durham's average annual sunshine over the 30-year period 1911–40 was just 1301 hours: the most recent 30-year mean (1991–2020) is 13 per cent greater. Every month of the year except June has shown an increase over the earlier period, the increase being greatest in the winter half-year and 30 per cent or more in the three winter months December, January, and February. The increase is mainly due to reduced pollution and aerosol loading following clean air legislation since the 1950s, although changes in atmospheric circulation have also played a part.

The annual average number of sunless days during the year in the period 1971–99 was 76. Until the daily record ceased at the end of 1999, the number of sunless days in a year has varied between 51 and 109 days since 1883 (some gaps in 1883–84; see Table 16.13). In January 1996, the dullest month on record (Table 16.12), 24 days remained sunless, and 23 in December 1882, December 1903, and December 1978. The frequency of sunless days had been decreasing in the two decades up to the end of the record.

The annual average number of days with 12 hours or more of sunshine in the period 1971–99 was 10.5, most likely in May and June with a little over three days apiece. The

Table 16.12 *Annual sunshine duration at Durham Observatory: extremes 1882–2021, sunniest days 1882–1999 only*

Annual mean sunshine duration 1473 hours, 4.0 hours per day (average 1991–2020).

Possible day length: 4492 hours (leap year 4500 hours). Mean sunshine duration as percentage of possible: 32.8

Sunniest years			Dullest years			Sunniest days 1882–1999	
Duration, hours	Per cent of possible	Year	Duration, hours	Per cent of possible	Year	Duration, hours	Date
1703	37.9	2003	982.2	21.9	1912	15.8	5 June 1940 and 15 June 1957
1646.5	36.6	1989	1065.8	23.7	1937	15.7	14 June 1959
1624.4	36.1	1995	1090.7	24.3	1917	15.6	26 June 1937 and 5 June 1962
1609	35.8	2018	1115.8	24.8	1888	15.5	13 July 1929 and 29 June 1931
1601	35.6	2019	1123.9	25.0	1966	15.4	Numerous
Last 50 years to 2021							
1703	37.9	2003	1139.6	25.4	1972	14.7	26 May 1989

Daily sunshine records ceased at the Observatory on 31 December 1999: monthly sunshine totals from January 2000 onwards are estimated (to the nearest hour) by regression from the Met Office East and North-east England sunshine series. See Appendix 6 for record metadata and missing months.

Table 16.13 *Annual frequency of sunless days at the Durham Observatory, 1883–1999, some gaps 1883–85*

Most sunless days	Year	Least sunless days	Year
109	1937	51	1988
108	1888	52	1999
107	1903, 1912	57	1949
105	1960	58	1995
100	1942, 1969	59	1989

annual frequency varied from just two such sunny days in 1958 and three in 1912, 1916, 1965, and 1998, to 26 in 1901 and 1989, 23 in 1921, and 21 in 1955 and 1990. In June 1940, the sunniest calendar month yet recorded, 11 days recorded at least 12 hours of sunshine; in June 1921, June 1933, and May 1980, nine such sunny days were recorded, and several other months have seen eight such days. Table 26.1 summarizes the sunniest and dullest calendar months on record (including estimated monthly totals after 1999, for which daily totals are not available).

Wind and weather

Snow cover

Information on snow cover is rather limited—the digital record does not begin until January 1960, and manual observations ended in late 1999. By convention, 'snow cover' is counted when more than half the ground representative of the observing site is covered with snow, regardless of its depth, while the usual statistical measure is 'mornings with snow cover', referring to a snow cover at the time of the 0900 GMT observation.

Over the period 1960 to 1999, there were an average of 17 mornings with snow cover at the Durham Observatory, and every winter had at least one morning with at 3 cm of lying snow (Table 25.9). For the period from 1971–99, the annual frequency decreased slightly to 14 mornings. By far the snowiest winter on the data available was the winter of 1962/63 (see also Chapter 22) which saw snow on the ground for 70 consecutive mornings from 27 December 1962 to 6 March 1963, inclusive. In all, that winter saw 76 mornings with a snow cover; the next greatest counts were 52 in 1978/79, 49 in 1965/66, and 38 in 1985/86. The least-snowy winters, in terms of snow cover, were those of 1960/61 and 1998/99, both of which witnessed only two snow-covered mornings.

In the period of record, 40 mornings experienced snow at least 30 cm deep, the greatest depths at 0900 GMT being 43 cm (on 15–18 February 1963) and 41 cm (on 31 December 1961, 8 and 9 February 1963, 20–26 February 1963, and 30 November 1965).

Fog

We are similarly limited to the period 1962–97 for information on the frequency of fog, which is determined from the visibility at the 0900 GMT observation.

Over the period 1962–97, fog at 0900 GMT (visibility below 1000 m) occurred on 22 mornings in a typical year, and six of these would be 'thick fog' (visibility below 200 m). The frequency of fog has been declining steadily since the early 1960s, following the introduction of the Clean Air Acts of the 1950s and the gradual phasing out of coal fires. Subsequently, during the late 1980s and into the 1990s some years recorded only half a dozen fogs.

Barometric pressure (1844–1960)

The highest pressure on record for Durham during this period occurred on 23 January 1907, when at 0900 GMT MSL pressure stood at 1052.5 hPa. This value appears slightly high (at 0800 GMT North Shields reported 1050.8 hPa): the true maximum probably lay between 1051 and 1051.5 hPa (see also Chapter 22). At the time of writing, this value has not since been exceeded in north-east England. The only other occasion of

1050 hPa on Durham's records occurred on 31 January 1902, when the barometer stood at 1050.8 hPa at 0900 GMT and 1050.7 hPa at the 2100 observation. That evening, the barometer rose to 1053.6 hPa in Aberdeen, a value which remains the highest on record within the British and Irish Isles [55]. Also notable was 1049.7 hPa at 0900 GMT on 26 January 1932.

At the other end of the scale, the MSL pressure at the 2100 GMT observation on 8 December 1886 stood at only 936.2 hPa, Durham's lowest on record (and not remotely approached since), although it may have been lower still a little earlier that evening. An exceptionally deep depression crossed the north of Ireland and southern Scotland during the afternoon and evening; at Belfast the barometer fell to 927.2 hPa at 1330h as the depression centre passed nearby [56] (Figure 22.12).

It should be noted that observations made just twice daily are unlikely to record the absolute highest or lowest barometric pressures: the extreme values quoted here, and in the other monthly and annual chapters, must be regarded as the extremes on the observational record, rather than the true absolute extremes, which could occur at any time of the day or night. From the observational record, however, we can show that the overall range in pressure at Durham in almost 180 years has been at least 115 hPa.

The highest mean MSL pressure for any month occurred in February 1932, with a remarkable 1034.6 hPa (see Figure 5.5). The closest approach to this during the period of available record has been 1030.4 hPa, in February 1959. The lowest monthly mean MSL pressure on record for any month occurred in January 1948, at 994.5 hPa and the wettest January on record; almost as low were 997.0 hPa in the Decembers of 1868 and 1959. The range in monthly mean pressures during 1959, from 1030.4 hPa in February to 997.0 hPa in December, remains unprecedented.

Wind speed and direction

Figure 16.9 shows the annual distribution of wind speed and direction at the 0900 GMT observation over the period 1961–97. The mean wind speed at 0900 GMT over the year as a whole was 6.9 knots. From an annual viewpoint, winds from between south and north-west are most frequent; winds from between north and east are most often experienced in spring, while winds from the south-east remain very infrequent at all times of the year.

The highest wind gusts known to have occurred at Durham include one of 80 knots on 3 November 1970 (during an hour in which the mean wind speed was westerly, 47 knots), 66 knots on 31 January 1953, and 65 knots on 13 January 1948. However, this list is based upon fragmentary notes and is certainly incomplete; other notable gusts may yet remain undiscovered in the manuscript archives. More recently, the passage of storm *Malik* on the morning of 29 January 2022 resulted in a gust of 56 knots being recorded at the AWS on the roof of the Geography Department on the University's Lower Mountjoy campus, about 800 m east of the Observatory.

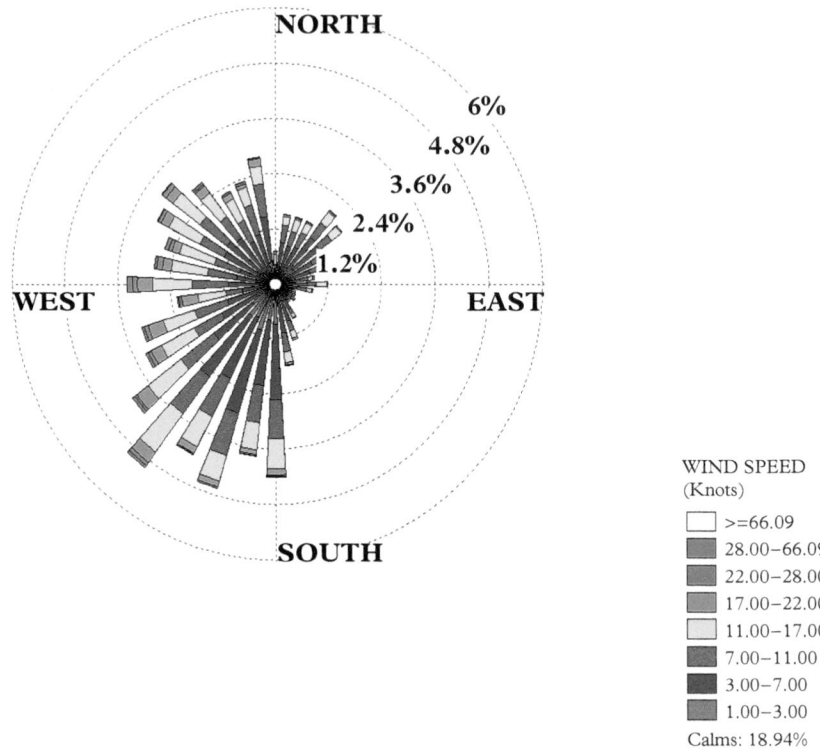

Figure 16.9 *Annual wind rose for 0900* GMT *observations at Durham, period 1961–97*

Durham temperature, precipitation, and sunshine in graphs—calendar years

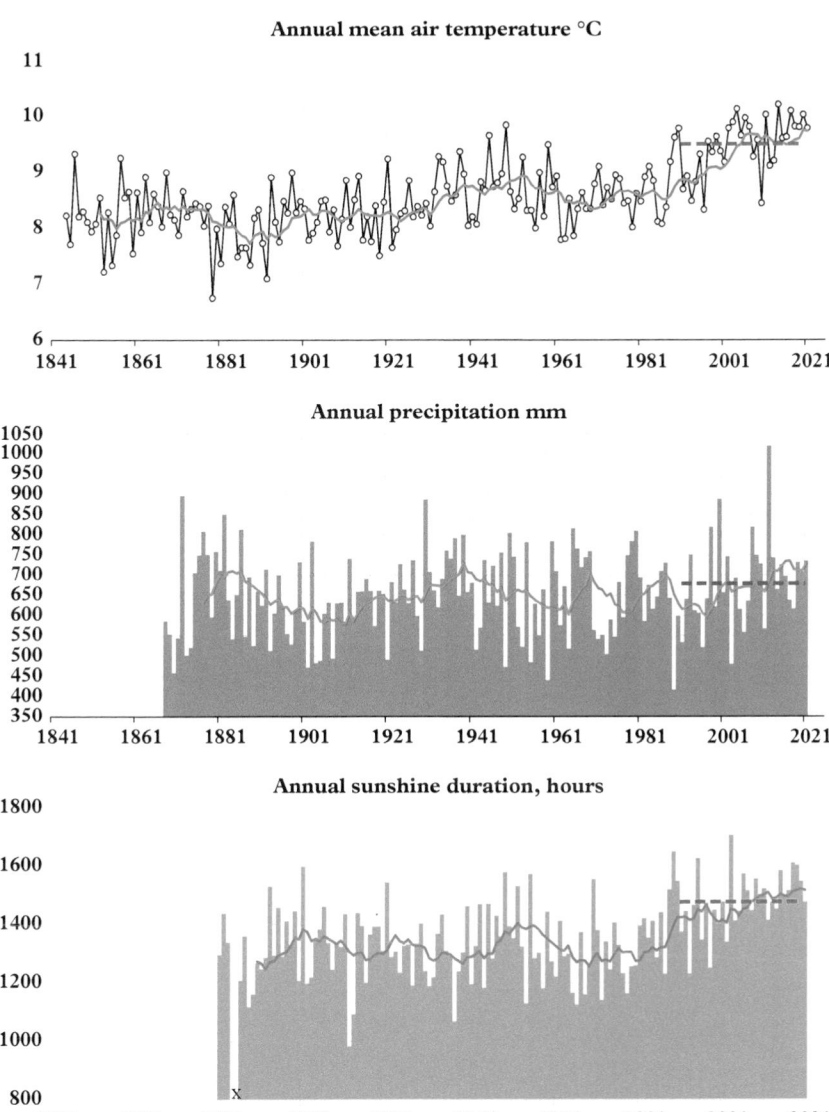

Annual mean air temperature °C

Annual precipitation mm

Annual sunshine duration, hours

Figure 16.10 *Annual values of (from top) mean temperature (°C, since 1844), total precipitation (mm, since 1868), and sunshine duration (hours, since 1881, incomplete data for 1884 and 1885) at Durham Observatory. The 1991–2020 averages are indicated by the thick blue dashed line, while the ten-year running mean ending at the year shown is indicated by the red line*

Part 3

Durham weather through the seasons

17

Winter

December, January, and February

By long convention, the year is divided meteorologically into four seasons, each comprising three calendar months: winter (December, January, February), spring (March, April, May), summer (June, July, August) and autumn (September, October, November). Of course, there are other possible divisions, including starting each season on the day of a solstice or equinox, but while this might make some climatic sense, using complete months makes data compilation a much simpler, more convenient task. Hubert Lamb [50] suggested dividing the year into five natural seasons, which would encompass numerous shorter episodes or 'singularities.' For statistical convenience, we will retain the three-month convention here, starting with winter. Throughout the chapter, winters are dated by the January year—thus, the winter of 1962/63 (December 1962 to February 1963 inclusive) is referred to as 'winter 1963'.

Temperature

Winter is the coldest season of the year: in the standard averaging period 1991–2020, winter's mean air temperature was 4.3 °C (mean maximum 7.2 °C, mean minimum 1.4 °C). There is only half a degree Celsius average difference between the mean monthly temperatures of the three winter months—December 4.2 °C, January 4.1 °C, and February 4.6 °C. Table 17.1 shows the individual 'average coldest' days of the year in the period 1991–2020, each value being the average of temperatures on that date over the 30 years. The coldest days (i.e. lowest mean maxima) come around the turn of the year, with the coldest nights (i.e. lowest mean minima) in mid- to late- December, late January, and a surprising inclusion of 2 March. The lowest mean air temperatures (the average of the daily maximum and minimum temperatures) again tend to cluster in late December and late January. Figure 3.1 shows the general pattern of temperature during the standard averaging period, with a seven-day running mean smoothing out the peaks and troughs. It might seem surprising that the coldest part of the winter is not in February, but the detail of which individual days are the coldest changes from decade to decade and between 30-year periods according to the characteristics of individual cold spells within the averaging period.

Table 17.1 *Coldest days of the year in Durham (daily averages over 1991–2020) by mean maximum, mean minimum, and mean temperature;* °*C. Leap Day (29 February) is excluded*

Lowest mean max °C	Date	Year day	Lowest mean min °C	Date	Year day	Lowest mean temp	Date	Year day
5.51	20 Dec	354	−0.10	29 Dec	363	3.00	20 Dec	354
5.83	27 Dec	361	0.30	30 Dec	364	3.06	29 Dec	363
5.89	28 Dec	362	0.31	30 Jan	30	3.25	30 Jan	30
6.03	3 Jan	3	0.34	2 Mar	61	3.28	27, 28 Dec	361, 362
6.07	4 Jan	4	0.49	20 Dec	354	3.51	22 Jan	22

Mildest and coldest winter days

Since the start of Durham's records in 1843, winter air temperatures have ranged from −18.0 °C on 8 February 1895 to 17.4 °C on 28 February 2012 (Table 17.2). Within the last 50 years (to 2021), the coldest winter night was −16.1 °C on 11 January 1982.

Figure 17.1 *Preparing the ice for winter sports on the frozen river just below Prebends Bridge, February 1895 (courtesy Gilesgate Archive). See also Chapter 22*

Figure 17.2 *Post box on Palace Green, Durham, 3 January 2010 (Tim Burt)*

Table 17.2 *Highest and lowest winter maximum and minimum temperatures at Durham Observatory, period 1844–2021*

Rank	Mildest days	Coldest days	Mildest nights	Coldest nights
1	17.4 °C, 28 Feb 2012	−6.6 °C, *13 Dec 1878*	11.9 °C, 18 Dec 1932	−18.0 °C, *8 Feb 1895*
2	16.8 °C, 7 Feb 1993 and 26 Feb 2019	−6.1 °C, *1 Jan 1871*	11.4 °C, 3 Jan 1932	−16.9 °C, *17 Jan 1881*
3	16.7 °C, 13 Feb 1998	−5.2 °C, *31 Dec 1870*	11.3 °C, 17 Dec 1904 and 17 Dec 2015	−16.5 °C, *10 Feb 1895*
4	16.6 °C, 14 Feb 1998	−4.7 °C, *25 Dec 1860 and 25 Jan 1881*	11.2 °C, 24 Dec 1843, 5 Dec 1898 and 19 Dec 2015	−16.4 °C, *25 Dec 1860*
5	16.5 °C, 27 Feb 2019	−4.5 °C, 7 Jan 1982	11.1 °C, 5 Feb 2004	−16.1 °C, 21 Jan 1940 and 11 Jan 1982
Last 50 years to 2021				
	17.4 °C, 28 Feb 2012	−4.5 °C, 7 Jan 1982	11.3 °C, 17 Dec 2015	−16.1 °C, 11 Jan 1982

Temperatures were recorded in different thermometer screens during the period of record; see Notes on page 411 and Appendix2 for more details.

The very coldest winter *nights* have tended to occur in January or February rather than December, although Christmas Day 1860, when the temperature fell to −16.4 °C, remains Durham's fourth-coldest night on record (see Chapter 22 for more on this cold spell). The coldest winter *days* tend to occur earlier in the winter, with shorter day lengths and minimum solar radiation. All of the five coldest days on record in Durham have occurred within three weeks of the winter solstice—most often after this date, but the coldest of all was 13 December 1878, when the maximum temperature reached only −6.6 °C, a week before the solstice (Figure 15.1). In contrast, the mildest winter *nights* are more likely in December, when some of the warmth of autumn lingers in the sea and the soil. As temperatures fall lower during late December and into January, mild nights are less likely, and only one of the five mildest winter nights has occurred in February.

Table 17.3 *Winter mean air temperatures at Durham Observatory, 1844–2021*

Winter mean air temperature 4.3 °C (average 1991–2020)

Mildest winters			Coldest winters		
Mean temperature, °C	Departure from 1991–2020 normal degC	Year	Mean temperature, °C	Departure from 1991–2020 normal degC	Year
6.3	+2.0	1989	−0.1	−4.4	1963
5.6	+1.3	1998, 2016, 2020	*0.1*	*−4.2*	*1879*
5.5	+1.2	2007	0.9	−3.4	*1895, 1979*
5.4	+1.1	2014	*1.0*	*−3.3*	*1855*
5.3	+1.0	1935, 1975, 2019	1.1	−3.2	*1881, 1940*
Last 50 years to 2021					
6.3	+2.0	1989	0.9	−3.4	1979

Winter comprises December, January, and February, and is dated by the January year.

Mild days can occur in any winter month, but the mildest of all tend to be in February as the Sun's power increases and the days begin to lengthen. Durham's mildest winter day occurred on 28 February 2012, when the temperature reached 17.4 °C—about the normal for mid-June. The last nine days of that month were generally very mild with a south-west to westerly airflow around an anticyclone centred over Brittany: in eastern Scotland and north-east England, in the lee of high ground, the temperature exceeded 14 °C almost daily.

Mildest and coldest winters

Table 17.3 lists Durham's mildest and coldest winters, by mean air temperature. Winter 1963 heads the list, at more than four degrees below average. On the Central England Temperature (CET) series, which begins in 1659 [2], 1963 is ranked third coldest, the only colder winters being 1684 and 1740. At Durham, the mean minimum temperature in January 1963 was −3.2 °C, and in February that year −3.9 °C: the coldest night of the winter was 23 January, when the air temperature fell to −13.3 °C. January and February 1963 (Figure 17.3) both feature in Durham's 'Top 20' coldest months (Table 24.8). Further details about this very cold winter are given in Chapter 22.

Figure 17.3 *A temporary ski tow rigged up by a Durham School housemaster on Observatory Hill during the very cold and snowy winter of 1962/63 (Copyright © Ward Philipson Collection)*

Of the next four coldest winters, the most recent is 1979, Durham's third-equal coldest winter, and aside from 1963 the only other winter from the twentieth century listed in Table 17.3 is 1940. The very wet December of 1978 was on the cold side, but not notably so until the closing days; this was followed by the third-equal coldest January (mean temperature −0.5 °C, the same as January 1963), and although February was less severe (mean temperature +0.9 °C) only the winters of 1879 and 1963 have been colder in Durham. Snowfalls were frequent from late December into March (snow lay 18 cm deep at the morning observation on New Year's Day) and lay on the ground on 51 mornings in all.

Table 17.4 *Most and least frosty 'winter seasons' at the Durham Observatory*

	Most frosty				Least frosty			
Rank	Air frosts	Winter season	Ground frosts	Winter season	Air frosts	Winter season	Ground frosts	Winter season
1	112	1917	180	1892	25	2014	65	2017
2	108	1879, 1881	163	1888, 1920	27	1961, 2007, 2020	69	1935
3	99	1888	160	1886	31	1939, 1948, 1989	70	2020
4	96	1924, 1963	158	1887	32	1998	73	2000, 2014
5	86	1880	155	1902, 1922	33	1859, 1884, 2016, 2019	76	2012
6	84	1942						

The 'winter season' is defined as the period 1 July to 30 June and dated by the January year. Periods of record: air frosts 1856 to 2021 (pre-Stevenson screen records prior to November 1899 italicized), ground frosts March 1874 to December 2021. Grass minimum records are inhomogeneous post July 2015 (see Appendix 3) and may be missing during snowfalls.

Winter 1989 remains the mildest on record at Durham by some margin: all the mildest winters listed in Table 17.3 date from 1989 onwards, including the recent winter of 2020, the equal-second mildest on record (the mean temperatures during the winters of 1998, 2016, and 2020 differ by only one-fiftieth of a degree). Full accounts of winters 1998 and 2020 are provided in Chapter 22.

Sometimes the winter season starts early or continues late into spring. When considering the 'extended winter', which we can take as the season 1 July to 30 June and dated according to the January, Table 17.4 shows the most and least frosty seasons on Durham's records since 1856/57.

Durham winters pre-1843

Using the Eglise extension to the Manley series (Chapter 2, and reference [45]), we can produce a monthly 'Durham temperature series' from January 1784 to December 1849, albeit with a gap from October 1791 to December 1794 where no suitable records from north-east England could be found. Manley had always hoped to extend back

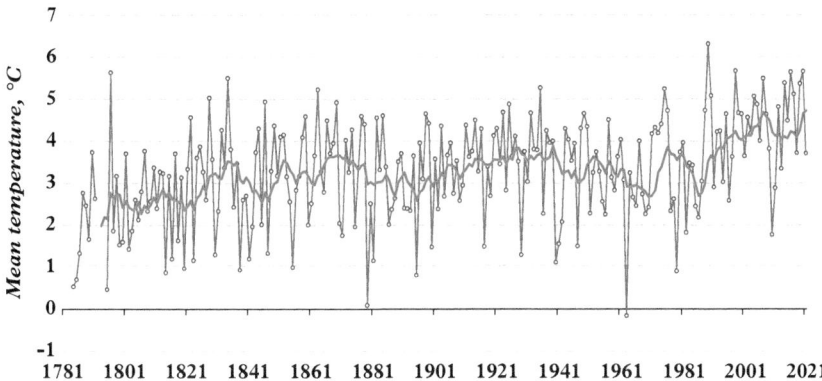

Figure 17.4 *Mean winter temperatures, °C, at Durham since 1785, using the 'Durham Eglise' series (see Chapter 2) for the period before the commencement of the Observatory's instrumental record in 1844. The red line is a ten-year running mean, plotted at year ending. Winters are dated by the January year*

as far as the winter of 1795, a notoriously severe winter throughout England and Scotland. Unfortunately, because the Eglise series recommences in January 1795, to obtain a value for the winter's mean temperature it was necessary to estimate December 1794's mean temperature using a regression equation between the CET series and the post-1844 Durham temperature record. The mean air temperature for winter 1795 is thus estimated at +0.5°C, making this the third coldest winter in the extended series.

Figure 17.4 shows other severe winters in the early part of the extended series including 1785, 1814 (mean temperature 0.9 °C), 1816 (1.2 °C), and 1820 (1.0 °C). There are also two other very cold winters in the pre-1844 period: 1838 and 1841 (0.9 °C and 1.2 °C, respectively). It is noteworthy that none of these winters from the end of the eighteenth century and the first half of the nineteenth century are as cold as either 1879 or 1963 (Table 17.3). In stark contrast, winter 1796 was exceptionally mild (mean temperature 5.6 °C), ranking equal second mildest in the entire extended series; 1834 was also very mild (mean 5.5 °C). Table 17.5 replicates Table 17.3 for the extended series.

Table 17.5 *Winter mean temperatures at Durham Observatory, 1785–2021, with gap from October 1791 to November 1794, December 1794 estimated from CET*

Winter mean air temperature 4.3°C (average 1991–2020)

Mildest winters			Coldest winters		
Mean temperature, °C	Departure from 1991–2020 normal degC	Year	Mean temperature, °C	Departure from 1991–2020 normal degC	Year
6.3	+2.0	1989	−0.1	−4.4	1963
5.6	+1.3	1796, 1998, 2016, 2020	*0.1*	*−4.2*	*1879*
5.5	+1.2	1834, 2007	0.5	−3.8	1795
5.4	+1.1	2014	0.7	−3.6	1785
5.3	+1.0	1935, 1975, 2019	0.9	−3.4	1814, 1838, 1895, 1979
Last 50 years to 2021					
6.3	+2.0	1989	0.9	−3.4	1979

Winter comprises December, January, and February, and is dated by the January year.

Precipitation

Durham's mean winter precipitation during the standard averaging period 1991–2020 was 160 mm. December is the wettest month (61 mm), followed by January (53 mm), and February (45 mm). Table 17.6 lists the driest and wettest winters on Durham's record, together with the wettest winter days.

Dry winters

Durham's driest winter was in 1926/27 (Table 17.6). The outstanding feature of the weather of December 1926 was its abnormal dryness, due to persistent high pressure (*Monthly Weather Report*); Durham received just under 18 mm rainfall that month, falling on just eight days. January was the wettest of the winter months, with 27 mm in Durham, just over half normal, and February's precipitation of 17 mm brought the winter total up to 61 mm. Winter 1904/05 was almost as dry—the totals for the three months being just 35 mm, 9 mm, and 18 mm, respectively—despite an unsettled start to

Table 17.6 *Winter precipitation at Durham Observatory 1868–2021*

Winter mean precipitation 160 mm (average 1991–2020)

Wettest winters			Driest winters			Wettest days	
Total fall, mm	Per cent of normal	Year	Total fall, mm	Per cent of normal	Year	Daily fall, mm	Date
328.6	205	2021	61.1	38	1927	50.0	19 Feb 1941
297.6	186	1877	62.6	39	1905	37.4	16 Dec 1872
291.0	182	1941	65.3	41	1889	34.3	27 Dec 1978
275.7	172	1979	65.4	41	1973	33.6	25 Dec 2015
267.0	167	2016	67.5	42	1909	32.8	12 Jan 1921
Last 50 years to 2021							
328.6	205	2021	65.4	41	1973	34.3	27 Dec 1978

December, with nine rain days in the first fifteen. January 1905 was notably anticyclonic, with a mean MSL pressure at Durham of 1021.5 hPa. Rain fell on only six days, in what remained the driest January on Durham's records until 1989. Anticyclonic conditions persisted for the first 18 days of February, after which the weather became more unsettled.

The most recent dry winter in the 'Bottom 5' of Table 17.6 occurred in 1972/73, with only 65 mm in the three months; December and January were both about half normal rainfall, while February was the driest of the three, with just 12 mm in all.

Wet winters

Durham's wettest winter occurred while this book was being written—the winter of 2020/21 surpassed any previous winter's precipitation total by more than 10 per cent, amounting to almost 330 mm (Table 17.6). This was the first winter on the Durham Observatory record to surpass twice normal rainfall; the previous wettest was back in 1877 at 298 mm. All three months were wet, January 2021 especially so: the month's total of 137 mm was the wettest January since 1948 and the second-wettest on record back to 1868. The east and north-east of England were particularly wet that winter, as the Met Office rainfall anomaly map in Figure 17.5 shows very clearly. On the Met Office regional and county series, winter 2021 was the wettest on the Tyne and Wear series, but on the County Durham series winter 2016 was slightly wetter, and 1877 considerably wetter (Mike Kendon, Met Office NCIC, pers. comm., July 2021). Both series commence in 1862.

The winter of 1876/77 remained the wettest at Durham Observatory for almost 150 years, and was a major contributor to a very wet decade [76]. Indeed, the three

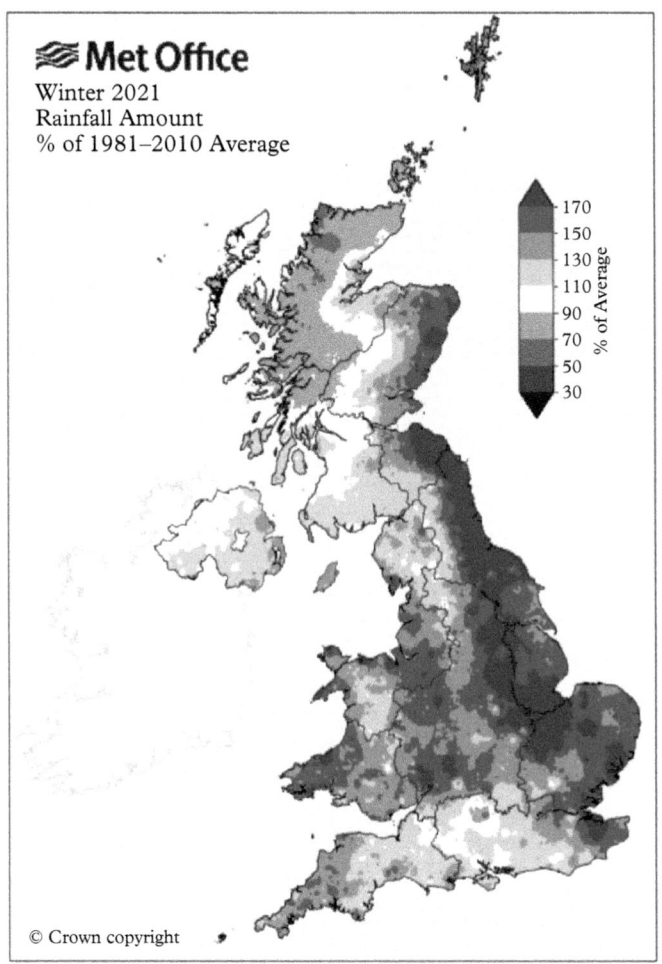

Figure 17.5 *Precipitation winter 2020/21 (December 2020 to February 2021) as a percentage of the 1981–2010 normal, showing how wet the east and north-east of England were. This was Durham's wettest-ever winter. This map is based upon pre-quality-controlled data: note that the dry 'bullseye' in central southern England is almost certainly erroneous. (Based upon the Met Office 1 km gridded rainfall dataset [121], courtesy of National Climate Information Centre, Met Office Exeter: © Crown Copyright, Met Office)*

months ending January 1877 remains the third-highest three-month precipitation total of any in the series: 356 mm, fully 52 per cent of the annual average, a value only exceeded in 1930 (375 mm, ending in September) and 1976 (359 mm, ending in November). December 1876's total precipitation, 165 mm, remains one of Durham's 20 wettest months, and there was no wetter December for over a century, until surpassed in 1978 (195 mm; Table 15.4). January 1877, with 89 mm, was wet, but not exceptionally so, while February 1877's 44 mm was very close to normal. More recent inclusions in the list are winter 1979, Durham's coldest winter in the last 50 years (Table 17.3) with several heavy snowfalls; and winter 2016, one of the mildest.

The 'wettest' winter day in Table 17.6 was due to a notable snowfall in February 1941; more details are given in Chapter 5 and in Chapter 22.

Sunshine

Mean winter sunshine during the standard averaging period 1991–2020 was 204 hours, equivalent to 2.3 hours per day on average and just 27 per cent of possible. As would be expected in relation to day length, December (57 hours, 1.8 hours per day) has the least sunshine, followed by January (62 hours, two hours per day), and then February (85 hours, three hours per day) as the sunniest. In the period 1971–99 there were an average of 33 days with no sunshine each winter, December and January accounting for 12 or 13 each, almost one day in two.

Sunniest winters

Winters are becoming sunnier; the top four sunniest in Table 17.7 all come from the present century, with 2015 (268 hours, an average of just under three hours per day) coming out just ahead of 2000 and 2019. Even so, this sunshine total represents barely 36 per cent of the possible duration in the cloudy winter months. December 2014, with 84 hours of bright sunshine, became the sunniest on record, while January 2015 continued the trend, with 85 hours, the third-sunniest January; February that year was sunnier still (99 hours), although this was closer to the monthly normal than the preceding two months. The recent very mild winter of 2020 just misses a place in Table 17.7, with 235 hours of bright sunshine, followed by a remarkably sunny spring.

Dullest winters

The dullest winter on record to date was 1972 with just 94 hours of bright sunshine, barely one hour per day on average and only 12.6 per cent of possible sunshine. All three months were dull; 19 days in January 1972 remained sunless, and February (33 hours sunshine) was the dullest since 1940. Winter 1913 was the only other winter to receive less than 100 hours of sunshine and must have seemed interminably dull following the

Table 17.7 *Winter sunshine duration at Durham Observatory. Observed extremes 1881–2021, sunniest days 1882–1999 only*

Winter mean sunshine duration 204 hours, 2.3 hours per day (average 1991–2020). Possible day length: 747 hours. Mean sunshine duration as percentage of possible: 27.3

Sunniest winters			Dullest winters			Sunniest days 1882–1999	
Duration, hours	Per cent of possible	Year	Duration, hours	Per cent of possible	Year	Duration, hours	Date
268	35.9	2015	94.1	12.6	1972	10.2	26 Feb 1995
264	35.4	2000	98.8	13.2	1913	9.5	27 Feb 1996
264	35.3	2019	100.7	13.5	1929	9.4	27 Feb 1946
254	34.0	2022	101.0	13.5	1904	9.3	27 Feb 1943 and 28 Feb 1996
237	31.7	2001	102.1	13.7	1935	9.2	22 Feb 1999
Last 50 years to 2021							
268	35.9	2015	94.1	12.6	1972	10.2	26 Feb 1995

Daily sunshine records ceased at the Observatory on 31 December 1999: monthly sunshine totals from January 2000 onwards are estimated (to the nearest hour) by regression from the Met Office East and North-east England sunshine series. See Appendix 6 for record metadata and missing months.

very low sunshine totals the previous year. In some of the larger industrial cities, sunshine receipts were as low as just 15 minutes per day in January 1913—no doubt owing to industrial pollution, and included among these were Bradford, Glasgow, Manchester, and Hull. Durham managed exactly 30 minutes per day that month, in what was then the dullest January on record, and to this day only the Januarys of 1942 and 1996 have been less sunny.

Wind and weather

Snow cover 1960–99

Over the period 1960 to 1999, the three winter months accounted for the majority of mornings with snow cover—averaging 13 out of the 17 annual mean: for the period from 1971–99, this decreased slightly to 11. By far the snowiest winter on the data available was the winter of 1962/63 (see also Chapter 22) which saw snow on the ground for 70 consecutive mornings from 27 December 1962 to 6 March 1963, inclusive. In all, that winter saw 76 mornings with a snow cover. The next highest winter totals were 52 in 1978/79 (41 between December and February), 49 in 1965/66 (33 between December and February), and 38 in 1985/86 (30 between December and February). The least snowy winters, in terms of snow cover, were those of 1960/61 and 1998/99, both of which witnessed only two snow-covered mornings, in the latter only one day between December and February, the other morning with snow cover being in April 1999.

In the period of record, 40 winter mornings experienced snow at least 30 cm deep, the greatest depths at 0900 GMT being 43 cm (on 15–18 February 1963) and 41 cm (on 31 December 1961, 8 and 9 February 1963, 20–26 February 1963, and 30 November 1965).

Fog 1962–97

We are similarly limited to the period 1962–97 for information on the frequency of fog, which is determined from the visibility at the 0900 GMT observation.

Over the period 1962–97, fog at 0900 GMT (visibility below 1000 m) occurred on nine winter mornings in a typical year, two or three of which would be 'thick fog' (visibility below 200 m). The frequency of fog has been declining steadily since the early 1960s, following the introduction of the Clean Air Acts of the 1950s and the gradual phasing out of coal fires, and during the late 1980s and into the 1990s some winters have remained almost fog-free. The reduction in fogs has also had a beneficial effect on sunshine totals (see Figure 17.7).

Barometric pressure (1844–1960)

Both the highest and lowest barometric pressure on record for Durham have occurred during the winter months, when the range in barometric pressure is at its greatest owing to the more disturbed north Atlantic atmospheric circulation at this time of year. The maximum pressure occurred on 23 January 1907, when at 0900 GMT MSL pressure stood at 1052.5 hPa, although this value appears slightly high, and the true maximum probably lay between 1051 and 1051.5 hPa (see also Chapter 22). At the 2100 GMT observation on 8 December 1886 the barometer stood at only 936.2 hPa as an exceptionally deep depression crossed the north of Ireland and southern Scotland during the afternoon and evening; this remains Durham's lowest barometer reading on record (and has not been remotely approached since). There are more details of this event in Chapter 15 and in Chapter 22.

The highest mean MSL pressure for any month occurred in February 1932, with a remarkable 1034.6 hPa (see Figure 5.6), and the lowest 994.5 hPa in January 1948.

Wind speed and direction

Figure 17.6 shows the distribution of wind speed and direction at the 0900 GMT observation during the three winter months over the period 1961–97. Winds from between south and north-west are most frequent, and those from between north and south-east distinctly infrequent at this time of year.

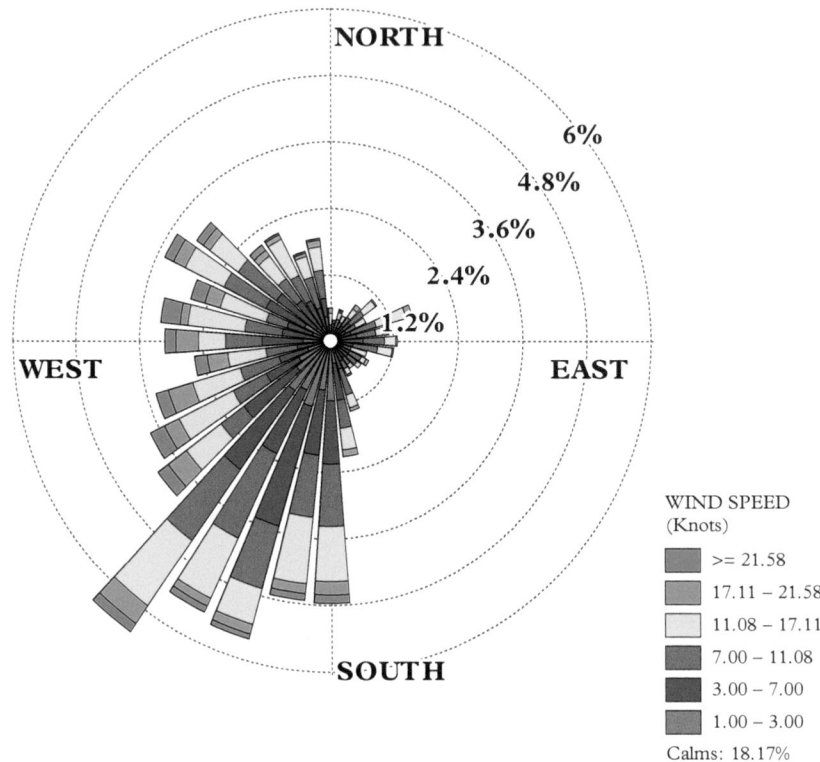

Figure 17.6 *Winter wind rose for 0900 GMT observations at Durham, period 1961–97*

Durham temperature, precipitation, and sunshine in graphs—winter

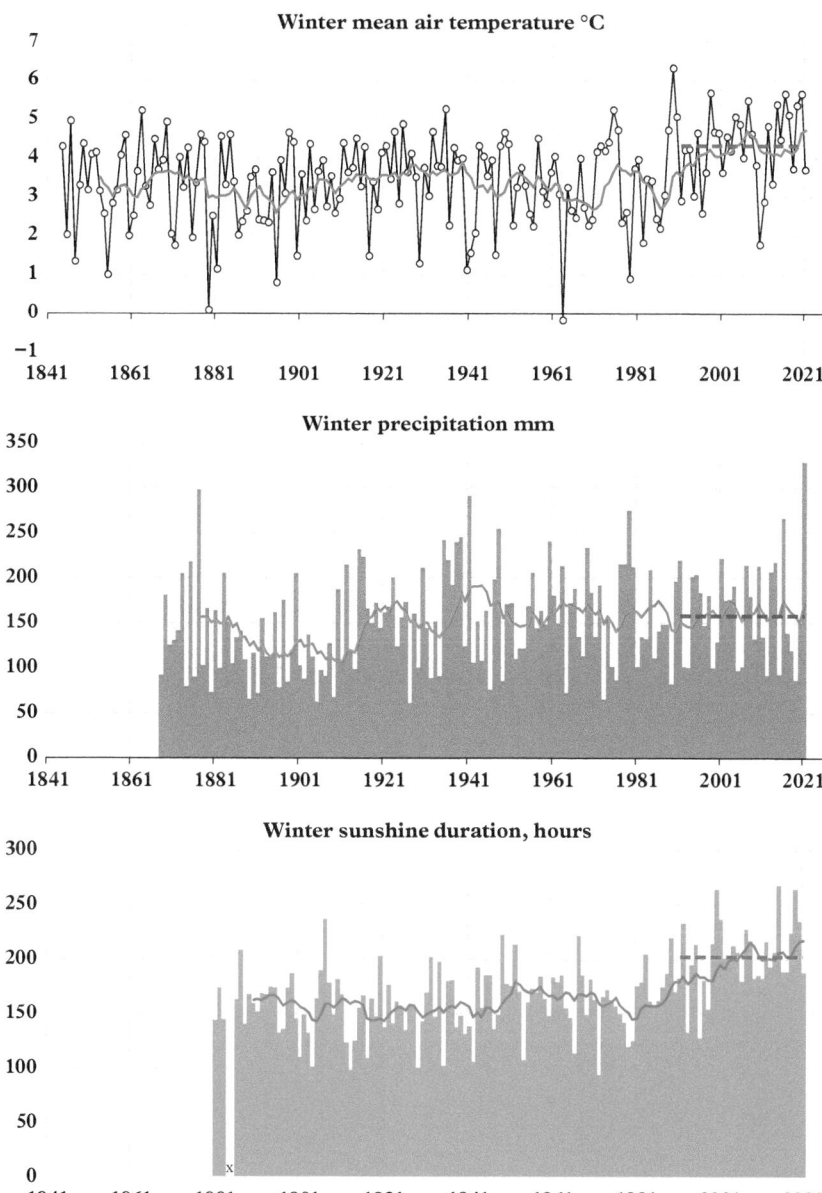

Figure 17.7 *Seasonal values of (from top) mean temperature (° C, since 1844), total precipitation (mm, since 1868), and sunshine duration (hours, since 1881, missing data 1884 and 1885) for winter at Durham Observatory. The 1991–2020 averages are indicated by the thick blue dashed line, while the ten-year running mean ending at the year shown is indicated by the red line*

18

Spring

March, April, and May

Spring is very much a season of transition. Like a longer version of March, spring comes in like a lion and goes out like a lamb, but the journey is sometimes a disappointing one: warm weather raises expectations, only for hopes to be dashed by spells of cool— even cold—weather. Spring marks the transition from snow and frost to the warmth of high summer, as the days lengthen, and the temperature rises. Figure 3.1 shows a pattern of more pronounced warming in late April after rather slower warming in March and early April. Average daytime temperatures rise by ten degrees during spring, from below 8 °C in early March to almost 18 °C by late May, while by night the range is from just above freezing to above 8 °C over the same period. Notwithstanding the sometimes-faltering rate of warming, spring's mean air temperature (8.5 °C) is four degrees warmer than winter, and since 1900, spring has seen the highest rate of warming of any season, equivalent to 1.4 degrees Celsius per century.

Spring is the driest of the seasons, with an average of 137 mm falling on 44 days, which compares to winter's 160 mm in 55 days. Taken together, the mean rainfall per rain day is slightly higher in spring (3.1 mm per spring rain-day against 2.9 mm in winter), reflecting the increased influence of convective rainfall generation. There are on average 462 hours of bright sunshine in spring compared to winter's 204, equivalent to five hours per day, not far short of summer's normal of 5.5 hours daily. Mid-May is, on average, the sunniest week of the year in Durham.

Temperature

Spring is the season with the greatest range in temperature in Durham, encompassing as it does a few very cold nights in early March with temperatures worthy of midwinter, to the occasional early summer heatwave towards the end of May. Since the Durham record commenced in 1843, spring temperatures have ranged from −15.0 °C on 4 March 1947, right at the end of a very cold winter, to 29.0 °C on 23 May 2001 (Table 18.1). May 2001 was a very dry month at Durham, and dry ground surfaces together with little cloud under anticyclonic conditions allowed for a large range in temperature on the

Table 18.1 *Highest and lowest maximum and minimum temperatures at Durham Observatory in spring, 1844–2021*

Rank	Warmest days	Coldest days	Mildest nights	Coldest nights
1	29.0 °C, 23 May 2001	−1.9 °C, 18 and 19 Mar 1846, 8 Mar 1858	14.6 °C, 19 May 2004	−15.0 °C, 4 Mar 1947
2	27.9 °C, 31 May 1922	−1.7 °C, 6 Mar 1942	14.3 °C, 13 May 1848	−12.2 °C, 3 Mar 1965
3	27.2 °C, 29 May 1947 and 23 May 2010	−1.1 °C, 1 Apr 1917	14.1 °C, 30 May 2003	−11.7 °C, 2 Mar 1963
4	26.7 °C, 29 May 1944	−0.8 °C, 18 Mar 1853	14.0 °C, 16 May 1845, 14 May 2004 and 26 May 2017	−11.6 °C, 9 Mar 1917
5	26.1 °C, 25 May 2017 and 7 May 2018	−0.7 °C, 4 Mar 1845 and 9 Mar 1931	13.9 °C, 28 May 2017	−11.3 °C, 4 Mar 1889
Last 50 years to 2021				
	29.0 °C, 23 May 2001	−0.3 °C, 17 Mar 1979	14.6 °C, 19 May 2004	−10.8 °C, 3 Mar 2001

Temperatures were recorded in different thermometer screens during the period of record; see Notes on page xvii and Appendix 3 for more details.

23rd, from a minimum of 10.0 °C to the maximum of 29.0 °C. Although the wind was a light north-easterly, Durham was clearly far enough inland not to be influenced by an onshore wind from the still-cool North Sea on the day in question.

Not surprisingly, almost all the daily 'cold' records are from March, and all the 'warm' records are from May, reflecting the pronounced rise in temperatures during the season, but the maximum temperature of −1.1 °C on 1 April 1917—the latest date in the year to have remained below zero (see Chapter 7 for details and the synoptic chart)—ranks third in the 'coldest days', while the minimum temperature the following night of −11.1 °C lies only just outside spring's 'coldest nights'. Notably, spring 1947 includes an entry under both 'coldest nights' and 'warmest days', with a huge seasonal range in temperature of 42.2 degrees Celsius.

Table 18.2 lists the warmest and coldest springs on Durham's records. The four coldest springs come from the nineteenth century, with only a few hundredths of a degree separating 1877, 1879, and 1891. In addition to ranking as (just) the coldest spring on Durham's records, 1877 was also the fourth wettest (Table 18.4), in the middle of a very wet twelve months (see Precipitation). Spring 1879 was as cold as 1877, but unlike 1877 was somewhat drier than normal; May 1879 recorded six air frosts—still a

Table 18.2 *Spring mean temperatures at Durham Observatory, 1844–2021*

Spring mean temperature 8.5 °C (average 1991–2020)

Warmest springs			Coldest springs		
Mean temperature, °C	Departure from 1991–2020	Year	Mean temperature, °C	Departure from 1991–2020	Year
9.9	+1.4	2004, 2017	5.5	−3.0	*1877, 1879, 1891*
9.7	+1.2	2011	5.6	−2.9	*1855*
9.4	+0.9	2007, 2014	5.7	−2.8	*1853*, 1917
9.3	+0.8	2003, 2020	5.7	−2.8	*1887, 1888*
9.1	+0.6	1945, 2009	5.8	−2.7	*1883*, 1941
Last 50 years to 2021					
9.9	+1.4	2004, 2017	6.4	−2.1	1979

record for the month. Both springs must have been a very trying time for gardeners and farmers. Spring 1917 makes it into the 'coldest springs' table and retains the record for the greatest frequency of air frosts in any spring—44; the current spring average is just ten.

It is very noticeable that almost all of the warm springs have occurred this century, spring 2004 and 2017 tying for warmest. The warmest spring of the twentieth century was in 1945 with a mean temperature 9.1 °C, but which now only ranks equal fifth. The very warm spring of 1893 followed a run of very cold springs between 1877 and 1892; its mean temperature of 8.6 °C was a full half-degree warmer than any other spring on record to that date. It remained the warmest for 40 years until (very slightly) surpassed by 1933 and then again by 1938, the latter at 8.7 °C; but 1893 now ranks as only 23rd warmest, with its mean temperature now only a tenth of a degree above today's spring average.

Spring 2004 became the warmest spring on record, supplanting the previous year's spring, which was itself the warmest since 1945. April 2004 was very warm, at that time the second-warmest on record behind 1949, and was followed by what remains the warmest May on record, surpassing May 1848 for the first time. Spring 2004 had the equal-second lowest total of air frosts on record, just five. Just over a decade later, spring 2017 equalled 2004's mean temperature as the equal-warmest on record; all three months were very warm, particularly May, which was the third-warmest on record, and included six consecutive days above 20 °C.

The most recent warm spring on Table 18.2, 2020, at fourth-equal warmest, also deserves particular mention: another very warm spring in the current century, and also exceptionally dry and sunny. The remarkable weather of the first COVID-19 lockdown is further described in Chapter 22.

Spring temperatures before 1843

Using the Eglise extension to the Manley series (Chapter 2, and reference [45]), two very cold springs and one very warm one stand out in Table 18.3: 1837 as the coldest spring on record (mean temperature just 4.9 °C, barely above today's winter average), and 1816 as the second-coldest (mean temperature 5.4 °C). This latter year was famously 'the year without a summer' following the eruption of the Tambora volcano in Indonesia in April 1815 (See Chapter 22). In contrast, spring 1811 was one of the warmest prior to 2000: only spring 1945 was equally warm in the twentieth century, while all the warmer springs are from the present century. The 1820s saw remarkably warm springs, and its decadal mean temperature of 8.0 °C remained the warmest until the 1990s: since then that decadal mean has been surpassed in every decade—1990s 8.1 °C, 2000s 8.6 °C, and 2010s 8.7 °C.

Table 18.3 *Spring mean temperatures at Durham Observatory, 1784–2021 excluding 1791–1794, from the extended Durham temperature series*

Spring mean temperature 8.5 °C (average 1991–2020)

Warmest springs			Coldest springs		
Mean temperature, °C	Departure from 1991–2020 normal degC	Year	Mean temperature, °C	Departure from 1991–2020 normal degC	Year
9.9	+1.4	2004, 2017	4.9	−3.6	1837
9.7	+1.2	2011	5.4	−3.1	1816
9.4	+0.9	2007, 2014	5.5	−3.0	1877, 1879, 1891
9.3	+0.8	2003, 2020	5.6	−2.9	1839, 1855
9.1	+0.6	1811, 1945, 2009	5.7	−2.8	1853, 1917
Last 50 years to 2021					
9.9	+1.4	2004, 2017	6.4	−2.1	1979

Precipitation

Spring is the driest of the four seasons, with a 1991–2020 average of 137 mm, falling on 44 rain days. Table 18.4 gives details of the extremes of spring precipitation at Durham since 1868. In contrast to the remaining three seasons, springs have become slightly drier during the 1991–2020 period, when compared with the 1981–2010 averages—137 mm

Table 18.4 *Spring precipitation at Durham Observatory 1868–2021*

Spring mean precipitation 137 mm (average 1991–2020)

| Wettest months | | | Driest months | | | Wettest days | |
Total fall, mm	Per cent of normal	Year	Total fall, mm	Per cent of normal	Year	Daily fall, mm	Date
307.2	224	1979	42.4	31	2020	70.2	31 May 1924
282.1	206	1998	45.3	33	1956	50.6	28 Mar 1979
263.6	192	1983	51.7	38	1929	43.0	1 Apr 1992
249.5	182	1877	54.1	39	1875	41.0	13 May 1993
235.2	172	1947	65.4	48	1870	39.8	15 April 2005
Last 50 years to 2021							
307.2	224	1979	42.4	31	2020	50.6	28 Mar 1979

vs 141 mm, a reduction of 3 per cent, although the average annual precipitation increased by almost 4 per cent between the two periods.

The wettest springs

The wettest springs are well distributed across the record, although three of the five wettest occurred in the 20 years from 1979. Spring 1979 was the wettest, in the bottom 20 per cent for temperature, too; to date, it is the only spring to exceed 300 mm precipitation at Durham. March that year was by far the wettest on record (166 mm, four times normal—Table 6.3) with heavy snowfall in the third week: snow lay every day during 17–24th, accumulating to 31 cm on the 18th and 19th, while 50.6 mm precipitation fell on the 28th, the second-wettest spring day on record. The heavy rain caused widespread flooding across the country, although the flood peak on the River Wear at Sunderland Bridge was a relatively modest 1.651 m. April 1979 was drier than average, but May was also wet, in the top 10 per cent of May rainfall totals.

Durham's second wettest spring was in 1998, helped in large measure by the wettest April on record; its 151 mm total alone, falling on 25 days, was well above spring's mean precipitation of 137 mm. April was also very dull and cool; unusually, the month was colder than both February and March that year. Easter was particularly cheerless, cold, and wet: Good Friday (10 April) received 15.8 mm precipitation, with the day's maximum a raw 5.7 °C. March and May were also wet, although not remarkably so. Spring 1983 was also very wet, amounting to the third wettest on record, with both April and May receiving more than 100 mm rainfall; an inauspicious start to the summer—one that was to become one of the hottest on record.

The driest springs

Durham's driest spring will be a vivid memory for many, coinciding as it did with the first national lockdown due to the coronavirus pandemic in 2020. The three-month rainfall total was less than one-third of normal, and rain fell on only 29 days instead of the normal 44. The weather was also mostly warm and sunny, the sunniest spring and the fifth-warmest on record. While March experienced a mixture of cyclonic and anticyclonic weather, April and May were dominated by settled, anticyclonic weather with the jet stream lodged to the north of the British Isles; April was the third driest on record, and, until a fall of 1.8 mm on the 29th during the passage of a weak cyclonic system, would have been the driest (Figure 18.1).

Until 2020, Durham's driest spring was in 1956, with just 45 mm in all, one-third of normal. Except for a changeable first week, the weather in March was dry and sunny, dominated by an anticyclone over European Russia and an equally persistent depression over south-west Ireland, and winds were mainly from the south and east. April saw little cyclonic activity with high pressure centred over or near the British Isles, while May was warm, sunny, and very dry with rain on only eight days during the month.

Spring 1875 was an unusual extended dry spell in the middle of a very wet decade with a preponderance of cyclonic weather [76]. Both March and April that year were dominated by anticyclonic conditions: Lamb Weather Types show 18 and 23 anticyclonic days, respectively, for the two months, and 16 in May.

The wettest spring days

Wet spring days tend to transition in character from March to May, from long-duration, persistent rain associated with slow-moving fronts or cyclonic systems more

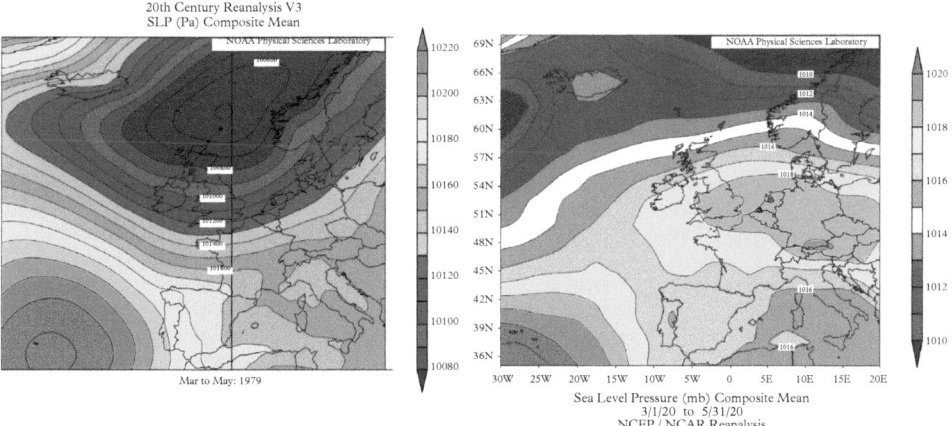

Figure 18.1 *Mean MSL pressure distribution during Durham's wettest spring (1979, left) and driest (2020, right)—the former recording seven times as much precipitation as the latter. Isobars are in Pascals (divide by 100 for hPa = millibars), at 100 Pa intervals. (Spring 1979 plot from the 20th Century Reanalysis version 3 (20CRv3) plot [58], spring 2020 plot provided by the NOAA/ESRL Physical Sciences Laboratory, Boulder Colorado from their website at http://psl.noaa.gov/ [84]*

typical of the winter months, to shorter, more intense convective rainfall falling as showers or thunderstorms more typical of the summer half-year. Spring's wettest days (Table 18.4) reflect this transition, with persistent cyclonic rainfall events (28 March 1979, 1 April 1992) appearing in the same list as heavy thundery rain (31 May 1924). The latter was Durham's wettest spring day, when 70.2 mm fell; the event is discussed in more detail in May's chapter. April 1992 very wet; it was also a little wintry with a cold start. There was 3 cm of snow lying on the morning of the 2nd and no doubt this contributed to the rainfall total of 43 mm thrown back to the day before. May 1993 had some very wet weather with a series of slow-moving complex fronts associated with a succession of low-pressure systems from the Atlantic; the fall of 41 mm on the 13th was the wettest day of the year and the fourth-wettest daily spring total. On the same day 62 mm fell at Nunraw Abbey in the Scottish Borders, the result of a southward-moving cold front.

Sunshine

During the standard averaging period 1991–2020, Durham's average sunshine receipt in spring was 462 hours, equivalent to five hours per day; this is 36 per cent of possible, the highest percentage of any season, and perhaps not surprising given that spring is the driest of the seasons. Sunshine in spring is not far short of the summer amount, despite

Figure 18.2 *Cricket on the Racecourse, 26 May 2005. Durham's relatively dry springs allow more days of cricket than at most places in the UK (Tim Burt)*

Table 18.5 *Spring sunshine duration at Durham Observatory. Extremes 1881 to 2021, sunniest days 1882–1999 only*

Spring mean sunshine duration 462 hours, 5.0 hours per day (average 1991–2020)
Possible day length: 1292 hours. Mean sunshine duration as percentage of possible: 35.8

Sunniest springs			Dullest springs			Sunniest days 1882–1999	
Duration, hours	Per cent of possible	Year	Duration, hours	Per cent of possible	Year	Duration, hours	Date
630	48.8	2020	293.8	22.7	1937	15.4	27 May 1883
548.0	42.4	1990	294.2	22.8	1928	15.3	29 May 1952
547.0	42.4	1921	306.3	23.7	1886	15.0	21 May 1921, 25 May 1929 and 18 May 1940
546	42.3	2011	313.7	24.3	1941	14.9	21 May 1888 and 25 May 1921
545.4	42.2	1982	316.6	24.5	1983	14.8	*Numerous occasions*
Last 50 years to 2021							
630	48.8	2020	316.6	24.5	1983	14.7	26 May 1989

Daily sunshine records ceased at the Observatory on 31 December 1999: monthly sunshine totals from January 2000 onwards are estimated (to the nearest hour) by regression from the Met Office East and North-east England sunshine series. See Appendix 6 for record metadata and missing months.

the fact that summer has an additional 200 hours of daylight. May is, on average, the sunniest month of the year in Durham, and mid-May the most reliable time of year for sunny weather.

The sunniest springs

Spring 2020 was truly remarkable—not only the sunniest spring on record with 630 hours bright sunshine (Table 18.5), it would also have ranked as the fourth-sunniest *summer*, not far behind the memorable summers of 1976 and 1989 and ahead of summer 1995. April and May both ranked within the top-five sunniest of those months. Predominantly anticyclonic conditions were responsible for the clear skies, although as noted in Chapter 22, easterly winds off the North Sea did bring a few cool, overcast days. Spring 2020 was also the driest on record (Table 18.4) and one of the warmest (Table 18.3).

   Durham's sunniest spring prior to 2020 was in 1990, its total of 548 hours of sunshine just surpassing spring 1921's 547 hours, itself only just ahead of spring 2011 (546 hours) and spring 1982 (545 hours). Spring 1921 was for nearly 70 years the sunniest on record, a significant contribution to another year well known for its fine weather; it now ranks third. March was close to normal, but it was the exceptional April and May that account for its placing in Table 18.5. The *Monthly Weather Report* noted that the outstanding feature of April's weather was the abundant sunshine, while May was another fine and dry month for most of the country except the north-west; Durham averaged a fraction under eight hours of bright sunshine daily.

The dullest springs

The dullest spring on record was 1937, just 24 minutes less sunshine over the three months than the previous dullest, nine years previously (Table 18.5). The coronation of King George VI and Queen Elizabeth took place at Westminster Abbey, London, on 12 May 1937, and in keeping with the dullest spring on record, Coronation Day was sunless at Durham. March that year was very dull with barely half normal sunshine, while April remains by some margin the dullest on record (Table 7.4); 13 days remained sunless, including eight consecutive days from the 10th to the 17th. The *Monthly Weather Report* stated that April was distinguished by 'a general and exceptional deficiency of sunshine', although May saw sunshine totals return closer to normal.

Figure 18.3 *Spring colours, 7 May 2003 (Tim Burt)*

Spring 1928 got off to a poor start with the dullest March on record to that date, a record not beaten until 1964. Fourteen days had no sunshine at all but there were a few bright days, with 9.5 hours on the 26th. April and May were both duller than normal.

March 1886 was a month of two halves: cold, wintry, anticyclonic weather in the first half of the month followed by mild, wet weather from the south-west: neither weather type permitted much in the way of sunshine in Durham. April was also dull but May was exceptionally so, and remains the third-dullest on record today: ten days remained sunless, including a bleak spell 8th to 13th.

After the sunny spring in 1982, that of 1983 was a disappointment, with only 317 hours sunshine—almost half of the amount registered in its immediate predecessor (Table 18.5). All three months managed barely 100 hours of sunshine; dull for March, very dull for an April, and extremely dull for a May month. April and May were both cold and wet. No spring has been as dull since 1983, although 1996 came close, with only 330 hours.

Wind and weather

Snow cover 1960–99

Snow cover is not uncommon in early spring, March averaging two mornings with snow on the ground, while about one April in three during this period recorded a snow cover. The snowiest spring was in 1979, with ten days, all in March, and there were eight days with snow cover in March 1965. The Aprils of 1966 and 1973 both saw a snow cover on three mornings, although snow cover in April has become less frequent within the last two decades. The deepest spring snowfalls during the above period were 38 cm on 1 March 1963, at the end of that long, cold winter, 28 cm on the same date in 1965, and 31 cm on 18 and 19 March 1979. Perhaps most notable was 25 cm snow depth on 5 April 1975.

The latest date with a snow cover during the period 1960–99 was 25 April 1981, when snow lay 7 cm deep.

Fog 1962–97

Fog is similarly more likely in March than in April or May, although considerably less frequent in both as the season advances and in more recent years.

Barometric pressure (1844–1960)

The highest spring barometric pressure on record for Durham occurred on 6 March 1852, when the MSL pressure at 0900 was 1044.9 hPa; this was equalled at the 0900 GMT observation on 9 March 1935. The spring lowest stands at 962.3 hPa, at 1000 on 10 March 1876. The highest mean MSL pressure for any spring month occurred in

March 1953, when the mean was 1029.4 hPa; the lowest mean was 997.6 hPa, in March 1876.

Wind speed and direction

Figure 18.4 shows the distribution of wind speed and direction at the 0900 GMT observation during the three spring months over the period 1961–97. March is the windiest month of the year in Durham, although wind speeds progressively reduce as the season advances. Winds from between south and north-west remain most frequent, although those from between north and east are at their most frequent at this time of year.

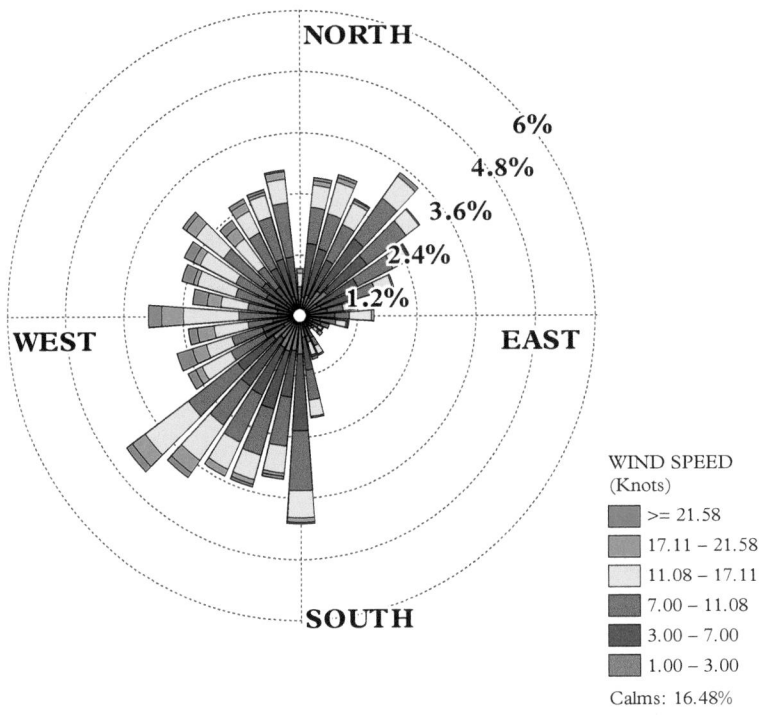

Figure 18.4 *Spring months wind rose for 0900 GMT observations at Durham, period 1961–97*

Durham temperature, precipitation, and sunshine in graphs—spring

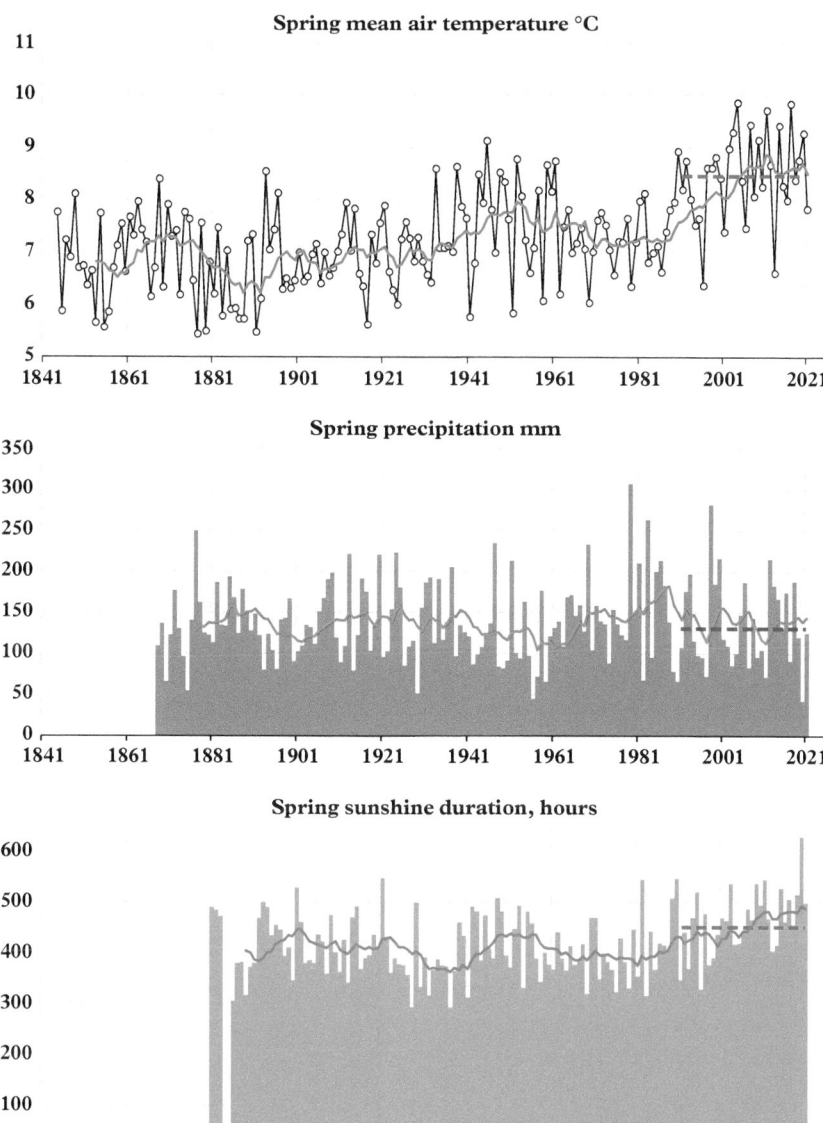

Figure 18.5 *Seasonal values of (from top) mean temperature (° C, since 1844), total precipitation (mm, since 1868), and sunshine duration (hours, since 1881, missing data 1884 and 1885) for spring at Durham Observatory. The 1991–2020 averages are indicated by the thick blue dashed line, while the ten-year running mean ending at the year shown is indicated by the red line*

19

Summer

June, July, and August

Summer is the warmest and sunniest season of the year, although not the wettest. The mean air temperature for Durham over the standard 1991–2020 averaging period was 15.0 °C (mean daily maximum temperature 19.3 °C, mean daily minimum temperature 10.6 °C). There is little to choose between July and August in mean temperatures as they differ by only 0.2 degrees Celsius, July being the warmer of the two at 15.8 °C: June's mean air temperature (13.6 °C) is almost identical to September's (13.3 °C). In the 178 years between 1844 and 2021, July was the warmest month of the year 96 times and August 75 times: June was the warmest in only seven of those years. Temperatures typically reach their annual peak at the end of July and the beginning of August (Figure 3.1). Only one summer air frost occurred during the 30-year averaging period 1991–2020, although ground frosts do occur occasionally.

Summer's mean precipitation total is 189 mm, wetter than spring but drier than autumn, and rain can be expected to fall on 45 or 46 days in an average summer—about one day in two. Average monthly rainfall totals vary little between the three months, although there is of course considerable variation from year to year, and the occasional heavy thundery downpour can be expected most years. Summer is also the sunniest season, averaging 508 hours in total, equivalent to 5.5 hours of bright sunshine per day.

Temperature

Durham's summer temperatures have ranged from −0.8 °C on 15 June 1927 (grass minimum −4.4 °C), to 32.9 °C on 25 July 2019 (Table 19.1). Not surprisingly, the lowest summer temperatures tend to be recorded early in the season: air frosts in June are not unknown given clear skies, dry ground conditions and a polar airmass, even as late as the closing week (−0.5 °C on 28 June 1974 is the latest air frost on record). At the other end of summer, the temperature fell to just 0.5 °C on 22 August 1864.

Durham's coldest summer days also tend to occur in early June; notable among these are 2 and 3 June 1953, the former being Queen Elizabeth II's Coronation Day, when the maximum temperature attained just 7.2 °C; it was an atrocious day in the middle of

Table 19.1 *Highest and lowest maximum and minimum temperatures at Durham Observatory in summer, 1844–2021*

Rank	Hottest days	Coldest days	Warmest nights	Coldest nights
1	32.9 °C, 25 July 2019	7.2 °C, 2 June 1953	18.9 °C, 20 July 2016	−0.8 °C, 15 June 1927
2	32.5 °C, 3 Aug 1990	*7.4 °C, 2 June 1888*	17.9 °C, 14 Aug 2001, 6 Aug 2006 and 26 July 2019	−0.7 °C, 19 June 1915
3	31.9 °C, 2 Aug 1990	*7.6 °C, 1 June 1886*	17.8 °C, 26 Aug 1959 and 5 June 1980	−0.6 °C, 8 June 1920 and 8 June 1991
4	31.3 °C, 16 July 1876	7.8 °C, 3 June 1953	17.7 °C, 3 Aug 2018 and 24 July 2019	−0.5 °C, 28 June 1974
5	31.0 °C, 1 July 2015	8.0 °C, 1 June 1902	17.6 °C, 24 July 1900	0.3 °C, 3 June 1924 and 14 June 1924
Last 50 years to 2021				
	32.9 °C, 25 July 2019	8.3 °C, 5 June 2009	18.9 °C, 20 July 2016	−0.6 °C, 8 June 1991

Temperatures were recorded in different thermometer screens during the period of record; see Notes on page xvii and Appendix 3 for more details

a spell of atrocious weather. A very cool and moist northerly airflow (Figure 9.1) gave conditions more like November: overcast skies, a chill wind and persistent cold rain—20 mm fell on 2 June and 13 mm the following day. There was no sunshine on the 2nd, 3rd, or 4th. Durham's outdoor celebrations must have been sadly curtailed. Only slightly less cold were 2 June 1888 (maximum temperature 7.4 °C) and 1 June 1886 (7.6 °C). In late August, the maximum temperature of 9.4 °C on 28 August 1919 was also extremely low for the time of year.

Until 1990, Durham's hottest day on record was 16 July 1876, when the Stevenson screen equivalent maximum temperature was 31.3 °C (the maximum in the Glaisher stand was 33.6°C, and in the 'North Shed' 30.1 °C). This extreme stood for well over a hundred years before finally being surpassed on 2 August 1990, when 31.9 °C was reached. This new record lasted only a single day, for on the following day (3 August 1990), the maximum temperature reached 32.5 °C (the synoptic situation at the time is shown in Figure 11.1). This short-lived but intense heatwave demonstrated an intensity and spatial extent of high temperatures without precedent on reliable instrumental records at that time [61]. New absolute maximum temperatures were established in England (37.1 °C at Cheltenham, also on 3 August, surpassing the previous UK record of 36.7 °C in August 1911) and in Wales (Hawarden Bridge 35.2 °C, on 2 August), while the all-time record for Scotland was closely approached. Warm air reached north-east England after a long land track on a strengthening south-westerly wind of sufficient strength to inhibit sea-breeze development from the much colder North Sea. Sea temperatures are never very high off north-east England, even in summer and, as a consequence, high surface temperatures are very rare [85]. At Hartburn Grange in Cleveland the temperature reached 33.9 °C, well in excess of the previous station highest (records began 1962) of 30.0 °C on 29 June 1976 and equalling the highest on record in north-east England (at Ushaw College on 1 September 1906). At Sunderland Poly-technic the maximum of 33.0 °C was well in excess of anything previously recorded in Tyne and Wear (previous county record 30.6 °C). Even right on the coast at Tynemouth (where records began in 1912) the temperature rose to an unprecedented 31.9 °C—the effect of the suppression of the sea-breeze being seen in an extraordinary rise from the previous day's maximum of 20.2 °C. Minimum temperatures were also exceptional during this spell, also establishing new record levels for the British Isles, although Durham's warmest night (on 3 August) was only 16.3°C, well below extreme levels.

The August 1990 records survived almost 20 years, until they in turn were surpassed on 25 July 2019, when Durham's maximum temperature reached 32.9 °C. This date also saw a new UK highest maximum temperature record set, this time at the Cambridge Botanic Garden where 38.7 °C was recorded. Previous site records were widely broken, including 36.5 °C at Oxford's Radcliffe Observatory [86]. A plume of hot air pushed northwards from the continent with large parts of central and western Europe, including Belgium, Germany, and the Netherlands also seeing their highest ever temperatures, in the low to mid-40s Celsius. Figure 10.2 shows the synoptic situation on this date. The nights were also very warm, with a minimum temperature of 17.7 °C on 24 July and 17.9 °C following the record heat on the morning of 26 July, Durham's second-warmest night on record (Figure 10.1 and Table 19.1).

Durham's warmest night saw the temperature fall only to 18.9 °C on 20 July 2016; this broke the record by a whole degree Celsius (Table 19.1), despite the lack of any exceptional heatwave—the maximum temperature the previous day was 29.2 °C, the only '25' recorded that month.

Only seven air frosts have been recorded in summer at Durham since 1844, all in June, all of them since 1900 (Table 19.1). June 1927 was very cool with much heavy rain and high winds in the last two weeks. The *Monthly Weather Report* notes some fine weather in eastern districts but with some notably low minima on 14th and 15th, including −3.3 °C at Castleton in Yorkshire; at Durham the screen minimum was −0.8°C, and the grass minimum as low as −4.4 °C. At the time of writing, no summer air frost has occurred in 30 years—most recently on 8 June 1991, when the screen minimum fell to −0.6 °C.

Figure 19.1 *Durham Regatta, 10 June 2007. Held on the second weekend of June, at its best fine weather allows many spectators to enjoy the racing. At its worst, wet and miserable, there are few people on the riverbank! June is very much a transition month into high summer (Tim Burt)*

Hottest and coldest summers

Table 19.2 sets out Durham's hottest and coldest summers since the Observatory's records commenced in 1844 (summer 1843 is incomplete, since continuous records did not begin until 23 July that year).

Durham's coldest summer, more than three degrees below average, was in 1888. The *Monthly Weather Report* states that June 1888 was 'rather cold', especially along the east coast owing to the prevalence of north-easterlies, while July was 'most unseasonable' with frequent heavy rainfall, and even a few snowfalls—July 1888 remains Durham's coldest and wettest July on record. August was less unseasonable but still cooler than normal. Given these three dreadful summer months, it is hardly surprising that summer 1888 comes bottom of the list. The contrast with the prolonged and brilliant summer of the previous year, Queen Victoria's Diamond Jubilee year (subsequently known for a generation as the 'Queen's summer'), must have been striking indeed.

Summer 1862 appears in second place in Table 19.2's coldest summers. We have less information on 1862 than later summers (sunshine records did not begin until 1880, for example) but the available statistics tell their own story of a dull and cold summer, slightly drier than normal. The year's highest temperature, just 21.1 °C, was reached on 17 May. Thereafter, no other day that year reached 20 °C, the highest in June and August (and October) being 19.8 °C: July's warmest day, the 13th, struggled to 18.8 °C.

All three months experienced below-average temperature in summer 1902, although rainfall was a little below normal. June was relatively the best of the three, although both June and July were in the bottom 10 per cent in terms of mean temperature, and in August, only two days reached 20 °C. Finally, summer 1912 was by far the dullest summer yet recorded in Durham, so it is not surprising that it was also one of the coolest; this summer is discussed in more detail later in this chapter and in Chapter 22.

There is just one hundredth of a degree between Durham's hottest summers, 1846 and 2018, with 2018 just the warmer, although it is difficult of course to be sure of the validity of instrumental readings for the former year, just three years into the Durham record. The summer of 1846 was dominated by very hot weather in June—which remains to this day the warmest on record, by a remarkable 1.7 degrees Celsius (see Table 9.2). Gordon Manley referred to June 1846 as 'phenomenally hot' [19]. Although the observatory registers for the month were not available to him in his lifetime, we now know that maximum temperatures reached 25 °C on 11 days in the month, still a record for any June. The rest of the summer was rather disappointing, however, as only one day in July and none in August reached this level, and of course this is 35 years before sunshine records commenced so we have little idea of how sunny or otherwise the summer actually was.

Table 19.2 *Summer mean temperatures at Durham Observatory, 1844–2021*

Summer mean temperature 15.0 °C (average 1991–2020)

Warmest summers			Coldest summers		
Mean temperature, °C	Departure from 1991–2020 normal degC	Year	Mean temperature, °C	Departure from 1991–2020 normal degC	Year
16.3	+1.3	*1846, 2018*	*11.9*	−3.1	*1888*
16.2	+1.2	2003, 2006	*12.2*	−2.8	*1862*
16.1	+1.1	1933	12.4	−2.6	1902
15.9	+0.9	1947, 1976, 2013	12.5	−2.5	1907
15.7	+0.7	2019	12.7	−2.3	1912
Last 50 years to 2021					
16.3	+1.3	2018	13.4	−1.6	1985

We have much more extensive and reliable records for 2018, of course, the warmest summer on record (Table 19.2). Although no new high temperature records were set, all three summer months were warm (July particularly so) and dry, while June and July (and May) were all very sunny.

The summers of 2003 and 2006 tie for second place, just a tenth of a degree below 2018's level. Summer 2003 became Durham's hottest in 70 years. June was remarkably warm, the fifth-warmest on record: July and August were also in the top 10 per cent warmest of those months. The summer was characterised by persistent anticyclonic conditions and frequent hot weather over much of western Europe: existing national temperature records were broken in France, Germany, Belgium, Switzerland, and Portugal. In the United Kingdom, the first half of August saw a ten-day hot spell during the course of which a new UK maximum temperature record was established: 38.1 °C at Kew on 10 August, surpassing the previous highest UK temperature record of 37.1 °C at Cheltenham on 3 August 1990 [87]. A new all-time temperature record for Scotland was also set, with 32.9 °C recorded at Greycrook, Borders Region, on 9 August. This was also Durham's hottest day of the year, although somewhat cooler at 28.5 °C. This came as the culmination of a noteworthy hot spell for the week 4–10 August in which 24 °C was exceeded every day. The heatwave broke down in violent thunderstorms in the north-east on 10 August; at Carlton-in-Cleveland on the edge of the North York Moors, about 50 km south-east of Durham, an astonishing 49 mm of rain fell in 13 minutes, the most intense fall yet reliably recorded instrumentally anywhere in the British Isles [88].

Just three years later, in 2006, came an equally warm summer. June that year was warm, the ninth warmest on record, and was followed by Durham's hottest and sunniest July on record (Tables 10.2 and 10.4). July's mean maximum temperature (24.2 °C) surpassed the previous highest monthly mean maximum temperature (23.1 °C in August 1995) by more than a degree Celsius; it has not since been exceeded, although July 2013 (23.2 °C) has come closest in the intervening years. After that excitement, August returned to near-normal conditions: had August's mean maximum temperature been less than one degree higher, 2006 would have become the warmest summer on Durham's records.

The warmest summer of the twentieth century in Durham was 1933, a record that lasted 70 years. All three months were well above average temperature: the *Monthly Weather Report* described July as notably warm, with much sunshine, and very dry in the east. Summer 1947 was the other summer from the first half of the last century that deserves a mention, equally as hot as 1976 and 2013. June and August 1947—the latter the warmest on record to that date—were the exceptional months. August was hot, dry, and sunny: although the absolute maximum was only 26.7 °C, temperatures were very consistent through the month and only three days failed to reach 20 °C.

Summer 2013 ranks fourth-equal warmest in mean temperature, along with 1947 and 1976. July was the exceptional month—very sunny and the second-warmest of any month on Durham's records—with a mean temperature of 18.0 °C (behind only July 2006's 18.3 °C). June and July were also warm, although less so than July.

Finally, only brief mention here of the hot summer of 1976, the year of a major drought which is discussed in more detail in Chapter 22. This remains the sunniest summer on record, as well as the third driest and the fourth-equal hottest.

More information on summer heatwaves, and the frequency of both 'hot days' (maximum temperature 25 °C or more) and 'warm nights' (minimum temperature 15 °C or more) is given in Chapter 16.

Summer temperatures before 1844

Extending the mean air temperature series back to 1784 [45] changes Table 19.3 only slightly. In the 'coldest' column, we add summer 1841 in third place, with a mean temperature of just 12.3 °C; only 1888 and 1862 have been colder. Just outside the 'Bottom 5' are the summers of 1799 (mean 12.6 °C), and 1795 and 1816 (mean 12.7 °C). The latter was famously the 'year with no summer' and followed a notably cold spring. The impact of the global dust veil from the Tambora volcanic explosion in April 1815 is discussed in Chapter 22. In contrast, summer 1826, with a mean temperature of 17.3 °C, surpasses every summer since, while summer 1818 (mean temperature 15.7 °C) falls just outside the 'Top 5'. The warmth of summer 1818 must have been particularly welcome after the very cool summers of 1816 and 1817.

Table 19.3 *Summer mean temperatures at Durham Observatory, 1784–2021 excluding 1791–1794, from the extended Durham temperature series*

Summer mean temperature 15.0 °C (average 1991–2020)

Warmest summers			Coldest summers		
Mean temperature, °C	Departure from 1991–2020 normal degC	Year	Mean temperature, °C	Departure from 1991–2020 normal degC	Year
17.3	+2.3	1826	11.9	−3.1	1888
16.3	+1.3	1846, 2018	12.2	−2.8	1862
16.2	+1.2	2003, 2006	12.3	−2.7	1841
16.1	+1.1	1933	12.4	−2.6	1902
15.9	+0.9	1947, 1976, 2013	12.5	−2.5	1907
Last 50 years to 2021					
16.3	+1.3	2018	13.4	−1.6	1985

Precipitation

During the standard averaging period 1991–2020, summer's average rainfall in Durham was 189 mm, falling on 46 days. Summer has the highest mean rainfall per rain day for any season, at just over 4 mm per rain day (40 per cent greater than in winter), a reflection of increased convective precipitation during the warmer months. The wettest summer day (69.1 mm) was 25 August 1986, associated with the passage of ex-hurricane *Charley* (see Chapter 11 and Chapter 22).

Wettest summers

Durham's wettest summer was in 1980 (Table 19.4)—and more than half of that summer's rainfall fell in less than a month. Between 5 June and 3 July that year, inclusive, 192.8 mm fell in Durham; that period included eight days with 10 mm or more, and three with 20 mm or more, the wettest day of the summer being 33.4 mm on 11 June. June's monthly fall was 191 mm, which at 313 per cent of normal remains Durham's wettest June on record. The *Monthly Weather Report* headline for the month read 'mostly unsettled, cool and thundery', and noted that rainfall exceeded three times normal in parts of northern and north-eastern England, including Durham. Thunder was heard on eight days at Low Etherley in County Durham and at Hartburn Grange in Cleveland. Thereafter, July was slightly drier than normal, while August was also wet, although not making it into the top 20 per cent of wettest Augusts.

Summer 2012 was cool and unusually wet, like several other summers in the first two decades of the present century—the 'Top 20 wettest' includes seven summers since 2000. After a dry start to 2012 (only 56 mm had been recorded by the end of March),

Table 19.4 *Summer precipitation at Durham Observatory 1868–2021*

Summer mean precipitation 189 mm (average 1991–2020)

Wettest summers			Driest summers			Wettest days	
Total fall, mm	Per cent of normal	Year	Total fall, mm	Per cent of normal	Year	Daily fall, mm	Date
339.8	181	1980	54.9	29	1995	69.1	25 Aug 1986
338.0	180	2012	65.2	35	1887	61.4	8 July 1893
331.8	176	1927	73.3	39	1976	57.1	25 July 1888
326.3	174	1930	80.4	43	2006	54.0	3 June 2000
307.1	162	1912	89.7	47	1868	52.2	9 Aug 2004
Last 50 years to 2021							
339.8	181	1980	54.9	29	1995	69.1	25 Aug 1986

the second half of spring was wet, April alone receiving 134 mm. Summer continued the theme with a very wet June (137 mm), and July and August were both wet (98 mm and 103 mm, respectively). The wet conditions continued throughout the autumn, making 2012 the wettest year on record, despite the dry start (see Chapter 16). Summer's total of 338 mm was the greatest in any of the four seasons that year. The turnaround was extraordinary—the 12 months ending March 2012 saw just 503 mm of rainfall. Exactly one year later, the 12-month total at the end of March 2013 had reached 1130 mm—more than twice as much as the previous 12 months—and the wettest such period on Durham's records back to at least 1868.

The third-wettest August on record contributed much to 1927's summer total. It was at the time Durham's wettest August on record, a fact mentioned in the *Monthly Weather Report*, which described the month as unsettled and wet.

Driest summers

Durham's driest summer came in 1995, when just 55 mm fell (in only 31 rain days) during the three months (Table 19.4). June was sunny and dry, if a little cool overall, July sunny with near-normal temperature and rainfall, but August was an exceptional month—the sunniest on record, equal-second hottest with 1947 (behind only 1975) and third driest. Overall, summer 1995 remains not only the driest on record, but to date the fourth sunniest and one of warmest on Durham's records, although since surpassed in warmth by the summers of 2003, 2006, 2013, 2018, and 2019.

The summer of 1887, Queen Victoria's Golden Jubilee year, was one of the finest summers of the nineteenth century. In a time when more people depended upon springs and wells for their water supply, the year was marked by widespread and at times severe drought and resulting water shortages. The *Monthly Weather Report* stated that:

> June was unusually fine, warm, and dry, especially over England and the greater part of Ireland, where no rain fell after the 8th. Pressure was considerably in excess of the normal with anticyclonic conditions dominating for most of the month. The amount of rainfall was in most instances excessively small and bright sunshine was unusually abundant.

In Durham, only 6 mm of rain fell in June (still the fourth-driest on record); July and August were also dry, although less remarkably so. The summer's rainfall total amounted to only 65 mm, falling on 29 rain days, and retained the title of Durham's driest summer for over 100 years, until 1995.

The third-, fourth-, and fifth-driest summers, namely 1976, 2006, and 1868, respectively, were also hot summers; summer 2006 remains Durham's second-hottest summer, while summer 1976 follows closely behind (Table 19.2).

Sunshine

During the current standard averaging period 1991–2020, Durham averaged 508 hours of bright sunshine in total, equivalent to 5½ hours per day. However, this is only an hour each day more than in spring, showing the impact of increased summer cloudiness despite the extra daylight. Table 19.5 shows the sunniest and dullest summers since 1881.

Summer 1976 remains Durham's sunniest summer; it was also the third driest and fourth-equal hottest. All but six days during summer 1976 recorded sunshine. A major drought affected north-west Europe from May 1975 to August 1976, culminating in the summer of 1976 before its dramatic reversal during the autumn. Durham was not at the heart of the drought, which was more severe further south, but it was nevertheless an exceptional summer (see Chapter 22 for further details about the drought and the 1976 summer).

Table 19.5 *Summer sunshine duration at Durham Observatory. Seasonal totals 1881 to 2021, sunniest days 1882–1999 only*

Summer mean sunshine duration 508 hours, 5.5 hours per day (average 1991–2020)

Possible day length: 1495 hours. Mean sunshine duration as percentage of possible: 34.0

Sunniest summers			Dullest summers			Sunniest days 1882–1999	
Duration, hours	Per cent of possible	Year	Duration, hours	Per cent of possible	Year	Duration, hours	Date
669.8	44.8	1976	222.4	14.9	1912	15.8	5 June 1940 and 15 June 1957
659.9	44.1	1989	350.2	23.4	1987	15.7	14 June 1959
647.4	43.3	1901	356.1	23.8	1931	15.6	26 June 1937 and 5 June 1962
627.4	42.0	1995	361.8	24.2	1954	15.5	13 July 1929 and 29 June 1931
617.0	41.3	1911	362.8	24.3	1888	15.4	7 July 1905, 6 June 1940 and 30 June 1940
Last 50 years to 2021							
669.8	44.8	1976	350.2	23.4	1987	14.6	19 June 1989 and 27 June 1995

Daily sunshine records ceased at the Observatory on 31 December 1999: monthly sunshine totals from January 2000 onwards are estimated (to the nearest hour) by regression from the Met Office East and North-east England sunshine series. See Appendix 6 for record metadata and missing months.

The second-sunniest summer came in 1989, another warm, dry summer; May, June, and July all exceeded 200 hours of bright sunshine, although August fell slightly short with 199 hours. Until 1976, summer 1901 held the title of Durham's sunniest on record—June and July were very sunny, August less so, but only July was very warm, and the summer rainfall, while below normal, was not exceptionally so. The summer of 1911 was also sunny and hot, although not particularly dry—in fact, June was wet.

It is worth commenting at this point that the very sunny *spring* of 2020, with 630 hours of sunshine in Durham, would outrank all but three *summers* in Table 19.5.

Gordon Manley attributed the low sunshine total in 1912 to the volcanic eruption of Mount Katmai in Alaska on 6 June 1912 [19] (see also Chapter 16). It was an exceptionally dull summer, well below any other summer before or since, with only one-third of the sunshine recorded in the sunniest of summers (Table 19.5). Of the summer's 92 days, 22 remained entirely sunless, and remarkably only one day (15 July, with 11.3 hours) managed even ten hours of bright sunshine. Not surprisingly, it remains also one of the coolest (Table 19.1) and wettest (Table 19.4) summers on record.

Summer 1987 was the dullest summer in Durham since 1912 (Figure 19.2), although even so sunshine duration was 50 per cent greater than in 1912. The summer was cool, although not exceptionally so (both of the previous summers were cooler), and a little wetter than usual, but again, nothing particularly notable. The other dull summers, 1888, 1931, and 1954, were all wet and cool.

Table 19.5 shows Durham's sunniest days on the record from 1882 to 1999. Almost all are in June; proximity to the summer solstice and long summer daylight hours are clearly important, together with lower cloud amounts and often clearer air before the peak of midsummer warmth.

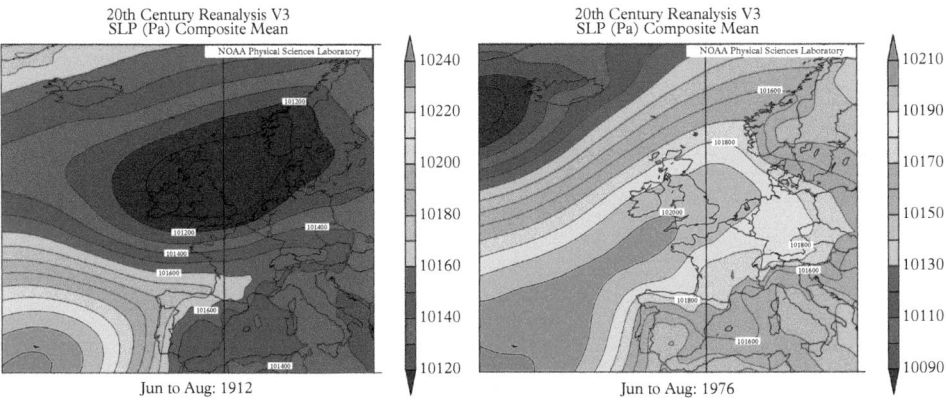

Figure 19.2 *The mean distribution of barometric pressure during the cold, dull, and wet summer of 1912 (left) and the hot, dry, and very sunny summer of 1976 (right). Isobars are in Pascals (divide by 100 for hPa = millibars), at 100 Pa intervals. (Both plots from the 20th Century Reanalysis version 3 (20CRv3) website [58])*

The finest summers

Climatologists have long been keen to summarize just how 'good' or 'bad' a particular summer season has been, typically more the perspective of the general public on holiday rather than a farmer looking for bumper crop yields. The optimum summer weather index (I) of Davis [171] is one of the most frequently quoted:

$$I = 18T_{max} + 20S_d - 0.267R + 320$$

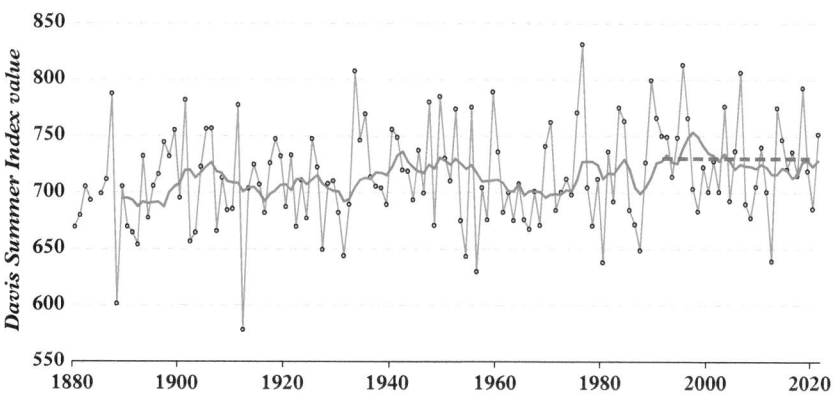

Figure 19.3 *Values of the Davis Summer Index at Durham 1880 to 2021, with a decadal running mean plotted at the year ending. The 1991–2020 average (728) is shown by the dashed line*

Table 19.6 *Best and worst summers at Durham 1880–2021, excluding 1884, using the Davis Summer Index*

Best summers		Worst summers	
Davis Index	Year	Davis Index	Year
832	1976	579	1912
814	1995	601	1888
808	1933	630	1956
807	2006	638	1981
800	1989	639	2012
Last 50 years to 2021			
832	1976	638	1981

See notes in temperature and sunshine tables for periods of records available.

where T_{max} is the mean daily maximum temperature (°C), S_d is the mean daily sunshine (hours), and R is the total rainfall (mm).

Annual values of the Davis Summer Index from 1880 to 2021 (excluding 1884, for which sunshine data are missing) are plotted in Figure 19.3. Summers gradually improved, at least in their index rating, through the twentieth century, mainly the result of improved temperatures but also more sunshine later in the century—probably as a result of changes in atmospheric circulation patterns and perhaps some reductions in air pollution too. The twenty-first century has seen several warm, dry, and sunny summers (2003, 2006, 2013, and 2018 in particular), but also the very wet summer in 2012.

Under the Davis Index classification over the period 1881–2021, Durham's worst summer was in 1912, and its best was in 1976 (Table 19.6).

Wind and weather

Snow cover 1960–99

Snow has occasionally fallen in June, and rarely even in July, but the latest date with a snow cover during the period 1960–99 was 25 April in 1981.

Fog 1962–97

Fog is infrequent in summer, at least the morning observation time of 0900 GMT, although occasional fog can be expected earlier in the morning during the summer months.

Barometric pressure (1844–1960)

The highest summer barometric pressure on record for Durham occurred on 13 June 1959, when the MSL pressure at 0900 GMT was 1040.7 hPa. The summer lowest stands at 968.5 hPa, at 0900 GMT on 28 August 1917. The highest mean MSL pressure for any summer month occurred in June 1865, when the mean was 1024.1 hPa; the lowest monthly mean was 1003.4 hPa, in August 1860.

Wind speed and direction

Compared to spring, summer sees an increase in frequency of winds from the south-west and west and a lower frequency from the north, north-east, and east. Wind speeds tend to be lower in summer with fewer instances of strong winds. Figure 19.4 shows the distribution of wind speed and direction at the 0900 GMT observation during the three summer months over the period 1961–97.

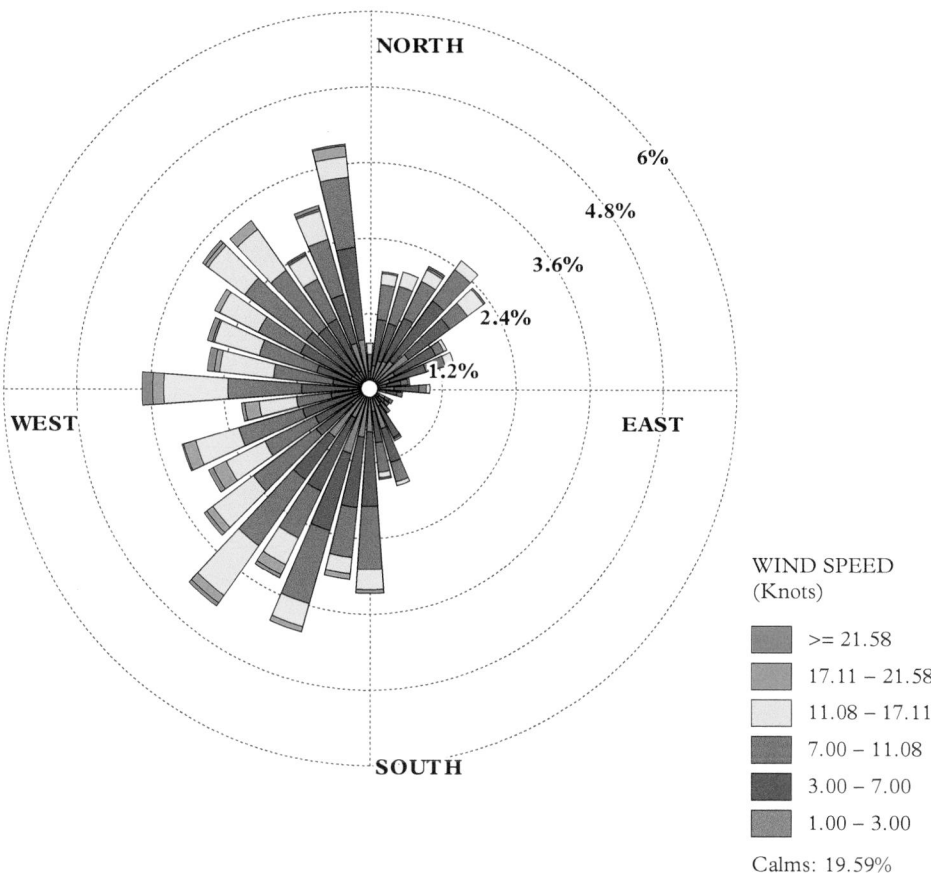

Figure 19.4 *Summer months wind rose for 0900* GMT *observations at Durham, period 1961–97*

Durham temperature, precipitation, and sunshine in graphs—summer

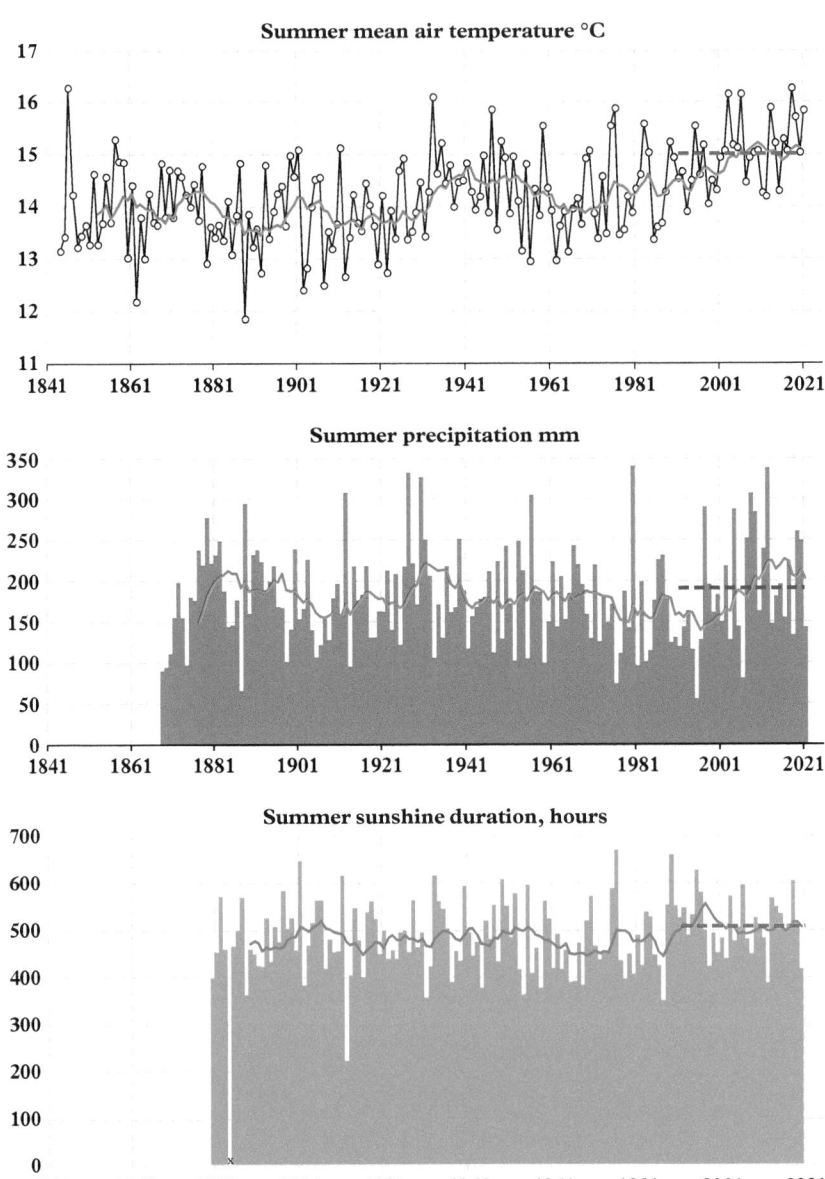

Figure 19.5 *Seasonal values of (from top) mean temperature (°C, since 1844), total precipitation (mm, since 1868), and sunshine duration (hours, since 1880, missing data 1884) for summer at Durham Observatory. The 1991–2020 averages are indicated by the thick blue dashed line, while the ten-year running mean ending at the year shown is indicated by the red line*

20

Autumn

September, October and November

Just as spring, autumn is a season of transition. It often starts with hot, summery days but can end with snow; strong winds and rainstorms are likely at some stage in between. But there can be calm conditions and fog, too, redolent of John Keats' 'season of mists and mellow fruitfulness'. Temperatures fall more steeply in autumn than they rise in spring (Figure 3.1); mean daily maximum temperatures in Durham fall from a little over 19 °C at the beginning of September to below 8 °C by the end of November, while the change in mean daily minimum temperatures is somewhat less, from 11 °C to 2 °C over the same period. The overall seasonal mean is exactly 10 °C in Durham. Autumns have become warmer in recent decades—only spring has shown a greater rate of warming over the last 170 years.

Rainfall amounts and frequency increase throughout the autumn months: November is, on average, the wettest month of the year, and autumn the wettest of the four seasons, with seasonal average (1991–2020 period) precipitation total of 194 mm, falling on 52 rain days: only summer has a higher average amount of rainfall per rain day. Sunshine amounts also decrease quickly at this time of year through a combination of increased cloud amounts and shorter days (Figure 3.3); the daily expectation reduces from almost five hours per day in early September (about 35 per cent of the possible duration) to fewer than two hours per day by the last week of November (25 per cent of possible). Autumn's average for the season is 299 hours.

Temperature

Table 20.1 summarizes the warmest and coldest autumn nights on Durham's records. It should not come as any surprise that all of the highest temperatures occur in September, and all of the lowest in November, although the minimum temperature of 15.9 °C on 1 October 1985 only just escapes inclusion in the 'warmest nights' column. The overall range in temperature during autumn, 42.0 degrees Celsius, is second only to spring (44.0 degrees).

Figure 20.1 *Autumn colours along the Elvet Waterside stretch of the River Wear during autumn 2003 (Tim Burt)*

The fine summer of 1906 ended with a flourish in the form of a remarkably late intense heatwave at the end of August. An anticyclone centred over central Europe on 30th and 31st moved slowly northwards to become centred over Sweden by 2 September. A depression lay off north-west Scotland, and between the two the British Isles lay in a broad southerly flow, although winds fell light and variable on the 2nd and 3rd. Temperatures reached or exceeded 32 °C widely on four consecutive days from 31 August, including 35.6 °C at Bawtry, South Yorkshire on the 2nd (this remains the highest September temperature on record in the United Kingdom) [87]. Durham reached 30.0 °C on 1 September, the Observatory's autumn record, although the following day was only a little cooler at 29.3 °C (Table 20.1). The *Monthly Weather Report* noted that the hot spell passed away as quickly as it appeared, and the maximum at Durham on 4 September was 'only' 18.7 °C, down 8 degrees on the previous day but otherwise a very respectable September day in most circumstances.

No autumn day since has surpassed 28 °C, the most recent very warm autumn day being 8 September 2021 (maximum temperature 28.0 °C), in a remarkably warm September—the second warmest on record (Table 20.2). Very mild nights occurred early in September 2016 (Table 20.1) including the equal warmest autumn night on 7 September that year at 17.2 °C.

Late October can produce hard frosts and very cold days, and minima below −5° C have been recorded towards the end of the month, but the earliest date in the autumn to

Table 20.1 *Highest and lowest maximum and minimum temperatures at Durham Observatory in autumn, 1843–2021*

Rank	Warmest days	Coldest days	Warmest nights	Coldest nights
1	30.0 °C, 1 Sept 1906	−1.4 °C, 23 Nov 1853	17.2 °C, 12 Sept 1945, 5 Sept 1949, and 7 Sept 2016	−12.0 °C, 24 Nov 1993
2	29.3 °C, 2 Sept 1906	−1.1 °C, 12 Nov 1919 and 28 Nov 1969	16.4 °C, 3 Sept 1898	−9.8 °C, 21 Nov 1880
3	28.0 °C, 8 Sept 2021	−0.8 °C, 29 Nov 1912	16.3 °C, 2 Sept 1999	−9.3 °C, 25 Nov 1993
4	27.5 °C, 4 Sept 1898	−0.6 °C, 29 Nov 1856, and 29 Nov 1952	16.2 °C, 8 Sept 2016	−8.9 °C, 16 Nov 1919
5	27.3 °C, 13 Sept 2016	−0.5 °C, 27 Nov 1874	16.1 °C, 5 Sept 1951	−8.7 °C, 28 Nov 2010
Last 50 years to 2021				
	28.0 °C, 8 Sept 2021	0.2 °C, 27 Nov 2010	17.2 °C, 7 Sept 2016	−12.0 °C, 24 Nov 1993

Temperatures were recorded in different thermometer screens during the period of record; see Notes on page xvii and Appendix 3 for more details.

remain below 0° C throughout the day was 12 November in 1919, when the maximum temperature reached only −1.1 °C, following a minimum temperature overnight of −6.7 °C. The lowest air temperature yet recorded in any autumn was −12.0 °C on 24 November 1993, the coldest November for over 70 years and the coldest month of the calendar year, a very unusual position usually held by one of the winter months. Snow fell several times during the last ten days of November 1993, reaching a maximum depth of 8 cm on the 23rd. November 2010 saw a very wide range in temperatures, from a maximum temperature of 16.1 °C on the 4th to the month's minimum of −8.7 °C on the 28th, following a maximum temperature of just 0.2 °C the previous afternoon, the coldest autumn day within the last 50 years.

All of the warmest autumns (Table 20.2) are recent, within the last 50 years, and all but one in the present century; 1978 was the mildest autumn on record at the time, just surpassing 1949 by less than one tenth of a degree, but four autumns have been warmer since, with autumn 2006 more than a degree warmer than autumn 1978. Autumn has shown the second-highest rate of warming of any season since 1900, just marginally less than spring, at 1.25 degrees Celsius per century.

Table 20.2 *Autumn mean air temperatures at Durham Observatory, 1843–2021*

Autumn mean temperature 10.0 °C (average 1991–2020)

Warmest autumns			Coldest autumns		
Mean temperature, °C	Departure from 1991–2020 normal degC	Year	Mean temperature, °C	Departure from 1991–2020 normal degC	Year
11.9	+1.9	2006	7.2	−2.8	1887
11.6	+1.6	2021	7.3	−2.7	1919
11.4	+1.4	2011	7.4	−2.6	1896, 1993
11.1	+1.1	2001	7.5	−2.5	1915
11.0	+1.0	2014	7.6	−2.4	1886, 1905, 1952
Last 50 years to 2021					
11.9	+1.9	2006	7.4	−2.6	1993

Autumn 2006 was the warmest on record, at almost two degrees above normal. All three months were exceptional: easily the warmest September on record (the first time since 1956 when September was warmer than August, and only the third time in the previous 100 years), followed by the fourth warmest October and the seventh mildest November. Both mean maximum and mean minimum temperatures established new records for September, with the mean maximum exceeding 20 °C for the first (and so far, only) time. In many ways the most remarkable aspect of temperatures in October was the small range, with neither very high nor very low temperatures recorded. The relatively constant, mild weather was reflected in the fact that both rainfall and sunshine duration were below normal, thanks to a mild maritime air flow, without the extremes associated with high-pressure systems. November was another mild month, sunnier and slightly drier than normal.

Autumn 2021 became Durham's second-warmest, with a mean temperature 1.6 degrees above normal. All three months were warm, particularly September, which featured Durham's warmest autumn day since 1906 with a maximum temperature of 28.0 °C on 8th.

Autumn 2011's most remarkable feature was the hot spell at the end of September and into early October, described earlier in the September and October chapters; the last three days of September 2011 all exceeded 24 °C, and the warm weather continued into October, when on 1 October the maximum temperature of 25.3 °C established a new record high for the month. Temperatures remained well above normal into November, which became the mildest on Durham's records. Only two air frosts were recorded in autumn 2011, both in early November. Almost inevitably, the warmest autumns benefit unduly from a mild November at the end of the season.

Aside from 1952 and 1993, all of the coldest autumns listed in Table 20.2 date from at least 100 years ago. The cold weather in November 1993 has already been described in Chapter 14, but September and October were also cool. Ironically, 1 September that year was the warmest day of the year, at 25.1 °C: until that date there had been only six previous years since 1844 when autumn (September) produced the warmest day of the year (1880, 1891, 1898, 1906, 1954, and 1956), although this has since occurred in 2004. Thereafter, temperatures declined steadily: September's mean air temperature was 1.8 degrees colder than normal, October 2.4 degrees below, and November 3.3 degrees below average.

Durham's coldest autumn on record was in 1887, almost three degrees below the current mean temperature. This must have come as something of a shock to the system after the warm, dry, and sunny summer of that year, one of the finest summers of the late nineteenth century. September was disappointingly cool, dull, and wet, and both October and November were cold; there were 17 air frosts, the earliest on 28 September.

Precipitation

Table 20.3 lists the wettest and driest autumns on Durham's record since 1868, together with the wettest autumn days. Curiously, the driest and wettest autumns in the near 150-year record occurred only five years apart, in 1971 and 1976, respectively.

The two driest autumns on Durham's rainfall record occurred in the consecutive years 1971 and 1972. September 1971 was the driest September in Durham since

Table 20.3 *Autumn precipitation at Durham Observatory 1868–2021*

Autumn mean precipitation 194 mm (average 1991–2020)

Wettest autumns			Driest autumns			Wettest days	
Total fall, mm	Per cent of normal	Year	Total fall, mm	Per cent of normal	Year	Daily fall, mm	Date
358.6	183	1976	81.4	42	1971	87.8	11 Sept 1976
351.8	180	2000	88.4	45	1972	72.8	12 Nov 1901
340.1	174	1944	89.1	46	1904	63.7	8 Oct 1903
330.2	169	1935	90.2	46	1958	57.4	24 Sept 2012
324.6	167	2012	95.0	49	1989	55.8	21 Sept 1935
Last 50 years to 2021							
358.6	183	1976	81.4	42	1971	87.8	11 Sept 1976

1910, although the amount was equalled in 1941: anticyclones or ridges of high pressure dominated the circulation until very near the end of the month. October and November were also dry in 1971. Autumn 1971 marks the start of the driest 60-month period on Durham's records, ending in August 1976; 2564 mm of rain fell in these 60 months—an average of just 513 mm/year. Autumn 1972 was only slightly wetter than its predecessor; this time October was the most notable month, the third driest on record with just 12 mm, 19 per cent of the normal amount.

The famous *summer* of 1976 was Durham's driest since 1887, with only 73 mm falling on just 17 rain days in the three months ending August. What a change during the *autumn* of 1976: 359 mm fell in 56 rain days (27 in October alone), Durham's wettest autumn on record (Table 20.3), and in fact Durham's wettest-ever season of any name. Autumn 1976 includes Durham's wettest-ever day, 11 September, when 87.8 mm was recorded (see also Chapter 12). This was associated with the passage of a very active depression, which had also provided 32.3 mm the day before, adding up to the largest two-day total on record and contributing substantially to the wettest September on record (193 mm), more than three times the average amount. October 1976 was also very wet (143 mm, the equal fourth-wettest on record); both months were dominated by a succession of deep depressions. November 1976 was a much quieter month, although depressions continued to pass across southern Britain; Durham was considerably drier than normal, with only 23 mm in all. There is more on the prolonged drought of 1976, and its rapid recovery, in Chapter 22.

Autumn 2000 was almost as wet as 1976, and the season's total of 352 mm represents the second-wettest season of any name in Durham's records. September and October were both very wet, while November (148 mm) was, at that date, the third wettest on record (Figure 20.2). The wet autumn of 2000 contributed substantially to that year's ranking, at that time Durham's wettest year since 1872, with an annual fall of 887 mm—since exceeded by 2012 (Table 16.9). In complete contrast, 1976's calendar-year rainfall total was only just above average (683 mm), comprising 273 mm for the first eight months (only 64 per cent of the normal fall for this period) followed by 410 mm for the remaining four months (161 per cent of normal).

Sunshine

In the standard averaging period 1991–2020, autumn received an average of 299 hours of bright sunshine, equivalent to 3.3 hours per day. Of course, the sunniest days are in September, when day length is longer than in the other two autumn months.

Not surprisingly, given how wet it was, autumn 1976 became the dullest autumn on record at Durham, with just less than 20 per cent of the possible duration (Table 20.4). September was both the wettest and dullest on record; remarkably, September's sunshine duration (59.4 hours, less than half normal and less than a normal January) was the lowest of the three autumn months (October's total was 70.2 hours, and November 61.2 hours). Twenty-nine days remained sunless that autumn.

Figure 20.2 *Double rainbow over the College of St Hild and St Bede, 28 November 2004 (Tim Burt)*

Autumn 1960, the ninth-wettest autumn, was also a very dull season (194 hours of bright sunshine), with only autumn 1976 less sunny; 33 days that autumn remained sunless, more than one in three. October was both the dullest October yet recorded (with only 41 hours of bright sunshine, less than a normal December) and the second-wettest (Table 13.3).

More recently, autumn 1993 was another very dull season, and very cool, too—equal third-coolest on Durham's records (Table 20.2). September and October could both be described as cool, wet, and dull, while November was dull, very cold, and wintry; snowfalls dominated the last third of the month.

Durham's sunniest autumn came in 1986, getting off to a good start with the sunniest September for over a century with 193 hours of bright sunshine, a cool and dry month dominated by anticyclonic conditions in which every day saw some sunshine—three days with ten hours or more. The *Monthly Weather Report* noted that the UK's highest temperature of the month (25 °C) was recorded at Newcastle on the last day of the month; the maximum temperature at Durham that day was 23.5 °C, the warmest final day of September for almost 80 years. October was also sunny and remains Durham's third-sunniest on record with 139 hours of bright sunshine: six days saw more than eight hours of sunshine; November was also sunny, at that date the fifth-sunniest on record with just under 90 hours, although four Novembers since have been sunnier.

The next-sunniest autumn was in 1981: September and November 1981 were both sunny, but October became the chief reason for such a sunny autumn, for, with 146 hours of sunshine, it became the sunniest October in Durham since 1893 and remains the second sunniest on record (Table 13.4).

Autumn 1895 remained Durham's sunniest autumn for over 80 years, until being slightly surpassed by 1981. The autumn was sunny mainly by reason of a very sunny September, which, with 204 hours of bright sunshine (greater than the average for any summer month), remains the sunniest September on record (Table 12.4): 14 days received more than eight hours of bright sunshine, and only one day remained sunless. October 1895 was also sunny, but November was by contrast a dull month, with just 46 hours in all.

The most recent sunny autumn in Table 20.4 is that of 2003, a dry and fine autumn following the hot summer of that year. All three months were sunny, albeit not exceptionally so, while October was on the cool side.

Table 20.4 *Autumn sunshine duration at Durham Observatory. Extremes 1880 to 2021, sunniest days 1882–1999 only*

Autumn mean sunshine duration 299 hours, 3.30 hours per day (average 1991–2020)
Possible day length: 960 hours. Mean sunshine duration as percentage of possible: 31.1

Sunniest autumns			Dullest autumns			Sunniest days 1882–1999	
Duration, hours	Per cent of possible	Year	Duration, hours	Per cent of possible	Year	Duration, hours	Date
420.8	43.8	1986	190.8	19.9	1976	12.7	2 Sept 1994
369.8	38.5	1981	194.0	20.2	1960	12.6	1 Sept 1951
361.7	37.7	1895	196.1	20.4	1946	12.1	1 Sept 1906, 3 Sept 1941, and 7 Sept 1991
359	37.4	2003	198.9	20.7	1993	12.0	2 Sept 1906
355.6	37.0	1947	200.8	20.9	1894	11.9	2 Sept 1987
Last 50 years to 2021							
420.8	43.8	1986	190.8	19.9	1976	12.7	2 Sept 1994

Daily sunshine records ceased at the Observatory on 31 December 1999: monthly sunshine totals from January 2000 onwards are estimated (to the nearest hour) by regression from the Met Office East and North-east England sunshine series. See Appendix 6 for record metadata and missing months.

Wind and weather

Snow cover 1960–99

Snow has occasionally fallen in September, but the earliest date with a snow cover during the period 1960–99 was 14 November in 1965, when snow lay 5 cm deep at the 0900 GMT observation. A morning snow cover can be expected on one day in November.

Fog 1962–97

Fog becomes more obvious during the autumn and can be expected on seven or eight mornings in a typical season; October is the foggiest month of the year, averaging 3.4 mornings with fog over the period 1962–97. Qualitative impressions in Durham more recently suggest that air quality can sometimes be poor due to fine particulates from vehicular traffic but, sadly, evidence of something like PM10 concentration together with visual observations of visibility are these days lacking in Durham.

Barometric pressure (1844–1960)

The highest autumn barometric pressure on record for Durham occurred on 15 November 1922, when the MSL pressure at 0900 GMT reached 1042.9 hPa. The autumn lowest stands at 958.5 hPa, at 0900 GMT on 27 October 1959. The highest mean MSL pressure for any autumn month occurred in November 1867, when the mean was 1026.2 hPa; the lowest monthly mean was 999.4 hPa, in November 1877.

Wind speed and direction

Autumn sees an increase in frequency of winds from the south-west and west and a reduced frequency from the north, north-east, and east. Wind speeds increase in autumn, and strong winds can be expected occasionally almost every year. A gust of 80 knots was recorded at 1325 GMT on 3 November 1970, at that time the highest gust on the Observatory's anemograph record, which commenced in 1937. For the hour ended 1400 GMT the mean wind speed was westerly at 47 knots. Figure 20.3 shows the distribution of wind speed and direction at the 0900 GMT observation during the three autumn months over the period 1961–97.

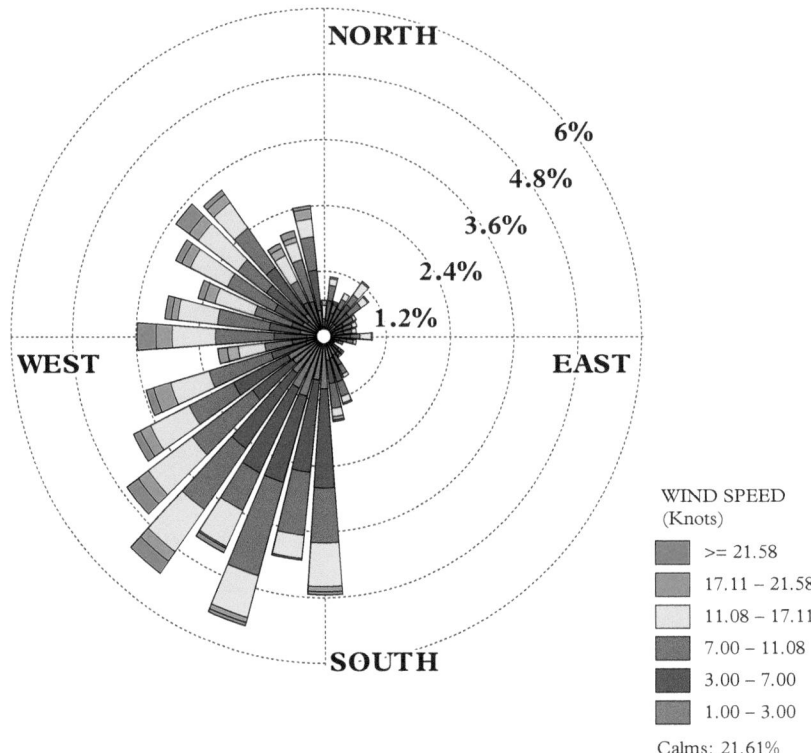

Figure 20.3 *Autumn months wind rose for 0900* GMT *observations at Durham, period 1961–97*

Figure 20.4 *One of the trees blown down in Durham's Botanic Garden by storm Arwen on 26-27 November 2021 (Courtesy of Michele Allan)*

Durham temperature, precipitation, and sunshine in graphs—autumn

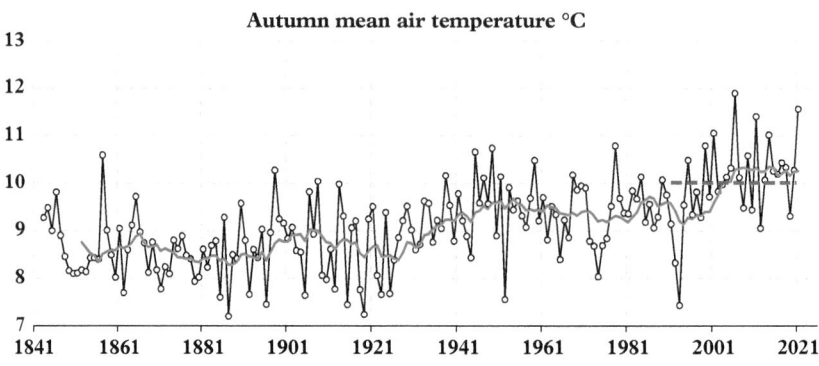

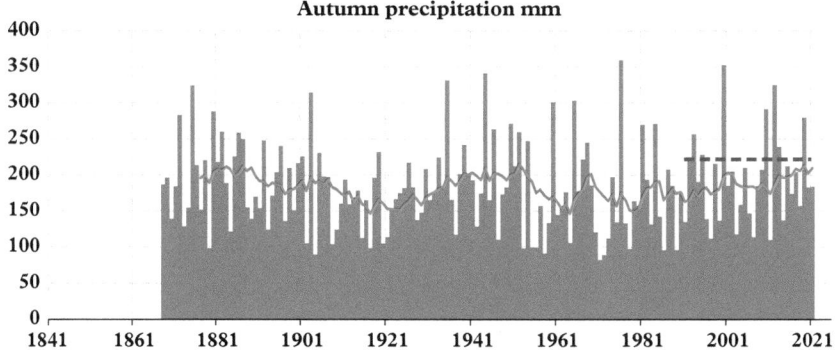

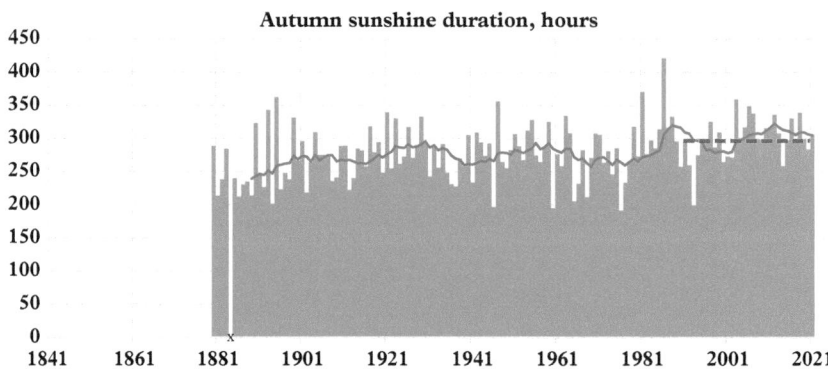

Figure 20.5 *Seasonal values of (from top) mean temperature (° C, since 1843), total precipitation (mm, since 1868), and sunshine duration (hours, since 1880, missing data 1884) for autumn at Durham Observatory. The 1991–2020 averages are indicated by the thick blue dashed line, while the ten-year running mean ending at the year shown is indicated by the red line*

Part 4

Long-term climate change in Durham

21

Climate change in Durham

The Intergovernmental Panel on Climate Change (IPCC) in its 2018 report [89] concluded with high confidence that human-induced global warming had reached approximately 1 degree Celsius (*likely* between 0.8 and 1.2 degrees Celsius) above pre-industrial levels by 2017, increasing at 0.2 degrees Celsius (*likely* between 0.1 and 0.3 degrees Celsius) per decade. It also concluded with high confidence that global warming is likely to reach 1.5 degrees Celsius between 2030 and 2052 if it continues to increase at the current rate, due to past and ongoing emissions of greenhouse gases. Reflecting the long-term warming trend since pre-industrial times, observed global mean surface temperature (GMST) for the decade 2006–2015 was 0.87 degrees Celsius (*likely* between 0.75 and 0.99 degrees Celsius) higher than the average over the 1850–1900 period (*very high confidence*). Estimated anthropogenic global warming matches the level of observed warming to within ±20 per cent (*likely* range; IPCC, 2018). This is the context within which we analyse the Durham Observatory temperature series from 1843, a series meticulously assembled by Gordon Manley from 1850 and continued to this day (and extended back to 1784 by Matthew Eglise [45] following Manley's provisional reconstruction from 1801). Of course, there are other aspects of climate change we can consider at Durham, most especially the precipitation and sunshine series: other data series are less satisfactory for long-term analysis, given much shorter periods of observation and changing methodology.

Temperature

Mean air temperature 1843 to present

Table 21.1 shows decadal mean air temperature (°C) at the Durham Observatory 1851–2020; the 1850s is the first full decade in the period of instrumental observation (from 1843; see Chapter 2). The 1880s was the coldest decade on Durham's records. While that decade contains neither 1879 nor 1892 (the two coldest years on Durham's instrumental record; see Table 16.2) it does include several other years—such as 1881, 1886, and 1888—which were almost as cold: 1888 was the coldest summer on the record (see Chapter 19). Significant warming with respect to the 1851–60 mean did not take place

Table 21.1 *Decadal mean air temperature (°C) at the Durham Observatory 1784–2020 in date order*

Decade	Mean air temperature (°C)	Difference from 1851–1860, degC	Decade	Mean air temperature (°C)	Difference from 1851–1860, degC
1784–90	*7.58*	*−0.55*	1901–10	8.12	−0.01
1791–1800	*7.90*	*−0.23*	1911–20	8.24	+0.11
1801–10	*8.15*	*+0.02*	1921–30	8.35	+0.22
1811–20	*8.24*	*+0.11*	1931–40	8.74	+0.61
1821–30	*8.79*	*+0.66*	1941–50	8.84	+0.71
1831–40	*8.22*	*+0.09*	1951–60	8.62	+0.49
1841–50	*8.17*	*+0.04*	1961–70	8.34	+0.21
1851–60	8.13		1971–80	8.62	+0.48
1861–70	8.40	+0.27	1981–90	8.85	+0.72
1871–80	8.11	−0.02	1991–2000	9.05	+0.92
1881–90	7.91	−0.22	2001–10	9.57	+1.44
1891–1900	8.22	+0.09	2011–20	9.76	+1.62

All differences are relative to the 1851–1860 decade. Pre-1844 data from Eglise [45] in italics.

until the twentieth century with a warm period from the 1930s to the 1950s, followed by a cooler decade in the 1960s. Thereafter, warming has accelerated since the turn of the millennium.

Figure 21.1 shows annual mean air temperature from 1844 (the first complete year of observations) together with annual mean maximum and annual mean minimum temperatures. All three plots have an 11-year unweighted running mean fitted, centred on the year plotted, to smooth out the noisy annual record. There is modest variability through to the late 1920s after which there is a warmer period peaking in 1949, a very warm year, followed by progressive cooling through to the mid-1960s. The remainder of the twentieth century was marked by substantial, rapid warming, which has continued and accelerated through to today. Maximum and minimum temperatures show a slightly different pattern in recent decades: mean maximum temperature show a steep rate of warming from the late 1980s, whereas mean minimum temperatures showed their fastest rate of warming about ten years later. Note that Figure 16.3 shows annual mean air temperatures as a 'climate stripe', where every year's mean is plotted as a thin vertical stripe; the stripe's colour palette represents grades from cold (blue) to warm (red): the recent warming is immediately obvious.

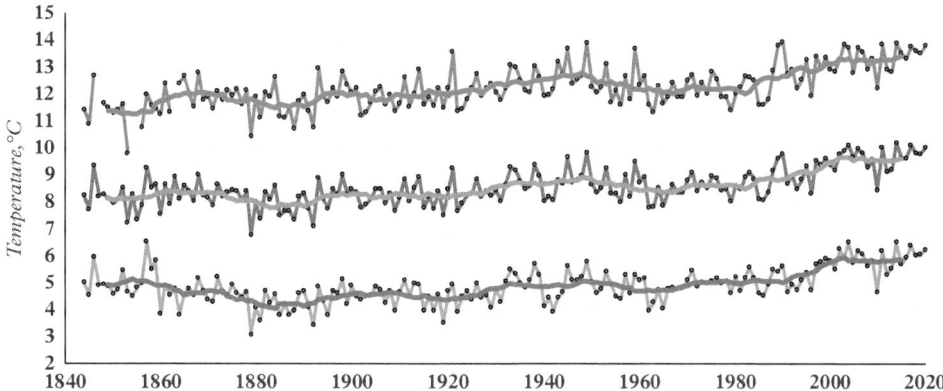

Figure 21.1 *Annual mean maximum (red line), minimum (blue line) and mean temperature (green line), in °C, at Durham, 1844–2020, together with an eleven-year unweighted running mean for each, centred on the year plotted.*

Figure 21.2 shows the annual mean air temperature record in a different way. First, the data are normalized relative to the mean of 1851–1900; this mirrors the approach of Hawkins et al. [90]. Then a 30-year running mean is fitted but this time, each mean is plotted at the *end* of the period, rather than centred; this allows mean air temperature for any 30-year period to be easily identified from its end point. In the recent period of warming, the running mean reaches one standard deviation above the 1851–1900 mean in 1953 and again in 1995, and two standard deviations above the mean in 2011. Adopting the language of Frame and colleagues [91] and Hawkins and colleagues [90] to describe how the climate has changed from being familiar, to being 'unusual', 'unknown', or even 'inconceivable' relative to lived experience, we can assess Durham's climate in the twenty-first century relative to the second half of the nineteenth century. Note that Figure 21.2 shows eleven years in the twenty-first century with standard deviations greater than +3 above the 1851–1900 mean. If we assume the annual mean air temperatures follow a Gaussian distribution and that each annual mean is independent, then the probability of any *one* year exceeding 3 standard deviations is 0.00135, approximately once in 750 years. We can therefore calculate the chances of having *eleven* years exceed 3 standard deviations in a period of 20 years: this is the infinitesimally tiny probability of 1×10^{-27}. It is clear that these repeated extreme deviations of twenty-first century annual mean air temperatures indicate that a significant regime change has occurred. We can conclude that Durham's current temperature regime would have been *inconceivable* in the second half of the nineteenth century, incontrovertible evidence of more recent global warming.

We have already made some reference in this chapter's introduction to the causes of global warming, but a little more detail is called for here. Based on a large amount of evidence, Wuebbles and colleagues [92] conclude that it is extremely likely that human activities, emissions of greenhouse gases in particular, are the dominant cause of

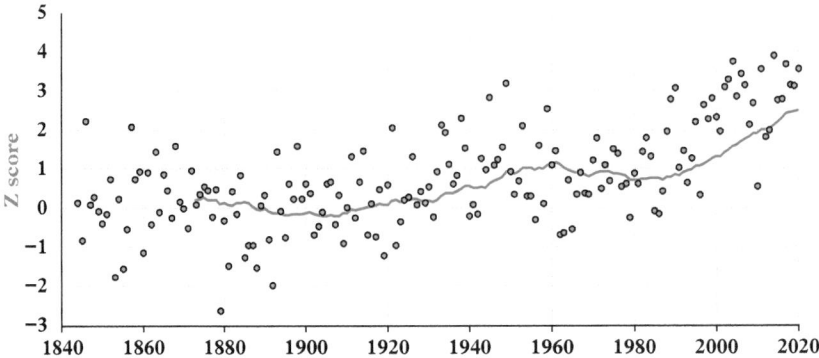

Figure 21.2 *Annual mean air temperature (°C) at Durham since 1844 with an unweighted 30-year running mean plotted at the end of each 30-year period. The temperature scale is plotted as a z-score based on the mean (8.13°C) and standard deviation (0.53 degC) of the 1851–1900 data.*

the observed warming since the mid-twentieth century—there is no convincing alternative explanation. They argue that radiative forcing due to human activities has been evident since about 1870 and has grown at an accelerating rate since about 1970. The resultant increase in global mean surface temperature accords well with what has been observed at Durham, with a clear increase in the rate of increase of mean air temperature from the 1970s onwards (Table 21.1 and Figure 21.1). At the same time, there is high confidence that contributions of natural radiative forcing (the impact of volcanic eruptions, variations in solar radiation) and the impact of atmosphere–ocean interactions such as El Niño are minor over timescales of several decades. However, it must be remembered that Durham does show a very small urban heat island effect, but probably only accounting for the equivalent of about 0.2 degrees Celsius of warming since the 1860s (see Chapter 2).

Mean air temperature since 1784

Table 21.1 includes data from the Eglise extension [45] back to 1784 (see Appendix 3). It is clear that in Durham the first half of the nineteenth century was a little warmer than the 1850s, even though there were individual cold years in the late 1830s and early 1840s. This period includes the very warm years of 1826, 1828, and 1846 (see notes on June 1846 in Chapter 21). It also includes several very cold years, including 1816, 'the year with no summer', and winter 1795, an extremely cold season that was Manley's target when he started his extension of the 'Durham' record back from 1850 during his retirement [28]. The end of the eighteenth century is the coldest period in this extended record. The NASA Earth Observatory glossary [93] notes three minima during the Little Ice Age: approximately 1650, 1770, and 1850. The second of these cold periods coincides approximately with the start of the Eglise extension in 1784 and the last with the start of the observed temperature record in 1843. A decadal running mean for the Central England Temperature (CET) series record shows a minimum in 1787; 1784 was the equal ninth-coldest year in CET (mean 7.9°C), together with 1799, 1816, and 1860. The last of these ranks ninth-coldest year in the Durham instrumental record from 1844.

Figures 16.1 and 16.2 show the annual extremes in air temperature at Durham (the lowest minimum and highest maximum respectively) by year since 1844. The plots do not show quite the same trends as seen for mean maximum and mean minimum (Figure 21.1) although the increases seen in recent decades are much the same. The decadal running mean of *annual maximum temperature* (Figure 16.2) first reaches a peak of 28.6°C in the ten years ending 1877, before falling away to a decadal minimum of 25.2°C ending in 1891. Thereafter there is an irregular increase until 28.6°C is reached once more in the decade ending 1949/50, before a sharp fall during the 1960s to a minimum of 25.6°C in the ten years ending 1971 and 1972. Since then there has been a further irregular rise almost reaching the late 1940s peak, namely 28.3°C for the ten years ending 2020 (this decade including two consecutive years reaching 30°C in 2019 and 2020). It is not thought that the changes within the last 90 years or more are influenced by any change in instruments: Gordon Manley added a new Stevenson screen in 1934, with new thermometers, but this predates the 1940s peak (see Appendix 3).

The decadal running mean plot for *annual minimum temperatures* (Figure 16.1) remains between about −9°C and −11°C until the beginning of the twentieth century, with decadal mean annual minima of −11.5°C in the ten years ending 1882 and −11.4°C ending 1895. Thereafter, there is a rise to almost −7°C during the 1930s followed by a decline into the 1940s, falling below −10°C once more before an irregular rise to the 1970s and a strong continuing upward trend since then, to a mean for the decade ending 2020 of −5.4°C, the highest in the entire series. The average annual minimum temperature at Durham Observatory for the 50 years ending 1900 was −9.9°C. At the time of writing in 2021, no minimum of −11°C or lower has been observed since November 1993, while only one minimum below −8°C has been recorded since December 2010, namely −9.4°C on 12 February 2021.

Other than a few years either side of 1940, the annual temperature range (Figure 21.3) shows little change since the early years of the twentieth century, after making allowances for instrumental changes (described more fully in Appendix 3), although obviously the year-to-year variations are considerable. Clearly, an above-average range in temperature in any year can only occur when the lowest minimum is below normal and/or the highest maximum is above normal, and the greatest annual ranges occur when both occur, such as in 1940 when the annual range reached 46 degrees (coldest night −16.1°C on 21 January, hottest day 30.0°C on 8 June). Taken together with the points considered here, there is clear evidence that very low minima have become much less common whereas very high maxima have become slightly more common, albeit still relatively unusual. There is a slight downward trend in annual temperature range, also evident at Oxford ([49], Figure 24.2), which can be understood partly as a result of the decreasing frequency of cold winter nights, offset to some extent by an increase in high summer temperatures. This downward trend is, however, a very gentle one, for within the last

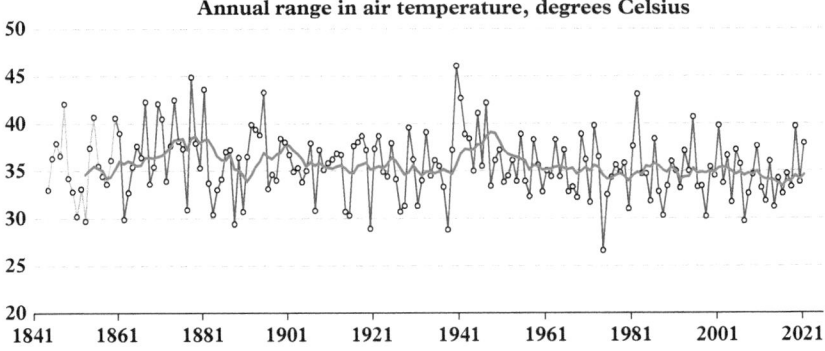

Figure 21.3 *Annual temperature range (highest daily maximum temperature minus lowest daily minimum temperature, in degrees Celsius) at Durham since 1844, with a superimposed ten-year running mean plotted at year ending. Data prior to 1860 are from the uncorrected North Wall exposures and are shown in pale grey; data during 1860–1899 are corrected to Stevenson screen equivalents from Glaisher Stand records; since 1900 records are from readings made in Stevenson screens—for more details on the observational record and exposure corrections applied refer to Appendix 3.*

60 years or so the decadal mean range has varied only from 35.7 degrees Celsius for the ten years ending 1986, to 33.3 degrees ending 2016, although the latter is the lowest on the entire record. Changes in observing practices may account for some of this reduction; minimum temperatures were recorded as 2100–2100 GMT until December 1960, and of course the automation of the site in late 1999 may also have affected the record slightly (see Appendix 2 and 3).

Figure 3.1 shows the annual temperature cycle for two different long-period averages: 1871–1900 and 1991–2020, clearly illustrating how the pattern of warming has changed across the year.

Extending the Durham mean air temperature record back to 1659

The correlation between annual mean air temperatures for CET and Durham is very strong: based upon 100 years data 1900–1999, $R^2 = 0.9378$ and the linear regression equation is

$$Durham\,°C = 0.9464 \star CET\,°C - 0.4245$$

with standard error = 0.035. On this basis it is possible to hindcast Durham temperatures between 1659 and 1784 (when the 'Durham Eglise' extension begins, as explained in Chapter 2), assuming that the relative difference in temperature between central and north-east England was the same in the seventeenth and eighteenth centuries as in the twentieth century. Table 21.2 shows long-period averages from 1659 to present. Durham's faltering emergence from the Little Ice Age is clear, with warmer periods in the first halves of the eighteenth and nineteenth centuries followed by cooler half-centuries.

Table 21.2 *Long-period (mostly half-century) means of air temperature at Durham from 1659, in °C, including estimates from CET to 1784, the Eglise series from 1784 and the instrumental record from 1844—see text for explanation*

Period	Mean air temperature (°C)
1659–1700	7.8
1701–50	8.3
1751–1800	8.1
1801–50	8.3
1851–1900	8.1
1901–50	8.4
1951–2000	8.7
2001–20	9.6

This is not repeated in the twentieth century, however, with global warming in the second half of the twentieth century accelerating into the first part of the twenty-first century. The two coldest years on the CET series are 1695 and 1740; we lack documentary evidence about the weather in Durham in 1695 and only have scant information about 1740 in north-east England (see Chapter 22), so we can only speculate that these years were as bitterly cold in Durham as more widely across Central England. From the period of instrumental record, only 1879 (Durham's coldest year on record, which ranks third coldest in CET) comes close to being so cold (again, see Chapter 22); in CET the mean air temperature in 1879 is 7.4°C compared to Durham's mean 6.8°C. For 1740, the estimated mean air temperature in Durham is 6.0°C (CET 6.8°C) and for 1695 the estimated mean air temperature for Durham is 6.5°C (CET 7.3°C).

Temperature-related indices

For gardeners, the incidence of frost is always a concern, especially during the growing season. For those going to work, by car, bike, or on foot, icy conditions are also worrying. Figure 21.4 shows the annual incidence of air frosts since 1844 and ground frosts since 1875 (the air frost plot also appears alone as Figure 16.5). Both show a high frequency of frost in the last 25 years of the nineteenth century, with a secondary peak in the 1910s and into the 1920s. After a decline in the 1930s, both showed an irregular increase up to the 1980s, with a steady decline since then. The annual total of air frosts averaged 68 in the 1880s but had fallen to 38 by the decade ending in 2020. Variations in ground frost frequency are broadly similar, with a peak of 145 per annum in the 1890s, down to 80 in the decade ending 2020, but it should be noted that episodes of cold springs or cold autumns, for example in 1917 and 1919, and more recently 2021, produce sharply higher ground frost counts without necessarily being cold enough for air frosts to occur.

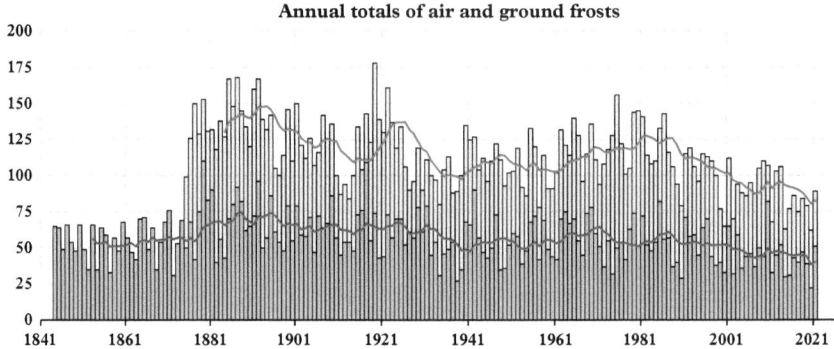

Figure 21.4 *Annual frequency of air frosts (blue columns) and ground frosts (green columns) at Durham Observatory 1844–2021, with ten-year running means shown at year ending. See Appendix 3 for variations in instrumentation during the period of this time series, in particular the introduction of the AWS in autumn 1999 and the use of artificial turf from July 2015, which will particularly affect the homogeneity of the ground frost record*

However, the introduction of the automatic weather station in autumn 1999, and in particular the installation of the grass minimum sensor above artificial turf in July 2015, have significantly affected the homogeneity of this record, so that the annual number of ground frosts has fallen a little more quickly than for air frosts during the present century.

The number of ice days (daily maxima below 0°C) shows a similar pattern: Table 21.3 shows the total number of ice days per decade since 1901; 1901–1910 is the first full decade after the installation of the Stevenson screen. Through to the 1960s, the number

Table 21.3 *Total number of ice days at Durham* per decade *since 1901*

Decade	Total ice days	Decade	Total ice days
1901–10	25	1961–70	31
1911–20	23	1971–80	13
1921–30	24	1981–90	27
1931–40	26	1991–2000	7
1941–50	49	2001–10	8
1951–60	31	2011–20	7

averaged about two or three per year, but this has fallen remarkably in the last three decades to less than one per year on average; half the years since 2000 have had none at all, while even the relatively colder winter of 2021 saw only one such day.

Just as gardeners are concerned about frost, so they take a keen interest in the length of the growing season. The Growing Day-Degree (GDD) index can be used to predict when a crop will reach maturity. Each day's index value is calculated by subtracting a threshold value from the daily mean air temperature, negative values being ignored. GDDs are then summed to provide a measure of the accumulated warmth of the growing season. The record can also be used to identify the start and finish date of the growing season in each year, that is, the period when mean air temperature remains above the threshold. In the UK, interest has traditionally been in the growing season for grass, for which a threshold value of 5.5 °C is used (GDD5.5). Figure 21.5 shows that annual totals of GDD5.5 at Durham have increased by about one-third since 1844, from an average at the start of the record of about 1320 degree-days, to close to 1750 by the 2020s; much of that increase has come since 1990. Figure 21.6 shows the start and end of the growing season. Linear regression shows that there is now an extra month of growth in spring, starting now on average on 23 March rather than 23 April, while there has been a lengthening of the growing season in autumn of only fifteen days, from 31 October to 15 November.

For some crops a different threshold is appropriate. For vines, a threshold of 10°C is used (GDD10). In the Oxford region GDD10 has remained above 900 since about 1990, allowing the development of commercial vineyards in recent years [49]. At Durham, a good deal cooler than Oxford, the best year on record for GDD10 was 2006 (867) and the average remains around 700. Thus, north-east England remains too cold to allow commercial vineyards to operate, although no doubt vines may be grown successfully where slope and aspect produce a favourable microclimate.

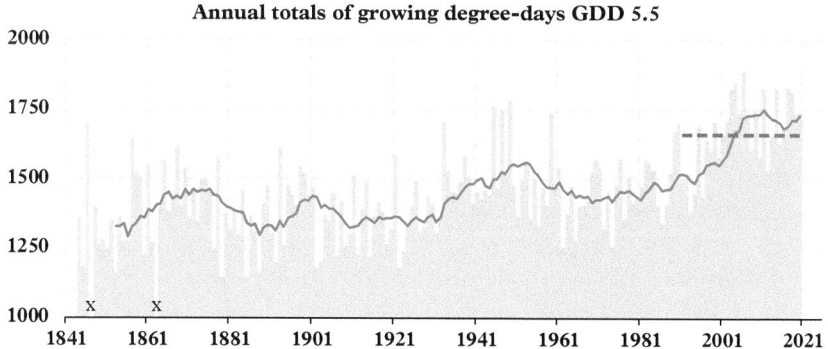

Figure 21.5 *Annual growing degree-days above 5.5° C (GDD5.5) at Durham, 1844–2020, together with an unweighted ten-year running mean plotted at year ending. The current (1991–2020) average GDD5.5 is 1670, shown by the dashed blue line. Daily data incomplete for some early years, shown by x.*

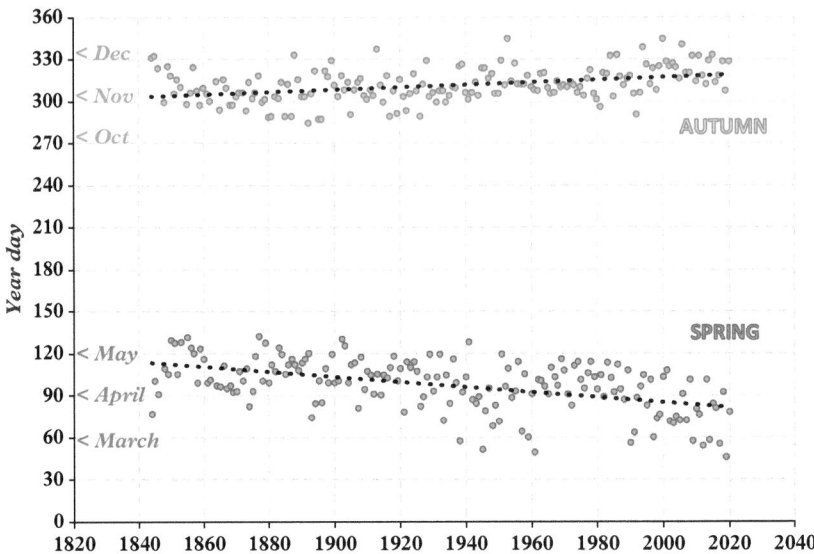

Figure 21.6 *An indication of the length of the growing season in Durham every year from 1844 to 2020. The start and end of the growing season (green and orange circles, respectively, for spring and autumn) are defined by GDD5.5 thresholds and are shown in Julian Days (days from 1 January); day 120 is the end of April and day 270 the end of September. Least squares regression linear trends over the period are shown by the black dotted lines.*

Discussion of growing seasons leads naturally to consideration of soil temperatures. Soil thermometer data are available from 12 August 1948, originally at 1 foot (30 cm) and 4 feet (120 cm) depths, the latter re-sited at the metric equivalent of 100 cm on 1 January 1970. Figure 21.7 shows the annual mean soil temperatures at 30 cm and 120 or 100 cm from 1949 to present.[*] After a couple of decades without much change in mean temperatures, both exhibited a steady increase through the 1970s followed by a faster rate of warming through to the early 2000s, since when the running mean has flattened out somewhat. Given variations in mean air temperature, the pattern of soil temperatures over 70 years is very much as expected. Monthly mean soil temperatures at the end of the record remain above 5.5°C from March to November on average, very similar to the GDD5.5 pattern discussed (not shown). In 2020, only February saw the average soil temperature at 30 cm fall slightly below the 5.5°C threshold.

---

[*] It might be expected that this record would show a discontinuity in 1970 when the thermometer depth was raised by 20 cm. Since the data plotted are annual means, the change of depth apparently makes very little difference, but this would not be the case if the annual range in soil temperature were plotted instead, with a smaller range at greater depth.

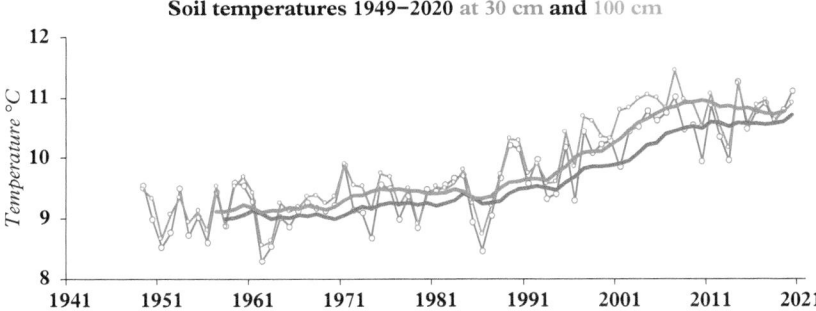

Figure 21.7 *Mean soil temperatures at 0900* GMT *at depths of 30 cm (red) and 100 cm (green, note 120 cm to 1969) since 1949. Annual means are indicated by the thinner line with markers, while ten-year running means ending at the year plotted are shown by continuous thicker lines. The 1991–2020 averages are 10.36° C and 10.60° C, at 30 cm and 100 cm depth, respectively*

Precipitation

Considerable attention has been paid to the Durham precipitation record [94] (see also Appendix 4) so we can be fairly confident that the record is reliable and homogenous, at least from 1868. Figure 16.10 shows that the Durham precipitation record is rather different to that for mean air temperature, exhibiting a slow upward trend among a series of wetter and drier periods. The plotted ten-year running mean smooths a typically noisy record. The notably wet decade of the 1870s [76] is very evident in the 712 mm ten-year mean ending in 1886, followed by the well-known drier period at the end of the nineteenth and beginning of the twentieth century.

Burt and colleagues [76] analysed the very wet 1870s across the British Isles, with particular attention being paid to the Durham record. Two wet periods in the 1870s were identified; the calendar year 1872 and the 12-month period September 1876 to August 1877 inclusive; comparisons were made with the wettest calendar year on record, 2012. For Durham, there is a strong and highly significant correlation between annual rainfall and the frequency of the cyclonic Lamb weather type ($r = 0.76$, $p < 0.001$, $n = 30$—see also Table 3.3). Since cyclonic weather is inversely correlated with the North Atlantic Oscillation (NAO) index [76, Table 1], this confirms that high annual precipitation totals at Durham result from cyclonic circulation together with a weak NAO. (Note that the correlation between Durham annual precipitation totals and NAO is not significant: $r = -0.32$, $n = 30$ but there is a significant inverse correlation in winter: $r = -0.43$). This is rather different to a much wetter upland location like Malham Tarn where there is a highly significant positive correlation between annual precipitation totals and the NAO (r = 0.35, p < 0.0001, n = 100) although Malham Tarn precipitation totals also have a significant correlation with cyclonic weather (r = 0.39, p < 0.0001, n = 100).

In relation to the 1870s, Burt et al [76] concluded that the very high rainfall totals of 1872, 1876/77 and 2012 were associated with a high frequency of cyclonic weather; at the same time, the frequency of anticyclonic weather and westerly winds tended to be very low. Places that are normally wettest in the British Isles, i.e., the uplands of the north and west, were not unusually wet at these times, whereas locations with extremely high rainfall totals relative to mean annual rainfall tended to be further east, in the lowlands.

The driest decade is the 1900s, which included Durham's longest period of protracted drought (see Drought section below); the ten-year mean ending in 1906 was only 584 mm, 14 per cent below the current 1991–2020 normal.[*] A second wet period is shown in the 1930s (ten-year mean 731 mm ending in 1939), and others in the cool, wet 1960s and in the 1980s, both separated by drier periods—notably the early 1970s, with six consecutive years 1970–75 all receiving less than 600 mm and the ten-year average ending in 1977 down to 607 mm. The early 1990s were even drier, with the ten-year mean ending in 1997 only 597 mm. Since then, the trend has been fairly sharply upwards, such that the wettest point on the record is the ten years ending 2017, which averaged 734 mm. The wettest decade on record at Durham is the most recent (Table 21.4), which at 710 mm slightly surpassed the average for the 1930s.

Table 21.4 *Mean annual precipitation at Durham by decade from 1871 to 2020; units mm*

Decade	Mean annual precipitation (mm)	Decade	Mean annual precipitation (mm)
1871–80	683	1951–60	617
1881–90	663	1961–70	684
1891–1900	620	1971–80	635
1901–10	580	1981–90	631
1911–20	648	1991–2000	663
1921–30	656	2001–10	668
1931–40	708	2011–20	710
1941–50	651		

[*] The rain gauge rim height was higher prior to 1906—at about 1.2 m, reduced thereafter to 30 cm—and this may also have had led to a slight reduction in catch; see Appendix 4 for details.

Notably, Durham's driest 12-month precipitation total occurred, not in one of the well-known dry years such as 1921 or 1976, but back in 1904/05, when the 12-month total commencing 25 August 1904 amounted to just 350 mm. The next-driest 12-month period was that commencing on 22 October 1958, with 356 mm. The wettest 12-month period commenced on 2 April 2012, when 1131 mm fell (Table 25.2).

Trends in annual rainfall

The annual rainfall series shown in Figure 16.10 exhibits a slow but clear upwards trend. A linear trendline (not shown) suggests an increase from about 625 mm in the late 1860s to about 675 mm in recent decades, amounting to about 7.5 per cent in 150 years. This is very close to the expected theoretical increase arising from the increase in atmospheric water vapour resulting from the observed 1.2 degrees rise in air temperatures over the same period (average 1851–80 8.21°C, 1991–2020 9.44°C).

Annual maximum daily rainfall

The annual maximum series for daily rainfall is shown on Figure 16.7. The linear trend-line also suggests a slight increase, about 10 per cent, over the whole series, also likely to be commensurate with the increase in atmospheric water vapour resulting from a warmer atmosphere.

The five highest daily rainfall totals at Durham since 1868 are given in Table 16.10 (a longer list is given in Table 25.5). The wettest day on Durham's records was 11 September 1976, which was also one of the coldest September days on record (Table 12.1); see also Chapter 22. The only other daily total over 70 mm within the last 100 years has been the 70.2 mm which fell on 31 May 1924 (Figure 21.8; Figure 21.10; see also Chapter 8).

Trends in rainfall frequency

The frequency of rain days has apparently increased markedly since 2000; although possibly an artefact of the AWS record (the rainfall record from which is now from a 0.2 mm tipping bucket sensor, rather than the five-inch manual checkgauge used until December 1999), the increase is mirrored in other sites in north-east England, although less strongly. A recent increase is much less apparent in the frequency of wet days (precipitation 1.0 mm or more), for which see Figure 16.6.

The average number of rain days for the 30 years ending 1999—the final period of manual records—was 177; for the 21 years ending 2020—all AWS records—that number rose to 204, an increase of 15 per cent. Of particular note is that the rainiest year prior to 2000, 1872 with 216 rain days, has been surpassed four times since and including the year 2000 (2000, 218 days; 2010, 217; 2014, 219; and 2015, 225 days). Suspicion immediately rests upon the change in observing method in late 1999, when the manually-read rain gauge was replaced by the AWS, as this increase in rain-days since 2000 is only partially confirmed in neighbouring records—for example at Esh,

Figure 21.8 *The old Baths Bridge during the flood of 1 June 1924, following heavy rainfall the day before. The daily rainfall total on 31 May 1924 remains the third-highest on Durham's post-1868 record (Table 25.5). (Reproduced by permission of Durham University Library and Collections and the MacFarlane-Grieve family)*

7 km west-north-west (Figure 1.1 and Appendix 4) where a daily record is available from May 1989, and whose average annual rainfall of 804 mm 1991–2020 is 18 per cent greater than at the Durham Observatory. Here, the annual average number of rain days over the 10 years 1990–99 (including a few estimates for missing days in 1996) was 188, and over the 21 years 2000–2020 was 197, an increase of 5 per cent. At Durham, the corresponding figures over the same periods are 176 and 204, up 15 per cent. In 2015, Durham's 225 rain days compared to 209 at Esh, the wetter site (annual total 972 mm at Esh, 725 mm at Durham). As a tipping-bucket rain gauge of the type in use at Durham would normally be expected to give a slightly *reduced* frequency of rain days, owing to evaporation between falls, this increase is difficult to explain, barring a substantial calibration error in the tipping-bucket rain gauge itself, a greater propensity to catch dewfall, or perhaps spurious tips resulting from software or communication glitches.

Exploring further, we can include records from the site at Cockle Park, near Morpeth (Figure 1.1) to examine the 'mean rainfall per rain day' at all three sites before and after the installation of the Durham AWS (Figure 21.9). The Cockle Park record extends only to 2013 as that site was itself converted to an AWS in 2014. Visually at least, the plot certainly suggests that some instrumental artefact, rather than climate change, is responsible for the observed increase in frequency of rainfall at the Durham Observatory site since 1999. Further comparisons are hindered by the dearth of long records at unchanged sites in and around the Durham area, but a detailed investigation is beyond the scope of this book.

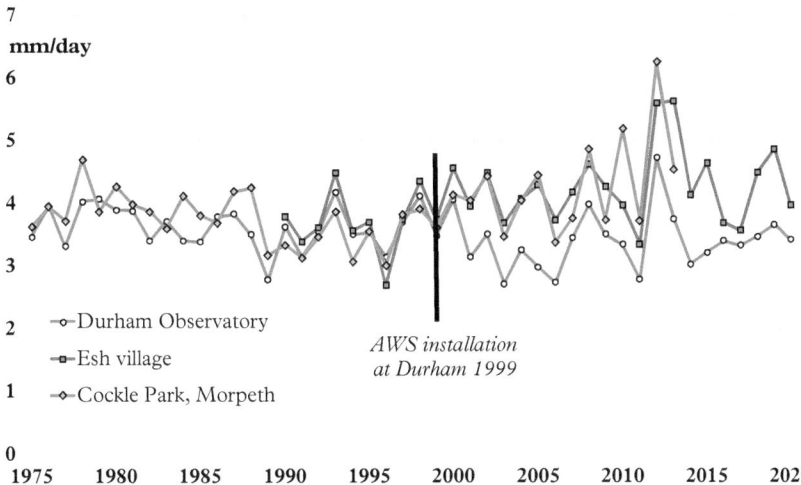

Figure 21.9 *Mean rainfall per rainday, in millimetres, at Durham Observatory 1975–2020, Esh Village 1990–2020, and Cockle Park 1975–2013, suggesting a significant discontinuity following the commencement of the AWS record at Durham in October 1999*

Drought

A drought is defined as a prolonged period of abnormally low rainfall, leading to a shortage of water. The lack of adequate precipitation, either rain or snow, can cause dry soil, diminished river flow, crop damage, and in the most severe cases, restricted water supply to the local population. Drought has been quantified in a number of ways in the past: historically, and with agriculture particularly in mind, the focus was on runs of days with little or no measurable rainfall. Two definitions were introduced by G. J. Symons in *British Rainfall 1887* (itself a notable drought year) and are still sometimes
used:

> *Absolute drought* - A period of at least 15 consecutive days, during which no day receives as much as 0.2 mm of precipitation; and
> *Partial drought*—A period of at least 29 consecutive days, whose mean daily precipitation does not exceed 0.2 mm.

These definitions ceased to be used officially more than fifty years ago, but they remain useful ways of summarizing short-term rainfall deficits. Today, longer-period totals are more relevant as water supplies are usually well buffered in the shorter term but can become scarce if the drought is prolonged. Any definition of 'prolonged' will depend on local conditions: in Durham, water supply comes from the Pennine Hills—rivers here are flashy in nature, with rapid production of flood runoff but little groundwater supply. To overcome the lack of sustained baseflow, many reservoirs have been built to maintain water supply through periods of water shortage (see Chapter 1).

To identify longer, potentially more severe, droughts, which may last for many months, we use accumulated precipitation totals. Standardization based on the mean and standard deviation allows comparison with other records; accordingly, results are reported here as z-scores, multiples of standard deviation (σ). A standardized precipitation index (SPI) may be calculated for any series of accumulated totals; we start with 12-month accumulated totals (SPI-12).

Following Noone et al [95] and Barker et al [96], droughts may be characterized according to four event characteristics, namely duration, accumulated deficit, mean deficit and maximum intensity. Severe droughts must last for more than three months with every month in that period having a z-score below -1.5σ. Table 21.5 shows major droughts in the Durham precipitation record identified using this approach. The droughts of 1959 and 1973 are just a little more extreme than the drought of 1904/05, which was the most protracted in the record with a z-score below -1.5 for 13 months (and produced the lowest 12-month annual precipitation total on Durham's instrumental record—350 mm in the 12 months commencing August 1904, Table 25.2). It is noticeable that the drought of 1975/76 does not register at Durham using this definition

Table 21.5 *Severe drought events in the Durham precipitation series, 1868 to 2020 (see text for details). Units are z-score (standard deviation multiples)*

Start month	Duration months	mean	max intensity
February 1874	9	−2.09	−2.43
October 1904	13	−2.12	−2.76
June 1929	6	−1.72	−1.98
August 1949	5	−1.95	−2.49
March 1956	5	−1.69	−1.91
May 1959	8	−2.13	−2.77
January 1973	8	−2.13	−2.57
July 1989	7	−2.04	−2.59
March 1990	4	−1.68	−1.93
October 2003	8	−1.92	−2.47

of severe drought: in fact, the SPI-12 index was negative at Durham from August 1972 to September 1976 inclusive but only one month, the 12-month total to August 1976, saw a SPI-12 score below the severe drought threshold (−2.09). The period 1989/90 saw another protracted drought; 1989 (Durham's driest calendar year on record—see also Chapter 22) appears in Table 21.5 but two months in 1990 with SPI-12 scoring above −1.5 (February 1990, −1.47; July 1990, −1.45) mean that the full 14-month period does not appear.

Table 21.6 presents a 12-month standardized streamflow index (SSI-12) for the River Wear gauging station at Sunderland Bridge (see location map in Chapter 1), where the record commences in October 1957; the method of calculation is identical to that used for the SPI, here using accumulated gauged daily flows over 12-month periods. A somewhat different pattern of droughts emerges for river flow, indicating that the Durham precipitation record is not an ideal location to use in relation to river flow in the Wear catchment, which is much more influenced by rainfall over the Pennine Hills in the upper catchment. Unlike Durham precipitation, the 1975/76 drought was a notable drought in the River Wear record, with ten consecutive z-scores below −1.5 and a maximum intensity equalling that recorded in 1973. The drought events of 1992 and 2017 are not replicated in the Durham precipitation series: in 1992, the SPI-12 only fell below −1.5 in two months at Durham (February and March), and in 2017 there were no negative SPI-12 values at Durham at all. In 2018/19, Durham did experience a sustained period negative SPI-12 scores from July 2018 to September 2019 inclusive, but only the 12-month total to April 2019 gets down to −1.5, whereas a severe drought was registered on the Wear for five months (Table 21.6).

To understand better periods of very low flow in the River Wear, Table 21.7 lists severe drought events for the rainfall record at Burnhope Reservoir in the headwaters of the catchment since 1957 (see Chapter 1 for location map). Unlike Durham, the list includes the 1975/76 drought, which clearly had much more impact in the Pennine hills than in

Table 21.6 *Severe droughts in the River Wear catchment since 1957 (Data source: National Rivers Flow Archive)*

Start month	Duration, months	mean	max intensity
July 1959	6	−1.92	−2.34
February 1973	12	−2.12	−2.37
December 1975	10	−2.01	−2.37
February 1992	7	−1.65	−2.12
November 2003	9	−2.00	−2.35
January 2017	12	−1.96	−2.08
March 2019	5	−1.95	−2.22

Data source: National Rivers Flow Archive.

Table 21.7 *Severe droughts at Burnhope Reservoir since 1931. See Figure 1.1 for location*

Start month	Duration, months	mean	max intensity
February 1942	5	−1.90	−1.71
November 1955	9	−1.90	−1.77
May 1959	7	−2.22	−1.77
October 1971	7	−1.87	−1.54
February 1973	7	−1.83	−1.57
December 1975	10	−1.87	−1.60
February 1996	12	−2.05	−1.62
August 2003	8	−1.92	−1.53
January 2006	11	−1.74	−1.59
April 2017	6	−1.82	−1.62
January 2019	6	−1.87	−1.57

the lowlands to the east. The Burnhope record also includes the two recent droughts in 2017 and 2019, which again were less severe at Durham, as noted earlier. The most prolonged and severe drought at Burnhope was the drought of 1996/97, although the drought maximum intensity was observed there in the earlier droughts of the 1940s and 1950s. Why droughts are generally less severe as measured by precipitation totals in the wet uplands than those indicated by river flow from the catchment is unclear and seems to be a matter that deserves further research.

Sunshine

Records from a Campbell–Stokes sunshine recorder commenced at Durham in late May 1880, one of the initial batch of sunshine stations established by the Meteorological Office that year following trials of the prototype instruments at the Kew and Greenwich Observatories in London [97] and the adoption of the modified Stokes version of the instrument [98] (see also Appendix 6). No trace has been found of the daily records prior to January 1882, but fortunately monthly totals for Durham up to 1884 were given by Scott [81]. After some problems lasting from December 1883 to May 1885 for which period there are no data, the Campbell–Stokes record is complete to the end of 1999. The cessation of this long and homogeneous sunshine record, second only to Oxford's Radcliffe Observatory [49] in its length at that time, is greatly to be regretted. The site's current AWS includes a pyranometer to provide approximate daily global solar radiation data, but these cannot be simply converted into a homogeneous 'hours of sunshine' record to continue the Campbell–Stokes record, particularly as the growth of trees around the site has reduced the previously excellent exposure of an unbroken horizon throughout the year. Since 2000, the *monthly* sunshine record has been estimated from areal values (as set out in Appendix 6), and no attempt has been made to estimate daily values.

Figure 16.10 shows the annual record of sunshine hours for Durham from 1881, the first complete year of record (1884 and 1885 are incomplete). Although there are wide variations from year to year, the decadal running mean shows little variation, remaining mostly within 50 hours of 1300 hours, until the late 1980s. The three very sunny years of 1988–1989–1990 saw the ten-year average rise above 1400 hours for the first time by 1990 and, making allowances for the estimated record since 1999, appears to have continued at a lesser rate over most of the time since. The annual average for the ten years ending 2020 stands at 1520 hours, the highest on the entire record and more than 20 per cent greater than the lowest ten-year average duration. Exceptionally sunny years occurred in 1989, 1995, and 2003, the latter the sunniest on record to date with 1703 hours (Table 16.13).

Comparing monthly averages for the 30-year period 1951–80 with those for the current period 1991-2020—noting of course that 20 of these latter 30 years are based upon areal estimates, rather than an instrumental record as in the 1951–80 period—an increase is apparent in every month except June, which declined by about 2 per cent. The largest percentage increases are all in the winter months—up 37 per cent in February and 26 per cent in December—although five months of the year (February, March, April, July, and August) show increases in monthly sunshine averages of between 20 and 25 hours between the two periods, contributing between them 112 hours of the 174 hours increase in annual duration. The reason for the increases appears mainly due to improvements in air quality, and the consequent reduction in the frequency, thickness, and duration of fogs particularly during the winter months, although also partly due to changes in the frequency of particular weather types and sequences over the British Isles.

The importance of maintaining long records

It is not surprising that a record as long as that at the Durham Observatory shows clear evidence of climate change, global warming in particular. The importance of long-term observation of the natural environment has long been recognized, and yet 'monitoring' is too often dismissed as low-grade science [99]. Long time series allow slow, subtle changes within noisy records to be identified, provided that the quality of the measurements remains consistently high. Climate systems are notoriously variable and there is no way to distinguish a normal extreme from a new trajectory without a long, well-maintained record. Nor is computer modelling the full answer: for all its great merits, model output requires verification—comparison with past or present data sets as necessary. A very long record allows present climate to be placed in its historical context. As we showed earlier in this chapter, Durham's current temperature regime, including eleven extremely warm years since 2000, would have been *inconceivable* in the second half of the nineteenth century, irrefutable evidence of global warming. Long series are also likely to include rare events, such as the dismal lack of sunshine in 1912, the coldest winter for over 200 years in 1962/63, the extreme transition from severe drought to high flood during autumn 1976, and the remarkably sunny spring of 2020. They also indicate limits to past and present climate: as extreme weather becomes more usual, it is good to know what is the best or worst that can be expected, although of course records are always there to be broken!

Like all other very long, reliable records, those at the Durham Observatory provide three important benefits [99]: they reveal important patterns for scientists to explain; they allow testing of new hypotheses, questions not imagined of at the time the monitoring was started; and they are a means of identifying significant changes in the natural environment that may eventually be harmful to the natural environment and to people themselves. Of course, not all aspects of long-term change are harmful, and while we may be genuinely cautious about global warming, decreases in very low winter temperatures and the incidence of air and ground frost may be welcomed by farmers and gardeners, while even a modest increase in winter sunshine duration will surely be welcomed by nearly everyone. Thankfully, Durham University and the UK Met Office are committed to maintaining meteorological observations at the Durham University as long as there is a need to do so. As we argued in relation to the Radcliffe Observatory meteorological station in Oxford [49], we see that ongoing necessity extending far into the future.

22

Chronology of notable weather events in and around Durham

1740—Durham's coldest year

The coldest year in the Central England Temperature (CET) series, which commenced in 1659, was 1740, with mean air temperature 6.8 °C, 3.4 degrees below the current (1991–2020) normal. The year includes the second-coldest winter on the series (mean temperature −0.4 °C, behind only 1684's −1.2 °C), the ninth-equal coldest spring (mean 6.3 °C), the 27th-equal coldest summer (mean 14.3 °C) and the second-coldest autumn (mean 7.5 °C). Both January (mean temperature −2.8 °C) and February (mean −1.6 °C) remain the fourth coldest on record, while October of that year remains more than a degree colder than anything before or since (mean 5.3 °C, 5.6 degrees below today's October normal).

There is little archival information available referring to the impact of the severe winter of 1740 in the Durham region. David Archer (pers. comm. via Deborah Smith) kindly provided three brief descriptions of conditions in Newcastle-upon-Tyne:

> A deep frost arrived in late December and by January rivers froze. Ships could not break through the ice, and so desperate were coal ships to break out of the frozen Tyne that hundreds of men hacked a channel through the ice to reach open waters. Shortages of fuel sent the price of coal rising eight-fold.
>
> *The Times*, Paul Simons' Weather Eye, 24 January 2013

> The frost continues still so strong and in all appearance likely to hold that most of the poor tradesmen are deprived of following their employments. The famous River Tyne on which depends most of our trade is frozen so hard that all manner of diversions are pursued upon as sliding, running jumping football, dressing of meat and selling all kinds of liquor in tents and is like a fair by the concourse of people of both sexes.
>
> *Caledonian Magazine*, 22 January 1740

> Immense masses of ice heaped up in the River Tyne which was completely frozen up. On 29 February a great quantity of ice was brought down the Tyne which with the weight of the water broke away a great part of the dam at Bywell. About 11 March, the rivers which had been blocked up were cleared of their ice by a gentle thaw without damage.
>
> *Sykes Local Records*

There were exceptional snowfalls over Scotland, especially in January—often with marked drifting. Further south, in the London area, snowfall was recorded on 39 days between November 1739 and May 1740. Deep snow fell around Christmas in Norwich, which remained on the ground until March. The streets of London were clogged with snow and ice, the Thames was frozen for about eight weeks: Thames shipping and London Bridge were damaged considerably by the ice. There were 'great shortages' of food and other essential supplies for the first seven weeks of 1740 due to the difficulty of shipping negotiating the ice. Moreover, this followed a violent easterly gale on 29 and 30 December 1739 [Old Style date: New Style 10–11 January 1740], accompanied by snow, which had already done considerable damage to ships. The gale and large blocks of drifting ice played havoc with shipping on the Thames; many ships were driven ashore and dashed to pieces. The snow finally melted in early March after which 'prodigious' floods were reported. More information on 'The Great Winter of 1740' can be found in references [100–103].

Using the relationship between CET and Durham for mean annual temperature (see Chapter 21), the estimated annual mean temperature for 1740 in Durham would be close to 6.1 °C, below that for 1695 (6.4 °C). The coldest year on Durham's instrumental record remains 1879, with a mean temperature of 6.8 °C, which thus ranks third lowest on the extended series back to 1659.

The great flood of 1771

On Saturday and Sunday, 16–17 November 1771, the north-east of England was subjected to a series of disastrous floods which damaged or destroyed many of the bridges on the rivers Tyne, Wear, and Tees. On the Tyne, all the bridges with the exception of Corbridge were destroyed along its entire length, from Alston in the headwaters to the Tyne Bridge in Newcastle (Figure 22.1). On the Wear, bridges had to be rebuilt at Frosterley, Wolsingham, Witton-le-Wear, and in Durham City [104]. Prebends Bridge used to stand some 100 m upstream of the present bridge; the Dean and Chapter moved the bridge to its present location when it was rebuilt in order to provide a more romantic view of the valley with cathedral and castle above. The foundation stone was laid on 17 August 1772 and it was opened on 11 April 1778. Engraved stone plaques show that flood levels on the Tyne in 1771 were considerably higher than anything recorded since, including the very large floods of 1955 and 2005. While precise flood discharges cannot be calculated given changes to the river channel since 1771, notably gravel extraction, the supremacy of the 1771 flood on the Tyne cannot be doubted [105]. Quite probably, the same can be said of the River Wear at Durham.

To the Right Worshipful John Erasmus Blackett Esq. Mayor of Newcastle upon Tyne, This VIEW of the RUINS of the BRIDGE of that TOWN; as they appeared after the Fall thereof in November 1771, Is most respectfully Inscribed, by his very obliged and most devoted faithful humble Servant John Bruno October 27 1772

Figure 22.1 *Destruction of the Tyne Bridge in Newcastle following the November 1771 flood (reproduced with permission from Durham University Library and Heritage Collections)*

The very cold winters of 1784 and 1795

Gordon Manley was well aware that January 1795 was the coldest month in his CET series, and his aim during retirement was to extend the Durham record back to that severe winter. Manley wrote to Joan Kenworthy [28] saying 'There is sufficient material to extend "Durham" back to 1795, a task now in progress' [36].

Manley was unable to achieve his aim before he died, but the Eglise extension [45] takes us back to the beginning of 1784 and, following a gap, starts again from January 1795. Two options are available to analyse these two very cold winters: either use CET to estimate the mean air temperature in December 1783 and December 1794 (to enable the usual winter average) or use a January/February average; in either case, some comparison with more recent winters is then possible. Table 22.1 shows the eight coldest winters and combined January/February periods at Durham since 1784. The third-coldest winter is 1784, with 1795 the fourth coldest; both appear in the January/February list, albeit in reverse order.

Table 22.1 *The eight coldest winters (Dec–Jan–Feb) and January/*
February at Durham since 1784; based upon Eglise [45] to 1844, then the
Durham Observatory temperature record. Years are dated to the January
year: units °*C*

Year	Winter (DJF)	Year	Jan–Feb
1963	−0.2	1963	−0.9
1879	−0.1	1795	−0.9
1784	0.5	1895	−0.8
1795	0.5	1838	−0.8
1895	0.7	1784	−0.4
1785	0.7	1814	−0.4
1814	0.9	1855	−0.2
1979	0.9	1940	−0.1

Winter 1784 seems to be related to the Laki eruption in Iceland in July 1783, the largest basaltic fissure eruption in the last thousand years [106–108]. Using tree-ring densities, Briffa and colleagues [109] identify 1783 as a year with significant negative anomaly, but do not include the Laki eruption in their list of the largest explosive volcanic eruptions since 1400 AD, as it only has a volcanic explosivity index (VEI) of 4 [107]. However, Ward [107] argues that gases were emitted high enough in the troposphere to be caught in the jet stream headed for Europe. For Iceland itself, the following winter (1783/84) was known as the 'Famine Winter': a quarter of the population died, many from wet and dry deposition of acidic pollutants. Because of the lack of major stratospheric impact, there is still some argument as to whether this led to changes to the regional/European climate, and by how much. In Scotland, the period around and after Christmas was bitterly cold with a 'violent' easterly storm accompanied by heavy snowfall and significant drifting on 25–26 December, which caused havoc along the Scottish east coast [110]. This was followed by a second severe snowstorm on 2 and 3 January, in which drifts were reported to have reached 5–6 metres deep in Aberdeenshire, seriously dislocating travel. Houses all down the eastern side of Scotland were unroofed, rocks were blown into harbours on the east coast, and stacks of corn and hay were carried away. Winter 1783/84 was the 14th-equal coldest winter in CET (mean temperature 1.2 °C).

Manley's target, winter 1795, was equally as cold in Durham as winter 1784 (Table 22.1) although even more extreme in CET (fifth coldest, mean temperature 0.5 °C). This winter was exceptionally severe, with the very cold conditions setting in on Christmas Eve 1794 [102]. The frost then lasted, with some breaks, until late March. The cold was most intense during January, the coldest month of any on the CET series (since 1659) and on the extended Eglise series in Durham (mean temperature −2.4 °C).

All but five days in the month saw mean daily CET below 0 °C, including −8.9 °C on 25 January 1795, quickly followed by a short mild spell (CET daily mean +6.0 °C on 27 January). By this time both the Severn and the Thames were frozen, and 'frost fair' festivities established on both rivers. A rapid but temporary thaw accompanied by heavy rain began on 7 February, which again resulted in much flooding across large areas of England with extensive damage to bridges. On 11 February, ice on the Tees at Stockton (Co. Durham), which had been frozen for five weeks, broke up, and the river rose 'higher than ever before remembered', 20 cm higher than the flood in 1771 [102]. The severe cold returned after 12 February and continued well into a cold, dry, late spring. Frequent and heavy snowfalls were accompanied by easterly winds with an anticyclone positioned to the north. February 1795 became the fourth-coldest February on the pre-Observatory record at Durham (mean temperature +0.7 °C), with only 1784 (0.3 °C), 1838 (−0.4 °C), and 1785 (−0.6 °C) being colder. For records since 1843, see Table 5.2.

January 1814—severe frost

> On 14 January, a hard frost was experienced on Tyneside. The Tyne, from St Peter's upwards, was completely frozen over, and the people of Newcastle held high carnival upon the ice. Tents, according to the local historian, were erected upon it, fires lighted, and donkey races and other sports held.

> *Newcastle Daily Chronicle*, reported in
> *Symons's Meteorological Magazine*, Vol. 23, 1888, p. 36

1816—the year without a summer

The Eglise extension of the Durham temperature record back to 1784 allows us to include the very cold year of 1816 in this chronology. This followed the April 1815 eruption of the Indonesian volcano Tambora—probably the largest caldera-forming eruption of the last few centuries [111] with a VEI of 7. The eruption injected approximately 60 Mt of sulphur into the atmosphere, forming a global sulphate aerosol in the atmosphere and resulting in pronounced climatic perturbations [109, 112, 113]. Anomalously cold weather affected the north-eastern USA, the maritime provinces of Canada, and Europe the following year [114, 115]. The year 1816 came to be known as 'the year without a summer' in these regions. Crop failures were widespread in the UK and the resulting famine was followed by a widespread typhus epidemic that affected almost every town and village in England. Post [116] characterized the period 1816–19 as the last great subsistence crisis to affect the western world. In the UK, the economy was already stagnating, with 400,000 men from the armed services having re-entered the job market at the end of the Napoleonic War. Mass reaction to the grim conditions included food riots, looting of granaries, and arson. Veale and Endfield [117] provide a detailed account of both the weather and the social conditions: they summarize 1816 as cold, very wet, and sunless, with frequent strong winds and storms. Unusually cold weather continued well into spring with snow in the Midlands and Wales on 12 May, while winter snowdrifts remained on Helvellyn in the Lake District as late as 30 July [102].

In the extended 'Eglise' temperature series for Durham, 1816 ranks as the second-coldest year (mean temperature 7.0 °C), with only 1879 colder (mean 6.8 °C). Summer 1816 at Durham remains the seventh-equal coldest (mean 12.7 °C), one of five summers with that average including 1912, which was affected by the eruption of Mount Katmai in Alaska in June [83, 109], an extremely dull summer in an exceedingly dull year (see *1912: another year with no summer*). Spring 1816 was also cold, with the second-lowest spring temperature (mean 5.4 °C; Table 18.3). Figure 22.2 shows the rank of each month in the extended Durham temperature series during the four-year period 1814 to 1817, where 1 represents the coldest of the 234-year analysis period. From November 1815 to the end of 1816, every month was below average temperature except for October 1816. For 1816 as a whole, such a cluster of months all ranking in the bottom quartile is not what would be expected in a normal year. Note that winter 1814 was also very cold (see *January 1814—severe frost*); Raible and colleagues [111] speculate about an unknown eruption in 1808/09 on the basis of evidence from ice cores, followed by a period of negative temperature anomaly. Although this event was not listed by Briffa and colleagues [109], it was included in Ward [107]. Guevara-Murua and colleagues [118] suggested a possible South American volcanic event in December 1808 based upon contemporary accounts.

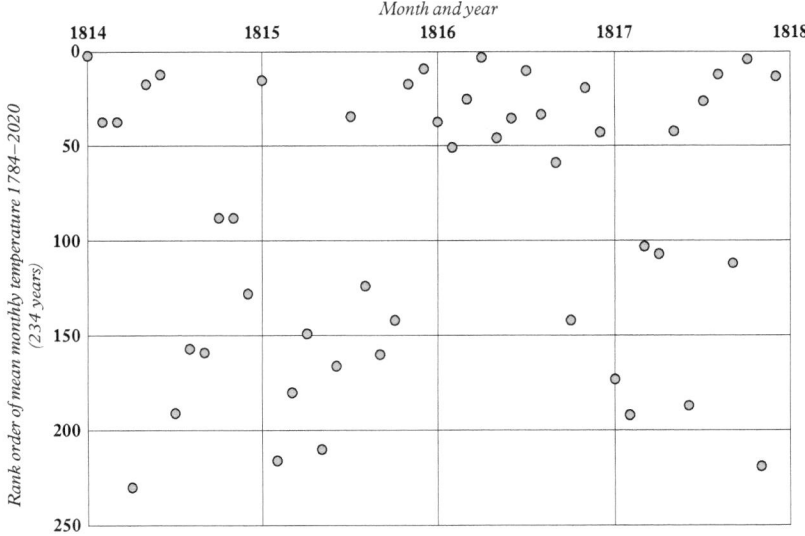

Figure 22.2 *Monthly temperature rankings in 234 years for the period of 1814 to 1817 at Durham; low numbers indicate the coldest months of that name in the record*

February 1823—snowstorm

What is described by the chroniclers of the day as being a dreadful snowstorm was raging in the North of England on 4 February 1823. The local roads, such as they were, became entirely blocked, and communication between villages, hamlets, and large towns was thoroughly cut off. Stagecoaches and mail coaches were then the conveyances most familiar to the public, and at the Turf Hotel, Collingwood Street, and at one or two other old hostelries in Newcastle, exciting scenes were witnessed. Travellers who had come to the town a day or two before were compelled to remain, and what was perhaps of more importance to the public, both inside and outside of Newcastle, all postal communication was, for the time being, brought to an end. For a whole week, the north and west mail coaches neither reached nor were despatched from Newcastle, and it was only when on one day thirteen saddle horses with messengers were despatched northwards that communication with the outer world was again established.

Newcastle Daily Chronicle, reported in
Symons's Meteorological Magazine, Vol. 23, 1888, p. 36

June 1846

This remains by some margin the warmest June on Durham's instrumental record. Manley was clearly aware of the 'phenomenally hot June' of 1846 (letter to Joan Kenworthy, 8 August 1979). An entry in the *Durham Chronicle*, Friday, 19 June 1846, confirms that conditions that month were exceptional:

It is so melting hot this week,
That, though our readers flout us,
The truth we must in conscience speak,
We've scarce our wits about us.

The Observatory's unscreened thermometer at the time was exposed outside the window on the Transit Room (above a little jutting wing) 4.6 m above the ground [19]. Figure 22.3 shows Durham's daily maximum, minimum, and mean daily temperatures for the month; the hottest days were at the start of the month and then on the 18th and 19th. The maximum temperature of 28.5 °C on 6 June 1846 remains the tenth-equal hottest June day on Durham's instrumental record, although of course we should be cautious of its accuracy given the difference in exposure from current standards; similarly, the maximum temperatures recorded on 18 and 19 June, almost as high at 28.4 °C. We should note that there is a comment in the ledger alongside the evening reading on the 6th:

This is the highest in 1843–1847, 84.5 [°F = 29.1 °C] after correcting to Standard Thermometer.

Were we to accept this 'correction', 6 June 1846 would then rank as the seventh-highest June maximum temperature on Durham's records. The minimum temperature of 15.1 °C on 19 June also remains one of the warmest June nights on record.

The reason for the hot weather is evident from Figure 22.4—a blocking anticyclone over Germany feeding warm or hot continental air towards the British Isles

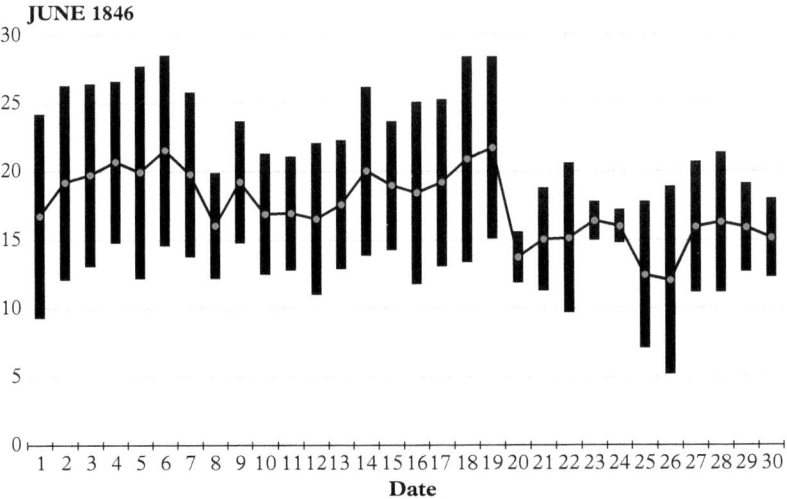

Figure 22.3 *Daily maximum, minimum, and mean daily temperatures recorded by the unscreened thermometers on the north wall of the Durham Observatory in June 1846*

with a very short sea track. In northern England, the dominant wind direction was southerly or south-westerly rather than south-easterly, thus minimizing the cooling effect of the North Sea. An easterly or south-easterly mean flow would have reduced mean temperatures considerably on facing coasts.

There are several possible methods to determine the monthly mean temperature of this exceptional month, which remains the warmest June on Durham's records by a large margin (Table 9.2). Manley [19] assumed that the north wall readings were well protected from direct radiation, and accordingly that a combination of the fixed-hour observations at 0900 and 2100 local time taken together with the means of the observed maximum and minimum temperatures would give a mean temperature similar to that derived from the mean of the daily maximum and minimum temperatures observed in a Stevenson screen. However, as already noted, he was unable to attempt to derive a mean for June 1846 as the observations ledgers pre-1850 were not available to him at the time. For this series, for consistency and to align with normal practice, we have throughout taken monthly mean temperatures as the average of the mean daily maximum and minimum temperatures for the month (corrected for exposure or calibration offsets where possible, as described in Appendix 3). For June 1846, accepting the caveats regarding thermometer exposure as mentioned, the mean daily maximum temperature was 22.6 °C and the mean daily minimum temperature 12.3 °C; the derived mean temperature for the month is therefore 17.45 °C, which we have rounded to 17.5 °C (see Chapter 9 and Table 9.2). (This monthly mean temperature was not surpassed until August 1975.) Eglise [45] used the Manley method of adopted means based on the 0900 and 2100 readings as well as the maximum and minimum to derive a monthly mean of 18.0°C. Manley's CET mean for the month was 18.2 °C—still the highest June mean temperature on the CET record. The average difference between

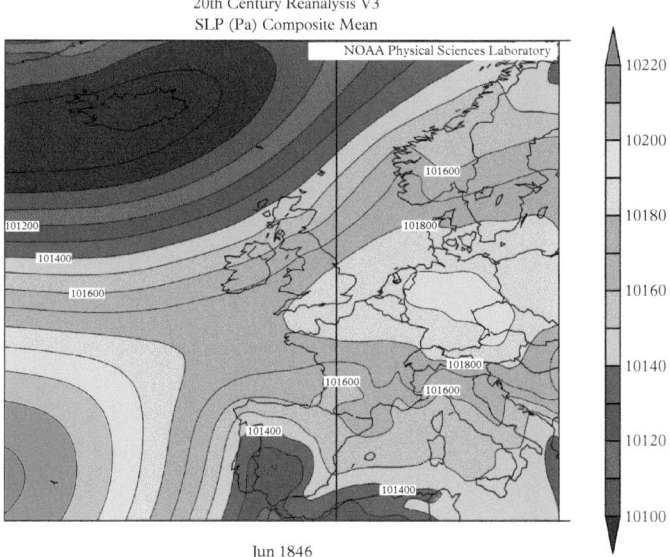

20th Century Reanalysis V3
SLP (Pa) Composite Mean

Jun 1846

Figure 22.4 *Composite mean MSL pressure distribution in June 1846, from the 20th Century Reanalysis version 3. Values are in Pascals (Pa) at 100 Pa contour interval: divide by 100 for hPa (millibars). Image provided by the NOAA/ESRL Physical Sciences Laboratory, Boulder Colorado from their website at http://psl.noaa.gov/*

CET and Durham mean temperatures in June over the period of the north wall readings 1844–59 was 1.3 degrees (Durham the lower of the two), suggesting a Durham mean closer to 16.9 °C for June 1846. This is still more than a degree greater than the next warmest June (Table 9.2). It was indeed a phenomenally hot month!

The severe frost in February 1855

February 1855 remains the coldest month of any on Durham's instrumental records; the monthly mean temperature of −2.2 °C (Table 24.8) was a partial estimate by Manley [19] as records have survived for only half of the month. On earlier estimates (the Eglise series, from 1784 with gaps 1791–94 [45]), the only colder months were January 1795 (estimated mean temperature in Durham −2.4 °C) and January 1814 (estimated mean −2.3 °C).

　The monthly mean pressure distribution (Figure 22.5) for February 1855 shows low pressure over France from the Gulf of Genoa to the Bay of Biscay, and high pressure over Iceland, resulting in persistent easterly winds which drew very cold air from the continent and Scandinavia. A succession of depressions passed over northern France during the first three weeks of the month—no doubt accompanied by heavy snowfalls in the cyclonic easterly flow to the north of the centre; Durham may have escaped the heaviest

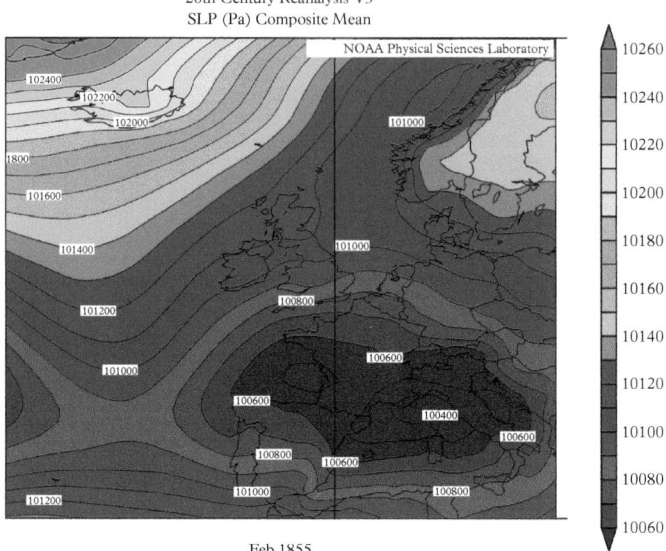

20th Century Reanalysis V3
SLP (Pa) Composite Mean

Feb 1855

Figure 22.5 *Composite mean* MSL *pressure distribution in February 1855, from the 20th Century Reanalysis version 3. Values are in Pascals (Pa) at 100 Pa contour interval: divide by 100 for hPa (millibars). Image provided by the NOAA/ESRL Physical Sciences Laboratory, Boulder Colorado from their website at http://psl.noaa.gov/*

of the snowfalls but remained in cold easterlies almost throughout. Unfortunately, the Durham temperature record is missing from 27 December 1854 to 14 February 1855 inclusive, and the monthly mean temperature has been estimated using daily CET data together with the two weeks of available record (Figure 22.6). The coldest day from the available record was 16 February, when a minimum temperature of −14.6 °C in the 'North Shed' was followed by a maximum of −4.3 °C, giving a mean temperature for the day of −9.5 °C. This remains sixth in the list of Durham's all-time coldest days (Table 24.6), unequalled in the twentieth century or thus far into the twenty-first.

The following contemporary account gives an indication of conditions in the north-east that month:

> The Tyne down to St Lawrence was firmly ice-bound, tents were pitched, vendors of all sorts of commodities did a thriving trade, and hundreds of persons disported themselves upon the ice. At that time the only weak parts in the glassy sheet were near the Quay edge, and keelmen and steamboatmen who were, for the time being, idle took advantage of that circumstance, and by making bridges of planks over their craft, and by levying toll, they turned to profitable account the inclemency of the weather. There are yet in the land of the living many persons who will recollect not only the stirring pastimes of the pleasure-seekers, but the exciting work performed by the steamers afterwards employed to break up the ice as the thaw began. These vessels

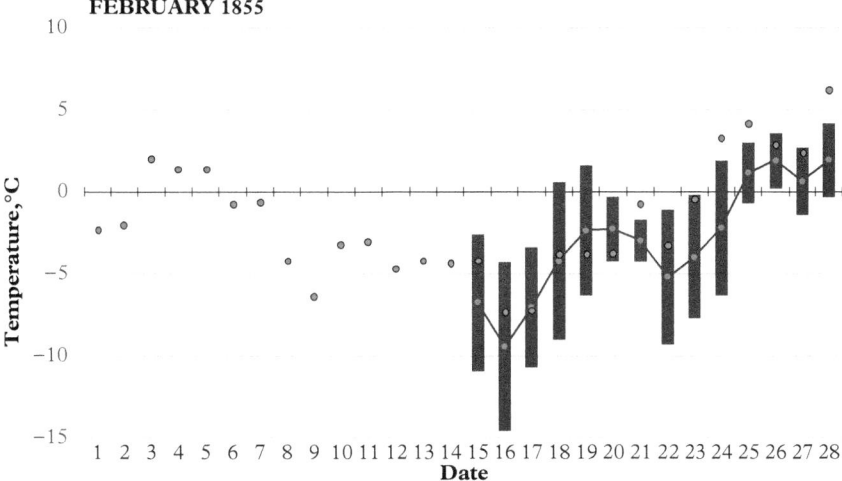

Figure 22.6 *Daily mean temperatures for February 1855—the coldest month on the Durham instrumental record. The Durham record is missing from 27 December 1854 to 14 February 1855 inclusive; daily mean CET are also plotted (blue circles), from the Met Office Hadley Centre website https://www.metoffice.gov.uk/hadobs/hadcet/data/download. html*

were strengthened at the stem for the work, and as they slowly moved up the river and cut their way through the resisting barrier of ice, hundreds of towns-people assembled on the quays and bridges to witness the work going on. That work was, however, but slowly performed, and it was a long time before full traffic on the river could be resumed.

Newcastle Daily Chronicle, reported in
Symons's Meteorological Magazine, Vol. 23, 1888, p. 36.

Durham's coldest-ever Christmas—1860

The very severe frost during Christmas 1860 remains one of the coldest spells on Durham's records (Table 24.12), indeed on many extant records in both Scotland and England, and as such the synoptic conditions are worth closer examination. Table 22.2 lists the main observational data from the daily 1000 and 2200 observations at Durham Observatory during the second half of December 1860, together with maximum and minimum temperatures and precipitation. Daily maximum and minimum temperatures are plotted in Figure 22.7. At this time, air temperatures were from a Glaisher stand and precipitation from a rain gauge of 30 cm diameter exposed at about 1.2 m above ground; the former have been corrected to approximate Stevenson screen equivalents (see Appendix 3 for details), while the latter are given for comparison only, as the accuracy of precipitation measurements prior to 1868 is open to question. Unfortunately no

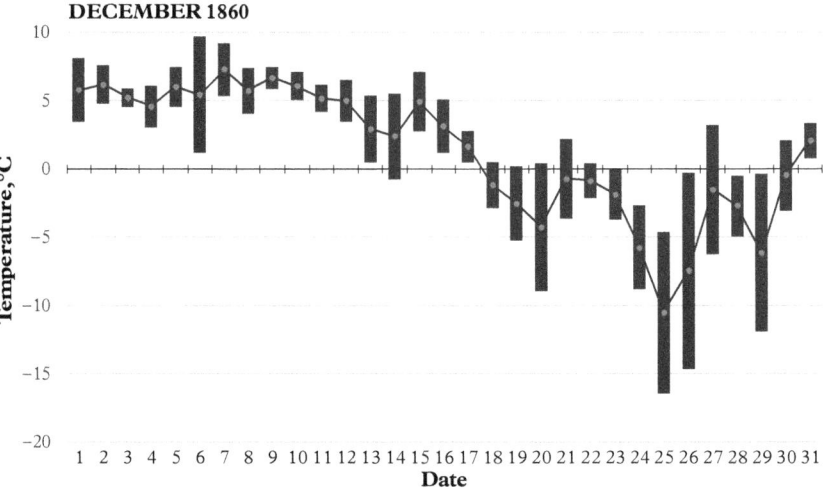

Figure 22.7 *Daily maximum, minimum, and mean daily temperatures from the Glaisher stand thermometers at Durham Observatory, corrected to approximate Stevenson screen equivalents, during December 1860*

detailed weather diary for the period has survived. We have no records of snow depths in Durham from so long ago, and accordingly some of the following sequence of events necessarily draws upon contemporary newspaper accounts and inferred synoptic developments together with the existing observational record and a previous account of the event by Pike [119].

The sequence of synoptic conditions has been derived from six-hourly reanalysis plots of MSL pressure from the 20th Century Reanalysis version 3 (20CRv3) provided by the NOAA/ESRL Physical Sciences Laboratory, Boulder Colorado from their website at http://psl.noaa.gov/ [84].

Relatively mild and changeable conditions prevailed until mid-December; there was no air frost recorded in Durham between 20 November and 14 December. A broad ridge from the Azores anticyclone extended into north-west Europe, and the maximum temperature on 15 December reached 7.1 °C in Durham; this was to be the mildest day for the coming five weeks. Thereafter, winds quickly veered towards the north and strengthened as a depression deepened over Scandinavia, engaging very cold Arctic air in its circulation; at Durham the wind was north-north-easterly Beaufort Force 6 at 2200 on the 16th, with a falling barometer (Table 22.2). Rain falling on the 16th began to turn to snow on the 17th as temperatures fell and the cold air pool advanced over Denmark and the Low Countries. As this very cold air crossed the relatively warm North Sea, the surface layers became extremely unstable and prolonged and heavy snow showers resulted, widely accompanied by thunder and lightning. The eastern coasts of Scotland and northern and north-east coasts in England were worst affected; in Aberdeen 30–40 cm of snow fell overnight, and up to 60 cm in rural Fife. On the morning of

Table 22.2 *Meteorological records from Durham Observatory, 15–31 December 1860. See Appendices for instrumental details.*

Date 1860	Cloud cover tenths	Wind dir'n	Wind force Bft	Temp °C	MSLP hPa	Cloud cover tenths	Wind dir'n	Wind force Bft	Temp °C	MSLP hPa	Tmin °C	Tmax °C	Rain mm
	At 1000 h					**At 2200 h**					**24 hours**		
15 Dec	4	NW	1	5.0	1026.8	4	NW	2	1.7	1023.9	2.8	7.1	–
16 Dec	9	NNW	1	3.3	1020.1	10	NNE	6	1.3	1015.9	1.2	5.1	9.1
17 Dec	10	NW	3	1.6	1011.3	2	NNE	1	–1.6	1004.8	0.5	2.8	4.8
18 Dec	9	NNW	1	–0.7	1002.0	10	NW	1	–3.2	997.5	–2.8	0.5	9.1
19 Dec	10	CALM		–1.6	997.0	10	CALM		–6.0	1006.5	–5.2	0.2	7.9
20 Dec	10	CALM		–0.6	1014.3	10	NNW	1	–3.2	1018.4	–8.9	0.4	8.9
21 Dec	10	CALM		–0.8	1020.2	10	NNW	1	–1.1	1018.7	–3.6	2.2	8.4
22 Dec	9	NNW	1	–1.4	1017.0	10	NNW	1	–1.7	1011.8	–2.1	0.4	3.3
23 Dec	10	NNW	2	–1.6	1004.8	10	NNW	2	–6.5	1000.1	–3.7	0.0	1.3
24 Dec	2	CALM		–7.8	998.2	3	NNW	1	–13.1	999.6	–8.8	–2.7	–
25 Dec	3	CALM		–11.0	1001.5	10	CALM		–11.9	1002.5	–16.4	–4.6	–
26 Dec	8	CALM		–6.9	1004.7	8	CALM		–4.4	1005.1	–14.6	–0.3	1.0
27 Dec	8	SSW	2	–0.6	1005.6	9	SSE	1	–1.7	1009.6	–6.2	3.2	–
28 Dec	7	CALM		–4.1	1017.5	2	WNW	2	–10.3	1025.6	–4.9	–0.5	–
29 Dec	8	WNW	2	–7.2	1026.3	10	SSE	5	–3.1	1016.9	–11.9	–0.4	2.3
30 Dec	10	SE	3	0.9	1004.4	10	ESE	1	1.2	1003.4	–3.0	2.1	–
31 Dec	10	SE	1	1.7	1008.8	10	SE	1	0.2	1000.4	0.8	3.4	7.1

19 December the depression, by now probably a full-fledged polar low, lay over northern England; at 10 a.m. the barometer at Durham stood at 997 hPa while the wind was calm, suggesting proximity to the centre. As quoted by Pike:

> Snow fell more or less the whole of yesterday and a hard frost has set in. During last evening, the atmosphere was highly charged with electricity and frequent discharges were observed. Shortly after one o'clock this morning it commenced to thunder, and again at the time we went to press.
>
> *Newcastle Chronicle*, 19 December 1860

The Times of 20 December reported:

> On Wednesday morning [19 December] the northern counties from coast to coast appeared in a covering of snow. At Liverpool Manchester and at Bradford there was a uniform depth of 5 or 6 inches [13-15 cm] and locomotion was suddenly and unexpectedly impeded. From Liverpool to Manchester there was little variation in the depth of the snow which had fallen evenly and without drifting. Between Manchester and Sheffield the depth of snow gradually diminished and in some places south of Sheffield, the ground was barely covered. Towards the Midlands, there had been a heavier fall.

The depression continued moving steadily southwards close to the east coast of England on 19–20 December. By 20 December, most of the British Isles had been blanketed in snow, although amounts were small in the south-east of England. Barometric pressure rose during 20th and 21st as an anticyclone developed to the south of Iceland, although persistent and heavy snow showers or longer spells of snow continued in the resulting very cold northerly or north-easterly flow—fresh snowfall would have been greatest on eastward-facing coasts. In Durham, cumulative precipitation amounts throughout 16–23 December suggest upwards of 30 cm of snow had fallen. Meanwhile, an intensely cold pool of continental air developed over continental Europe, subsequently reinforced by a further cold plunge from the Arctic on 22–23 December, and as this again moved over the warmer waters of the North Sea a second polar low developed. Despite a falling barometer (998 hPa at Durham on Christmas Eve morning), skies had begun to clear, for that morning the minimum temperature at Durham fell to −8.8 °C with just 2/10 cloud cover at 1000. (Pike postulates the formation of a weak anticyclone over the Irish Sea, and this development may have been just sufficient to assist cloud clearance [119]). At 10 a.m. the temperature in Durham had risen only slightly to −7.8 °C, and that day it rose no higher than −2.7 °C. In a slack pressure field and well-broken cloud cover (3/10 at 10 p.m.) temperatures fell rapidly overnight: at 10 p.m. the temperature was already down to −13.1 °C. On Christmas morning the minimum temperature at Durham Observatory was −16.4 °C, a value which has only been surpassed twice since, in January 1881 and in February 1895 (Table 24.5). At 10 a.m. the temperature was −11.0 °C with 3/10 cloud cover and almost certainly a deep snow cover—surely a picture-perfect Christmas morning.

Figure 22.8 shows the synoptic situation at 0900 GMT on 25 December 1860, based on 20CRv3 [84]. By day, the temperature reached only −4.6 °C in Durham, and by 10 p.m. was down to −11.9 °C, on the way to the minimum temperature on Boxing

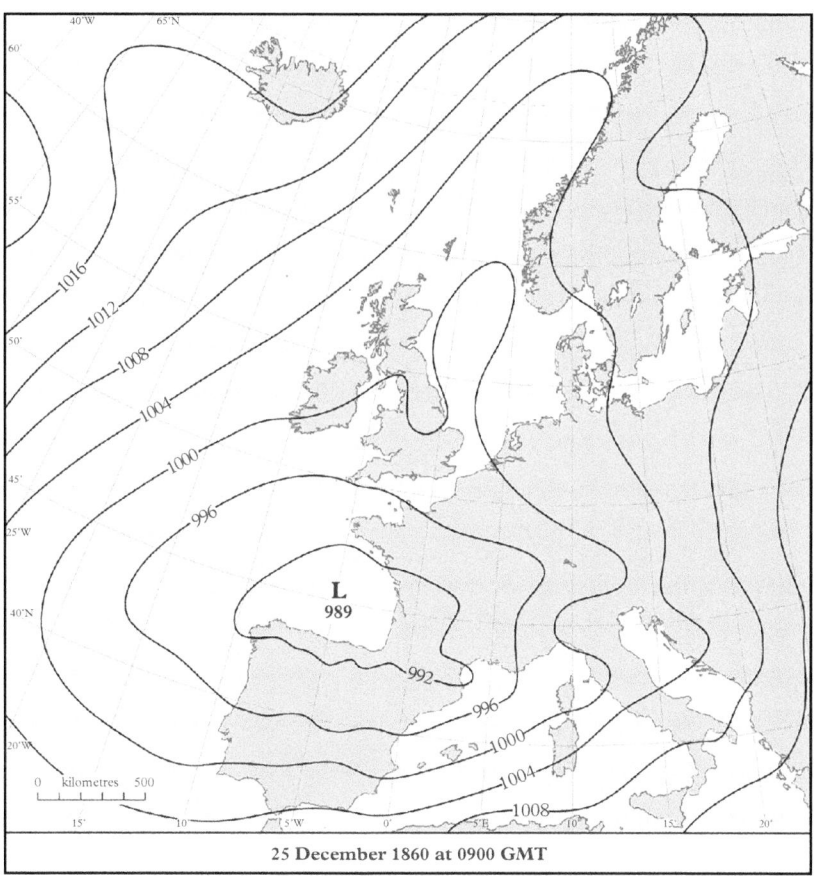

25 December 1860 at 0900 GMT

Figure 22.8 *The synoptic situation at 0900* GMT *on 25 December 1860, Durham's coldest-ever Christmas: the maximum temperature reached only −4.7 °C following an overnight minimum of −16.4 °C. Isobars are in hPa (= millibars). (Based upon the 20th Century Reanalysis version 3 (20CRv3) plot [120], drawn by Chris Orton, Durham University)*

Day of −14.6 °C. The 24-hour mean temperature on 25 December, namely −10.5 °C, remains Durham's coldest day on record (Table 24.6), equalled on 17 January 1881 but not approached within a degree by anything in the twentieth or, to date, twenty-first centuries.

Elsewhere in the north-east, Bywell (about 30 km north-west of Durham) fell to −15.8 °C on Christmas morning, while even at North Shields on the coast the minimum was −14 °C. In southern Scotland, Hawick recorded −19 °C and Kelso −21 °C [119]. Temperatures below −15 °C were widespread over the deep and fresh snow cover as far south as the English Midlands; Christmas Eve 1860 remains the coldest morning on

Oxford's records since they began in 1813, with a minimum temperature of −17.8 °C [49]. At Beeston in Nottinghamshire −22 °C was recorded, and −20.5 °C at the Cambridge Botanical Gardens.* Queen Victoria, in Windsor, noted in her diary on Christmas Eve:

> Already this dear Festival returned again, & this year with true Xmas weather, snow on the ground & sharp frost.

Conditions began to ease a little as a depression formed to the south-west of the British Isles on 26–27th; winds veered south or south-easterly for a time and Durham reached 3 °C for the first time in 11 days on the 27th. Pressure rose quickly as an anticyclone developed over the British Isles on the 28th, moving eastwards into Europe on the 29th. As skies cleared once more during 28 December, temperatures once more fell rapidly after dark, down to −10.3 °C at 10 p.m. and to the minimum of −11.9 °C by next morning. By New Year's Eve a strengthening southerly flow in the wake of the anticyclone brought a temporary respite with slightly milder conditions.

It is very likely that substantial snow cover from December persisted throughout the first three weeks of January 1861, and although not quite as severe, further snowfalls and occasional severe frosts occurred until mild weather finally arrived on 21 January, ending one of the most remarkable cold spells on the English meteorological record. The mean air temperature at Durham for the 28 days 18 December 1860 to 14 January 1861 was −2.6 °C; only six days in this period saw the mean daily temperature surpass 0 °C. Had this period been confined within a calendar month, rather than being split across two mid-month to mid-month, it would have outranked February 1855 (mean temperature −2.2 °C) as the coldest on Durham's records.

September 1865—a warm and dry month, but not the driest

What we originally believed to be the driest month in the Durham record came in September 1865, when the total rainfall was apparently just 0.2 mm. Reference to the original ledgers shows that 0.003 inch (< 0.1 mm) was recorded on the first day of the month, and a further 0.005 inch (0.1 mm) on the 9th. The two falls of 0.1 mm were the only rain gauge entries in a 46-day period from 23 August–7 October 1865. However, while evidence for September 1865 being a very dry month is not in doubt, Durham's monthly total of 0.2 mm was much lower than other gauges in the north-east. The total rainfall that month at nearby Ushaw, for example, was 9.1 mm, at Sedgefield 7.6 mm, and at Darlington 8.9 mm. Thomas Backhouse in Sunderland recorded 16.3 mm, while the lowest total in the north-east excluding Durham was 4.1 mm at Barnard Castle. The Met Office Hadley Centre 'County Durham' rainfall series, which

---

* Some caution is advised before accepting reported temperatures from this period as comparable to modern records, as this event predated the invention of the Stevenson screen (in 1864) and its subsequent widespread adoption by the end of the nineteenth century.

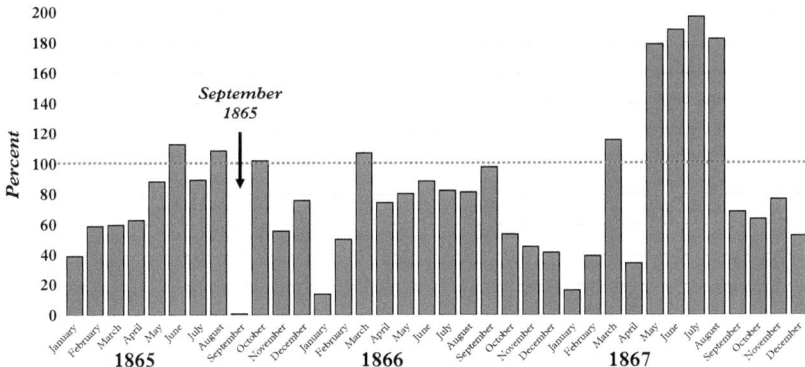

Figure 22.9 *A comparison of Durham Observatory's recorded monthly rainfall, expressed as a percentage of 75 per cent of the Met Office 'County Durham' gridded rainfall series, by month January 1865 to December 1867. On average, the Observatory's monthly totals should lie close to the green dashed line; prior to 1868 observed records are simply too erratic to be considered reliable. See also Appendix 4 for possible explanations. September 1865's monthly total can be seen to be unrealistically low, and summer 1867 unrealistically high*

extends back to 1862 [121], lists 21 mm as the monthly total for September 1865. Over the long term, Durham Observatory averages about 75 per cent of the 'County Durham' series, and although wide variations about this figure will inevitably occur from month to month, this would suggest the true rainfall total for September 1865 at Durham Observatory lay closer to 15 mm. When we examined the other remaining months in 1865, the discrepancies became even clearer (Figure 22.9). Durham's register shows 92.9 mm falling in the 13 days immediately preceding the commencement of the apparent dry spell (including 36.6 mm on 20 August), while 149.2 mm fell in the 20 days following the end of the dry spell. This pattern is entirely lacking in the 'County Durham' series. These discrepancies, together with many others in 1866 and 1867, led us to believe the record prior to 1868 was simply too unreliable to be used (see also Appendix 4).

1872—a very wet and rainy year

Durham's annual rainfall total in 1872 amounted to 895 mm, 32 per cent greater than the current annual average of 680 mm. This remained the wettest calendar year for 140 years, until surpassed in 2012 (Table 16.10). January, May, and June 1872 were slightly drier than the current monthly average for those months, but March (69 mm, 167 per cent), October (105 mm, 165 per cent), and December (125 mm, 204 per cent) were all 50 per cent (or more) wetter than normal.

This was a year dominated by cyclonic conditions; 1872 had by far the highest frequency of cyclonic weather in the Lamb Weather type series, with a very low frequency

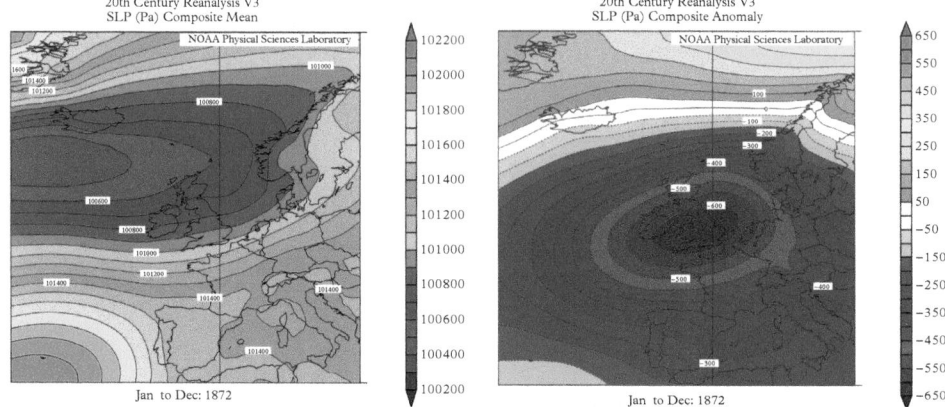

Figure 22.10 *Monthly mean* MSL *pressure (left, units Pa, interval 100 Pa = 1 millibar) and anomaly from 1981–2010 normal (right, units Pa, interval 50 hPa) over the north-east Atlantic and western Europe for the year 1872. Plots from 20th Century Reanalysis data provided by the NOAA/OAR/ESRL PSL, Boulder, Colorado, USA, from their website at https://psl.noaa.gov/*

of anticyclonic and westerly air flow (Figure 22.10). The normally wettest locations in the British Isles, i.e. the uplands in the north and west, were not unusually wet, whereas locations with extremely high rainfall totals (relative to mean annual rainfall) such as Durham tended to be further south and east in the lowlands. What was important in 1872 was not a few, extreme events, but rather the high frequency of rainfall of moderate extent (216 rain days, a value not exceeded until 2000, and 58 days with 5 mm or more). The 1870s were a wet decade, with only the 1930s, 1960s, and 2010s being wetter at Durham (Table 21.4; [76]).

1879—cold, cold, cold

This remains the coldest year on Durham's instrumental records, with a mean temperature of 6.8 °C, 2.6 degrees below the current average (Table 16.2). In the CET series back to 1659, 1879 ranks third coldest (mean temperature 7.4 °C) with only 1695 (7.3 °C) and 1740 (6.9 °C) being colder. If we include the Eglise extension for 'Durham' back to 1795, only 1816 comes close with an annual mean of 7.0 °C, followed on the instrumental record by 1892 at 7.1 °C.

As Table 22.3 shows, every month in 1879 was below the current average temperature. Winter 1879 at Durham includes December 1878, the coldest December on Durham's instrumental record with a mean temperature of −0.7 °C, almost 5 degrees

Table 22.3 *Monthly mean air temperatures (° C) at Durham, 1879, with departure from 1991–2020 mean temperature, and rank order in 178 years of record★*

	Jan	Feb	Mar	Apr	May	Jun	Jul	Aug	Sep	Oct	Nov	Dec	1879
Mean temperature °C	−0.5	1.5	3.8	4.9	7.7	12.1	13.0	13.7	11.5	8.2	4.2	0.7	**6.8**
Diff from 1991–2020 mean degC	−4.6	−3.1	−2.4	−3.3	−3.3	−1.5	−2.8	−1.9	−1.8	−1.8	−2.5	−3.5	**−2.7**
Rank; 1 = coldest	6	24	46	4	10	32	8	36	37	35	19	6	1

★ *In this table the months are ranked strictly to two decimal places of mean temperature. Rankings in monthly, seasonal, and annual tables are to one decimal place, and shown equal if mean temperatures differ by less than 0.1 degC.*

below normal; winter as a whole (December 1878 to February 1879) remains the second coldest on record, behind 1963 (Table 17.3). There was little improvement in spring, which ranked equal coldest on record, with the mean temperature 3.0 degrees below normal (Table 18.2), April ranking as one of the coldest on record (Table 7.2). Summer 1879 was little better, ranking only tenth coolest at 2.1 degrees below normal; the highest temperature that summer was just 23.3 °C, recorded on 11 and 12 August. Only 13 days reached even 20 °C, the longest such 'warm spell' (if indeed that is the correct term for such low summer temperatures) being just three consecutive days 28–30 July. We have no sunshine records for this summer, of course, as Durham's records did not commence until June 1880, but it was almost certainly a very cloudy summer too. Autumn 1879 fared a little better; the mean temperature for the season was 2.1 degrees below the current normal, but there have been 16 colder autumns on record. Another shock came at the end of the year; December 1879 remains the equal-fourth coldest on record (Table 15.2), at 3.5 degrees colder than normal. Although the month as a whole was not quite as cold as the previous December, two very severe frosts very early in the month (−14.0 °C on the 3rd and −14.6 °C on the 4th; Table 15.1) remain among the lowest on record for the month and are unprecedented for so early in the winter. The month became milder towards the end of the year, however, and the temperature reached 11.6 °C on the 28th in near gale-force west-south-westerly winds: later that evening, the Tay rail bridge collapsed in a violent westerly storm, derailing a train and costing 75 lives [74].

The annual mean MSL pressure and anomaly charts for 1879 are shown in Figure 22.11.

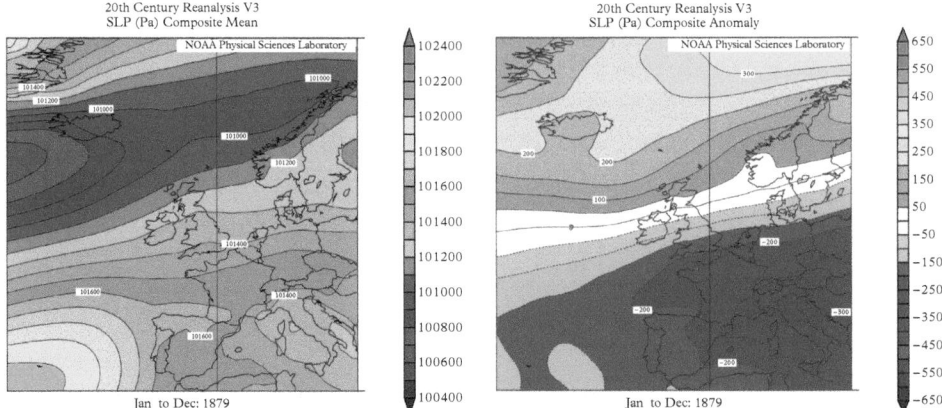

Figure 22.11 *Monthly mean* MSL *pressure (left, units Pa, interval 100 Pa = 1 millibar) and anomaly from 1981–2010 normal (right, units Pa, interval 50 hPa) over the north-east Atlantic and western Europe for the year 1879. Plots from 20th Century Reanalysis data provided by the NOAA/OAR/ESRL PSL, Boulder, Colorado, USA, from their website at https://psl.noaa.gov/*

March 1886 snowstorm

The following letter from Thomas Backhouse of Sunderland [122] was published in *Symons's Meteorological Magazine* in April 1866 (old units have been converted):

SIR, — The great snowstorm of March 1st and 2nd was in some respects the most severe I ever knew. It snowed here incessantly, as far as I observed, for 50 hours, up to 9 a.m. on the 3rd—though at times very slightly, and it may have ceased at some time. The snow drifted very much, so that it was impossible to tell the exact depth, but as near as I could judge it was 5 cm at 2 p.m., 25 cm at 10.15 p.m. of the 1st; and 32 cm deep at 8 a.m. on the 2nd; and it reached a maximum of about 35 cm on the 3rd. The amounts caught in the rain-gauges, when melted, gave approximately 21.8 mm and 20.3 mm on the 1st, in my new and old gauges respectively; and 9.1 mm and 14.0 mm on the 2nd; but owing to the drifting these amounts would be very inaccurate. The most probable amount, judged by taking the average of the snow gathered and melted from five different spots, is 27.2 mm for the 1st, and 13.5 mm for the 2nd; but, as the five quantities obtained differed widely, it is impossible to judge exactly.

To your list of snowed-up trains and impediments to traffic there should be added that on the 2nd inst. a train became imbedded in the snow between Seaham Harbour and Sunderland; two engines sent to its assistance were also snowed up, and railway communication between the two towns became wholly suspended, as well as between this town and Hartlepool. The deepest drift on road or railway in this district was probably in the railway cutting at Fulwell, a short way out of Sunderland, where the snow was said to be 5.5 m deep, probably somewhat of an exaggeration, which, with accumulations at other points, completely stopped the train service between Sunderland and Newcastle, and Sunderland and Shields, for many hours. A train becoming fixed in this drift at night on the 2nd (Tuesday) blocked one line of rails till

(Saturday) the 6th, when the last carriages were extricated. The Newcastle and Jarrow line was also blocked, and I believe others. I doubt whether such a stoppage to railway communication ever occurred before in this district. The snowstorm most resembling the late one of any I have known was that of 1881, March 4th and 5th, on which occasion the total yield was heavier, but as it was partly sleet the depth here was much less. The weather has continued very snowy ever since the great storm, snow or sleet falling with little interruption, though at times very slightly, from the morning of the 15th to last evening; it mostly did not lie, and the total fall when melted was only 11.7 mm.

T. W. BACKHOUSE
Sunderland
March 18th, 1886

It is a shame that no information was recorded at Durham about snowfall at that time, neither depth nor 'snow lying'. The equivalent precipitation amounts for the 1st and 2nd are 4.2 mm and 11.9 mm, respectively, somewhat less than in Sunderland; for both days the maximum temperature was near 0 °C so most (if not all) would have been snow. At 0900 GMT on 2 March, Durham recorded 10/10 cloud cover, wind north-east 15 knots, air temperature −1.2 °C, and MSL barometer 991.4 hPa. Conditions the following morning were little changed, the temperature −1.9 °C and the wind north-north-east at 24 knots; the barometer had begun rising, however, and stood at 1005.0 hPa.

Durham's lowest barometric pressure: 8 December 1886

An extremely disturbed spell of weather in early December 1886 saw a succession of intense depressions cross north-west Europe. A depression forming in an area of marked thermal contrast in mid-Atlantic on 7 December merged with a depression that had formed near south-west Iceland, and the combined storm deepened rapidly to form a single system with a central pressure close to 930 hPa off north-west Ireland on the morning of 8 December [56, 69]. The storm continued to deepen and to slow as it crossed the north of Ireland, the barometer falling to 927.2 hPa at Belfast at 1330h as the depression centre passed nearby (Figure 22.12). The central pressure may have fallen as low as 924 hPa during the afternoon. At the height of the storm, winds exceeded gale force on all windward coasts between southern Norway and northern and western Spain. At Fleetwood in Lancashire the mean wind speed remained above 26 knots for 65 consecutive hours, the highest hourly mean reaching 69 knots; near the peak of the storm, the lifeboats from Southport and St Anne's were both lost, and 27 of the 29 crew members perished.

An early account of the storm was given by Harding [123]. The synoptic situation at 1400 GMT (Figure 22.12) is based upon Harding's account, together with observations and maps published in the *Daily Weather Report*. This storm established new low barometric pressure records widely across the north of Ireland, southern Scotland, and northern England; they have not been closely approached at any time since. At Stonyhurst in Lancashire, where the barometer fell to 940.4 mbar at MSL, the reading was

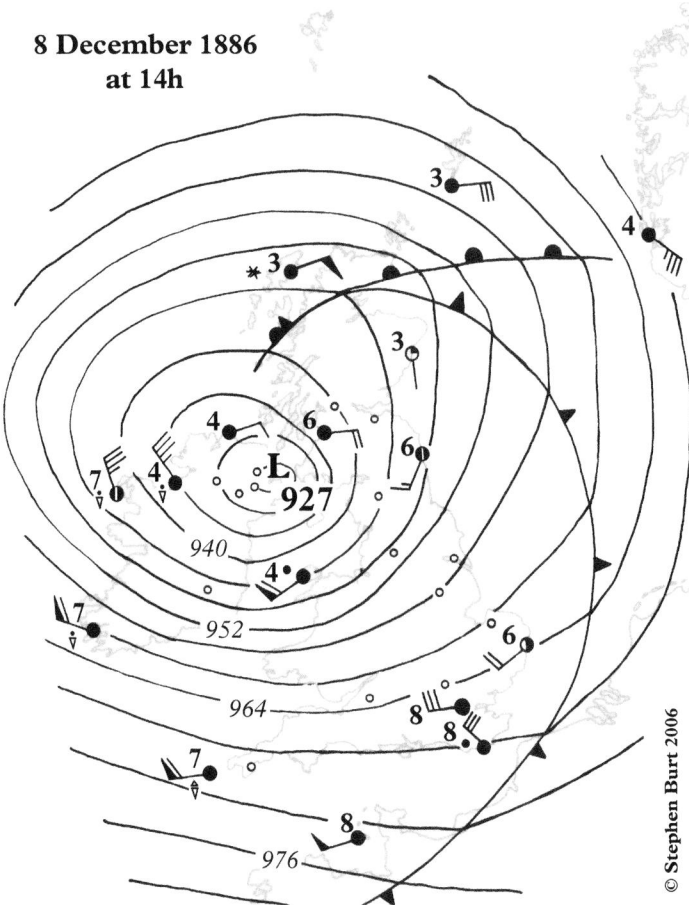

8 December 1886
at 14h

© Stephen Burt 2006

Figure 22.12 *Synoptic chart for 1400* GMT *on 8 December 1886, based upon observations in the Met Office* Daily Weather Report *and in Harding [123]. Sites with pressure observations only are shown as small open circles. This exceptionally deep depression passed close to Durham later that evening (Stephen Burt [56])*

'lower than at any time in the last 40 years'. At Durham, the MSL pressure at 2100 GMT stood at 936.2 hPa, almost 15 hPa lower than anything recorded before or since. At that time, the wind at the Observatory was southerly at 16 knots, indicating that the depression centre still lay to the west or north-west, and as a result the true minimum pressure was probably a little lower than the 2100 reading. By next morning, with the wind west-north-westerly 18 knots at the 0900 GMT observation, the barometer had risen 17 hPa overnight and the depression had moved out across the North Sea.

July 1888: Cold, dull, and extremely wet—Durham's wettest month

July 1888 was both the coldest and wettest July on Durham's instrumental records, and for good measure the sixth dullest, too. With a monthly rainfall of 209.7 mm, it also remains Durham's wettest-ever month, one of only two months to surpass 200 mm (Table 25.3). The *Monthly Weather Report* commented that 'the weather was most unseasonable, pressure was low, depressions passed over England, producing very heavy and constant rains, with frequent thunderstorms'. The absolute maximum for the British Isles was just 25.0 °C at Lairg in northern Scotland; the *Monthly Weather Report* added that 31 °C had been reached at Cambridge in June which made the 'remarkable chilliness of July more apparent'.

The month's statistics make depressing reading for a summer month: four of the five coldest days on record for July (Table 10.1); the coldest July on record, with a mean temperature of just 12.1 °C, 3.7 degrees below the current normal (Table 10.2); the wettest July on record, with more than three times normal rainfall, together with the second-wettest July day (57.1 mm on 25th, Table 10.3); and the third-highest two-day total (95.1 mm 25–26 July) on record. The mean maximum temperature for the month was just 15.6 °C, while only four days reached 20 °C during the month, the warmest day of the month (and indeed of the entire year) being 23.0 °C on the 20th. Figure 22.13 shows the daily temperature, rainfall, and sunshine conditions in Durham during this very cold, very wet, and very dull summer month.

The reason for the dreadful weather can be found in the stream of depressions that crossed the country throughout the month. Many passed to the south of Durham and thus exposed north-east England to persistent and cold cyclonic precipitation. Figure 22.14 shows the monthly mean MSL pressure; a mean low-pressure area lay over south Norway, with an elongated trough westwards to the British Isles.

February–March 1891: very dry, followed by heavy snowfall

Another note from Thomas Backhouse in Sunderland (see also *March 1886 snowstorm)* regarding conditions in February–March 1891 (old units have been converted):

> **Sunderland, West Hendon House**: The rainfall in February was 1.0 mm, part of which may have been condensed fog. The driest month I have recorded previously in my 31 years' observation was June, 1889, when the fall was 6.1 mm. There was no rain from the end of February until March 6th, on which date there was 0.8 mm. The total fall from January 21st till March 6th inclusive was 5.3 mm.
>
> We have now, however, had a sudden change, as it commenced to snow on the evening of the 7th [March], and by 8 a.m. on the 8th the snow was 37 cm deep. This is the deepest fall I have ever known in one day, though not the heaviest, as the snow was very light, being equal to about 24 mm of water up to 8 a.m. On the 8th it snowed nearly all day,

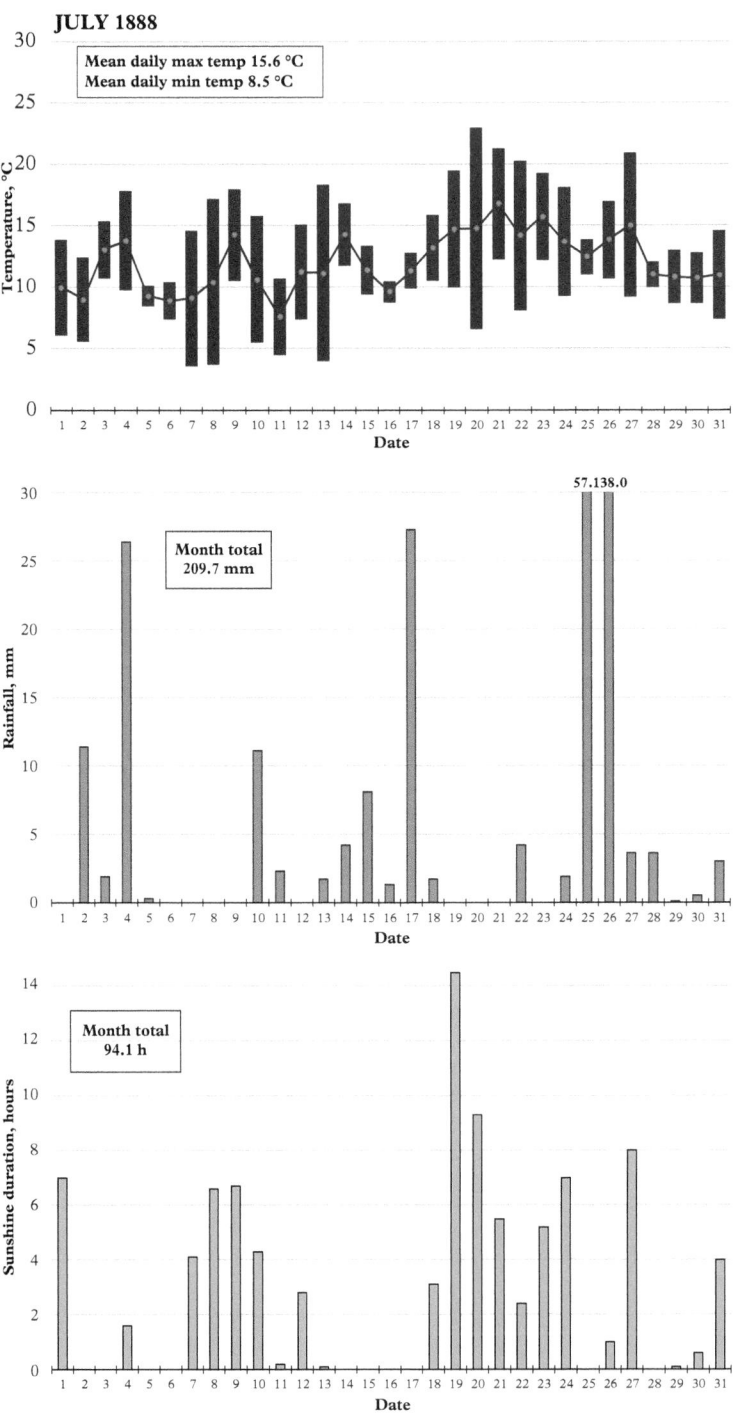

Figure 22.13 *Daily temperature, rainfall, and sunshine data for Durham Observatory for July 1888: (top) daily maximum and minimum temperatures, °C, derived 'Stevenson screen equivalent' from the original Glaisher stand records; (middle) daily rainfall, 24 hours from 0900* GMT, *in mm; (bottom) daily sunshine duration, in hours*

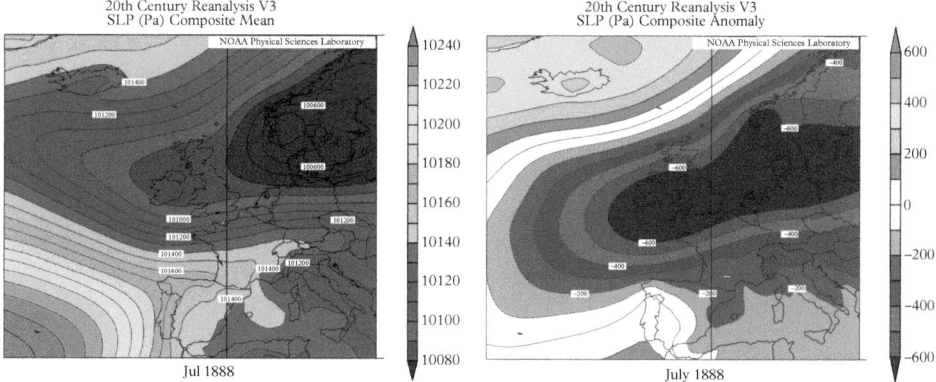

Figure 22.14 *Monthly mean* MSL *pressure (left, units Pa, interval 100 Pa = 1 millibar) and anomaly from 1981–2010 normal (right, units Pa, interval 100 hPa) over the north-east Atlantic and western Europe for July 1888. Plots from 20th Century Reanalysis data provided by the NOAA/OAR/ESRL PSL, Boulder, Colorado, USA, from their website at https://psl.noaa.gov/*

with I am told a little rain, but the depth of snow diminished. In the 24 hours ending 8 p.m. on the 8th, the total fall would be about 36 mm of water, and in the 26 hours up to 10 p.m., when it practically ceased, 37 mm, which is more than the total fall of the previous 71 days. The amount of melted snow in the 24 hours has been exceeded once before in my experience—viz., Nov. 12th, 1878, when the fall (which was 27 cm) gave 41 mm of water.

THOMAS BACKHOUSE
Observer's Notes in *Symons's Meteorological Magazine*, Vol. 26, 1891, pp. 24–25

At Durham, 8.2 mm precipitation was recorded in the 45 days from 21 January to 6 March inclusive. We do not have any records of snow depth in Durham during this spell, but 25.7 mm precipitation fell during 7–8 March, almost evenly divided between the two days. The maximum temperature on 8 March was just 0.5 °C, so most if not all of that day's precipitation would have fallen as snow, and probably that during the previous night, too. At 0900 on 8 March, cloud cover was 10/10, temperature −1.1 °C, and wind north-north-east 10 knots.

20–22 May 1894—sharp May frosts

Few persons perhaps remember so destructive and disastrous a May frost as that which occurred on the 21st of May in this year. Following upon a week of abnormally cold weather, we had a slight frost on Saturday night and Sunday morning, but it was left to the night of the latter day, and yesterday, to complete the work of destruction. A frost of any sort in May is always dangerous, and often destructive; but when it comes to 22 °Fahrenheit [−5.6 °C] then, indeed, is the disaster complete, and the hopes of the year are absolutely and entirely destroyed.

The morning of the 21st broke bright and sunny after the overnight frost, and disclosed a sorrowful sight. Ice a quarter of an inch in thickness [6 mm] was plentiful, and on the north side of walls and buildings the ground remained frost-bound far into the forenoon. In most gardens, and in many fields, potatoes were well above the ground; by noon there was nothing but the blackened remains of the haulms above the ground level, so effectually are they killed. Seeding turnips, instead of being green, are turned to a sickly purple hue. According to Mr W. Thompson, the thermometer at the Deanery Gardens registered 22 °F [−5.6 °C], or 10° of frost.

Newcastle Daily Chronicle, reproduced in *Symons's Monthly Meteorological Magazine*, June 1894, pp. 69–70

The minimum temperatures at Ushaw College for the three nights 20–21–22 May 1894 were −1.3 °C, −0.8 °C, and −0.7 °C, respectively. At Durham, the equivalent values were −0.4 °C, −1.7 °C, and −1.1 °C respectively, and −1.3 °C the following night (23rd). Grass minimum temperatures for the respective four nights were −2.8 °C, −5.6 °C (the 21st), −5.3 °C, and −4.7 °C, among the lowest on record so late in the year at Durham.

The winter of 1895

The weather of early 1895 was quite unlike the previous autumn, when sustained heavy rainfall generated the greatest recorded flood on the River Thames at Teddington since records began in 1883 [49, 124]. During January and February 1895, a 'blocking' Siberian anticyclone extended westwards over Scandinavia and Scotland, producing extremely cold weather over most of the British Isles.

The winter of 1895 remains the third coldest on record at Durham, with a mean air temperature of 0.9 °C; only the winters of 1963 and 1879 have been colder (Table 17.3). In the long CET record, winter 1895 is ranked eleventh coldest, with a mean CET 1.2 °C, ranking equal with 1709 and 1784, still some way behind the coldest of all on the CET record, which were (ranked, coldest first) 1684, 1740, 1963, 1814, and 1795.

December 1894 was relatively mild, but cold weather set in just before New Year. Between 29 December and 27 February, inclusive, Durham's highest temperature was 6.3 °C, and the mean −0.8 °C. The most severe cold occurred 10–12 January (three consecutive nights below −12 °C) and 7–11 February (five consecutive nights below −12 °C). On three days in January and five in February, air temperature failed to rise above zero, so-called 'ice days'. The coldest days were 6 and 7 February, both days with a Stevenson-screen equivalent maximum of −3.3 °C, while the coldest night was 8 February, when the minimum temperature was −18.0 °C, the coldest night on Durham's records (see February's chapter, and Table 24.5). During the same spell, the −16.5 °C recorded on 10 February remains Durham's third-lowest minimum temperature, with nothing lower since (Figure 22.15).

January 1895's mean temperature was −0.6 °C, 4.7 degrees below the current normal and Durham's third-coldest January (Table 4.2); only 1881 and 1940 have been colder. February remains the fourth coldest by mean monthly temperature (−1.0 °C, 5.6 degrees

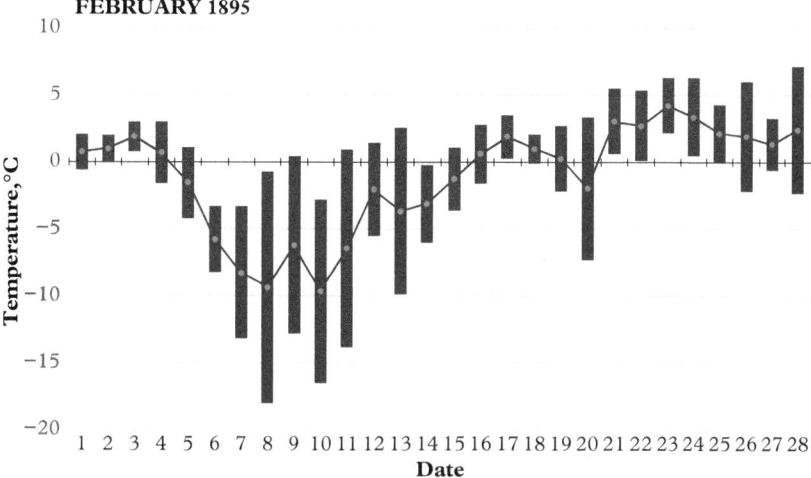

Figure 22.15 *Daily maximum, minimum, and mean daily temperatures from the Glaisher stand thermometers at Durham Observatory, corrected to approximate Stevenson screen equivalents, during February 1895*

below normal); only the Februarys of 1855, 1947, and 1963 were colder in Durham (Table 5.2). Air frost occurred on 25 nights in January and 20 in February. No surprise, then, that the River Wear froze over during this prolonged cold spell, permitting curling and skating to take place (Figures 5.2 and 17.4).

Further west, the severe cold caused lakes to freeze. Windermere was frozen from end to end by early February and thousands came to skate and sledge—an early example of mass tourism by train. On Saturday 16 February special trains from Lancaster, Manchester, and Liverpool brought an estimated 2000 skaters, while the following Saturday the crowds were estimated at 15,000 to 20,000 [125, 126]. Arthur Ransome was at school in Bowness at the time and was able to use his memories to produce the Swallows and Amazons story *Winter Holiday*.

Durham's highest barometric pressure: 23 January 1907

Contrary perhaps to expectation, most intense midwinter anticyclones that affect the British Isles are 'Atlantic' rather than 'Siberian' in origin. However, Durham's record highest barometric pressure was without doubt merely a sideshow to the very intense anticyclone that developed over western Russia in late January 1907. On 23 January 1907 at 1015 GMT, the barometer at Aberdeen reached 1051.8 hPa, and during the morning 1050 hPa was exceeded over all of Scotland except the far north-west, northern England, and the extreme north-east of Ireland (Figure 22.16). On this occasion the

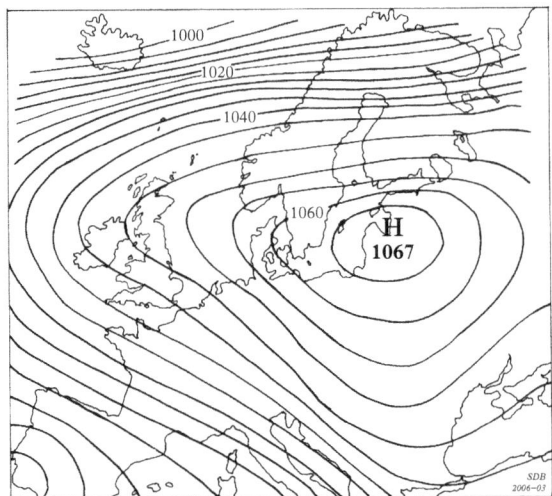

Figure 22.16 *Synoptic chart for the morning of 23 January 1907, when the barometric pressure at Durham reached its highest level on record. Isobars are in hPa (= millibars), at 5 hPa interval. (Stephen Burt [55])*

barometer reached 1067 hPa at Pernau in Russia (now Pärnu, Estonia) on the Gulf of Riga, at the evening observation on 22 January, and also at Riga (now in Latvia) on the morning of 23 January [55, 127]. These observations are notable as both cities are on the Baltic coast and therefore the value of the barometric correction to mean sea level is very much smaller than at stations on the higher Siberian or Mongolian plateaux much further east. At Durham the MSL pressure at 0900 GMT on 23 January 1907 was calculated as 1052.5 hPa, although this appears to be slightly too high; at North Shields the MSL pressure at 0800 GMT was 1050.8 hPa, while a little further south at Meltham in West Yorkshire the maximum reached on this occasion was 1050.1 mbar. The true maximum at Durham on this occasion was probably between 1051 and 1051.5 hPa—a value that has not been attained before or since.

Heavy snowfall in County Durham, 28–29 January 1910

The note in *British Rainfall* for 28 January 1910 quotes *The Times*:

Reports brought in by drivers of express trains north and south show that snow had generally fallen over the entire country between London and Aberdeen. The snow plough was used on the Waverley route. Over the Lammermuir area there has been no communication since the beginning of the week, and as renewed drifting occurred yesterday morning, the promise of getting in touch with many farm places is still further removed. There has been heavy loss of flocks.

British Rainfall goes on to say that roads and railways were much blocked in Northumberland and County Durham.

This date is too early in the Durham Observatory records for observations of snow depth, but we can infer weather conditions from the twice-daily observations. On 28 January, temperatures at the Durham Observatory ranged between −5.4 °C (minimum) and +0.6 °C (maximum), and things were little better the next day with temperatures between −2.2 °C and +1.1 °C. Given the low temperature, we can be sure that the 22 mm of precipitation recorded on 28 January fell as snow. Wind speeds were high in this period too: 18 knots from the north-east at 0900 GMT on 28 January, increasing to near-gale 31 knots from the north at 2100; it was still 18 knots from the north-west at 0900 GMT next morning. No surprise then that drifting was a widespread problem.

Figure 22.17 shows Rowley Station, just south of Consett, some 18 km west-north-west from Durham. The Beamish Museum has the following (slightly edited) description for the two images, which are entitled *Rowley Station after a heavy fall of snow*:

The 6.20 p.m. Blackhill to Darlington train was held up at Rowley from 6.53 p.m. on Friday 28 January to 4.30 p.m. on 30 January because the line ahead was blocked by snow. This information was provided by Mr G.W. Bainton whose father, Mr G. R. Bainton, the stationmaster, is seen on the right of one of the photographs. This particular incident started on Friday 28 January a day of heavy snow: difficulty was experienced in keeping the line open to Burn Hill. The 'ploughs' derailed at Burn Hill on starting the last run for the last Blackhill to Saltburn train. A 'pilot' from Consett shed (the 0-6-0) was leading, but it was hopeless to go forward and impossible to go back to Blackhill. This was about 6.30 p.m. and we were really cut off. It was decided to keep the steam heating on by shovelling snow into the loco tanks. There were about 17 people on the train including the loco crew, but only one lady. She spent the night in the station house, the others remained in the coaches. When daylight came, it was found that the drifts were up to 12 ft [4 metres]. My mother had a great deal of food in stock, but that soon went down. A huge joint for a farmer living 3 miles [5 km] away was cooked. The lady passenger helped in the home whilst stranded. By Sunday afternoon the road was cut through to Castleside, this is how the photos were able to be taken. Late that afternoon the plough made it. The approach to Rowley Station was down the side of a north-facing hill and the view of the ploughs coming down was marvellous, with snow being sprayed quite a distance. Of course the line could only be opened as single line working. In view of any possible troubles, both locos set off with only one coach each. At the time I was only 9 years old, but it was something never to be forgotten.

Figure 22.17 *Rowley Station in County Durham following the heavy snowfall on 28 January 1910. (Courtesy of Beamish Museum)*

Rowley Station (NGR NZ 087 479) opened in 1845 and was situated on the west side of the A68 on what is now the Waskerley Way cycle track, which passes through the station site at about 268 m above MSL. By the 1970s, the station buildings had fallen into disrepair; however, in 1972 they were dismantled for reassembly at Beamish Museum.

1912: another year with no summer

Gordon Manley was well aware of the impact of the Tambora volcanic eruption on global climate and clearly had no hesitation in attributing the poor weather in Durham (which lasted from summer 1912 through to winter 1913) to the eruption of Mount Katmai in Alaska on 6–9 June 1912 [19]. Notwithstanding that this eruption is commonly referred to as the Katmai eruption, there was in fact another, larger eruption on the same day from Novarupta—a second nearby volcano—which is located approximately 10 km west of Katmai, the volcano responsible for the majority of the SO_2 injected into the stratosphere. A large amount of the magma from Novarupta drained from under Katmai, which caused its summit to collapse, forming a caldera more than a kilometre deep [83].

If it really was Katmai that was largely responsible for the dreadful summer of 1912—one which must have come as an unpleasant shock after the prolonged and hot summer of 1911 [82]—then its effects were surprisingly immediate.* All three summer months were exceptionally dull: June (92 hours of sunshine, 18 per cent of possible) was the dullest on record until 1987 (Table 9.4), while both July (74 hours of sunshine, 14 per cent of possible) and August (just 57 hours of sunshine, 12 per cent of possible and equalling *December's* average sunshine) remain the dullest of those months today (Tables 10.4 and 11.4, respectively). It is hardly a surprise then that 1912 was, by a huge margin, Durham's dullest summer with just 222 hours of sunshine in all (the next-dullest summer, 1987, had 350 hours—Table 19.4). Not even one day managed more than twelve hours of bright sunshine during the whole summer, while in dismal August the sunniest day, the 8th, recorded only 6.1 hours—many midwinter days have surpassed this duration of sunshine, and it remained easily the lowest 'sunniest day' for any August on Durham's records until daily records ceased in 1999. As a result of the cloudy and wet conditions, all three months were cool or very cool—August 1912 is still the coldest on Durham's records, with a mean temperature of just 11.7 °C, almost 4 degrees below normal (Table 11.2). As well as gloomy and cool, it was a wet summer, still the fifth wettest on record—Table 19.3: the wettest periods in June and August coincided with the hay and grain harvests, with calamitous impact on agriculture and crops rotting in the fields.

For the year as a whole, the only month with above-average sunshine was April (the month of the *Titanic's* ill-fated maiden voyage), which received 190 hours, at the time Durham's third-sunniest April—and rain fell on only three days. Not surprisingly, 1912 remains Durham's dullest year by far, and the only one to date to have recorded less than 1000 hours of bright sunshine: the annual total was just 982 hours, 22 per cent of possible sunshine, and barely half that of the sunniest year (2003, with 1703 hours—Table 16.13). In 1912, much of Scotland except the west and south-west coasts, most of northern England, and parts of the Midlands also recorded less than 1000 hours sunshine during 1912; among the lowest totals being 779 hours in Manchester (City) and 822 hours at Manchester (Whitworth Park), 827 hours in Hull, 868 hours in Glasgow, 926 hours at Stonyhurst, 933 hours in Birmingham, and 944 hours in York (*Monthly*

---

* Popular media of the day attributed the poor summer to the huge increase in use of 'wireless telegraphy' (i.e. radio) following on from the sinking of the *Titanic* in April that year.

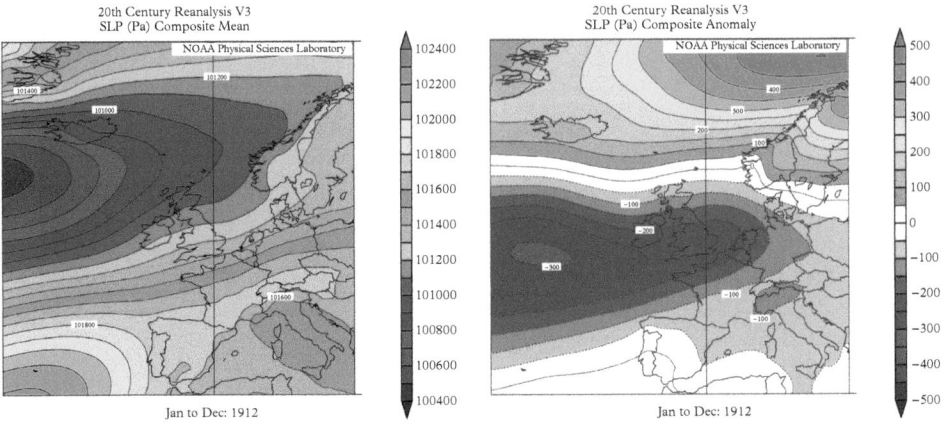

Figure 22.18 *Monthly mean* MSL *pressure (left, units Pa, interval 100 Pa = 1 millibar) and anomaly from 1981–2010 normal (right, units Pa, interval 50 hPa) over the north-east Atlantic and western Europe for the year 1912. Plots from 20th Century Reanalysis data provided by the NOAA/OAR/ESRL PSL, Boulder, Colorado, USA, from their website at https://psl.noaa.gov/*

Weather Report). Figure 22.18 shows the mean MSL pressure for the year over the North Atlantic region, together with the anomaly from the 1981–2020 normal.

The coming of the New Year made little difference—winter 1912/13 remained resolutely dull, December managing just 29 hours of sunshine, while January was duller still, with just 15.5 hours of sunshine, or 6 per cent of possible; there have been only two duller Januarys since (Table 4.4).

While we remember 1816 as 'the year with no summer', 1912 was very similar in that respect. However, the eruption did not have a global impact. Aerosols from tropical volcanic eruptions such as Tambora tend to spread over the entire globe, whereas those ejected from high-latitude eruptions typically remain in the hemisphere in which they were injected. The largest radiative forcing effects thus tend to happen in extratropical regions, and the climate response differs to that from tropical eruptions [83]. Tambora was also a much bigger eruption than Katmai, accentuating differences in climatic impact. Nevertheless, the Katmai–Novarupta eruption apparently made a major impression on the Durham climate record, as Gordon Manley recognized, and is still notable in the records more than 100 years later.

As a postscript to the preceding tale, the following perceptive observations by Thomas Backhouse in Sunderland during 1912 certainly confirm the presence of atmospheric aerosols, presumably of volcanic origin. This note was written late in December 1912:

The aspect of the sky in 1912

Excess of cloud has often been noted this year, but I have not seen any mention of, what has appeared to me very remarkable, the want of blueness in the sky, especially during the summer. With this seems to be connected a phenomenon of the nature of 'Bishop's

Ring'. It may be remembered that after the eruption of Krakatoa in 1883, there appeared a large corona round the sun, the outer part of which was pink or salmon-coloured, and was called 'Bishop's Ring'; the interior was light blue or green; it was best seen near sunrise and sunset. In the course of many years this gradually faded, and ultimately no trace of it remained; whether there was any feeble manifestation of the kind before the eruption, I am unable to say, as one's attention was naturally not directed particularly to the point. Since then, however, there has at times been a recrudescence of a large corona, which no doubt is capable of being formed by other kinds of dust; and this year it has been more marked, though of very feeble colours, the reddish being replaced by a dull brownish hue, and the interior being bluish or greenish white, the result being an excessive glare round the sun. The corona is rather larger than in the case of the Krakatoa dust, the distance of the brownest part being perhaps 40° from the sun.

It would appear from these observations that there must be some kind of film high up in the atmosphere—not so high up as the Krakatoa dust was—and that this film still remains, as the corona continues, though the sky is sometimes fairly blue.

It is not only at Sunderland that these remarks apply, but also in other places where I have travelled.

There has been, however, a singular scarcity of true cirrus, and also of halos.

T. W. BACKHOUSE
West Hendon House, Sunderland
21st December, 1912
From *Symons's Meteorological Magazine*, Vol. 47, 1912, pp. 244–245

The following comments regarding atmospheric conditions that year also appeared in the Annual Summary for 1912 of the *Monthly Weather Report*, dated 'Meteorological Office, South Kensington, March 11th, 1913':

> **16. Summer Sky.**—In this and in other countries on both sides of the Atlantic there was, during the summer months, a peculiar haze-like veil over the sky, 'producing a grey whiteness,' and resulting in a marked diminution in the brightness of the sunshine. There is, as yet, no adequate explanation of the phenomenon, which has been attributed to such widely different circumstances as the large quantity of ice in the Western Atlantic, and an Alaskan volcanic eruption.

25 June 1930: Gordon Manley marries the girl next door!

Gordon Manley married Audrey Fairfax Robinson in Durham Cathedral on Wednesday 25 June 1930 (Figure 22.19). Audrey was the daughter of Arthur Robinson, Master of Hatfield College. Since Gordon Manley was a resident don in the College, it seemed reasonable to assume that they met in Hatfield, but their marriage certificate raises another intriguing and climatically romantic possibility: that they met at the Observatory.

Since the head of a college is usually resident in their college, it might too easily be assumed that the Robinson family was resident in Hatfield College, but the marriage certificate shows that they were actually living in Observatory House, a large property

Figure 22.19 *The wedding of Gordon Manley and Audrey Robinson, 25 June 1930 (reproduced with permission from Durham University Library and Heritage Collections)*

within the Observatory grounds. They were living there because the Master's house in College was still occupied by Robinson's predecessor [128]. Robinson had moved into Observatory House when he was Vice-Chancellor 1922–24 (a rotating two-year appointment in those days); he was appointed Master of Hatfield College during his tenure as Vice-Chancellor in 1923. Clearly, the Robinsons stayed put in Observatory House at least until 1930. Thus, Gordon Manley, the man who put the Durham Observatory temperature records on the map, literally married the girl next door!

The wedding day was pleasantly warm, although not as warm as might have been hoped. It was already 15 °C at 0900 but the daytime maximum reached only 17.8 °C; the day was dry, with 10.8 hours of bright sunshine. Cloud cover was 4/10 at 0900 GMT and 3/10 at 2100 GMT; winds were westerly, a brisk Beaufort Force 4 at 0900, moderating to a light Force 1 from the south-west by 2100. A pleasant summer's day, if perhaps a little on the cool side.

21 January 1940—Durham's coldest night of the twentieth century

The *Monthly Weather Report* commented that January 1940 'will long be remembered for the intense cold'. The majority of the month was dominated by high pressure over Scandinavia (Figure 22.20), and winds from between east and south-east dominated the month. The coldest conditions occurred between the 17th and 23rd. At Durham Observatory, the temperature fell to −16.1 °C on 21 January (minimum temperature above snow −19.4 °C), the coldest night since February 1895 (Table 24.5) and the coldest night of the twentieth century—later equalled on 11 January 1982. With clear skies and light winds over a snow cover overnight 20/21 January, conditions were ideal for katabatic drainage into the Wear Valley: the minimum temperature at Houghall on the valley floor (Figure 1.1) that morning was −20.0 °C (Table 1.4). That same morning

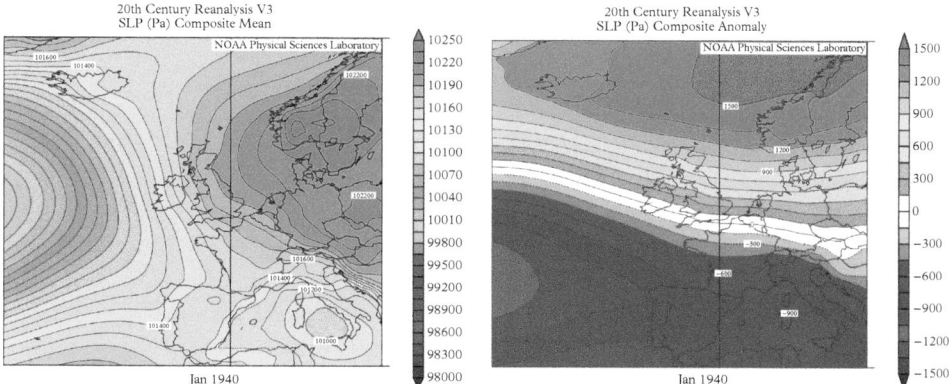

Figure 22.20 *Monthly mean* MSL *pressure (left, units Pa, interval 100 Pa = 1 millibar) and anomaly from 1981–2010 normal (right, units Pa, interval 100 hPa) over the north-east Atlantic and western Europe for January 1940. Plots from 20th Century Reanalysis data provided by the NOAA/OAR/ESRL PSL, Boulder, Colorado, USA, from their website at https://psl.noaa.gov/*

the temperature fell to −23.3 °C at Rhayader in central Wales, the lowest temperature anywhere in the British Isles that month.

The mean temperature for January 1940 at Durham was −1.5 °C, 5.6 degrees below the current average and, with the sole exception of 1881, the coldest January on the instrumental record (Table 4.2). On the Eglise series prior to the commencement of the instrumental record at Durham Observatory, only the Januarys of 1795 (mean temperature −2.4 °C) and 1814 (mean −2.3 °C) were colder.

The great north-east blizzard of 18–19 February 1941

In Durham, snow started to fall around midday on 18 February 1941 and did not stop until 7 a.m. on 21 February; this period of 67 hours of continuous snowfall might well be the longest in the Durham record. By the time it stopped, snow lay to a depth of 107 cm, a value noted in both *British Rainfall* and the *Monthly Weather Report*. Wheeler [129] reports that nothing comparable could be found in newspaper reports going back to 1814 in Durham and 1831 in Sunderland.

On 15 February a deep depression formed at an unusually southerly location near the Azores. It then moved north-east to Portugal while deepening, then towards the British Isles, with its centre at 976 hPa west of Cornwall at 0700 GMT on 17 February. The depression then began to fill, and was located near Lyme Bay at 984 hPa 24 hours later, and offshore from The Wash by noon GMT on 19 February with central pressure little changed (Figure 22.21) [129]. Very cold air was drawn into the depression from as far afield as Russia; conditions there were so cold that the German military advance came to a halt. The easterly airstream's passage over the relatively warm North Sea allowed it to gain water vapour and heat from the sea surface, becoming highly unstable in the process. As a result, air temperatures at ground level did not fall much below 0 °C while copious quantities of 'wet' snow fell in coastal districts. The deep snow caused

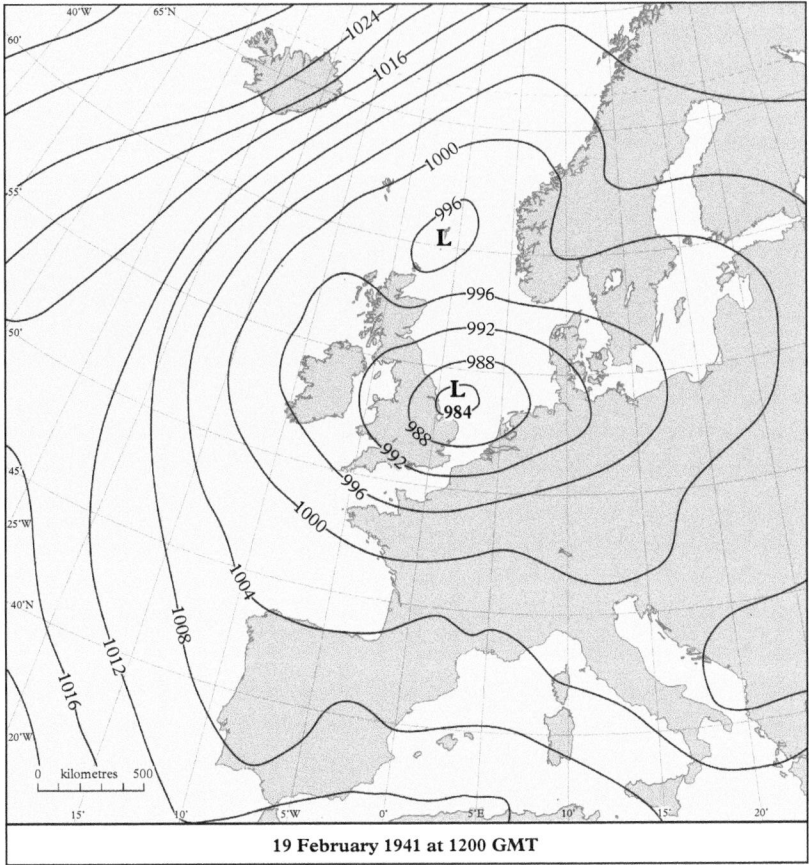

19 February 1941 at 1200 GMT

Figure 22.21 *The synoptic situation at 1200* GMT *on 19 February 1941, during one of Durham's heaviest snowfalls. (Based upon the 20th Century Reanalysis version 3 (20CRv3) plot [58] and the Met Office* Daily Weather Report: *drawn by Chris Orton, Durham University)*

considerable disruption, not least to the mainline railway to Scotland, with six trains buried in drifts just north of Newcastle. Sunderland was cut off for three days.

Following the snowfall, the development of a weak anticyclone resulted in clear, calm conditions, which with the deep, fresh snow cover favoured very low night-time temperatures. Table 22.4 shows conditions in Durham during the period 15–28 February 1941. Rainfall equivalents during the blizzard on 18–19 February amounted to 75 mm at Durham, although this is certainly something of an estimate considering the depth of the snowfall[*] (a standard rain gauge rim is only 30 cm above ground level); the fall of 50 mm on 19 February remains Durham's highest precipitation for any February day

[*] There is a letter from the Meteorological Office in the Observatory files in the Durham University Library, DU reference 1326 dated 7 March 1941, which confirms that the rainfall equivalents for these two days of the blizzard were estimated based on snow depths.

Table 22.4 *Durham Observatory weather observations for the second half of February 1941*

	24 HOURS					0900 GMT			
Date	Max temp °C	Min temp °C	Grass or snow min temp °C	Precip'n mm	Sunshine hours	Cloud cover tenths	Wind direction	Wind force Bft Force	MSL pressure hPa
15	7.2	1.1	1.7	0.3	2.8	10	CALM		986.0
16	5.6	−1.7	−5.0	13.8	0	9	NE	1	991.5
17	3.3	0.0	1.1	17.5	0	10	NE	4	989.3
18	1.1	0.0	0.0	25.0	0	10	N	1	990.1
19	0.6	−0.6	0.0	50.0	0	10	N	2	987.8
20	0.0	−1.1	*no data*	15.0	0	10	N	2	987.2
21	1.7	−1.7	*no data*	–	4.0	8	W	1	991.8
22	3.9	−6.1	−6.7	–	7.5	0	CALM		997.4
23	2.2	−7.2	−6.7	–	4.2	8	WNW	2	996.7
24	2.8	−7.8	−8.3	2.5	1.1	8	NW	2	997.0
25	3.9	−5.6	−5.0	–	5.0	6	NW	1	1011.4
26	1.1	−10.0	−11.1	2.0	7.5	0	CALM		1014.5
27	7.2	−4.4	−1.7	1.0	0	10	SSW	4	994.8
28	7.8	3.3	0.0	1.0	7.2	2	SW	6	988.5

(Table 5.3) and is supported by 70 mm reported at Ushaw. Durham's coldest night during this spell was on 26 February, when the minimum temperature fell to −10.0 °C. At the Houghall site in the Wear valley, the screen minimum was −15.6 °C on 24 February, with a snow surface minimum of −21.1 °C, while on 26 February the screen minimum was −16.1 °C with a snow surface minimum of −21.7 °C [129]. A thaw set in from the 27th as winds backed to south-west and the barometer fell sharply with the approach of a deep Atlantic depression. Widespread flooding resulted from heavy frontal rainfall over the Pennine hills combined with snowmelt, but details are scant given wartime censorship of information regarded as sensitive at the time, and in the period prior to river gauging records on north-eastern rivers.

The snowy winter of 1947

The winter of 1947 only became severe towards the end of January (Figure 22.22). February was a bitterly cold and snowy month—Durham's only colder February to that time was in 1855, although February 1963 was fractionally colder (Table 5.2). Heavy snowfalls blocked roads and railways, particularly in eastern and north-eastern districts, causing problems transporting coal to coal-fired power stations. As a result, many were forced to shut down, imposing major restrictions in power supply, including limiting domestic electricity supply to nineteen hours per day and cutting some industrial supplies completely. The harsh winter caused severe hardship in both economic terms and living conditions in a country still recovering from the Second World War.

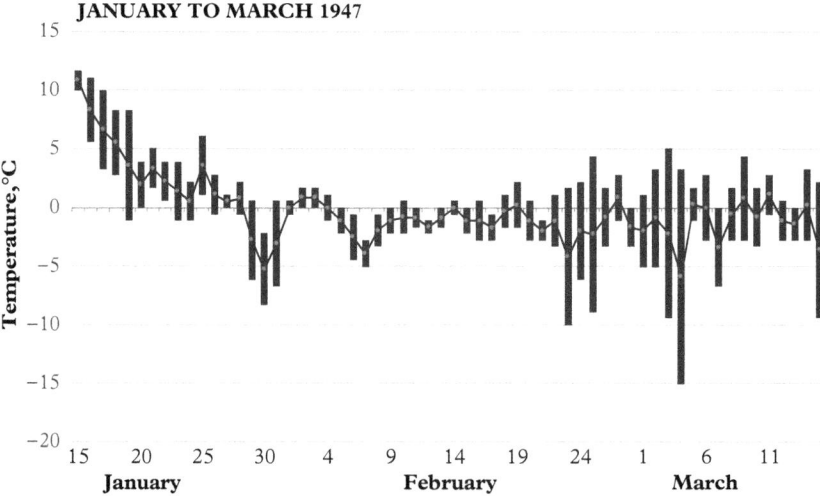

Figure 22.22 *Daily maximum, minimum, and mean daily temperatures at Durham Observatory between 15 January and 15 March 1947*

At Durham, snow fell on ten days in January, and lay on six, almost certainly the last six days of the month. Snow lay throughout February, and snow fell on 18 days; in March, snow lay on 21 days, and fell on ten. For the period late January to the end of March, snow lay on 55 mornings, probably continuously or nearly so: in Durham, March 1947 remains both the equal coldest on record (with 1883; Table 6.2) and one of the five wettest (Table 6.3), for which read 'snowiest'. We have records of a few snow depths from this notorious winter: at Ushaw, level snow lay 38 cm deep on 5 February, 56 cm on the 15th, 71 cm on the 28th, and 86 cm on 14 March, on which date it lay 43 cm deep at Durham Observatory (*Monthly Weather Report*). Snow depths were even greater in the upland villages—at Forrest-in-Teesdale in County Durham, snow lay 112 cm deep on 6 February and 135 cm deep on 18 February.

The coldest night of the winter came in early March, when under clearing skies and deep snow cover the minimum temperature at Durham fell to −15.0 °C on the 4th—still the lowest temperature on record for the month (Table 6.1). As in January 1940, conditions were ideal for exceptionally low temperatures in the Wear Valley at Houghall, which that night recorded its lowest minimum on record, −21.1 °C, a value equalled once previously on 5 January 1941 [20, 130]. The Observatory occupies an open summit on a plateau surface into which the River Wear is entrenched by some 60 m: Manley [20] remarked upon the exceptional effect of outward radiation on clear, calm nights with a deep snow cover impeding the conduction of heat from the soil below, emphasizing the considerable aggregate effect of shallow, gentle cold-air drainage, often only a few metres deep at most. As the cold air accumulates on the valley floor, it continues to lose heat to the clear sky above, and, again, the coldest air sinks to the bottom. Manley estimated that the difference in temperature between plateau and valley floor on a calm clear night with deep snow was three times greater than it otherwise would have been. On this occasion, the difference in screen minima was 6 degrees, whereas the mean difference for March over the period 1925–45 was only 0.3 degrees [130].

When milder conditions finally returned in mid-March, heavy rainfall combined with snowmelt to cause extensive flooding across the country.

The very cold winter of 1962/63

The coldest winter on Durham's instrumental record—and probably since 1740—began abruptly just before Christmas in 1962, when on 22 December a high-pressure system moved to the north-east of the British Isles, dragging bitterly cold winds across the country. This situation, with only slight variations, was to persist much of the winter (Figure 22.23). At Durham, the first air frost of the long, cold spell was on 22 December, with the first ice day on the 23rd. Air frost occurred every night thereafter until 5 March (74 consecutive nights), with the exception of three nights at the end of January (27th–29th). Daytime temperatures struggled to get much above zero and remained at or below 0 °C all day on five days in both January and February 1963: the mean temperature for the winter (December 1962 to February 1963) was −0.1 °C (Table 17.4). Most of the precipitation in January and February thus fell as snow (and both months were 'wetter'

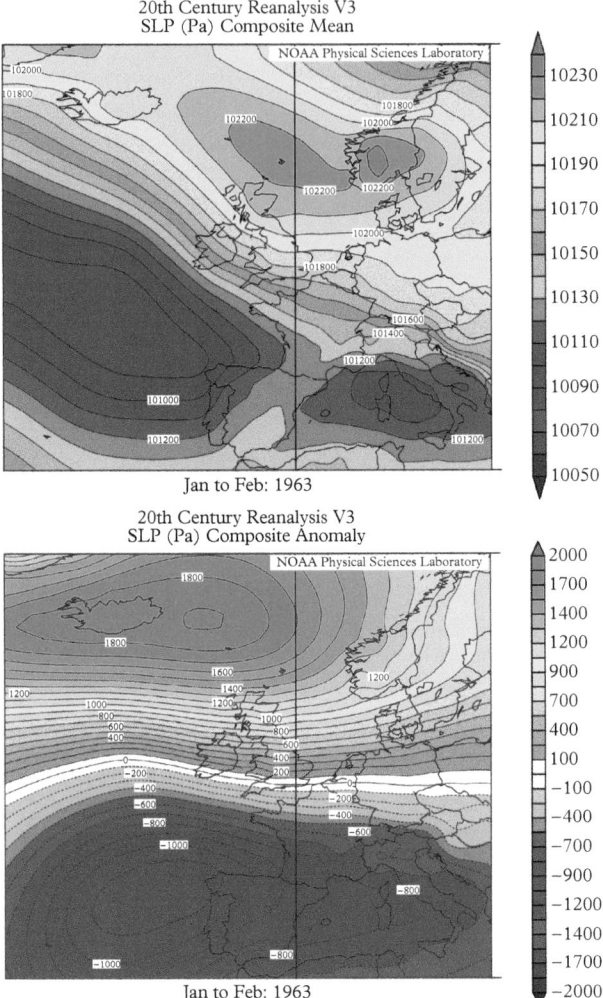

20th Century Reanalysis V3
SLP (Pa) Composite Mean

Jan to Feb: 1963

20th Century Reanalysis V3
SLP (Pa) Composite Anomaly

Jan to Feb: 1963

Figure 22.23 *Mean MSL pressure (top, units Pa, interval 100 Pa = 1 millibar) and anomaly from 1981–2010 normal (bottom, units Pa, interval 100 hPa) over the north-east Atlantic and western Europe for January–February 1963. Plots from 20th Century Reanalysis data provided by the NOAA/OAR/ESRL PSL, Boulder, Colorado, USA, from their website at https://psl.noaa.gov/*

than normal, with 75 and 72 mm, respectively). At Ushaw College, snow fell on 16 days in January and 17 in February; at Durham Observatory, snow lay on the ground for 70 consecutive mornings from 27 December to 6 March inclusive; in all, that winter recorded 76 mornings with snow cover.

Figure 22.24 shows minimum temperatures at four locations from 23 December 1962 to 6 March 1963. For Durham, there are data from both the Observatory and Houghall

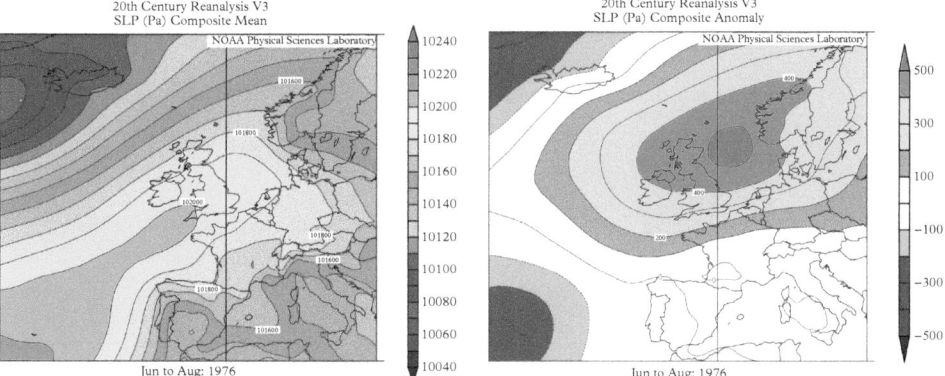

Figure 22.26 *Seasonal mean* MSL *pressure (left, units Pa, interval 100 Pa = 1 millibar) and anomaly from 1981–2010 normal (right, units Pa, interval 100 hPa) over the north-east Atlantic and western Europe for summer 1976 (June to August). Plots from 20th Century Reanalysis data provided by the NOAA/OAR/ESRL PSL, Boulder, Colorado, USA, from their website at https://psl.noaa.gov/*

Autumn 1976, and Durham's wettest day on 11 September

Coming at the end of the long drought of 1975/76, this event was a decisive end to that long, dry period. There were heavy thunderstorms at the end of August over the Bank Holiday weekend, followed in early September by two quite deep depressions, which crossed the country from west to east bringing the drought to an end. Rainfall amounts from the first depression were modest, but the second system produced many hours of heavy cyclonic rainfall to the north of the depression centre (Figures 12.1 and 22.27), amounting to 32.3 mm on 10 September and 87.8 mm on 11 September. The latter remains Durham's greatest one-day fall, while the combined total of 120.1 mm holds the same rank in the two-day falls (Tables 28.3 and 29.7). Notwithstanding very dry antecedent conditions, the River Wear recorded what is now its 15th-highest flood peak since late 1957, 255.5 cumecs; at the time only the peak of 6 November 1967 had been larger.

The hydrological and synoptic contrasts between the hot, dry, and sunny summer and the cold, wet, and dull autumn of 1976 could hardly be greater. Figure 22.28 shows the mean pressure distribution for autumn 1976, together with the anomaly from the 1981–2010 normal. Compare with Figure 22.26; the anomaly centre of +5 hPa over the North Sea in summer was replaced by one of −4 hPa over the English Channel during autumn. Autumn 1976 remains the wettest on Durham's records (Table 20.3); its total of 359 mm was almost five times the summer total. It also remains the dullest autumn on record, with less than half normal sunshine (Table 20.4), and without doubt one of the most startling reversals of seasonal weather in Durham's long climate record.

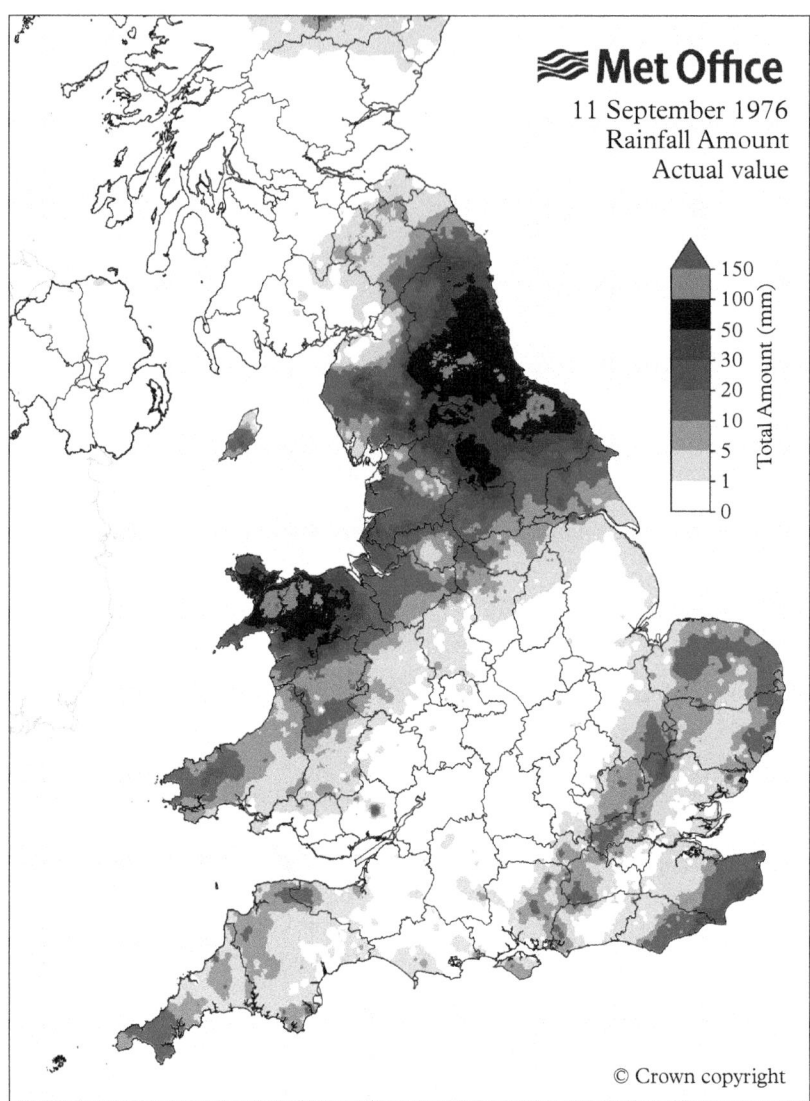

Figure 22.27 *Daily rainfall totals for the 24 hours commencing 0900 GMT on 11 September 1976. Durham recorded 87.8 mm on this date, the wettest day in 150 years of rainfall records. (Based upon the Met Office 1 km gridded rainfall dataset [121], courtesy of National Climate Information Centre, Met Office Exeter: © Crown Copyright, Met Office)*

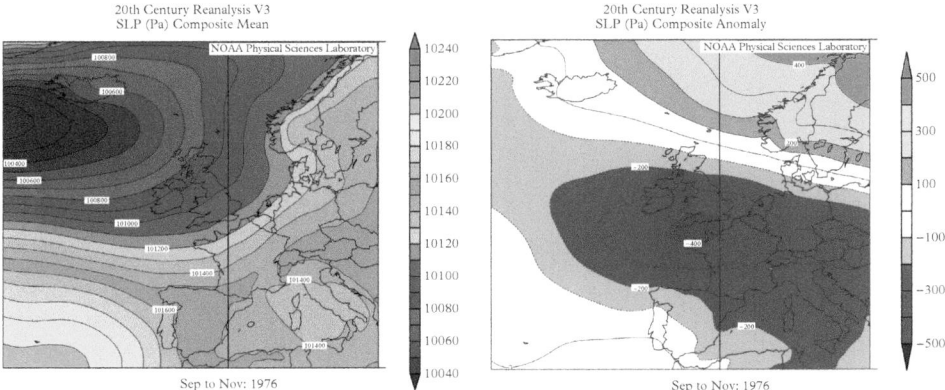

Figure 22.28 *Seasonal mean* MSL *pressure (left, units Pa, interval 200 Pa = 1 millibar) and anomaly from 1981–2010 normal (right, units Pa, interval 200 hPa) over the north-east Atlantic and western Europe for autumn 1976 (September to November). Plots from 20th Century Reanalysis data provided by the NOAA/OAR/ESRL PSL, Boulder, Colorado, USA, from their website at https://psl.noaa.gov/*

Winter 1981/82—two very severe cold spells

The winter of 1981/82 featured two exceptionally severe spells of cold weather, separated by a much milder interval. The opening days of December were mild, but frontal systems gradually introduced very cold arctic air southwards to all areas between the 4th and 8th (Figure 22.29). Vigorous depressions on 8 and 13 December brought widespread snow-falls, and in Durham, 17 cm snow lay on the morning of 13 December. Clearing skies and cold, dry, polar air allowed temperatures to fall to very low levels in many places, including in Durham where the temperature fell to −11.4 °C on 17th and −12.5 °C on 18th, the lowest in December since 1879 (Table 15.2). Daytime temperatures re-mained below 0 °C on both days. Cold weather persisted until early January, making this the coldest December in Durham since 1878: the mean temperature of −0.3 °C was 4.5 degrees below normal, one of only ten sub-freezing months during the twentieth century. Snow lay on 19 consecutive mornings 12–30 December.

Milder conditions became established in early January, and the temperature reached 9 °C on the 2nd and 3rd, before colder weather returned with the establishment of a ridge of high pressure over northern districts; this intensified and advanced south-east to central Europe, from where a ridge of high pressure to Ireland persisted until the 15th. The ten-day spell 6–15 January 1982 remains one of the most severe on record, and in Durham included nine consecutive ice days 6–14 January, including Durham's coldest day in any month since 1881 when on 7 January the maximum temperature reached only −4.5 °C (Tables 4.1, 22.5, and 29.1). Despite modest amounts of snow cover, much less than in the previous month, six nights of these ten fell below −10 °C, including a screen

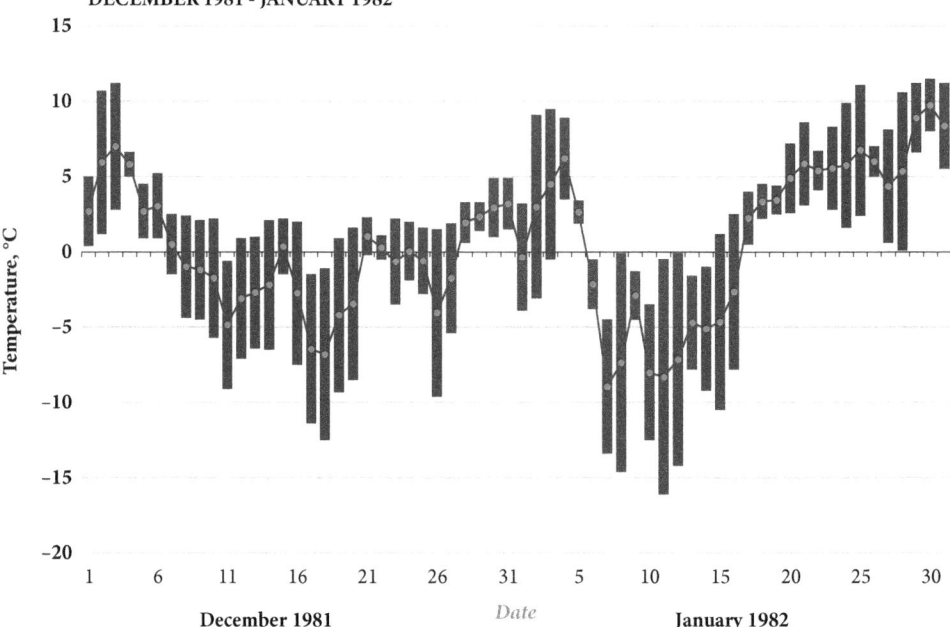

Figure 22.29 *Daily maximum, minimum, and mean daily temperatures at Durham Observatory during December 1981 and January 1982*

minimum of −16.1 °C on 11 January (Durham's coldest January night since 1881, and equalling 21 January 1940 as the coldest night of the twentieth century). At 0900 GMT the temperature was only −14.0 °C. Record or near-record low temperatures occurred widely, including a minimum of −27.2 °C at Braemar on 10 January, which remains the equal-lowest temperature reliably observed anywhere in the United Kingdom [75, 135]. Based upon interpretations of satellite observations of near-surface temperatures, the coldest areas of all were in the Welsh border valleys, where surface temperatures below −27 °C were evident, probably falling to −30 °C by the end of the night [135]. Milder conditions returned during the second half of the month, leaving January 1982's mean temperature of 1.3 °C almost 3 degrees below normal, but well outside the coldest Januarys on record (Table 4.2).

Hurricane *Charley* 22–26 August 1986

Hurricane *Charley* was the second hurricane of the 1986 season [136]. *Charley* formed as a subtropical low on 13 August along the Florida panhandle. After moving north off the coast of South Carolina, the system intensified into a tropical storm on 15 August and attained hurricane status before moving across eastern North Carolina. It gradually weakened over the North Atlantic Ocean and was downgraded to an extratropical cyclone on 20 August. *Charley*'s remnants remained identifiable for over a week until, after crossing the British Isles, it finally dissipated on 30 August [136]. *Charley* brought heavy rainfall and strong winds to the British Isles, causing at least 11 deaths. In eastern

Ireland, Phoenix Park, Dublin recorded its highest 24-hour rainfall total (85.1 mm) in a record stretching back to 1885, and there was widespread flooding in Dublin as a result. In the UK, the storm also caused rivers to flood, with severe floods in Cumbria and Gloucestershire, and brought down trees and power lines.

Charley arrived on Bank Holiday Monday. At 0600 GMT the central pressure was 990 hPa at 49°N 12°W, the centre reaching 52°N 3°W by 0600 on 26 August and deepening to 981 hPa, notably deep for August. Many places in England and Wales experienced more than 12 hours of continuous heavy rain and this was reflected in the totals, with 135 mm falling in 24 hours Aber College Farm, Gwynedd; a map of the rainfall distribution appears as Figure 22.30. At Durham, 69.1 mm fell on 25 August with 27.7 mm the following day. The former remains Durham's fourth-wettest day on record (Table 25.5), and the two-day total of 96.8 mm the second highest two-day total, exceeded only by the two-day total for 10–11 September 1976. Looking at rainfall totals across northern England (Table 22.6), the highest totals were in the northern Pennines rather than the Lake District; Durham on the eastern Pennine dip-slope had the highest one-day and two-day totals of all, closely followed by Burnhope Reservoir to the north west, Widdybank Fell in upper Teesdale, and Cockle Park, near Morpeth, much closer to the coast. Carlisle and Edinburgh were clearly too far north to have had significant rainfall from the storm, although Carlisle would certainly have experienced very strong winds.

Table 22.5 *Durham Observatory weather observations during the first half of January 1982*

	24 HOURS					0900 GMT			
Date	Max temp °C	Min temp °C	Grass or snow min temp °C	Precip'n mm	Sunshine hours	Cloud cover oktas	Wind direction	Wind speed knots	Snow depth cm
1	3.2	−3.9	−6.4	0.1	3.6	1	CALM		0
2	9.1	−3.1	−5.2	9.4	0	8	SSW	6	0
3	9.5	−0.5	−0.6	19.1	0	7	SW	1	0
4	8.9	3.5	3.6	6.7	0.2	8	NW	2	0
5	3.4	1.9	1.5	12.7	0	8	ENE	7	0
6	−0.5	−3.8	−1.0	–	3.7	4	NNW	4	8
7	−4.5	−13.4	−14.1	2.7	3.5	3	CALM		6
8	−0.1	−14.6	−14.0	1.5	0	8	CALM		9
9	−1.3	−4.5	−2.5	0.6	1.3	4	E	10	5
10	−3.5	−12.5	*no data*	–	4.5	4	W	1	8
11	−0.5	−16.1	−12.5	–	2.5	1	WSW	1	5
12	−0.1	−14.2	−10.5	–	3.9	2	WSW	1	5
13	−1.6	−7.8	−12.3	–	2.6	3	SSW	4	5
14	−1.0	−9.2	−10.7	–	3.4	4	WSW	2	5
15	1.2	−10.5	−11.0	–	0	8	SSW	6	5

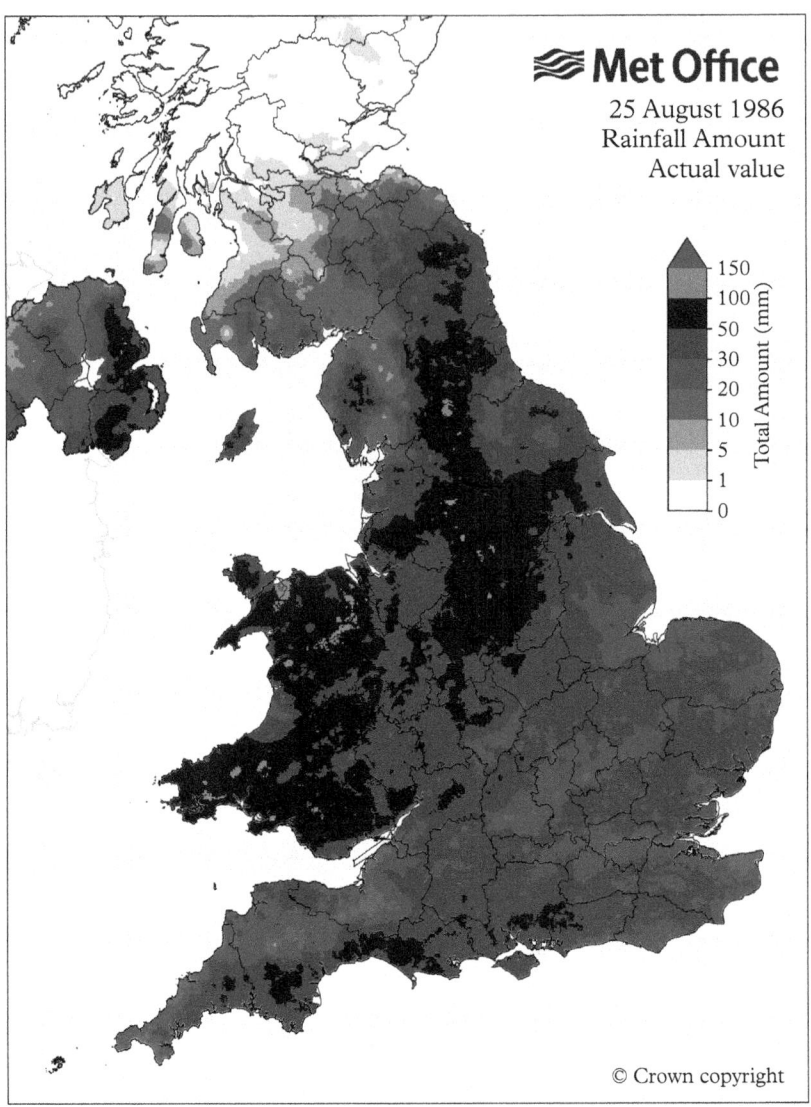

Figure 22.30 *Daily rainfall totals on 25 August 1986; Durham recorded 69 mm on this date. For the UK considered as a whole, this remained the wettest day on record until 3 October 2020 [63]. (Based upon the Met Office 1 km gridded rainfall dataset [121], courtesy of National Climate Information Centre, Met Office Exeter: © Crown Copyright, Met Office)*

Table 22.6 *Rainfall at selected locations during the passage of Hurricane* Charley, *August 1986. Units mm*

Station	24th	25th	26th	27th	25th+26th
Carlisle	–	12.4	1.1	3.0	13.5
Barrow	–	25.1	0.0	0.6	25.1
Coniston Holywath	–	27.3	1.9	3.4	29.2
Ambleside	–	27.8	5.7	4.2	33.5
Holehird	–	28.7	5.6	3.4	34.3
Appleby	–	34.5	2.8	2.0	37.3
Malham Tarn	–	36.5	10.4	2.7	46.9
Widdybank Fell	–	63.6	16.1	4.4	79.7
Burnhope Res.	–	42.9	39.0	10.5	81.9
Durham	–	**69.1**	**27.7**	**9.5**	**96.8**
Cockle Park	–	33.8	38.8	14.4	72.6
Edinburgh	–	4.9	0.5	0.3	5.4

The River Wear at Sunderland Bridge recorded its tenth highest peak discharge since records began in 1957: 304 cumecs on 26 August at 1930 GMT. The mean flow for the whole day was 196 cumecs, the 17th highest daily mean on record.

Durham's mildest winter—1988/89

Durham's mildest winter was in 1988/89, with a mean temperature of 6.3 °C, 2.0 degrees above normal and a full degree milder than the previous mildest winter on the instrumental record to that time (Table 17.3). December 1988 was an unusually mild month dominated by south-west winds, with cold spells from the north or east conspicuous by their absence (Figure 22.31). It was the equal-mildest December on record at the time, along with 1843, with a mean temperature 3.1 degrees above normal (Table 15.2); since then, only December 2015 has been milder. The month remained free of air frost for only the second time since the record began in 1843 (the others were 1857 and 1928); there were only six ground frosts, the fewest in December since 1934. Unusually, this mild winter month was also dry (28 mm, less than half normal) and fairly sunny (63 hours, 10 per cent above normal). January 1989 was an outstandingly mild and windy month almost everywhere, with very little frost or snowfall. In Durham, this was the second-mildest January on record, behind only 1916 (Table 4.2),

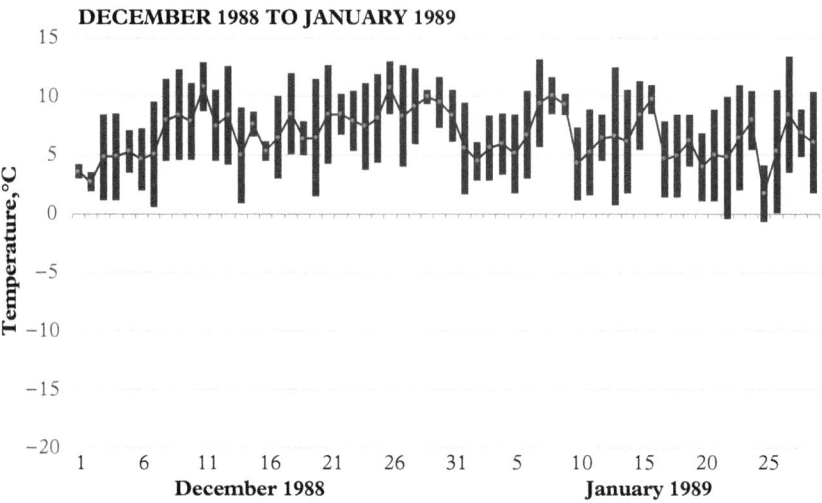

Figure 22.31 *Daily maximum, minimum, and mean daily temperatures at Durham Observatory during December 1988 and January 1989; compare with Figure 22.29 (same scale)*

while precipitation totalled just 7.2 mm during the month, the driest January on record (Table 4.3). There were just two air frosts, equalling 1938's total as the lowest on record for any January. As with December, the month was also a little sunnier than normal.

The mild and dry conditions during December and January were due to a persistent anticyclone over Europe, and these continued into the second week of February, which although also milder than normal ranked some way below the levels of the previous two months. In all, only ten air frosts were recorded during the winter (December to February), a record which still stands at the time of writing. Prior to 1989, the lowest air frost count for any winter had been 14, in 1859, 1869, and 1925, although since 1989 the winters of 2014 and 2020 have recorded just 11. The total of air frosts for the 'extended winter season' (the 12 months July to June) of 31 in 1989 was lower in 1961 (27) and has since been lower in 2007, 2014, and 2020 (Table 16.7). Only one morning in winter 1989 (25 February) had a very slight covering of snow.

Durham's driest and sunniest year—1989

Table 16.10 lists Durham's driest years on record; 1989 heads this list, with an annual total of just 416 mm, 61 per cent of the current 1991–2020 average. Appendix 11 shows monthly rainfall totals for 1989, and their percentage of normal. The year got off to a very dry start—January was, most unusually, the driest month of that year and the driest on Durham's record (Table 4.3), with just 7.2 mm precipitation in all, falling on eight

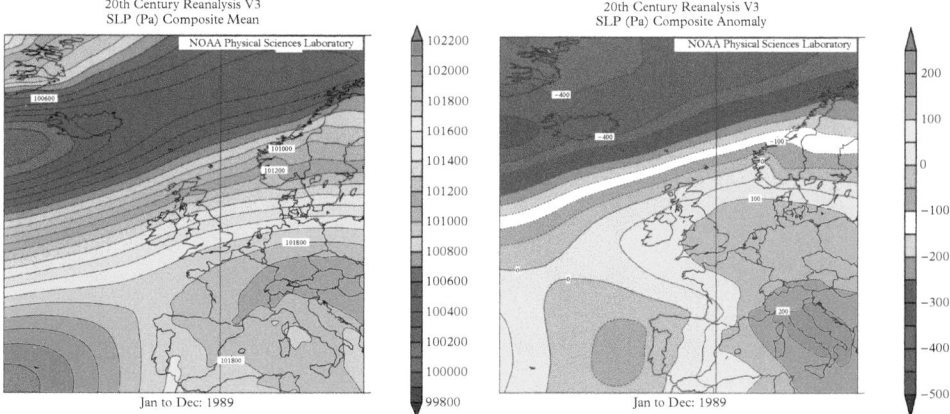

Figure 22.33 *Monthly mean* MSL *pressure (left, units Pa, interval 100 Pa = 1 millibar) and anomaly from 1981–2010 normal (right, units Pa, interval 50 hPa) over the north-east Atlantic and western Europe for the year 1989. Plots from 20th Century Reanalysis data provided by the NOAA/OAR/ESRL PSL, Boulder, Colorado, USA, from their website at https://psl.noaa.gov/*

days. Only two months during the year received above-normal rainfall, February and December, and both within 5 per cent of the expected value. May (7.7 mm, 17 per cent of normal), July (10.1 mm, 16 per cent), and September (11.7 mm, 21 per cent) were all extremely dry, while March and November both received only half the normal accumulation. Rain fell on 149 days in 1989, well below the average of 196.

Figure 22.32 shows annual rainfall totals for 1989 as a percentage of the 1961-90 averages, from which it can be seen that parts of north-east England received less than 65 per cent of normal during the year. Among the lowest annual totals were 342 mm at Hartburn Grange on Teesside (59 per cent of 1941–70 average), 346 mm at South Shields' South Pier (55 per cent of normal), 350 mm at West Hartlepool Sewage Works, and 362 mm at Tynemouth Coastguard. The annual distribution of MSL mean pressure and the anomaly from the 1981–2010 normal are shown in Figure 22.33, which makes an interesting comparison with the plots for 1872 (Figure 22.10).

The year was also exceptionally sunny—at that time, the sunniest on record with 1646 hours of sunshine, 12 per cent above normal and 36.6 per cent of the possible amount, although 2003's estimated annual sunshine duration of 1703 hours has since surpassed this figure—Table 16.13. May, June, and July were all exceptionally sunny, each about one-third sunnier than normal, July being at that time second only to 1955 and 1911 in sunshine duration (Table 10.4) with 237 hours. The summer of 1989— June, July, and August—enjoyed 660 hours of sunshine, second only to summer 1976's 670 hours (Table 19.4). Only a very dull December, just 25 hours of sunshine, less than half normal, was to spoil 1989's sparkling record.

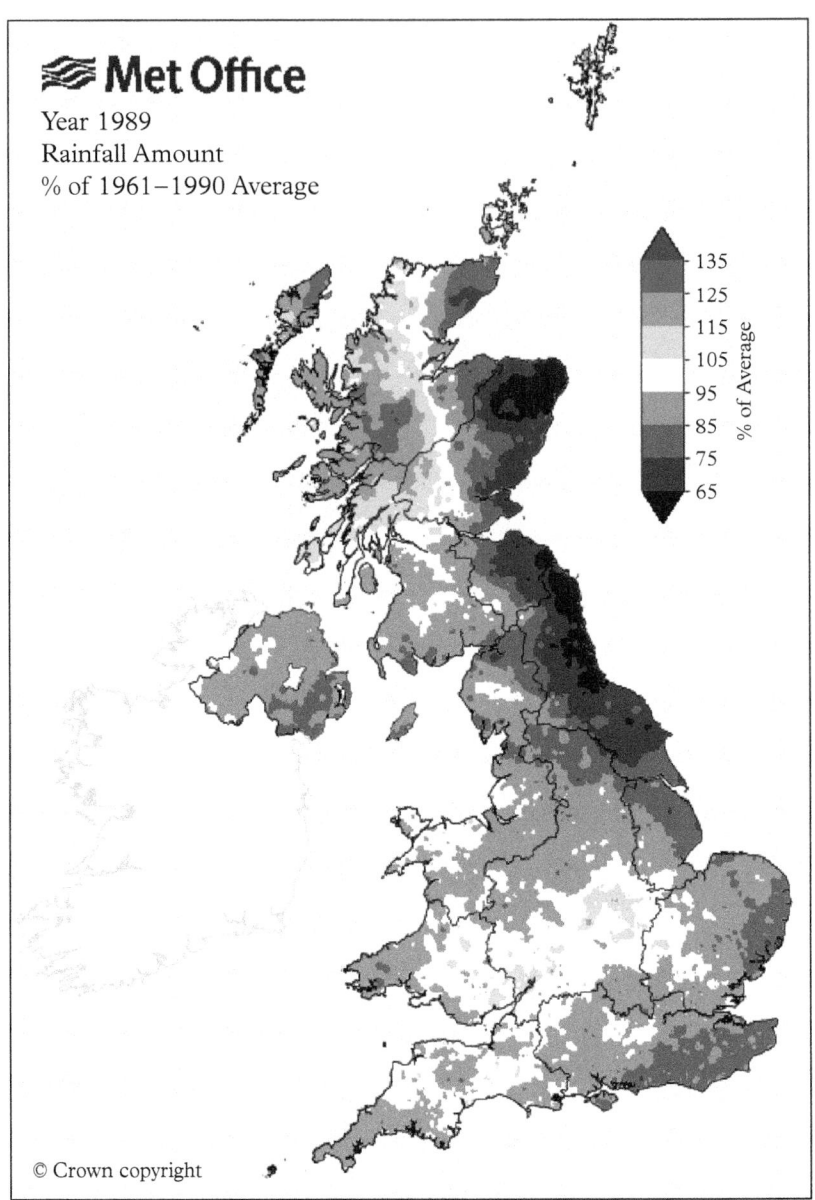

Figure 22.32 *Annual rainfall totals across the United Kingdom for 1989, expressed as a percentage of the 1961–90 average. © Crown Copyright, Met Office*

Another very mild winter—1998

Winter 1998 was another very mild winter in Durham, ranking second behind 1989 (and since equalled by winters 2016 and 2020)—Table 17.3—with a mean temperature of 5.6 °C, 1.3 degrees above normal. December 1997 and January 1998 were both mild and wet, but February 1998 became the mildest on record by over a degree on the previous record in 1945 (Table 5.2), with a mean temperature 3.0 degrees above normal; both mean daily maximum and mean daily minimum established also new records for the month. Much of this warmth was due to the period 8–27 February which was exceptionally mild; during this 20-day period, only two days failed to reach 10 °C, and then only by a few tenths of a degree. The mild conditions resulted from fresh to strong south-westerly winds across the UK that introduced a constant flow of sub-tropical maritime air from the Azores or the Canary Islands, between high pressure centred over Europe and low pressure to the north-west of the British Isles. On 13 February, a day with almost seven hours of sunshine, the temperature reached 16.7 °C, only a tenth of a degree short of the all-time February record at that time (Table 5.1): the following day was only a tenth of a degree cooler. These are about the maximum temperatures expected in late May rather than mid-February. Frosts were infrequent, with only 15 air frosts during the entire winter. To add to a record-breaking month, February was also very dry, the month's precipitation total of just 6.1 mm remains the fourth-driest February on Durham's records (Table 5.3).

Durham's wettest June day on record: 3 June 2000

A slow-moving occluded front lying from western Ireland across to Poland produced widespread prolonged rainfall over north Wales, the Midlands and northern England on 3 June 2000. In Durham, 54.0 mm fell in the 24 hours commencing 0900 GMT, Durham's wettest June day on record (Table 9.3), and to date the only June day to receive more than 50 mm in the day. Figure 9.3 shows the hourly distribution of rainfall during this event. Other high falls in the north-east included 72.8 mm at Carlton-in-Coverdale (about 60 km south-west of Durham), 67.6 mm at Malham Tarn, 65.2 mm at Copley (with 15.8 mm the day before), and 47.7 mm at Burnhope Reservoir. Cockle Park in Morpeth, to the north of Durham, recorded 29.6 mm. Flooded roads were reported widely in West Yorkshire, North Yorkshire, Teesside, and Tyne and Wear. About 700 people were evacuated as houses in Bishop Auckland and Todmorden were flooded. In Durham, the River Wear peaked at 1300 GMT at 3.48 m (375 cumecs) on 4 June, the second-highest water level on record (Figure 22.34). This caused extensive flooding in Durham the next day, most notably at the University's Maiden Castle sports centre, where examinations were disrupted the following week.

(a)

(b)

Figure 22.34   *a: The River Wear in flood, 4 June 2000; b: the same view just a week later at the Durham regatta (Tim Burt)*

July 2006: Durham's hottest month

With a mean monthly temperature of 18.3 °C, 2.5 degrees above normal, this was—and remains—the hottest month on Durham's instrumental record. The previous hottest month to that date, July 1983, was surpassed by 0.6 degrees, a large difference when it is considered that the difference between the next five ranked entries in that particular list was only 0.5 degrees at that time (Table 24.7). Since 2006, though, the Julys of 2013 (mean temperature 18.0 °C) and 2018 (mean 17.9 °C) have come closer. July 2006 was also the hottest month of any name on the CET record, with a mean temperature of 19.7 °C, just ahead of July 1983's 19.5 °C.

July 2006's mean maximum temperature in Durham, 24.3 °C, remains the highest on record for any month, and one of only three months to date to have exceeded 23 °C (the others being August 1995, at 23.1 °C, and more recently July 2013 with 23.2 °C). The hottest day was 17 July, maximum temperature 30.5 °C, Durham's first '30' since August 1995, and Durham's eighth-hottest day on record to date (Table 24.4). Seventeen days during the month surpassed 25 °C, including nine consecutive 21–29 July (Figure 22.35). Night temperatures were also above normal, but largely clear skies allowed nocturnal cooling and only three nights remained at or above 16 °C. The month was also very dry, the monthly rainfall total of 10.0 mm being the fourth-lowest on record for the month (Table 10.3), and extremely sunny, the estimated 273 hours of bright sunshine establishing a new July Record (Table 10.4), second only to June 1940's 297 hours (Table 26.1).

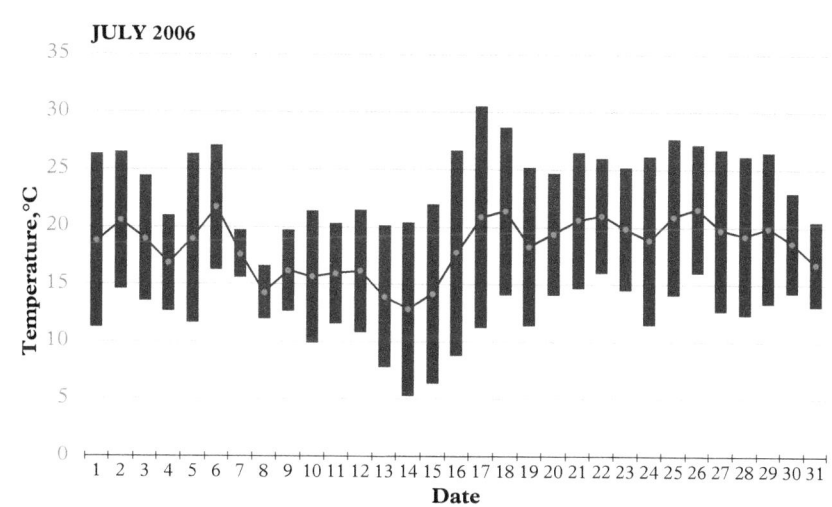

Figure 22.35 *Daily maximum, minimum, and mean daily temperatures at Durham Observatory during July 2006*

The flood of 18 July 2009

Durham's largest recorded flood occurred on 18 July 2009, with an estimated discharge at Sunderland Bridge of 410 cumecs at 0730 GMT on 18 July 2009. At the Durham Observatory, 44.4 mm of rain fell in the 24 hours commencing 0900 GMT on 16 July, followed by a further 39.8 mm in the following 24 hours, in all 84.2 mm within 48 hours—the eighth highest such fall since 1868. Ground conditions were already close to saturation with three days receiving more than 10 mm earlier in the month, thus encouraging a high rate of runoff.

On this occasion, as on so many other very wet days in the north-east of England, a depression moving slowly north-east across England to lie just offshore in the North Sea brought persistent cyclonic rainfall. Between the afternoon of 16 July and the early hours of 18 July, only four hours remained completely dry. On the 17th 15.8 mm fell in two hours between 1400 and 1600 GMT, resulting in severe flooding in Durham. The University's Maiden Castle sports centre was inundated for only the third time since being built in the early 1960s. Just across the road at Houghall College, there was dramatic erosion of the River Wear's floodplain, resulting in a deep gully— soon nicknamed the 'Durham canyon' (Figure 22.36). The gully was about 30 m across, 5 m deep, and 200 m long; it was estimated that the water carried into the river up to 12,000 cubic metres of soil (Jeff

Figure 22.36 *The 'Durham Canyon' resulting from flooding due to heavy rainfall on 16–17 July 2009. This photograph was taken on 25 August 2009. Compare with the photograph taken a month earlier (Figure 10.4): in the meantime, a temporary embankment was constructed to support the pipework (Tim Burt)*

Warburton, pers. comm.). Part of the gully needed to be filled in quickly to prevent a water pipe from collapsing. The rest was infilled later in the autumn.

2012: an extraordinary year for rainfall

The first three months of 2012 were mild and dry; by 1 April only 56 mm of rainfall had fallen, only 40 per cent of the normal over that period. The next nine months received 962 mm, an amount in itself sufficient to surpass Durham's previous annual record fall, back in 1872 (895 mm). The annual fall of 1018 mm represented 150 per cent of normal, surpassing 1872's record by more than 10 per cent.

The wet weather began early in April, with 26.2 mm on the 3rd; in all, five days surpassed 10 mm, including a two-day total of 39 mm over 25–26 April. April's total of 134 mm ranked the third wettest on record (Table 7.3), following other recent wet Aprils in 1998 and 2000. May was less wet, although still wetter than average, but the 11-day dry spell that ended on the 29th was to remain the longest dry spell for the next 15 months. June was also exceptionally wet, the third wettest on record with 137 mm (Table 9.3), more than twice normal. Seven days recorded more than 10 mm of rainfall, including 26.6 mm on 28 June, 18.6 mm of which fell between 1600 and 1800 GMT during a very intense thunderstorm following a relatively warm afternoon. Most of this probably fell within half an hour or so.

The first half of July continued very wet, with 79 mm falling by the 15th, but the second half of the month and the beginning of August were mainly dry. August saw four days with more than 10 mm of rainfall, including 23.8 mm on the 15th; its monthly total of 103 mm was already the third month in the year to surpass 100 mm.

After a dry start to September, more exceptionally heavy rainfall, and resultant flooding, occurred towards the end of the month, when 90.2 mm fell in the three days 23–25 September, including 57.4 mm on the 24th—16 mm of which fell in three hours. The latter became Durham's second-wettest September day, behind only 11 September 1976 (Table 12.3), and its seventh-wettest day on record to date (Table 25.5). October was also wetter than normal, although not exceptionally so, and included a fall of 27.4 mm on the 11th. The first half of November was mostly dry, but the second half was very wet, including a three-day fall amounting to 88.6 mm 24–26 November; 48.0 mm fell on the 26th, including 15.6 mm in four hours 0900–1300 GMT. Significantly, this fall also saw the year-to-date rainfall total already surpass 1872's record 895 mm as the wettest calendar year yet recorded in Durham, with more than a month to go. December continued the trend; although wet, it was not exceptionally so, and brought the annual fall up to 1018 mm, the first time 1000 mm had been recorded within a calendar year in Durham since reliable rainfall records commenced in 1868. A notable year for rainfall included the second-wettest summer on record (Table 19.3) and the fifth-wettest autumn (Table 20.3). In all, rain fell on 215 days during 2012, 19 more than average, compared with 216 in 1872.

The first three months of 2013 were considerably wetter than the first three months of 2012, as a result of which the 12-month rainfall accumulation did not reach its maximum

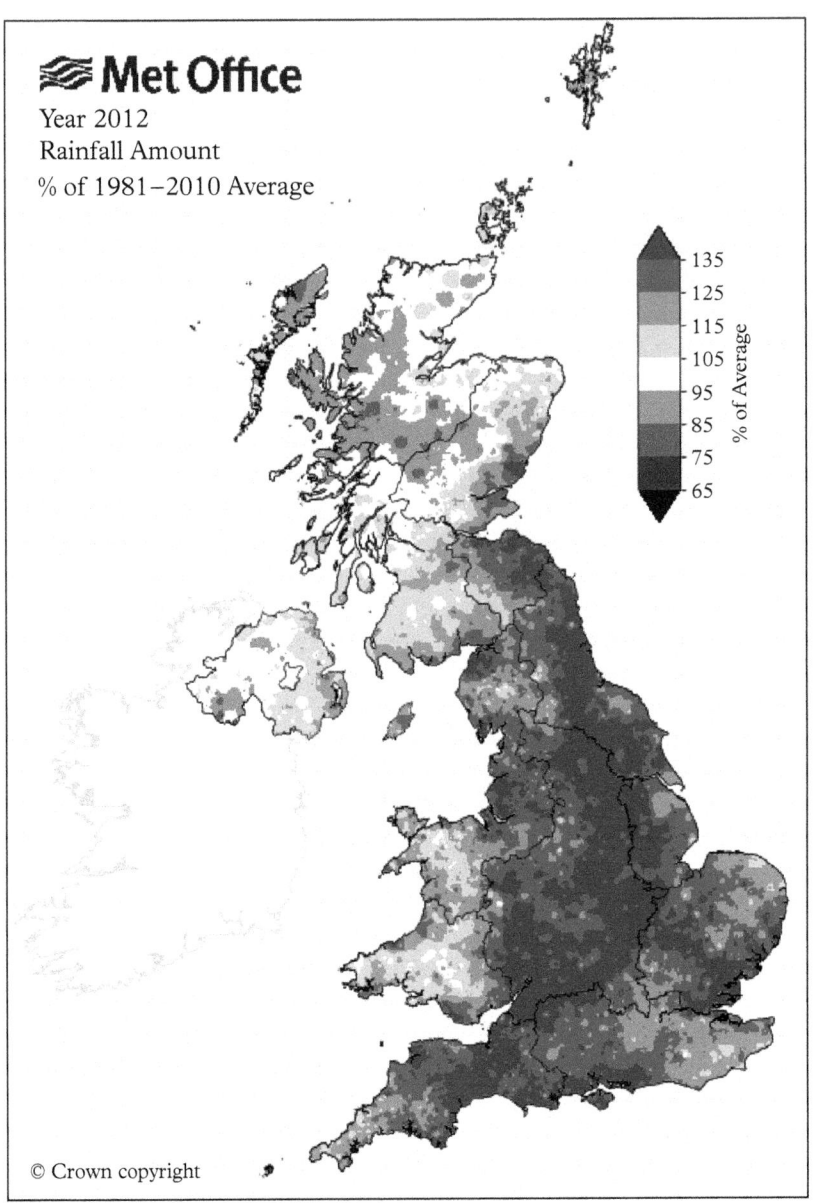

Figure 22.37 *Annual rainfall totals across the United Kingdom for 2012, expressed as a percentage of the 1981–2010 average. Met Office UK Monthly Climate Summary— Annual 2012 [137] © Crown Copyright, Met Office*

until April 2013. The total fall for the 365 days 2 April 2012 to 1 April 2013 amounted to 1131 mm, a remarkable 166 per cent of the annual average of 680 mm, and by a large margin the greatest such fall on Durham's record (Table 25.2). Table 25.2 also shows that 2012 holds the record for total rainfall over any 120-day and 180-day periods.

Figure 22.37 shows 2012's annual rainfall totals as a percentage of the 1981–2010 average. It was a very wet year across most of England and Wales, the wettest year in this series since 1872—only 1872 and 1768 have been wetter since the series began in 1766 [137]. A year which began with ongoing concerns over long-term drought ended with widespread and damaging flooding.

The Pelaw Wood landslide: 26 January 2013

The very wet weather of 2012 (Figure 22.38) extended into the new year. On 26 January, an accumulation of 40 mm of rainfall over the previous week caused a small landslide on the riverbank in Pelaw Wood, just upstream of Durham city, opposite the racecourse (Figure 22.39). Two trees toppled into the river blocking the footpath and cycle route. The response to this seemed sensible: remove the trees blocking the river and bulldoze soil away from the blocked path, which is a main cycle and pedestrian route into the city centre from the east. Unfortunately, the slope was so wet that removing the failed material from the toe of the slope was like removing a retaining wall, weakening the slope still further. This started a sequence of retrogressive landslides, gradually extending back up the slope and, because the slope was so wet, small mudflows were formed too. Soon after the path was cleared, a large tree fell over and it was clear that just removing the

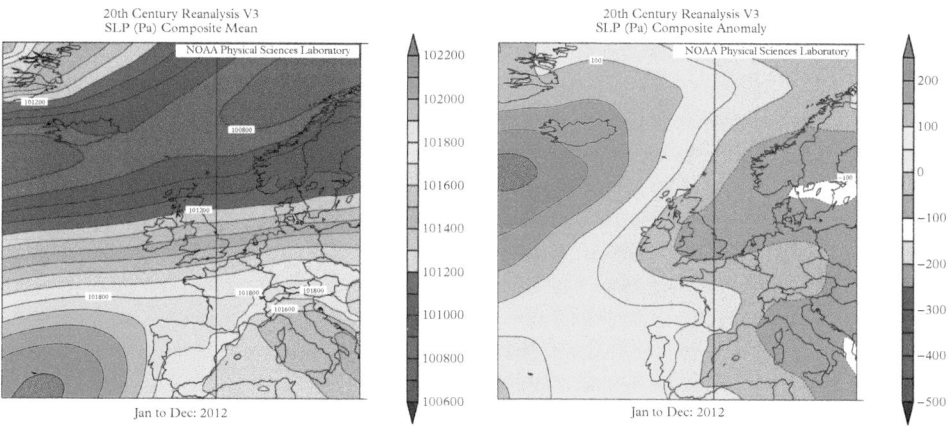

Figure 22.38 *Monthly mean* MSL *pressure (left, units Pa, interval 100 Pa = 1 millibar) and anomaly from 1981–2010 normal (right, units Pa, interval 50 hPa) over the north-east Atlantic and western Europe for the year 2012. Plots from 20th Century Reanalysis data provided by the NOAA/OAR/ESRL PSL, Boulder, Colorado, USA, from their website at https://psl.noaa.gov/*

Figure 22.39 *Top, The Pelaw Wood landslide, 27 January 2013, the day after the slope failed; bottom, The heavily engineered landslide, 11 September 2016 (Tim Burt)*

failed material would not be enough, since this would simply unload the toe of the slope and lead to still more landslides. The path remained closed for several years and in the end costly, major engineering works were required to stabilize the slope and allow the path to be re-opened: drainage was installed, and major retaining walls built.

Just a little earlier that winter, the complete collapse of an entire section of river cliff happened just downstream of Kingsgate Bridge on Christmas Eve 2012 (Figure 22.40). No one is quite sure how the Durham meander formed but it may have been related to the temporary damming of the river by glacial ice and then sudden drainage of the lake, causing rapid downcutting. Probably the entire incised meander once had outer banks as steep as the one shown, but quarrying has widened the left-hand bank downstream towards Prebends Bridge. This section of river cliff survived because the nearby St Oswald's church meant it was never quarried. The slab failure of the cliff resulted

Figure 22.40 *Collapsed river cliff as seen from Kingsgate Bridge, 28 December 2012 (Tim Burt)*

from a combination of undercutting during successive flood events plus the influence of very wet bedrock. The loss of a complete slab, perhaps only a metre thick, reminds us how the Durham 'gorge' would have gradually widened after the initial incision [138].

Durham's hottest day: 25 July 2019

The maximum temperature of 32.9 °C recorded at the Durham Observatory site on 25 July 2019 became the highest temperature on record there, surpassing 32.5 °C on 3 August 1990 (Table 24.4).* On the same day, the highest temperature ever recorded in the British Isles, 38.7 °C, was measured at the Cambridge Botanic Garden, together with a new record maximum temperature of 36.5 °C at the Radcliffe Observatory, Oxford— less than two months after the publication of *Oxford Weather and Climate since 1767* [49], illustrating the maxim that meteorological records are bound to go out of date almost as soon as they are published!

Figure 22.41 shows maximum and minimum temperatures at Durham during July 2019. Temperatures rose steadily through the month; four consecutive days surpassed 25 °C from the 23rd. There were also several notably warm nights, two of which are included in Durham's all-time warmest nights (Table 24.5), notably the minimum temperature of 17.9 °C on 26 July. Figure 10.1 shows the hourly variation in temperature during this short hot spell.

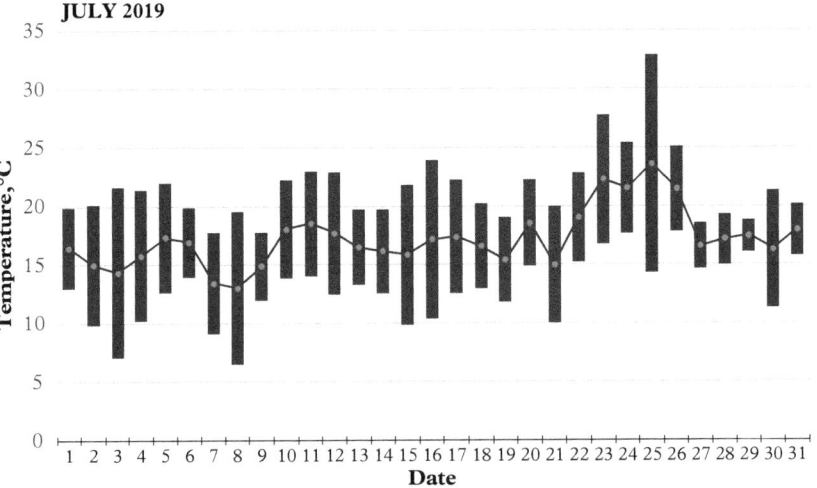

Figure 22.41 *Daily maximum and minimum temperatures at Durham during July 2019, which included Durham's hottest day (32.9° C on the 25th) and third-warmest night (17.8° C on the 26th)*

* Note that we discount the maximum of 33.6 °C recorded at Durham in a Glaisher stand on 16 July 1876, for which the equivalent Stevenson screen temperature is assessed to be close to 31.3 °C (see Chapter 10 and Appendix 3 for more details).

The heatwave experienced in late July 2019 was short but exceptional. The synoptic chart at 1200 GMT on 25 July 2019 (Figure 10.2) shows much of the British Isles in a light southerly flow drawing exceptionally hot air from the near continent, where temperatures reached the low 40s Celsius. New national temperature records were set in Belgium, the Netherlands, and Germany.

Winter 2020—another mild winter

Winter 2020 became the equal second-mildest winter on Durham's records, behind 1989 and alongside 1998 and 2016; all four mildest winters to date have occurred since and including 1989 (Table 17.3). December 2019 was mild and dry, and a little sunnier than normal. January 2020 became the fourth-equal mildest January on record with a mean temperature of 6.1 °C, 2.0 degrees above normal (Table 4.2); the mean minimum temperature of 3.2 °C was second only to January 1916's 3.7 °C. The maximum temperature reached 13.7 °C on the 7th and 14.0 °C on the 22nd, both more typical of early May, while only a single slight air frost occurred during the month (−0.5 °C on the 19th). February was a mild, wet, and windy month, with just three slight air frosts; the winter as a whole (December to February) recorded just 11 air frosts, fewer than were recorded in the frosty *April* of 2021 (12).

Spring 2020: anticyclonic lockdown

As April 2020 started, the first full calendar month of the coronavirus lockdown, fortunately the weather improved to give some welcome pleasure at an otherwise very difficult time. For much of the month, anticyclonic conditions dominated and, depending exactly where the anticyclone was located, Durham either basked in unseasonable warmth or was chilled by winds coming off the cold North Sea, still only about 8 °C, often with full cloud cover, too. This contrast was first obvious on 8–9 April, when the maximum of 18.2 °C on the 8th was followed by 10.7 °C on the 9th as the winds swung round to the east. The coldest day of the month was on the 13th, with a chilly 7.8 °C maximum, accompanied by a brisk east-north-easterly breeze and only a glimmer of sunshine. Only two days later, the temperature reached 21.2 °C on a day with lighter winds from the north-west and upwards of ten hours of sunshine. Sunshine was indeed a major feature of April's weather, the estimated monthly total of 218 hours, an average of 7¼ hours daily, ranking as the third-sunniest April on record to that date, behind only 1914 (238 hours) and 2015 (219 hours) (Table 7.4).

Until the very end of April it appeared as if this could become Durham's driest month on record, for only 0.8 mm of precipitation had fallen by the 25th. Thereafter, a little rain fell on 26th, 29th, and 30th to bring the month's total up to 3.8 mm, still a very dry month (the driest month since March 1953) but now only the third-driest April, behind 2.2 mm in 1938 and 2.4 mm in 1912. Rain fell on five days in all, and only four years have seen fewer rain days in April: three in 1912 and four in 1938, 1974, and 2011. The fine weather continued into May, which, with an estimated 256 hours of sunshine, ranked equal with 1881 and second only to May 1971, which recorded 259 hours. As in April, anticyclonic conditions dominated the weather and wind direction determined whether Durham was warm or cool. There was a notably chilly day on the 10th (maximum 8.6 °C with northerly winds), providing a sharp contrast to the three immediately preceding days, all of which had reached 20 °C. The second half of the month saw seven days surpass 20 °C, including 24.9 °C on the 20th and 24.0 °C on the 29th. May was also dry, with less than half normal rainfall, but not as dry as April.

For the spring as a whole, 2020 became the equal fourth-warmest on record, with a mean temperature of 9.3 °C (Table 18.2), surpassed only by 2003, 2004, 2007, 2011, 2014, and 2017: notably, all these warm springs have been this century. It was the driest spring on record at Durham, with just 42 mm (31 per cent of normal), 3 mm drier than the previous record driest spring in 1956 (Table 18.4). It was also by far the sunniest spring on record, with an estimated 630 hours of bright sunshine between March and May (Table 18.5). Not only did this extraordinary season eclipse the previous sunniest spring (1990) by more than 80 hours, but perhaps its most lasting claim to fame is that it would have ranked as the fourth-sunniest *summer* on record, behind only the memorable summers of 1976 and 1989 and, to an earlier generation, 1901.

On the national scale, too, spring 2020 became the UK's sunniest spring on record by a very wide margin, and fifth driest, as a result of the jet stream lodging itself to the north of the British Isles for many weeks, allowing high pressure and settled weather to dominate [139]. However, from early June this blocking high gradually broke down, allowing Atlantic weather systems to introduce increasingly unsettled conditions.

Summer 2020: the second consecutive summer to reach 30 °C

An isolated day of great heat saw the temperature reach 37.8 °C at London's Heathrow Airport on 31 July, and 30.8 °C at Durham—the previous day's maximum was just 21.2 °C, and the following day 22.5 °C. This was the first instance since the record commenced almost 180 years previously when two consecutive years have reached 30 °C in Durham.

A notable heatwave in southern and south-east England during the first half of August saw temperatures reach 34 °C for six consecutive days 7th to 12th [139], although on this occasion temperatures were less extreme in the north-east—Durham's highest maximum temperature being 26.9 °C on 11 August. In all, eight days during summer 2020 reached or exceeded 25 °C, equalling summer 2019's total but some way below summer 2018 (14 days) and 2006 (20). The average frequency over 1991–2020 is just five days per annum.

Weather highlights in 2021

The year began with the wettest winter on Durham's records, the total of 328 mm surpassing the previous wettest in 1877 by more than 10 per cent. Rain fell on 70 days, a frequency only exceeded once previously, in winter 2010.

After a warm March, with the temperature reaching 20.8 °C on the last day of the month, the remainder of spring was colder than normal. April was the coldest since 1989, with a mean temperature lower than March, and with 12 air frosts - the highest in any April since 1922. It was also a very sunny month – the sunniest April since 1914 (Table 7.4). There were few noteworthy features during the summer months: the year's highest temperature, 28.5 °C on 17 July, was lower than recent years, although five consecutive days in July reached 25 °C. The beginning of September saw Durham's highest autumn temperatures since 1906, when 27.2 °C was reached on 7th followed by 28.0 °C the following day. The autumn as a whole became the second-mildest on record, the mean temperature exceeding the normal by 1.6 degrees; only 2006 has been milder. Later in the year, a rare red weather warning was issued for north-east and south-east Scotland and the north and north-east of England with the approach and passage of storm *Arwen* on 26-27 November. Although Durham escaped the worst of the winds, the maximum gust at the rooftop AWS on the Geography Department being 'only' 49 knots from the north-east, a gust of 85 knots was recorded at Brizlee Wood in Northumberland late on 26 November.

Figure 22.42 *A fallen tree in Durham Botanic Gardens following storm* Arwen *on 26-27 November 2021 (Michele Allan)*

A sunny start to 2022

As we completed our work on this book, 2022 started with exceptional sunshine. After a dull December, January 2022 became the sunniest January on Durham's records extending back to 1881 (Table 4.4), the duration estimate (based on the Met Office East and North-east regional series) of 112 hours being almost 20 per cent in excess of the previous sunniest, 94 hours in 1959. February was similarly sunny, the preliminary estimate of 104 hours surpassing even January's daily average, although still some way behind February 2019's record 128 hours (Table 5.4). Despite the dull December, winter 2022 became the fourth-sunniest winter on record, with 254 hours — almost 25 per cent more than normal (Table 17.7).

Part 5

Durham weather averages and extremes

23

Warmest, driest, sunniest . . .

The following analysis (Table 23.1) is based upon daily statistics for the Durham Observatory over the standard 30-year period 1991–2020 (1970–99 for sunshine) and excludes 29 February. Different analysis periods would probably show slightly different results.

The most reliably warm week of the year in Durham is the first week of August, while mid-to late-March, late August, and mid-September share honours as—on average—the driest weeks of the year, although of course there is much year-to-year variability. Mid-May is the most reliably sunny time of the year. Mid-December, just before the winter solstice, is the least sunny week of the year, while late December tends to be the coldest. See also 'annual cycle' plots, Figure 3.1.

Table 23.1 *Details of the warmest, coldest, driest, wettest, sunniest and dullest days and weeks in an average year, based upon Durham's daily averages 1991–2020 (sunshine 1970–99)*

		Daily values		Weekly values	
		Value	Date	Value	Week centred on
Warmest	Highest mean daily maximum temperature	21.3 °C	5 Aug	20.8 °C	2 and 3 Aug
	Highest mean daily minimum temperature	12.4 °C	1 Aug	12.1 °C	4 Aug
	Highest mean daily temperature	16.8 °C	5 Aug	16.4 °C	3 and 4 Aug
Coldest	Lowest mean daily maximum temperature	5.5 °C	20 Dec	6.3 °C	29 Dec
	Lowest mean daily minimum temperature	−0.1 °C	29 Dec	0.6 °C	29 Dec
	Lowest mean daily temperature	3.0 °C	20 Dec	3.5 °C	29 Dec
Wettest	Greatest mean daily rainfall	4.1 mm	24 Sept	2.81 mm/day	19 Nov
	Greatest frequency of rain days	25 days in 30 years (83%)	20 Oct	3.1 days/week	10 Nov
Driest	Lowest mean daily rainfall	0.52 mm 0.53 mm	18 Sept 13 March	0.86 mm/day	16 March
	Lowest frequency of rain days	7 days in 30 years (23%)	4 May	1.71 days/week 1.73 days/week	30 and 31 Aug 4 and 22 May
Sunniest	Greatest mean daily sunshine duration	7.18 hours	29 June	6.45 hours/day	16 May
Dullest	Lowest mean daily sunshine duration	0.83 hours	17 Dec	1.24 hours/day	14 and 15 Dec

24

Temperature extremes in Durham

Temperatures recorded prior to the introduction of a Stevenson screen in November 1899 are shown in italics; these are regression estimates from Glaisher stand readings from January 1860 to October 1899, as read from the 'North Shed' October 1851 to December 1859 and as read from the north wall exposure prior to October 1851. Daily data are missing for various periods up to February 1855 (Table A3.5) and for much of 1863, and for these periods only monthly mean temperatures—some estimated—are available. Temperatures were recorded in degrees Fahrenheit until October 1971, read to a precision of 0.1 °F during the period 1843 to 1918 and 1 °F for the remainder of the record. All values in degrees Fahrenheit have been converted to degrees Celsius, but those for the 1 °F precision period are subject to approximately 0.2 degrees Celsius uncertainty as a result. Since November 1971 temperatures have been recorded to a precision of 0.1 °C. See Appendix 2 and 3 for more details.

Earliest and latest dates

Table 24.1 *Earliest and latest dates of occurrence of various maximum air temperatures, °C, period August 1843 to December 2021*

Maximum temperature	Earliest date	Value, °C	Latest date	Value, °C
≤ 0 °C	12 Nov 1919	−1.1	1 Apr 1917	−1.1
	23 Nov 1853	*−1.4*	*23 Mar 1899*	*−0.2*
	27 Nov 1874	−0.4	23 Mar 1916	0.0
	28 Nov 1969	−1.1		
≥ 20 °C	8 Mar 1929	20.4	16 Oct 1951	20.6
	18 Mar 1990	20.0	15 Oct 1945	20.6
			13 Oct 2018	21.1
≥ 25 °C	7 May 2018	26.1	1 Oct 2011	25.3
≥ 27 °C (80 °F)	23 May 2001	29.0	13 Sep 2016	27.3
	23 May 2010	27.2	8 Sep 1959	27.2
			8 Sep 2021	28.0
≥ 30 °C	6 Jun 1939	30.0	21 Aug 1995	30.3
	8 Jun 1940	30.0	12 Aug 1953	30.0
≥ 32 °C (90 °F)	25 Jul 2019	32.9	3 Aug 1990	32.5
≥ 35 °C	Not attained			

Table 24.2 *Earliest and latest dates of occurrence of various low minimum air temperatures, °C, period August 1843 to December 2021*

Threshold temperature	Earliest date	Value, °C	Latest date	Value, °C
Dates within the winter season, starting 1 July				
≤ 0 °C	4 Sep 1907	−1.1	28 Jun 1974	−0.5
	11 Sep 1931	−0.3	19 Jun 1915	−0.7
	12 Sep 1991	−0.5	15 Jun 1927	−0.8
≤ −5 °C	25 Oct 1926	−5.3	24 Apr 1908	−8.0
	31 Oct 1881	*−5.0*	22 Apr 1929	−5.3
	31 Oct 1909	−5.2	*16 Apr 1892*	*−6.5*
≤ −10 °C	24 Nov 1993	−12.0	2 Apr 1917	−11.1
	3 Dec 1879	*−14.0*	18 Mar 1900	−10.7
	3 Dec 2010	−10.4		
≤ −15 °C	*14 Dec 1878*	*−15.4*	4 Mar 1947	−15.0

Table 24.3 *Earliest and latest dates of occurrence of various high minimum air temperatures, °C, period August 1843 to December 2021*

Threshold temperature	Earliest date	Value, °C	Latest date	Value, °C
Dates within the calendar year				
≥ 15 °C	2 Jun 2003	16.1	14 Oct 2017	15.1
	5 Jun 1980	17.8	3 Oct 2011	15.2
	7 Jun 1950	15.0	1 Oct 1953	15.0
			1 Oct 1985	15.9
≥ 20 °C	Not attained			

Hottest and coldest days

Table 24.4 *Hottest and coldest days at the Durham Observatory, by maximum temperature, period August 1843 to December 2021*

Rank	Hottest days Maximum temperature, °C	Date	Coldest days Maximum temperature, °C	Date
1	32.9	25 Jul 2019	−6.6	*13 Dec 1878*
2	32.5	3 Aug 1990	−6.2	*1 Jan 1871*
3	31.9	2 Aug 1990	−5.1	*31 Dec 1870*
4	*31.3*	*16 Jul 1876*	−4.7	*25 Jan 1881*
5	31.0	1 Jul 2015	−4.6	*25 Dec 1860*
6	30.8	31 Jul 2020	−4.5	7 Jan 1982
7	30.6	10 Jul 1921	−4.4	5 Jan 1941
		31 Jul 1943		
8	30.5	17 Jul 2006	−4.3	*2 Jan 1854*
				16 Feb 1855
9	30.4	10 Jun 1925	−4.2	*19 Dec 1859*
10	30.3	21 Aug 1995	−4.1	*17 Jan 1881*
				26 Jan 1881
Last 50 years to 2021				
	32.9	25 Jul 2019	−4.5	7 Jan 1982

Table 24.5 *Warmest and coldest nights at the Durham Observatory, by minimum temperature, period August 1843 to December 2021*

| Rank | Warmest nights | | Coldest nights | | |
	Minimum temperature, °C	Date	Minimum temperature, °C	Date	Grass minimum temperature, °C*
1	18.9	20 Jul 2016	−18.0	*8 Feb 1895*	−13.8
2	*18.1*	*12 Jul 1859*	−16.9	*17 Jan 1881*	−19.4
3	17.9	14 Aug 2001	−16.5	*10 Feb 1895*	−13.3
		6 Aug 2006			
		26 Jul 2019			
4	17.8	26 Aug 1959	−16.4	*25 Dec 1860*	N/A
		5 Jun 1980			
5	17.7	3 Aug 2018	−16.1	21 Jan 1940	−19.4
		24 Jul 2019		11 Jan 1982	−12.5
6	*17.6*	*27 Jun 1856*	−15.8	*26 Jan 1881*	−16.7
		24 Jul 1900			
7	*17.5*	*27 Jun 1878*	−15.7	*29 Jan 1848*	−20.0
		12 Jul 1926			
		3 Aug 1933			
		3 Jul 2001			
8	17.4	24 Jul 2012	−15.4	*30 Dec 1874*	−16.2
				14 Dec 1878	−12.2
9	17.3	20 Jul 1995	−15.0	4 Mar 1947	−19.4
10	17.2	Six occasions	−14.7	*19 Feb 1892*	−13.9
Last 50 years to 2021					
	18.9	20 Jul 2016	−16.1	11 Jan 1982	−12.5

* *Records of grass minimum temperature commenced on 1 March 1874. The record from 29 January 1848 is one of a handful of 'lowest temperature on the surface of grass' entries noted in manuscript in the observation register prior to 1874. Note that the nominal 'grass minimum temperature' in very cold weather is more than likely to be above a snow surface rather than grass. Where the 'grass/snow' minimum is shown as higher than the air minimum, the grass thermometer may have been buried in snow and read higher than the actual minimum temperature as a result.*

In July 2015, the siting of the grass temperature sensor was changed such that it was mounted 35 mm above an artificial grass surface ('AstroTurf') rather than natural grass, unfortunately without any overlap of record: this will have significantly affected the homogeneity of the grass minimum series.

Other low grass minimum temperatures not included above, in date order:

Date	Air minimum, °C	Grass or snow minimum, °C
4 Dec 1879	−14.6	−16.1
14 Jan 1881	−12.0	−16.6
21 Jan 1881	−12.5	−16.1
25 Jan 1881	−14.1	−15.7
6 Jan 1894	−12.6	−16.3
2 Apr 1917	−11.1	−17.8
16 Feb 1929	−11.4	−16.7
17 Feb 1929	−10.8	−15.6
23 Feb 1947	−10.0	−16.1

Since 1970, the lowest grass (or snow) minima have been:

Date	Air minimum, °C	Grass or snow minimum, °C
1 Feb 1972	−13.1	−13.6
5 Jan 1979	−10.9	−13.6
18 Dec 1981	−12.5	−14.5
7 Jan 1982	−13.4	−14.1
8 Jan 1982	−14.6	−14.0
11 Jan 1982	−16.1	−12.5
24 Nov 1993	−12.0	−13.5
29 Dec 1995	−9.6	−12.6
21 Dec 2010	−10.4	−12.7

Table 24.6 *Hottest and coldest days at the Durham Observatory, by mean daily temperature, period August 1843 to December 2021*

Rank	Hottest days Mean daily temperature, °C	Date	Minimum temperature, °C	Maximum temperature, °C
1	24.4	3 Aug 1990	16.3	32.5
2	23.6	25 Jul 2019	14.3	32.9
3	23.5	2 Aug 1990	15.0	31.9
4	23.2	19 Jul 1901	17.2	29.1
5	*23.0*	*27 Jun 1878*	*17.5*	*28.5*
		22 Jul 1873	*16.2*	*29.7*
6	22.7	*19 Jul 1878*	*15.8*	*29.5*
		9 Aug 1911	15.3	30.1
		19 Aug 1996	16.6	28.8
		12 Aug 1997	15.9	29.5
7	22.6	1 Jul 2015	14.1	31.0
8	22.5	2 Sep 1906	15.7	29.3
		10 Jul 1921	14.4	30.6
		22 Jun 1941	15.6	29.4
9	*22.4*	*5 Jul 1852*	*15.8*	*29.0*
10	22.3	9 Jul 1921	16.7	27.8
		3 Aug 1933	17.5	27.1
		23 Jul 2019	16.8	27.8
Last 50 years to 2021				
1	24.4	3 Aug 1990	16.3	32.5

Continued

Table 24.6 *Continued*

	Coldest days			
Rank	Mean daily temperature, °C	Date	Minimum temperature, °C	Maximum temperature, °C
1	−10.5	*25 Dec 1860*	−16.4	−4.6
		17 Jan 1881	−16.9	−4.1
2	−10.1	*13 Dec 1878*	−13.5	−6.6
3	−10.0	*1 Jan 1871*	−13.7	−6.2
		26 Jan 1881	−15.8	−4.1
4	−9.7	*10 Feb 1895*	−16.5	−2.8
5	−9.6	*31 Dec 1870*	−14.0	−5.1
6	−9.5	*16 Feb 1855*	−14.6	−4.3
7	−9.4	*25 Jan 1881*	−14.1	−4.7
		8 Feb 1895	−18.0	−0.7
8	−9.0	*8 Jan 1861*	−14.2	−3.8
		7 Jan 1982	−13.4	−4.5
9	−8.9	5 Jan 1941	−13.3	−4.4
10	−8.7	*7 Jan 1894*	−13.8	−3.5
Last 50 years to 2021				
8	−9.0	7 Jan 1982	−13.4	−4.5

By standard meteorological convention, the 'mean daily temperature' is defined as the average of the daily maximum and minimum temperatures.

Table 24.7 *Hottest calendar months at the Durham Observatory, by mean daily temperature, period August 1843 to December 2021*

Rank	Hottest months Mean temperature, °C	Departure from 1991–2020 normal degC	Month and year	Mean daily max °C	Mean daily min °C
1	18.3	+2.5	Jul 2006	24.2	12.4
2	18.0	+2.2	Jul 2013	23.2	12.8
3	17.9	+2.1	Jul 2018	22.9	12.9
4	17.7	+1.9	Jul 1983	22.8	12.5
5	17.6	+2.0	Aug 1975	22.8	12.3
6	*17.5*	*+3.9*	*Jun 1846*	*22.6*	*12.3*
7	*17.4*	*+1.6*	*Jul 1852*	*21.6*	*13.2*
8	17.2	+1.6	Aug 1995	23.1	11.3
9	17.2	+1.4	Jul 2019	21.6	12.9
10	17.2	+1.6	Aug 1947	23.0	11.4
Last 50 years to 2021					
1	18.3	+2.5	Jul 2006	24.2	12.4

By standard meteorological convention, the 'mean daily temperature' is defined as the average of the daily maximum and minimum temperatures: the monthly mean temperature is taken as the average of all available mean daily temperatures within that month.

Table 24.8 *Coldest calendar months at the Durham Observatory, by mean temperature, period August 1843 to December 2021*

| Rank | Coldest months | | | | |
	Mean temperature, °C	Departure from 1991–2020 normal degC	Month and year	Mean daily max °C	Mean daily min °C
1	−2.2	−6.8	*Feb 1855*	See notes	
2	−1.7	−5.8	*Jan 1881*	2.0	−5.3
3	−1.5	−5.6	Jan 1940	1.6	−4.6
4	−1.2	−5.8	Feb 1963	1.5	−3.9
5	−1.1	−5.7	Feb 1947	0.6	−2.9
6	−1.1	−5.7	*Feb 1895*	2.4	−4.5
7	−0.7	−4.9	*Dec 1878*	2.3	−3.8
8	−0.6	−5.2	Feb 1986	1.8	−3.0
9	−0.5	−4.6	*Jan 1879*	2.4	−3.4
			Jan 1895	2.3	−3.4
			Jan 1963	2.1	−3.2
			Jan 1979	2.4	−3.5
Last 50 years to 2021					
8	−0.6	−5.2	Feb 1986	1.8	−3.0

NOTE: February 1855 has only 14 days' daily data (15–28th) in the surviving register and the monthly mean is an estimate: see Chapter 22 for more details. The mean daily maximum and minimum temperature for the available data are +0.3 and −6.1 °C, respectively.

Table 24.9 *Greatest daily ranges in temperature (in degrees Celsius) at the Durham Observatory, period August 1843 to December 2021*

Rank	Daily range, degC	Date	Minimum temperature, °C	Maximum temperature, °C
Greatest daily ranges				
1	23.7	5 Aug 1916	5.1	28.8
2	22.9	8 Mar 1929	−2.5	20.4
3	22.8	9 Jun 1953	2.8	25.6
		14 Jun 1959	3.9	26.7
4	*22.3*	*28 Jul 1868*	*6.0*	*28.3*
5	22.2	28 Jun 1921	1.7	23.9
		3 Jun 1939	1.7	23.9
6	21.7	6 May 1938	−2.8	18.9
		6 Jun 1939	8.3	30.0
7	21.5	7 Jun 1911	4.6	26.1
8	21.3	Five occasions		
Last 50 years to 2021				
	20.6	7 Sep 1991	3.5	24.1

The daily temperature range is the difference between the maximum and minimum temperatures over the period 0900–0900 GMT.

Prolonged heatwaves

The Met Office definition of a 'heatwave' is met when a location records a period of at least three consecutive days with maximum temperatures meeting or exceeding a heat-wave temperature threshold [140]. The threshold varies by UK county in the range 25–28 °C; for Durham the threshold is 25 °C. Table 24.10 lists all spells of at least three consecutive days reaching or exceeding 25.0 °C at the Durham Observatory since August 1844 (some months are missing in the early years of the record; see Appendix 2).

To and including summer 2021, there have been 73 such spells in almost 180 years, an average recurrence period of two to three years. There were no heatwaves for 11 years between 1846 and 1857, and ten years between 1959 and 1969. More than one-third of the 72 heatwave spells (26) have occurred since 1975, in just one-quarter of the record.

Only two years have recorded three or more heatwaves—1933 with four, and 1995 with three. The longest such unbroken heatwave was in August 1975 with 13 consecutive days at or above 25 °C, although there have been longer periods where the spell has been interrupted by only slightly cooler conditions for a day or two.

Table 24.11 lists the hottest seven-day spells by mean temperature over the same period.

Table 24.10 *The longest heatwaves (three consecutive days all reaching at least 25 °C) on Durham's records, between August 1843 and December 2021*

Year	Dates in spell	Spell length	Highest max reached, °C	Hottest day in spell
1846	2–7 Jun	6	28.5	6 Jun
	16–19 Jun	4	28.4	18 and 19 Jun
1857	24–26 Jun	3	25.6	24 Jun
1866	11–13 Jul	3	29.2	12 Jul
1868	1–6 Aug	6	29.0	2 Aug
	5–7 Sep	3	26.2	6 Sep
1870	22–24 Jul	3	28.1	24 Jul
1871	9–12 Aug	4	26.8	12 Aug
1873	21–23 Jul	3	29.7	22 Jul
1876	13–16 Jul	4	31.3	16 Jul
1878	17–20 Jul	4	29.5	19 Jul
1886	2–4 Jul	3	26.6	2 Jul

Continued

Table 24.10 *Continued*

Year	Dates in spell	Spell length	Highest max reached, °C	Hottest day in spell
1887	*1–3 Jul*	*3*	*26.3*	*2 Jul*
1896	*13–16 Jun*	*4*	*26.5*	*16 Jun*
1897	*2–5 Aug*	*4*	*28.0*	*5 Aug*
1899	*30 Jul–2 Aug*	*4*	*28.6*	*1 Aug*
1901	17–20 Jul	4	29.1	19 Jul
1905	8–10 Jul	3	27.4	10 Jul
1906	31 Aug–3 Sep	4	30.0	1 Sep
1911	5–7 Jul	3	25.8	5 Jul
	12–14 Jul	3	29.4	14 Jul
1917	22–24 Jul	3	26.4	22 Jul
1919	10–12 Aug	3	27.2	10 Aug
1921	8–11 Jul	4	30.6	10 Jul
1926	12–14 Jul	3	28.8	14 Jul
1930	26–29 Aug	4	29.3	27 Aug
1933	4–7 Jun	4	26.6	4 and 5 Jun
	2–5 Jul	4	29.9	3 Jul
	3–6 Aug	4	27.1	3 Aug
	26–28 Aug	3	27.7	27 and 28 Aug
1934	5–7 Jul	3	28.9	7 Jul
	9–11 Jul	3	27.2	9 Jul
1935	6–8 Aug	3	26.7	8 Aug
	20–22 Aug	3	27.2	21 and 22 Aug
1939	4–6 Jun	3	30.0	6 Jun
1940	3–10 Jun	8	30.0	8 Jun
1942	3–6 Jun	4	28.9	6 Jun
1945	2–4 Aug	3	28.3	2 and 3 Aug
1946	11–13 Jul	3	27.2	12 Jul
1948	27–31 Jul	5	27.8	31 Jul
1949	10–12 Jul	3	30.0	11 Jul
1952	20–22 Jul	3	27.8	21 and 22 Jul
1955	12–16 Jul	5	26.7	22 Jul

Continued

Table 24.10 *Continued*

Year	Dates in spell	Spell length	Highest max reached, °C	Hottest day in spell
	23–25 Aug	3	25.6	23 Aug
1959	18–22 August	5	28.9	20 Aug
	8–12 Sep	5	27.2	8 Sep
1969	14–16 Jul 1969	3	27.2	16 Jul
1975	**2–14 Aug**	**13**	**28.7**	**9 Aug**
	26–28 Aug	3	26.3	27 Aug
1976	25 Jun–3 Jul	9	29.3	29 Jun
1982	1–3 Jun	3	25.5	3 Jun
1984	6–8 Jul	3	26.2	7 Jul
1989	19–25 Jul	7	27.5	22 Jul
1990	18–20 Jul	3	28.0	20 Jul
	1–3 Aug	3	32.5	3 Aug
1995	29 Jul–2 Aug	5	27.6	1 Aug
	10–12 Aug	3	26.8	11 Aug
	15–22 Aug	8	30.3	21 Aug
1996	20–22 Jul	3	27.3	21 Jul
1999	31 Jul–3 Aug	4	28.0	2 Aug
2000	17–19 Jun	3	28.1	19 Jun
2003	4–6 Aug	3	27.8	6 Aug
2005	10–12 Jul	3	28.2	12 Jul
2006	16–19 Jul	4	30.5	17 Jul
	21–29 Jul	9	27.7	25 Jul
2013	15–18 Jul	4	28.2	18 Jul
2017	24–26 May	3	26.1	25 May
2018	24–26 Jun	3	26.6	25 Jun
	5–8 Jul	4	27.6	7 Jul
2019	23–26 Jul	4	32.9	25 Jul
	24–26 Aug	3	28.1	25 Aug
2020	24–26 Jun	3	27.1	24 Jun
2021	16–20 Jul	5	28.5	17 Jul

*The longest spell is highlighted in **bold**. Temperatures recorded prior to the introduction of the Stevenson screen in November 1899 are shown in italics; see Notes on p xvii and Appendix 3 for more details.*

Table 24.11 *The warmest seven-day periods on Durham's records, August 1844 to December 2021, in date order. The warmest such spell is shown in **bold** font.*

Dates	Mean temp, °C
1–7 Jun 1846	*19.7*
13–19 Jun 1846	*19.6*
4–10 Jul 1852	*19.7*
16–22 Jul 1878	*19.9*
15–22 Jul 1901	19.6
9–15 Jul 1905	19.6
1–7 Aug 1933	20.2
26 Jul–1 Aug 1948	19.9
21–27 Jul 1949	19.6
4–10 Aug 1975	**20.8**
20–26 Jul 1989	19.6
30 Jul–5 Aug 1990	19.8
27 Jul–2 Aug 1995	20.5
16–22 Aug 1995	20.3
8–14 Aug 1997	20.1
4–10 Aug 2003	20.4
21–27 Jul 2006	20.4
13–19 Jul 2013	19.8
17–23 Jul 2016	19.6
21–27 Jul 2018	19.7
22–28 Jul 2019	20.3
17–23 Jul 2021	19.7

Prolonged cold spells

Table 24.12 lists all spells since 1844 of at least three consecutive days when the maximum temperature did not exceed 0 °C at Durham Observatory, although again there are some gaps in the early years of the record.

Up to and including winter 2021/22, there have been 44 such spells in almost 180 years, an average recurrence period of four years, although at the time of writing none have been recorded for over 30 years. In contrast, in less than 30 years (between 1853 and 1881 inclusive), at least 18 such spells are known, occurring more often than every other year on average, although the thermometer exposure prior to 1900 rendered such events more likely to occur than with modern instrumentation and exposure standards. Several years have recorded two such spells, sometimes separated by only a day or two above freezing. The longest continuously sub-zero spell on Durham's record is nine days, 6–15 January 1982.

Table 24.13 shows the coldest spells of seven- and 14-day duration recorded in Durham since August 1844, with some gaps in the early years (including only 14 days of records during the very cold February in 1855). These are arranged in date order with the lowest mean temperature in each spell length shown in **bold**. See also Table 24.8.

Table 24.12 *The longest spells of sub-freezing days (consecutive days not exceeding 0.0 °C) on Durham's records, August 1843 to December 2021*

Year	Dates in spell	Spell length, days	Lowest max, °C and date	Coldest night, °C and date
1846	18–20 Mar	3	−1.9 °C, 18 and 19 Mar	−6.7 °C, 19 Mar
	13–15 Dec	3	−2.8 °C, 14 Dec	−5.9 °C, 14 Dec
1848	18–20 Jan	3	−2.2 °C, 18 Jan	−4.8 °C, 20 Jan
1853	11–14 Feb	4	−1.2 °C, 12 Feb	−6.0 °C, 11 Feb
	16–18 Feb	3	−0.8 °C, 18 Feb	−5.7 °C, 18 Feb
1853/54	31 Dec–4 Jan	5	−4.3 °C, 2 Jan	−11.7 °C, 3 Jan
1855	20–23 Feb*	4	−1.7 °C, 21 Feb	−9.3 °C, 22 Feb
	19–21 Dec	3	−1.3 °C, 19 Dec	−5.4 °C, 21 Dec
1856	29 Nov–1 Dec	3	−0.7 °C, 1 Dec	−4.7 °C, 1 Dec
1857	27–30 Jan	4	−1.5 °C, 29 Jan	−6.8 °C, 20 Jan
1859	14–19 Dec	6	−4.2 °C, 19 Dec	−9.1 °C, 19 Dec
1860	23–26 Dec	4	−4.6 °C, 25 Dec	−16.4 °C, 25 Dec
1862	17–19 Jan	3	−0.8 °C, 18 and 19 Jan	−5.6 °C, 19 Jan
1867	11–15 Jan	5	−2.2 °C, 12 Jan	−7.1 °C, 12 Jan
1870	12–14 Feb	3	−1.0 °C, 13 Feb	−4.4 °C, 12 Feb
	22–24 Dec	3	−2.3 °C, 24 Dec	−10.0 °C, 24 Dec
1870/71	31 Dec–4 Jan	5	−6.2 °C, 1 Jan	−14.0 °C, 31 Dec
1874/75	28 Dec–1 Jan	5	−3.5 °C, 29 Dec	−15.4 °C, 30 Dec
1879	3–5 Dec	3	−2.4 °C, 4 Dec	−14.6 °C, 4 Dec

* *Records incomplete for February 1855, spells probably longer than shown.*

Continued

Table 24.12 *Continued*

Year	Dates in spell	Spell length, days	Lowest max, °C and date	Coldest night, °C and date
1881	*14–17 Jan*	*4*	*−4.1 °C, 17 Jan*	*−16.9 °C, 17 Jan*
1894	*5–8 Jan*	*4*	*−3.5 °C, 7 Jan*	*−13.8 °C, 7 Jan*
1895	*6–8 Feb*	*3*	*−3.3 °C, 6 and 7 Feb*	*−18.0 °C, 8 Feb*
1904	20–22 Dec	3	−1.8 °C, 21 Dec	−7.6 °C, 22 Dec
1908	28–30 Dec	3	−3.9 °C, 29 Dec	−9.4 °C, 30 Dec
1912	2–5 Feb	4	−2.2 °C, 4 Feb	−12.4 °C, 4 and 5 Feb
1922	14–17 Jan	4	−0.4 °C, all dates	−5.3 °C, all dates
1927	16–20 Dec	5	−1.8 °C, 19 Dec	−7.5 °C, 17 Dec
1929	12–16 Feb	5	−3.4 °C, 12, 13, and 15 Feb	−11.9 °C, 14 Feb
	26–28 Feb	3	−1.8 °C, 26 Feb	−11.9 °C, 28 Feb
1938	19–21 Dec	3	−1.1 °C, 20 and 21 Dec	−3.9 °C, 20 Dec
1940	19–22 Jan	4	−1.7 °C, 20 Jan	−16.1 °C, 21 Jan
	28 Jan–1 Feb	5	−1.1 °C, 30 Jan	−3.3 °C, 30 Jan
1941	3–5 Jan	3	−4.4 °C, 5 Jan	−13.3 °C, 5 Jan
	16–19 Jan	4	−1.7 °C, 16 Jan	−12.2 °C, 18 Jan
1942	18–22 Jan	5	−2.8 °C, 20 and 21 Jan	−9.4 °C, 21 Jan
1945	23–26 Jan	4	−2.2 °C, 23 Jan	−12.8 °C, 24 Jan
1947	5–9 Feb	5	−2.8 °C, 7 Feb	−5.0 °C, 7 Feb
	11–13 Feb	3	−1.1 °C, 12 Feb	−2.2 °C, 12 Feb
1955	13–17 Jan	5	−1.7 °C, 13 Jan	−12.2 °C, 14 Jan
1982	**6–15 Jan**	**9**	**−4.5 °C, 7 Jan**	**−16.1 °C, 11 Jan**
1985	9–11 Feb	3	−0.8 °C, 11 Feb	−4.1 °C, 11 Feb
	27–29 Dec	3	−1.8 °C, 27 Dec	−5.6 °C, 29 Dec
1986	10–12 Feb	3	−3.5 °C, 10 Feb	−11.8 °C, 11 Feb
1987	11–13 Jan	3	−4.0 °C, 12 Jan	−7.1 °C, 13 Jan

*The longest spell is highlighted in **bold**. Prior to November 1899, records are not from a Stevenson screen, and this is reflected by showing pre-1899 values in italics. Details of the thermometer exposure throughout the record are given in Appendix 2 and 3.*

Table 24.13 *The coldest seven- and 14-day periods on Durham's records, August 1844 to December 2021, in chronological order*

Dates Seven-day spells	Mean temperature, °C
15–21 Feb 1855	*−5.0*
23–29 Dec 1860	*−5.1*
30 Dec 1870–5 Jan 1871	*−5.2*
1–7 Dec 1879	*−5.1*
12–18 Jan 1881	*−5.3*
20–26 Jan 1881	*−5.8*
6–12 Feb 1895	***−6.8***
11–17 Feb 1929	−5.3
7–13 Jan 1982	**−6.8**

14-day spells	
15–28 Feb 1855 (Month incomplete)	*−2.9*
17–30 Dec 1860	*−3.2*
22 Dec–3 Jan 1871	*−3.6*
21 Dec–2 Jan 1875	*−3.2*
12–25 Dec 1878	*−3.4*
14–27 Jan 1881	***−5.5***
24 Dec–6 Jan 1893	*−3.1*
3–16 Feb 1895	*−3.9*
5–18 Jan 1982	−3.8

Within the last 50 years to 2021, the coldest seven-day spell has been 17–23 December 2010, mean temperature −3.5 °C, and the coldest 14-day spell 16–29 December 2010, mean temperature −1.9 °C.

Records prior to the introduction of the Stevenson screen in November 1899 are shown in italic (see Appendix 2). In the case of multiple candidates within a particular range of dates, only the spell with the lowest mean temperature is quoted. The coldest spell in each duration is highlighted in **bold**.

25

Precipitation extremes in Durham

Droughts

Prolonged spells of deficient rainfall have been covered in Chapter 21. Here we itemize only the longest spells of 'absolute drought', which, as defined by Symons in *British Rainfall 1887*, consists of a period of at least 15 consecutive days during which no day receives as much as 0.2 mm of precipitation. Table 25.1 shows the longest absolute droughts recorded at the Durham Observatory since 1868. Prior to 1938, the longest such spells were the 22 days ending 2 July 1887 and two spells of 24 days, ending 18 June 1899 and 4 May 1912, respectively.

Table 25.1 *Longest spells of 'absolute drought' (no day with > 0.1 mm of precipitation, 25 days or longer) at Durham Observatory, January 1868 to December 2021*

Year	Start and end dates	Length
1938	27 Mar–21 Apr	26 days
1947	5 Aug–5 Sep (**Longest dry spell**)	32
1953	27 Feb–29 Mar	31
1955	4–31 Jul	28
1961	3–28 Mar	26
1976	3–27 Aug	25
2018	20 Jun–15 Jul	26

Precipitation depth-duration extremes

Table 25.2 sets out the driest and wettest spells of various durations from one day (35 days upwards for dry spells) to 365 days on the Durham record from 1868 to 2021.

Table 25.2 *Precipitation depth-duration extremes for Durham Observatory, January 1868 to June 2021 The ten wettest days on Durham's records are listed in Table 25.3.*

Period length	Wettest spells			Driest spells		
	Amount mm	mm/day	Start and end dates	Amount mm	mm/day	Start and end dates
1 day	87.8	87.8	11 Sep 1976			
2 days	120.1	60.1	10–11 Sep 1976			
	96.8	48.4	25–26 Aug 1986			
	95.1	47.5	25–26 Jul 1888			
3 days	127.1	42.4	9–11 Sep 1976			
	106.3	35.4	25–27 Aug 1986			
	98.7	32.9	25–27 Jul 1888			
4 days	133.7	33.4	10–13 Sep 1976			
5 days	140.7	28.1	9–13 Sep 1976			
7 days	146.9	21.0	8–14 Sep 1976			
	126.4	18.1	14–20 Feb 1941			
	123.9	17.7	6–12 Oct 1903			
10 days	150.5	15.1	8–17 Sep 1976			
	142.8	14.3	29 Oct–7 Nov 2000			
	142.1	14.2	25 Aug–3 Sept 1986			
14 days (2 weeks)	167.0	11.9	20 Jul–2 Aug 1930			
	160.2	11.4	26 Oct–8 Nov 2000			
	154.0	11.0	2–15 Jan 1948			
21 days (3 weeks)	199.6	9.5	13 Jul–2 Aug 1930			
	184.5	8.8	8–28 Sep 1976			
28 days (4 weeks)	230.9	8.2	8 Sep–5 Oct 1976			
	216.1	7.7	13 Jul–9 August 1930			
35 days (5 weeks)	231.1	6.6	1 Sep–5 Oct 1976	0.2	< 0.1	18 Feb–24 Mar 1953
				0.3	< 0.1	5 Aug–8 Sep 1947
40 days	301.5	7.5	8 Sep–17 Oct 1976	0.2	< 0.1	18 Feb–29 Mar 1953
				0.6	< 0.1	3 Aug–11 Sep 1947

Continued

Table 25.2 *Continued*

Period length	Wettest spells			Driest spells		
	Amount mm	mm/day	Start and end dates	Amount mm	mm/day	Start and end dates
45 days	301.5	6.7	3 Sep–17 Oct 1976	3.6	0.1	25 Mar–8 May 1938
				3.8	0.1	18 Jun–1 Aug 1878
50 days	325.0	6.5	8 Sep–27 Oct 1976	5.9	0.1	7 Mar–25 Apr 1938
				6.5	0.1	13 Jun–1 Aug 1878
60 days *(2 months)*	339.9	5.7	8 Sep–6 Nov 1976	7.5	0.1	7 Mar–5 May 1938
	300.5	5.0	7 Nov 1876–5 Jan 1877	13.7	0.2	18 Jun–16 Aug 1977
75 days	356.5	4.8	25 Aug—7 Nov 1976	18.0	0.2	27 Feb–12 May 1938
	348.1	4.6	15 Jul–27 Sep 1930	23.1	0.3	2 Mar–15 May 1893
90 days *(3 months)*	390.9	4.3	11 Jul–8 Oct 1930	34.2	0.4	5 Jul–2 Oct 1955
	376.3	4.2	25 Jun–22 Sep 1927	35.5	0.4	15 Feb–15 May 1938
120 days *(4 months)*	461.4	3.8	31 May–27 Sep 2012	59.8	0.5	17 Jan–16 May 1938
	443.1	3.7	3 Dec 1978–1 Apr 1979	69.7	0.6	5 Jul–1 Nov 1955
180 days *(6 months)*	661.8	3.7	1 Jun–27 Nov 2012	130.6	0.7	26 Feb–24 Aug 1870
	590.5	3.3	6 Nov 1876–4 May 1877	133.5	0.7	21 Sep 1972–19 Mar 1973
365 days *(1 year)*	1130.9	3.1	2 Apr 2012–1 Apr 2013	350.1	1.0	25 Aug 1904–24 Aug 1905
	1017.6	2.8	8 Apr 2000–7 Apr 2001	355.9	1.0	22 Oct 1958–21 Oct 1959

The period totals are exact day totals, the monthly equivalents shown are an approximation to the number of days shown. Only the driest or wettest spell in each named period is listed.

Wettest and driest calendar months

Table 25.3 *Wettest and driest calendar months at the Durham Observatory, by total precipitation, period January 1868 to December 2021*

	Wettest months			Driest months		
Rank	Total pptn, mm	Per cent of monthly 1991–2020 normal	Month and year	Total pptn, mm	Per cent of monthly 1991–2020 normal	Month and year
1	209.7	337	Jul 1888	1.3	3	Mar 1953
2	201.8	323	Oct 1903	1.7	3	Jun 1925
3	195.5	329	Dec 1978	2.1	5	Feb 1891
4	193.2	340	Sep 1976	2.2	4	Apr 1938
5	191.4	313	Jun 1980	2.4	5	Apr 1912
6	188.0	352	Jan 1948	3.6	6	Jul 1878
7	186.1	244	Nov 1965	3.7	8	Feb 1921
8	183.9	296	Jul 1930	3.8	7	Apr 2020
9	175.7	272	Aug 1956	4.1	7	Jun 1891
10	174.4	285	Jun 1997	4.2	10	Mar 1929
Last 50 years to 2021						
	195.5	329	Dec 1978	3.8	7	Apr 2020

Table 25.4 *Distribution by month of the wettest and driest calendar months at the Durham Observatory, period January 1868 to December 2021 (from Table 25.3)*

	J	F	M	A	M	J	J	A	S	O	N	D
Wettest	1					2	2	1	1	1	1	1
Driest		2	2	3		2	1					

No single calendar year features two months in the 'Top 10 wettest' category, although 1891 has two months in the driest—February and June.
See also Table 25.2 for details of other periods from one day to one year.

Wettest days

Table 25.5 *Wettest days recorded at the Durham Observatory, January 1868 to June 2021*

	Wettest days	
Rank	Total precipitation, mm	Date
1	87.8	11 Sep 1976
2	72.8	12 Nov 1901
3	70.2	31 May 1924
4	69.1	25 Aug 1986
5	63.7	8 Oct 1903
6	61.4	8 Jul 1893
7	57.4	24 Sep 2012
8	57.1	25 Jul 1888
9	55.8	21 Sep 1935
10	55.6	3 Sep 1948
Last 50 years to 2021		
	87.8	11 Sept 1976

For a heavy fall of rain to appear in this table, the timing of heavy rain needs to somewhat fortuitously fall within the 0900–0900 GMT standard 'rain day'. There are other examples in the record where the fall has been split across two rain days and therefore does not appear in the table.

Table 25.6 *Distribution by month of the wettest days at the Durham Observatory, period January 1868 to December 2021 (from Table 25.5)*

	J	F	M	A	M	J	J	A	S	O	N	D
Wettest days					1		2	1	4	1	1	

Snowfalls

Table 25.7 lists the earliest and latest dates with snow cover at Durham during the period 1960–99. Table 25.8 ranks the greatest snow depths during the period January 1960 to December 1999. Cold winters tend to produce a large number of days with similar snow depths, often a slow reduction after one or two major snowfalls, so only the greatest depth in each winter is shown. Table 25.9 lists the greatest single snow depth at 0900 GMT, with date, for every winter between 1959/60 and 1998/99. The winters are dated by the January.

Table 25.7 *Earliest and latest dates of snow cover, and various depths of snow, at Durham: period January 1960 to December 1999, based on 0900 GMT observations*

Depth threshold	Earliest date/depth	Latest date/depth
> 50% cover, > 0 cm	14 Nov 1965, 5 cm	25 Apr 1981, 7 cm
≥ 5 cm	14 Nov 1965, 5 cm	25 Apr 1981, 7 cm
≥ 10 cm	15 Nov 1965, 10 cm	7 Apr 1977, 10 cm
≥ 15 cm	26 Nov 1965, 20 cm	5 Apr 1975, 25 cm
≥ 20 cm	26 Nov 1965, 20 cm	5 Apr 1975, 25 cm
≥ 30 cm	27 Nov 1965, 36 cm	20 Mar 1979, 30 cm
≥ 40 cm	30 Nov 1965, 41 cm	26 Feb 1963, 41 cm

Table 25.8 *Greatest snow depths at 0900 GMT, period January 1960 to December 1999*

Greatest snow depths

Rank	Snow depth, cm	Date(s)
1	43	15–18 Feb 1963
2	41	31 Dec 1961
		30 Nov 1965
3	37	11 Feb 1991
4	31	18–19 Mar 1979
5	28	1 Mar 1965
6	25	5 Apr 1975

Table 25.9 *Greatest snow depths at 0900* GMT *in each winter, period January 1960 to December 1999. Winters dated by January year*

Winter	Snow depth, cm	Date	Winter	Snow depth, cm	Date
1961	13	3 Jan	1981	7	25 Apr
1962	41	31 Dec 1961	1982	17	14 Dec 1981
1963	43	15–18 Feb	1983	18	11 Feb
1964	10	18–19 Feb	1984	16	24–25 Jan
1965	28	1 Mar	1985	13	17 Jan
1966	41	30 Nov 1965	1986	15	25 Feb
1967	8	5 Jan	1987	12	14 Jan
1968	18	5 Feb	1988	4	24 Jan and 24 Feb
1969	23	21 Feb	1989	7	21 Nov 1988
1970	18	18 Feb	1990	8	28 Jan
1971	5	31 Dec and 28 Feb	1991	37	11 Feb
1972	14	2 Feb	1992	4	20 Dec 1991
1973	5	20 Jan	1993	3	27 Feb–1 Mar
1974	5	2 Dec 1973	1994	13	26 Feb
1975	25	5 Apr	1995	6	26 Jan
1976	8	25–26 Jan	1996	14	27 Jan
1977	10	7 Apr	1997	7	1 Jan
1978	24	14 Feb	1998	6	20 Jan
1979	31	18–19 Mar	1999	3	12 Jan
1980	6	2 and 5 Feb			

We have very few measurements of snow depths prior to 1960, and none since late 1999, and for this reason it is difficult to be sure of the greatest depth of snow recorded in Durham. One candidate would be 21 February 1941, when following a period of 67 hours of continuous snowfall the snow lay 107 cm deep. Wheeler [129] reported that nothing comparable could be found in newspaper reports going back to 1814 in Durham and 1831 in Sunderland; more on this event appears in Chapter 22.

26

Sunshine extremes in Durham

Table 26.1 lists Durham's sunniest and dullest calendar months by total sunshine duration, period June 1880 to December 2021, with some gaps in the early years—see Appendix 6 for more details.

Table 26.2 list Durham's sunniest and dullest months on record by percentage of possible duration of sunshine, over the same period as Table 26.1.

Data from the commencement of the record to and including December 1999 are monthly totals from the daily Campbell–Stokes sunshine recorder record, but from January 2000 to date, monthly totals (only) are derived by regression from the East and North-east England monthly sunshine regional value published by the Met Office to a precision of one hour. Some early monthly totals are also available only to a precision of one hour.

Table 26.1 *The sunniest and dullest months on record at Durham Observatory, by monthly duration of sunshine, June 1880 to December 2021*

	Sunniest months				Dullest months			
Rank	Total sunshine, hours	Per cent of 1991–2020 monthly normal	Per cent of possible	Month and year	Total sunshine, hours	Per cent of 1991–2020 monthly normal	Per cent of possible	Month and year
1	297.0	179	57.6	Jun 1940	7.6	12	3.1	Jan 1996
2	273	156	52.9	Jul 2006	9.7	17	4.3	Dec 1978
3	258.8	138	51.8	May 1971	10.4	18	4.6	Dec 1903
4	256	137	51.2	May 1881	10.8	19	4.8	Dec 1890
	256	137	51.1	May 2020				
5	254.5	145	49.3	Jul 1955	12.5	20	5.1	Jan 1942
6	253.5	153	49.1	Jun 1957	15.5	25	6.3	Jan 1913
7	252.0	144	48.8	Jul 1911	16.5	29	7.3	Dec 1927
8	250.9	151	48.6	Jun 1949	17.8	31	7.8	Dec 1934
9	249.3	148	54.1	Aug 1995	19.5	31	7.9	Jan 1917
10	248	133	49.6	May 2018	21.4	38	9.4	Dec 1882
					21.4	35	8.7	Jan 1929
Last 50 years to 2021								
	273	156	52.9	Jul 2006	7.6	12	3.1	Jan 1996

Table 26.2 *The sunniest and dullest months on record at Durham Observatory, by percentage of possible duration of sunshine, June 1880 to December 2021*

Rank	Sunniest months as % possible			Dullest months as % possible		
	Per cent of possible	Total sunshine, hours	Month and year	Per cent of possible	Total sunshine, hours	Month and year
1	57.6	297.0	Jun 1940	3.1	7.6	Jan 1996
2	56.2	238.1	Apr 1914	4.3	9.7	Dec 1978
3	54.1	249.3	Aug 1995	4.6	10.4	Dec 1903
4	54.0	229	Apr 2021	4.8	10.8	Dec 1890
5	53.5	203.9	Sep 1895	5.1	12.5	Jan 1942
6	52.9	273	Jul 2006	6.3	15.5	Jan 1913
7	52.0	191.4	Mar 1894	7.3	16.5	Dec 1927
8	51.8	258.8	May 1971	7.8	17.8	Dec 1934
9	51.7	219	Apr 2015	7.9	19.5	Jan 1917
10	51.5	218	Apr 2020	8.0	29.3	Mar 1996
Last 50 years to 2021						
	54.1	249.3	Aug 1995	3.1	7.6	Jan 1996

27

Barometric pressure extremes in Durham

Appendix 5 provides details of Durham's barometric pressure record.

Daily extremes

Table 27.1 lists the highest and lowest barometric pressure recorded at Durham Observatory by month (with date and time) and by year, from the *corrected* series (details in Appendix 5; none of these records are from gap-filled missing data). It should be noted that as readings were taken only twice per day, these are certainly under-estimates of the range of barometric pressure at the site, since extremes may occur at any hour of the day.

Table 27.1 *Monthly and annual extremes of MSL pressure at Durham Observatory, from morning and evening observations only (usually 0900 and 2100 GMT), period July 1843 to December 1960. Units—hPa, omitting initial digit (i.e. 10 for the maximum values, and 9 for the minimum)*

Period	J	F	M	A	M	J	J	A	S	O	N	D	Year
Maximum	52.5	48.2	44.9	42.5	42.3	40.7	37.7	35.0	39.3	40.5	42.9	47.4	52.5
Year	*1907*	*1902*	*1935*	*1906*	*1881*	*1959*	*1911*	*1874*	*1851*	*1956*	*1922*	*1926*	*1907*
Date	*23*	*1*	*9*	*8*	*10*	*13*	*10*	*20*	*16*	*31*	*15*	*23*	*23/1*
Hour	*09*	*09*	*09*	*21**	*10*	*09*	*09*	*22*	*09*	*09*†	*09*	*21*	*09*
Minimum	50.7	59.8	62.3	70.2	72.0	82.4	77.2	68.5	71.9	58.5	64.2	36.2	36.2
Year	*1884*	*1951*	*1876*	*1948*	*1943*	*1938*	*1922*	*1917*	*1935*	*1959*	*1881*	*1886*	*1886*
Date	*26*	*5*	*10*	*1*	*8*	*28*	*6*	*28*	*17*	*27*	*27*	*8*	*8/12*
Hour	*22*	*09*	*10*	*09*	*21*	*21*	*09*	*09*	*09*	*09*	*10*	*21*	*21*
Monthly pressure range	101.8	88.4	82.6	72.3	70.3	58.3	60.5	66.5	67.4	82.0	78.7	111.2	116.3

* *April highest value 1042.5 hPa equalled at 21h on 11 April 1938.*
† *October highest value 1040.5 hPa equalled at 09h on 23 October 1958*

The ten highest and ten lowest MSL pressure readings at Durham over the period 1843–1960 are listed in Table 27.2, together with the equivalent 20CRv3 grid point ensemble mean value closest to that observation time (see Appendix 5 for details).

Details of the circumstances of many of the events listed in Table 27.2, together with synoptic descriptions, can be found in Burt [55, 56].

Table 27.2 *The ten highest and ten lowest barometric pressures on the Durham Observatory record 1843–1960, with the date and observation time, in rank order, together with the 20CRv3 grid point ensemble mean for that date and time. Units hPa.*

Date	Time	Durham QC MSLP hPa	20CRv3 grid point ensemble mean hPa	Notes
Highest MSL pressures				
23 Jan 1907	0900	1052.5★	1050.8	
31 Jan 1902	0900	1050.8	1047.8	
31 Jan 1902	2100	1050.7	1047.7	
22 Jan 1907	2100	1049.9	1047.5	
26 Jan 1932	0900	1049.7	1046.5	
9 Jan 1896	0900	1048.8	1047.1	'Raw' value 1051.2 hPa appears too high
26 Jan 1932	2100	1048.7	1045.1	
1 Feb 1902	0900	1048.2	1046.7	
23 Jan 1907	2100	1048.1	1047.1	
25 Jan 1932	2100	1047.7	1045.4	
Lowest MSL pressures				
8 Dec 1886	2100	936.2	952.8	Reanalysis in error—depth and timing. See also Chapter 22
26 Jan 1884	2200	950.7	945.1	Reanalysis system speed too slow
9 Dec 1886	0900	953.7	958.3	
3 Dec 1909	0900	954.7	954.2	
4 Feb 1951	2100	955.4	954.3	
1 Jan 1949	2100	955.7	956.3	
8 Dec 1886	0900	956.3	973.1	
4 Feb 1951	2100	956.4	954.3	
6 Dec 1847	2100	956.9	965.8	Minimum noted as 956.8 hPa at 2025 common time
19 Feb 1900	2100	957.0	962.5	
1 Jan 1949	0900	957.7	957.1	

★ Slightly high; North Shields 1050.8 hPa at 0800 and 1049.1 hPa at 1800.
True maximum probably between 1051 and 1051.5 hPa. See also Chapter 22

Monthly and annual extremes

Table 27.3 lists the highest and lowest monthly barometric pressure means recorded at Durham Observatory, using the *corrected* series including any gap-fills from 20CRv3 as necessary. The anomaly from the 1931–60 monthly average (from Table A5.5) is also shown for each entry. The highest monthly mean pressure on the Durham record was in February 1932, when the mean was 1034.6 hPa; only two other months have averaged in excess of 1030 hPa, namely February 1959 (1030.4 hPa) and February 1891 (1030.3 hPa); not surprisingly, all three months were very dry in Durham. The lowest monthly mean pressure on the record was in January 1948, when the mean was 994.5 hPa. Other notably cyclonic months include the Decembers of 1868 and 1959, with a monthly mean of 997.0 hPa.

Table 27.3 *Highest and lowest monthly mean MSL pressure (hPa) at Durham Observatory, and the extreme range in monthly means, over the period 1843–1960. The monthly mean is the average of the morning and evening observations (usually 0900 and 2100* GMT*). Anomaly from 1931–60 normal also stated. Values omit initial digit (i.e. 10 for values > 1000 hPa, and 9 < 1000 hPa). Highest and lowest values shown in* **bold***.*

	J	F	M	A	M	J	J	A	S	O	N	D	Year
Maximum	27.4	**34.6**	29.4	27.0	25.8	24.1	21.8	22.2	24.7	22.2	26.2	24.8	17.0
Year	1880	1932	1953	1938	1896	1865	1955	1947	1865	1947	1867	1926	1921
Anomaly (+)	15.5	20.9	14.6	12.3	9.4	8.7	8.7	8.8	10.3	8.7	14.7	13.7	+3.3
Minimum	**94.5**	98.8	97.6	05.1	07.7	04.9	05.1	03.4	03.6	00.0	99.4	97.0	07.9
Year	1948	1937	1876	1920	1925	1852	1861	1860	1918	1903	1877	1868, 1959	1872
Anomaly (−)	17.4	14.9	17.2	9.6	8.7	10.5	8.0	10.0	10.8	13.5	12.1	14.1	−5.8
Monthly range	32.9	35.8	31.8	21.9	18.1	19.3	16.7	18.8	21.1	22.2	26.8	27.8	9.1

Appendices

1. Observers at Durham University Observatory, 1840–1999
2. Metadata: meteorological records from Durham Observatory, 1843 to date
3. Temperature records made at Durham Observatory, 1843 to date
4. Precipitation records made at Durham Observatory, 1843 to date
5. Barometric pressure records made at Durham Observatory, 1843 to 1960
6. Sunshine duration records made at Durham Observatory, 1880–1999, and estimates to date
7. Wind speed and direction records made at Durham Observatory, 1866–2011
8. Climatological averages and extremes for Durham Observatory, period 1981–2010
9. Climatological averages and extremes for Durham Observatory, period 1991–2020
10. Records of other meteorological elements at Durham Observatory
11. Monthly and annual summaries of Durham's weather by year, 1843–2021

Durham from Observatory Hill, by John Wilson Carmichael. Oil on canvas.

It is thought this painting was completed in the early 1840s before the artist moved to London, at much the same time as the Observatory was built.

Reproduced with permission from Durham University Library and Heritage Collections

Appendix 1
Observers at Durham University Observatory, 1840–1999

List of observers

The following list was taken from Kenworthy and colleagues [44]. Note that the day of the month when a new Observer took over was not always clear in the record, but observations took place on the dates not covered, presumably by Professor Temple Chevallier in the early years. The weather observations were made by astronomers until 1938, after which conventional astronomical observations ceased.

Dates of service	Name	Notes
16 June 1840 to 1841, then Director until his retirement in 1871	Temple Chevallier, Observer	
From 16 June 1840	John Steward Browne, Assistant Observer, later Observer	
18 February 1842 to 3 February 1846	Arthur Beanlands	
3 February 1846 to 25 June 1849	Robert Anchor Thompson	
Part of June 1849	W. A. LeJeune	
June to October 1849	R. H. Blakey (acting)	
October 1849 to March 1852	Richard C. Carrington	Note 1
March 1852 to May 1853	William Ellis	Note 2
1854 to 1855	George F.W. Rümker	
1856 to January 1863	Albert Marth	Note 3
1863 to 1864	E. G. Marshall	
February 1865 to April 1867	Mondeford Reginald Dolman	
November 1867 to February 1874	John J. Plummer	Note 4
June 1874 to September 1885	Gabriel Alphonsus Goldney	Note 5
September 1885 to 24 September 1900	Henry James Carpenter	
1900 to 1919	Frederick Charles Hampshire Carpenter *(son of the above)*	
1919 to 10 December 1938	Frank Sargent	Note 6
December 1938 to December 1939	Dr E. Gluckauf	Note 7
May 1940 to 21 August 1945	A. Beecroft	
22 August 1945 to 9 January 1948	L. S. Joyce	
1 January 1949 to 6 January 1951	K. F. and G. A. Chackett	
January 1951 to March 1957	J. Musgrave	
24 March 1957 to 1968	F. and D. Glockling	
1969 to September 1999	Mrs Audrey Warner	Note 8

Compiled from Observatory records, the University Calendar and the Observatory plaque (Figure 2.9).

Notes

1. **Richard Carrington** (1826–75) was at Durham for only 2½ years; his time at Durham is covered in Chapter 2. He remains well-known in astronomical circles, for it was Carrington who discovered the differential rotation of the Sun, and for his observation of, and account of, the first recorded solar flare on 1 September 1859 [25, 26]. This solar storm is still known today as the 'Carrington Event'.

2. **William Ellis** (1828–1916) was the son of Thomas Ellis, an astronomer at Greenwich Observatory, and upon leaving school at the age of 13 he joined the junior computing staff at Greenwich, where he remained until taking up the post of Observer at Durham upon Carrington's departure, almost certainly at the command of George Airy, Astronomer Royal. He remained at Durham only little over a year, returning to Greenwich in 1853. He subsequently took charge of Greenwich's Time Signals for the following 20 years, until appointed Superintendent of the Magnetic and Meteorological Department at Greenwich upon James Glaisher's departure in 1874. His interests were as much meteorological as astronomical in nature, and he published more than 20 papers in the *Quarterly Journal of the [Royal] Meteorological Society* during his lifetime. He was elected a Fellow of the Meteorological Society (later the Royal Meteorological Society) in 1875 and served as the Society's President during 1886 and 1887. He was elected FRS in 1893 [141].

3. **Albert Marth** (1828–1897) was a noted German astronomer, publishing over 50 papers on astronomical subjects in his lifetime, several during his period at Durham; he was subsequently awarded an honorary M.A. by the University. From Durham he moved to an observatory in Malta from 1862 to 1865, eventually returning to the north-east of England in 1868 at the Gateshead Observatory. He later became Director of the Markree Observatory in Sligo, Ireland where he remained until his death. His obituary appeared in *Monthly Notices of the Royal Astronomical Society*, Vol. 58, p. 139. He has craters named after him on both the Moon and Mars.

4. **John Plummer** (Observer November 1867 to February 1874) wrote one of the early descriptions of the Observatory's meteorological record [27], and conducted extensive comparisons between the thermometer readings in the open Glaisher Stand and the so-called 'North Shed' (see Appendix 3 for more details). It was probably Plummer's appointment that led to the adoption of more consistent and reliable rainfall measurements (see Appendix 4 for more details), and this is the main reason why we are confident in the rainfall record from 1868.

5. **Gabriel Goldney** was Durham's Observer from June 1874 to September 1885. His letter of application for the post of Observer at Durham, dated 15 May 1874 together with a letter of recommendation from William Christie of the Royal Observatory, Greenwich (later Astronomer Royal) and a testimonial from Sir George Airy, then Astronomer Royal, to the trustees of the Observatory, survive in the Durham University Library archive ([24], items OBS331-333). It was during Goldney's time as Observer that the new Campbell–Stokes sunshine recorder was introduced; Durham received one of the very first instruments to be issued on loan by the Meteorological Office in spring 1880 [81]. Records continued satisfactorily until December 1883 (although we have not yet managed to trace the daily records for 1880 and 1881), but the record ceased entirely at the end of January 1884. The Meteorological Office (then at 116 Victoria Street, London SW) apparently wrote to the Observatory several times during 1884 to enquire of the sunshine records, but seemingly received no reply. On 19 March 1885, a letter from the Meteorological Office went to Mr Goldney asking him to return the sunshine recorder and all unused cards 'at once, in order that it might be placed elsewhere'. When no reply was received, the recently retired Head of Westminster School Dr Charles Broderick Scott wrote directly to the Dean of the University on the matter in April 1885. He was the brother of the Director of the Meteorological Office, Robert H Scott, and was presumably personally acquainted with the Dean and hoped that this approach might move things along. Sure enough it did, for

on 18 April 1885 Robert Scott himself wrote to the Dean, referencing his brother's letter and formally requesting the return of the sunshine recorder if it was no longer being used. (Durham University Library archive, op. cit. items OBS338 and 341). He stated that no sunshine records from Durham had been received by the Meteorological Office since July 1883, apart from the (presumably un-analysed) sunshine cards from February 1885 which had since been forwarded. Scott's letters are quite critical of Goldney's performance in the role, referring to him as 'impracticable' and commenting that previous efforts to re-solve the sunshine recorder issue 'had not resulted in any improvements' and seeking to 'get him to attend to our instructions; this he has not done'. The letter makes clear that the sunshine recorder was provided on loan to the Observatory only on condition that the results were sent weekly to the Meteorological Office, that the lack of compliance was becoming a 'serious inconvenience', and that they seek to 'make other arrangements'.

On 25 April, Goldney wrote to the head of the Observatory Trustees, a Professor Pearce, apologizing for not having called that morning and saying he will see him early in the week. Just three days later, on 28 April, Pearce wrote to Goldney formally dismissing him from the post of Observer as of 28 July. A second personal letter from Pearce to Goldney, of the same date, urged him to make the best use of the remainder of his time in Durham, 'so that by regularity and attention you may regain a little of the position you have forfeited by your neglect of duty' (both letters OBS340). On 17 July, Goldney wrote to Pearce pledging himself to discharge his duties in 'a most scrupulously careful manner' for the remainder of his time at Durham (OBS349). Meanwhile, on 29 April, Robert Scott wrote again to Pearce, stating that the sunshine recorder may remain at Durham provided it is attended to (OBS341). Sure enough, the sunshine record recommenced on 29 May 1885, and the daily record remained unbroken until terminated on 31 December 1999.

Goldney's post was very quickly advertised, probably via word of mouth within the observatory network, and was soon offered to Henry James Carpenter, then of the Radcliffe Observatory at Oxford (OBS342-343, 345–346). Carpenter was unable to start at Durham before the end of August owing to existing work commitments at the Radcliffe, as he set out in a letter to Pearce (OBS347, 29 May 1885) accepting the appointment. He also explained that, as he had formerly worked with the present Observer at Greenwich, 'he would in the circumstances prefer to delay his arrival in Durham until the other had left'. He arrived in Durham to take up his appointment on 24 September 1885, remaining in post for 15 years. (Goldney had presumably left the Observatory shortly before his arrival.) The post was for unmarried men only; Carpenter had been married, and had a family of boys, but his wife had died some years previously. He brought the two eldest boys with him to Durham 'so that he can give more attention to their needs and education'. One of his sons succeeded him in post, eventually gaining a B.Sc. and M.A. and being appointed lecturer in astron-omy and optics at Durham (Source: Met Office inspection report, 1912). Aside from War Service from 1916–18, where he was wounded at the Front, he continued as meteorological observer until 1919.

6. **Frank Sargent** left the Observatory in 1938. In a letter to John Glasspoole at the Meteorological Office dated 15 January 1940 (in the Met Office site file), he states 'I should not have left Durham, had it not been that what I confess was and is my major interest, astronomical work, was so seriously neglected, interest as I judged it being entirely lacking, and limited to practically confining me to utterly useless observations, even though I had to use my own telescopes to do it.' He was at the time employed in the Filton Aeroplane Works at Filton.

7. Since **Dr Gluckauf** was a registered foreign national, responsibility for the observations was given in June 1940, and subsequently for much of the Second World War, to his two deputies R. B. Jacobi and G. O'Neill. A letter in the Met Office site file from the Board of Curators of the Observatory to the Meteorological Office dated 6 June 1940 clarifies this arrangement: 'The new police regulations made it impossible for Dr Gluckauf, being an alien, to continue his work, and that the Board of Curators has asked the two men mentioned above to carry on provisionally. I suppose that a more definite arrangement will shortly be made, on which I shall inform you in due course.' This letter was acknowledged by John Glasspoole on 17 June 1940.

8. **Mrs Audrey Warner**'s time as meteorological observer is covered in Chapter 2. She was in the post for 30 years, the longest of any observer, until retiring in September 1999. Her position was not replaced, and an AWS was installed instead.

Appendix 2
Summary metadata: meteorological records from Durham Observatory, 1843 to date

This Appendix contains common details to the record, including site description, location details and site photographs, data sources, observing practices, observation times and terminal hours, and a summary of the digital dataset. The Appendices that follow set out details by element (temperature, precipitation, barometric pressure, sunshine and wind), of the instruments and exposures specific to that element, noting original data sources, units, gaps/missing data, and any adjustments or quality control procedures applied to the record.

The observatory site

The Durham Observatory site is located at 54.768°N, 1.584°W, altitude 102 m above MSL.

Some details of the site have previously been given in Chapter 1 but are repeated or provided in more detail here in order to provide a seamless metadata source. Manley [19] described the site as follows (updating original units):

> The University Observatory is a small stone building of two stories in 0.4 ha of garden, 102 m above sea level, at the top of the sharply rising slopes a kilometre to the south-west of the Cathedral and 2 km south-west of the centre of the town. From the garden the ground falls away immediately on the north-east and east to the River Wear, 500 m distant and over 60 m below. To the west the ground is nearly level for about 400 metres and then falls sharply to the Browney valley. To the south the ground again falls noticeably over a distance of 200 metres; to north-west and north the ground is undulating with a prevailing level of 90–105 m. The Observatory therefore stands on a decidedly well-exposed ridge open to winds from all quarters. The soil in the immediate neighbourhood is generally sandy loam but is interspersed with patches of heavier clay. Trees and shrubs have been planted on the north, west and east of the Observatory; on the north and east they are close to the building and some of them reach the level of the roof, 8–10 m above the ground. Thermometers and rain-gauges are now exposed on the lawn about 20 m south of the front of the building; the sunshine recorder is on the roof. Some smaller out-buildings erected in recent years are not likely to affect the exposure.

To this was added the following by Kenworthy [28] in 1985:

> The qualities of the site and its representativeness are as described [in Manley [19]]. The local exposure of the observatory is good, although the basin-like character of the Wear valley causes extreme minima, even on the observatory ridge (336 feet, 102 m), to be

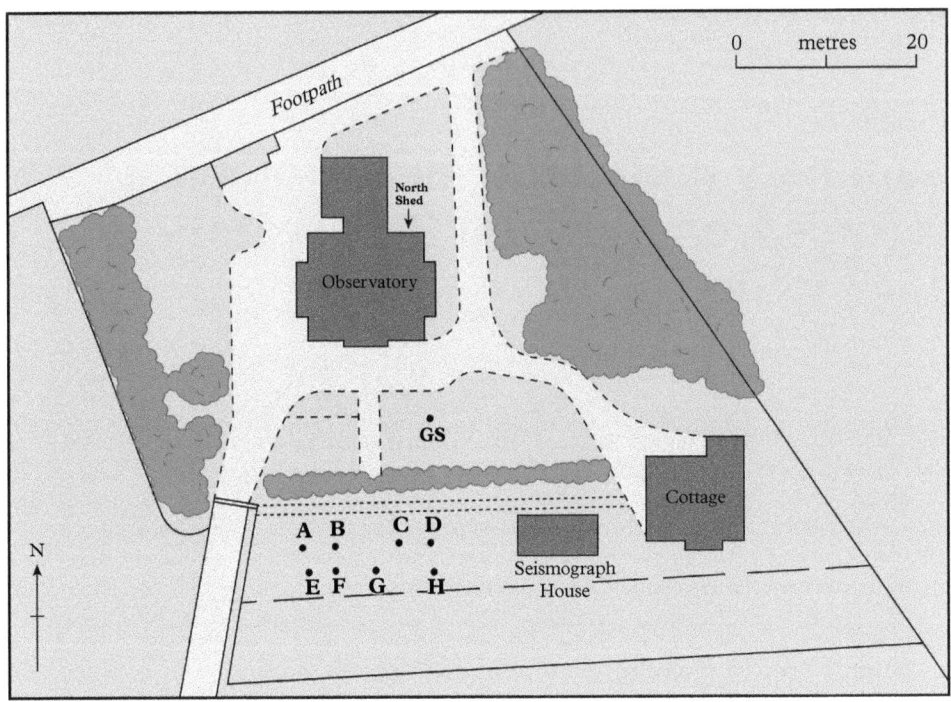

Figure A2.1 *Plan of instruments, past and present, at the Durham Observatory site. Redrawn from Baxter (1956) with additional information from Manley [19]. GS Glaisher stand 1860–1900; A and B Stevenson screens; C, five-inch rain gauge; D, recording rain gauge (since discontinued); E and F, earth thermometers; G, grass minimum thermometer; H, solar radiation thermometer (discontinued mid-1950s). Today, the Metspec Stevenson screen is located close to position C, with grass minimum, 30 cm, and 100 cm earth temperature sensors nearby, and the tipping-bucket rain gauge at position H (as shown in Figure A3.7). The location of the 'North Shed' on the north balcony 1851–1880 is also indicated. The Campbell–Stokes sunshine recorder was located on the front parapet of the Observatory building. (Drawn by Chris Orton, Durham University)*

'somewhat lower than the exposure would at first suggest'. The fact that very little building had taken place within the vicinity of the observatory added to the value of the record and, as it seemed likely that this state of affairs would continue, it was 'very desirable that the meteorological observations on this site should be carefully maintained'. This point holds good today.

A map of the past and present instrument locations is shown in Figure A2.1. Past and present photographs of the site are shown in Figures A2.2 to A2.4; see also Figure A3.7.

Data sources

The commencement of the Durham Observatory record has been described in Chapter 2. Most of the original observation registers or logbooks have been retained within the Durham University Library ([24], items M1–M16). The two original manuscript registers, covering the period July

Figure A2.2 *Durham Observatory from the south, c. 1955—from Baxter 1956 [35], showing the two Stevenson screens (one housing a thermograph), the Observatory building, and the vane of the Dines pressure tube anemograph rising above the roof*

1843 to April 1850, were scanned by the library and are available online (items M1, M2; [29]). The observation registers from January 1850 to August 1997 were digitized by Andrew Joyce as part of a Leverhulme Trust grant to Joan Kenworthy and Nicholas Cox [44]. These include twice-daily observations up to December 1960 of most elements (including barometric pressure, dry- and wet-bulb temperatures, precipitation, sunshine, wind direction and speed, etc.) and the once-daily (0900 GMT) observations thereafter. Some further 'manual' observation data, such as visibility, are available in annual reports published by the University's Department of Geography up to the end of manual observations on 30 September 1999, when Audrey Warner retired. The final volume includes most observations to the end of the calendar year 1999, providing overlap with the automatic weather station (AWS) that commenced records from 10 October 1999, technicians in the Geography Department providing these observations. The early temperature records for 1843–1850 were digitized by Joan Kenworthy in 1985 [28], and the twice-daily pressure records for these years by the authors during the preparation of this book in 2021 using scanned copies of the logbook pages kindly made available by Durham University Library. The first page of the logbook, starting on 23 July 1843, is illustrated as Figure 2.3.

From January 1880 to date, the Met Office MIDAS dataset [142] includes daily values of maximum and minimum temperatures (grass minimum temperatures from December 1930), precipitation and sunshine (from January 1882) for Durham Observatory. For most of the record up to 1999 we have more than one source for each record, and generally we have adopted the readings given in the original register unless there are good reasons to doubt those. Most of the

Figure A2.3 *Meteorological instruments at Durham Observatory from the east, c. 1955—from Baxter 1956 [35], showing the two Stevenson screens, the grass and soil thermometers, the standard (daily-read) five-inch rain gauge, and the recording rain gauge*

temperature observations up to 1918, in degrees Fahrenheit at that time, were read to a precision of 0.1 °F and in most cases, we have used these more precise values after conversion to Celsius in place of the MIDAS values which are conversions from truncations to 1 °F precision. In places the original Durham records show clear evidence of 5 °F or 10 °F reading errors that have been corrected on MIDAS, and in these circumstances the MIDAS record was preferred; see also Appendix 3 for more details on the temperature records and the quality control checks applied.

Observation times

For most of the period from the first surviving manuscript observations in July 1843 to December 1960, observations were made twice daily, most often at 9 a.m. and 9 p.m. local time (about six minutes behind Greenwich time), eventually 0900 and 2100 GMT following the introduction of standardized Greenwich Mean Time in the 1880s. The main exceptions are the period February 1855 to June 1858 (observations at 1000 and 1400 local time) and July 1858 to September 1885 (1000 and 2200 local time), and some minor variations resulting from both staff shortages and the introduction of Daylight Savings Time during the First World War. (Of course, as an astronomical

Figure A2.4 *Durham Observatory looking south-south-east, 1991. Visible are the large Stevenson screen (the smaller screen was not in use), the grass and soil thermometers and several rain gauges. Photographed on 6 September 1991, by Stephen Burt*

observatory, 9 p.m./2100 was a more convenient 'main hour' for meteorological observations than the usual 9 a.m./0900 convention.)

The 2100 GMT observations ceased after 31 December 1960, and from then until 31 December 1999 only a single daily observation was made at 0900 GMT (the observational content was reduced after September 1999).

Table A2.1 provides an overview of the digital data available for Durham Observatory; these brief details are expanded in the metadata sections by element which follow. From 10 October 1999 to date, the AWS on the lawn on the south side of the Observatory provides hourly and once-daily observations of variables shown in Table A2.2. Temperature and humidity sensors are sampled every 15 seconds, and the pyranometer every 30 seconds, with one-minute means being logged and summarized every hour; hourly values are transmitted to the Met Office via automated telephone data links, and the additional daily 'climate' values, such as maximum and minimum temperatures, are transmitted following the 0900 GMT observation data.

Terminal hours

In this context, 'terminal hours' refer to the period over which a daily (24-hour) variable is measured. For daily precipitation measurements for example, the convention within the United Kingdom is for the 24 hours to refer to the period 0900–0900 GMT, in which case 0900 GMT is the terminal hour (and the measurement is 'thrown back' to the previous day, because 15 of the 24 hours in that measurement occurred the previous day, even if all of the rainfall fell five minutes before the 0900 GMT measurement). As with maximum temperature observations, rainfall records are not 'thrown back' on MIDAS.

Knowledge of the terminal hours to which measurements refer is important for consistent and homogeneous datasets, but they are not always obvious with the Durham records. Accordingly, we have had to make reasonable assumptions based upon the time of the observation when the entry was made in the observation registers, as in Table A2.3. The assumptions made define the date of the occurrence—for example, a maximum temperature observed at 2100 could normally be expected to have occurred that day, whereas one observed at 0900 would be 'thrown back' to the previous day. During the winter months, it is not uncommon for the highest temperature to be reached overnight: for 0900 observations and 0900–0900 terminal hours, such a record value would be credited to the day *prior* to the observation, whereas a 2100 observation/2100–2100 terminal hour record would be credited to the *day of* the 2100 observation. Note, however, that both maximum and minimum temperatures refer throughout to 24-hour periods as far as we can tell, not 12-hour periods (typically 0900–2100 for maximum temperature and 2100–0900 for minimum), which result in the omission of some daily extremes, particularly during the winter months.

For maximum and minimum temperatures in particular, there are minor differences between using records based upon 0900–0900 or 2100–2100 terminal hours, particularly during the winter months, and in this sense the records presented here are not completely homogeneous. However, the differences are expected to be slight, and no corrections or adjustments have been made for differing terminal hours.

Dataset location

The entire Durham series is available online at **durhamweather.webspace.durham.ac.uk**

Table A2.1 *Inventory of Durham Observatory digitized meteorological data, to and including 2021: see also Table A2.2 for details of AWS data from October 1999*

Element and parameter	Frequency of data	Record commences	Record terminates	Notes
TEMPERATURE				
Maximum and minimum air temperatures	Daily	23 July 1843	Current	See Appendix 3 metadata for detail
Dry- and wet-bulb temperatures	Twice daily, mostly 0900 and 2100	Dry-bulb 23 July 1843 Wet-bulb 1 Oct 1846	31 Dec 1960	See Appendix 3 metadata for detail
	0900 GMT daily	1 Jan 1961	31 Aug 1997	Hourly dry-bulb from AWS late 1999 onwards
Grass minimum temperature	Daily	1 March 1874 (occasional earlier records)	Current	See Appendix 3 metadata for detail

Continued

Table A2.1 *Continued*

Element and parameter	Frequency of data	Record commences	Record terminates	Notes
EARTH TEMPERATURES				
At 30 cm depth	0900 GMT daily	12 Aug 1948	Current	See Appendix 3 metadata for detail
At 120 cm depth (4 ft)	0900 GMT daily	12 Aug 1948	31 Dec 1969	
At 100 cm depth	0900 GMT daily	1 Jan 1970	Current	
BAROMETRIC PRESSURE				
Station level pressure and barometer temperature	Twice daily, mostly 0900 and 2100	23 July 1843	31 Dec 1960	See Appendix 5 metadata for detail. MSL pressure retrospectively calculated, see metadata for method
PRECIPITATION				
Daily rainfall totals	Daily, mostly read at 0900	23 July 1843	Current	Regarded as unreliable prior to 1868 and not used. See Appendix 4 metadata for detail
SUNSHINE DURATION				
	Daily	June 1880	31 December 1999	See Appendix 6 metadata for detail
CLOUD COVER				
	Twice daily, mostly 0900 and 2100	1 April 1847	31 Dec 1960	Largely complete; see Appendix 10 metadata for detail
	Daily 0900 GMT	1 Jan 1961	31 Aug 1997	

Continued

Table **A2.1** *Continued*

Element and parameter	Frequency of data	Record commences	Record terminates	Notes
WIND DIRECTION AND SPEED				
	Twice daily, mostly 0900 and 2100	23 July 1843	31 Dec 1960	See Appendix 7 metadata for detail
	Daily 0900 GMT	1 Jan 1961	31 Aug 1997	Some gaps
	Daily run of wind (mean speed)	1 Aug 1885	1 Nov 1916	2100–2100 only
STATE OF GROUND				
	Daily 0900 GMT	1 Jan 1961	30 Sept 1999	See Appendix 10 metadata for detail and coding
SNOW DEPTH				
	Daily 0900 GMT	1 Jan 1961	30 Sept 1999	See Appendix 10 metadata for detail
PRESENT WEATHER				
	Daily 0900 GMT	1 Jan 1961	30 Sept 1999	WMO ww code; see Appendix 10 metadata for detail
VISIBILITY				
	Daily 0900 GMT	1 Jan 1961	30 Sept 1999	See Appendix 10 metadata for detail and coding

Table A2.2 *Meteorological variables observed by the Met Office automatic weather station (AWS) at Durham Observatory, since 1 October 1999*

Hourly observations	Dry-bulb temperature	Platinum resistance temperature sensor in Stevenson screen; maximum and minimum temperatures are highest and lowest one-minute means, respectively
	Dew point temperature	Calculated from air temperature and humidity sensor values
	Grass temperature (for daily grass minimum)	Thermistor temperature sensor, originally above grass but since July 2015 35 mm above artificial turf
	Concrete slab temperature (for daily concrete minimum)	Thermistor temperature sensor just touching the surface of a 90 × 90-cm concrete slab
	Soil temperatures at 10 cm, 30 cm, and 100 cm depths	Buried thermistor temperature sensors
	Cumulative rainfall total since 0900 GMT	0.2 mm resolution tipping bucket rain gauge
	Global solar radiation cumulative total since 0900 GMT	Pyranometer
	Daily sunshine total since 0900 GMT	Estimated from pyranometer values: not used in this analysis owing to inhomogeneity with records from Campbell–Stokes recorder, and over-shelter by trees on the northern side of the Observatory given the AWS location at ground level—see Appendix 6
	Relative humidity	Humidity sensor in Stevenson screen
Once-daily 'climate' observations	Maximum and minimum temperatures	0900–2100 GMT, 2100–0900 GMT, and 0900–0900 GMT
	Precipitation	Totals over the periods 0900–2100 GMT, 2100–0900 GMT, and 0900–0900 GMT
	Grass minimum and concrete minimum temperatures	Period 1800–0900 GMT
	Soil temperatures at 0900 GMT at 10 cm, 30 cm and 100 cm depth	
	Global solar radiation total 00–00 GMT previous day	
	Sunshine duration 00–00 GMT previous day	Estimated from pyranometer values: not used in this analysis

Table A2.3 *Terminal hours of the main daily variables in the Durham climatological record. The morning (nominally 0900) and evening (nominally 2100) observations were made at 1000 and 2200 between July 1858 and September 1885, and there were minor variations in morning and evening observation times during 1916–19.*

Element	Terminal hours
Daily maximum temperature	0900–2100 July 1843 to September 1850 2100–2100 October 1850 to December 1960, except February 1855 to June 1858 when they are probably 1000–1000
	0900–0900 January 1961 to date, including AWS record
Daily minimum temperature	0900–0900 July 1843 to February 1900
	2100–2100 March 1900 to December 1960
	0900–0900 January 1961 to date, including AWS record
Daily grass minimum temperature	2100–0900 1 March 1874 to 28 December 1884 2100–2100 29 December 1884 to 31 December 1960
	0900–0900 January 1961 to October 1999
	1800–0900 since AWS record commenced in October 1999
Daily precipitation	Period accumulations to January 1850, sometimes subsequently; morning to morning (mostly 0900–0900) January 1850 to date (but not used prior to January 1868 as unreliable; see Appendix 4 for details)
Daily sunshine	*Probably* 0900–0900 January 1882 to January 1884 (no data February 1884 to May 1885)
	2100–2100 or civil day June 1885 to December 1960
	00–00h civil day January 1961 to September 1999 *(terminated)*

Appendix 3
Temperature records made at Durham Observatory, 1843 to date

Gordon Manley's greatest contribution to the Durham record was to attempt the first homogenized record of mean air temperature at Durham, covering the period 1850–1940 [19]. His aim was to place the Durham records on a similar basis to those at Oxford, but he noted immediately the stark contrast between the Oxford records, where abundant metadata exist as context to the observations, and the 'scanty records beyond the figures themselves' at Durham.

Since Manley's death in 1980, two other significant sources have become available that have shed useful additional light on the 'scanty' Durham records, particularly the early series. Several hundred items relating to the Observatory have recently been catalogued by Durham University Library [24], although only a minority refer specifically to the meteorological record or the observers. The relevant documents consist of letters or memoranda in addition to the original meteorological registers and observations summaries. Some have been digitized and made available online, including the first two volumes of the Observatory's meteorological logbook covering 1843–47 and 1848–50 [29], which were thought lost in Manley's lifetime but which were re-discovered in the University Library in 1982 [28]. An extremely useful complement to Durham's archive has been the contents of the Met Office Archive 'station file' for Durham Observatory covering the period from 1902 to 1960, kindly scanned and made available by the Met Office Library and Archive to assist in the preparation of this book. The file includes numerous inspection reports and some site photographs, together with copies of correspondence with the Observatory. Both sources have added considerably to the 'scanty records' identified by Manley. From a careful examination and compilation of the available metadata we summarize here what we know about Durham's air temperature records and set out the adjustments we have made in attempting to provide as consistent a basis as possible for examination, comparison, and analysis of the long series.

Air temperature—instruments and exposure

Table A3.1 briefly summarizes the exposure of the thermometers at Durham, updated to 2021. Where known, more detail is provided in the following starting with records of maximum and minimum air temperatures; specific additional details regarding the records of dry- and wet-bulb temperatures follow.

Units

All air and grass temperatures were recorded in degrees Fahrenheit from the commencement of the record until 1 November 1971, when Celsius thermometers were introduced. Maximum and minimum air temperatures and grass minimum temperatures were usually noted to 0.1 degrees Fahrenheit until June 1918, after which they were recorded only to the nearest

Table A3.1 *Summary of the Durham Observatory air temperature records and exposure. Based originally upon Manley [19] with other details added from additional sources since, and updated to 2021*

Period of record	Air temperature instruments and exposure
?1841–30 Sep 1851	NORTH WALL. Unscreened 'standard' thermometer and maximum-minimum thermometer, on the north wall of the observatory.
1 Oct 1851–31 Dec 1859	'NORTH SHED'. A louvred screen structure, later called the 'penthouse' or 'north shed' was built to screen the above thermometers, sheltered by a buttress from any westward sunshine.
1 Jan 1860–28 Feb 1900	GLAISHER STAND. A Glaisher stand on the lawn in front of the building.
1 Apr 1868–4 Aug 1880	OVERLAP—'NORTH SHED' The 'North Shed' readings were maintained for maximum and minimum (and for additional 2 p.m. and 6 p.m. observations of the dry-bulb thermometer Apr–Jul 1868), to compare with the readings from the Glaisher stand.
1 Nov 1899–31 Dec 1999	STEVENSON SCREEN. A Stevenson screen, 20 m from the front of the building on the lawn, here sloping very slightly to the south. Thermograph installed Oct 1934.
1 Oct 1999 to date	MET OFFICE AWS—STEVENSON SCREEN. An automatic weather station continues the record in the same location as previously.

degree Fahrenheit until October 1971.* Other temperature measurements, such as dry- and wet-bulb, soil thermometers, etc., were usually noted to a precision of 0.1 degrees Fahrenheit until October 1971, and 0.1 degrees Celsius thereafter. Since November 1971, all temperatures have been recorded to 0.1 degrees Celsius. All temperatures in this book and in the accompanying database are in degrees Celsius, converted from degrees Fahrenheit as required, except where the Fahrenheit value is quoted for a particular reason.

North Wall exposure, June 1841 to September 1851

The earliest records of air temperature at Durham Observatory were from June 1841; we know of an incomplete series of daily observations, in weekly summary form, published in the *Durham Advertiser*, from about this date, but as previously described the earliest surviving manuscript observation register dates from 23 July 1843. The first page of the first register is reproduced as Figure 2.3; included are twice-daily measurements of the barometer temperature ('Att. Thm'— see Appendix 5) and external air temperature ('Ext. Thm', or dry-bulb), with daily minimum and

---

* For the period where air and grass minima are known only to the nearest 1°F, air and ground frosts are counted when the air or grass minimum temperature respectively was recorded as 32 °F (which could thus represent temperatures between 31.6 °F / −0.2 °C and 32.4 °F / +0.2 °C). Doing so will result in a slight over-estimation of true air and ground frost frequencies between July 1918 and October 1971.

maximum and temperatures noted at 9 a.m. and 9 p.m. local time, respectively. Wet-bulb records commenced on 1 October 1846.

Very few details of the instruments themselves or their exposure have survived. However, based upon brief manuscript entries in the register and knowledge of the practice of other leading observatories of the time (such as those at the Radcliffe Observatory at Oxford ([49], Appendix 1), and other leading observers—for example, Symons's account of meteorological instruments in 1837 [143]) we can surmise that the external 'standard thermometer' was probably a high-quality unit from a reputable manufacturer with a calibration certificate, as this was the practice at other similar observatories at the time. Alongside it would have been (probably) a Six's maximum-minimum thermometer very similar to that shown in Figure A3.1. James Six (1731–93) invented the self-registering maximum-minimum thermometer in 1780 [144], and one was brought into use at the Radcliffe Observatory in Oxford from 9 April 1815 ([49], Appendix 1). Many variants of this type of thermometer can still be found in greenhouses and other domestic settings today, although they have not been used for accurate meteorological purposes for over 150 years. Confirmation that the maximum-minimum thermometer at Durham was almost certainly of Six's pattern is afforded by numerous references in the register to 'float in contact' or 'float stuck'. A note in the register for 28 March 1847 reads as follows:

> For some cause unknown the float of the maximum Thermr stuck & would not move. In trying to put it right the tube was broken. Months elapse before a new one is supplied.

Maximum temperature readings did not recommence until 15 July of that year, but minimum readings continued throughout. Although it was very likely that the instruments were

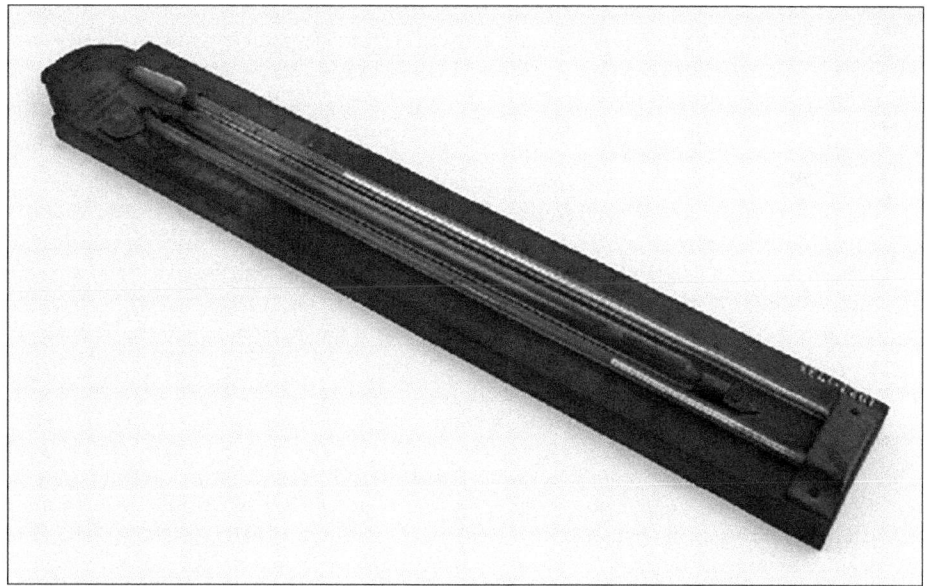

Figure A3.1 *A Six's maximum-minimum self-registering thermometer by Newman, manufactured 1834–1857, and probably similar to that in use at Durham Observatory until the 1860s. London Science Museum collection, item number 1927-1733, licensed under Public Domain Dedication (Creative Commons CC0 1.0 Universal)*

obtained from a leading maker, these details have not survived, with one exception, when on 3 May 1849 there is a note stating 'Adie's maxm thermr replaced'.

The thermometers were exposed on the north wall of the Observatory, 5.2 m above the ground and 1.2 m above the flat roof, which forms a wide balcony outside the first-floor window on the north wall, and were accessible from the room inside [9]. From the limited descriptions we have, it seems probable that the thermometers were not screened in any way. Their position on the north wall of a substantial stone building, some distance above ground level but out of direct sunshine at any time of day or year, shows in a smaller daily and annual range in temperature than would be expected from a well-exposed modern Stevenson screen at 1.25 m above ground level over short grass. As no direct comparisons between the unscreened north wall thermometers and either the Glaisher stand or Stevenson screens were made at Durham, we have not attempted to derive 'Stevenson screen equivalents' (SSEs) for these early records, although for the sake of completeness, all air temperature records (dry- and wet-bulb at 9 a.m. and 9 p.m. and daily maximum and minimum temperatures) at Durham from July 1843 are included in the online source datafiles listed in Appendix 2.

As stated previously, the maximum thermometer was read and noted at the 9 p.m. observation, and the minimum thermometer at 9 a.m. From this it is assumed that minimum temperatures refer to the 24-hour period 0900–0900, while the maximum temperature appears to refer to 0900–2100 (see also Table A2.3), at least for the first few years of the record. There are occasional instances during the winter months at this time of the dry-bulb or standard thermometer at 0900 being higher than the maximum as read at 2100, which would therefore appear to rule out a 2100–2100 period for the maximum temperature in this period. Within a few years, however, the record appears to have become 24-hour (i.e. 2100–2100), although this appears to have been a gradual evolution rather than a distinct dated change.

North Wall 'Shed', 1 October 1851–31 December 1859

In October 1851, the north wall thermometers were screened, when a ventilated penthouse or 'North Shed' was constructed on the north wall of the Observatory (located on Figure A2.1). In 1873, John Plummer (Observer November 1867 to February 1874), described the structure as follows [27]:

> The penthouse, or, as it is usually designated, the 'North Shed,' consists of a large louvre-boarded enclosure, solidly protected from above, and open upon the south side, where the thermometers are sheltered by the north wall of the building. Upon the west side there is also a partial protection from a projecting angle of the wall, but there is everywhere ample space for a free current of air. The thermometers are 3 feet 6 inches [115 cm] from the walls of the building, 17 feet [5.2 m] above the ground, and 4 feet 6 inches [1.4 m] above a balcony floored with lead, upon which, of course, the shed is erected. The position is thus fair, if not unexceptionable, and may perhaps be considered as representing tolerably well this system of exposure.

We know from published records and notes that the professional astronomical community of the time maintained close links, exchanging notes and comments regarding instruments, observations, and observing methods. It is a reasonable assumption that such 'networking' probably included sharing of best meteorological practices, and it is entirely possible that Durham's 'penthouse' was along similar lines to the 'penthouse' brought into use to screen the thermometry at Oxford's Radcliffe Observatory in August 1849. Unfortunately, no sketches or photographs of either structure have survived, but from the description we may envisage something similar to the

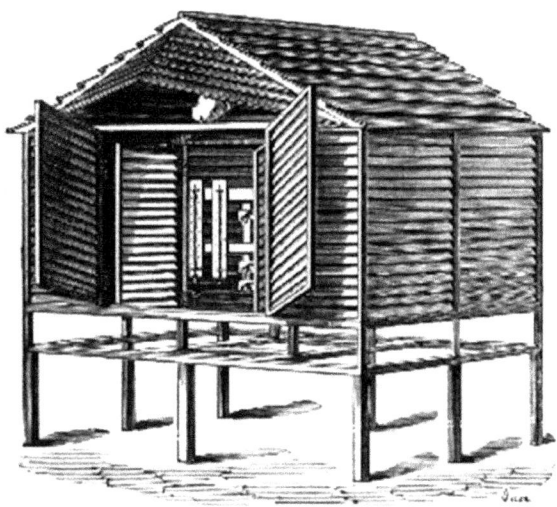

Figure A3.2 *The 'Kew screen', an early thermometer shelter from 1853, probably similar in appearance to the 'North Shed' structure erected at Durham Observatory in October 1851, and in use until at least August 1880. From the Frontispiece to* Symons's Meteorological Magazine *for 1869*

louvred penthouse erected at Kew Observatory in 1853 (Figure A3.2). It is probably significant that all three sites used the same term.

Plummer described the 'North Shed' exposure as producing 'decided low maxima' (at least compared to the Glaisher stand). The average difference varied seasonally, from about 0.5 degrees Celsius cooler than the Glaisher stand in midwinter to 2.5 degrees cooler in midsummer, occasionally more than 4 degrees on hot days. Minimum temperatures were lower all year round by between 0.5 and 1.0 degrees, the greatest differences again occurring in summer. As we have no direct comparisons between the 'North Shed' and a Stevenson screen, we have no reliable basis on which to attempt reduction of these early records to an SSE as we have attempted with the subsequent Glaisher stand records from January 1860. Accordingly, where used within the text these records are quoted 'as is', with a cautionary note. As with the 'north wall' exposure, it must of course be stressed that the exposure of Durham Observatory's thermometers prior to the introduction of a Stevenson screen in November 1899 differed considerably from modern standards. Accordingly, these early records must be treated with caution as they are *not* directly comparable with modern methods of measuring air temperatures.

Glaisher stand on front lawn: 1 January 1860–31 October 1899

On or about 1 January 1860, a Glaisher stand was installed on the lawn in front of the Observatory building, as shown in Figure A2.1 in Appendix 2. (It is not clear from extant records whether the thermometer or thermometers in use were relocated from the 'North Shed', or whether they were duplicates, for the 'North Shed' was brought back into use in April 1868 when an overlap period comparing maximum and minimum temperatures in both exposures commenced.) The Glaisher stand remained in use at Durham until at least the end of November 1900, although as referred to subsequently we have chosen in this analysis to use the Stevenson screen records from when they began on 1 November 1899.

The Glaisher stand was simply a vertical board to which the frame carrying the thermometers was fixed [145–147], and is illustrated in Figure A3.3[*]. The stand was double at the back, had a small canopy to reduce the chance of the thermometers getting wet during rainfall, and was free to pivot on a vertical post such that the thermometers were fixed about 1.2 m above the ground. The frame was raised from the back-board by blocks allowing air to circulate freely round the thermometers. The original form of the stand required it to be manually turned twice per day so that the thermometers were protected from the direct rays of the Sun, although some later versions had side screens which reduced or eliminated the necessity of turning. This requirement was presumably easy enough to organize at observatories staffed throughout the 24 hours, such as Greenwich, but the smaller astronomical observatories such as Durham would be less likely to have staff available to turn the stand during daylight hours.

The Glaisher stand suffers from two major disadvantages [146]: firstly, that an observer must be on the alert to turn it regularly during the day to avoid direct (short-wave) solar radiation affecting the thermometers, and secondly that the open nature of the stand also permits considerable indirect or reflected (terrestrial or long-wave) radiation to affect the thermometers. Glaisher stands tended to record higher maximum temperatures, especially in summer (longer days), and lower

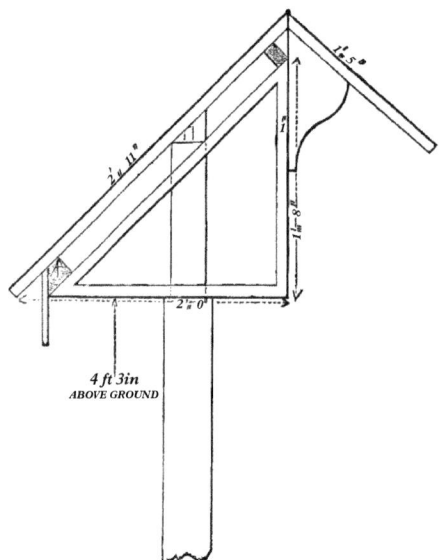

Figure A3.3 *The Glaisher stand—from Glaisher's 1868 description [145]*

[*] The stand was originally designed by George Airy, then Astronomer-Royal, although commonly known as the Glaisher stand after James Glaisher of the Royal Observatory Greenwich, who did much to popularize its adoption in the 1850s (*Symons's Meteorological Magazine*, 35, pp. 90–92). The design was first brought into use at Greenwich in March 1841. See also Margary, I. D., 1924: Glaisher stand versus Stevenson screen. A comparison of forty years' observations of maximum and minimum temperature as recorded in both screens at Camden Square, London. *Quarterly Journal of the Royal Meteorological Society*, 50 (211): pp. 209–226.

minima, more so during winter (longer nights). From the available records, the Glaisher stand in use at Durham appears to have had the usual characteristics of such a stand in the cooler months, but in summer the minima may have been slightly higher following warm summer days owing to re-radiation from the building and perhaps from the slightly sloping ground.

It was probably at an early stage during this period that the Six's maximum-minimum thermometer used in the 'North Shed' was replaced with separate maximum and minimum self-registering instruments in the Glaisher stand; this may have happened at about the time the latter was introduced, but unfortunately the observations register makes no reference to this. The comparison between maximum and minimum temperatures measured in the Glaisher stand and the 'North Shed' referred to here began in April 1868 (when the first specific mention was made of separate maximum and minimum thermometers) and the results of five years' comparison was published by Plummer in 1873 [27] and commented upon by Manley in 1941 [19]. A duplicate daily series survives for the two screens from April 1868 to August 1880 ([24], item OBS_230). Although this comparison confirmed the significant reduction in daily temperature range in the 'North Shed' exposure, it is of course of only limited benefit for understanding how the Glaisher stand records at Durham compare to those from a Stevenson screen.

Manley's monthly and annual mean temperatures 1850–1899

For the period 1850–1899, Manley produced an 'adopted' mean (A) for each month as follows:

$$A = 0.5 \star M + 0.5 \star (F + K)$$

where M is the mean of mean maximum and mean minimum temperature, F is the mean of all the fixed hour (0900 and 2100) observations, and K is a constant, the average amount by which the mean of the fixed time observations was thought to require correction in order to bring it to the mean of the maxima and minima (both measurements made at that time mostly at 2100). For the period 1850–1899, the average difference between A and M are given in Table A3.2, M always being the higher value through the period March to October.

Manley concluded that his means of temperature given for the years 1850–99 must be regarded as 'representative of Durham' rather than those which were actually observed, and we concur. With a desire to move to be coherent with modern climatological practice throughout this volume and in our online database, including Appendix 11, we sought to update Manley's monthly means 1850–1940 with a new series which is consistent throughout in using the mean of the available daily maximum and minimum temperatures for that month, season or year. To do so, we needed an improved statistical representation of the Glaisher stand/Stevenson screen differences.

Table A3.2 *The average difference (°C) between the adopted mean (A) and the mean of mean maximum and mean minimum temperatures (M), Durham 1850–1899—after Manley [19].*

Jan	Feb	Mar	April	May	June	July	Aug	Sep	Oct	Nov	Dec
0.1	0.0	−0.1	−0.3	−0.4	−0.5	−0.5	−0.4	−0.3	−0.1	0.0	0.0

Deriving Stevenson screen equivalent (SSE) temperatures from the Glaisher series

Rather than simply write off the 40 years of daily Glaisher stand records prior to the introduction of the Stevenson screen in late 1899, we sought to derive approximate daily maximum and minimum SSE temperatures from the Glaisher series in an attempt to bring the temperature readings to a common standard, and thereby make better use of them. Manley also attempted to do so in his 1941 paper [19]. A 13-month overlap (November 1899 to November 1900) exists of manuscript daily maximum and minimum temperatures at Durham as observed in the Glaisher stand and the new Stevenson screen ([24], item OBS_230, *Comparison of thermometers on Glaisher's screen with similar ones in Stevenson's*), but Manley was critical of this:

> A brief comparison of the Glaisher–Stevenson readings was made over twelve months; but this is very difficult to interpret on account of the uncertain thermometer corrections, and reluctantly has had to be abandoned.

Looking at the document 80 years on from Manley, even with the benefit of personal computers and spreadsheets to simplify the data processing involved, we must reluctantly agree with him; the document is far from clear even which record is which. The period with apparently the most reliable data is February to July inclusive, where there are just a few gaps and corrections ($n = 178$ days). The use of Student's t test shows the expected differences in maximum and minimum temperatures and daily range (all in degrees Fahrenheit):

	Glaisher	*Stevenson*	*t*	*P*
Mean maximum	54.91	54.36	5.22	4.52E-08
Mean minimum	39.94	40.12	−4.97	1.54E-06
Mean daily range	14.98	14.24	7.26	1.2E-11

The differences are all highly significant in statistical terms. However, as Manley commented, there are too many corrections and uncertainties to allow comparison over a complete annual cycle, and of course, ideally such comparisons need to be made over many years not just one.

Instead, Manley made use of monthly Glaisher stand/Stevenson screen corrections derived from previous multi-year comparisons, such as those undertaken at Greenwich (Preface to the *Monthly Weather Report*, 1938) and over 40 years at Camden Square in north London [147], together with more local observations from Ushaw and Rounton. He also examined differences in mean diurnal range in both screens over 40 years of data from each (i.e. Glaisher 1860–99 and Stevenson 1900–40), and we return to this approach shortly. The output was *monthly mean* corrections to Glaisher maximum and minimum temperatures 1860–99 to approximate to those from a Stevenson screen ([19], p. 369). Note that maximum and minimum temperatures require different corrections because of their very different radiation balance characteristics, by day and by night respectively.

Rather than simply applying Manley's monthly corrections as they stood to the daily Glaisher stand maximum and minimum temperatures over the January 1860 to October 1899 period, instead we used Manley's data to construct empirical regression plots separately for both maximum and minimum temperatures, as doing so enabled us to construct daily corrections without discontinuities at month boundaries, and also to extend the adjustments out to climatological extremes. To do this we related Manley's monthly mean correction to the mean maximum or minimum for that month over the period 1911–40 (as this would have been the most recent

Table A3.3 *Regression equations used to derive 'Stevenson screen equivalent' maximum and minimum temperatures from the observed Glaisher stand equivalents.*

Maximum temperature	
February to July	$T_x SS°F = T_x GS°F + (T_x GS°F \star -0.0794 + 3.2594)$
August to January	$T_x SS°F = T_x GS°F + (T_x GS°F \star -0.0772 + 3.6343)$
Minimum temperature	
All year	$T_n SS°F = T_n GS°F + (T_n GS°F \star -0.0332 + 1.3935)$

T_x and T_n represent maximum and minimum temperatures respectively, and SS and GS the Stevenson screen and Glaisher stand values. Units are °F, the conversion to °C being undertaken only as the final step to avoid rounding errors.

30 years average available to Manley at that time), as shown in Figure A3.4. For maximum temperatures, Figure A3.4(top) shows that there is a clear difference between the 'warming' first half of the year (February to July) and the 'cooling' second half (August to January) owing to thermal lag in ground temperatures, and accordingly slightly different regression equations were used for each half-year as shown in Table A3.3. For minimum temperatures, Figure A3.4b shows how the seasonal variations were less marked and a single regression equation could be applied to all months (Table A3.4). For clarity, although Figure A3.4 presents the correction to the observed Glaisher stand maximum or minimum temperature in temperature terms, in fact the temperature axis here represents only a proxy for the varying seasonal influences on the performance of the Glaisher stand, from shorter days and weaker solar radiation by day and longer nights with greater outgoing radiation at midwinter, to the opposite at midsummer. It should not be interpreted as a simple calibration adjustment for the thermometers involved.

We then applied the derived empirical regression equations to each observed Glaisher stand daily maximum and minimum temperature (in degrees Fahrenheit) to calculate the approximate SSE, also in degrees Fahrenheit—a process which is, of course, much very quicker and easier to do with 40 years of twice-daily data using modern computers and spreadsheets than it would have been to do by hand in Manley's day! Note that although original readings were recorded to no better than 0.1 degree Fahrenheit, and the inputs to the regression plots to adjust Glaisher stand values to SSE were also no better than 0.1 degree Fahrenheit, to avoid rounding errors the calculations from the regression equations were worked to several decimal places prior to being rounded to 0.1 degree Celsius to derive the final database entries used in the preparation of this volume.

While it is unfortunate that there is no reliable overlapping series at Durham with which to assess the likely range of error in doing so—which may be substantial on individual days—we believe that this approach provides an acceptable statistical method to make best use of the long Glaisher stand series.

One test whether our daily regression method could provide acceptable values—even at climatological extremes of temperature—was to evaluate and compare numerous individual extreme values on the Durham series with SSEs estimated by Manley ([19], p. 377); these comparisons (covering ten extreme values) are given in Table A3.4. The average root-mean-square difference between the two is acceptably close at 0.59 degrees Fahrenheit (0.33 degrees Celsius). While on its own this is hardly a robust statistical verification of our method, it does at

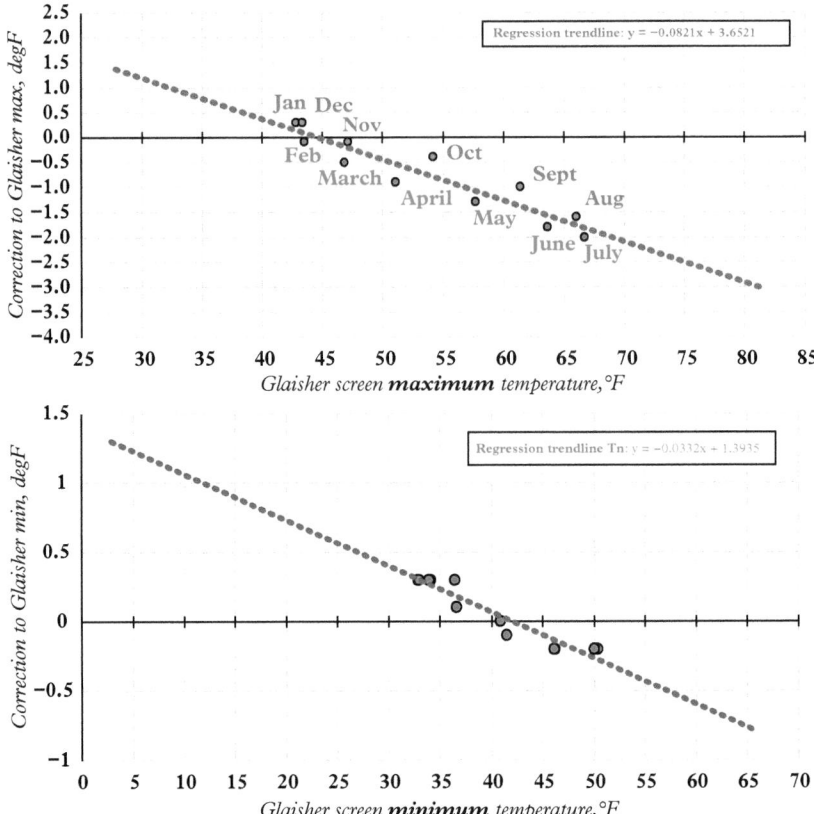

Figure A3.4 *Regression plots of corrections (degrees Fahrenheit) to observed Glaisher stand temperatures to derive an approximate Stevenson screen equivalent temperature SSE, based upon Manley [19]. Upper plot refers to maximum temperatures (the trend line shown is an annual least-squares fit, slightly different fits were used for the first and second half of the year as described in the text): lower plot refers to minimum temperatures. Note scales differ*

least demonstrate that our regression-based approach produced results very similar to Manley's intentions.

Next, we calculated the daily range in temperature (maximum minus minimum) for all dates where both maximum and minimum temperatures were available (for the few dates missing one or both values were excluded) for both the observed Glaisher stand values and the derived SSE maximum and minimum temperatures. Thorne and colleagues [148] found that measures of daily temperature range were often more effective in determining breakpoints in long-term temperature series than using other temperature elements. Accordingly, a time series plot was prepared showing both Glaisher and SSE annual mean daily ranges over the period 1860–1899 together with the annual mean daily range from Stevenson screen observations over the period

Table A3.4 *A comparison of extreme temperatures on the Durham record recorded in the Glaisher screen, the SSE estimated by Manley to 1 °F precision ([19], p. 377), and the SSE derived using the separate regression equations for maximum and minimum temperatures as described in the text*

MAXIMUM TEMPERATURES

Date	Glaisher max °F	Manley SSE °F	Regression SSE °F	Regression SSE °C
28 Aug 1869	88.8	86	85.6	29.8
22 July 1873	89.3	87	85.5	29.7
16 July 1876	92.5	89	88.4	31.3

MINIMUM TEMPERATURES

Date	Glaisher min °F	Manley SSE °F	Regression SSE °F	Regression SSE °C
25 Dec 1860	1.2	2	2.5	−16.4
30 Dec 1874	2.9	3	4.2	−15.4
14 Dec 1878	3.0	4	4.3	−15.4
17 Jan 1881	0.2	1	1.6	−16.9
26 Jan 1881	2.3	3	3.6	−15.8
8 Feb 1895	−1.8	−1	−0.3	−18.0
11 Feb 1895	0.9	2	2.3	−16.5

1900–2020 (Figure A3.5). If the SSE method as outlined here provides a good approximation to the Stevenson screen values, there should be no significant difference between the 'Glaisher SSE' means of annual mean daily range pre-1900 and the 'true Stevenson' means in the decades post 1900, assuming no long-term trends are present (this is an extended version of the second of Manley's comparisons).

This is indeed so: the annual mean daily range using the SSE regression method over the period 1860–1899, excluding 1863/64, was 7.42 degrees Celsius, statistically indistinguishable ($p = 0.13$, $n = 38$ records vs 121) from the annual mean daily range of 7.46 degrees for the Stevenson screen record for the whole period 1901–2020, suggesting that there is little if any long-term trend in the mean daily range over the period of record (Figure A3.5). In contrast, the observed Glaisher stand annual mean daily range over the period 1860–1899 excluding 1863/64 was 7.90 degrees Celsius (grey plot on Figure A3.5) and is highly significantly different from the Stevenson screen mean range 1900–2020 of 7.46 degrees ($p = 1.3 \times 10^{-7}$, $n = 38$ records vs 121).

A similar analysis and conclusion—not shown here—was reached from an analysis of mean diurnal temperature ranges during the summer months (June, July, and August), which differed by only 0.1 degrees Celsius for the 40 years before and after 1900. This is, of course, well within the range of expected thermometer error, particularly more than a hundred years since. Daily temperature ranges are greater in summer, and thus this test is more sensitive to inconsistencies in the statistical adjustment methods.

The Glaisher stand values for 1863 and part of 1864 were excluded, both from this analysis and from the online database, because the maximum temperatures are erratically high during these years in comparison with both the rest of the record and neighbouring records. Although it is difficult to be sure of the cause after more than 150 years, it seems probable that the Glaisher stand was not always turned during the day, perhaps due to an inexperienced or careless new observer (see Appendix 1). As a result, the thermometers were probably affected by direct sunshine for at least a part of the day. Gordon Manley also shared this opinion ([19], p. 368) and made greater use of 0900 and 2100 dry-bulb temperatures in assessing Durham's true monthly and annual mean temperatures for these years.

Also plotted for comparison as the pale blue series on Figure A3.5 is the mean daily range from the north wall exposure (1844–1851) and the North Shed (1852–59 and 1868–1879). As expected, this exposure considerably under-recorded the true diurnal temperature range. In time, it might be possible to use this relationship to derive a similar SSE for the earlier exposures, but with a greater uncertainty than for the Glaisher stand corrections.

Stevenson screen: 1 November 1899–31 December 1999

A Stevenson screen was installed at the Observatory towards the end of October 1899, sited 20 m from the front of the building on the lawn (Figure A2.1). For 13 months temperatures were read in both Stevenson screen and Glaisher stand, but from 1 March 1900 records from the Stevenson screen became the Observatory's standard (we have used them from the commencement of the record, five months beforehand). The location of the Stevenson screen has not changed by more

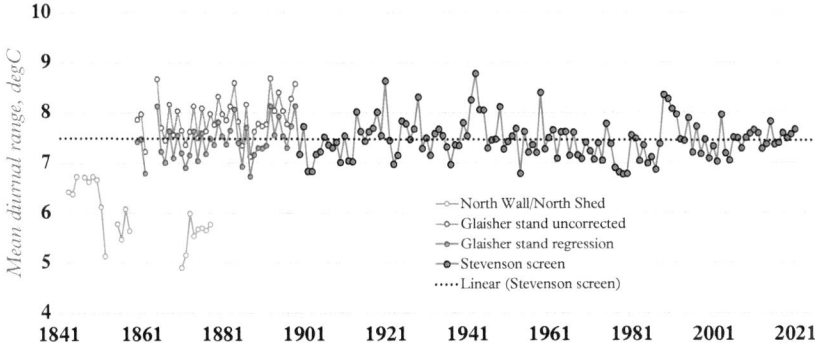

Figure A3.5 *Annual mean daily range in temperature at Durham since 1844, comparing observed North wall/North Shed records 1844–59 and 1868–79 (pale blue plot), the uncorrected Glaisher stand records 1860–1899 (grey plot), together with the derived SSE values for the same period, and observed Stevenson screen values from 1901–2020 (red plot). The black dotted line shows the linear trendline for the combined derived SSE and Stevenson screen instrumental series over the whole period of record 1860–2020*

than a few metres since 1899, although the screen itself has of course been replaced numerous times since.

The original screen was mounted on a single post until 1903, and vibration in strong winds tended to 'shake down' the maximum and minimum thermometers (Met Office inspection report, 16 September 1903); shortly afterwards the screen was remounted on four wooden legs in the more normal arrangement. A new Stevenson screen was erected beside the older screen and equipped with sheathed maximum and minimum thermometers on 1 January 1934. A second screen also housed a thermograph (installed October 1934), and later a thermohygrograph, recording on weekly paper charts (Figure A2.2 in Appendix 2). At some later date both screens were replaced by a single large-pattern Stevenson screen (nominal dimensions W × D × H approximately 120 × 53 × 56 cm) in the same location, although the date of changeover was not noted locally. Photographs of the site (Figures A2.4, A3.6) show this large-pattern screen in use throughout the 1980s and 1990s, and the automatic weather station (AWS) modules were installed in and on this screen in October 1999.

Manley was dissatisfied with the Observatory's 1922–33 temperature data, believing there were long-term systematic errors in both maximum and minimum thermometers (of about 0.5 degrees Fahrenheit [~ 0.3 degrees Celsius] in the minimum thermometer from 1922–33 and 0.8 degrees Fahrenheit [~ 0.4 degrees Celsius] in the maximum thermometer from 1924–33, based upon comparisons with reliable local sites such as Ushaw ([19], p. 372). Both thermometers were quickly replaced in late 1933, while Manley estimated corrections for this period. We have used Manley's corrections to adjust the individual daily readings (again, taking care to minimize rounding errors while doing so by applying the corrections to the original Fahrenheit values prior to conversion to Celsius), and the monthly means for maximum, minimum, and mean air temperature are based on the average of these corrected daily values.

Stevenson-screen based AWS: 10 October 1999 to date

A Met Office AWS was installed on 10 October 1999 and continues the record in the same location as previously. The existing large-pattern wooden screen was replaced by a new plastic and aluminium standard size Metspec screen (nominal dimensions W × D × H 57 × 39 × 58 cm) on 6 June 2007. The large Stevenson screen is shown in Figure A3.6, while the current screen is shown in a July 2021 photograph in Figure A3.7.

Significant gaps in Durham's daily temperature record

There are occasional gaps in the Durham temperature record, but fortunately most are short, rarely more than two or three consecutive days. The only periods in the record where maximum and/or minimum temperatures are missing for five consecutive days or more are shown in Table A3.5, although in addition to this numerous maximum temperatures for several months in 1863/64 are suspect and have been omitted in summary tables, extremes lists, etc.; monthly means for affected months during this period have been estimated using Manley's calculations, which were based more heavily on 0900 and 2100 temperatures [19].

The record is largely unbroken between January 1884 and September 1999, although unfortunately multi-day gaps have become more frequent since the introduction of the AWS in October 1999: in 2018, 40 days record were lost—the largest break in records since 1854. It is to be hoped that consideration can be given by the Met Office observing team to provide duplicate backup AWS logging equipment at the site to minimize the risk of such record loss continuing to damage the integrity of the long Durham series going forward.

Figure A3.6 *Audrey Warner reading the thermometers in the large Stevenson screen at Durham, 1990s (Department of Geography, Durham University)*

In most cases we have not attempted to fill gaps with estimates where daily data are missing, and missing values are left blank in the online database. The exception has been from April 2017 to June 2021, when we were able to use records from the AWS on the roof of the West Building about 800 m to the east to derive a reasonably satisfactory regression estimate for missing maximum and/or minimum temperatures from the observatory AWS. These estimates are included in the dataset and flagged accordingly by source code.

Monthly means are the average of all available dates within that month unless six or more days are missing, in which case the monthly mean is omitted or, where estimates from other sources are available, the monthly mean is shown bracketed thus (10.5) in the tables in Appendix 11.

Figure A3.7 *The Durham Observatory site on 27 July 2021, with the Stevenson screen fitted with the Met Office AWS logging equipment in centre frame (Michele Allan)*

Quality control of maximum and minimum temperatures

For maximum and minimum temperatures, the (digitized) original source between July 1843 and August 1997 was used in preference to MIDAS: this was the only source of data prior to January 1880. For records dated January 1880 onwards (when the MIDAS record commences), all daily values were then checked against the MIDAS value for that date, after allowing for the MIDAS date issue (maximum temperatures are not thrown back on MIDAS). Most values agreed closely, although there were minor discrepancies between values digitized to 0.1 °F or 1 °F precision, as expected.[*] Every entry where the two sources disagreed by 3 degC or more was individually checked against the digitized original registers to make an informed assessment wherever possible of which source was most likely to be correct, taking into account other sources of information in the observation registers, such as 0900 and 2100 temperatures, sunshine duration, wind direction and speed, and cloud cover. Some digitized register entries were found to be incorrect (typically, thermometers not reset, occasional 5 or 10 degree Fahrenheit reading errors, and the like), and some MIDAS entries too (most often mistyped entries, such as 45 °F instead of 65 °F). Some

---

[*] For the calendar year 1912, for some unknown reason, all MIDAS max and min entries are shown only to the nearest 1 degree Celsius; the digitized register values, which are mostly to 0.1 degrees Fahrenheit precision, are used in preference throughout.

Table A3.5 *Periods of five consecutive days or more where maximum and/or minimum temperatures are missing from the Durham Observatory record, July 1843 to December 2021*

Year	Dates (inclusive)	No of days missing	Maximum temperature missing	Minimum temperature missing
1845	4–8 Aug	5	x	x
1845	30 Dec–3 Jan 1846	5		x
1847	27 Mar–14 Jul	110	x	
1849	24–31 Dec	7	x	x
1854	1 Jul–14 Aug	45	x	x
1854	27 Sep–7 Nov	42	x	x
1854	7–12 Dec	6	x	x
1854	27 Dec–14 Feb 1855	50	x	x
1863	Many dates suspect Jan to Dec	> 20	x	
1883	24–31 Dec	8	x	x
2005	15 Aug–6 Sep	23	x	x
2017	31 Jan–6 Feb	7	x	x
2018	22 Apr–10 May	19	x	x
2018	18–30 May	13	x	x
2018	12–19 Jul	8	x	x

MIDAS values were available for dates missing in the digitized registers, and although these are themselves quite possibly 'best guess' estimates by Met Office quality control, they were used in place of missing values wherever possible. Each day's entries were then subjected to range checks (for example, maximum temperature above minimum temperature, daily range within monthly norms, changes from day-to-day remaining within climatological limits, and so on); these items were then flagged for individual checking once more. Finally, the most likely daily value was written back to the dataset for use in preparing the book's contents, the final 'source file' thus (we hope) combining the best of both digital sources. Additionally, each daily entry in the master dataset is linked back to a source code whereby its exact provenance can be quickly determined, and if necessary, the original source re-examined in case of doubt.

Dry- and wet-bulb temperatures

Instruments and exposure

Twice-daily (9 a.m. and 9 p.m.) dry-bulb readings commenced on 23 July 1843. Initially these were almost certainly from the Observatory's certified 'standard thermometer' exposed on the north wall, similar to the documented practice at Oxford's Radcliffe Observatory ([49], Appendix 1). In the absence of any information to the contrary, it has simply been assumed that the instrument exposure at the time and subsequently was the same as the maximum and minimum thermometers as described above. Wet-bulb readings commenced on 1 October 1846, and about this time a separate 'dry-bulb' reading also appears, in addition to the 'external (or standard) thermometer', suggesting the introduction of a second thermometer. Readings of both 'standard thermometer' and 'dry-bulb' are given in the meteorological registers until October 1849, after which the heading 'Air temperature' replaces 'standard thermometer', although these probably refer to the same physical thermometer. The 'air temperature' column is dropped after March 1861, leaving only the dry-bulb record. The limited space on a Glaisher stand (see Figure A3.3) was normally sufficient to permit only two thermometers to be mounted, and if separate maximum and minimum thermometers were fitted to the stand then the dry- and wet-bulb may well have remained in the 'North Shed' for an indeterminate time. The 'North Shed' remained in use until at least August 1880, but as the Stevenson screen, with sufficient room for dry- and wet-bulb as well as maximum and minimum thermometers, was not installed until November 1899 then it is possible the dry- and wet-bulb remained exposed in the 'North Shed' for many years, or it may be that a second Glaisher stand was used for the dry- and wet-bulb thermometers. Unfortunately, other than knowing from Plummer's records [27] that dry- and wet-bulb records were kept in the North Shed for four months in 1868, the surviving records do not provide clarity on this point.

The twice-daily thermometer readings are useful to provide a check upon the observed daily maximum and minimum temperatures and are also required for the calculation of barometer corrections to mean sea level (1843–1960 only; see Appendix 5). However, from at least September 1902 (when a calibration check was undertaken during the Met Office inspection, that for September 1902 being the first surviving inspection record in the Durham Observatory site file in the Met Office Archives) until August 1911 (when both thermometers were finally replaced with new units), both dry- and wet-bulb thermometers read between 0.6–1.0 degrees Fahrenheit (0.4–0.5 degrees Celsius) too high. Checks were repeated at subsequent inspections in 1903, 1906 and 1909 with similar results. Because we are unsure of the period prior to September 1902 to which these corrections could be applied, we have not made any corrections to the digitized dry- and wet-bulb records which are included in the online database which accompanies this volume, but this error should be noted by future users. When compiling his original series of monthly mean temperatures for Durham from 1847 to 1940, Manley [19] placed as much if not more credence in the twice-daily dry-bulb readings than in the records from the north wall and Glaisher exposures. Manley was almost certainly unaware of these thermometer errors; as stated previously, he used the morning and evening dry-bulb temperature along with the mean of the daily maximum and minimum temperatures as part of his method of determining Durham's mean monthly temperatures up to 1940. As a result, his published monthly means may therefore be 0.2 or 0.3 degrees Celsius too high for this period (and possibly for an unknown period prior to 1902). Throughout this volume and in the accompanying online database, we have derived daily, monthly, and annual mean temperatures from the mean of the daily maximum and minimum temperatures, following modern standard conventions, applying corrections for SSE as defined in the earlier sections within this Appendix for the period of records made in the Glaisher stand.

Dry- and wet-bulb readings were read and noted to a precision of 0.1 degrees Fahrenheit until October 1971 (except for the period 29 Jun 1918–31 Aug 1921, when they were noted only to the nearest degree): all have been subsequently converted to Celsius to a precision of 0.1 degrees. From November 1971, all readings are to 0.1 degrees Celsius.

Since 1999, hourly dry-bulb readings have been logged by the automatic weather station; a wet-bulb is no longer in use, although if necessary wet-bulb temperatures can still be derived from the AWS dry-bulb and humidity records.

Significant gaps in the record

Most of the gaps in the record are prior to 1884, although there is an eight-day gap in 0900 GMT readings in May 1961. Between 23 July 1843 and 31 August 1997 (56,288 days), morning observation dry-bulb records are available for 56,039 days (99.5 per cent); morning wet-bulb records are available on 54,296 out of 54,709 days (99.2 per cent) from their commencement on 1 October 1846 until 31 August 1997. Capillary action cannot maintain a flow of water to the wet-bulb wick in sub-freezing conditions, and as a result wet-bulb readings are more likely to be missing or incorrect under such conditions.

Grass minimum temperatures

Instruments and exposure

The grass minimum temperature is measured by a thermometer (or more recently an electronic sensor, such as a thermistor or platinum resistance thermometer, or PRT) placed horizontally with its bulb or sensing chip just above the tips of short grass; this has presumably always been sited on the lawn in front of the Observatory (the thermometer can just be seen in Figure A2.3). Regular grass minimum records (then referred to as 'terrestrial radiation minimum') commenced on 1 March 1874, but there are occasional references to earlier records, the earliest being on 30 December 1845 when a manuscript note in the observations register records 'Minimum thermometer placed upon the grass fell below 28 degrees [°F, or −2 °C]'. This is the earliest known reference to the reading of a grass minimum thermometer for any site in the United Kingdom.

The thermometer used until late 1999 was a self-registering minimum thermometer, usually with alcohol as its working fluid; but following the installation of the AWS in October 1999, a thermistor replaced the traditional alcohol thermometer (Table A2.2). Since July 2015, the 'grass minimum' sensor has been mounted 35 mm above an artificial turf surface rather than over natural grass. As the radiative properties of the two surfaces differ considerably, this change has introduced a significant inhomogeneity in the 'grass minimum' series, the artificial surface usually being somewhat warmer.

The grass minimum thermometer was read and reset at the morning observation from 1 March 1874 to 28 December 1884, after which it was read and reset at the evening observation until 31 December 1960. Thereafter, it was read and reset at the 0900 GMT observation until December 1999. From October 1999 the AWS records the grass minimum as the lowest temperature on the grass surface in the period 1800 to 0900 GMT (this can, and does, lead to missing some 'early evening' ground frosts during the winter months). The period of measurement of the grass minimum record is therefore not strictly homogeneous throughout the record period, for during the period when the grass minimum thermometer was read and reset at 0900 GMT, a mild night following a cold night would 'inherit' the grass temperature as it was at 0900 GMT the previous morning. During the March 1874 to December 1884 period, the different times of reading of the grass and screen minimum thermometers sometimes lead to confusion regarding exactly when the grass minimum temperature actually occurred, although for consistency the date credited is that of the date of reading in all cases. This does leads to some inconsistency in the treatment of ground frost frequencies (see Ground frosts). The re-siting of the AWS sensor above artificial turf from July 2015, as stated earlier, has also led to a significant inhomogeneity in 'grass' minimum temperatures and thus the ground frost frequencies derived from them.

Data sources and quality control

A continuous record of daily grass minimum temperatures is available from 1 March 1874 to date. From March 1874 to August 1999 the digital records are mostly from the original sources [44]. A daily grass minimum record is available on MIDAS from 1 December 1930 to date. For the over-lapping period (December 1930 to August 1997) both records were compared, and discrepancies identified. Where the two values differed by more than 2.5 degrees Celsius, the most probable value was entered into the database, taking into account the screen minimum temperature. A source code was also written to identify from which dataset the value or the correction originated. Although many of the errors found were again 5, 10, or even 20 degrees Fahrenheit mis-readings of the thermometer, or quite possibly mis-reading the 'wrong end' of the thermometer index, the meteorological context was also checked to ensure that the grass minimum reading was the value at fault and not the screen minimum. In a few such cases, the check identified that the grass min-imum was more likely to be the correct value and it was probably the screen minimum that was erroneous, which then led to a correction to the latter. However, the quality control objective for grass minima was simply to identify and remove only the most obvious errors in order to improve the accuracy of the ground frost totals, rather than attempting to verify every datapoint, and this should be borne in mind when using the records.

Ground frosts

The definition of a 'ground frost' is a day when the grass minimum temperature falls below 0 °C, and ground frost frequencies in this book have been derived from the daily grass minimum temper-ature records using that definition, starting in March 1874. Prior to 1961, the threshold definition of a ground frost was 30.4 °F (−0.9 °C) or below, or 30 °F when the thermometer was read only to the nearest degree Fahrenheit. The change in definition means that published values of ground frost frequency at Durham in, for example, the *Monthly Weather Report* prior to 1961 will differ from those in this volume.

Where grass minimum readings are missing but the screen minimum is available, a ground frost is counted if the screen minimum fell below 3 °C, as this is the average difference at Durham between the two minima. The count of ground frosts is likely to be slightly higher for 0900–0900 GMT measurement periods than for 1800–0900 GMT. During the period June 1918 to October 1971, when grass minima were ready only to 1 °F, however, '32 °F' has necessarily been counted as a ground frost, some of these readings would have been slightly above 32.0 °F/0.0 °C had the thermometer been read and noted to better than one degree.

Significant gaps in the record

The grass minimum records are almost complete, with very few observations missing—only 124 dates out of 45,109 days (99.7 per cent availability) between March 1874 and August 1997. The longest period of broken records prior to December 1999 are in 1917 (11 Jun–29 Aug inclusive) and in 1927 (4–24 Jan and 27 Oct–18 Nov)—both presumably owing to a broken thermometer. Occasionally a 'grass minimum' record will be lost when snowfall covers the thermometer: by convention the reading is ignored unless it is known that the lowest temperature was reached prior to snowfall. Where the observations were made manually, the observer would sweep the thermometer free of snow and replace it resting on the snow surface, but at an AWS site the ground sensors may remain buried in the snow for as long as a snow cover persists. Unfortunately, loss of record, and discrepancies in daily grass minimum temperatures when compared to screen minima, has been much greater during the period of AWS record.

Soil or earth temperatures

Instruments and exposure

Earth temperatures have been measured at Durham since 12 August 1948, when records of the temperature at 0900 GMT at 30 cm depth (1 ft) and 120 cm (4 ft) commenced using lagged thermometers inserted in steel tubes sited on the lawn in front of the Observatory (the tips of the tubes can be seen in Figure A2.2 and A2.3). The 120 cm earth thermometer record was replaced by observations of the metric 100 cm depth thermometer on 1 January 1970. When the AWS was installed in October 1999, thermistor-based sensors located within the existing steel tubes replaced the traditional mercury thermometers, and a buried 10 cm soil sensor (under grass) was also added; since October 1999, soil and earth temperatures at 10 cm, 30 cm, and 100 cm have been logged hourly, rather than a single daily 0900 GMT observation as heretofore.

Significant gaps in the record

The earth temperature records are almost complete, with very few observations missing—the longest gap being for the 120 cm thermometer between 1 May and 12 June 1969—presumably again owing to a broken thermometer.

Quality control

Other than the deletion of a few gross errors, no quality control has been applied to earth temperature records in the online datasets.

Barometer temperatures

Barometer temperatures are required to correct observed barometric pressure readings to a specified temperature, 0 °C by international standard since 1955, to correct for expansion of the mercury column. More details on the barometer temperature record (the 'Attached Thermometer' in the observations register) are presented in Appendix 5.

Appendix 4
Precipitation records made at Durham observatory, 1843 to date

We know that precipitation records were being kept at the Durham Observatory as early as June 1841, because weekly weather summaries published in the *Durham Advertiser* at the time included rainfall data.[*] The first surviving manuscript meteorological register in Durham University Library [29], commencing on 23 July 1843 (Figure 2.3), contains what appears to be daily rainfall totals, measured at the 9 a.m. observation, although on closer inspection up to September 1851 the daily values sometimes turn out to be accumulations over several days, or cumulative totals during the month or part of the month,. A page in the first ledger lists some monthly totals from as early as May 1842, raising the tantalizing possibility that an earlier ledger might yet be discovered. Unfortunately, we have very little information on the type or types of rain gauge in use in the early years of the record, and what few comparisons we are able to make with other local records before the late 1860s leave us in considerable doubt about their reliability and accuracy. For this reason we decided not to publish the rainfall record prior to 1868 in this book, other than in this Appendix.

From 1868 onwards we possess an increasing amount of information about the Observatory's rain gauges and their exposures, and we can confidently compile an almost unbroken daily precipitation record for the site up to the present day. This Appendix describes the sources of the precipitation records, the instruments and exposures, identifying and filling gaps in the record, and the various quality control measures which have been applied to generate, as far as possible, a complete and reasonably homogeneous daily rainfall record now extending over 150 years.

Data sources

Daily totals were extracted by the authors from the first two volumes of the Durham Observatory registers, covering the period July 1843 to December 1849 [29]. From January 1850 to December 1997, daily precipitation totals from the manuscript observation registers were digitized alongside the other entries as part of the Leverhulme Trust project [44], and these records were subsequently updated to the cessation of manual observations in December 1999 as described in Appendix 2. Since the installation of the automatic weather station (AWS) in October 1999, a sub-daily precipitation record has been maintained by the tipping bucket rain gauge that forms part of the AWS. Daily rainfall totals from the site are also available on the Met Office MIDAS dataset [142] from January 1880, continuing to date.

---

[*] 'Precipitation' includes rainfall, snowfall, drizzle, and hail, as well as small amounts from dew and fog. Generally we use the terms 'precipitation' and 'rainfall' synonymously unless the context requires the more specific term (such as when referring to winter snowfall).

Monthly totals—sometimes from more than one gauge at the Observatory site, as described subsequently—are also given on the Met Office ten-year sheets [149], where the Durham record is first included in 1850: a note on this sheet in what appears to be a contemporary hand states 'Observations before 1850 not reliable'. The publication and digitization of the ten-year sheets [150] has been incredibly useful in helping to track down both additional record metadata for the Durham site as well as identifying numerous historical rainfall records from elsewhere in north-east England, which can be used to provide some quantitative measure of comparison against Durham's records as discussed subsequently. In this context, we were also able to make objective comparisons of the Durham Observatory record against those of the recently-constructed County Durham monthly and annual rainfall series constructed by the Met Office Hadley Centre (part of the HadUK-Grid dataset [121]). This series commences in 1862 and is referred to in more detail in a later section.

A more detailed chronological description of the various periods of record follows below, broadly arranged in chronological order.

Units

Daily rainfall totals were measured in inches, usually to three decimal places, up to June 1915. From July 1915 to December 1959 measurements were in millimetres, to one decimal place, then from January 1960 to December 1970 the measurements reverted to inches, to two decimal places. Since January 1971 rainfall has been measured in millimetres, the manual measurements to 0.1 mm precision while the automatic weather station, used from October 1999, has a resolution of 0.2 mm.

All inch measurements have been converted to millimetres, rounding to one decimal place; amounts below 0.05 mm have been rounded down to zero.

The Durham daily precipitation series

There are very few substantive references in the Durham records to the types and exposure of rain gauges in use at the Observatory until the late 1860s. It should of course be borne in mind that standards for the optimum pattern and exposure of British rain gauges were still evolving at this time, and were only defined in the early 1870s following numerous field trials [151]. Manley was dismissive of the worth of the Durham rainfall record prior to 1886, and there is a note from him dated December 1938 to this effect on the Durham ten-year rainfall sheets held in the Met Office Archives, although in fairness his main research area concerned the temperature series (Appendix 3). A published account in 1985 [152] used only the record from 1886, presumably for this reason, but since then more details of the various gauges used at the Observatory have been unearthed in various archives. This new information enabled comparisons to be made between overlapping periods of record of the gauges in use, and permitted extension of a robust series back to the early 1870s [94].

More recently, we have been able to make objective comparisons of the Durham Observatory record back to 1862 against those of the recently-constructed County Durham monthly and annual rainfall series [121].* Figure A4.1 shows Durham Observatory annual precipitation totals expressed as a percentage of the County Durham series 1862–2020, with a ten-year unweighted

---

* Strictly speaking, the two series are not independent, since the long Durham Observatory series is a constituent record in the gridded County series in some years; however, it is but one record among many over a large area and its influence is not dominant in the compilation.

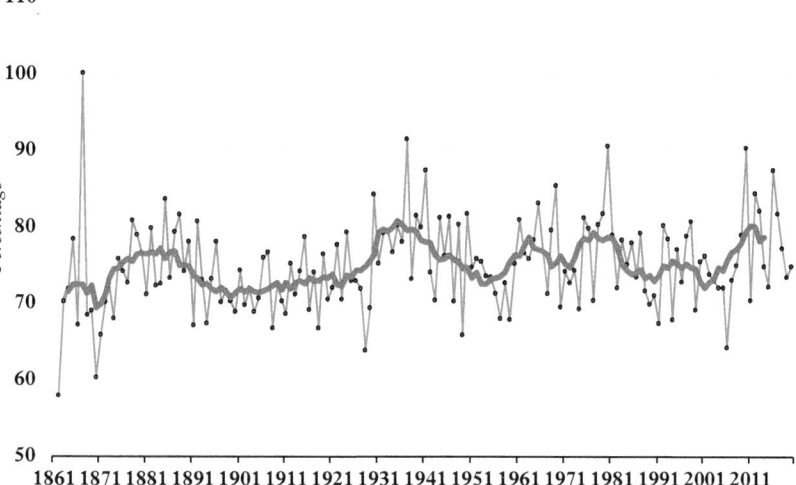

Figure A4.1 *Annual precipitation totals at Durham Observatory 1862–2020 expressed as a percentage of the County Durham series (green line and marker points from 1868, less reliable records prior to 1868 shown in orange). The thicker red line shows a nine-year centred running mean*

running mean plotted at the year ending, while Figure A4.2 shows the same but for monthly data, together with a 12-month running mean, for the period 1862–1880. From these comparisons it is clear that significant inhomogeneities remain in the early Durham series, likely due to a combination of differing instruments and exposure and variable observational diligence, and accordingly we reluctantly decided to exclude the Durham Observatory daily precipitation series prior to and including 1867 from this book. A similar exercise shown in Figure A4.3 compares ten-year running means of annual precipitation totals since 1868 at Durham Observatory alongside individual long-period sites at Ushaw College (record available 1861 to 1972), Tunstall Reservoir and Waskerley Reservoir (both 1895 to date), Cockle Park (1898 to 2013), and the climatological station at Houghall (1925 to 1977), all expressed as a percentage of the County Durham value. The location of these sites is mapped in Figure 1.1.

Now that the entire collection of ten-year rainfall sheets in the Met Office Archives have been digitized and made available [150], including some records in County Durham extending back to 1835, it may yet be possible over time to prepare a representative 'Durham monthly precipitation series' back to the 1840s, using the Durham records as the core sequence wherever possible. In the meantime, evidence from Figure A4.1 and other comparisons suggest that the Durham Observatory rainfall record has remained reasonably homogeneous from at least 1868 to the present day, at 75 per cent ± 5 per cent of the County Durham gridded total. Fluctuations above and below this level probably reflect variations in atmospheric circulation over time.

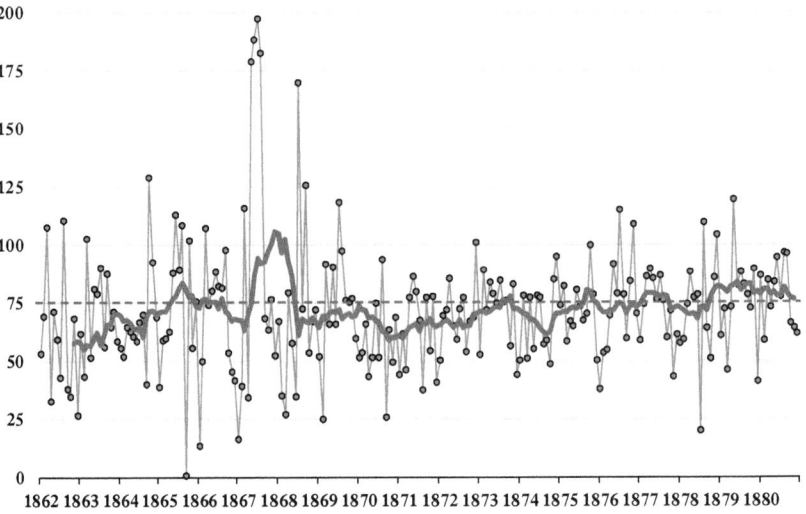

Figure A4.2 *As Figure A4.1, but for monthly totals and including a 12-month running mean, over the period 1862–1880 only; the dashed line represents the average relationship since 1868. The series can only be regarded as reasonably stable from January 1868, and earlier records have not been utilized further in this volume*

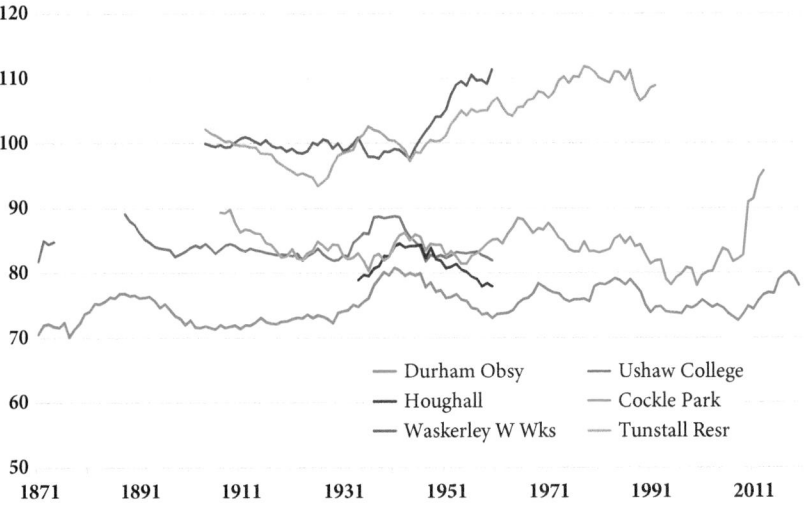

Figure A4.3 *Ten-year running means of annual precipitation totals since 1868, expressed as a percentage of the Hadley Centre County Durham series, for Durham Observatory, Ushaw College (record available 1861 to 1972), Tunstall Reservoir and Waskerley Reservoir (both 1895 to date), Cockle Park (1898 to 2013), and Houghall (1925 to 1977)*

Cataloguing the Durham Observatory rain gauges

Documentation regarding instruments in use at various periods has been compiled from three main sources as follows:

- The Observatory's meteorological observation registers, both manuscript (the 1843–47 and 1848–50 records from the Durham University Library [29]) and the digitized records 1850–1997 [44]). There is, unfortunately, very little information in the original ledgers concerning the design or location of the Observatory's rain gauges, or indeed instrumental metadata in general, and what little we have is often contradictory and confusing.

- The British Rainfall Organization ten-year sheets held in the Met Office archive [149], covering each decade from the 1850s to the 1950s. In addition to monthly and annual rainfall totals, the ten-year sheets include details of the site and authority as well as rain gauge information (type, diameter, and rim height—summarized in the annual *British Rainfall* volumes), and occasional notes on site changes, inspection notes, and the like. The quality and quantity of the information increases with time. A very useful source, but until 1902, dependent upon information supplied by the Observatory in its annual rainfall returns, which may have been simply copied over from previous years.

- Met Office site inspection reports from 1902 to 1959, contained within the Durham Observatory site file in the Met Office Archives. These reports provide information on the site and exposure as well as details of the instrument(s) in use, and are the most comprehensive and reliable source.

Previous work on the Durham rainfall series [76, 94] was also consulted.

All available sources at the time of writing have been examined, but it is possible that additional information may yet come to light to supplement the meagre metadata for the early years of the Durham record.

The 1850s ten-year sheet for the Observatory brusquely states, in what appears to be a contemporary hand, 'Observations before 1850 not reliable'. We have digitized the surviving records from 1843 to 1849 from ledgers M1 and M2, but they appear much too low in comparison with the few other rain gauges known to be operating in north-east England at the time and are not felt to be worth publishing. Until September 1851, rainfall totals were accumulated, and often the gauge had to be emptied part way through the month. In some cases, marginal notes in the ledger give a calculation of the monthly total, but this is not always the case, and it is often not clear what sums need to be made in order to produce the final total.

Durham rain gauge metadata

For ease of reference, gauges are lettered in broadly chronological sequence, and summarized in Table A4.1.

The first information on rain gauges then in use comes in a short note in Richard Carrington's neat hand referring to the observations taken over the Christmas/New Year period 1849/50, when he was away for a few days. He was clearly a careful and conscientious observer, only recently appointed to the position, and his frustration and dismay when procedures were not carried out correctly are clear from his comments in the second (M2) observations ledger:

> The observations taken by Mr Cruddas watchmaker between the 24th and the 2nd of January were found full of mistakes and unfit to be retained. The rain this month has been taken by the inner gauge, and there is reason to believe that the indication is far too small. Between the 23 and 31 [December 1849] the outer gauge caught 0.28 in[ch] whilst the inner caught only 0.08 in[ch].

Table A4.1 *Summary details of rain gauges in use at the Durham Observatory from 1841, together with period(s) used in the compilation of the daily rainfall series used in this volume. See text for additional details.*

Gauge	Identifier	Period in use	Funnel diameter in/cm	Rim height ft/m	Period used in daily rainfall series
A	'Inner' ?	? 1841—?	Unknown	4 ft 6 in/1.37 m	*Not used*
B	'Outer' ?	? 1841—?	Unknown	4 ft 6 in/1.37 m	*Not used*
C		? 1850 to 9/1867	8 in/20 cm	4 ft 6 in/1.37 m	*Not used*
D	'Large gauge'	9/1867 to 12/1880	12 in/30 cm	4 ft 6 in/1.37 m, 4 ft 8 in/1.42 m from 1876	11/1872 to 2/1876, adjusted see text
E	'Small gauge'	9/1867 to 7/1929	8 in/20 cm	4 ft/1.22 m, 3 ft 10 in/ 1.17 m 1902 to 11 in/28 cm 1906	1/1868 to 11/1872, 2/1876 to 7/1929
F	'Roof gauge'	9/1867 to 12/1870	Unknown	Unknown	*Not used*
G	'8 inch'	7/1929 to 6/1946	8 in/20 cm	12 in/30 cm	7/1929 to 6/1946
H	'5 inch'	6/1946 to ? 1999	5 in/12.7 cm	12 in/30 cm	6/1946 to 9/1999
I	Auto	12/1935 to at least 12/1971	8 in/20 cm	12 in/30 cm ?	*Not used*
J	Munro AWS TBR	10/1999 to 7/2015	20 cm	30 or 32 cm	10/1999 to 7/2015
K	EML AWS TBR	7/2015 to date	25.2 cm	47 cm	7/2015 to date

It is not certain what 'inner' and 'outer' gauges refer to, although perhaps we might surmise that the 'inner' gauge could refer to an internal measure fed by a pipe from a funnel on the Observatory roof (as was the case at the Radcliffe Observatory in Oxford at the time), and perhaps the 'outer' gauge was exposed on the lawn in front of the Observatory. Let us tentatively identify these as *Gauge A* and *Gauge B*.

Carrington started a new ledger in 1850, copying across data from January to the end of April 1850 from Ledger M2. The commencement of the new ledger from January 1850, together with the loss for many years of the two earlier observation ledgers for 1843–47 and 1848–50, obviously explains why the Durham record (including rainfall) was believed to have started in 1850.

Burt [94] quotes a note in the observation ledgers regarding the 1852 measurements as follows: 'The rain gauge of which the record is here given is that one of two near the south fence, enclosed in a wooden casing. The receiving surface is 4 feet 6 inches [1.37 m] above the ground'.

Could one of these be Gauge A or B? Until September 1867, there is only one daily rainfall total entered in the observations register, and that is definitely *Gauge C*, an 8 in/20 cm diameter gauge. There can be no doubt that this was the gauge whose daily readings appear in the observations register, for almost all the monthly and annual totals on the Observatory's ten-year sheet are identical with monthly sums of the daily values in the register from January 1852 to April 1867 (prior to October 1851 the accumulated values in the registers render such comparisons more challenging). However, the gauge details on the Observatory's 1850s and 1860s ten-year sheets refer to a rim at 1 foot (30 cm) above ground. It is possible that the rim height of the gauge, originally at 4 ft 6 in/1.37 m, had been reduced to 1 ft/30 cm by the time the gauge details were provided to the British Rainfall Organization in the 1860s. However, it seems far more likely that the ten-year sheet rim height is in error, for we know that at least one gauge with its rim height at about 4 ft/1.2 m was documented in Observatory site inspection reports by the Met Office until 1906.

The ten-year sheet entry for this gauge terminates in December 1866, and a contemporary note on the ten-year sheet states:

> After 1866 adjacent shrubs vitiate the readings of this gauge and a new one was adopted as the true record. During 1866 both were used and the difference was inconsiderable.

Despite the second sentence, there are no 1866 totals for the 'new' gauge, which is shown as commencing in January 1867; however, the daily readings from the observations register would appear to suggest that this gauge remained in use until March or April 1867.

A later comment from Gordon Manley dated 7 December 1938 appears on the 1850s ten-year sheet, copied onto the 1860s sheet, as follows:

> I should distrust any record from Durham Observatory from 1865 to 1879 inclusive, although from February 1876 an 8 inch gauge was running, the totals from which are in better agreement with neighbouring stations.

From September 1867, the observation registers include columns for three rainfall measurements, labelled respectively as *Rainfall, large gauge*, *Rainfall, small gauge*, and *Rainfall, roof gauge*. It may not be coincidental that this expansion of the Observatory's meteorological equipment occurred shortly before the appointment of John Plummer as Observer in November 1867, for, like Carrington, he seems to have brought a higher level of care to the role of Observer (Appendix 1), and took greater interest in the Observatory's meteorological records [27].

The ten-year sheet states that a new gauge, *Gauge D*, was brought into use from January 1867. As its diameter is given as 12 in (30 cm) and its rim height as 4 ft 6 in (1.37 m), we can safely assume this was the 'Large gauge'. The rim height was subsequently amended slightly to 4 ft 8 in/1.42 m from 1876, but this may simply be a transcription error. The gauge is also identified as a 'side tube gauge', suggesting that rainfall to be measured was drawn off by a tap arrangement. However, the totals of daily entries in the register continue to agree with the ten-year sheet monthly totals until March or April 1867, suggesting that Gauge C remained in use until that time.

Thereafter, the total of the daily values from the register and the monthly totals given on the ten-year sheet differ for the remainder of 1867, very considerably in some months. For July 1867, for example, the total of the daily values in the observations register amounts to 7.28 in (185 mm), while the ten-year sheet value is 3.14 in (80 mm); the latter looks to be much more realistic when compared with the County Durham rainfall series monthly total (94 mm).[*] From January 1868 to December 1876 the monthly sums of the daily 'large gauge' record agree very closely with the ten-year sheet monthly totals for this gauge, confirming that the 'large gauge' is Gauge D. The ten-year sheet notes a minor site move at the beginning of 1877. The observations register contains daily records identified as being from this gauge until December 1877.

Burt [94] notes that 'There are a few clues in the original ledger that suggest possible problems with rainfall measurements during the 1870s. The normally used "large" (12 inch) gauge was "broken by frost" on 4 December 1871; a new gauge was in use from 6 January 1872'. The ten-year sheet records that 'the gauge being broken, an older gauge in a somewhat sheltered situation was recorded; the total is therefore somewhat too small'. Perhaps the 'older gauge' was Gauge C, still in its sheltered location. Gordon Manley's unpublished notes (held in the Cambridge University Library) show that he doubted whether the new gauge was accurate: 'the 12 inch gauge catches too much', he commented at one point. Manley concluded that the record was acceptable from August 1879, although he gives no indication what changed from that time; nor does the original ledger [94]. This gauge certainly read in excess, and corrections to the observed records have been proposed [76, 94] and have been adopted in assembling the Durham daily rainfall series presented here.

The 'small gauge' whose records appear in the register from September 1867 to November 1872, *Gauge E*, was apparently an 8 in/20 cm diameter unit, although there is no confirmation of an 8 inch gauge in the ten-year sheets until 1879. It is possible this was the original Gauge C, in use since at least 1850 and possibly 1841, although the note on the ten-year sheet concerning the damage to the 'large' gauge in December 1871 referred to earlier implies that Gauge C was still in use as a spare at that time, albeit in a less than ideal exposure. Instead, it seems more likely that this was a new gauge of the same or similar construction to the older unit. Gauge E was almost certainly the gauge that eventually continued in use as the Observatory's standard after January 1881, which we know from that time was an 8 in/20 cm gauge with rim height 4 ft/1.22 m; however, we have been unable to find unambiguous confirmation of the gauge details prior to January 1878.

The third gauge whose record ran from September 1867 to December 1870 only was *Gauge F*, a roof gauge. We know very little about this gauge, other than it was presumably exposed on the roof of the observatory, where wind speeds could be expected to be considerably greater than at ground level and catchment losses through wind eddies correspondingly larger. Not surprisingly, amounts measured were much lower than the other gauges. The annual average fall in the two years 1869–70 amounted to just 357 mm, compared to 505 mm in the 'small' gauge and 678 mm in the 'large'. This was presumably an experiment which had served its purpose after three years and was then terminated.

From January 1872 to December 1877, only the record from a single gauge is included in the digitized observations register, although there is a note in the original to confirm that two rainfall records continued into 1873. The monthly totals of the daily record agree very closely with the ten-year sheet monthly and annual totals identified as being from Gauge D (12 in/30 cm at about 4 ft 6 in or 4 ft 8 in/1.4 m) until December 1876 but differ somewhat for 1877.

---

[*] A useful and sensitive indicator of daily record quality is the number of rain days. The Durham record shows a rapid fall-off in rain day frequency between winter 1865 and autumn 1867, to well outside of climatological extremes for the remainder of the record (just 86 in calendar year 1866, for example), a clear signal of something seriously amiss with the daily record. The rain day count returns within its expected climatological range during the autumn of 1867 and appears reliable thereafter.

From January 1878 to December 1880, the register includes columns labelled as '12 in gauge' and '8 in gauge', and we can safely assume these refer to gauges D and E, respectively. Finally, the 1880s ten-year sheet notes that the record from the 12 in/30 cm Gauge D was discontinued in January 1881 'as the gauge was so much damaged by frost'. From January 1881 to December 1884 the observations register column is again headed 'Rainfall 8 inch gauge', and it is safe to assume that this is the continuation of the Gauge E record. From January 1885 only a single rainfall entry is included in the observations register, and this is presumably also Gauge E. However, the ten-year sheet contains a record commencing in January 1885 identified as from a 12 in/30 cm diameter gauge, with rim height 4 ft/1.22 m, and this record continues to be shown as such until 1918, when the details changed to 8 in/20 cm diameter gauge with rim height 10 in/25 cm. However, this is almost certainly a rare error on the ten-year sheet—presumably owing to lack of information to the contrary.

Unambiguous confirmation of details concerning the gauge in use at the time can be found in the Met Office site inspection reports for Durham Observatory, kindly made available through the Met Office Archives. The first surviving record, for the inspection conducted on 5 September 1902, lists an 8 in/20 cm diameter gauge with its rim at 3 ft 4 in/1.02 m above ground level, and this is presumably still Gauge E. The 16 September 1903 inspection report notes the rim as 3 ft 10 in/1.17 m, but three years later (inspection 20 September 1906) the rim has been reduced to 11 in/28 cm above ground level following a recommendation at the previous inspection to reduce the rim height to the standard 1 ft/30 cm. The 20 October 1909 inspection confirms the 8 in/20 cm gauge (now identified as a deep-funnel Snowdon pattern, by Hicks) with its rim about 11 in/28 cm above ground level. Today, we would have reservations about accepting the catch of a gauge with its orifice at 1.2 m above ground level, but until at least 1906 this was the case with the Durham gauge. A statistical t-test comparing annual rainfall at Durham Observatory with the County Durham series before and after the change in gauge height in or about 1906 (39-year averages 1868–1906 versus 1907–45; see Figure A4.3) showed a measurable but statistically borderline difference ($p = 0.0506$) between the two periods of record.

Gauge E was finally replaced after 62 years' service by a new 8 in/20 cm gauge in July 1929 (*Gauge G*), rim height 12 in/30 cm, before this was itself replaced by a Met Office pattern 5 in/127 mm copper gauge, rim height 12 in/30 cm above ground level, on 12 June 1946 (*Gauge H*). Gauge H remained in use until at least the 1960s and possibly to the end of the manual record in December 1999; if it was replaced during this period, it was by an identical model.

A Casella natural siphon autographic rain gauge (*Gauge I*) was installed in December 1935, in addition to the standard 'five-inch' checkgauge. The 1940s and 1950s ten-year sheets note that the charts from this rain recorder were microfilmed from July 1936 to December 1971.

A Munro 0.2 mm tipping bucket rain gauge (TBR), *Gauge J*, was installed as part of the AWS in October 1999. The funnel diameter was 20 cm and rim height 30 or 32 cm. Finally, this gauge was replaced by an Environmental Measurements Ltd SBS 500 gauge, also 0.2 mm tip resolution, on 23 July 2015, *Gauge K*. The rim height is 47 cm, and funnel diameter 25.2 cm (500 cm^2). At the time of writing this is the gauge currently in use, and it can be seen in the foreground of Figure A3.7 taken in July 2021.

As a digression of historical interest, the word 'trace'—denoting a rainfall amount under 0.005 inches (a little over 0.1 mm)—first appears in the Durham observation register on 6 October 1899. The instruction to record such small amounts as 'trace' rather than zero, as had previously been the official policy of the British Rainfall Organization, did not come into general use until about 1923, following the absorption of the British Rainfall Organization by the Meteorological Office in 1919.

Based upon the above, Table A4.1 provides a short tabular summary of the rain gauges in use.

Assembling the daily rainfall series, 1868 to 2021

As explained, we have discounted daily and monthly rainfall data prior to January 1868, given uncertainty about gauge catch and observational procedures. From January 1868 the sources of the Durham Observatory daily rainfall series are as set out in Table A4.2. Following standard convention, the measurement made at or close to 0900 GMT each morning is thrown back to the previous day—so that the rainfall read on the morning of, say, 20 June will be entered to the 'rain day' of 19 June, as 15 of the 24 hours of the record will refer to the day before. The majority of the record has been extracted from the Met Office MIDAS dataset [142], cross-checked against the digitized Durham observation registers, with minor adjustments where disagreements were evident; confusingly, the MIDAS record for rainfall amount is held against 'date of reading' and not thrown back. Measurements made in inches have been converted to millimetres and rounded to one decimal place.

The AWS record—from October 1999

With effect from 0900 GMT 9 October 1999, the majority of daily rainfall totals have been obtained by summing the sub-daily record of a 0.2 mm capacity TBR sited within a few metres of the location of the previous 5 inch manual gauge, on the same 0900–0900 GMT rainfall day basis. This gauge is telemetered (logged by telephone line) and sub-daily totals are available on MIDAS for most of the record—mostly hourly, but sometimes only 12-hourly totals. There are periods of missing data due to instrumental defects, gaps in telecommunications, or (during the winter half-year) snowfall accumulations. The longest periods without data to date have been a spell of more than three months from 19 November 1999 to 1 March 2000 when the AWS record starts

Table **A4.2** *Sources of the Durham Observatory daily rainfall series*

Period	Record source and gauge used	Exceptions or adjustments made
1 Jan 1868– 22 Nov 1872	Digitized record from original registers [44] using 'small' 8 inch Gauge E	12 inch Gauge D used 31 Jan–7 Feb 1868 as Gauge E data missing. Minor multi-day accumulations split using obs notes or other gauge data where available.
23 Nov 1872– 21 Feb 1876	Digitized record from original registers [44] using 'large' 12 inch Gauge D	Too high; daily totals reduced using ratios set out in [94]
22 Feb 1876– 31 Dec 1879	Digitized record from original registers [44] using 'small' 8 inch Gauge E	Other than unit conversion, no adjustments made
1 Jan 1880– 31 Dec 2021	Sourced from MIDAS—manual daily record to Dec 1999, then (mostly) hourly record from AWS	See notes below regarding filling gaps in AWS data, and record QC from 1997

again, and the four months July to October 2001, since when missing days are less frequent. Since and including 2008, only one year has had as many as ten days missing days in total—in 2018, a total of 51 days records were lost, including a long spell in April and May of that year.

There remains a concern with regard to the record homogeneity following the introduction of the AWS, particularly in the frequency of rainfall; see Chapter 21 and Figure 21.9 in particular.

Estimation of missing daily totals since October 1999

In order to ensure that, as far as possible, the continuity of the long Durham record was not adversely affected by gaps in the daily record, estimates have been used to infill the missing dates. Where the observed rainfall total has been flagged as incorrect by Met Office quality control (QC) and a QC estimate of the daily total entered on MIDAS, that figure has generally been used instead of the reported value/values. Otherwise, the daily total has been estimated from the records of three nearby gauges (Table A4.3).

Barkers Haugh Sewage Treatment Works lies 2.05 km north-east (035°) of the Observatory, alongside the banks of the River Wear at a lower altitude (30 m versus 102 m for the Observatory). Daily rainfall records began here in February 1975 and, aside from the occasional missing day and a few missing months, are mostly complete until November 2001. From that date onwards there are increasing gaps in the record, and accumulations of several days' rainfall, up to February 2008. Thereafter, the record quality improved, and is almost complete until February 2013. The record for 2014 is patchy and ceases after August of that year. The MIDAS record is extremely poor, with numerous duplicate entries, which made extraction and error checking on the record

Table A4.3 *Details of nearby rainfall sites used to infill missing daily rain gauge data for Durham Observatory; see also Figure 1.1 for locations*

Site	Distance	Met Office rainfall station no.	Latitude °N	Longitude °W	Altitude AMSL	Average Annual Rainfall 1991–2020, mm*
Durham Observatory	—	024726	54.768	1.584	102 m	680
Barkers Haugh Sewage Works	2.05 km	024734	54.783	1.568	30	688
Esh Village	7.25 km	023987	54.791	1.692	210	804
Durham University, West Building Davis Instruments AWS (rooftop site, 20 m AGL)	800 m	N/A	54.768	1.573	60	674

* Averages for 1991–2020 for Durham Observatory and Esh Village are based upon complete monthly totals within this period. For Barkers Haugh, the figure is scaled by reference to all common months overlapping record with Durham Observatory over the period 1975–2014. For West Building AWS the average is approximate, based upon only three years' data overlap.

time consuming. However, the Barkers Haugh rain gauge provided a useful 'backup' daily rainfall record from a close site for most of the period 1975–2013, i.e. both before and after the change to the AWS. Where no other record was available, it was used unadjusted as an estimate for Durham Observatory, but where the Esh Village record was also available for the same date, an estimate based on a combination of both records was used as set out below.

The rain gauge at Esh Village is located 7.25 km west-north-west (290°) of the Observatory and lies in the low hills to the west of the city at a greater altitude (210 m vs 102 m); there is likely a greater orographic component to the rainfall here. Daily rainfall records are available on MIDAS from May 1989 and are largely unbroken at the time of writing, with only five months during this period mostly or completely missing daily values (all monthly totals are complete). This record was also used to draw comparisons with the significant increase in rain days evident on the Durham record since about 1999, for which please refer to Chapter 21.

The third local record, available from 13 April 2017, is from a Davis Instruments AWS mounted on the roof of the West Building (Geography) on the Durham University Lower Mountjoy campus, located just over 800 metres east of the Observatory site. The AWS is mounted on the roof, approximately 20 metres above ground level; the TBR unit is mounted on a paving slab sitting on the roof with its rim height at approximately 35 cm. Daily totals (in mm) were obtained by summing the individual sub-daily rainfall totals (at 0.2 mm/15 minute resolution) over the standard 0900–0900 GMT rainfall day, thrown back in the usual fashion. Although the rooftop siting is not ideal, the rainfall records are very similar to those from the Observatory—for the dates on which both records were available, the average difference over the two years 2018–19 was less than 3 per cent, although individual monthly totals varied by up to 15 per cent (winter months apparently drier, perhaps reflecting greater wind losses on the rooftop site). The record is also valuable for being closer to the Observatory site than any other current rainfall record.

Estimation of missing daily totals since October 1999

Throughout the period of AWS record commencing October 1999, infill estimates of the daily precipitation total at Durham Observatory were prepared using one or (usually) two local rain gauge records when the Observatory total was missing or suspect. Of course, spatial variations in rainfall distribution mean that no method of estimation can ever be perfect, but the simple method employed (detailed later) provided satisfactory results when verified on days when the actual total at the Observatory was known, monthly and annual totals agreeing within ± 10 per cent in over 80 per cent of months. It should be remembered that in most years, estimates are required for less than 5 per cent of missing or suspect daily data, so the overall error contribution from estimates owing to missing data is likely to be very small.

The gauges used for infill necessarily changed over the period as record availability dictated. Between 1 October 1999 and 31 August 2014, a period of 5449 days, daily records for both Barkers Haugh and Esh Village are available on 4780 days (87.7 per cent, of which about half would be dry), thus enabling a reasonable estimate of the missing Durham Observatory fall from the combination of two local sites. The Barkers Haugh record was weighted more heavily (as it is nearer), while the Esh Village daily total was reduced by 20 per cent (reflecting the difference in average annual rainfall). The final algorithm used for infilling daily totals when both sites were available was therefore

$$Estimate \ = \ (((Esh \ Village \ \star \ 0.8) \ + \ (2 \ \star \ Barkers \ Haugh)) \, / 3) \ \star \ 1.03$$

The factor 1.03 is an empirical value, based on method comparison on dates when the Durham Observatory daily rainfall total was known. The resulting estimate was then rounded to one decimal place, and values less than 0.1 mm set to zero.

The Barkers Haugh record ended after August 2014, and from September 2014 to April 2017 inclusive the Esh Village record was used as a single-site estimate for Durham's daily fall (after a reduction of 20 per cent, again reflecting the difference in average annual rainfall). Only ten days during this 973-day period were missing, seven of them in February 2017 alone, a rate of just over 1 per cent.

From 14 April 2017 to the current date at the time of writing, the two-site method is used again, combining Esh Village and West Building AWS instead of Barkers Haugh, using the same formula as above, excluding only the factor 1.03.

The Durham daily precipitation dataset, 1868–2021

The Durham daily precipitation series is available at **durhamweather.webspace.durham.ac.uk**. If you refer to the series, please include a citation to this volume.

Appendix 5
Durham's barometric pressure records, 1843 to 1960

The records for Durham Observatory include a long twice-daily record of barometric pressure, commencing on 23 July 1843 and terminating on 31 December 1960. The record was first published in *Geoscience Data Journal* in October 2021 [153], and this Appendix is largely based upon that paper. The Durham record, which is 98.7 per cent complete, is by far the longest digital barometric pressure series in northern England and fills a very large temporal and spatial gap in the International Surface Pressure Database (ISPD: [154]). In what is believed to be the first study of its kind, the record was independently quality-controlled against the NOAA–CIRES–DOE Twentieth Century Reanalysis version 3 (20CRv3) [58, 120], which did not include the Durham records in its assimilation set.

This Appendix describes the instruments used and their exposure, the sources of the record, digitization work undertaken to generate the digital time series (including quality control assessments using 20CRv3), reduction to MSL pressure from station-level observations, and examines consistency over the period of record against 20CRv3. A summary of monthly and annual means and extremes over the 117-year series has been included as Chapter 27. Throughout this appendix, the terms millibar (mbar) and hectopascal (hPa) are used interchangeably. For consistency with original sources, the millibar unit has been retained where appropriate.

Instruments and exposure

The astronomical observatory at Durham University (Figure A5.1) opened in 1840 and commenced meteorological observations shortly afterwards. Astronomical observatories such as Durham required observations of barometric pressure and external air temperature to correct star positions for atmospheric refraction ([49], Chapter 5), and for this purpose high-quality instruments were usually procured from reputable manufacturers. Information on the barometers used, their calibrations, and their exposure within Durham Observatory was assembled from original archival records held in Durham University Library together with previously restricted Met Office site inspection reports and correspondence held in the Met Office Library and Archives in Exeter.

The earliest details we have of Durham's barometer are from the monthly climatological return to the Met Office for January 1877, when the barometer's serial number (209) was first stated. The earliest surviving Met Office site inspection record for Durham Observatory, for September 1902, adds the detail that barometer 209 was a Fortin-pattern instrument made by Browning. Brownings were a long-established optician and scientific instrument-maker, trading in London from the early nineteenth century; Brownings' shop has been suggested as Charles Dickens' model for that of Solomon Gills in *Dombey and Son* [155]. John Browning took over from his father

Figure A5.1 *Durham Observatory, photographed about 1955. The observatory's barometer was located behind the eastern window (arrowed). Photograph from Baxter [35]*

William in 1865 and the business advertised itself as 'Optical and physical instrument-maker to Her Majesty's Government, the Royal Observatory [Greenwich] and Kew Observatory' [156].

At this distance in time it is impossible to state with certainty whether the Browning barometer noted as being in use in 1877 and 1902 was the original station barometer in use when the Observatory opened in 1840/41, but it is certainly possible; it is equally possible that the Browning barometer replaced an earlier instrument in June 1867, as a very brief comment ('New barometer installed') appears in the observations register at that time, although no equipment receipts or other metadata confirming a possible change of instrument have yet been found [24].

Although there were occasional temporary substitutions by other instruments as detailed subsequently, this barometer remained in use between at least January 1877 and termination of the record in December 1960.

In 1910, and probably earlier, the barometer was located at the side of the eastern window on the first floor of the Observatory (Met Office inspection report, 1910) (Figure A5.1). The height of the barometer cistern above MSL was stated as 352 ft or 107.3 m throughout the record. Thereafter, only minor relocations of the instrument took place.

Until 1949, the barometer was checked against a standard mercury barometer carried by the inspector at every Met Office inspection (conducted at typically two- to three-year intervals); errors were found to be small, within ± 0.1 or 0.2 hPa.

From the beginning of the record the observed ('As Read') barometer readings, in inches of mercury, at the morning and evening observation hours are given in the original registers, together with the barometer temperature (the 'Attached Thermometer'), the latter in degrees Celsius to March 1867, thereafter in Fahrenheit until November 1948, after which the 'Att. Therm' entry was omitted from the registers. The 'corrected' reading of the barometer (here referring to correction for the thermal expansion of the mercury column from room temperature to the standard 0 °C, rather than the correction to MSL pressure), usually referred to as the 'station level pressure', is also included in the observation registers until November 1948. This correction was made by reference to a local table, not always accurately, and for consistency during the preparation of the series the correction to 0 °C was re-calculated using the observed Attached Thermometer reading, in preparation for reducing the 'station level pressure' so obtained to 'mean sea level (MSL) pressure' (as documented subsequently). The 1910 inspection report noted that a minor index correction from the Browning barometer's calibration at Kew Observatory was included within the table used within the observatory to reduce the barometer reading to 0 °C, and the re-calculation of 'station level' pressures adopted here will therefore not include this. However, the barometer was recertified on 14 January 1925 following minor cleaning and repair work, after which the corrections applied were all noted as 0.1 hPa or less across the working range of the instrument (1926 inspection report). The effect of the omission of the corrections is therefore probably insignificant.

From early 1948 the barometer readings appear to have given cause for concern, because the mean 0900 and 2100 GMT MSL pressures were not published in the *Monthly Weather Report* after February 1948. It is probably no coincidence that the Browning Fortin barometer was removed for cleaning and repair (by Negretti & Zambra) in December 1948. A Kew pattern barometer, on loan from the Met Office, was read in its place until 31 August 1949 (September 1949 inspection report refers). This barometer was graduated and read in millibars.

The original Browning Fortin barometer returned to Durham in September 1949 and remained in use until the cessation of the record on 31 December 1960, except for the five months 28 August 1957 to 28 February 1958 when it was again away for cleaning. Records during this period were from a temperature-compensated aneroid barometer on loan from the Met Office, and accordingly there are no 'Attached Thermometer' readings. The Fortin barometer was recertified by Negretti & Zambra on 5 February 1958 and brought back into use from 1 March 1958.

From September 1949 the barometer readings in the register continue in millibars, suggesting that the Fortin barometer had been refitted with a new scale during its cleaning and repair. The barometer at this date was at least 80 years old, and quite possibly over 100. The reading of the Attached Thermometer was omitted from the observation register from this point, as examined in more detail subsequently.

Unfortunately, there is no record of what happened to the Browning Fortin barometer following the discontinuation of readings in December 1960, and its current whereabouts are unknown.

Instrumental record summary

Observation hours

From the commencement of record until September 1885, hours were reckoned by local or 'common' time, about six minutes later than the Greenwich Meridian, but from October 1885 onwards Greenwich Mean Time (GMT) was adopted as the observatory's standard time. The barometer readings were noted for the majority of the period of record at 0900 and 2100 daily, with the following exceptions.

Morning observation:

- At 10h February 1855 to September 1885
- At 09h Summer Time (0700 GMT) during operation of Summer Time, 1916–18
- At 09h Summer Time (0800 GMT) during operation of Summer Time, April to September 1945 (letter in site file, Met Office Archives, dated 30 September 1945)

Evening observation:

- No evening observations were made between February 1855 and June 1858; instead, a 14h observation was made during this period
- At 22h July 1858 to September 1885

Occasional observations at other 'non-standard' times such as 2140 have been assumed to be 'late observations' from the intended hour and simply included without adjustment along with the other observations for the 'standard' hour.

Missing data

The record is remarkably complete; for the period July 1843 to December 1960, 117 years, 98.7 per cent of observations appear in the database. The main gaps (four or more consecutive missing observations) are shown in Table A5.1.

Units

Until November 1948, barometer readings were stated in inches of mercury (inHg), to two places of decimals until 23 March 1846 and thereafter to three places of decimals (0.001 inHg = 0.03 hPa). These have been converted throughout to hectopascals (hPa: 1 hPa = 1 millibar; 1 inHg = 33.86388 mbar) and rounded to 0.1 hPa after the application of any corrections. From December 1948 onwards, observations were noted in millibars. Attached

Table A5.1 *Missing data in the Durham Observatory pressure series, July 1843 to December 1960*

Period missing	No. of days	Period missing	No. of days
4–8 Aug 1845	5	24–31 Dec 1883	8
1 Jul–14 Aug 1854	46	6–8 Oct 1908	3
27 Sep–7 Nov 1854	43	5–16 Feb 1923	12
7–12 Dec 1854	6	9 Mar–17 Apr 1923	39
27 Dec 1854–14 Feb 1855	51	10 Oct 1934–28 Feb 1935	143
1–11 Feb 1856	11	1 Apr–31 May 1935	62
1–6 Jun 1865	6	1 Jun–31 Aug 1936	92
		12–15 Jun 1960	4

thermometer readings are stated in degrees Fahrenheit or Celsius at various times: all have been converted to degrees Celsius. On the database, both 'station level pressure reduced to 0 °C' and the calculated MSLP are given in hPa for each available observation. The method used to reduce station level pressures to MSL is detailed later.

Creation of the Durham pressure database

Digitization of the record

Most of the original manuscript meteorological records from Durham Observatory have been retained, either in the Department of Geography or in the Durham University library. An initiative funded by the Leverhulme Trust saw many of the manuscript instrumental records from 1850 to 1997 digitized [44], although until recently knowledge of this dataset remained almost entirely limited to Durham University. The barometric pressure observations (which terminated after December 1960) were included in this project—fortunately, for the Durham pressure record now represents by far the longest digital barometric pressure series in northern England, filling a very large gap in the International Surface Pressure Database (ISPD: [154]). The ISPD forms the majority of the input to the NOAA–CIRES–DOE 20CRv3; [58, 120], currently extended back to 1806. The remaining early records, from July 1843 to December 1849, were digitized by hand by the authors in 2021.

Errors in the series

Unfortunately, the 'Leverhulme Trust' barometric pressure record as originally digitized contained a significant number of major errors, some of which were due to the original observers and some due to mis-digitization. An understanding of the causes leading to typical errors enables the process of checking and correcting to be more efficiently undertaken. Pressure records in inch units tend to exhibit distinct and characteristic errors, which can be briefly summarized as follows:

1. *Observer errors in reading the instrument or noting the observation.* The typical scale of barometric pressure in the United Kingdom covers the pressure range from below 28 to above 30 inHg. It was very common in manuscript entries in inch pressure registers to omit the inch value until and unless the integer changed, by entering only the values following the decimal point. The transition from one-inch integer to another can easily be omitted by the observer, as Table A5.2 indicates.

   In line 6, the entry '30' has been omitted, implying that the observed pressure is 29.013 inHg rather than 30.013, and this leads to a 1 inHg error (33.9 hPa) in the digitized value. The '30' entry then appears correctly in line 7, but the '29' entry has been omitted from the following entry, leading to a large positive error in the converted inch value. This type of error is particularly common where the average station level pressure is close to 30 inHg (1015.9 hPa), and thus the changeover of leading digits from '29' to '30' or vice versa happens frequently. Where the observations are close together in time, such as hourly or three-hourly records, correction is usually a matter of simple continuity checks to flag as suspect any particularly rapid rises or falls which are immediately followed by a similar magnitude fall or rise. Where observations are spaced 6–12 hours or more apart, as is the case with Durham, great care is required to avoid false corrections as changes of

Table A5.2 *Example of common errors in barometric pressure entries in manuscript register*

Entry	Manuscript entry inHg	Digitized as inHg	Conversion to hPa	Correct inch values	Conversion to hPa	Error hPa
1	30 106	30.106	1019.5	30.106	1019.5	
2	062	30.062	1018.0	30.062	1018.0	
3	004	30.004	1016.1	30.004	1016.1	
4	29 994	29.994	1015.7	29.994	1015.7	
5	906	29.906	1012.7	29.906	1012.7	
6	013	29.013	982.5	**30.013**	1016.4	33.9
7	30 042	30.042	1017.3	30.042	1017.3	
8	981	30.981	1049.1	**29.981**	1015.3	-33.9

30 hPa in six hours or more, rises or falls, do occur from time to time, particularly during the winter months.

2. *Incorrect reading of the 0.5 inHg vernier scale.* Most inch barometers have a major division at 0.5 inHg intervals, and it can be easy to misread (say) 29.72 for 29.22 inHg, thereby introducing a 16–17 hPa error. These errors are much harder to spot on a twice-daily or daily record, but comparison with 20CRv3 and manual examination of the relevant synoptic chart, where available, often helped determine whether the observation value was correct or otherwise.

3. *Incorrect or unclear digits in the observations register.* There are many possible errors here—confusions between 0, 6, or 9; 1, 4, or 7; 2, 4, or 7; and 3, 5, and 8 are typical, and often mistaken for each other at the digitization stage if the written record is in any way unclear. Any such errors will generally only be obvious where they amount to at least several hPa from the 20CRv3 background, and thus are difficult to identify and correct.

4. *Incorrect entry in the register.* It is very easy to enter, say, 29.852 inHg in the register when the barometer was read as 29.582. Such errors are almost impossible to identify and correct unless the noted observation value differs by at least several hPa from the 20CRv3 background.

5. *Incorrect transcription of any of the above errors during digitization* will produce similar results. Some are easy to spot by out-of-range checks, especially errors in the inch value itself; digitized values of 20 inHg or 39 inHg are easy to flag and clearly incorrect, but whether the true value in such instances should be 29 inHg or 30 inHg may not be immediately obvious. Such cases require close consideration of the context of the record including, where possible, examination of the original register entry.

6. *Incorrect millibar entries.* For the millibar records (from December 1948 to end of series), occasional errors in reading the scale persist; most appear to be around ± 10 mbar (hPa), or multiples thereof.

Correction to mean sea level pressure (MSLP)

For the period 1843 to 1948, reduction of the observed or 'As read' barometer readings to MSL involved a two-stage process; firstly, to correct for the observed temperature of the barometer (the 'Attached thermometer' reading) to the standard 0 °C, this value then referred to as the 'station level pressure'; and secondly to correct the derived station level pressure to MSL, for which the outside air temperature (dry bulb temperature) is also required. From 1949 onwards, the value of the incremental MSL correction was derived at the observatory and noted in the observations register, and thus has been used to correct the 'As read' pressure to MSL. Further details are given in a subsequent section.

A detailed account of the rationale for, and the methods involved in, the reduction of barometric pressure to MSL is beyond the scope of this paper, and the reader is referred to specialist texts, such as the *Handbook of Meteorological Instruments, Volume 1: Measurement of atmospheric pressure* [159] and the World Meteorological Organization CIMO observing guide [160], Chapter 3, *Measurement of atmospheric pressure*. At the Durham Observatory, only the station level pressure was required for astronomical work and MSL pressure was not routinely evaluated until January 1949, and thus is not included in the manuscript observation registers prior to that date.

Throughout, calculations were performed in an Excel spreadsheet.

Correction to 0 °C

For a Fortin barometer, and for pressure in millibars, the barometer temperature correction in millibars per Kelvin is 0.163 mbar/K at 1000 mbar 'As read' pressure. The correction is proportional to the 'As read' pressure, and is negative for Attached Thermometer readings above 0 °C or 273 K ([161], Chapter 2, *Atmospheric Pressure*). Thus, for 'As read' 1020 mbar and Attached Thermometer 20 °C, the correction is −3.32 mbar and the station level pressure at 0 °C becomes 1016.68 mbar. The calculation is slightly different for a Kew barometer and is closer to 0.171 mbar/K at 1000 mbar. Normally no correction for barometer temperature is required for aneroid instruments, such as the temperature-compensated aneroid barometer in use at the Observatory for five months during 1957–58.

Prior to 1955, standard corrections for barometer temperature assumed a barometer temperature of 62 °F (17 °C) and standard gravity 9.8062 m/s^2; correction tables based upon these values (or even earlier standards) would most likely have been used for most of the Durham pressure record. To ensure the application of consistent corrections to modern standards, and to bypass occasional errors in the observation record and the digitized series, the entire Fortin 'As read' record prior to 1949 was corrected to 0 °C using the post-1955 standard.

Reduction to MSL

The reduction to MSL is computed in two stages. The first is to sum the barometer's index errors (calibration differences at various pressures) together with the correction for standard gravity of 9.806 65 m/s^2, which varies with latitude, then add this to the station level pressure as calculated above. This sum is then added to the MSL correction term to obtain the MSL pressure or MSLP.

The index errors for the barometer in use are not individually known, although inspection reports stated they were small, and accordingly they have been neglected in this calculation.

The correction for standard gravity at the Observatory's latitude (54.768°N) at 1000 mbar is +0.83 mbar (Table LIIA in [161], p. 439), and is in proportion to the station level pressure.

The method of reduction to MSL follows that set out by WMO in the CIMO guide, section 3.7 [160] as follows, assuming a constant lapse rate:

$$ p_{msl} = p_{stn}.exp\left(\frac{\frac{g}{R}.H}{T + \frac{L.H}{2} + e_s.C} \right) $$

where the terms are as follows:

p_{msl} is the MSL pressure (in hPa)

p_{stn} is the station level pressure (hPa)

g is the standard acceleration of gravity (9.806 65 m/s^2)

R is the gas constant of dry air (287.05 J/kg/k)

H is the station elevation (in geopotential metres—the error in using altitude in metres is insignificant and can be neglected below about 500 m AMSL. For Durham Observatory the barometer cistern height of 107.3 m was used)

T is the outside air temperature (in Kelvin), from observed dry-bulb temperature $T_C + 273$

L is the assumed lapse rate in the fictitious air column extending from sea level to the level of the barometer cistern, taken as 0.0065 K/gpm

e_s is the station vapour pressure (in hPa)

C is a coefficient (0.12 K/hPa).

In this calculation, the station vapour pressure e_s is taken as 85 per cent of the saturation vapour pressure e_{sat} at the outside air temperature T_C (in °C)—(i.e. the humidity is taken as 85 per cent, very close to a true mean for Durham).* e_{sat} is calculated using the formula due to Bolton [162], which is acceptably accurate between −30 °C and 35 °C:

$$ e_{sat}(T_C) = 6.112 \exp(\frac{17.67\ T_c}{T_c + 243.5}) $$

The value of the MSL pressure thus obtained was then subjected to quality control measures as set out below.

Applying reanalysis data to effect quality control of the Durham MSLP series

Without some form of independent record, it would be difficult to provide objective assessment of the accuracy and reliability of the new Durham series, particularly at the daily or sub-daily level. Fortunately, the increasing accuracy and lengthening timescale of reanalysis datasets provides an objective means to assess record quality; this is believed to be the first time an underpinning assessment using reanalysis has been applied to verify an independent long-period pressure

---

* While we do have dry- and wet-bulb temperatures for almost all observations, it would require a great deal of additional computation to calculate relative humidity and vapour pressure for each observation, for no significant benefit to the MSLP calculation. Averaged over a year, the difference between MSLP assuming RH 75 per cent and RH 95 per cent (accounting for the majority of diurnal and seasonal variations, and 0900 and 2100 observation times) is about 0.01 hPa, which is much lower than instrument and observer errors, and can therefore be safely neglected.

dataset. Alternative methods include comparisons against gridded MSLP series or weighted comparisons with other sites. Clearly there is the potential for circularity in the correction of the candidate series against the Twentieth Century Reanalysis (20CR) if and when the corrected series is subsequently assimilated into the reanalysis at a later date. However, the risk of circularity in this case is considered to be very small. Firstly, the reanalysis series was only used to flag potential errors, which were then followed up wherever possible by scrutiny of the original registers or other sources (such as the online Met Office *Daily Weather Report* and subsequent publications). Secondly, the number of flagged errors is very small, amounting to just 0.42 per cent of the entire series between 1843 and 1960 (Table A5.3). The published database also includes both 'raw' and 'post-quality control' values.

Reanalyses can provide complete and consistent atmospheric fields by objectively combining historical observations with modern numerical weather prediction model forecasts, while accounting for estimated errors in both [84]. The latest version of the 20CR has been generated by the University of Colorado Boulder's Cooperative Institute for Research in Environmental Sciences (CIRES) together with the National Oceanic and Atmospheric Administration (NOAA) and the U.S. Department of Energy (DOE). This NOAA–CIRES–DOE 20CRv3 uses a newer, higher-resolution model, assimilates a larger set of observations, and includes an improved data assimilation system relative to its predecessors. The 20CRv3 system further extends the reanalysis period to 1836–2015, with an experimental extension spanning 1806–35. Slivinski and colleagues [120] provide an in-depth description of the system that generated the 20CRv3 reanalysis product, which consists of a numerical weather prediction model, an observational dataset, and an assimilation method. Using an 80-member ensemble Kalman filter, the 20CRv3 system assimilates only surface pressure observations (a so-called 'sparse' reanalysis) from the open, unrestricted, and publicly available International Surface Pressure Databank (ISPD) version 4.7 [154], into the U.S. National Centers for Environmental Prediction (NCEP) Global Forecast System (GFS) model, with a horizontal resolution of about 60 km at the equator and a vertical atmospheric resolution of 64 levels. Sea surface temperature and sea ice, solar radiation and time-varying atmospheric constituents of volcanic aerosols, stratospheric ozone, and atmospheric carbon dioxide (CO_2) levels are also specified. Output fields are available at three-hourly resolution.

Until very recently, there were no pressure records from sites in England held on the ISPD (and thus available to the 20CRv3 reanalysis) prior to 1925. Before 1925, the only ISPD records within the British and Irish Isles are those from Armagh in Northern Ireland (pressure data from 1796–1826, 1833–1965; Figure A5.2), Aberdeen in Scotland (1871–1948, 1957–1988),

Figure A5.2 *Location map showing the other UK and Ireland long-period pressure observations contained within the International Surface Pressure Databank (see text for details)*

Table A5.3 *Durham pressure record QC summary, comparing 'raw' (original digital dataset) and 'Post QC' (after quality control measures as described in the text). A positive value in the 'Average arithmetic error' indicates that the Durham value is higher than the 20CRv3 gridpoint ensemble mean value, and vice versa*

Period and number of observations		No of corrected observation	Corrected %	Average arithmetical error 20CRv3 minus Durham, hPa	Error Std Dev, hPa	Average RMS error 20CRv3 minus Durham	MSLP average, hPa	% Values 20CRv3 ±2 hPa	% Values 20CRv3 ±10 hPa
Fortin barometer Jul 1843 to Nov 1948									
Raw	75 816 of 76 964 (98.5%)	None		+0.43	11.15	2.21	1013.35	64.6	99.1
Post QC	75 816	319	0.42	+0.34	2.71	1.97	1013.26	64.6	99.4
Kew barometer Dec 1948 to Aug 1949									
Raw	548 of 548 (100.0%)	None		+0.47	1.38	1.05	1017.53	89.2	99.6
Post QC	548	2	0.36	+0.43	1.22	1.01	1017.49	89.6	100.0
Fortin barometer Sep 1949 to Dec 1960 excl. 28 Aug 1957 to 28 Feb 1958									
Raw	7901 of 7910 (99.9%)	None		+0.59	1.85	1.23	1013.48	85.9	99.4
Post QC	7901	28	0.35	+0.63	1.29	1.13	1013.51	86.7	100.0
Aneroid barometer 28 Aug 1957 to 28 Feb 1958									
Raw	370 of 370 (100.0%)	None		+0.54	1.76	1.16	1012.66	85.4	99.5
Post qc	370	2	0.54	+0.45	1.34	1.08	1012.58	85.9	100.0
Combined record, Jul 1843 to Dec 1960									
Raw	84 635 of 85 792 (98.7%)	None		+0.45	10.56	2.11	1013.38	66.8	99.1
Post QC	84 635	351	0.41	+0.37	2.60	1.88	1013.30	66.9	99.5

and Valentia Observatory in the Republic of Ireland (1892 to date). Recent work by Hawkins and colleagues [158] rescued pressure data from Fort William and Ben Nevis from 1883–1904 and digitized records from multiple European sites published in the UK Met Office contemporary *Daily Weather Report* publication from 1860 [163, 164]; this work has been completed, although at the time of writing it had not yet been incorporated into ISPD. Consequently, the accuracy of atmospheric reanalyses over the north-eastern Atlantic prior to 1925 has been constrained owing to the dearth of reliable surface pressure records in and around the British and Irish Isles. This newly available record from Durham should therefore be helpful in assessing both likely errors in gridpoint pressure data from the reanalysis ensemble means, and changes over time, and in improving the accuracy/reducing ensemble spread in future reanalyses once the Durham data are eventually included into ISPD and a future version of 20CR.

Method

A time series of pressure values (ensemble mean pressures) at a given gridpoint can be obtained from the 20CRv3 reanalysis website (https://psl.noaa.gov/data/timeseries/hour/), for any period within the main dataset (currently 1836 to 2015). This output, in the form of one or more sub-daily gridpoint ensemble mean pressures at three-hourly intervals (00, 03, 06 GMT, etc.) for the chosen period, was used to provide an independent underpinning quality control measure to the Durham MSLP record (the two series can be regarded as independent since the Durham series was not included in the ISPD datasets from which this version of the reanalysis was built). The nearest 20CRv3 gridpoint to Durham Observatory is at 55°N 2°W, about 36 km north-west of the observatory (54.768°N, 1.584°W). Gridpoint ensemble mean pressure values were extracted for 0900 and 2100 GMT* throughout the period of record, except for February 1855 to June 1858 when an afternoon observation was made at 1400, for which 1500 GMT gridpoint values were used instead.

An example, from the first winter of the record in December 1843-January 1844 and using observed (i.e. pre-QC) data, is shown in Figure A5.3. The tendency for ensemble averaging to reduce the absolute range in extremes is evident in the plot. This is understandable when it is considered that the assimilation of this version of the reanalysis is based upon only a single site in the British and Irish Isles at this time (viz. Armagh, Figure A5.2), and consequently the spread of ensemble members is likely to be less constrained than in later periods with a denser spread of surface pressure observations.

When Durham's observations were made at 10h and 22h, 0900 and 2100 GMT gridpoints were also used as representing the closest point in time. The resulting 20CRv3 twice-daily gridpoint pressure series was then compared against the Durham MSLP record.

Of course, no quality control measure can ever render an imperfect record into a perfect one, and it is important to minimize changes to the original record commensurate with removing at least the most obvious errors. The 20CRv3 daily and sub-daily series are particularly useful where short spells of observations are missing or have been mis-coded with incorrect dates, for pattern-matching (by eye or by algorithm) can quickly suggest a fitting sequence. However, as Table A5.3 shows, the vast majority of the Durham record appears extremely reliable. In turn, the independent Durham dataset provides a useful benchmark to assess the likely accuracy of 20CRv3 within the north-east Atlantic region.

---

* Strictly, GMT as a defined time standard did not exist until 1885, but for convenience times are referred to as GMT for dates prior to 1885 unless otherwise stated.

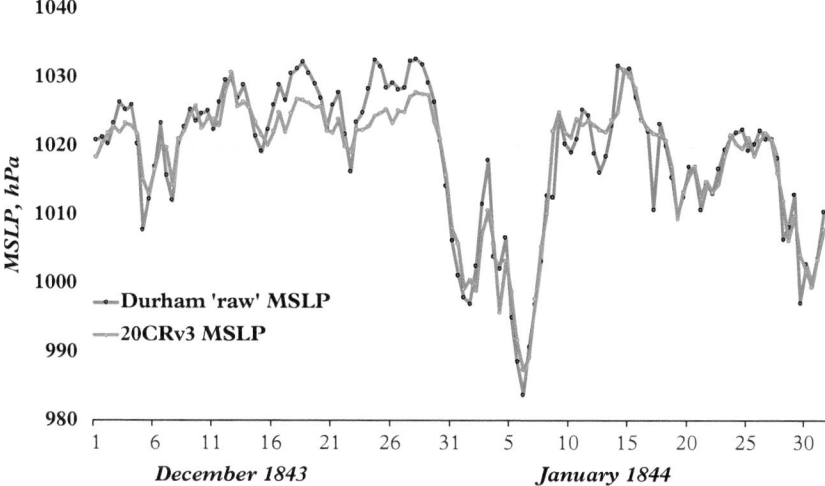

Figure A5.3 *Comparison of daily 'raw' (uncorrected) Durham* MSL *pressure observations at 0900 and 2100 solar time (six minutes later than Greenwich time) (red line) with 20CRv3 nearest gridpoint values at 0900 and 2100* GMT *(green line), for the months of December 1843 to January 1844, the first winter of the record. The very close agreement is evident, and the few significant differences are most likely due to uncertainties in the reanalysis—see text for details*

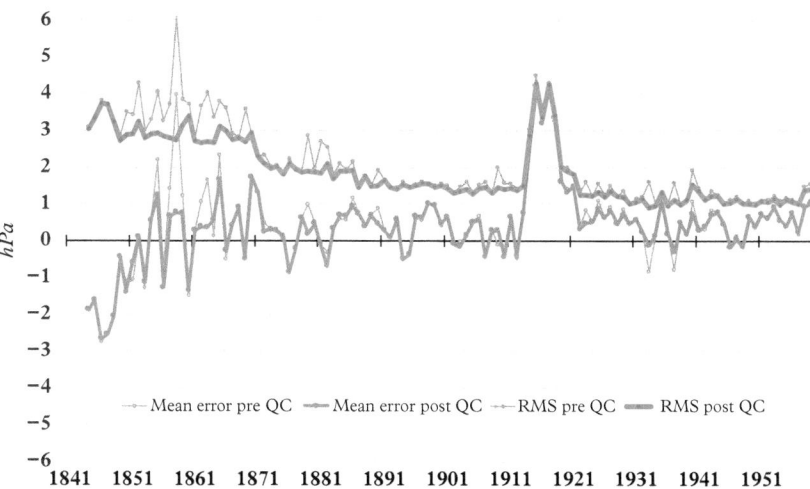

Figure A5.4 *Annual mean arithmetic error (green line) and root-mean-square RMS error (orange line) between the quality-controlled Durham pressure series and 20CRv3 nearest gridpoint value, 1844 to 1960. For comparison, the thin grey lines on both series show the arithmetic mean and root-mean-square errors from the uncorrected (pre-QC) series.*

Figure A5.4 shows both the annual mean root-mean-square (RMS) and absolute errors (the arithmetic difference 20CRv3 gridpoint minus Durham) over the period 1844–1960. From this it is evident that the relative accuracy of the reanalysis dataset usually increases over time, as would be expected with increasing density of assimilated surface pressure data. The one significant exception to this fairly smooth trend occurs between 1914 and 1919, when the Durham data provides clear evidence of a previously unknown bias in 20CRv3 over the North Atlantic. This anomaly has been confirmed by examination of other reanalysis products; at the time of writing, the reasons for it remain unclear and under investigation (Ed Hawkins, Clive Wilkinson and Gil Compo, personal communications February–March 2021).

The Durham pressure series is divided into four periods, and each is discussed in more detail below.

July 1843 to November 1948

At just over 105 years in length, this comprises the first and longest subdivision of the Durham pressure series. As far as can be ascertained, the record originates from the same instrument throughout, namely the Browning Fortin barometer. The series is 98.5 per cent complete; most of the gaps after 1883 resulted from periods when the Browning instrument was away for cleaning or repair.

Throughout this period, the twice-daily series appears extremely reliable when compared against the relevant 20CRv3 gridpoint value (Table A5.3 and Figure A5.4). Even prior to the introduction of quality control corrections, between July 1843 and November 1948 64.6 per cent of the available observations lay within ± 2 hPa of the corresponding 20CRv3 gridpoint value, and 99.1 per cent lay within ± 10 hPa. Comparisons against 20CRv3 suggested that 319 records (just 0.42 per cent of the total observations) were most likely to be incorrect.[*] Most of these (235, or 73 per cent) were found to be clustered close to 0.5 or 1.0 inHg multiples, thereby suggesting an error in the 'As Read' reading as the most likely cause: a few were 9 or 10 inHg errors (i.e. an entry of 20 or 39 inHg instead of 29 or 30 Hg). Such discrepancies are almost certainly due to one or a combination of the reasons outlined earlier. The distribution of errors was Gaussian and almost symmetrically distributed about zero, as is evident from the very minor change in mean MSL pressure (Table A5.3). The post-QC Durham mean pressure is slightly higher than the 20CRv3 gridpoint mean—for the whole 111-year period by 0.37 hPa—as would be expected with the gridpoint being 36 km north-west of the Observatory location and the climatological decrease in mean MSL pressure with increasing latitude in the British and Irish Isles.

An automated quality control check against 20CRv3 was successful in flagging most if not all of the major errors during this period, increasing the number of the Durham observations within ± 10 hPa of the corresponding 20CRv3 gridpoint value from 99.1 per cent to 99.4 per cent. However, objective automated error checking became progressively less reliable with smaller discrepancies. At errors below roughly 17 hPa/0.5 inHg difference, the very high quality of most of the Durham series would suggest that the difference is more likely to arise from errors in the reanalysis field, particularly prior to about 1925 for reasons referred to earlier. The two most likely causes of such errors are timing differences (almost always cyclonic storms moving or developing more

---

[*] The term 'error' in this sense refers to the difference (20CRv3 gridpoint value minus Durham). Such differences can arise only through an error in the Durham value (whether observer, transcription or subsequent digitization), or uncertainty in the 20CRv3 reanalysis field. The latter is larger in the early years of the record where fewer surface pressure observations have so far been assimilated.

quickly than suggested by the reanalysis model), and insufficient deepening of intense cyclonic systems within the reanalysis. In particular, the latter effect clearly resulted in some Durham records being incorrectly flagged as erroneous, when a manual check on the original manuscript records (at present, online files available for 1843–1850 only), or from observations published in the relevant *Daily Weather Report* from September 1860 (and, later, synoptic charts from March 1872), showed that they were almost certainly correct—in which case, the error flags were removed, and the original observations reinstated.

While it would be easy enough automatically to correct all discrepancies larger than, say, ± 10 hPa by arbitrarily assuming the Durham value was in error and flagging it as 'incorrect', and of course doing so would immediately reduce the number of errors larger than 10 hPa from 0.6 per cent to zero, it was felt that such actions would almost certainly result in some valid observations being wrongly flagged as incorrect. Accordingly, to avoid arbitrary corrections to the Durham series, for this period quality control checks were applied only where the discrepancy was greater than about 15 hPa, or where smaller errors could be investigated by manual checking against other nearby sites (usually from stations included in the *Daily Weather Report*) or by other methods, such as continuity checks in periods of settled weather conditions. Particular attention was paid to records close to or exceeding monthly long-term climatological extremes.

In time, it should be possible to check all discrepancies greater than, say, ± 5 hPa back against the original manuscript observation registers to identify and correct any mistyped digitization entries. Unfortunately, coronavirus travel and access restrictions made this impossible during the preparation of this book, but when this step becomes feasible then it will be possible to re-examine and corrected the Durham series as necessary.

In summary therefore: the first and longest part of the Durham pressure series appears impressively complete and shows a very high degree of accord with the 20CRv3 reanalysis. Most if not all of the larger errors have been identified and corrected, but it is likely that an unknown number of smaller errors remain in this period of record. While some uncertainty remains with regard to individual values, monthly and annual means over this 107-year period are believed to be correct to within a few tenths of a millibar.

December 1948 to December 1960

The barometer in use during the final 12 years of the long Durham record changed several times, as stand-in instruments were used in place of the Browning Fortin unit while away for cleaning and refurbishment in 1948–49 and 1957–58.

The period is sub-divided by barometer type in use; the distinction is important because the detail of the corrections applied to the observed value differ somewhat according to the type of barometer, and these have been allowed for in the MSL corrections applied. From December 1948, all barometer readings were made in millibars; with the change of graduation comes a change in the nature of 'major' errors, which are now more likely to be about ± 10 mbar, or multiples thereof.

Readings of the 'Attached Thermometer' ceased with effect from 1 December 1948. From January 1949 onwards, the entry in the Att Therm column in the observations register appears to be the MSL correction to the As Read value, presumably derived from a barometer correction card taking the barometer temperature into account in doing so (the method is explained in the reference [159]). Although a different MSL derivation scheme for the final years of the series is unfortunate, the absence of the observed Att Therm reading renders this a necessity and from January 1949—with exceptions listed below—the MSLP was taken as As Read + MSL correction as given in the observations register. For the few occasions when the correction was missing, estimates based on neighbouring values were used.

December 1948 to August 1949: Kew barometer

A loan Kew-pattern barometer replaced the Browning Fortin barometer while the latter was away for repair between December 1948 and August 1949. This barometer was graduated in millibars. Readings of the 'Attached Thermometer' ceased with effect from 1 December 1948. For December 1948, MSL values were prepared taking the Att Therm value as the average for December for the previous 10 years 1938–47 and using the calculation as set out previously. From January to August 1949, the incremental MSL correction is given in the observations register, and the MSLP has been taken as the As Read + MSL correction.

September 1949 To December 1960: Fortin barometer

The Browning Fortin barometer was re-introduced on 1 September 1949, and aside from the six months 28 August 1957 to 28 February 1958, when it was away for cleaning once more, it remained in use until barometric pressure observations were discontinued after 31 December 1960. The barometer was re-certified by Negretti & Zambra (certificate dated 5 February 1958), with errors no more than 0.4 hPa at any point within the calibration range. The August 1958 inspection report states that a handful of comparisons of Durham's MSL pressure against neighbouring synoptic stations around the date of the inspection indicated that the barometer read up to 1.9 hPa too *low*, although the average of the observatory's 0900 GMT readings for that month compared with nearby sites would suggest the error was about 1.0 hPa too *high* at that time.

28 August 1957 to 28 February 1958: Aneroid barometer

Daily observations were made with a temperature-compensated aneroid barometer loaned by the Met Office for this five-month period (note on Durham Observatory site file dated 28 October 1957). This type of instrument does not require temperature correction and thus no 'Attached Thermometer' readings are available for this period.

This instrument appears to have read 27–28 hPa too low—possibly as a result of the MSL adjustment being incorrectly set—but otherwise appears to have indicated daily pressure changes accurately and reliably. During this period, the MSLP was taken as As Read + 27.5 hPa. Two observations during this period required additional corrections for 10 mbar reading errors.

The Durham observatory digital pressure record

The entire Durham twice-daily pressure series 1843–1960 is available on the University of Reading Research Data Archive at http://dx.doi.org/10.17864/1947.328 as an open access dataset under a Creative Commons Attribution 4.0 International Licence. The file format is Comma Separated Variable (.csv), and the file size is 48 MB. The contents of the file are listed in Table A5.4, as a ReadMe file within the same location.

Three options are available for the Durham MSLP series—the 'raw' (as observed) record, including gaps where they occur; the corrected (post-QC) record, including gaps; and a corrected (post-QC) record, where gaps have been filled using the 20CRv3 gridpoint data +0.5 mbar to provide an unbroken series. Gaps in the record amount to 1.3 per cent of the record (Table A5.3) and corrections to 0.41 per cent, distributed fairly evenly throughout the record.

Table A5.4 *Details of the twice-daily Durham Observatory digital pressure series 1843–1960*

Column header	Cell contents
YYYY MM DD	Date as YYYY MM DD character string (two entries per day)
YYYY MM DD HHmm	Date and time as YYYY MM DD HHmm character string (one entry per observation)
DD	Date in month (1–31)
MM	Month (1–12)
YYYY	Year (1843–1960)
HHmm	Observation hour HHmm—mostly 0900 or 2100, GMT from Oct 1885
Missing	Flag: Barometer 'As Read' reading missing = 1, else 0
AsRead_inHg	Barometer 'As Read' in inches of mercury (inHg) to November 1948, blank thereafter. This is the barometer reading as observed and digitized and includes numerous errors
AsRead_hPa	Barometer 'As Read' in millibars (mbar or hPa) throughout—inHg converted by × 33.86388. This is the barometer reading as observed and digitized and includes numerous errors
AttTherm_C	Barometer 'Attached Thermometer' in degrees Celsius—converted from °F as necessary. The record runs from July 1843 to November 1948 only
SLP_hPa	Station level pressure—barometer 'As Read' reduced to 0 °C using the Attached Thermometer reading (see text for details), to November 1948 only
Tdry_C	Observed external air temperature (dry bulb in screen) in degrees Celsius. Some are missing; estimates were made using neighbouring values or, occasionally, monthly means over several years
MSLP_RAW_hPa	Calculated MSL pressure in millibars (hPa). MSL calculation details are given in the text. This is the RAW value, prior to quality control. Between 1843 and 1948 this is calculated from the observed As Read and Attached Thermometer with dry bulb temperature, as explained in the text; from 1949 to 1960 this is the As Read + MSL correction as given in the digitized observations register
MSLP_QC_hPa	Calculated MSL pressure in millibars (hPa). This is the CORRECTED value, post quality control (see text for QC details), based upon the SLP_hPa value 1843–1948, or the As Read + MSL correction 1949–60
MSLP_QC_gapfilled_hPa	This is identical to **MSLP_QC_hPa** except that gaps are filled using the 20CRv3 ensemble mean gridpoint value + 0.5 hPa to provide an unbroken series

Missing data are shown blank (empty cell) but note that there are no missing cells in the date/time headers, for the 20CRv3 ensemble mean gridpoint value or for the gap-filled pressure series.

Dataset details

Identifier: http://dx.doi.org/10.17864/1947.328

Creator: Burt, Stephen D.

Title: A twice-daily barometric pressure record from Durham Observatory in north-east England, 1843–1960

Publisher: Burt, Stephen D and University of Reading

Publication year: 2021

Resource type: Dataset

Version: 1.1

Record summary

Monthly and annual averages

Table A5.5 sets out monthly and annual averages of MSLP at Durham over various 30-year periods. These averages are of the *corrected* series, with gaps filled by 20CRv3 gridpoint values +0.5 hPa where necessary. Values are shown minus 1000 hPa.

Extremes on record

Daily and monthly extremes of barometric pressure recorded at Durham Observatory by month between 1843 and 1960 are given in Chapter 27.

Table A5.5 *Monthly and annual averages of MSL pressure at Durham Observatory, average of 0900 and 2100 GMT observations, various periods 1851–80 to 1931–60. Units—hPa less 1000.*

Period	J	F	M	A	M	J	J	A	S	O	N	D	Year
1851–80	10.8	13.5	12.7	14.6	15.2	14.5	13.8	13.4	13.7	11.2	12.9	12.0	13.17
1861–90	12.4	13.8	12.2	14.0	15.4	15.6	13.3	13.2	13.5	12.0	11.4	12.4	13.24
1871–1900	13.7	14.1	12.7	13.5	15.7	15.6	13.3	12.9	14.1	11.9	12.0	11.6	13.41
1881–1910	14.5	14.0	11.8	13.5	15.6	16.5	14.1	12.9	15.9	12.2	12.6	10.9	13.70
1891–1920	13.9	12.5	10.6	13.8	16.1	16.2	14.9	13.1	16.1	12.3	12.7	9.2	13.43
1901–30	12.9	12.2	11.4	12.8	15.2	15.9	14.4	12.6	16.0	12.3	11.6	9.4	13.06
1911–40	11.6	13.1	12.2	13.0	15.5	15.6	13.3	13.5	14.9	12.1	10.6	10.7	12.99
1921–50	11.7	13.9	14.8	13.0	15.3	15.4	13.0	13.3	14.4	12.6	10.9	12.2	13.36
1931–60	11.9	13.7	14.8	14.7	16.4	15.4	13.1	13.4	14.4	13.5	11.5	11.1	13.68

Appendix 6
Sunshine duration records made at Durham observatory, 1880 to 1999, and estimates to date

Durham was on the list of the first 30 sites to be allocated one of the new sunshine recorders, as recorded in the minutes of the Meteorological Committee of the Royal Society (the body that was appointed to set the strategy for the Meteorological Office following the death of Admiral Fitzroy in 1865) in its meeting on 29 November 1879.[*] This amounted to the remarkably rapid introduction of a completely new instrument, what we now know as the Campbell–Stokes sunshine recorders, for the Meteorological Committee's allocation list predated by more than three weeks the reading of the paper in the Meteorological Society's rooms in London on 17 December of that year by Professor George Gabriel Stokes FRS, the noted Cambridge mathematician, who described and announced the new instrument. Stokes' paper [98] did not even appear in the *Quarterly Journal of the Meteorological Society* (*Royal* only after 1882) until April 1880, by which time several of the 30 stations had already installed the instruments and commenced records (including Oxford, where an almost unbroken daily record using this new instrument commenced in February 1880, and continues today [49]). Notably, the serial number of Durham's unit was MO28, confirming its early origin; its appearance would have resembled Figure A6.1, taken from Stokes' 1880 paper. This original unit remained in use at the Observatory from its installation in May 1880 until May 1981, when after 101 years use it was replaced by an updated model of the same instrument, which can be seen in Figure 16.8.

Durham's sunshine record was therefore one of the earliest in the world. Daily records survive from January 1882 and are included in the digitized online observations database. Daily records continued until 5 December 1883, after which the rest of the month is missing; January 1884 is complete but thereafter there is no record in the surviving Durham registers until 29 May 1885. The reasons behind this long gap can be found in Appendix 1, note 5; it is probable that the letter from Scott to the Observatory in April 1885 referred to in that note came about as the result of his 1885 paper [81]. To date, we have not found the daily records from 1880 and 1881, although fortunately

---

[*] The complete list of the original sites was given as follows: Scotland, North—Stornoway, Sandwick; Scotland, East—Aberdeen, Glenalmond; England, North-East—Durham, York, Kelstern (Lincolnshire); England, East—Cambridge, Hillington (Norfolk); Midland Counties—Leicester, Oxford; England, South—London (2), Kew, Southampton, Jersey; Scotland, West—Glasgow, Douglas, Silloth; England, North-West—Stonyhurst, Llandudno, Churchstoke; England, South-West—St Ann's Head, Falmouth, Cirencester; Ireland, North—Armagh, Markree; Ireland, South—Dublin, Parsonstown, Valencia [Valentia]. Sunshine records had commenced by the end of 1880 at all of these sites except Glenalmond, which (for whatever reason) did not establish a record.

Figure A6.1 *The original version of the Campbell–Stokes sunshine recorder, from Stokes [98], Figure 3*

monthly totals for 43 sites for the five years ending 1885 including Durham were given in Scott's 1885 paper, while we tracked down a few missing months in *Ten Years Sunshine in the British Isles 1881–1890* (Meteorological Office, M.O. 98, 1891). With the exception of December 1883, April to July and October to December 1884, and March to May 1885, the *monthly* Campbell–Stokes sunshine record is complete from June 1880 to December 1999, although the totals for some months are known only to the nearest hour up to and including May 1885. The *daily* record is unbroken from 29 May 1885 to the cessation of the record on 31 December 1999.

Exposure

Throughout, the instrument was mounted on the front pediment of the Observatory. In its early years, the exposure was unobstructed but by 1939 tree growth on the north-east side of the building had begun to obstruct early-morning summer sunshine, necessitating some pruning [19]. Site photographs from the mid-1950s (Figure A2.2) show an open site, but subsequent tree growth had once more begun to obstruct early-morning summer sunshine from the 1960s, as indicated by the gradual reduction in the duration of sunshine on the sunniest day in the year, by about an hour from the early 1960s to the 1980s and 1990s.

Estimates from regional values, 2000 onwards

Looking back, it is an enormous pity that the record was terminated at the end of 1999, after almost 120 years record. The current AWS includes a pyranometer (a global solar radiation sensor), which does provide a very crude estimate of sunshine duration, but the record is incompatible with the record from a Campbell–Stokes instrument. Subsequent unchecked tree growth has caused further sheltering of the site reducing the effective exposure of the radiation sensor, and accordingly the post-1999 AWS record has not been used in this book. We are hopeful that the Met Office will consider installing an electronic sunshine recorder on the Observatory roof, assuming tree growth can be cut back from its levels in 2021/22 and managed thereafter.

Rather than omit sunshine figures entirely for the period from 2000 onwards, it was decided to estimate ongoing monthly sunshine totals for Durham based upon regional figures from the Met Office's East and North-East (E&NE) areal series, which covers Northumberland to Lincolnshire. These are adjusted slightly by regression comparisons with the existing period of record at Durham from 1929 (the E&NE sunshine series commenced in 1919) to 1998 (shortly before the Observatory record terminated), although it must of course be borne in mind that the Durham values

Table A6.1 *Regression coefficients used to multiply Met Office regional E&NE monthly sunshine totals to obtain an estimate for Durham. Results based on 1929–1998 data. All the correlation coefficients (r^2) were ~ 0.99.*

	J	F	M	A	M	J	J	A	S	O	N	D
Adjustment factor	1.113	1.043	1.025	0.976	0.949	0.947	0.936	0.961	0.983	1.000	1.069	1.087

were themselves a component of the calculations used to derive these areal values up to 1999. The derived regression adjustment factors by month are shown in Table A6.1.

The E&NE regional sunshine dataset is available from the Met Office website at https://www.metoffice.gov.uk/research/climate/maps-and-data/uk-and-regional-series. To obtain a nominal Durham monthly sunshine duration, the areal value (in hours) is adjusted using the regression factors in Table A6.1 to generate an approximate monthly sunshine duration, rounded to the nearest hour: seasonal and annual values are then summed from the monthly estimate. Such estimates have been used in this book for all sunshine data from 2000 onwards. However, they are only estimates and carry an uncertainty of ± 5–10 per cent at best, particularly as there are very few Campbell–Stokes type sunshine recorders still operating within north-east England. Note that 'daily sunshine totals' listed on the Durham Observatory website originate from the (partly-sheltered) solar radiation sensor on the AWS, and as a result monthly sunshine totals from that source will differ from the more reliable estimate based upon the regional values, particularly during the summer months where shading of the pyranometer by trees is more significant than in the winter months.

Appendix 7
Wind records from Durham Observatory, 1866 to 2011

Records of wind direction and speed were made at the Observatory from July 1843 to December 1999. The records are a mixture of eye estimates and instrumental readings.

Table A7.1 lists the contents of the wind records as identified in the observational registers. It is a reasonable assumption that, during the periods for which wind speeds in Beaufort Force are shown, that these were manual estimates of both wind direction and speed. Beaufort Force is assumed in the early registers, although not explicitly stated as such until the 1850s; a note heading the January 1850 register columns states that wind speeds were observed on a scale 0–10, which may of course have simply been a pragmatic shortening of the Beaufort scale as Force 11 and 12 were unlikely to occur at an inland site. Instrumental records, once they became available, were presumably obtained from examination of the wind speed (and, later, wind direction) indicator mechanism from the mast-mounted anemometer during the 0900 and 2100 observations, or from subsequent analysis of chart recordings covering the time of the observations. Tables of wind speed equivalents for Beaufort Forces 0–12 are available in WMO codes ([165], page A-377). Wind direction is conventionally the direction from which the wind is blowing. Note, however, that owing to the slow variation of magnetic declination with time, wind directions given as compass bearings may differ slightly from those based upon degrees True, in which the variation of compass bearing has been corrected for. Calm winds were usually shown in the observation registers by zero for both wind direction and speed.

It is very unlikely that wind speeds observed over the period of record are homogeneous, and time series must be treated with caution.

Instruments and exposure

A *Robinson anemometer* was installed in late summer 1866, following parallel-running comparisons with similar instruments at Greenwich Observatory from 5 June to 17 August ([24], item Papers 226). This would have been the original four-cup instrument as developed at Armagh Observatory by Thomas Romney Robinson in the 1840s ([166], Chapter 6). Each cup was 30 cm in diameter and fixed to an iron cross, mounted such that the centres of the cups were 58 cm from the vertical spindle. The whole apparatus was mounted on a vertical mast some distance above the Observatory roof. Robinson's original calibration tests suggested that the centres of the cups moved with one-third of the velocity of the wind, although by the 1880s this assumption was shown to be incorrect. The factor was eventually reduced to 2.2 [167–169] and all previously measured wind speeds shown to be too high as a result. One example concerns the gale of 28 November 1892, when the Durham anemometer indicated a mean speed of 75 mph ([24], item 426); reducing this by 2.2/3 produces a more realistic 55 mph or 48 knots. There are regular references in the following 50 years to maintenance and repairs to the Robinson unit. The demise of the

Table A7.1 *Details of wind records in the twice-daily observational records from Durham Observatory, 1843 to 2011*

Period	Wind direction	Wind speed
July 1843 to Aug 1867	Compass bearing—compass points (N, NNE, NE etc)	Beaufort Force estimate
Sept 1867 to March 1916	*As above*	Instrumental value, in miles per hour *(but see note below re Robinson's anemometer)*
April 1916 to Dec 1957	*As above*	Beaufort Force estimate
Jan to Dec 1958	Degrees, to nearest 10 degrees, North as 360°	Beaufort Force estimate
Jan to Dec 1959	Compass bearing—compass points (N, NNE, NE etc)	Instrumental value, in knots
Jan to Dec 1960	Degrees, to nearest 10 degrees, North as 360°	Beaufort Force estimate
Jan 1961 to Feb 2011 *(record terminates)*	*As above*	Instrumental value, in knots *(Dines PTA record terminated 2011)*

Source: Digitized version of the original records [44].

Robinson anemometer is not known with certainty, but wind speeds reverted to Beaufort Force estimates from April 1916, and a letter in the Durham University Library archive dated 25 June 1917 (Papers, 863A) refers to the need for the Observatory to be fitted with a new Dines Pressure Tube Anemometer. However, it was to be another 20 years before this would come about.

'Run of wind' observations were entered in the observations registers at the Observatory from August 1885 to October 1916; it is assumed that this was a separate instrument to the Robinson anemometer (they are usually exposed at 2 m above ground level, rather than on a 10 m mast or above the Observatory roof), but it is possible the values were derived from the Robinson chart record. The 'run of wind' is simply the nominal distance travelled by air past the anemometer in any 24-hour period (usually read once daily at the 0900 GMT observation), the anemometer spindle geared in similar fashion to the odometer in a car. The mean daily wind speed, in the units of record, is obtained by dividing the 'run of wind' by 24. The units for this early period of record are miles per day.

An *electric cup anemometer* was installed in July 1926, but during the gale of 8 February 1928 'the anemometer mast collapsed whilst the 9 p.m. reading was being taken', according to a later note in the 1926 Met Office site inspection report. The wind speed and direction were noted at the time—presumably by the mildly traumatized observer—as 'south-westerly, Force 8'.

Figure A7.1 *The recording section of the Dines Pressure Tube Anemometer within Durham Observatory, during the 1990s. (Department of Geography, Durham University)*

A *Dines pressure tube anemograph* (PTA) [170] was installed in July 1937 (later note in 1926 inspection report); the height of the wind vane was given as 389 ft (119 m) AMSL, 16 m above ground level with an effective height of 10 m. Hourly tabulations of the chart record from this instrument were prepared and submitted to the Met Office, and published in the *Monthly Weather Report*, until at least 1979. These records still exist in the Met Office archive, although other than noteworthy mean or gust wind speeds specifically extracted from the *Monthly Weather Report*, few have been utilized in this book as they are not readily accessible via MIDAS. Chart records from this instrument were in miles per hour until late 1949, and thereafter in knots (Figure A7.1). The Dines anemometer and wind vane can be seen on the September 1991 Observatory photograph (Figure A7.2).

Figure A7.2 *Durham Observatory from the south, 6 September 1991, showing the Dines Pressure Tube Anemometer mounted above the roof (Stephen Burt)*

Figure A7.3 *Durham Observatory from the east, 22 June 2011, showing the two anemometer masts, the Dines Pressure Tube Anemometer towards the rear. This was taken just a few months after some of the equipment was damaged in a storm. Unfortunately, funding to replace the masts was not available but the existing structures have remained in place, although no measurements are currently possible (compare with Figure 2.6) (Tim Burt)*

A second period of 'daily run of wind' records exists from October 1982 to August 1997, units this time in kilometres per day. There are many days with missing or zero record. No information is available on the instrument providing the record. Usually, a cup counter anemometer would be sited 2 m above ground, close to the other instruments (thermometer screen and rain gauge), but site photographs such as Figure A7.3 taken in 2011 suggest that this anemometer was mounted above the dome of the observatory, at a height a few metres below the main anemometer head, and presumably read from within the observatory.

Wind speed readings were available from the AWS between October 1999 and 9 February 2011, when a stanchion supporting the mast on the Observatory roof broke and was not repaired or replaced by the Met Office. At the time of writing, wind speed and direction are no longer recorded at the Observatory: limited wind data are available from the Davis Instruments AWS on the roof of the West Building, about 800 metres to the east.

Appendix 8
Records of other meteorological elements at Durham Observatory

Records of various other elements are available from the digitized meteorological registers, as detailed below with the period for which they are available.

Cloud cover

Digitized records of cloud cover are available from July 1843 to August 1997. Units are tenths up to December 1948, thereafter oktas (eighths). In oktas coding, 9 is used to denote 'sky obscured' (by fog, snow, etc.). If numerical averages of cloud amounts are prepared, days with cloud cover = 9 oktas should be omitted from the analysis to avoid introducing bias.

Present weather

Observations of present weather using WMO ww code 4677 are available from January 1961 to August 1997. Details of the ww code can be found in WMO code books [165], page A-356.

Visibility

Although there are clear references to visibility points in site inspection reports as far back as December 1927, eye estimates of visibility were not included in the observation registers until January 1961, but thereafter appear until the cessation of manual records in September 1999.

The code used is as given in Table A8.1. It is possible that the records for 1961 refer to a different code as otherwise fog and low visibility observations appear abnormally frequent in that year, a conclusion not supported by local sites. For this reason, fog statistics (visibility code 3 or less) have been taken only over the period 1962–97, taking 'fog' as reported code 3 or less, and 'thick fog' as code 1 or less.

An earlier version (dated 15 December 1927) can be found in the Met Office site inspection reports and is shown in Table A8.2. Although no visibility estimates were noted in the observation registers until 1961, it may have been used for supplementary observations recorded elsewhere, which have not yet been catalogued.

Table A8.1 *Visibility code objects and distances as used at Durham Observatory 1999*

Code	Distance	Description	Visibility point
X	0–19 m	Dense fog	Screen visible from Observatory
E	20–39 m	Dense fog	Observatory Cottage visible from Observatory
0	40–99 m	Thick fog	Trees in drive visible from Observatory
1	100–199 m	Thick fog	Gate to Observatory on Potters Bank visible
2	200–399 m	Fog	St Aidan's College visible from Observatory
3	400–999 m	Moderate fog	Grey College visible from Observatory
4	1000–1999 m	Very poor visibility	Whinney Hill School visible from Observatory
5	2000–3999 m	Poor visibility	Gilesgate Moor Estate visible
6	4–9 km	Moderate visibility	Littletown spoil heap visible
7	10–19 km	Good visibility	Burnhope Mast visible from Observatory

There were no objects or features beyond point 7 visible from the Observatory, and thus this was the maximum visibility reported.

Table A8.2 *Visibility code objects and distances—Durham Observatory, 1927*

Code	Standard distance	Actual distance	Description of object
A	27 yards	27 yards	Screen from front door
B	55 yards	55 yards	Group of trees in field
C	110 yards	110 yards	West side of Observatory House
D	220 yards	210 yards	West House, main building
E	550 yards	600 yards	Neville's Cross Mission (St John's Church)
F	1100 yards	1080 yards	Castle Keep (close to Palace Green)
G	1 ¼ miles	1 ¼ miles	St Giles Church
H	2 ½ miles	2 ½ miles	Stob House ? [sic]
i	*4 ⅔ miles*	-	*No object*
J	6 ¼ miles	5 ⅔ miles	Ferry Hill
K	12 ½ miles	11 miles	Gateshead Fell
l	*18 ⅔ miles*	-	*No object*
m	*31 miles*	-	*No object*

State of ground

Coded observations of 'state of ground' are included in the observations registers from January 1961 to August 1997. Two distinct codes appear to have been used, from 1961 to 1970 inclusive, and from 1971 to the end of the record in August 1997. The latter code is reproduced as Table A8.3. This element has not been analysed—occasions of snow lying have been flagged instead, using snow depth reports.

Snow depth

Snow depths at the 0900 GMT observation are included in the observations register from January 1960 to the cessation of observations in September 1999: units inches until December 1969, thence cm. In the online database, inch readings have been converted to cm. This element was used to define 'mornings with snow cover' (snow depth > 0) and analyse the distribution of snow depths over the available period of record.

Between January 1961 and December 1977 the observations registers contain a column headed 'Depth of fresh snowfall', with units inches to December 1969 and thence cm. It was never very clear what was intended for this column, and it was little used in the Durham records.

Table A8.3 *State of ground code in use at Durham, 1971–97*

Code	Description
0	Surface of ground dry (no appreciable amount of dust or loose sand)
1	Surface of ground moist
2	Surface of ground wet (standing water in small or large pools on surface)
3	Surface of ground frozen
4	Glaze or ice on ground (but no snow or melting snow)
5	Snow or melting snow, snow grains, snow pellets, or ice pellets (with or without ice on the ground) covering less than one half of the ground
6	Snow or melting snow, snow grains, snow pellets, or ice pellets (with or without ice on the ground) covering more than one half of the ground but not completely
7	Snow or melting snow, snow grains, snow pellets, or ice pellets (with or without ice on the ground) covering ground completely
8	Loose dry snow, dust, or sand, covering more than one half of ground but not completely
9	Loose dry snow, dust, or sand, covering ground completely

Codes 0–3 apply to the bare soil plot, while codes 4–9 refer to ground representative of the station. Source: Met Office Instructions for completing monthly returns on Metform 3208, January 1973 edition.

Appendix 9
Climatological averages and extremes for Durham Observatory, period 1981–2010

Durham Observatory

	Site details				
Averages for period 1981–2010	Record began 1841	Lat 54.77 °N	Long 1.59 °W	NGR NZ 267 415	Altitude AMSL 102 m
Extremes for period 1981–2010		Ob time AWS	Temperature Rainfall		0900–0900 GMT 0900–0900 GMT

TEMPERATURE °C	Jan	Feb	Mar	Apr	May	Jun	Jul	Aug	Sep	Oct	Nov	Dec	Annual	Yrs
Mean daily maximum	6.6	7.2	9.5	11.9	15.0	17.6	20.1	19.8	17.1	13.3	9.4	6.7	12.9	30
Mean daily minimum	0.9	0.9	2.3	3.7	6.2	9.0	11.1	11.0	9.0	6.3	3.4	1.1	5.4	30
Mean temperature	3.7	4.1	5.9	7.8	10.6	13.3	15.6	15.4	13.1	9.8	6.4	3.9	9.1	30
Highest maximum	14.4	16.8	20.0	24.1	29.0	29.2	30.5	32.5	26.7	24.1	17.1	15.1	32.5	30
Year	2003	1993	1990	2003	2001	1989	2006	1990	2006	1985	1983	1991	3 Aug 1990	
Lowest minimum	−16.1	−11.8	−10.8	−4.4	−2.5	−0.6	3.1	2.7	−0.5	−3.8	−12.0	−12.5	−16.1	30
Year	1982	1986	2001	1999	2010	1991	1990	1999	1991	1997	1993	1981	11 Jan 1982	
Lowest grass min	−14.1	−12.0	−11.1	−8.4	−6.3	−1.7	0.6	−0.9	−4.1	−9.1	−13.5	−14.5	−14.5	30
Year	1982	1986	2009	1996	1984	1987	1995	1986	1995	1997	1993	1981	18 Dec 1981	
Highest minimum	9.8	11.1	11.1	12.5	14.6	16.8	17.5	17.9	16.3	15.9	12.5	11.0	17.9	30
Year	2003	2004	1981	1994	2004	2005	2001	2001,06	1999	1985	2003,07	1994	2001,2006	
Lowest maximum	−4.5	−3.5	0.3	3.6	6.5	8.3	12.6	11.5	10.2	6.4	0.2	−1.8	−4.5	30
Year	1982	1986	1986	1986	1997	2009	1987	1986	1991	1992	2010	1985	7 Jan 1982	
Air frosts	11.3	11.1	6.8	4.0	0.8	0.0	0	0	0.0	1.6	4.8	11.0	51.5	30
Ground frosts	19.0	17.5	14.9	11.1	4.0	0.5	0	0.1	1.4	5.5	12.1	17.9	103.8	30

PRECIPITATION mm	Jan	Feb	Mar	Apr	May	Jun	Jul	Aug	Sept	Oct	Nov	Dec	Annual	Yrs
Monthly mean	53.8	42.8	44.6	52.7	44.2	55.5	54.9	60.7	54.7	61.4	69.2	59.4	653.8	30
Days ≥ 0.2mm	18.4	15.1	15.6	14.8	14.1	14.6	13.2	14.0	14.3	17.2	17.7	18.3	187.3	30
Days ≥ 1.0 mm	11.8	9.3	9.7	9.5	9.2	9.8	9.2	9.6	9.3	11.3	12.0	11.6	122.2	30
Wettest day	24.5	23.0	31.7	43.0	41.0	54.0	44.4	69.1	45.8	38.1	43.2	28.3	69.1	30
Year	1984	2002	1981	1992	1993	2000	2009	1986	1992	1981	1995	1983	25 Aug 1986	
Wettest month	113.4	103.6	113.5	151.0	102.4	174.4	168.8	169.1	117.6	120.2	153.3	98.1	887.3	30
Year	2008	2001	1981	1998	1983	1997	2009	1986	1993	2004	2010	1983	2000	
Driest month	7.2	5.7	10.1	5.8	7.7	8.6	10.0	14.2	11.3	13.0	18.2	19.8	415.9	30
Year	1989	1985	1993	1997	1989	1996	2006	1995	1997	2007	2004	2004	1989	

Continued

Appendix 9 (*Continued*)

Monthly and annual averages for the period 1981–2010 for Durham Observatory

SUNSHINE *hours*	Jan	Feb	Mar	Apr	May	June	July	Aug	Sept	Oct	Nov	Dec	Annual	Yrs
Monthly mean	60	80	115	147	180	164	174	168	132	102	68	52	1441	30
Daily mean	*1.93*	*2.84*	*3.70*	*4.89*	*5.82*	*5.45*	*5.61*	*5.41*	*4.41*	*3.29*	*2.27*	*1.67*	*3.95*	*30*
Daylight hours	*246*	*274*	*368*	*424*	*500*	*516*	*517*	*461*	*381*	*325*	*254*	*227*	*4494*	
% possible	*24.3*	*29.2*	*31.1*	*34.6*	*36.1*	*31.7*	*33.6*	*36.4*	*34.7*	*31.4*	*26.9*	*22.8*	*32.1*	
Days nil sunshine*	12.2	8.7	6.1	4.7	3.9	3.3	2.1	3.1	3.7	6.7	8.7	12.7	75.8	29
Sunniest day*	7.6	10.2	11.8	13.7	14.7	14.6	14.5	13.7	12.7	10.2	8.1	7.0	14.7	29
Year	1987,99	1995	1997	1990	1989	1989,95	1989	1995	1994	1980	1985	1997	26 May 1989	
Sunniest month	86	118	172	212.7	246.0	224	273	249.3	192.7	146.1	107	77.4	1703	30
Year	2000	2008	2003	1990	1989	1989	2006	1995	1986	1981	2005	1999	2003	
Dullest month	7.6	52.3	29.3	92.2	104.4	89.4	122	106	76.2	51.6	33.0	25.4	1229.5	30
Year	1996	1991	1996	1986	1983	1987	2000	2008	1993	1982	1997	1987	1987	

DAYS WITH …	Jan	Feb	Mar	Apr	May	June	July	Aug	Sept	Oct	Nov	Dec	Annual	Yrs
Snow/sleet falling	*No data available*													
Snow lying 0900 GMT*	5.0	4.3	1.6	0.3	0	0	0	0	0	0	1.1	2.0	14.3	30
Thunder heard	*No data available*													

Site and Instrument metadata

A well-exposed site, environment little changed since 1841 other than increasing shelter by trees. AWS records since October 1999 Sunshine records ceased Dec 1999, records since are estimated from regional series to 1h precision only.
* Period 1971–99 only.

Appendix 10
Climatological averages and extremes for Durham Observatory, period 1991–2020

Durham Observatory

Averages for period 1991–2020

Extremes for period 1991–2020

Site details

Record began 1841

Lat 54.77° N   Long 1.59° W

Altitude AMSL 102 m

NGR NZ 267 415

Ob time AWS

Temperature / Rainfall

0900–0900 GMT
0900–0900 GMT

TEMPERATURE °C	Jan	Feb	Mar	Apr	May	Jun	Jul	Aug	Sep	Oct	Nov	Dec	Annual	Yrs
Mean daily maximum	6.9	7.7	9.9	12.5	15.5	18.0	20.2	19.8	17.3	13.5	9.7	7.1	13.2	30
Mean daily minimum	1.3	1.4	2.5	4.0	6.5	9.3	11.3	11.3	9.2	6.5	3.6	1.4	5.7	30
Mean temperature	4.1	4.6	6.2	8.3	11.0	13.6	15.8	15.6	13.3	10.0	6.6	4.2	9.4	30
Highest maximum	14.4	17.4	21.8	24.1	29.0	28.1	32.9	30.3	27.3	25.3	18.5	15.9	32.9	30
Year	2008	2012	2012	2003	2001	2000	2019	1995	2016	2011	2020	2015	25 Jul 2019	
Lowest minimum	−8.9	−8.5	−10.8	−4.4	−2.5	−0.6	4.1	1.9	−0.5	−3.8	−12.0	−10.4	−12.0	30
Year	2010	1991 1994	2001	1999	2010	1991	1999	2012	1991	1997	1993	2010	24 Nov 1993	
Lowest grass min	−11.8	−11.1	−11.1	−8.4	−6.2	−1.6	0.6	−0.7	−4.1	−9.1	−13.5	−12.7	−13.5	30
Year	2009	1991	2009	1996	2010	1998	1995	1992	1995	1997	1993	2010	24 Nov 1993	
Highest minimum	10.6	11.1	10.8	12.5	14.6	16.8	18.9	17.9	17.2	15.2	13.7	11.3	18.9	30
Year	2016	2004	2017	1994	2004	2005	2016	2001 2006	2016	2011	2015	2015	20 Jul 2016	
Lowest maximum	−0.8	−0.8	−0.1	4.0	6.5	8.3	12.8	12.7	10.1	6.4	0.2	−1.5	−1.5	30
Year	2013	2018	2018	1999	1997	2009	1992	2020	2020	1992	2010	2010	17 Dec 2010	
Air frosts	10.2	9.7	6.5	3.2	0.8	0.03	0	0	0.03	1.2	4.6	10.4	46.6	30
Ground frosts	16.9	15.5	13.8	9.7	3.5	0.4	0	0.03	0.8	4.7	11.1	16.1	92.4	30

PRECIPITATION *mm*	Jan	Feb	Mar	Apr	May	Jun	Jul	Aug	Sep	Oct	Nov	Dec	Annual	Yrs
Monthly mean	53.4	45.3	41.3	51.2	44.5	61.2	62.0	66.2	56.8	63.7	73.4	61.2	680.2	30
Days ≥ 0.2 mm	19.3	16.2	15.3	14.2	14.2	14.9	15.0	15.6	14.9	18.1	19.1	19.2	196.1	30
Days ≥ 1.0 mm	11.9	9.9	8.6	9.2	8.6	10.0	10.7	10.2	9.5	11.9	12.0	12.1	124.5	30
Wettest day	28.6	23.0	20.0	43.0	41.0	54.0	44.4	52.2	57.4	27.4	48.0	33.6	57.4	30
Year	2016	2002	2005 2013	1992	1993	2000	2009	2004	2012	2012	2012	2015	24 Sep 2012	
Wettest month	116.6	103.6	92.7	151.0	100.8	174.4	168.8	156.8	117.6	120.2	153.3	120.2	1018.0	30
Year	2016	2001	1999	1998	2013	1997	2009	2004	1993	2004	2010	2015	2012	
Driest month	12.0	6.1	10.1	3.8	15.0	8.6	10.0	14.2	11.3	13.0	18.2	19.8	479.5	30
Year	2019	1998	1993	2020	2001	1996	2006	1995	1997	2007	2004	2004	2003	

Continued

Appendix 10 (*Continued*)

Monthly and annual averages for the period 1991–2020 for Durham Observatory

SUNSHINE *hours*	Jan	Feb	Mar	Apr	May	Jun	Jul	Aug	Sep	Oct	Nov	Dec	Annual	Yrs
Monthly mean	62	85	120	154	187	166	175	168	133	99	68	57	1473	30
Daily mean	*1.99*	*3.02*	*3.89*	*5.14*	*6.03*	*5.52*	*5.65*	*5.41*	*4.42*	*3.18*	*2.27*	*1.84*	*4.03*	*30*
Daylight hours	*246*	*274*	*368*	*424*	*500*	*516*	*517*	*461*	*381*	*325*	*254*	*227*	*4494*	
% possible	*25.1*	*31.1*	*32.7*	*36.4*	*37.4*	*32.1*	*33.9*	*36.3*	*34.7*	*30.3*	*26.8*	*25.0*	*32.8*	
Days nil sunshine*	12.2	8.7	6.1	4.7	3.9	3.3	2.1	3.1	3.7	6.7	8.7	12.7	75.8	29
Sunniest day*	7.6	10.2	11.8	13.7	14.7	14.6	14.5	13.7	12.7	10.2	8.1	7.0	14.7	29
Year	1987 1999	1995	1997	1990	1989	1989 1995	1989	1995	1994	1980	1985	1997	26 May 1989	
Sunniest month	86	128	180	219	256	217.0	273	249.3	171.2	129	107	84	1703	30
Year	2000	2019	2012	2015	2020	1996	2006	1995	1997	2003	2005	2014	2003	30
Dullest month	7.6	52.3	29.3	104	124.3	102	122	106	76.2	58	33.0	26.6	1230.0	30
Year	1996	1991	1996	2012	1991	2012	2000	2008	1993	2005	1997	1997	1993	30

DAYS WITH …	Jan	Feb	Mar	Apr	May	Jun	Jul	Aug	Sep	Oct	Nov	Dec	Annual	Yrs
Snow/sleet falling	No data available													
Snow lying 0900 GMT*	5.0	4.3	1.6	0.3	0	0	0	0	0	0	1.1	2.0	14.3	29
Fog at 0900 GMT‡	2.8	2.8	1.9	1.0	0.6	0.3	0.3	0.3	1.0	2.9	2.2	2.3	18.6	27

Site and Instrument metadata

A well-exposed site, environment little changed since 1841 other than increasing shelter from trees. Grass min exposure changed to 'AstroTurf' July 2015. AWS records since Oct 1999; sunshine records ceased Dec 1999, records since are estimated from reginal series to 1 h precision only.

* Period 1971–99 only; ‡ Period 1971–97 only.

Appendix 11
Monthly and annual summaries of Durham's weather by year, 1843–2021

This Appendix contains monthly and annual summaries of temperature, rainfall and sunshine for Durham Observatory for each month and year from 1843 to 2021 for which data are available.

Each table is headed with the year followed by monthly totals or means for various elements as set out in Table A11.1. Where no data are available (for example, prior to the start of sunshine records in June 1880) the relevant columns are greyed out or left blank. Details of the available period of record by element are given in Appendix 2. Metadata relating to each of the elements included in the tables (temperature, precipitation, and sunshine) are included in Appendices 3–6. Temperature records are from a north wall exposure until January 1860, approximate Stevenson screen equivalent values from records originally made in a Glaisher stand from January 1860 to October 1899, and since November 1899 in a Stevenson screen (Appendix 3). Manuscript daily rainfall records are available from 1843 [29], but the rainfall record is unreliable in comparison with the few neighbouring sites prior to 1868, and possibly as late as 1871, and are therefore not included until 1868 (Appendix 4). Sunshine data commenced in June 1880, with some gaps until 1885, but records ceased in December 1999 and monthly totals since have been estimated from areal values (Appendix 6).

Records are normally quoted to one decimal place. Sunshine records for 1880–81 are known only as monthly totals, to the nearest hour. Sunshine records from 2000 onwards are estimated from areal values, and thus are also given only to the nearest hour. If a value has been estimated, usually because of missing data or doubts about instrumental accuracy, the value is given in brackets. Monthly records are normally left blank where there are more than seven days of data missing for the month. Where possible, monthly extremes are quoted based on period of data available, although that may be less than the full month. Appendix 2 lists the main periods of missing data in the datasets used.

The period of the 'daily' values (the 'terminal hour') varies slightly across the dataset: details where known, and any corrections applied, are given by element in Appendix 2. Since 1961, daily temperature and precipitation measurements refer to the period 0900 to 0900 GMT. Sunshine duration refers to the 'civil day', midnight to midnight.

Monthly and annual totals and means presented here and elsewhere in this volume may disagree slightly owing to rounding errors. Continuing work on digitizing the Durham dataset (and re-digitizing to greater precision, for instance to 0.1 degF temperatures from 1 degF) may lead to slight inconsistencies over time between values in these tables and the daily, monthly, and annual values published online.

For annual values, the same terminology and terms apply, except that the totals and means relate to the calendar year, and the date of each of the extremes is replaced by the number of the month of occurrence (1 = January, 2 = February, etc.): reference back to the month line will give the date of the event within the month, from which the date of the annual extreme can easily be determined. In the case of two or more months having the same extreme value, only the first month value is shown.

Table A11.1 *Explanation of column headings in the monthly and annual tables*

Column abbreviation	Value	Unit
Month	Month name	
Mean max	Mean daily maximum temperature	°C
Mean min	Mean daily minimum temperature	°C
Mean temp	Mean daily temperature (average of mean daily max and mean daily min)	°C
Anom	Difference ('anomaly') of the mean daily temperature from the 1991–2020 normal (Appendix 10); negative value indicates a value below the current normal	°C
Highest max	The highest daily maximum temperature during the month (warmest day)	°C
Date	The date of the highest daily maximum during the month. Where the extreme occurred on more than one date during the month, only the first date is shown. (This is more likely between June 1918 and October 1971, when temperatures were read only to 1 degF resolution.)	
Lowest min	The lowest daily minimum temperature during the month (coldest night)	°C
Date	The date of the lowest daily minimum temperature during the month. Where the extreme occurred on more than one date during the month, only the first date is shown.	
Lowest max	The lowest daily maximum temperature during the month (coldest day)	°C
Date	The date of the lowest daily maximum during the month. Where the extreme occurred on more than one date during the month, only the first date is shown.	
Highest min	The highest daily minimum temperature during the month (warmest night)	°C
Date	The date of the highest daily minimum during the month. Where the extreme occurred on more than one date during the month, only the first date is shown.	
Air frost	Count of the number of days in the month when the minimum temperature fell to −0.1 °C or below (or to 0.0 °C/32 °F or below for the period when read only to 1 °F resolution)	Count
Days ≥ 25 °C	Count of the number of days in the month when the maximum temperature reached 25.0 °C or above ('hot days')	Count
Total pptn	Total precipitation during the month, from 1868 onwards; this is the sum of the daily values	mm
Anom	Total precipitation during the month as a percentage of the 1991–2020 normal (Appendix 10); > 100 indicates a wetter month than the current normal	%
Rain days	Count of the number of days in the month when 0.2 mm or more of rainfall was recorded, since 1868. See comments in Appendix 4 regarding the reliability of the rain day count in the early years of the record	Count
Wettest day	The amount of precipitation on the wettest day in the month	mm
Date	The date of the wettest day during the month. Where the extreme occurred on more than one date during the month, only the first date is shown.	
Total sunshine	Total duration of bright sunshine during the month; from 2000, this is an estimate to the nearest hour based upon the Met Office England East and North-East areal sunshine series (see Appendix 6 for details)	hours
Anom	Total sunshine duration during the month as a percentage of the 1991–2020 normal (Appendix 3); > 100 indicates a sunnier month than the current normal	
Sunniest day	The duration of bright sunshine on the sunniest day in the month (not available for 1880–81 or since 2000)	hours
Date	The date of the sunniest day during the month. Where the extreme occurred on more than one date during the month, only the first date is shown.	

DURHAM 1843

Month	Mean max	Mean min	Mean temp	Anom	Highest max	Date	Lowest min	Date	Lowest max	Date	Highest min	Date	Air frost	Days ≥ 25 °C	Total pptn	Anom	Rain days	Wettest day	Date	Total sunshine	Anom	Sunniest day	Date
Jan																							
Feb																							
Mar																							
Apr																							
May																							
June																							
July	*Record commenced 23 July 1843*																						
Aug	18.6	10.6	14.6	−1.0	23.4	17	5.7	22	13.9	22	15.3	8	0	0									
Sep	18.4	10.8	14.6	+1.4	24.0	8	2.4	29	10.7	26	16.0	3	0	0									
Oct	10.8	4.6	7.7	−2.3	18.9	1	−2.1	15	5.6	30	13.9	1	7	0									
Nov	8.3	2.7	5.5	−1.1	12.2	28	−2.1	25	3.6	9	8.2	27	6	0									
Dec	9.6	5.0	7.3	+3.1	13.3	7	−1.4	2	6.4	1	11.2	24	1	0									
1843																							

DURHAM 1844

Month	Mean max	Mean min	Mean temp	Anom	Highest max	Date	Lowest min	Date	Lowest max	Date	Highest min	Date	Air frost	Days ≥ 25 °C	Total pptn	Anom	Rain days	Wettest day	Date	Total sunshine	Anom	Sunniest day	Date
Jan	6.3	1.5	3.9	−0.1	11.4	29	−7.2	3	1.4	15	7.8	28	10	0									
Feb	4.7	−1.3	1.7	−2.9	9.4	15	−6.3	27	1.8	21	4.3	17	22	0									
Mar	8.1	1.0	4.6	−1.6	15.2	29	−5.6	18	2.3	16	6.4	27	11	0									
Apr	13.9	5.0	9.5	+1.2	17.9	9	0.8	28	9.8	4	8.9	15	0	0									
May	12.8	5.6	9.2	−1.8	19.2	9	−0.1	18	8.0	17	11.8	14	1	0									
June	16.0	9.4	12.7	−0.9	23.7	23	4.3	4	8.8	19	13.6	13	0	0									
July	17.5	10.0	13.7	−2.1	25.8	23	5.7	8	12.2	4	15.2	25	0	1									
Aug	16.7	9.4	13.1	−2.5	21.9	31	4.4	4	13.6	23	14.1	20	0	0									
Sep	16.1	9.7	12.9	−0.4	23.6	1	3.2	22	11.2	23	13.7	6	0	0									
Oct	11.9	5.7	8.8	−1.2	18.1	3	−0.9	23	8.3	19	12.3	13	2	0									
Nov	9.3	4.3	6.8	+0.2	18.4	2	−2.4	26	5.8	25	9.4	16	3	0									
Dec	3.7	−0.2	1.8	−2.5	7.6	31	−5.9	7	0.4	13	3.6	19	16	0									
1844	11.4	5.0	8.2	−1.2	25.8	7	−7.2	1	0.4	12	15.2	7	65	1									

DURHAM 1845

Month	Mean max	Mean min	Mean temp	Anom	Highest max	Date	Lowest min	Date	Lowest max	Date	Highest min	Date	Air frost	Days ≥ 25 °C	Total pptn	Anom	Rain days	Wettest day	Date	Total sunshine	Anom	Sunniest day	Date
Jan	5.5	0.7	3.1	−0.9	10.9	23	−10.0	30	−0.4	30	7.8	6	11	0									
Feb	4.3	−2.0	1.2	−3.4	8.5	21	−6.8	25	−0.8	12	3.0	5	20	0									
Mar	5.7	−0.6	2.5	−3.6	13.3	23	−9.6	13	−0.7	4	7.5	28	17	0									
Apr	11.3	2.6	6.9	−1.3	17.2	25	−2.9	6	5.7	10	8.1	26	2	0									
May	10.9	5.5	8.2	−2.8	18.6	16	2.4	9	6.9	24	14.0	16	0	0									
June	18.1	10.1	14.1	+0.5	26.3	16	5.6	30	13.2	28	15.4	9	0	2									
July	17.0	9.6	13.3	−2.5	21.7	7	5.0	1	11.9	22	12.9	27	0	0									
Aug	16.2	9.6	12.9	−2.7	20.6	30	6.2	22	13.3	16	13.1	31	0	0									
Sep	14.5	7.0	10.8	−2.5	20.6	1	0.8	23	10.4	3	13.1	2	0	0									
Oct	12.1	6.4	9.2	−0.8	15.8	19	−0.7	6	9.1	4	12.0	18	1	0									
Nov	9.3	4.7	7.0	+0.4	13.3	26	−0.7	24	3.0	24	9.4	8	3	0									
Dec	6.3	1.0	3.7	−0.6	10.0	29	−3.3	14	2.0	21	5.4	15	10	0									
1845	10.9	4.6	7.7	−1.7	26.3	6	−10.0	1	−0.8	2	15.4	6	64	2									

DURHAM 1846

Month	Mean max	Mean min	Mean temp	Anom	Highest max	Date	Lowest min	Date	Lowest max	Date	Highest min	Date	Air frost	Days ≥ 25 °C	Total pptn	Anom	Rain days	Wettest day	Date	Total sunshine	Anom	Sunniest day	Date
Jan	7.8	2.4	5.1	+1.0	11.7	25	−2.7	16	3.6	13	5.3	8	5	0									
Feb	8.5	3.7	6.1	+1.5	13.9	24	−2.9	10	1.1	9	9.4	25	4	0									
Mar	7.9	1.5	4.7	−1.5	12.9	10	−9.4	21	−1.9	18	9.1	4	9	0									
Apr	9.1	3.4	6.2	−2.0	16.3	12	−1.1	29	3.9	6	8.2	12	2	0									
May	14.9	6.8	10.8	−0.2	21.7	29	3.1	16	9.4	4	10.3	23	0	0									
June	22.6	12.3	17.5	+3.9	28.5	6	5.2	26	15.6	20	15.1	19	0	11									
July	19.6	12.4	16.0	+0.2	28.0	5	9.1	8	13.0	24	16.3	28	0	1									
Aug	19.3	11.5	15.4	−0.2	24.1	3	3.6	26	16.1	24	16.0	8	0	0									
Sep	18.4	9.2	13.8	+0.5	23.4	7	1.6	29	12.7	28	13.0	6	0	0									
Oct	12.4	5.7	9.0	−1.0	16.9	1	−1.8	29	7.3	28	12.2	6	3	0									
Nov	8.9	4.3	6.6	−0.0	13.3	3	−3.9	30	0.7	30	9.4	4	4	0									
Dec	2.9	−1.7	0.6	−3.6	7.4	8	−6.4	27	−2.8	14	4.4	9	22	0									
1846	12.7	6.0	9.3	−0.1	28.5	6	−9.4	3	−2.8	12	16.3	7	49	12									

DURHAM 1847

Month	Mean max	Mean min	Mean temp	Anom	Highest max	Date	Lowest min	Date	Lowest max	Date	Highest min	Date	Air frost	Days ≥ 25 °C	Total pptn	Anom	Rain days	Wettest day	Date	Total sunshine	Anom	Sunniest day	Date
Jan	3.0	0.0	1.5	−2.6	7.4	26	−4.2	21	−0.4	12	4.2	6	18	0									
Feb	4.2	−0.4	1.9	−2.7	10.4	21	−8.2	9	−1.8	8	7.8	18	17	0									
Mar	8.8	1.2	4.7	−1.5	15.8	18	−6.3	11	1.2	10	6.7	17	9	0									
Apr	–	1.7	(5.6)	(−2.7)	–	–	−2.3	2	–	–	5.9	12	10	–									
May	–	6.6	(10.4)	(−0.6)	–	–	0.9	3	–	–	11.6	29	0	–									
June	–	9.0	(13.1)	(−0.5)	–	–	4.9	7	–	–	14.5	29	0	–									
July	20.4	11.6	16.0	+0.2	28.4	15	5.6	18	15.3	17	16.8	11	0	2									
Aug	18.1	9.1	13.6	−1.9	23.6	1	1.1	24	12.8	23	14.7	12	0	0									
Sep	14.5	6.3	10.4	−2.9	19.9	22	−1.2	27	10.8	26	12.5	10	1	0									
Oct	11.8	6.1	9.0	−1.0	16.7	12	1.9	26	8.3	5	11.8	11	0	0									
Nov	10.0	4.7	7.4	+0.7	15.7	8	−3.8	27	4.1	17	12.0	8	3	0									
Dec	6.2	2.6	4.4	+0.2	12.7	9	−0.8	8	0.6	29	7.2	3	8	0									
1847		4.9	8.2	−1.2	28.4	7	−8.2	2	−1.8	2	16.8	7	66	2									

DURHAM 1848

Month	Mean max	Mean min	Mean temp	Anom	Highest max	Date	Lowest min	Date	Lowest max	Date	Highest min	Date	Air frost	Days ≥ 25 °C	Total pptn	Anom	Rain days	Wettest day	Date	Total sunshine	Anom	Sunniest day	Date
Jan	2.7	−1.3	0.7	−3.4	10.1	3	−15.7	29	−2.9	28	6.6	4	23	0									
Feb	7.4	2.2	4.8	+0.2	11.3	5	−4.4	1	2.7	24	8.7	5	5	0									
Mar	7.6	2.0	4.8	−1.4	14.0	30	−2.9	4	3.3	21	8.2	31	6	0									
Apr	10.1	2.9	6.5	−1.8	20.6	3	−4.2	10	5.6	7	9.3	3	5	0									
May	18.4	7.8	13.1	+2.1	24.7	13	−0.8	1	14.1	14	14.3	13	1	0									
June	17.0	8.8	12.9	−0.7	23.7	21	4.9	26	13.5	24	13.0	22	0	0									
July	19.0	10.2	14.6	−1.2	26.4	12	4.9	5	13.4	1	15.3	12	0	3									
Aug	16.6	7.9	12.2	−3.3	20.0	6	1.8	14	13.4	14	13.2	27	0	0									
Sep	15.6	8.0	11.8	−1.4	22.4	5	1.6	13	11.1	27	12.2	8	0	0									
Oct	11.5	5.8	8.6	−1.4	20.9	7	−1.3	30	3.2	18	12.6	5	3	0									
Nov	7.2	2.6	4.9	−1.7	11.9	28	−4.4	5	0.8	4	8.4	29	6	0									
Dec	6.7	2.3	4.5	+0.3	14.3	15	−5.9	22	0.0	21	10.2	13	5	0									
1848	11.6	4.9	8.3	−1.2	26.4	7	−15.7	1	−2.9	1	15.3	7	54	3									

DURHAM 1849

Month	Mean max	Mean min	Mean temp	Anom	Highest max	Date	Lowest min	Date	Lowest max	Date	Highest min	Date	Air frost	Days ≥ 25 °C	Total pptn	Anom	Rain days	Wettest day	Date	Total sunshine	Anom	Sunniest day	Date
Jan	5.6	0.6	3.1	−0.9	11.2	21	−9.4	3	−2.3	2	9.0	25	10	0									
Feb	8.6	2.3	5.5	+0.9	13.3	16	−5.0	27	3.3	1	8.2	19	8	0									
Mar	7.9	1.7	4.8	−1.3	13.1	15	−4.4	10	2.1	9	7.2	15	6	0									
Apr	8.2	2.1	5.2	−3.1	16.0	29	−4.3	17	1.1	17	7.6	30	7	0									
May	13.7	6.6	10.1	−0.9	20.7	29	0.0	12	7.1	10	10.7	31	0	0									
June	16.4	7.7	12.1	−1.5	21.8	4	1.7	10	10.4	10	11.9	5	0	0									
July	18.8	9.1	14.0	−1.8	24.8	7	5.8	1	15.7	23	12.8	23	0	0									
Aug	18.3	10.3	14.3	−1.2	24.2	8	4.0	19	12.4	3	15.0	8	0	0									
Sep	15.0	8.9	11.9	−1.3	22.0	3	3.1	19	9.5	30	14.2	1	0	0									
Oct	10.9	4.2	7.5	−2.5	16.7	19	−0.9	5	7.7	4	12.2	25	2	0									
Nov	7.5	2.6	5.0	−1.6	13.1	9	−5.2	29	0.6	28	11.1	11	8	0									
Dec	5.3	1.5	3.4	−0.9	9.5	16	−6.1	5	0.8	4	6.2	15	7	0									
1849	11.4	4.8	8.1	−1.4	24.8	7	−9.4	1	−2.3	1	15.0	8	48	0									

DURHAM 1850

Month	Mean max	Mean min	Mean temp	Anom	Highest max	Date	Lowest min	Date	Lowest max	Date	Highest min	Date	Air frost	Days ≥ 25 °C	Total pptn	Anom	Rain days	Wettest day	Date	Total sunshine	Anom	Sunniest day	Date
Jan	2.2	−2.0	0.1	−3.9	8.8	28	−8.0	22	−2.2	21	2.6	26	26	0									
Feb	8.5	3.3	5.9	+1.3	12.8	15	−2.2	13	1.3	13	8.3	9	3	0									
Mar	7.6	1.1	4.3	−1.8	14.1	7	−6.0	26	2.3	24	7.2	2	13	0									
Apr	10.4	4.3	7.3	−0.9	15.5	19	0.3	29	6.6	15	8.3	9	0	0									
May	12.7	4.5	8.6	−2.4	20.5	30	−2.7	7	5.8	9	10.2	28	5	0									
June	18.6	9.3	13.9	+0.3	24.8	24	2.1	16	10.6	15	13.7	25	0	0									
July	18.1	10.0	14.0	−1.7	24.8	22	4.9	10	13.5	28	13.9	13	0	0									
Aug	17.3	8.7	13.0	−2.5	21.8	2	1.4	22	11.9	30	14.4	16	0	0									
Sep	14.8	6.9	10.8	−2.4	19.7	1	1.2	9	11.1	8	13.2	2	0	0									
Oct	10.4	4.0	7.2	−2.8	15.1	3	−2.2	27	6.2	24	10.6	19	4	0									
Nov	8.7	3.9	6.3	−0.3	14.2	11	−7.7	30	0.5	30	12.4	2	6	0									
Dec	6.4	1.1	3.7	−0.5	12.2	31	−3.9	19	−2.3	10	7.3	6	9	0									
1850	11.3	4.6	7.9	−1.5	24.8	6	−8.0	1	−2.3	12	14.4	8	66	0									

DURHAM 1851

Month	Mean max	Mean min	Mean temp	Anom	Highest max	Date	Lowest min	Date	Lowest max	Date	Highest min	Date	Air frost	Days ≥ 25 °C	Total pptn	Anom	Rain days	Wettest day	Date	Total sunshine	Anom	Sunniest day	Date
Jan	7.0	2.0	4.5	+0.4	12.8	1	−3.4	7	2.8	26	10.3	1	9	0									
Feb	6.8	1.3	4.1	−0.5	12.1	19	−4.0	22	1.9	3	9.4	19	10	0									
Mar	7.6	1.6	4.6	−1.6	10.9	4	−2.7	7	3.1	12	5.5	26	7	0									
Apr	8.8	2.5	5.7	−2.6	13.9	21	−0.7	7	4.3	30	8.1	21	3	0									
May	13.1	4.7	8.9	−2.1	20.8	31	−0.6	1	5.9	3	11.5	22	5	0									
June	17.3	8.4	12.9	−0.7	26.2	28	2.4	4	12.2	4	12.8	29	0	2									
July	17.1	9.0	13.0	−2.7	22.8	1	4.4	5	11.1	10	13.2	14	0	0									
Aug	18.0	9.9	14.0	−1.6	22.7	12	2.4	31	11.8	24	16.1	3	0	0									
Sep	16.0	7.4	11.7	−1.6	23.1	3	1.6	28	9.3	26	15.5	3	0	0									
Oct	12.5	6.4	9.5	−0.5	16.6	19	1.4	31	6.4	30	13.7	19	0	0									
Nov	5.6	0.8	3.2	−3.4	9.9	21	−3.3	17	1.3	17	6.2	11	11	0									
Dec	6.7	2.7	4.7	+0.5	13.3	10	−2.7	25	1.7	24	11.0	10	4	0									
1851	11.4	4.7	8.0	−1.4	26.2	6	−4.0	2	1.3	11	16.1	8	49	2									

DURHAM 1852

Month	Mean max	Mean min	Mean temp	Anom	Highest max	Date	Lowest min	Date	Lowest max	Date	Highest min	Date	Air frost	Days ≥ 25 °C	Total pptn	Anom	Rain days	Wettest day	Date	Total sunshine	Anom	Sunniest day	Date
Jan	6.5	1.7	4.1	−0.0	9.7	31	−2.2	11	1.7	10	4.9	27	6	0									
Feb	6.1	1.3	3.7	−0.9	11.1	2	−3.9	20	0.8	13	5.9	2	9	0									
Mar	7.4	1.0	4.2	−2.0	18.6	23	−4.1	4	3.1	1	6.7	23	11	0									
Apr	11.1	2.4	6.8	−1.5	20.7	13	−0.8	20	5.4	18	7.8	29	4	0									
May	12.0	6.0	9.0	−2.0	17.6	17	0.2	3	8.2	22	10.6	9	0	0									
June	15.4	7.8	11.6	−2.0	20.3	20	0.6	2	10.8	14	12.6	26	0	0									
July	21.6	13.2	17.4	+1.6	29.0	5	7.1	11	17.1	1	15.8	5	0	4									
Aug	18.8	11.1	15.0	−0.6	21.7	27	7.6	16	15.8	12	14.9	17	0	0									
Sep	14.5	8.3	11.4	−1.9	20.6	3	2.6	18	7.2	27	14.6	9	0	0									
Oct	9.8	4.6	7.2	−2.8	13.3	19	0.8	8	5.6	28	9.0	23	0	0									
Nov	7.8	4.0	5.9	−0.7	15.6	8	−2.5	25	1.6	30	11.3	8	4	0									
Dec	8.1	4.1	6.1	+1.9	12.4	11	−1.9	1	2.1	22	10.1	12	1	0									
1852	11.6	5.5	8.5	−0.9	29.0	7	−4.1	3	0.8	2	15.8	7	35	4									

DURHAM 1853

Month	Mean max	Mean min	Mean temp	Anom	Highest max	Date	Lowest min	Date	Lowest max	Date	Highest min	Date	Air frost	Days ≥ 25 °C	Total pptn	Anom	Rain days	Wettest day	Date	Total sunshine	Anom	Sunniest day	Date
Jan	5.6	1.9	3.7	−0.3	11.5	4	−1.4	28	2.5	23	8.3	20	4	0									
Feb	1.6	−2.4	−0.4	−5.0	4.7	5	−6.3	19	−1.2	12	1.4	5	21	0									
Mar	4.8	−0.8	2.0	−4.2	10.2	10	−8.1	25	−0.8	18	5.1	10	17	0									
Apr	10.0	3.6	6.8	−1.5	17.0	18	−0.4	23	5.3	22	8.8	19	1	0									
May	11.9	4.4	8.2	−2.8	16.8	28	0.2	9	5.9	9	11.8	27	0	0									
June	16.6	9.4	13.0	−0.6	21.6	8	1.8	9	12.2	2	12.6	19	0	0									
July	16.4	11.0	13.7	−2.1	19.1	7	8.6	1	12.8	2	14.1	21	0	0									
Aug	15.9	10.5	13.2	−2.4	19.1	9	6.4	24	12.7	14	13.8	10	0	0									
Sep	13.4	9.0	11.2	−2.1	16.8	20	3.6	6	9.4	26	12.9	14	0	0									
Oct	10.4	6.6	8.5	−1.5	13.4	23	1.2	3	7.9	2	10.8	28	0	0									
Nov	6.6	2.8	4.7	−1.9	11.5	5	−3.9	23	−1.4	23	8.7	1	8	0									
Dec	3.3	−0.3	1.5	−2.7	9.5	1	−5.0	29	−2.7	31	4.8	14	15	0									
1853	9.7	4.6	7.2	−2.3	21.6	6	−8.1	3	−2.7	12	14.1	7	66	0									

DURHAM 1854

Month	Mean max	Mean min	Mean temp	Anom	Highest max	Date	Lowest min	Date	Lowest max	Date	Highest min	Date	Air frost	Days ≥ 25 °C	Total pptn	Anom	Rain days	Wettest day	Date	Total sunshine	Anom	Sunniest day	Date
Jan	4.4	−0.3	2.0	−2.1	12.1	30	−11.7	3	−4.3	2	6.5	18	13	0									
Feb	6.5	1.7	4.1	−0.4	14.3	6	−4.8	18	1.7	15	10.2	6	7	0									
Mar	10.2	3.4	6.8	+0.6	13.8	23	−2.0	6	4.2	19	8.9	10	4	0									
Apr	11.0	3.5	7.2	−1.0	15.0	2	−0.9	13	5.2	23	8.5	7	3	0									
May	13.2	5.4	9.3	−1.7	17.3	17	1.4	19	8.8	18	9.7	21	0	0									
June	15.3	8.6	12.0	−1.7	25.7	24	4.4	1	9.7	4	14.6	23	0	1									
July	–	–	(14.4)	(−1.4)	–	–	–	–	–	–	–	–	–	–									
Aug	19.0	11.0	(14.7)	(−0.9)	23.4	27	6.8	15	14.7	17	16.3	27	0	0									
Sep	17.9	9.7	13.8	+0.5	23.8	3	4.9	22	12.6	22	13.6	17	0	0									
Oct	–	–	(7.8)	(−2.2)	–	–	–	–	–	–	–	–	–	–									
Nov	5.6	1.8	3.7	−2.9	10.5	8	−3.1	27	2.1	26	6.9	8	0	0									
Dec	–	–	(3.3)	(−0.9)	12.8	22	−2.7	19	1.7	18	8.7	22	–	0									
1854	–	–	(8.3)	(−1.1)	25.7	6	−11.7	1	−4.3	1	16.3	8	–	–									

DURHAM 1855

Month	Mean max	Mean min	Mean temp	Anom	Highest max	Date	Lowest min	Date	Lowest max	Date	Highest min	Date	Air frost	Days ≥ 25 °C	Total pptn	Anom	Rain days	Wettest day	Date	Total sunshine	Anom	Sunniest day	Date
Jan	–	–	(1.9)	(–2.2)	–	–	–	–	–4.3	16	–	–	–	0									
Feb	–	–	(–2.2)	(–6.8)	4.2	28	–14.6	16	0.8	23	0.2	26	18	0									
Mar	5.2	–0.5	2.3	–3.8	7.8	6	–3.5	7	6.1	3	2.6	17	7	0									
Apr	11.3	2.4	6.8	–1.4	16.7	19	–2.3	1	4.2	4	9.7	16	3	0									
May	11.6	3.5	7.6	–3.4	21.4	26	–3.2	5	8.9	1	10.1	25	0	0									
June	17.2	8.5	12.9	–0.7	24.7	28	2.4	19	15.8	5	13.7	27	0	1									
July	19.8	11.7	15.7	–0.0	26.1	14	8.6	6	13.8	5	16.8	23	0	0									
Aug	18.8	11.5	15.2	–0.4	23.2	11	6.0	6	11.0	14	14.7	19	0	0									
Sep	15.3	8.3	11.8	–1.5	20.0	21	0.3	7	6.1	24	13.3	23	3	0									
Oct	11.5	5.5	8.5	–1.5	16.7	21	–2.3	29	4.4	16	11.6	21	5	0									
Nov	7.3	2.6	5.0	–1.7	11.9	11	–2.6	17	–1.7	12	6.8	12	15	0									
Dec	4.0	–0.4	1.8	–2.5	9.1	14	–6.0	23			6.0	15											
1855	–	–	(7.3)	(–2.1)	26.1	7	–14.6	2	–4.3	2	16.8	7	–	1									

DURHAM 1856

Month	Mean max	Mean min	Mean temp	Anom	Highest max	Date	Lowest min	Date	Lowest max	Date	Highest min	Date	Air frost	Days ≥ 25 °C	Total pptn	Anom	Rain days	Wettest day	Date	Total sunshine	Anom	Sunniest day	Date
Jan	4.0	0.0	2.0	–2.1	8.5	24	–8.8	14	–1.2	14	4.4	6	14	0									
Feb	6.9	2.5	4.7	+0.1	12.2	9	–4.3	3	0.7	18	8.6	9	9	0									
Mar	6.0	1.1	3.6	–2.6	11.9	31	–5.1	30	2.2	12	4.8	2	7	0									
Apr	10.0	3.3	6.6	–1.6	15.6	2	–0.8	1	4.8	16	7.8	12	1	0									
May	10.3	4.5	7.4	–3.6	15.5	26	–1.1	6	5.9	1	7.8	28	1	0									
June	15.9	9.4	12.7	–1.0	24.9	26	6.2	15	8.8	14	17.6	27	0	0									
July	18.0	9.7	13.9	–1.9	24.5	31	4.1	1	12.2	8	15.7	30	0	0									
Aug	17.9	11.3	14.6	–1.0	26.7	1	5.6	23	12.0	22	16.4	2	0	2									
Sep	14.4	7.8	11.1	–2.2	18.9	4	3.8	14	11.1	30	12.2	15	0	0									
Oct	12.2	7.4	9.8	–0.2	16.1	4	1.9	25	8.2	30	11.8	4	0	0									
Nov	7.0	1.6	4.3	–2.4	14.1	23	–4.7	28	–0.6	29	10.6	23	13	0									
Dec	6.3	1.2	3.7	–0.5	14.4	7	–8.6	3	–1.7	27	10.7	7	14	0									
1856	10.8	5.0	7.9	–1.6	26.7	8	–8.8	1	–1.7	12	17.6	6	59	2									

DURHAM 1857

Month	Mean max	Mean min	Mean temp	Anom	Highest max	Date	Lowest min	Date	Lowest max	Date	Highest min	Date	Air frost	Days ≥ 25 °C	Total pptn	Anom	Rain days	Wettest day	Date	Total sunshine	Anom	Sunniest day	Date
Jan	3.8	0.0	1.9	−2.2	9.9	1	−6.8	30	−1.5	29	6.5	1	15	0									
Feb	6.5	1.2	3.8	−0.7	12.3	28	−8.3	1	−0.6	4	7.2	22	6	0									
Mar	6.5	2.2	4.4	−1.8	12.4	18	−1.8	11	2.3	23	6.8	2	6	0									
Apr	9.1	3.3	6.2	−2.1	16.7	18	−2.6	28	4.7	26	7.9	19	3	0									
May	12.6	6.6	9.6	−1.4	18.3	26	0.6	6	6.6	3	12.6	20	0	0									
June	17.9	10.4	14.2	+0.6	26.1	26	3.9	13	10.6	11	16.2	27	0	3									
July	19.6	12.0	15.8	+0.0	24.1	14	5.5	2	14.2	1	16.1	24	0	0									
Aug	18.9	12.9	15.9	+0.3	23.3	3	9.4	5	12.6	7	16.7	20	0	0									
Sep	16.9	10.9	13.9	+0.6	20.8	15	5.1	20	13.3	19	15.4	16	0	0									
Oct	13.4	8.5	11.0	+0.9	18.8	1	2.3	7	8.9	23	13.5	14	0	0									
Nov	8.9	5.0	6.9	+0.3	12.7	1	−3.6	26	2.2	25	8.8	14	3	0									
Dec	9.2	4.8	7.0	+2.8	12.9	22	0.1	27	4.4	31	9.8	18	0	0									
1857	11.9	6.5	9.2	−0.2	26.1	6	−8.3	2	−1.5	1	16.7	8	33	3									

DURHAM 1858

Month	Mean max	Mean min	Mean temp	Anom	Highest max	Date	Lowest min	Date	Lowest max	Date	Highest min	Date	Air frost	Days ≥ 25 °C	Total pptn	Anom	Rain days	Wettest day	Date	Total sunshine	Anom	Sunniest day	Date
Jan	5.8	1.1	3.4	−0.7	11.2	30	−3.4	5	0.7	6	7.3	30	14	0									
Feb	4.3	−0.7	1.8	−2.7	8.9	6	−7.1	3	−1.3	1	3.9	6	13	0									
Mar	7.8	1.5	4.6	−1.5	16.9	23	−8.7	11	−1.9	8	7.2	31	13	0									
Apr	11.2	2.9	7.0	−1.2	21.8	22	−3.0	13	3.2	11	8.8	16	6	0									
May	13.2	6.3	9.7	−1.3	24.4	31	0.3	1	5.9	2	11.2	18	0	0									
June	20.1	11.5	15.8	+2.1	24.9	14	5.9	28	15.3	27	16.4	15	0	0									
July	17.4	10.4	13.9	−1.9	21.3	24	5.8	29	13.0	2	15.3	14	0	0									
Aug	18.6	11.4	15.0	−0.6	22.6	8	6.1	29	14.3	28	14.9	18	0	0									
Sep	16.8	10.3	13.5	+0.3	21.8	13	4.9	21	13.7	22	13.7	13	0	0									
Oct	11.1	6.2	8.7	−1.3	16.5	3	1.2	26	5.7	29	12.8	3	0	0									
Nov	6.8	2.8	4.8	−1.8	11.6	3	−5.7	24	0.9	24	7.9	2	7	0									
Dec	5.7	2.3	4.0	−0.2	11.2	21	−3.6	15	0.8	15	7.5	2	4	0									
1858	11.6	5.5	8.5	−0.9	24.9	6	−8.7	3	−1.9	3	16.4	6	57	0									

DURHAM 1859

Month	Mean max	Mean min	Mean temp	Anom	Highest max	Date	Lowest min	Date	Lowest max	Date	Highest min	Date	Air frost	Days ≥ 25 °C	Total pptn	Anom	Rain days	Wettest day	Date	Total sunshine	Anom	Sunniest day	Date
Jan	6.5	2.9	4.7	+0.6	11.4	18	−3.3	14	0.9	7	8.1	21	5	0									
Feb	7.4	2.7	5.1	+0.5	11.9	16	−2.6	8	1.7	3	9.9	16	5	0									
Mar	9.7	4.7	7.2	+1.0	14.7	4	−3.9	31	1.9	31	11.5	5	3	0									
Apr	8.8	3.0	5.9	−2.4	21.9	7	−2.9	1	4.4	15	10.8	5	9	0									
May	13.8	5.3	9.5	−1.5	19.7	29	0.7	2	6.5	4	12.1	31	0	0									
June	16.1	10.0	13.0	−0.6	22.7	28	4.4	17	11.8	2	14.4	26	0	0									
July	19.6	12.6	16.1	+0.3	27.0	11	8.4	1	12.9	3	18.1	12	0	2									
Aug	19.2	11.7	15.4	−0.1	24.5	25	6.2	9	14.8	8	16.1	18	0	0									
Sep	15.2	8.9	12.0	−1.2	18.6	25	4.5	15	11.3	14	14.4	25	0	0									
Oct	11.2	6.5	8.8	−1.2	20.3	3	−3.3	22	1.8	21	14.9	3	6	0									
Nov	6.8	2.5	4.6	−2.0	12.2	6	−2.3	14	0.8	21	7.8	6	5	0									
Dec	2.9	−1.0	1.0	−3.3	9.3	31	−9.1	19	−4.2	19	5.4	5	15	0									
1859	11.4	5.8	8.6	−0.8	27.0	7	−9.1	12	−4.2	12	18.1	7	48	2									

DURHAM 1860

Month	Mean max	Mean min	Mean temp	Anom	Highest max	Date	Lowest min	Date	Lowest max	Date	Highest min	Date	Air frost	Days ≥ 25 °C	Total pptn	Anom	Rain days	Wettest day	Date	Total sunshine	Anom	Sunniest day	Date
Jan	5.3	0.6	2.9	−1.2	11.7	1	−4.3	26	1.5	26	8.7	1	14	0									
Feb	5.1	−0.9	2.1	−2.5	9.4	26	−7.3	11	0.9	11	5.4	5	18	0									
Mar	7.1	0.8	3.9	−2.2	11.2	17	−3.9	11	3.5	12	5.4	31	12	0									
Apr	9.6	1.3	5.4	−2.8	16.3	29	−2.9	28	6.2	2	7.8	30	5	0									
May	15.4	5.8	10.6	−0.4	22.3	4	0.4	28	7.5	5	10.6	11	0	0									
June	16.0	7.7	11.8	−1.8	20.2	21	4.3	8	12.4	10	10.6	21	0	0									
July	19.3	9.0	14.2	−1.6	24.2	5	4.3	9	13.3	22	12.1	4	0	0									
Aug	17.4	8.9	13.1	−2.4	20.8	16	5.8	24	11.0	6	12.2	16	0	0									
Sep	15.5	5.9	10.7	−2.6	21.6	7	0.5	11	10.3	30	11.7	6	0	0									
Oct	12.0	5.6	8.8	−1.2	17.9	6	−1.2	12	5.5	10	10.6	23	1	0									
Nov	6.9	2.3	4.6	−2.0	11.0	3	−3.1	19	2.6	18	7.2	1	4	0									
Dec	3.8	−1.2	1.3	−2.9	9.7	6	−16.4	25	−4.6	25	5.9	9	14	0									
1860	11.1	3.8	7.5	−2.0	24.2	7	−16.4	12	−4.6	12	12.2	8	68	0									

DURHAM 1861

Month	Mean max	Mean min	Mean temp	Anom	Highest max	Date	Lowest max	Date	Lowest min	Date	Highest min	Date	Air frost	Days ≥ 25 °C	Total pptn	Anom	Rain days	Wettest day	Date	Total sunshine	Anom	Sunniest day	Date
Jan	4.9	−1.0	1.9	−2.2	11.6	25	−3.8	8	−14.2	8	7.4	28	17	0									
Feb	7.1	1.5	4.3	−0.3	10.5	1	2.0	14	−9.2	13	6.0	1	6	0									
Mar	9.5	2.6	6.1	−0.1	13.9	8	6.4	21	−1.6	14	7.7	6	3	0									
Apr	11.2	2.7	6.9	−1.3	17.6	24	5.6	1	−0.9	9	7.3	25	4	0									
May	15.1	5.0	10.1	−1.0	22.8	29	6.8	12	−2.8	10	11.9	16	4	0									
June	18.3	9.4	13.9	+0.2	24.8	21	10.3	8	3.6	7	13.8	21	0	0									
July	18.7	9.9	14.3	−1.4	22.5	15	13.8	11	6.4	4	12.8	25	0	0									
Aug	19.1	11.0	15.1	−0.5	23.6	28	14.9	22	7.6	14	15.3	12	0	0									
Sep	16.7	8.4	12.6	−0.7	21.0	1	11.3	25	3.5	27	12.6	2	0	0									
Oct	14.0	6.7	10.3	+0.3	19.0	8	7.8	31	−0.2	17	13.5	8	2	0									
Nov	7.5	1.0	4.3	−2.4	12.6	29	1.7	24	−5.6	24	7.9	21	11	0									
Dec	5.9	1.3	3.6	−0.6	11.8	12	1.9	21	−5.6	29	8.0	13	10	0									
1861	12.3	4.9	8.6	−0.8	24.8	6	−3.8	1	−14.2	1	15.3	8	57	0									

DURHAM 1862

Month	Mean max	Mean min	Mean temp	Anom	Highest max	Date	Lowest max	Date	Lowest min	Date	Highest min	Date	Air frost	Days ≥ 25 °C	Total pptn	Anom	Rain days	Wettest day	Date	Total sunshine	Anom	Sunniest day	Date
Jan	5.1	0.9	3.0	−1.1	10.4	9	−0.8	18	−5.6	19	5.6	29	9	0									
Feb	6.5	2.2	4.3	−0.2	12.9	4	2.6	7	−7.0	10	7.8	5	4	0									
Mar	6.5	1.5	4.0	−2.1	12.7	12	3.0	3	−8.8	4	8.0	7	7	0									
Apr	11.1	3.2	7.1	−1.1	20.0	30	3.8	11	−3.5	12	8.6	2	4	0									
May	15.2	6.5	10.9	−0.1	21.1	17	8.7	13	−0.3	3	11.2	30	1	0									
June	15.0	7.8	11.4	−2.2	19.8	2	11.9	22	2.4	2	11.0	26	0	0									
July	16.2	8.2	12.2	−3.6	18.8	13	12.6	7	4.8	5	12.1	25	0	0									
Aug	16.6	9.4	13.0	−2.6	19.8	4	11.9	7	5.9	26	13.7	1	0	0									
Sep	15.4	6.9	11.2	−2.1	19.2	6	11.6	22	0.6	16	12.1	13	0	0									
Oct	13.2	4.8	9.0	−1.0	19.8	3	8.4	30	−4.1	30	13.2	3	2	0									
Nov	6.3	−0.5	2.9	−3.7	11.7	2	2.5	25	−6.9	25	7.9	1	19	0									
Dec	8.3	3.2	5.8	+1.5	11.9	6	5.3	21	−0.2	27	7.2	28	1	0									
1862	11.3	4.5	7.9	−1.5	21.1	5	−0.8	1	−8.8	3	13.7	8	47	0									

DURHAM 1863

Month	Mean max	Mean min	Mean temp	Anom	Highest max	Date	Lowest min	Date	Lowest max	Date	Highest min	Date	Air frost	Days ≥ 25 °C	Total pptn	Anom	Rain days	Wettest day	Date	Total sunshine	Anom	Sunniest day	Date
Jan	–	1.7	(4.3)	+0.2	13.2	31	-1.7	8	3.1	20	6.7	23	8	0									
Feb	–	2.0	(5.6)	+1.3	13.3	4	-2.6	16	6.1	9	7.2	27	8	0									
Mar	–	2.6	(6.4)	+2.2	17.2	29	-5.5	12	10.0	1	8.6	3	6	0									
Apr	–	2.8	(8.0)	+1.6	18.3	5	-1.6	2	15.7	6	7.6	26	6	0									
May	–	4.8	(9.5)	+0.5	19.9	9	0.8	6	16.8	6	10.3	29	0	0									
June	–	8.1	(12.9)	+0.2	24.2	14	4.3	5	17.5	6	12.2	14	0	0									
July	–	7.7	(14.2)	-2.1	27.2	10	2.8	26	13.1	22	13.6	12	0	3									
Aug	–	9.4	(14.3)	-1.3	22.8	3	3.8	21	13.8	27	15.7	8	0	0									
Sep	–	7.2	(10.9)	-1.6	20.1	4	4.4	11	11.5	22	11.2	19	0	0									
Oct	–	5.3	(8.5)	-0.6	17.2	1	-0.4	29	10.5	25	10.0	11	2	0									
Nov	–	3.4	(6.4)	+1.1	15.4	17	-3.5	7	6.5	7	9.6	17	6	0									
Dec	–	2.4	(5.7)	+4.1	–	–	-3.8	28	4.0	27	8.4	8	6	0									
1863	–	4.8	(8.9)	+0.5	27.2	7	-5.5	3	3.1	1	15.7	8	42	3									

DURHAM 1864

Month	Mean max	Mean min	Mean temp	Anom	Highest max	Date	Lowest min	Date	Lowest max	Date	Highest min	Date	Air frost	Days ≥ 25 °C	Total pptn	Anom	Rain days	Wettest day	Date	Total sunshine	Anom	Sunniest day	Date
Jan	5.4	-1.7	1.9	-2.2	12.3	31	-8.5	5	0.5	7	4.8	22	21	0									
Feb	5.7	-1.2	2.2	-2.3	12.3	13	-7.3	23	2.4	6	5.8	1	19	0									
Mar	7.9	-0.1	3.9	-2.3	14.7	24	-6.2	10	4.4	9	5.7	14	13	0									
Apr	12.8	3.1	8.0	-0.3	19.8	21	-0.1	1	6.4	6	8.3	21	1	0									
May	15.8	5.0	10.4	-0.6	25.3	18	-0.6	30	8.7	9	11.0	15	1	1									
June	17.2	7.9	12.6	-1.0	21.5	8	1.8	1	12.1	23	11.1	25	0	0									
July	18.4	8.8	13.6	-2.2	25.1	19	4.7	27	12.8	9	11.7	21	0	1									
Aug	18.6	7.1	12.9	-2.7	26.9	16	0.5	22	13.9	19	13.1	5	0	1									
Sep	17.2	7.0	12.1	-1.2	20.3	9	1.8	3	12.5	30	12.6	8	0	0									
Oct	12.5	5.8	9.1	-0.9	17.0	12	-0.1	21	8.9	21	9.1	29	1	0									
Nov	9.7	2.5	6.1	-0.5	12.6	28	-3.0	12	6.4	12	6.6	5	5	0									
Dec	6.9	1.3	4.1	-0.1	13.6	4	-6.2	25	-0.2	25	9.7	5	9	0									
1864	12.4	3.8	8.1	-1.4	26.9	8	-8.5	1	-0.2	12	13.1	8	70	3									

DURHAM 1865

Month	Mean max	Mean min	Mean temp	Anom	Highest max	Date	Lowest min	Date	Lowest max	Date	Highest min	Date	Air frost	Days ≥ 25 °C	Total pptn	Anom	Rain days	Wettest day	Date	Total sunshine	Anom	Sunniest day	Date
Jan	5.7	−0.9	2.4	−1.7	11.9	6	−10.6	26	0.7	25	8.4	10	16	0									
Feb	5.1	−1.3	1.9	−2.7	12.6	24	−10.2	15	0.0	20	5.2	23	19	0									
Mar	7.0	−0.7	3.1	−3.0	12.2	30	−3.9	28	4.1	26	3.4	2	19	0									
Apr	12.5	3.8	8.1	−0.1	19.8	25	−0.5	11	5.5	14	10.1	26	2	0									
May	14.5	6.1	10.3	−0.7	23.2	23	−0.1	1	5.7	12	12.1	27	1	0									
June	19.3	8.6	14.0	+0.3	27.0	20	1.1	12	8.6	2	13.2	6	0	3									
July	20.3	9.8	15.1	−0.7	25.4	25	3.3	2	14.5	31	13.2	16	0	2									
Aug	18.2	9.3	13.7	−1.8	22.3	7	5.0	5	12.4	2	13.5	27	0	0									
Sep	20.2	9.8	15.0	+1.7	24.1	8	3.7	24	13.0	29	14.4	3	0	0									
Oct	12.1	4.9	8.5	−1.5	18.6	3	−0.5	7	6.8	28	11.0	12	4	0									
Nov	8.9	2.4	5.7	−0.9	12.4	17	−2.0	27	4.9	27	6.7	21	4	0									
Dec	7.4	2.5	5.0	+0.7	12.2	7	−0.9	23	3.6	24	7.9	7	6	0									
1865	12.6	4.5	8.6	−0.9	27.0	6	−10.6	1	0.0	2	14.4	9	71	5									

DURHAM 1866

Month	Mean max	Mean min	Mean temp	Anom	Highest max	Date	Lowest min	Date	Lowest max	Date	Highest min	Date	Air frost	Days ≥ 25 °C	Total pptn	Anom	Rain days	Wettest day	Date	Total sunshine	Anom	Sunniest day	Date
Jan	7.6	2.0	4.8	+0.7	12.5	25	−4.7	13	0.6	11	7.4	14	5	0									
Feb	6.6	0.9	3.7	−0.8	11.3	6	−3.7	15	2.5	14	7.1	23	11	0									
Mar	7.2	−0.4	3.4	−2.8	13.8	29	−7.2	1	2.2	6	5.7	28	15	0									
Apr	10.5	2.7	6.6	−1.7	20.3	27	−2.4	30	5.6	3	6.8	17	4	0									
May	13.8	3.2	8.5	−2.5	20.8	19	−0.8	16	5.6	2	7.8	9	3	0									
June	18.3	8.7	13.5	−0.1	26.8	27	3.2	18	9.2	1	13.3	28	0	2									
July	18.5	9.8	14.1	−1.7	29.2	12	4.2	7	11.7	29	15.3	13	0	3									
Aug	17.4	9.6	13.5	−2.1	21.7	26	5.1	10	13.5	29	16.1	26	0	0									
Sep	14.7	7.9	11.3	−2.0	18.2	6	2.9	24	10.4	22	13.2	10	0	0									
Oct	12.1	7.3	9.7	−0.3	16.8	7	0.3	15	8.3	25	12.5	22	0	0									
Nov	8.8	3.0	5.9	−0.7	15.0	1	−3.3	29	2.7	18	9.5	1	5	0									
Dec	7.7	2.5	5.1	+0.9	12.2	4	−3.6	9	2.2	2	10.1	18	6	0									
1866	11.9	4.8	8.3	−1.1	29.2	7	−7.2	3	0.6	1	16.1	8	49	5									

DURHAM 1867

Month	Mean max	Mean min	Mean temp	Anom	Highest max	Date	Lowest min	Date	Lowest max	Date	Highest min	Date	Air frost	Days ≥ 25 °C	Total pptn	Anom	Rain days	Wettest day	Date	Total sunshine	Anom	Sunniest day	Date
Jan	3.1	−2.3	0.4	−3.7	10.9	27	−13.5	2	−2.3	1	8.3	28	21	0									
Feb	8.1	3.1	5.6	+1.1	11.8	21	−5.0	28	4.6	28	7.4	21	3	0									
Mar	5.1	−0.6	2.3	−3.9	13.1	24	−6.7	17	0.5	13	7.3	26	16	0									
Apr	12.0	4.4	8.2	−0.1	15.6	23	0.2	1	7.1	25	9.2	19	0	0									
May	13.6	5.8	9.7	−1.3	21.4	6	0.2	22	6.9	26	12.3	7	0	0									
June	17.3	8.4	12.8	−0.8	25.4	27	2.4	8	8.2	7	13.6	12	0	1									
July	17.8	9.1	13.4	−2.3	23.9	9	4.4	28	11.7	26	12.2	14	0	0									
Aug	19.7	9.8	14.7	−0.8	28.8	14	5.7	11	12.1	1	14.1	13	0	2									
Sep	16.2	9.1	12.6	−0.6	20.7	9	2.8	26	9.6	18	14.0	4	0	0									
Oct	11.9	4.8	8.4	−1.6	17.7	22	−0.7	6	5.7	5	12.0	23	3	0									
Nov	8.2	2.3	5.3	−1.4	13.4	8	−1.6	24	4.9	27	6.7	4	5	0									
Dec	5.6	0.6	3.1	−1.1	12.2	16	−5.2	4	1.0	20	7.6	12	16	0									
1867	11.5	4.5	8.0	−1.4	28.8	8	−13.5	1	−2.3	1	14.1	8	64	3									

DURHAM 1868

Month	Mean max	Mean min	Mean temp	Anom	Highest max	Date	Lowest min	Date	Lowest max	Date	Highest min	Date	Air frost	Days ≥ 25 °C	Total pptn	Anom	Rain days	Wettest day	Date	Total sunshine	Anom	Sunniest day	Date
Jan	5.2	1.0	3.1	−1.0	11.7	14	−4.2	24	0.8	3	8.7	17	11	0	72.2	135	20	24.0	31				
Feb	8.1	3.2	5.6	+1.1	13.3	25	−2.9	9	3.9	3	9.0	25	3	0	20.2	45	10	5.1	7				
Mar	9.7	2.0	5.8	−0.3	14.6	30	−3.9	25	4.9	24	7.3	22	5	0	20.7	50	10	6.0	2				
Apr	11.7	3.5	7.6	−0.7	18.0	16	−2.3	12	5.8	8	10.2	16	2	0	63.7	124	17	19.1	7				
May	16.8	6.7	11.7	+0.7	23.3	19	−3.3	6	9.9	5	11.5	20	1	0	24.6	55	11	5.7	24				
June	19.5	8.3	13.9	+0.3	26.1	21	3.7	5	14.8	4	13.5	21	0	1	8.3	14	4	3.9	11				
July	21.2	10.4	15.8	+0.0	29.4	15	3.9	25	14.6	3	16.2	21	0	5	21.7	35	7	14.9	16				
Aug	20.1	9.7	14.9	−0.7	29.0	2	5.8	31	13.7	19	12.4	11	0	6	59.7	90	9	19.1	11				
Sep	16.5	9.0	12.8	−0.5	27.1	7	3.8	9	11.3	12	13.2	5	0	3	100.2	177	18	22.9	18				
Oct	11.2	3.5	7.3	−2.7	15.3	11	−2.2	19	7.7	19	9.2	9	3	0	41.3	65	16	13.4	3				
Nov	6.4	2.2	4.3	−2.3	14.2	1	−3.8	7	2.1	29	11.4	1	7	0	44.6	61	21	11.4	22				
Dec	7.2	2.7	5.0	+0.7	12.0	4	−2.6	31	1.6	30	9.5	5	3	0	109.0	178	27	13.0	20				
1868	12.8	5.2	9.0	−0.5	29.4	7	−4.2	1	0.8	1	16.2	7	35	15	586.2	86	170	24.0	1				

DURHAM 1869

Month	Mean max	Mean min	Mean temp	Anom	Highest max	Date	Lowest min	Date	Lowest max	Date	Highest min	Date	Air frost	Days ≥ 25 °C	Total pptn	Anom	Rain days	Wettest day	Date	Total sunshine	Anom	Sunniest day	Date
Jan	6.4	1.6	4.0	-0.1	10.9	31	-4.0	1	1.2	24	5.6	10	8	0	46.4	87	13	9.4	13				
Feb	8.4	3.3	5.9	+1.3	11.6	6	-1.1	13	2.9	28	8.8	8	3	0	24.6	54	14	6.1	1				
Mar	6.0	0.3	3.1	-3.0	9.6	26	-2.9	3	3.4	12	4.4	26	12	0	39.7	96	18	15.6	11				
Apr	13.2	3.9	8.6	+0.3	21.9	11	-1.2	1	6.8	9	10.4	12	1	0	30.4	59	10	11.2	14				
May	10.5	4.1	7.3	-3.7	16.9	1	-1.2	14	5.1	3	8.6	25	1	0	66.3	149	13	13.3	18				
June	16.5	7.1	11.8	-1.8	22.2	7	0.4	1	12.1	16	11.8	6	0	0	21.0	34	5	16.6	15				
July	21.4	10.5	15.9	+0.1	26.8	16	4.3	1	15.0	19	16.6	8	0	3	22.1	36	6	15.5	26				
Aug	18.6	8.8	13.7	-1.9	29.8	28	2.7	31	12.4	29	13.9	28	0	3	51.0	77	13	8.9	3				
Sep	15.8	9.1	12.5	-0.8	19.6	25	5.3	21	11.3	12	14.4	8	0	0	88.6	156	18	18.5	12				
Oct	11.8	5.4	8.6	-1.4	19.2	9	-1.9	27	3.7	27	12.6	10	4	0	47.5	75	15	11.2	28				
Nov	8.1	2.3	5.2	-1.4	13.2	13	-3.4	30	2.1	30	9.0	2	10	0	59.7	81	12	16.0	23				
Dec	4.7	-0.1	2.3	-2.0	11.5	18	-5.6	28	-0.7	28	4.1	11	15	0	56.2	92	21	8.7	26				
1869	11.8	4.7	8.2	-1.2	29.8	8	-5.6	12	-0.7	12	16.6	7	54	6	553.5	81	158	18.5	9				

DURHAM 1870

Month	Mean max	Mean min	Mean temp	Anom	Highest max	Date	Lowest min	Date	Lowest max	Date	Highest min	Date	Air frost	Days ≥ 25 °C	Total pptn	Anom	Rain days	Wettest day	Date	Total sunshine	Anom	Sunniest day	Date
Jan	4.0	-0.1	2.0	-2.1	7.8	8	-6.2	27	0.6	29	3.6	7	13	0	36.6	69	17	7.6	6				
Feb	4.2	-0.4	1.9	-2.7	10.2	28	-4.4	12	-1.0	13	5.0	3	13	0	32.5	72	20	4.5	6				
Mar	7.4	1.0	4.2	-1.9	14.0	20	-4.4	13	2.3	12	6.9	17	10	0	28.9	70	12	8.3	2				
Apr	13.7	3.9	8.8	+0.6	22.2	20	-0.9	6	7.9	28	10.3	25	3	0	10.1	20	6	5.4	9				
May	15.2	6.2	10.7	-0.3	21.2	29	0.6	4	8.7	4	11.3	20	0	0	26.4	59	9	13.8	11				
June	18.6	8.9	13.7	+0.1	25.3	21	3.6	28	13.4	12	13.2	22	0	1	43.8	72	10	16.9	24				
July	20.8	11.2	16.0	+0.2	28.1	24	7.2	2	14.9	3	15.5	21	0	3	14.6	24	7	4.2	3				
Aug	19.5	9.2	14.4	-1.2	25.3	2	4.4	27	11.4	28	13.6	1	0	0	52.0	79	9	22.7	27				
Sep	17.2	7.4	12.3	-1.0	20.1	20	0.1	15	13.2	14	11.4	5	0	0	10.3	18	9	2.2	12				
Oct	11.8	5.0	8.4	-1.6	18.1	4	-0.8	15	7.4	9	8.5	22	1	0	86.5	136	17	25.8	8				
Nov	6.6	1.1	3.9	-2.8	11.9	3	-2.2	15	2.6	11	6.0	1	10	0	41.4	56	14	12.3	24				
Dec	3.5	-1.3	1.1	-3.1	10.4	19	-14.0	31	-5.1	31	7.3	19	18	0	74.6	122	14	13.5	8				
1870	11.9	4.3	8.1	-1.3	28.1	7	-14.0	7	-5.1	12	15.5	7	68	5	457.7	67	144	25.8	10				

DURHAM 1871

Month	Mean max	Mean min	Mean temp	Anom	Highest max	Date	Lowest min	Date	Lowest max	Date	Highest min	Date	Air frost	Days ≥ 25 °C	Total pptn	Anom	Rain days	Wettest day	Date	Total sunshine	Anom	Sunniest day	Date
Jan	2.3	-3.1	-0.4	-4.5	8.2	14	-13.7	1	-6.2	1	2.8	14	24	0	18.3	34	12	5.8	15				
Feb	6.9	2.2	4.5	-0.0	11.4	25	-4.1	12	-0.2	11	7.4	19	7	0	37.9	84	15	5.3	4				
Mar	10.1	2.4	6.3	+0.1	17.9	25	-4.4	15	4.0	28	7.3	12	7	0	15.4	37	9	4.5	15				
Apr	10.4	2.3	6.4	-1.9	15.7	27	-2.8	8	4.3	18	6.8	13	7	0	66.4	130	13	15.3	19				
May	14.2	4.3	9.3	-1.7	23.4	24	-2.0	17	7.7	12	9.2	31	2	0	40.3	91	6	14.9	26				
June	15.0	7.2	11.1	-2.5	21.0	16	2.8	27	10.0	23	12.8	15	0	0	65.1	106	10	32.9	15				
July	18.7	9.9	14.3	-1.5	22.8	16	7.5	1	14.0	3	15.1	14	0	0	73.4	118	21	11.5	4				
Aug	21.2	10.8	16.0	+0.5	26.8	12	4.9	22	17.0	26	15.3	11	0	4	16.0	24	4	8.8	20				
Sep	14.9	8.1	11.5	-1.8	22.0	1	2.4	29	8.9	27	13.3	1	0	0	107.9	190	14	31.8	27				
Oct	11.9	5.4	8.7	-1.3	16.1	18	-1.8	10	8.8	29	11.7	19	3	0	40.6	64	13	11.5	28				
Nov	5.5	0.8	3.2	-3.5	9.4	15	-5.1	19	0.8	22	7.1	1	12	0	35.2	48	14	9.8	25				
Dec	5.9	0.8	3.3	-0.9	10.8	18	-5.9	8	1.8	4	5.7	25	14	0	27.7	45	13	9.3	22				
1871	11.4	4.3	7.9	-1.6	26.8	8	-13.7	1	-6.2	1	15.3	8	76	4	544.2	80	144	32.9	6				

DURHAM 1872

Month	Mean max	Mean min	Mean temp	Anom	Highest max	Date	Lowest min	Date	Lowest max	Date	Highest min	Date	Air frost	Days ≥ 25 °C	Total pptn	Anom	Rain days	Wettest day	Date	Total sunshine	Anom	Sunniest day	Date
Jan	6.2	1.2	3.7	-0.4	11.3	30	-4.2	21	1.7	10	6.4	31	11	0	53.1	99	24	8.0	23				
Feb	7.2	2.8	5.0	+0.5	10.9	29	-1.5	8	4.1	16	6.9	11	2	0	60.6	134	25	16.1	24				
Mar	8.6	2.6	5.6	-0.6	14.2	6	-3.0	26	2.4	22	8.1	30	8	0	69.1	167	18	8.1	22				
Apr	11.3	3.7	7.5	-0.8	17.3	12	0.1	20	5.0	2	9.0	8	0	0	67.7	132	14	33.8	21				
May	13.3	5.0	9.2	-1.8	20.0	26	0.1	20	7.1	11	9.6	28	0	0	40.0	90	11	11.9	11				
June	18.4	8.6	13.5	-0.1	26.0	18	3.0	7	13.3	6	14.3	19	0	1	50.0	82	17	8.3	9				
July	21.2	11.1	16.1	+0.4	29.7	21	6.6	10	12.5	14	15.1	27	0	6	79.1	128	14	25.1	26				
Aug	18.4	10.4	14.4	-1.2	24.4	18	4.2	4	15.0	5	13.7	22	0	0	68.8	104	13	20.0	10				
Sep	15.3	7.9	11.6	-1.6	21.7	13	1.0	21	7.9	24	15.4	12	0	0	78.8	139	18	14.9	25				
Oct	11.2	4.2	7.7	-2.3	15.1	2	-2.1	15	7.7	11	10.0	2	2	0	104.6	164	19	19.7	21				
Nov	8.2	2.6	5.4	-1.2	15.6	6	-0.6	19	2.9	13	7.0	6	2	0	98.7	134	23	20.2	15				
Dec	5.9	1.9	3.9	-0.4	10.4	23	-3.9	12	1.4	12	7.0	28	6	0	124.8	204	20	37.4	16				
1872	12.1	5.2	8.6	-0.8	29.7	7	-4.2	1	1.4	12	15.4	9	31	7	895.3	132	216	37.4	12				

DURHAM 1873

Month	Mean max	Mean min	Mean temp	Anom	Highest max	Date	Lowest min	Date	Lowest max	Date	Highest min	Date	Air frost	Days ≥ 25 °C	Total pptn	Anom	Rain days	Wettest day	Date	Total sunshine	Anom	Sunniest day	Date
Jan	6.4	1.9	4.1	+0.1	11.7	13	-5.1	21	0.8	20	8.6	14	9	0	43.8	82	13	7.6	18				
Feb	4.3	-0.8	1.8	-2.8	8.6	20	-7.9	24	-1.6	23	3.2	16	15	0	36.2	80	9	13.9	3				
Mar	6.9	1.0	4.0	-2.2	14.5	30	-3.5	13	1.8	1	3.2	5	8	0	48.9	118	17	11.2	6				
Apr	10.0	2.6	6.3	-1.9	14.3	30	-2.3	20	6.1	24	5.8	3	4	0	26.2	51	16	6.8	24				
May	12.2	4.3	8.3	-2.7	17.3	25	-0.7	20	4.4	16	9.4	1	1	0	54.9	123	17	19.1	5				
June	18.3	9.1	13.7	+0.1	22.6	21	5.7	7	10.3	1	13.9	22	0	0	35.7	58	12	12.7	15				
July	20.4	10.7	15.6	-0.2	29.7	22	7.0	5	16.9	19	16.2	22	0	3	59.6	96	12	18.1	3				
Aug	18.7	10.2	14.4	-1.1	23.4	7	5.1	11	14.9	25	14.8	8	0	0	58.9	89	16	12.7	31				
Sep	15.4	6.9	11.1	-2.1	20.1	27	0.6	30	10.5	30	12.6	1	0	0	39.6	70	13	7.0	30				
Oct	11.4	3.5	7.5	-2.6	17.2	3	-3.5	25	6.8	26	10.7	2	6	0	43.5	68	14	16.2	2				
Nov	8.4	3.1	5.7	-0.9	12.7	23	-1.9	4	5.4	16	7.4	1	4	0	44.6	61	15	11.7	7				
Dec	8.2	2.2	5.2	+1.0	12.4	16	-3.4	29	1.7	29	8.6	2	6	0	9.3	15	7	3.3	21				
1873	11.7	4.6	8.1	-1.3	29.7	7	-7.9	2	-1.6	2	16.2	7	53	3	501.2	74	161	19.1	5				

DURHAM 1874

Month	Mean max	Mean min	Mean temp	Anom	Highest max	Date	Lowest min	Date	Lowest max	Date	Highest min	Date	Air frost	Days ≥ 25 °C	Total pptn	Anom	Rain days	Wettest day	Date	Total sunshine	Anom	Sunniest day	Date
Jan	7.6	1.6	4.6	+0.5	10.3	15	-2.1	18	2.6	4	6.3	16	10	0	36.5	68	11	9.6	3				
Feb	6.4	-0.3	3.0	-1.5	9.9	15	-6.6	5	-0.7	11	5.2	3	13	0	33.0	73	13	15.8	26				
Mar	10.2	2.6	6.4	+0.2	14.8	23	-6.9	11	1.8	10	8.0	17	7	0	24.8	60	14	5.8	19				
Apr	13.7	3.9	8.8	+0.6	21.5	26	-1.1	11	5.4	13	9.4	26	2	0	29.2	57	10	8.9	13				
May	11.6	4.6	8.1	-2.9	19.0	27	-0.7	17	7.7	5	11.7	31	2	0	42.2	95	16	18.6	23				
June	17.6	7.7	12.6	-1.0	21.6	19	3.0	12	10.5	17	13.9	8	0	0	22.6	37	12	4.5	24				
July	21.1	11.0	16.1	+0.3	27.1	18	5.0	7	15.8	12	14.1	14	0	3	29.8	48	11	13.4	23				
Aug	18.7	9.4	14.0	-1.5	26.5	19	3.8	24	12.0	4	13.0	19	0	1	44.4	67	16	11.2	30				
Sep	16.1	8.7	12.4	-0.8	19.1	27	3.7	14	13.5	16	14.5	29	0	0	40.6	72	18	7.5	11				
Oct	12.3	5.7	9.0	-1.0	16.6	27	2.0	24	8.6	29	10.7	13	0	0	38.6	61	13	9.8	6				
Nov	7.6	2.4	5.0	-1.7	14.1	5	-3.4	28	-0.4	27	9.7	4	11	0	74.2	101	20	20.4	28				
Dec	2.6	-3.3	-0.3	-4.6	8.2	5	-15.4	30	-3.5	29	4.4	5	24	0	104.5	171	19	26.7	8				
1874	12.1	4.5	8.3	-1.1	27.1	7	-15.4	12	-3.5	12	14.5	9	69	4	520.4	77	173	26.7	12				

DURHAM 1875

Month	Mean max	Mean min	Mean temp	Anom	Highest max	Date	Lowest min	Date	Lowest max	Date	Highest min	Date	Air frost	Days ≥ 25 °C	Total pptn	Anom	Rain days	Wettest day	Date	Total sunshine	Anom	Sunniest day	Date
Jan	6.7	1.5	4.1	+0.0	11.5	18	−13.1	1	−1.5	1	7.0	28	5	0	80.4	151	20	13.7	19				
Feb	4.1	0.1	2.1	−2.5	10.7	1	−4.1	6	−0.5	8	5.4	1	13	0	32.9	73	15	7.5	17				
Mar	7.2	1.4	4.3	−1.8	13.8	30	−2.2	5	1.7	3	6.8	31	11	0	13.2	32	10	2.5	6				
Apr	12.1	3.2	7.7	−0.6	20.9	20	−1.8	14	5.3	10	7.2	27	1	0	19.2	38	7	4.9	8				
May	16.0	5.8	10.9	−0.1	23.3	14	0.9	8	9.1	1	10.2	12	0	0	21.7	49	12	4.8	28				
June	17.8	7.9	12.8	−0.8	23.9	25	4.9	19	12.7	13	11.8	29	0	0	49.4	81	17	9.3	12				
July	18.7	9.7	14.2	−1.6	23.7	7	5.8	13	14.1	3	13.7	19	0	0	81.7	132	15	21.0	10				
Aug	18.9	11.0	15.0	−0.6	25.0	15	5.7	2	14.1	8	14.7	16	0	1	49.0	74	14	11.1	7				
Sep	16.8	9.3	13.0	−0.2	20.7	7	5.1	13	11.9	21	13.3	5	0	0	61.5	108	19	18.0	2				
Oct	11.4	5.4	8.4	−1.6	19.1	1	−2.1	12	6.9	31	10.0	7	1	0	126.9	199	20	34.9	20				
Nov	7.2	1.8	4.5	−2.2	14.5	5	−1.7	9	2.6	30	8.8	18	8	0	135.0	184	23	51.7	13				
Dec	6.0	1.7	3.8	−0.4	11.4	22	−3.4	15	0.0	15	5.8	26	11	0	34.8	57	13	5.3	3				
1875	11.9	4.9	8.4	−1.0	25.0	8	−13.1	1	−1.5	1	14.7	8	50	1	705.7	104	185	51.7	11				

DURHAM 1876

Month	Mean max	Mean min	Mean temp	Anom	Highest max	Date	Lowest min	Date	Lowest max	Date	Highest min	Date	Air frost	Days ≥ 25 °C	Total pptn	Anom	Rain days	Wettest day	Date	Total sunshine	Anom	Sunniest day	Date
Jan	5.6	0.0	2.8	−1.3	12.4	31	−6.0	8	0.5	9	7.3	31	16	0	9.9	19	9	2.5	12				
Feb	6.3	0.6	3.4	−1.1	11.1	29	−5.5	10	1.8	9	5.9	22	13	0	45.2	100	25	5.8	3				
Mar	7.2	0.3	3.8	−2.4	15.4	31	−4.3	20	2.3	18	3.6	31	16	0	54.0	131	18	16.0	29				
Apr	11.0	3.0	7.0	−1.3	18.2	8	−2.9	13	4.1	13	9.4	7	6	0	59.3	116	19	8.3	10				
May	12.9	4.4	8.7	−2.3	19.3	29	−1.1	3	7.4	1	9.5	30	2	0	26.9	60	11	7.1	23				
June	18.2	7.7	13.0	−0.6	26.3	20	3.4	11	13.2	15	11.5	22	0	1	61.4	100	14	32.5	22				
July	21.5	10.0	15.8	−0.0	31.3	16	6.2	3	15.6	23	13.4	13	0	6	63.5	102	10	16.6	26				
Aug	19.9	9.2	14.6	−1.0	29.9	14	2.8	24	14.4	24	14.0	7	0	1	51.1	77	11	14.0	2				
Sep	15.2	7.8	11.5	−1.8	19.8	22	4.7	12	10.4	30	9.7	6	0	0	63.7	112	19	8.0	3				
Oct	13.3	7.0	10.1	+0.1	18.2	4	−0.2	31	5.9	31	12.5	7	1	0	46.3	73	12	12.9	8				
Nov	7.9	2.1	5.0	−1.6	12.9	15	−3.8	10	2.7	11	6.6	4	7	0	102.8	140	18	16.7	26				
Dec	6.9	2.7	4.8	+0.6	11.2	28	−2.2	23	0.5	26	6.6	4	7	0	164.9	269	21	24.4	19				
1876	12.2	4.6	8.4	−1.1	31.3	7	−6.0	1	0.5	1	14.0	8	68	8	749.0	110	187	32.5	6				

DURHAM 1877

Month	Mean max	Mean min	Mean temp	Anom	Highest max	Date	Lowest min	Date	Lowest max	Date	Highest min	Date	Air frost	Days ≥ 25 °C	Total pptn	Anom	Rain days	Wettest day	Date	Total sunshine	Anom	Sunniest day	Date
Jan	7.0	1.1	4.0	-0.1	12.9	19	-4.0	2	3.1	3	4.7	7	5	0	88.2	165	20	21.5	3				
Feb	8.1	1.8	5.0	+0.4	11.6	24	-5.6	27	0.5	27	5.3	6	4	0	44.5	98	14	8.4	3				
Mar	7.3	0.0	3.6	-2.5	13.6	30	-6.5	1	2.0	10	5.6	31	15	0	83.3	201	18	28.1	25				
Apr	8.4	2.4	5.4	-2.9	14.4	4	-1.8	20	4.6	13	7.5	3	1	0	106.5	208	20	25.5	9				
May	11.0	3.7	7.3	-3.7	16.3	16	-2.6	4	5.3	1	8.2	27	4	0	59.7	134	16	9.7	17				
June	18.3	8.6	13.4	-0.2	23.2	28	3.7	16	12.3	12	12.1	3	0	0	35.8	59	12	8.9	22				
July	18.3	9.9	14.1	-1.7	24.4	30	4.1	5	15.3	17	15.3	30	0	0	76.5	123	14	20.8	14				
Aug	17.5	9.8	13.7	-1.9	21.9	7	3.1	24	12.6	22	14.2	7	0	0	125.4	189	20	19.7	19				
Sep	14.5	6.6	10.6	-2.7	19.8	11	2.5	25	10.7	20	11.8	14	0	0	39.0	69	12	12.9	6				
Oct	12.6	4.6	8.6	-1.4	16.5	7	-1.9	18	8.5	16	11.0	14	2	0	68.7	108	19	16.3	19				
Nov	9.4	3.3	6.3	-0.3	14.4	6	-1.2	24	4.9	24	10.1	6	4	0	42.9	58	16	8.4	11				
Dec	6.9	1.6	4.3	+0.0	10.9	12	-3.6	25	1.2	25	6.8	21	7	0	37.2	61	14	14.7	30				
1877	11.6	4.4	8.0	-1.4	24.4	7	-6.5	3	0.5	2	15.3	7	42	0	807.7	119	195	28.1	3				

DURHAM 1878

Month	Mean max	Mean min	Mean temp	Anom	Highest max	Date	Lowest min	Date	Lowest max	Date	Highest min	Date	Air frost	Days ≥ 25 °C	Total pptn	Anom	Rain days	Wettest day	Date	Total sunshine	Anom	Sunniest day	Date
Jan	6.8	1.0	3.9	-0.2	11.5	22	-5.3	25	2.8	9	8.2	21	13	0	42.5	80	14	8.6	27				
Feb	8.2	1.9	5.1	+0.5	13.4	17	-2.6	9	4.0	6	7.4	17	6	0	23.0	51	9	12.0	14				
Mar	8.6	1.5	5.0	-1.1	15.0	18	-4.0	24	2.3	31	7.8	1	14	0	21.7	52	12	5.8	24				
Apr	11.8	3.3	7.6	-0.7	17.2	14	-2.6	6	7.1	10	7.9	15	5	0	52.8	103	14	15.5	20				
May	14.3	5.9	10.1	-0.9	20.2	5	0.3	21	8.2	8	10.1	17	0	0	87.5	197	22	16.1	7				
June	18.0	8.4	13.2	-0.4	28.5	27	1.7	1	8.9	4	17.5	27	0	3	64.2	105	13	24.1	11				
July	20.4	11.2	15.8	+0.0	29.5	19	4.7	4	13.4	2	16.5	18	0	0	3.6	6	6	1.2	28				
Aug	19.4	11.3	15.4	-0.2	22.4	7	5.9	1	15.1	23	13.9	3	0	0	151.3	229	19	35.3	3				
Sep	16.8	8.4	12.6	-0.6	22.1	6	2.5	24	12.3	20	14.8	3	0	0	38.1	67	12	8.6	22				
Oct	12.9	6.1	9.5	-0.5	19.0	5	-1.2	29	6.3	29	13.4	7	2	0	32.8	52	18	8.7	30				
Nov	5.8	0.5	3.1	-3.5	9.4	10	-2.9	22	2.3	21	4.1	17	10	0	148.8	203	23	30.9	14				
Dec	2.3	-3.8	-0.7	-5.0	10.0	31	-15.4	14	-6.6	13	7.9	31	25	0	83.9	137	22	24.6	8				
1878	12.1	4.6	8.4	-1.1	29.5	7	-15.4	12	-6.6	12	17.5	6	75	7	750.2	110	184	35.3	8				

Month	Mean max	Mean min	Mean temp	Anom	Highest max	Date	Lowest min	Date	Lowest max	Date	Highest min	Date	Air frost	Days ≥ 25 °C	Total pptn	Anom	Rain days	Wettest day	Date	Total sunshine	Anom	Sunniest day	Date
Jan	2.4	−3.4	−0.5	−4.6	8.2	1	−12.9	27	−0.5	11	1.2	15	28	0	21.7	41	13	4.3	11				
Feb	3.9	−0.9	1.5	−3.1	7.7	27	−7.7	22	1.0	1	4.1	7	14	0	59.8	132	23	6.9	23				
Mar	7.1	0.5	3.8	−2.4	13.1	19	−8.7	14	0.4	24	4.7	10	13	0	20.7	50	16	4.3	15				
Apr	8.5	1.4	4.9	−3.3	13.3	26	−2.3	13	4.5	12	4.5	6	10	0	47.7	93	20	7.3	23				
May	12.2	3.3	7.7	−3.3	19.9	5	−1.5	3	5.7	9	9.7	21	6	0	56.4	127	17	12.9	21				
June	16.4	7.8	12.1	−1.5	21.1	19	3.6	2	11.5	7	11.4	20	0	0	91.2	149	21	23.6	7				
July	16.3	9.7	13.0	−2.8	22.5	30	7.4	12	11.1	22	14.5	29	0	0	101.1	163	20	23.7	21				
Aug	17.7	9.7	13.7	−1.9	23.3	11	5.9	10	13.2	16	12.3	20	0	0	84.8	128	18	17.2	6				
Sep	15.7	7.3	11.5	−1.8	20.3	7	−0.2	30	12.4	19	12.7	3	1	0	31.0	55	11	10.7	8				
Oct	12.8	3.6	8.2	−1.8	19.2	9	−2.0	26	7.3	15	10.6	4	4	0	17.3	27	12	7.9	14				
Nov	7.3	1.0	4.2	−2.5	14.0	18	−3.7	30	1.2	30	6.8	18	12	0	48.9	67	16	9.3	9				
Dec	4.6	−3.2	0.7	−3.5	11.6	28	−14.6	4	−2.4	4	5.9	28	22	0	15.0	25	8	3.2	5				
1879	10.4	3.1	6.8	−2.7	23.3	8	−14.6	12	−2.4	12	14.5	7	110	0	595.6	88	195	23.7	7				

Month	Mean max	Mean min	Mean temp	Anom	Highest max	Date	Lowest min	Date	Lowest max	Date	Highest min	Date	Air frost	Days ≥ 25 °C	Total pptn	Anom	Rain days	Wettest day	Date	Total sunshine	Anom	Sunniest day	Date
Jan	4.6	−1.2	1.7	−2.4	12.0	1	−7.7	19	0.1	13	9.1	1	22	0	13.5	25	5	6.1	16				
Feb	8.4	1.9	5.2	+0.6	13.8	29	−2.9	6	4.3	9	6.5	29	9	0	44.3	98	13	10.9	18				
Mar	8.9	1.1	5.0	−1.2	14.7	5	−4.6	20	4.0	16	7.4	6	11	0	32.7	79	10	13.7	31				
Apr	11.1	2.8	6.9	−1.3	15.3	19	−0.7	30	6.5	12	6.7	19	2	0	51.5	101	16	9.3	25				
May	13.4	3.7	8.5	−2.5	19.2	19	−1.7	8	7.6	7	7.3	22	3	0	38.1	86	8	18.5	26				
June	16.7	7.6	12.1	−1.5	22.6	30	2.1	10	10.7	4	14.3	28	0	0	66.6	109	15	23.4	15	140	75		
July	18.1	9.8	14.0	−1.8	23.1	25	6.3	10	13.0	15	13.1	6	0	0	105.0	169	19	16.0	12	118	71		
Aug	18.9	10.7	14.8	−0.7	23.3	10	5.6	20	14.5	18	13.8	5	0	0	49.4	75	14	17.8	5	139	79		
Sep	18.1	8.7	13.4	+0.1	25.5	3	4.1	9	14.1	16	15.3	3	0	0	128.3	226	12	30.2	14	143	85		
Oct	10.0	2.9	6.4	−3.6	17.5	1	−4.3	22	4.7	20	8.0	7	7	0	102.0	160	20	22.6	5	74	56		
Nov	8.2	0.2	4.2	−2.4	15.2	13	−9.8	21	2.8	21	7.8	13	15	0	56.7	77	13	11.1	16	71	72		
Dec	6.5	0.6	3.6	−0.7	11.3	6	−6.5	17	1.0	27	9.1	6	14	0	70.2	115	18	27.0	29	55	81		
1880	11.9	4.1	8.0	−1.5	25.5	1	−9.8	9	0.1	1	15.3	9	83	1	758.3	111	163	30.2	9				

DURHAM 1881

Month	Mean max	Mean min	Mean temp	Anom	Highest max	Date	Lowest min	Date	Lowest max	Date	Highest min	Date	Air frost	Days ≥ 25 °C	Total pptn	Anom	Rain days	Wettest day	Date	Total sunshine	Anom	Sunniest day	Date
Jan	2.0	−5.3	−1.7	−5.8	7.4	2	−16.9	17	−4.7	25	2.3	1	23	0	22.3	42	10	10.4	12	51	83		
Feb	4.1	−0.9	1.6	−3.0	9.1	5	−5.6	6	1.6	22	3.7	4	17	0	70.4	155	19	14.0	10	38	45		
Mar	7.6	−0.4	3.6	−2.6	15.8	17	−7.4	2	1.5	4	8.6	10	18	0	56.2	136	17	19.7	6	123	102		
Apr	9.3	1.2	5.3	−3.0	17.5	13	−5.9	4	2.6	2	7.6	13	13	0	24.9	49	11	8.1	11	110	71		
May	15.0	4.6	9.8	−1.2	23.7	30	−0.9	11	7.0	2	8.1	26	1	0	31.7	71	12	15.4	24	256	137		
June	17.3	7.2	12.3	−1.3	26.7	2	1.0	10	9.4	9	13.1	4	0	1	54.8	90	18	10.3	17	140	85		
July	19.6	10.4	15.0	−0.8	26.2	5	5.8	21	15.2	31	15.5	14	0	1	78.5	127	12	31.0	5	168	96		
Aug	16.9	9.0	13.0	−2.6	24.6	5	5.3	2	11.8	31	13.1	8	0	0	97.9	148	15	21.2	25	145	86		
Sep	14.8	8.3	11.5	−1.7	19.8	18	4.0	28	11.0	23	11.6	21	0	0	81.5	144	17	17.7	21	66	50		
Oct	10.1	3.6	6.9	−3.1	14.9	11	−5.0	31	4.6	30	8.5	3	6	0	90.9	143	18	14.1	22	94	95		
Nov	10.3	4.5	7.4	+0.8	15.2	14	−2.5	18	1.8	1	11.1	11	2	0	44.4	60	16	8.9	23	53	78		
Dec	6.1	0.9	3.5	−0.8	10.0	29	−3.6	15	2.5	15	5.1	29	10	0	57.9	95	18	13.3	11	51	90		
1881	11.1	3.6	7.3	−2.1	26.7	6	−16.9	1	−4.7	1	15.5	7	90	2	711.4	105	183	31.0	7	1295	88		

DURHAM 1882

Month	Mean max	Mean min	Mean temp	Anom	Highest max	Date	Lowest min	Date	Lowest max	Date	Highest min	Date	Air frost	Days ≥ 25 °C	Total pptn	Anom	Rain days	Wettest day	Date	Total sunshine	Anom	Sunniest day	Date
Jan	7.4	2.1	4.7	+0.7	11.3	5	−0.8	26	4.0	26	5.2	11	3	0	18.4	34	8	8.0	29	55.6	90	5.8	21
Feb	8.6	2.3	5.5	+0.9	13.8	21	−2.3	2	3.4	1	7.6	26	4	0	23.3	51	7	8.5	26	66.8	78	7.8	19
Mar	10.5	2.3	6.4	+0.2	17.5	16	−1.6	22	4.7	1	6.6	10	6	0	37.6	91	13	10.8	1	127.1	106	10.1	26
Apr	9.8	3.0	6.4	−1.8	15.2	21	−2.8	16	5.9	3	8.0	21	4	0	91.4	179	17	16.6	14	118.2	77	11.7	25
May	14.7	4.4	9.6	−1.4	19.5	18	−0.7	17	7.6	8	9.4	27	2	0	57.7	130	11	14.9	25	238.9	128	14.4	12
June	16.6	7.7	12.2	−1.4	19.7	20	3.5	17	12.3	10	10.8	28	0	0	84.5	138	18	23.6	22	193.2	117	14.7	1
July	19.0	10.0	14.5	−1.3	22.0	29	5.9	27	14.8	13	12.4	14	0	0	90.6	146	25	13.4	21	194.9	111	12.0	19
Aug	19.0	9.6	14.3	−1.2	24.4	11	5.8	16	15.0	31	13.5	17	0	0	73.3	111	17	32.5	22	182.1	109	14.0	3
Sep	15.2	6.8	11.0	−2.3	19.5	3	1.3	12	10.3	14	11.4	2	0	0	53.4	94	16	15.7	14	104.3	79	10.7	7
Oct	12.0	6.2	9.1	−0.9	19.2	1	−1.2	26	7.2	24	11.7	13	1	0	121.8	191	28	22.9	27	72.9	74	8.2	2
Nov	7.4	1.9	4.6	−2.0	13.4	5	−3.9	18	1.8	18	8.0	23	5	0	84.0	114	27	17.5	10	60.4	89	6.5	9
Dec	4.2	−0.2	2.0	−2.3	11.4	28	−9.3	11	−3.9	12	8.1	28	15	0	114.2	187	25	17.5	8	21.4	38	5.3	23
1882	12.0	4.7	8.4	−1.1	24.4	8	−9.3	12	−3.9	12	13.5	8	40	0	850.2	125	212	32.5	8	1435.8	97	14.7	6

DURHAM 1883

Month	Mean max	Mean min	Mean temp	Anom	Highest max	Date	Lowest min	Date	Lowest max	Date	Highest min	Date	Air frost	Days ≥25°C	Total pptn	Anom	Rain days	Wettest day	Date	Total sunshine	Anom	Sunniest day	Date
Jan	5.8	0.8	3.3	−0.8	11.6	1	−4.1	31	1.8	7	4.8	18	9	0	47.9	90	20	8.1	14	40.7	66	6.3	25
Feb	7.8	1.6	4.7	+0.1	12.1	28	−2.4	1	0.6	1	7.9	21	4	0	42.4	94	17	11.4	1	82.3	96	8.3	25
Mar	5.5	−2.1	1.7	−4.5	13.2	4	−6.9	15	1.3	22	3.5	1	25	0	47.7	115	20	11.7	7	131.6	109	11.3	31
Apr	11.6	2.0	6.8	−1.5	17.4	4	−2.1	26	6.6	23	5.8	28	8	0	40.5	79	11	10.9	28	130.8	85	10.0	6
May	13.5	4.2	8.9	−2.2	19.9	22	0.3	7	4.0	8	9.6	13	0	0	46.1	104	13	13.5	8	210.3	112	15.4	27
June	16.5	7.8	12.2	−1.4	22.9	29	2.7	22	9.5	5	13.4	29	0	0	62.3	102	17	25.1	25	131.6	79	13.2	3
July	18.1	9.2	13.6	−2.2	22.6	1	4.6	20	10.0	21	14.0	3	0	0	94.7	153	20	35.0	21	157.9	90	13.2	1
Aug	19.3	9.3	14.3	−1.2	23.5	24	6.2	24	14.7	29	13.0	14	0	0	30.4	46	15	11.2	12	168.9	101	13.5	18
Sep	15.8	8.8	12.3	−1.0	21.2	17	4.2	9	10.5	30	12.1	24	0	0	104.5	184	21	25.5	20	99.6	75	8.5	9
Oct	12.1	4.7	8.4	−1.6	18.8	8	−0.8	22	9.2	24	10.2	14	1	0	60.0	94	15	17.5	16	103.0	105	9.9	8
Nov	8.3	2.4	5.4	−1.3	13.3	28	−1.2	11	5.3	15	9.2	28	5	0	23.7	32	19	4.1	15	81.1	119	6.9	12
Dec	6.9	1.6	4.2	−0.0	13.1	13	−1.7	6	2.6	7	5.4	22	4	0	37.1	61	10	10.1	15	–	–	3.3	4
1883	11.8	4.2	8.0	−1.5	23.5	8	−6.9	3	0.6	2	14.0	7	56	0	637.3	94	198	35.0	7	–	–	15.4	5

DURHAM 1884

Month	Mean max	Mean min	Mean temp	Anom	Highest max	Date	Lowest min	Date	Lowest max	Date	Highest min	Date	Air frost	Days ≥25°C	Total pptn	Anom	Rain days	Wettest day	Date	Total sunshine	Anom	Sunniest day	Date
Jan	8.1	2.7	5.4	+1.3	11.5	9	−2.3	26	2.0	2	7.9	9	6	0	80.6	151	17	18.5	3	25.7	42	5.3	15
Feb	7.1	1.3	4.2	−0.4	11.6	4	−3.6	2	2.6	28	7.0	4	8	0	32.4	72	16	8.1	23	43	–	–	–
Mar	9.1	1.8	5.4	−0.7	17.4	16	−2.4	3	3.9	2	6.6	16	4	0	64.4	156	12	23.9	4	83	–	–	–
Apr	10.0	2.1	6.0	−2.2	15.4	2	−2.4	26	5.5	17	7.4	5	5	0	51.9	101	20	7.1	1	–	–	–	–
May	14.7	4.5	9.6	−1.4	23.0	11	0.5	7	9.1	29	9.5	9	0	0	17.0	38	10	3.5	4	–	–	–	–
June	17.6	7.3	12.4	−1.2	26.5	28	2.0	1	11.0	8	10.8	27	0	1	25.8	42	9	8.8	6	–	–	–	–
July	19.7	9.6	14.7	−1.1	25.2	5	5.6	28	13.1	26	12.7	8	0	1	98.8	159	16	27.4	5	–	–	–	–
Aug	21.0	9.5	15.2	−0.3	26.2	12	5.2	29	12.4	27	14.0	11	0	4	18.6	28	7	6.1	27	181	–	–	–
Sep	17.0	8.5	12.7	−0.5	22.8	17	3.4	4	13.5	12	13.6	10	0	0	35.4	62	12	12.5	21	107	–	–	–
Oct	12.4	4.4	8.4	−1.6	16.3	15	−0.7	11	5.8	10	10.3	31	2	0	38.2	60	12	15.5	11	–	–	–	–
Nov	8.4	2.0	5.2	−1.4	13.8	1	−6.5	30	1.1	30	8.5	1	6	0	46.8	64	12	11.7	4	–	–	–	–
Dec	6.1	0.9	3.5	−0.7	12.2	13	−2.9	27	1.1	29	7.3	13	12	0	32.3	53	13	7.6	19	–	–	–	–
1884	12.6	4.6	8.6	−0.9	26.5	6	−6.5	11	1.1	11	14.0	8	43	6	542.2	80	156	27.4	7	–	–	–	–

DURHAM 1885

Month	Mean max	Mean min	Mean temp	Anom	Highest max	Date	Lowest min	Date	Lowest max	Date	Highest min	Date	Air frost	Days ≥ 25 °C	Total pptn	Anom	Rain days	Wettest day	Date	Total sunshine	Anom	Sunniest day	Date
Jan	4.5	0.2	2.4	–1.7	11.4	29	–3.2	22	1.3	25	5.2	30	13	0	50.1	94	16	9.7	13	17	–	–	–
Feb	7.4	1.2	4.3	–0.3	12.8	27	–4.9	21	2.4	21	7.4	13	11	0	22.7	50	14	9.9	8	41	–	–	–
Mar	7.3	0.1	3.7	–2.5	12.6	11	–2.6	14	3.5	5	4.6	17	17	0	56.9	138	12	14.9	4	–	–	–	–
Apr	10.9	2.5	6.7	–1.6	19.2	19	–3.5	3	5.1	9	9.3	21	6	0	38.7	76	13	10.4	9	–	–	–	–
May	11.7	3.1	7.4	–3.6	19.6	28	–1.2	12	5.1	4	10.1	28	5	0	98.1	220	23	17.3	3	–	–	–	–
June	16.8	7.6	12.2	–1.4	23.9	4	2.6	27	8.7	8	12.2	19	0	0	43.0	70	7	18.5	24	185.9	112	14.3	18
July	20.1	9.6	14.9	–0.9	27.2	25	4.6	13	14.5	27	15.0	26	0	3	32.9	53	14	9.9	3	171.8	98	13.3	7
Aug	15.9	8.6	12.2	–3.3	22.8	16	2.9	30	11.5	3	13.1	10	0	0	69.5	105	16	33.1	5	107.8	64	11.4	12
Sep	15.6	6.1	10.9	–2.4	19.8	15	0.5	1	10.2	26	10.7	15	0	0	61.3	108	16	9.7	4	133.2	101	10.6	10
Oct	9.7	3.7	6.7	–3.3	14.0	8	–0.1	12	5.1	22	7.3	15	1	0	115.2	181	23	19.2	10	74.9	76	9.4	5
Nov	8.0	2.5	5.2	–1.4	14.1	4	–5.5	18	4.3	18	9.2	3	5	0	48.2	66	14	17.5	26	30.8	45	5.1	17
Dec	6.4	0.2	3.3	–1.0	12.7	16	–6.9	8	0.1	8	6.6	31	12	0	14.3	23	11	5.0	9	51.3	90	5.5	1
1885	11.2	3.8	7.5	–2.0	27.2	7	–6.9	12	0.1	12	15.0	7	70	3	650.9	96	179	33.1	8	–	–	14.3	6

DURHAM 1886

Month	Mean max	Mean min	Mean temp	Anom	Highest max	Date	Lowest min	Date	Lowest max	Date	Highest min	Date	Air frost	Days ≥ 25 °C	Total pptn	Anom	Rain days	Wettest day	Date	Total sunshine	Anom	Sunniest day	Date
Jan	4.2	–1.1	1.5	–2.6	11.1	1	–9.8	20	0.6	10	8.2	1	18	0	96.4	181	19	18.2	25	50.8	82	5.5	8
Feb	4.1	–1.5	1.3	–3.3	8.0	10	–6.5	20	0.9	19	1.3	11	22	0	23.1	51	15	4.2	22	60.8	71	7.2	2
Mar	6.4	0.2	3.3	–2.9	15.9	24	–10.0	10	–0.1	2	8.7	23	18	0	62.8	152	18	12.6	2	82.8	69	9.0	9
Apr	9.5	2.5	6.0	–2.2	18.2	27	–2.7	30	5.2	18	5.8	4	2	0	38.4	75	20	10.5	18	124.6	81	11.6	27
May	12.3	4.7	8.5	–2.5	21.9	6	–2.7	2	5.1	12	8.4	6	1	0	67.0	151	18	14.9	12	98.9	53	9.8	2
June	16.5	7.6	12.1	–1.6	24.0	28	2.5	4	7.6	1	12.5	29	0	0	23.7	39	11	16.9	1	155.3	94	13.8	15
July	19.6	10.2	14.9	–0.9	26.6	2	4.7	28	12.5	26	14.6	19	0	3	129.2	208	13	37.3	26	178.9	102	14.4	5
Aug	19.0	10.2	14.6	–1.0	23.8	30	4.9	4	13.9	22	14.1	26	0	0	23.5	36	12	7.2	18	165.2	99	13.1	1
Sep	15.8	8.5	12.2	–1.1	22.4	1	2.6	16	10.3	21	13.4	30	0	0	38.5	68	13	14.1	9	73.2	55	8.6	4
Oct	12.6	7.2	9.9	–0.1	17.2	8	2.0	23	9.0	27	11.7	6	0	0	128.4	202	22	29.2	1	54.8	56	8.6	2
Nov	8.9	2.6	5.8	–0.9	12.6	1	–2.1	22	5.9	30	8.0	20	3	0	90.8	124	15	38.9	5	83.4	122	6.9	16
Dec	4.0	–1.1	1.5	–2.8	11.1	6	–10.4	21	–0.6	21	4.5	6	16	0	90.3	148	17	26.8	15	78.8	138	6.1	1
1886	11.1	4.2	7.6	–1.8	26.6	7	–10.4	7	–0.6	12	14.6	7	80	3	812.1	119	193	38.9	11	1207.5	82	14.4	7

DURHAM 1887

Month	Mean max	Mean min	Mean temp	Anom	Highest max	Date	Lowest min	Date	Lowest max	Date	Highest min	Date	Air frost	Days ≥ 25 °C	Total pptn	Anom	Rain days	Wettest day	Date	Total sunshine	Anom	Sunniest day	Date
Jan	4.7	−0.8	1.9	−2.1	11.4	30	−9.6	7	−1.5	17	6.5	29	18	0	37.1	69	17	8.1	7	41.5	67	5.2	12
Feb	7.2	0.3	3.7	−0.8	12.9	27	−6.9	10	0.0	9	7.0	23	13	0	13.6	30	7	3.4	24	87.9	103	8.5	28
Mar	7.5	−0.3	3.6	−2.6	13.4	2	−7.3	17	1.1	13	4.9	30	16	0	52.8	128	16	21.6	11	137.8	114	9.9	13
Apr	10.0	1.1	5.5	−2.7	16.0	18	−2.6	15	5.2	14	6.8	19	9	0	19.6	38	17	3.1	26	145.2	94	12.9	28
May	11.8	4.4	8.1	−2.9	18.8	8	−1.1	4	6.0	3	8.0	16	1	0	51.8	116	16	10.6	19	97.5	52	12.1	31
June	19.1	9.0	14.0	+0.4	27.6	18	3.7	21	8.8	2	13.6	13	0	4	6.4	10	5	5.1	3	179.0	108	14.6	21
July	21.6	11.0	16.3	+0.5	27.0	21	3.6	6	14.7	5	15.5	13	0	5	20.7	33	12	7.0	5	216.0	123	13.8	20
Aug	19.5	9.0	14.2	−1.3	24.0	5	3.4	14	11.8	13	15.0	7	0	0	38.1	58	12	9.0	12	174.5	104	13.0	3
Sep	14.5	6.9	10.7	−2.6	18.6	6	−0.7	28	10.6	14	11.9	2	1	0	114.6	202	17	24.6	6	94.9	72	10.5	10
Oct	10.2	3.4	6.8	−3.2	14.9	3	−2.9	26	6.8	24	8.7	6	7	0	58.2	91	14	14.4	9	94.9	96	8.2	17
Nov	6.7	1.6	4.1	−2.5	11.2	26	−5.0	16	2.6	29	6.3	8	9	0	75.9	103	18	15.9	6	40.0	59	5.7	27
Dec	5.0	0.1	2.5	−1.7	10.6	1	−3.8	22	1.1	27	7.4	2	18	0	58.3	95	21	17.7	20	49.2	86	5.7	3
1887	11.5	3.8	7.6	−1.8	27.6	6	−9.6	6	−1.5	1	15.5	7	92	9	547.1	80	172	24.6	9	1358.4	92	14.6	6

DURHAM 1888

Month	Mean max	Mean min	Mean temp	Anom	Highest max	Date	Lowest min	Date	Lowest max	Date	Highest min	Date	Air frost	Days ≥ 25 °C	Total pptn	Anom	Rain days	Wettest day	Date	Total sunshine	Anom	Sunniest day	Date
Jan	6.2	0.5	3.4	−0.7	16.3	9	−6.4	30	1.1	19	7.1	8	12	0	18.6	35	17	2.4	27	36.2	59	6.0	9
Feb	4.3	−0.2	2.0	−2.5	10.9	8	−6.3	14	0.9	17	7.2	6	20	0	31.8	70	20	5.8	18	54.7	64	8.0	12
Mar	4.9	−0.2	2.4	−3.8	12.8	8	−3.8	26	0.5	12	7.4	9	21	0	103.4	250	20	35.8	29	83.5	69	10.0	24
Apr	8.9	1.9	5.4	−2.8	15.7	14	−1.8	7	3.7	24	6.7	29	10	0	61.8	121	20	21.0	20	96.9	63	11.0	16
May	14.2	4.6	9.4	−1.6	22.0	19	−1.1	10	8.7	28	9.9	19	1	0	13.3	30	7	7.5	29	202.1	108	14.9	21
June	14.3	7.0	10.7	−3.0	20.4	26	3.2	5	7.4	2	11.5	28	0	0	40.7	67	15	6.6	2	140.1	85	13.7	25
July	15.6	8.5	12.1	−3.7	23.0	20	3.6	7	10.1	5	12.3	21	0	0	209.7	338	20	57.1	25	94.1	54	14.5	19
Aug	16.9	8.8	12.9	−2.7	22.7	9	3.0	19	10.5	16	13.1	24	0	0	43.5	66	19	7.8	28	128.6	77	12.7	14
Sep	14.9	6.6	10.7	−2.5	19.1	2	1.9	30	9.2	30	11.9	2	0	0	32.8	58	17	11.6	1	117.8	89	10.2	14
Oct	11.8	4.2	8.0	−2.0	17.2	27	−2.3	2	7.3	6	14.4	27	6	0	17.2	27	9	5.0	3	78.5	80	7.4	14
Nov	9.1	4.4	6.7	+0.1	14.4	16	0.5	20	5.4	8	8.3	19	0	0	104.0	142	20	36.6	2	38.0	56	6.0	6
Dec	7.0	1.4	4.2	−0.0	13.1	4	−4.5	18	2.1	13	8.6	4	12	0	18.6	30	15	4.2	25	45.3	80	6.3	9
1888	10.7	4.0	7.3	−2.1	23.0	7	−6.4	7	0.5	3	14.4	10	82	0	695.4	102	199	57.1	7	1115.8	76	14.9	5

DURHAM 1889

Month	Mean max	Mean min	Mean temp	Anom	Highest max	Date	Lowest min	Date	Lowest max	Date	Highest min	Date	Air frost	Days ≥ 25 °C	Total pptn	Anom	Rain days	Wettest day	Date	Total sunshine	Anom	Sunniest day	Date
Jan	6.5	0.8	3.7	−0.4	11.8	31	−3.4	3	1.5	16	6.4	18	11	0	20.7	39	14	5.5	9	47.8	77	6.9	27
Feb	5.6	−0.3	2.6	−1.9	12.2	1	−5.4	10	0.5	10	7.6	18	18	0	26.0	57	15	6.0	3	74.8	88	7.7	9
Mar	7.7	0.5	4.1	−2.0	14.6	24	−11.3	4	1.9	2	8.0	29	11	0	57.5	139	10	22.7	8	94.8	79	9.6	10
Apr	9.3	2.7	6.0	−2.3	17.6	18	−0.7	3	4.5	6	7.3	20	1	0	54.5	106	18	8.9	4	78.5	51	9.9	29
May	16.0	7.0	11.5	+0.5	22.3	22	3.0	26	8.6	13	10.7	23	0	0	38.0	85	12	8.5	27	144.5	77	13.4	21
June	18.9	8.9	13.9	+0.3	24.8	26	4.2	11	10.8	9	11.2	3	0	0	9.5	16	5	5.9	9	214.0	129	15.2	21
July	18.3	9.4	13.8	−2.0	25.1	6	5.0	8	14.7	11	13.5	10	0	1	68.1	110	15	16.8	21	125.6	72	14.8	4
Aug	17.5	10.2	13.9	−1.7	22.3	1	6.0	25	13.9	24	13.0	1	0	0	81.9	124	19	10.6	19	120.7	72	10.4	17
Sep	15.2	7.7	11.5	−1.8	20.7	10	0.7	22	7.7	24	13.3	12	0	0	21.5	38	11	6.7	24	117.4	89	9.7	25
Oct	11.0	5.2	8.1	−1.9	14.3	11	1.4	11	7.0	27	8.0	16	0	0	94.0	148	24	14.6	22	42.1	43	7.4	8
Nov	8.7	2.7	5.7	−0.9	14.4	10	−2.2	28	2.3	26	9.6	9	7	0	23.0	31	13	7.2	14	53.6	79	7.1	10
Dec	5.9	0.1	3.0	−1.2	11.9	18	−3.8	29	−0.2	3	5.6	17	14	0	30.1	49	12	13.8	6	47.1	83	5.1	11
1889	11.7	4.6	8.1	−1.3	25.1	7	−11.3	3	−0.2	12	13.5	7	62	1	524.8	77	168	22.7	3	1160.9	79	15.2	6

DURHAM 1890

Month	Mean max	Mean min	Mean temp	Anom	Highest max	Date	Lowest min	Date	Lowest max	Date	Highest min	Date	Air frost	Days ≥ 25 °C	Total pptn	Anom	Rain days	Wettest day	Date	Total sunshine	Anom	Sunniest day	Date
Jan	8.0	2.0	5.0	+0.9	13.3	31	−6.4	3	−1.3	3	7.6	7	7	0	52.5	98	21	5.5	9	55.9	91	6.0	27
Feb	5.8	0.5	3.1	−1.5	10.6	1	−2.8	28	1.8	20	6.4	1	12	0	33.5	74	14	19.0	15	57.0	67	7.5	5
Mar	9.4	2.6	6.0	−0.2	15.9	27	−3.1	4	1.7	2	8.8	12	6	0	51.4	124	16	10.2	23	98.2	82	9.2	30
Apr	10.2	1.5	5.9	−2.4	16.7	30	−2.0	11	4.4	13	6.4	30	11	0	16.9	33	16	3.5	25	120.9	78	10.4	3
May	14.4	5.9	10.2	−0.8	22.2	22	1.5	4	9.2	11	9.9	22	0	0	59.1	133	14	11.3	7	152.8	82	14.2	14
June	17.2	8.4	12.8	−0.8	21.4	23	2.1	8	10.6	13	12.1	16	0	0	59.9	98	16	24.7	30	126.9	77	9.4	1
July	17.6	9.5	13.5	−2.3	21.9	21	5.9	5	12.1	1	14.5	31	0	0	45.3	73	16	9.8	2	159.1	91	12.5	24
Aug	17.6	9.3	13.4	−2.1	24.0	5	3.7	31	12.5	30	13.5	5	0	0	126.5	191	21	45.8	12	163.0	97	12.2	8
Sep	18.9	10.1	14.5	+1.2	23.4	8	2.1	1	15.2	29	14.8	27	0	0	46.3	82	12	11.1	21	143.6	108	10.9	7
Oct	12.8	6.0	9.4	−0.6	18.2	12	−1.9	28	4.5	26	13.4	4	2	0	37.5	59	15	8.4	26	122.0	124	8.9	12
Nov	8.5	1.3	4.9	−1.8	15.5	19	−4.7	29	2.5	29	6.7	19	8	0	85.6	117	24	21.0	23	56.4	83	6.5	1
Dec	3.1	−1.2	1.0	−3.3	9.3	1	−6.7	22	−1.3	19	4.9	1	19	0	44.1	72	15	6.3	25	10.8	19	2.3	20
1890	11.9	4.7	8.3	−1.1	24.0	8	−6.7	12	−1.3	1	14.8	9	65	0	658.6	97	200	45.8	8	1266.6	86	14.2	5

DURHAM 1891

Month	Mean max	Mean min	Mean temp	Anom	Highest max	Date	Lowest min	Date	Lowest max	Date	Highest min	Date	Air frost	Days ≥ 25 °C	Total pptn	Anom	Rain days	Wettest day	Date	Total sunshine	Anom	Sunniest day	Date
Jan	4.9	-1.5	1.7	-2.4	9.1	29	-12.2	18	-2.3	18	4.7	26	18	0	25.7	48	15	5.7	15	49.7	80	6.0	22
Feb	8.9	0.2	4.6	-0.0	14.3	15	-4.7	20	1.7	21	7.1	3	13	0	2.1	5	3	0.5	13	91.6	107	7.1	27
Mar	6.6	0.2	3.4	-2.8	15.9	1	-8.8	10	0.5	8	8.4	5	12	0	69.0	167	22	17.4	15	111.5	93	9.8	12
Apr	8.6	1.4	5.0	-3.3	16.2	30	-2.4	1	4.0	6	6.6	30	6	0	41.9	82	17	14.2	7	107.5	70	10.0	12
May	12.5	3.6	8.1	-2.9	21.9	13	-4.8	18	6.9	25	9.2	13	4	0	36.9	83	14	9.2	25	161.5	86	12.8	12
June	17.1	8.7	12.9	-0.7	23.6	19	4.2	3	9.9	7	13.2	29	0	0	4.1	7	9	0.8	15	175.1	106	13.4	20
July	18.3	10.4	14.4	-1.4	23.6	17	7.5	28	12.8	9	13.3	17	0	0	106.6	172	16	22.3	22	151.1	86	12.7	28
Aug	17.1	9.8	13.5	-2.1	21.7	14	5.1	1	13.4	23	13.9	8	0	0	126.7	191	21	25.2	21	98.5	59	9.5	13
Sep	16.9	9.0	12.9	-0.4	24.3	10	4.7	8	10.1	21	12.8	17	0	0	47.4	84	13	9.0	22	128.4	97	11.3	7
Oct	12.0	4.9	8.5	-1.5	15.4	4	-1.6	31	7.8	29	10.2	6	2	0	48.9	77	20	8.7	6	73.8	75	6.3	12
Nov	7.6	2.5	5.0	-1.6	11.4	19	-3.3	28	2.9	25	7.5	2	6	0	56.9	77	20	18.2	10	42.0	62	6.3	27
Dec	5.9	-0.2	2.8	-1.4	13.3	3	-9.3	25	-3.8	25	7.3	5	11	0	57.5	94	18	16.1	7	56.2	99	5.8	6
1891	11.4	4.1	7.7	-1.7	24.3	9	-12.2	1	-3.8	12	13.9	8	72	0	623.7	92	188	25.2	8	1246.9	85	13.4	6

DURHAM 1892

Month	Mean max	Mean min	Mean temp	Anom	Highest max	Date	Lowest min	Date	Lowest max	Date	Highest min	Date	Air frost	Days ≥ 25 °C	Total pptn	Anom	Rain days	Wettest day	Date	Total sunshine	Anom	Sunniest day	Date
Jan	4.7	-1.1	1.8	-2.3	12.5	29	-8.1	7	-0.9	16	6.5	29	19	0	48.6	91	15	25.0	9	47.6	77	5.9	3
Feb	5.6	-0.4	2.6	-2.0	11.9	7	-14.7	19	0.4	18	5.0	8	9	0	49.2	109	20	10.9	14	64.6	76	8.5	12
Mar	6.5	-1.7	2.4	-3.7	17.6	31	-6.6	30	0.7	27	7.5	18	23	0	28.9	70	13	6.3	10	139.1	115	11.5	30
Apr	10.9	1.2	6.0	-2.2	20.2	1	-6.5	16	4.7	14	7.4	22	11	0	33.3	65	17	18.0	27	172.8	112	11.6	30
May	14.4	5.4	9.9	-1.1	22.0	31	-1.2	2	6.8	4	11.9	27	2	1	59.5	134	17	15.9	27	156.8	84	12.8	1
June	16.2	7.4	11.8	-1.8	25.2	9	2.9	12	8.6	11	13.1	27	0	0	80.2	131	19	14.6	28	176.4	107	14.8	8
July	16.4	9.1	12.7	-3.1	21.2	24	4.1	21	11.4	14	12.7	31	0	0	42.9	69	15	10.3	19	123.0	70	13.4	7
Aug	17.7	9.7	13.7	-1.8	23.3	23	1.6	10	11.8	29	15.3	22	0	0	99.7	151	17	21.9	7	121.7	73	11.7	20
Sep	14.7	7.5	11.1	-2.2	18.6	10	2.5	5	10.5	30	11.8	19	0	0	74.2	131	17	19.3	1	116.6	88	10.7	8
Oct	9.6	3.0	6.3	-3.7	15.0	28	-4.0	26	5.4	25	7.5	13	6	0	141.7	223	26	45.8	14	71.8	73	7.3	26
Nov	8.5	2.7	5.6	-1.0	12.1	5	-4.4	2	1.3	18	8.1	28	5	0	31.5	43	20	9.1	6	37.6	55	6.3	30
Dec	4.0	-1.7	1.1	-3.1	12.3	18	-9.8	26	-1.8	30	7.3	17	21	0	24.4	40	12	4.6	8	57.6	101	6.0	4
1892	10.8	3.4	7.1	-2.4	25.2	6	-14.7	2	-1.8	12	15.3	8	96	1	714.1	105	208	45.8	10	1285.6	87	14.8	6

DURHAM 1893

Month	Mean max	Mean min	Mean temp	Anom	Highest max	Date	Lowest min	Date	Lowest max	Date	Highest min	Date	Air frost	Days ≥ 25 °C	Total pptn	Anom	Rain days	Wettest day	Date	Total sunshine	Anom	Sunniest day	Date
Jan	5.0	-0.2	2.4	-1.7	10.5	23	-10.9	6	-2.3	5	7.3	23	9	0	33.5	63	20	9.7	31	37.1	60	5.5	25
Feb	6.3	0.7	3.5	-1.0	11.7	19	-4.4	28	2.2	26	6.3	19	11	0	54.0	119	19	13.7	26	72.6	85	7.8	27
Mar	11.5	1.8	6.7	+0.5	18.4	24	-5.2	19	4.5	18	6.9	5	11	0	11.1	27	10	6.0	1	166.3	138	10.2	19
Apr	13.2	2.6	7.9	-0.3	20.9	20	-4.1	12	6.1	17	7.4	20	2	0	9.4	18	9	4.5	16	172.7	112	13.5	30
May	15.8	6.3	11.1	+0.1	22.9	14	0.8	10	9.4	30	10.7	20	0	0	59.0	133	13	18.6	17	160.4	86	13.3	10
June	18.4	9.3	13.8	+0.2	27.9	18	4.6	1	11.8	23	13.4	28	0	2	20.7	34	11	5.6	26	194.2	117	13.6	5
July	18.8	10.1	14.4	-1.4	25.7	7	4.9	28	12.7	13	13.6	24	0	2	114.1	184	19	61.4	8	135.9	78	12.2	7
Aug	21.0	11.4	16.2	+0.6	28.5	15	6.4	28	13.8	27	16.2	18	0	3	54.1	82	17	7.9	2	194.8	116	12.0	12
Sep	16.0	7.3	11.7	-1.6	22.5	5	0.8	23	7.8	23	12.9	14	0	0	45.1	79	16	11.2	25	139.5	105	9.9	20
Oct	13.3	4.9	9.1	-0.9	17.9	16	-2.8	31	5.4	30	14.0	15	2	0	18.1	28	10	9.0	7	150.6	153	10.0	1
Nov	8.3	2.0	5.1	-1.5	13.6	28	-3.4	1	2.3	23	9.5	28	9	0	60.1	82	23	8.4	7	51.9	76	6.3	13
Dec	7.6	1.6	4.6	+0.4	11.1	16	-5.0	2	2.1	2	8.9	16	6	0	32.9	54	16	7.6	12	51.9	91	6.2	2
1893	12.9	4.8	8.9	-0.6	28.5	8	-10.9	1	-2.3	1	16.2	8	50	7	512.1	75	183	61.4	7	1527.9	104	13.6	6

DURHAM 1894

Month	Mean max	Mean min	Mean temp	Anom	Highest max	Date	Lowest min	Date	Lowest max	Date	Highest min	Date	Air frost	Days ≥ 25 °C	Total pptn	Anom	Rain days	Wettest day	Date	Total sunshine	Anom	Sunniest day	Date
Jan	5.4	-0.4	2.5	-1.6	10.6	11	-13.8	11	-3.5	7	4.9	11	13	0	27.3	51	20	3.5	8	46.4	75	7.0	23
Feb	7.3	0.3	3.8	-0.8	12.0	7	-6.3	20	0.5	19	5.9	2	10	0	52.1	115	14	23.6	16	76.2	89	7.4	14
Mar	11.0	0.8	5.9	-0.3	16.7	30	-3.4	17	7.1	5	5.4	6	9	0	22.9	55	17	12.1	12	191.4	159	10.9	26
Apr	12.4	3.3	7.8	-0.4	17.9	10	-1.7	20	6.7	13	6.6	15	3	0	28.0	55	18	6.4	16	125.8	82	11.9	11
May	11.4	3.4	7.4	-3.6	14.9	11	-1.7	21	6.5	19	6.8	15	5	0	72.3	162	22	16.3	28	172.7	92	14.2	24
June	16.0	7.7	11.8	-1.8	25.0	30	2.1	1	8.8	5	12.9	26	0	1	94.1	154	16	23.7	4	146.3	88	14.1	21
July	19.5	10.8	15.1	-0.7	24.5	2	7.0	16	15.8	23	13.3	25	0	0	40.4	65	15	10.9	21	177.3	101	14.0	12
Aug	17.3	9.2	13.2	-2.3	21.3	30	4.9	23	11.9	27	12.4	1	0	0	64.0	97	20	11.3	9	108.4	65	11.9	22
Sep	14.0	6.8	10.4	-2.8	19.7	11	2.5	13	11.3	26	11.6	15	0	0	22.1	39	14	5.6	1	90.8	69	9.7	27
Oct	11.0	4.4	7.7	-2.3	16.8	13	-4.1	23	6.2	19	9.8	12	3	0	128.9	202	25	17.5	20	53.1	54	8.8	2
Nov	9.9	4.5	7.2	+0.6	15.3	2	0.8	28	6.2	27	11.7	1	0	0	19.8	27	15	4.2	12	56.9	84	6.2	30
Dec	7.1	1.0	4.1	-0.2	12.0	13	-3.1	2	1.9	31	6.7	11	14	0	32.7	53	18	8.7	14	48.1	84	5.6	15
1894	11.9	4.3	8.1	-1.4	25.0	6	-13.8	1	-3.5	1	13.3	7	57	1	604.6	89	214	23.7	6	1293.4	88	14.2	5

DURHAM 1895

Month	Mean max	Mean min	Mean temp	Anom	Highest max	Date	Lowest min	Date	Lowest max	Date	Highest min	Date	Air frost	Days ≥ 25 °C	Total pptn	Anom	Rain days	Wettest day	Date	Total sunshine	Anom	Sunniest day	Date
Jan	2.3	-3.4	-0.6	-4.6	5.2	2	-12.5	11	-1.6	11	0.7	20	25	0	94.6	177	24	14.4	19	43.4	70	6.3	18
Feb	2.4	-4.5	-1.1	-5.6	7.2	28	-18.0	8	-3.3	6	2.2	23	20	0	34.0	75	14	7.6	24	81.9	96	8.3	16
Mar	7.9	1.0	4.4	-1.7	14.8	21	-5.0	3	1.3	3	4.3	22	9	0	48.9	118	19	11.2	9	91.1	76	9.1	16
Apr	11.4	3.2	7.3	-1.0	16.5	21	-2.8	8	5.0	2	8.6	23	7	0	33.8	66	12	16.8	25	144.8	94	12.5	14
May	15.6	5.5	10.6	-0.4	23.2	30	1.2	9	5.3	17	12.4	31	0	0	21.0	47	12	5.1	9	197.5	106	14.2	7
June	18.4	7.4	12.9	-0.7	25.3	26	0.7	13	11.2	14	12.8	23	0	1	32.5	53	12	9.4	27	222.9	135	15.1	7
July	18.1	9.7	13.9	-1.9	24.0	8	6.9	16	12.5	25	13.8	18	0	0	126.1	203	18	39.6	25	143.8	82	15.0	15
Aug	19.2	10.7	15.0	-0.6	24.4	17	6.5	1	13.9	4	16.5	17	0	0	59.0	89	19	11.1	4	139.3	83	10.0	19
Sep	19.6	9.2	14.4	+1.1	24.3	26	2.5	22	14.9	5	13.7	27	0	0	15.1	27	12	4.3	3	203.9	154	10.8	1
Oct	10.2	2.5	6.4	-3.6	18.7	1	-3.5	29	4.3	26	10.5	1	9	0	125.4	197	25	26.2	15	112.2	114	8.8	4
Nov	9.4	3.3	6.3	-0.3	14.6	16	-1.6	2	6.6	24	7.9	6	4	0	62.7	85	23	8.7	5	45.6	67	5.5	8
Dec	5.7	0.4	3.0	-1.2	12.1	5	-6.2	22	1.5	28	4.2	31	12	0	47.1	77	20	6.3	29	29.8	52	5.8	10
1895	11.7	3.8	7.7	-1.7	25.3	6	-18.0	2	-3.3	2	16.5	8	86	1	700.2	103	210	39.6	7	1456.2	99	15.1	6

DURHAM 1896

Month	Mean max	Mean min	Mean temp	Anom	Highest max	Date	Lowest min	Date	Lowest max	Date	Highest min	Date	Air frost	Days ≥ 25 °C	Total pptn	Anom	Rain days	Wettest day	Date	Total sunshine	Anom	Sunniest day	Date
Jan	7.3	1.6	4.5	+0.4	10.9	2	-6.5	23	2.1	21	8.1	17	11	0	17.9	34	13	6.7	13	62.8	102	7.0	28
Feb	7.4	1.3	4.4	-0.2	12.1	8	-4.4	18	0.9	25	7.5	11	11	0	12.8	28	8	6.4	20	39.4	46	6.6	21
Mar	9.7	1.7	5.7	-0.5	17.3	25	-2.1	13	2.9	14	7.3	25	8	0	46.7	113	19	15.7	13	111.1	92	8.6	27
Apr	12.8	3.9	8.3	+0.1	18.3	24	-1.5	15	5.5	14	8.6	8	1	0	20.6	40	15	4.1	10	153.8	100	10.6	18
May	15.5	5.2	10.3	-0.7	25.0	31	-1.7	4	9.2	1	8.9	18	1	1	13.6	31	9	5.9	21	190.6	102	14.4	11
June	19.7	9.7	14.7	+1.1	26.5	16	4.5	1	14.1	6	13.5	16	0	4	50.8	83	16	16.0	4	198.3	120	14.2	21
July	19.7	10.0	14.9	-0.9	26.6	20	5.6	28	14.4	27	14.7	19	0	2	64.1	103	11	19.2	24	179.1	102	13.9	28
Aug	16.9	9.5	13.2	-2.3	21.9	11	4.8	17	13.7	5	12.9	13	0	0	53.3	81	24	13.7	19	97.1	58	10.6	26
Sep	15.1	8.4	11.8	-1.5	19.5	10	1.2	21	10.4	25	13.1	2	0	0	56.7	100	22	15.9	25	90.7	68	9.5	15
Oct	9.3	2.8	6.1	-4.0	16.3	2	-2.9	26	4.2	29	10.7	3	6	0	148.8	234	29	26.8	10	79.9	81	8.3	9
Nov	7.6	1.6	4.6	-2.1	11.5	11	-5.0	30	2.3	30	6.8	11	10	0	34.1	46	16	9.9	7	52.1	77	7.2	4
Dec	5.8	0.8	3.3	-0.9	11.6	26	-4.5	23	1.8	23	6.2	31	13	0	103.9	170	23	10.7	14	35.0	61	5.6	27
1896	12.2	4.7	8.5	-1.0	26.6	7	-6.5	1	0.9	2	14.7	7	61	7	623.3	92	205	26.8	10	1289.9	88	14.4	5

DURHAM 1897

Month	Mean max	Mean min	Mean temp	Anom	Highest max	Date	Lowest min	Date	Lowest max	Date	Highest min	Date	Air frost	Days ≥ 25 °C	Total pptn	Anom	Rain days	Wettest day	Date	Total sunshine	Anom	Sunniest day	Date
Jan	3.7	-0.9	1.4	-2.7	9.1	1	-6.6	25	1.3	26	2.6	8	17	0	47.2	88	20	7.1	7	40.7	66	6.4	18
Feb	7.3	1.9	4.6	+0.0	13.9	22	-3.7	12	2.3	4	8.0	25	10	0	24.0	53	13	6.7	5	59.8	70	8.2	27
Mar	8.4	2.0	5.2	-1.0	14.2	21	-3.5	30	4.2	31	7.9	22	6	0	61.6	149	23	9.9	14	95.9	80	8.4	27
Apr	9.5	1.8	5.6	-2.6	17.2	28	-3.5	1	4.2	2	5.9	27	7	0	42.2	82	19	6.7	4	152.0	99	11.4	18
May	12.3	3.9	8.1	-2.9	18.7	31	0.0	27	7.7	12	8.3	31	0	0	37.1	83	12	11.3	27	197.9	106	14.8	17
June	16.9	8.9	12.9	-0.7	25.7	12	3.9	10	8.8	8	14.4	12	0	2	74.7	122	17	18.2	18	144.7	87	12.6	22
July	19.7	10.0	14.9	-0.9	24.7	23	5.4	8	13.7	20	14.8	24	0	0	31.3	50	8	19.7	20	233.3	133	14.8	12
Aug	20.4	10.5	15.5	-0.1	28.0	5	7.2	2	17.0	21	15.2	6	0	4	60.7	92	18	15.5	5	206.0	123	12.1	27
Sep	15.0	6.5	10.8	-2.5	19.4	12	0.5	9	11.5	8	11.2	14	0	0	54.4	96	15	13.0	1	112.7	85	10.5	25
Oct	12.9	5.6	9.2	-0.8	17.5	17	-3.4	14	6.4	14	9.1	18	2	0	36.1	57	4	16.7	14	105.2	107	9.2	13
Nov	9.6	4.2	6.9	+0.3	13.9	13	-3.3	16	4.3	15	9.8	13	4	0	44.5	61	17	18.3	14	29.3	43	5.3	22
Dec	7.0	1.0	4.0	-0.2	12.3	16	-5.1	23	1.2	24	8.3	17	8	0	40.4	66	19	5.6	15	31.9	56	4.3	11
1897	11.9	4.6	8.2	-1.2	28.0	8	-6.6	8	1.2	12	15.2	8	54	6	554.2	81	185	19.7	7	1409.4	96	14.8	5

DURHAM 1898

Month	Mean max	Mean min	Mean temp	Anom	Highest max	Date	Lowest min	Date	Lowest max	Date	Highest min	Date	Air frost	Days ≥ 25 °C	Total pptn	Anom	Rain days	Wettest day	Date	Total sunshine	Anom	Sunniest day	Date
Jan	9.2	3.1	6.2	+2.1	13.8	30	-2.0	25	6.3	3	10.1	30	6	0	25.4	48	11	10.1	3	40.9	66	6.7	7
Feb	7.2	0.4	3.8	-0.8	14.7	1	-6.3	21	2.5	21	5.4	10	13	0	19.1	42	18	2.3	20	100.8	118	7.7	7
Mar	7.7	0.1	3.9	-2.2	13.4	18	-4.7	8	2.6	24	8.3	17	17	0	49.7	120	15	15.5	26	118.1	98	10.1	20
Apr	11.5	3.3	7.4	-0.8	18.0	8	-3.1	5	6.6	27	8.7	7	2	0	51.8	101	17	12.8	28	128.5	83	12.0	17
May	12.2	4.1	8.2	-2.8	17.0	7	-0.8	15	7.7	2	7.4	8	1	0	41.4	93	21	9.5	21	148.1	79	11.6	18
June	16.6	7.8	12.2	-1.4	23.2	17	1.8	1	10.2	13	12.1	18	0	0	38.8	63	15	7.5	6	136.7	83	13.4	9
July	18.8	8.8	13.8	-2.0	22.6	15	2.5	20	12.5	29	14.7	7	0	0	16.4	26	9	5.6	22	208.9	119	14.3	24
Aug	19.6	10.3	14.9	-0.6	25.8	12	5.2	25	13.7	29	15.8	11	0	2	45.5	69	16	13.2	5	157.1	94	11.3	14
Sep	18.6	9.7	14.1	+0.9	27.7	4	2.2	24	11.1	24	16.4	3	0	4	8.1	14	6	6.3	29	106.6	80	11.8	4
Oct	13.7	7.4	10.6	+0.6	20.2	3	4.5	11	8.5	16	11.3	22	0	0	123.6	194	23	29.8	17	75.9	77	9.0	4
Nov	9.3	3.0	6.1	-0.5	15.1	2	-4.2	29	0.9	29	7.0	12	5	0	77.5	106	23	19.6	28	55.2	81	6.8	3
Dec	9.5	3.1	6.3	+2.0	14.1	5	-4.9	31	1.9	31	11.2	5	4	0	30.5	50	16	10.0	29	51.0	90	5.8	30
1898	12.8	5.1	9.0	-0.5	27.7	9	-6.3	2	0.9	11	16.4	9	48	6	527.8	78	190	29.8	10	1327.8	90	14.3	7

DURHAM 1899

Month	Mean max	Mean min	Mean temp	Anom	Highest max	Date	Lowest min	Date	Lowest max	Date	Highest min	Date	Air frost	Days ≥ 25 °C	Total pptn	Anom	Rain days	Wettest day	Date	Total sunshine	Anom	Sunniest day	Date
Jan	6.5	0.8	3.6	−0.5	11.1	21	−6.3	28	2.7	27	6.6	21	9	0	56.0	105	23	10.9	22	48.3	78	6.2	14
Feb	7.1	−0.2	3.4	−1.1	13.4	10	−6.2	4	2.0	4	4.7	12	17	0	33.6	74	12	9.8	1	88.0	103	7.6	23
Mar	8.8	0.4	4.6	−1.6	18.1	17	−6.0	21	−0.2	23	7.1	11	15	0	31.5	76	15	4.5	23	120.6	100	9.8	15
Apr	10.6	2.7	6.6	−1.6	16.1	28	−2.7	17	4.5	15	8.5	2	5	0	55.3	108	19	10.6	14	115.6	75	11.2	16
May	11.7	3.7	7.7	−3.3	22.1	31	−2.1	5	6.5	12	8.7	31	2	0	80.3	180	18	20.5	11	174.5	93	14.4	27
June	19.3	8.1	13.7	+0.1	25.8	6	1.4	14	12.8	8	11.9	29	0	2	34.3	56	9	12.2	29	210.2	127	14.7	14
July	20.2	11.4	15.8	+0.0	25.8	30	7.9	9	14.9	2	15.1	29	0	2	77.0	124	18	16.5	2	152.5	87	14.5	30
Aug	20.3	10.7	15.5	−0.1	28.6	1	6.2	11	15.2	6	13.8	3	0	3	29.0	44	11	10.8	28	163.3	97	13.7	1
Sep	16.5	7.8	12.2	−1.1	23.5	5	−0.2	29	11.0	23	14.1	5	1	0	61.5	108	19	22.1	29	139.6	105	10.2	7
Oct	12.6	3.8	8.2	−1.8	15.8	19	−1.5	20	8.6	14	9.6	11	3	0	48.8	77	11	32.0	1	124.0	126	8.6	4
Nov	11.1	3.8	7.4	+0.8	16.2	5	−3.2	17	3.6	18	10.0	27	5	0	39.7	54	14	10.4	2	66.9	98	7.7	9
Dec	3.4	−2.7	0.3	−3.9	11.6	4	−9.8	13	−1.8	12	4.9	6	22	0	63.7	104	19	12.6	28	40.0	70	6.3	31
1899	12.3	4.2	8.3	−1.2	28.6	8	−9.8	12	−1.8	12	15.1	7	79	7	610.7	90	188	32.0	10	1443.5	98	14.7	6

DURHAM 1900

Month	Mean max	Mean min	Mean temp	Anom	Highest max	Date	Lowest min	Date	Lowest max	Date	Highest min	Date	Air frost	Days ≥ 25 °C	Total pptn	Anom	Rain days	Wettest day	Date	Total sunshine	Anom	Sunniest day	Date
Jan	6.0	0.9	3.5	−0.6	11.6	23	−4.6	28	2.4	14	8.9	23	9	0	47.3	89	19	7.7	1	41.0	66	6.4	18
Feb	3.5	−2.2	0.6	−3.9	11.0	23	−8.6	13	−1.9	9	3.4	24	18	0	93.8	207	21	27.7	26	64.5	76	8.4	22
Mar	5.7	−0.1	2.8	−3.4	12.7	11	−10.7	18	0.1	18	3.6	15	12	0	35.7	86	13	7.4	25	79.7	66	9.2	16
Apr	12.3	2.7	7.5	−0.8	23.1	21	−3.1	26	6.9	8	6.0	15	6	0	24.3	47	14	6.0	3	146.5	95	11.7	28
May	13.1	5.1	9.1	−1.9	19.8	27	0.2	12	6.7	11	10.8	28	0	0	30.3	68	13	7.1	8	119.5	64	12.5	16
June	18.0	8.8	13.4	−0.2	26.7	12	2.1	9	8.3	1	13.6	11	0	2	79.9	131	18	19.8	24	128.1	77	11.2	17
July	21.1	12.3	16.7	+0.9	27.3	11	4.5	8	13.4	4	17.6	24	0	5	28.5	46	14	6.5	16	202.3	115	14.8	18
Aug	17.5	9.9	13.7	−1.9	24.2	13	4.9	9	10.9	8	13.7	19	0	0	130.3	197	15	36.7	3	126.6	76	12.6	13
Sep	17.3	8.4	12.8	−0.4	21.1	12	3.9	3	12.9	7	13.1	22	0	0	17.2	30	10	9.6	7	138.1	104	10.9	12
Oct	12.3	4.6	8.5	−1.5	18.4	7	−2.1	16	6.2	26	11.3	7	2	0	137.7	216	20	50.1	26	98.4	100	9.4	14
Nov	8.6	3.8	6.2	−0.5	13.0	1	−1.7	24	4.7	19	10.0	1	5	0	60.8	83	21	12.8	3	29.3	43	6.3	9
Dec	8.8	4.0	6.4	+2.2	13.9	20	−1.5	30	3.9	23	10.2	12	3	0	46.3	76	19	15.0	30	32.5	57	5.3	17
1900	12.0	4.8	8.4	−1.0	27.3	7	−10.7	7	−1.9	2	17.6	7	55	7	732.1	108	197	50.1	10	1206.5	82	14.8	7

DURHAM 1901

Month	Mean max	Mean min	Mean temp	Anom	Highest max	Date	Lowest min	Date	Lowest max	Date	Highest min	Date	Air frost	Days ≥ 25 °C	Total pptn	Anom	Rain days	Wettest day	Date	Total sunshine	Anom	Sunniest day	Date
Jan	5.0	-0.3	2.3	-1.8	10.3	21	-7.6	9	-1.1	9	3.7	12	17	0	27.2	51	15	4.3	19	30.0	49	4.5	27
Feb	4.5	-0.5	2.0	-2.6	7.9	25	-4.7	7	1.9	1	4.2	25	17	0	28.9	64	17	6.8	5	47.8	56	5.7	13
Mar	7.0	0.4	3.7	-2.5	13.7	10	-6.1	26	1.8	27	5.8	12	11	0	35.4	86	19	6.3	16	114.7	95	10.3	27
Apr	12.4	2.4	7.4	-0.8	21.8	22	-2.1	2	7.6	2	7.7	23	5	0	24.6	48	15	5.2	29	194.1	126	12.7	21
May	14.6	5.0	9.8	-1.2	19.8	18	-0.3	22	6.8	9	10.7	31	1	0	42.2	95	13	12.1	26	218.6	117	14.1	13
June	17.8	7.6	12.7	-0.9	23.4	8	1.8	19	13.4	14	13.3	21	0	0	31.1	51	10	7.5	22	230.0	139	14.8	11
July	22.4	12.0	17.2	+1.4	29.1	19	6.6	4	16.2	25	17.2	19	0	9	65.0	105	13	26.6	21	219.2	125	15.1	8
Aug	20.2	10.6	15.4	-0.2	26.8	22	3.8	20	14.9	17	16.3	8	0	3	56.6	86	10	17.1	10	198.2	118	12.8	16
Sep	16.4	9.4	12.9	-0.3	20.1	25	0.8	2	12.9	30	15.7	28	0	0	35.0	62	15	11.4	8	124.4	94	11.2	5
Oct	12.4	4.5	8.5	-1.6	16.3	1	-2.4	26	8.0	30	12.0	28	3	0	60.2	95	19	18.9	21	92.6	94	8.9	7
Nov	8.2	2.0	5.1	-1.5	11.7	20	-4.9	16	2.2	14	9.1	19	10	0	129.9	177	12	72.8	12	78.0	115	6.9	15
Dec	5.0	0.2	2.6	-1.7	12.2	7	-5.9	17	0.0	16	8.2	1	15	0	48.1	79	15	8.5	13	48.8	86	5.8	27
1901	12.2	4.5	8.3	-1.1	29.1	7	-7.6	1	-1.1	1	17.2	7	79	12	584.2	86	173	72.8	11	1596.4	108	15.1	7

DURHAM 1902

Month	Mean max	Mean min	Mean temp	Anom	Highest max	Date	Lowest min	Date	Lowest max	Date	Highest min	Date	Air frost	Days ≥ 25 °C	Total pptn	Anom	Rain days	Wettest day	Date	Total sunshine	Anom	Sunniest day	Date
Jan	6.6	1.4	4.0	-0.1	13.2	4	-5.2	31	-0.9	27	7.6	21	14	0	13.3	25	9	3.9	1	36.1	58	7.2	26
Feb	3.3	-2.1	0.6	-4.0	10.7	28	-10.7	14	-0.4	11	5.3	24	16	0	26.4	58	14	6.2	26	64.3	75	7.9	10
Mar	9.5	1.7	5.6	-0.6	16.2	17	-3.4	24	5.8	11	5.7	18	5	0	14.1	34	12	4.1	24	109.9	91	10.5	20
Apr	10.8	2.0	6.4	-1.9	15.9	21	-3.7	10	6.2	6	9.5	22	6	0	37.3	73	11	16.5	15	187.6	122	12.3	28
May	11.0	3.7	7.3	-3.7	22.3	23	-0.8	9	7.5	11	10.1	23	3	0	57.9	130	19	13.2	30	164.3	88	12.3	29
June	15.6	7.5	11.6	-2.0	24.2	27	3.0	11	8.0	1	12.1	23	0	0	50.1	82	13	8.2	1	132.0	80	15.1	28
July	17.6	9.0	13.3	-2.5	23.6	14	4.1	12	11.2	20	13.6	14	0	0	69.9	113	20	22.2	26	134.7	77	11.9	5
Aug	16.4	8.4	12.4	-3.2	20.3	23	3.1	11	11.4	6	13.3	23	0	0	45.1	68	15	8.1	17	116.0	69	11.7	26
Sep	15.7	8.2	12.0	-1.3	22.2	8	2.1	25	10.9	11	12.6	4	0	0	23.1	41	14	4.9	15	101.7	77	8.9	24
Oct	11.7	5.9	8.8	-1.2	16.0	13	0.2	12	9.1	10	10.1	13	0	0	53.2	84	21	12.6	10	68.9	70	8.2	16
Nov	8.9	4.0	6.5	-0.2	14.2	1	-1.6	24	1.1	22	9.2	6	4	0	28.3	39	9	10.7	8	47.0	69	6.5	10
Dec	6.8	2.2	4.5	+0.3	12.6	14	-2.8	6	0.6	7	8.1	27	11	0	52.1	85	12	17.0	1	34.9	61	5.4	6
1902	11.2	4.3	7.7	-1.7	24.2	6	-10.7	2	-0.9	1	13.6	7	59	0	470.8	69	169	22.2	7	1197.4	81	15.1	6

DURHAM 1903

Month	Mean max	Mean min	Mean temp	Anom	Highest max	Date	Lowest min	Date	Lowest max	Date	Highest min	Date	Air frost	Days ≥25°C	Total pptn	Anom	Rain days	Wettest day	Date	Total sunshine	Anom	Sunniest day	Date
Jan	5.3	0.1	2.7	-1.4	11.7	26	-9.9	13	-0.6	14	7.3	26	14	0	57.9	108	14	11.0	5	42.3	68	5.7	23
Feb	8.9	2.9	5.9	+1.3	12.2	9	-2.1	13	4.2	2	10.7	8	6	0	26.7	59	10	7.9	26	54.9	64	7.0	26
Mar	9.4	2.2	5.8	-0.4	14.3	25	-0.9	4	4.3	2	7.2	22	5	0	30.6	74	17	5.0	17	103.8	86	8.7	6
Apr	9.2	0.6	4.9	-3.4	15.1	10	-4.1	23	4.0	17	5.2	6	13	0	35.2	69	17	9.0	14	134.1	87	9.7	25
May	12.8	5.1	8.9	-2.1	22.5	22	-1.5	12	6.1	9	9.8	21	1	0	68.6	154	14	12.6	5	143.0	76	13.8	26
June	15.4	7.4	11.4	-2.2	22.2	4	0.7	21	8.9	13	15.3	28	0	0	60.8	99	12	19.0	15	156.3	94	13.1	19
July	17.9	10.1	14.0	-1.8	25.4	9	3.7	8	11.9	19	14.9	10	0	2	122.8	198	18	19.8	14	142.5	81	13.4	31
Aug	17.2	9.1	13.1	-2.4	20.3	1	3.4	23	13.9	24	13.4	27	0	0	42.0	63	16	7.3	24	168.4	100	12.3	22
Sep	15.2	7.9	11.6	-1.7	19.1	26	0.2	16	11.2	14	12.9	24	0	0	59.5	105	17	16.6	2	120.1	91	9.3	18
Oct	11.8	6.0	8.9	-1.1	16.6	3	1.7	24	8.9	18	9.6	12	0	0	201.8	317	28	63.7	8	69.8	71	8.4	4
Nov	8.4	2.2	5.3	-1.3	12.7	12	-4.4	19	0.5	30	8.0	2	8	0	52.1	71	14	24.1	30	72.6	107	7.3	3
Dec	4.4	0.3	2.3	-1.9	9.1	22	-5.8	6	0.8	1	4.8	22	11	0	25.6	42	13	5.8	24	10.4	18	3.5	4
1903	11.3	4.5	7.9	-1.6	25.4	7	-9.9	1	-0.6	1	15.3	6	58	2	783.6	115	190	63.7	10	1218.2	83	13.8	5

DURHAM 1904

Month	Mean max	Mean min	Mean temp	Anom	Highest max	Date	Lowest min	Date	Lowest max	Date	Highest min	Date	Air frost	Days ≥25°C	Total pptn	Anom	Rain days	Wettest day	Date	Total sunshine	Anom	Sunniest day	Date
Jan	6.0	0.9	3.4	-0.6	10.7	27	-3.8	25	1.7	2	6.8	27	13	0	44.8	84	16	8.9	31	35.7	58	5.1	16
Feb	4.6	-0.1	2.3	-2.3	10.0	20	-4.6	29	0.6	26	4.9	20	15	0	42.0	93	20	8.4	3	54.9	64	6.6	18
Mar	6.5	0.4	3.4	-2.7	12.9	20	-3.9	27	1.2	4	9.1	20	13	0	31.2	75	15	5.1	7	85.8	71	9.8	30
Apr	12.1	4.0	8.1	-0.2	18.8	24	0.1	12	6.3	1	9.0	29	0	0	25.6	50	14	5.3	14	163.0	106	11.8	9
May	13.3	5.3	9.3	-1.7	20.0	26	-0.1	20	6.2	7	9.7	13	1	0	74.9	168	16	23.2	27	137.6	74	12.1	18
June	16.5	8.4	12.5	-1.1	22.0	29	3.5	3	10.1	6	11.9	16	0	0	33.1	54	8	20.4	24	180.2	109	14.1	4
July	19.7	11.0	15.4	-0.4	23.8	11	5.9	18	15.3	27	14.1	15	0	0	43.3	70	12	18.5	30	172.3	98	14.7	11
Aug	18.4	9.9	14.2	-1.4	26.2	29	2.5	21	10.7	22	16.2	4	0	2	63.1	95	18	26.3	22	162.4	97	12.0	1
Sep	16.1	7.4	11.8	-1.5	20.4	5	1.7	21	11.6	25	11.3	2	0	0	25.0	44	8	12.1	24	137.3	104	10.6	18
Oct	12.6	4.9	8.7	-1.3	17.4	20	-1.8	13	8.6	16	11.0	19	4	0	18.5	29	7	8.1	6	111.3	113	9.7	2
Nov	8.3	2.1	5.2	-1.5	15.2	3	-5.8	26	1.1	22	9.4	4	9	0	45.6	62	13	20.8	21	60.1	88	5.7	8
Dec	5.7	-0.1	2.8	-1.5	14.0	4	-7.6	22	-1.8	21	11.3	17	16	0	35.2	58	12	10.6	9	50.4	89	5.8	8
1904	11.7	4.5	8.1	-1.4	26.2	8	-7.6	12	-1.8	12	16.2	8	71	2	482.3	71	159	26.3	8	1351.0	92	14.7	7

DURHAM 1905

Month	Mean max	Mean min	Mean temp	Anom	Highest max	Date	Lowest min	Date	Lowest max	Date	Highest min	Date	Air frost	Days ≥ 25 °C	Total pptn	Anom	Rain days	Wettest day	Date	Total sunshine	Anom	Sunniest day	Date
Jan	6.6	1.2	3.9	−0.2	11.3	6	−4.8	19	0.5	16	7.8	8	13	0	9.1	17	6	2.4	23	42.3	68	6.1	25
Feb	7.3	1.4	4.3	−0.2	12.3	5	−4.4	12	2.7	24	6.1	4	9	0	18.3	40	12	3.0	1	71.1	83	7.6	17
Mar	10.2	2.2	6.2	+0.0	13.4	23	−1.7	3	5.5	1	5.2	23	4	0	30.8	75	11	9.1	25	132.5	110	9.7	19
Apr	9.7	2.2	5.9	−2.3	14.1	27	−4.8	8	1.7	7	8.4	28	5	0	72.5	142	21	10.5	14	98.8	64	11.3	8
May	13.8	5.0	9.4	−1.6	21.4	28	−2.6	5	7.4	21	13.5	28	2	0	8.1	18	5	4.2	2	148.5	79	11.8	18
June	17.9	8.8	13.3	−0.3	25.0	27	5.3	5	11.1	7	13.6	28	0	1	17.5	29	7	6.7	18	220.5	133	14.7	25
July	21.6	11.4	16.5	+0.7	28.4	13	4.8	19	13.8	1	16.1	12	0	5	19.1	31	5	11.9	22	202.7	116	15.4	7
Aug	17.6	9.9	13.7	−1.8	20.6	8	3.7	9	13.4	26	13.3	4	0	0	69.3	105	17	14.8	3	139.6	83	12.4	24
Sep	15.3	8.1	11.7	−1.5	21.2	4	0.7	21	10.1	30	13.3	3	0	0	56.4	99	15	10.2	9	116.7	88	10.8	11
Oct	10.4	3.2	6.8	−3.3	17.1	9	−1.2	17	6.6	16	9.6	9	4	0	56.8	89	16	14.4	14	99.0	100	8.3	17
Nov	7.1	1.9	4.5	−2.2	10.9	23	−6.6	19	2.3	19	6.3	2	9	0	116.8	159	19	17.2	1	58.5	86	6.7	27
Dec	7.3	3.0	5.2	+0.9	11.2	7	−2.2	31	1.1	31	8.7	21	1	0	13.1	21	5	6.0	28	50.8	89	6.0	9
1905	12.1	4.8	8.4	−1.0	28.4	7	−6.6	11	0.5	1	16.1	7	47	6	487.8	72	139	17.2	11	1381.0	94	15.4	7

DURHAM 1906

Month	Mean max	Mean min	Mean temp	Anom	Highest max	Date	Lowest min	Date	Lowest max	Date	Highest min	Date	Air frost	Days ≥ 25 °C	Total pptn	Anom	Rain days	Wettest day	Date	Total sunshine	Anom	Sunniest day	Date
Jan	6.8	2.1	4.5	+0.4	11.1	26	−2.3	1	1.7	1	8.1	28	4	0	57.7	108	17	14.3	18	35.7	58	6.5	22
Feb	5.4	−0.9	2.2	−2.3	7.9	7	−4.2	24	0.8	13	3.2	1	17	0	27.0	60	12	9.3	19	103.2	121	7.8	20
Mar	7.6	0.9	4.3	−1.9	13.0	15	−7.9	14	2.2	12	8.4	7	11	0	34.2	83	14	6.4	23	132.8	110	9.3	14
Apr	11.6	0.2	5.9	−2.4	18.4	12	−3.5	3	5.7	26	5.3	16	15	0	22.1	43	10	4.3	25	195.5	127	12.2	10
May	12.6	5.6	9.1	−1.9	19.4	28	−1.8	22	7.1	20	12.6	7	3	0	94.6	213	21	38.9	19	107.3	57	10.0	22
June	18.1	8.8	13.4	−0.2	24.7	12	1.2	5	11.7	16	14.7	24	0	0	21.1	34	10	4.8	16	194.1	117	13.3	19
July	19.7	9.7	14.7	−1.1	23.4	22	5.2	13	15.1	1	14.6	18	0	0	35.1	57	10	21.8	31	211.4	121	15.2	20
Aug	20.0	11.1	15.6	−0.0	28.7	31	8.3	26	15.1	18	16.1	22	0	1	64.9	98	19	13.4	7	156.3	93	12.9	6
Sep	17.5	8.0	12.8	−0.5	30.0	1	0.5	27	11.5	23	15.7	2	0	3	11.7	21	6	4.6	13	163.1	123	12.1	1
Oct	12.8	6.5	9.6	−0.4	19.6	5	0.1	14	6.0	30	13.4	7	0	0	113.4	178	26	36.0	18	71.9	73	9.0	17
Nov	9.5	4.6	7.1	+0.4	15.3	22	−1.7	14	4.8	19	9.9	29	3	0	72.3	98	17	18.1	2	39.1	57	7.0	6
Dec	5.1	−0.1	2.5	−1.8	12.4	5	−6.5	26	−2.2	30	7.9	3	19	0	50.3	82	16	11.5	5	48.5	85	5.4	1
1906	12.2	4.7	8.5	−1.0	30.0	9	−7.9	3	−2.2	12	16.1	8	72	4	604.4	89	178	38.9	5	1458.9	99	15.2	7

DURHAM 1907

Month	Mean max	Mean min	Mean temp	Anom	Highest max	Date	Lowest min	Date	Lowest max	Date	Highest min	Date	Air frost	Days ≥ 25 °C	Total pptn	Anom	Rain days	Wettest day	Date	Total sunshine	Anom	Sunniest day	Date
Jan	5.9	0.5	3.2	-0.9	11.1	7	-6.9	24	-0.6	23	6.7	14	14	0	13.8	26	8	5.2	1	72.5	117	6.8	16
Feb	5.8	-0.6	2.6	-2.0	13.6	27	-6.9	3	0.8	8	5.1	25	17	0	26.4	58	11	7.7	12	115.8	136	8.8	22
Mar	10.6	0.6	5.6	-0.6	18.2	27	-4.3	12	3.9	4	6.4	16	10	0	21.6	52	13	7.0	9	187.1	155	11.8	28
Apr	10.7	2.8	6.8	-1.5	17.9	24	-3.3	19	6.0	12	9.1	23	4	0	51.4	100	14	16.7	3	130.9	85	12.0	24
May	12.1	5.2	8.6	-2.4	18.8	11	0.4	23	6.9	21	10.9	11	0	0	94.6	213	19	21.7	14	105.5	56	13.3	3
June	15.2	7.7	11.5	-2.1	21.2	9	1.7	4	10.6	1	11.2	15	0	0	77.0	126	19	14.4	20	143.2	86	12.0	17
July	16.8	8.7	12.8	-3.0	23.9	16	3.6	2	10.9	9	13.8	15	0	0	39.7	64	16	9.7	4	134.8	77	14.0	16
Aug	17.3	9.2	13.3	-2.3	21.5	4	4.9	31	12.4	15	12.8	4	0	0	38.0	57	15	10.3	14	140.4	84	12.2	30
Sep	17.4	7.2	12.3	-1.0	22.7	19	-1.1	4	11.3	30	12.1	7	1	0	6.3	11	6	2.7	2	139.5	105	10.0	1
Oct	12.5	5.6	9.0	-1.0	15.6	3	-2.2	8	8.8	27	11.2	6	2	0	142.0	223	22	20.9	30	80.3	81	7.0	4
Nov	8.4	2.5	5.5	-1.2	15.2	2	-3.1	25	3.9	25	8.4	3	8	0	48.2	66	19	9.4	26	54.3	80	6.7	13
Dec	6.3	1.2	3.8	-0.5	10.8	8	-1.8	24	1.7	29	7.3	21	8	0	72.6	119	19	10.5	4	31.6	56	5.3	6
1907	11.6	4.2	7.9	-1.5	23.9	7	-6.9	1	-0.6	1	13.8	7	64	0	631.6	93	181	21.7	5	1335.9	91	14.0	7

DURHAM 1908

Month	Mean max	Mean min	Mean temp	Anom	Highest max	Date	Lowest min	Date	Lowest max	Date	Highest min	Date	Air frost	Days ≥ 25 °C	Total pptn	Anom	Rain days	Wettest day	Date	Total sunshine	Anom	Sunniest day	Date
Jan	5.1	-0.8	2.1	-2.0	12.4	27	-7.6	4	-1.7	13	6.9	17	19	0	27.3	51	9	12.4	7	63.9	103	6.3	20
Feb	8.1	1.3	4.7	+0.1	12.8	6	-1.9	2	1.9	28	5.3	11	8	0	27.7	61	13	9.3	29	82.6	97	6.1	15
Mar	6.5	0.3	3.4	-2.8	12.4	23	-4.8	5	2.2	4	3.9	25	13	0	74.5	180	18	16.5	6	90.5	75	9.3	28
Apr	8.4	1.6	5.0	-3.2	13.9	17	-8.0	24	3.0	24	6.6	30	8	0	70.7	138	17	13.1	26	90.8	59	9.6	3
May	15.6	6.9	11.2	+0.2	23.7	27	2.6	22	5.9	3	11.4	18	0	1	45.4	102	17	8.1	5	177.2	95	14.8	27
June	17.5	7.9	12.7	-0.9	26.7	25	0.9	21	11.1	6	14.1	26	0	2	11.8	19	8	2.8	16	180.4	109	14.9	21
July	19.0	10.2	14.6	-1.2	27.8	2	6.1	29	13.3	18	15.3	25	0	2	63.7	103	13	17.9	8	150.7	86	14.5	2
Aug	17.4	9.2	13.3	-2.2	24.9	3	4.2	12	13.3	20	14.8	4	0	0	51.7	78	18	15.0	11	148.2	88	10.1	30
Sep	16.3	9.0	12.7	-0.6	24.6	30	1.8	5	11.8	16	15.9	29	0	0	42.0	74	20	10.5	20	91.6	69	10.5	2
Oct	15.1	7.7	11.4	+1.4	24.9	4	1.4	25	8.3	24	13.8	1	1	0	33.8	53	16	7.0	27	82.7	84	9.2	4
Nov	9.3	2.8	6.1	-0.5	12.4	12	-4.8	8	2.6	10	10.0	4	6	0	27.4	37	13	6.5	13	61.0	90	7.2	7
Dec	5.3	-0.2	2.6	-1.7	11.0	8	-9.4	30	-3.9	29	6.4	22	13	0	17.1	28	16	4.7	10	25.3	44	3.1	6
1908	12.0	4.7	8.3	-1.1	27.8	7	-9.4	12	-3.9	12	15.9	9	67	3	493.1	72	178	17.9	7	1244.9	84	14.9	6

DURHAM 1909

Month	Mean max	Mean min	Mean temp	Anom	Highest max	Date	Lowest min	Date	Lowest max	Date	Highest min	Date	Air frost	Days ≥ 25 °C	Total pptn	Anom	Rain days	Wettest day	Date	Total sunshine	Anom	Sunniest day	Date
Jan	5.6	-0.2	2.7	-1.4	11.1	17	-7.9	27	-2.2	27	4.4	2	16	0	15.1	28	10	4.7	14	74.0	120	5.5	4
Feb	5.6	-0.6	2.5	-2.1	12.5	3	-4.1	7	0.6	7	6.0	3	18	0	35.3	78	10	13.3	4	49.6	58	6.1	21
Mar	5.6	-0.2	2.7	-3.5	12.5	20	-7.6	6	1.6	4	6.1	20	14	0	108.1	261	22	14.1	28	62.6	52	9.0	5
Apr	13.1	2.6	7.8	-0.4	20.3	10	-2.3	7	6.6	1	6.8	23	5	0	56.8	111	16	13.0	19	188.8	122	11.5	5
May	14.7	4.5	9.6	-1.4	22.9	22	-1.7	1	6.2	17	11.5	23	4	0	33.5	75	11	19.1	25	222.1	119	13.7	5
June	15.2	6.6	10.9	-2.7	20.4	19	0.6	7	10.2	26	10.7	21	0	0	48.4	79	11	17.1	26	128.9	78	15.0	13
July	18.0	10.0	14.0	-1.8	22.3	16	2.4	2	12.6	1	13.8	3	0	0	55.7	90	17	13.0	15	151.8	87	12.5	23
Aug	19.2	10.2	14.7	-0.8	27.2	9	6.3	22	13.4	2	15.7	15	0	4	74.5	113	17	16.6	16	171.5	102	13.2	8
Sep	14.1	7.3	10.7	-2.6	20.3	19	2.2	9	10.3	21	11.4	24	0	0	50.4	89	16	12.8	27	79.6	60	10.4	2
Oct	12.5	5.5	9.0	-1.0	17.8	4	-5.2	31	4.8	29	12.4	4	4	0	53.3	84	21	15.8	23	99.1	101	8.5	30
Nov	7.6	1.4	4.5	-2.1	15.3	3	-4.9	14	2.8	15	10.1	3	12	0	19.3	26	11	3.9	22	61.6	90	7.5	13
Dec	5.3	0.3	2.8	-1.4	12.1	10	-7.7	21	-1.4	19	7.8	11	13	0	79.4	130	17	17.1	2	47.2	83	6.6	5
1909	11.4	4.0	7.7	-1.8	27.2	8	-7.9	1	-2.2	1	15.7	8	86	4	629.8	93	179	19.1	5	1336.8	91	15.0	6

DURHAM 1910

Month	Mean max	Mean min	Mean temp	Anom	Highest max	Date	Lowest min	Date	Lowest max	Date	Highest min	Date	Air frost	Days ≥ 25 °C	Total pptn	Anom	Rain days	Wettest day	Date	Total sunshine	Anom	Sunniest day	Date
Jan	5.5	-0.3	2.6	-1.5	12.7	2	-12.4	27	-1.1	27	8.9	2	15	0	55.0	103	18	22.0	28	60.4	98	6.8	26
Feb	6.3	0.6	3.4	-1.1	10.2	6	-2.7	9	1.6	26	6.7	6	12	0	53.0	117	21	10.2	7	74.5	87	7.7	15
Mar	9.6	2.0	5.8	-0.4	15.7	20	-2.5	29	4.8	18	6.0	9	6	0	16.0	39	8	7.0	9	124.1	103	11.3	28
Apr	10.0	2.1	6.0	-2.2	14.6	13	-3.8	2	5.7	5	8.3	13	4	0	61.1	119	15	15.1	4	115.2	75	12.4	14
May	13.3	5.1	9.2	-1.8	21.3	26	-0.4	11	9.1	8	11.3	27	2	0	42.1	95	17	10.0	15	162.7	87	11.7	14
June	17.8	8.5	13.2	-0.4	23.3	12	4.3	17	11.1	5	12.7	11	0	0	60.8	99	13	20.5	2	184.2	111	14.7	16
July	17.0	9.6	13.3	-2.5	20.4	13	4.9	4	13.2	16	13.9	29	0	0	73.4	118	14	22.9	25	126.9	72	11.3	14
Aug	18.4	10.7	14.6	-1.0	23.4	13	5.2	23	14.2	22	15.2	17	0	0	62.2	94	16	15.3	28	143.3	85	10.1	1
Sep	15.6	7.2	11.4	-1.8	20.4	28	1.6	17	11.6	14	11.4	11	0	0	9.5	17	7	4.3	10	112.0	85	10.0	3
Oct	12.9	7.1	10.0	+0.0	19.8	7	2.7	20	9.1	29	13.6	2	0	0	73.8	116	16	12.1	20	83.3	85	10.0	5
Nov	5.4	-0.4	2.5	-4.2	10.7	1	-4.9	23	1.7	23	3.3	1	16	0	75.7	103	19	19.4	13	92.7	136	7.9	2
Dec	7.8	3.1	5.4	+1.2	12.9	24	-3.9	28	4.7	3	6.8	10	2	0	48.7	80	21	12.5	3	43.3	76	6.0	18
1910	11.6	4.6	8.1	-1.3	23.4	8	-12.4	1	-1.1	1	15.2	8	57	0	631.3	93	185	22.9	7	1322.6	90	14.7	6

DURHAM 1911

Month	Mean max	Mean min	Mean temp	Anom	Highest max	Date	Lowest min	Date	Lowest max	Date	Highest min	Date	Air frost	Days ≥ 25 °C	Total pptn	Anom	Rain days	Wettest day	Date	Total sunshine	Anom	Sunniest day	Date
Jan	6.4	0.6	3.5	−0.6	12.1	25	−6.1	21	1.9	13	7.0	26	9	0	37.1	69	10	11.4	11	55.1	89	6.0	18
Feb	7.3	1.1	4.2	−0.3	12.3	18	−6.1	1	2.2	6	7.5	17	11	0	21.1	47	10	4.9	23	69.5	81	8.7	26
Mar	6.5	1.5	4.0	−2.1	14.1	3	−1.3	8	3.9	13	7.9	3	4	0	43.8	106	18	5.1	14	83.8	70	9.8	11
Apr	9.8	3.4	6.6	−1.7	14.2	21	−2.4	2	1.7	5	9.4	22	6	0	27.6	54	15	6.7	28	106.2	69	12.0	14
May	16.0	6.9	11.4	+0.4	21.6	25	1.3	6	8.9	20	12.4	25	0	0	17.7	40	10	5.2	14	173.4	93	14.4	29
June	18.2	7.8	13.0	−0.6	26.1	5	1.1	15	12.3	25	12.3	22	0	3	90.7	148	12	30.8	24	196.2	118	14.4	2
July	22.2	10.4	16.3	+0.5	29.4	14	2.2	10	16.7	1	15.2	30	0	9	16.4	26	8	6.7	20	252.0	144	15.3	11
Aug	20.6	11.7	16.1	+0.6	30.1	9	6.3	17	15.0	22	15.3	9	0	5	50.5	76	12	13.7	21	168.8	101	13.4	10
Sep	17.4	7.5	12.5	−0.8	25.1	8	0.6	22	9.6	30	14.4	1	1	1	41.2	73	10	18.4	12	167.8	127	11.3	7
Oct	11.2	5.0	8.1	−1.9	14.9	21	−3.1	29	5.4	29	9.9	20	4	0	61.9	97	17	21.1	26	64.7	66	7.6	10
Nov	8.0	2.5	5.3	−1.4	14.7	14	−1.8	21	3.2	21	6.5	15	7	0	89.9	122	19	25.6	19	55.9	82	6.4	13
Dec	7.5	2.3	4.9	+0.6	10.7	28	−1.1	25	3.4	8	7.1	18	4	0	86.5	141	23	12.2	10	40.2	71	4.8	7
1911	12.6	5.1	8.8	−0.6	30.1	8	−6.1	1	1.7	4	15.3	8	45	18	584.4	86	164	30.8	6	1433.6	97	15.3	7

DURHAM 1912

Month	Mean max	Mean min	Mean temp	Anom	Highest max	Date	Lowest min	Date	Lowest max	Date	Highest min	Date	Air frost	Days ≥ 25 °C	Total pptn	Anom	Rain days	Wettest day	Date	Total sunshine	Anom	Sunniest day	Date
Jan	4.9	−0.2	2.3	−1.7	11.1	1	−8.7	29	0.6	8	6.2	2	16	0	79.0	148	19	13.2	17	33.4	54	5.6	29
Feb	6.4	1.0	3.7	−0.9	12.4	28	−12.4	4	−2.2	4	8.6	28	10	0	49.2	109	16	13.2	19	50.2	59	7.3	24
Mar	9.5	3.1	6.3	+0.2	14.6	28	−2.6	21	4.3	21	8.4	26	4	0	49.0	119	17	6.5	29	107.1	89	9.2	30
Apr	12.5	2.5	7.5	−0.8	20.3	21	−3.6	12	6.6	11	9.4	5	5	0	2.4	5	3	1.5	10	190.2	123	13.0	22
May	14.4	5.6	10.0	−1.0	19.6	8	−0.6	14	7.4	22	11.8	9	2	0	57.5	129	14	16.7	15	128.4	69	12.3	25
June	16.1	8.8	12.4	−1.2	22.8	22	5.4	17	9.4	2	13.7	23	0	0	111.7	183	24	13.3	3	91.7	55	9.3	22
July	17.3	10.4	13.8	−2.0	24.4	12	6.7	6	11.1	2	15.6	11	0	0	89.5	144	16	14.2	31	73.9	42	11.3	15
Aug	15.3	8.1	11.7	−3.8	19.1	8	1.9	3	11.8	26	11.2	16	0	0	105.9	160	23	14.7	23	56.8	34	6.1	8
Sep	14.0	6.4	10.2	−3.1	20.0	13	1.0	25	10.0	11	11.2	16	0	0	42.6	75	8	10.8	30	94.2	71	8.9	13
Oct	11.5	3.8	7.7	−2.4	16.1	7	−2.4	26	7.2	31	8.7	16	4	0	65.3	103	18	18.2	26	83.6	85	7.5	17
Nov	8.2	2.8	5.5	−1.1	12.3	6	−5.8	30	−0.8	29	9.2	8	8	0	50.5	69	18	13.2	12	43.8	64	7.1	1
Dec	8.1	1.6	4.8	+0.6	12.8	14	−6.9	1	−0.6	1	8.6	20	5	0	38.2	62	16	9.9	15	28.9	51	4.9	16
1912	11.5	4.5	8.0	−1.4	24.4	7	−12.4	2	−2.2	2	15.6	7	54	0	740.8	109	192	18.2	10	982.2	67	13.0	4

DURHAM 1913

Month	Mean max	Mean min	Mean temp	Anom	Highest max	Date	Lowest min	Date	Lowest max	Date	Highest min	Date	Air frost	Days ≥ 25 °C	Total pptn	Anom	Rain days	Wettest day	Date	Total sunshine	Anom	Sunniest day	Date
Jan	4.8	0.1	2.4	-1.7	10.9	7	-7.1	13	0.6	12	4.8	8	15	0	64.3	120	19	22.6	11	15.5	25	5.8	25
Feb	7.3	0.7	4.0	-0.6	11.6	13	-4.1	22	3.5	23	4.8	5	13	0	16.7	37	9	4.9	6	54.4	64	6.4	12
Mar	8.1	1.1	4.6	-1.5	11.7	4	-4.8	18	3.3	18	5.5	30	7	0	56.0	135	20	11.4	22	118.1	98	9.2	16
Apr	10.1	3.3	6.7	-1.5	16.2	22	-1.9	13	2.9	11	7.7	23	3	0	101.6	198	18	19.7	26	100.4	65	12.6	20
May	13.7	5.8	9.7	-1.3	22.7	30	-0.6	16	7.3	6	12.6	30	1	0	63.6	143	16	17.7	8	123.0	66	13.4	25
June	17.6	8.5	13.0	-0.6	22.6	18	2.7	6	11.9	9	11.4	29	0	0	25.4	42	13	6.8	23	177.3	107	14.2	1
July	17.2	9.4	13.3	-2.5	21.0	29	3.8	9	12.7	23	12.9	18	0	0	22.5	36	8	10.0	6	93.3	53	12.1	29
Aug	18.8	9.1	14.0	-1.6	27.8	3	3.3	5	13.3	5	13.6	30	0	2	46.8	71	11	25.0	29	132.9	79	11.7	3
Sep	16.9	9.3	13.1	-0.2	21.5	11	1.8	8	11.6	1	13.6	27	0	0	71.1	125	16	22.4	26	98.9	75	9.0	28
Oct	12.6	6.5	9.5	-0.5	15.8	17	-2.2	24	6.0	23	11.6	1	3	0	67.4	106	16	38.4	7	74.8	76	8.2	17
Nov	10.4	4.2	7.3	+0.7	14.3	17	-1.9	8	4.5	6	9.9	28	3	0	30.1	41	16	7.7	20	65.8	97	6.5	3
Dec	6.6	1.6	4.1	-0.1	12.2	9	-8.9	31	-0.6	6	9.4	3	9	0	28.8	47	7	14.6	29	36.3	64	5.5	28
1913	12.0	5.0	8.5	-1.0	27.8	8	-8.9	12	-0.6	12	13.6	8	54	2	594.3	87	169	38.4	10	1090.7	74	14.2	6

DURHAM 1914

Month	Mean max	Mean min	Mean temp	Anom	Highest max	Date	Lowest min	Date	Lowest max	Date	Highest min	Date	Air frost	Days ≥ 25 °C	Total pptn	Anom	Rain days	Wettest day	Date	Total sunshine	Anom	Sunniest day	Date
Jan	6.0	0.7	3.4	-0.7	12.2	31	-4.6	24	0.8	12	5.9	30	13	0	36.2	68	20	8.2	9	32.6	53	6.4	26
Feb	9.1	3.0	6.0	+1.5	12.3	1	-2.2	25	5.7	17	9.8	2	2	0	33.2	73	15	10.4	21	56.6	66	6.7	21
Mar	8.4	1.6	5.0	-1.2	14.6	30	-3.0	11	3.4	8	9.2	31	8	0	34.6	84	20	6.0	20	97.0	81	9.0	6
Apr	14.8	3.2	9.0	+0.7	22.2	22	-1.7	19	9.0	9	7.3	13	3	0	8.7	17	8	2.3	5	238.1	154	12.3	18
May	14.1	4.9	9.5	-1.5	24.0	18	-3.5	2	7.8	1	10.4	22	3	0	35.7	80	13	8.0	23	136.0	73	12.7	15
June	18.9	7.7	13.3	-0.3	26.1	29	1.8	3	9.9	8	15.2	29	0	2	113.8	186	11	44.1	9	222.1	134	15.0	30
July	19.0	10.3	14.7	-1.1	24.0	13	5.9	7	13.7	29	15.4	14	0	0	66.9	108	13	24.8	16	134.4	77	13.9	4
Aug	19.6	10.0	14.8	-0.8	23.4	14	5.6	18	15.9	30	14.7	24	0	0	36.6	55	12	11.9	2	188.8	113	12.7	3
Sep	17.4	7.2	12.3	-1.0	25.2	3	-0.5	30	12.2	17	13.9	9	2	1	41.7	73	10	15.0	17	167.1	126	10.6	18
Oct	13.0	6.2	9.6	-0.4	18.4	3	-0.1	29	9.2	31	9.8	3	1	0	48.5	76	16	19.8	29	67.1	68	7.7	26
Nov	8.9	3.1	6.0	-0.6	13.9	30	-2.8	15	4.0	14	11.7	9	6	0	87.4	119	26	11.3	15	49.7	73	7.1	14
Dec	6.0	0.9	3.4	-0.8	9.8	2	-3.9	25	1.3	25	5.3	14	10	0	115.3	188	25	23.3	12	47.5	83	5.0	29
1914	12.9	4.9	8.9	-0.5	26.1	6	-4.6	1	0.8	1	15.4	7	48	3	658.6	97	189	44.1	6	1437.0	98	15.0	6

DURHAM 1915

Month	Mean max	Mean min	Mean temp	Anom	Highest max	Date	Lowest min	Date	Lowest max	Date	Highest min	Date	Air frost	Days ≥ 25°C	Total pptn	Anom	Rain days	Wettest day	Date	Total sunshine	Anom	Sunniest day	Date
Jan	5.5	0.9	3.2	-0.9	12.3	13	-4.8	26	1.6	28	8.2	14	11	0	52.9	99	21	10.3	3	40.5	66	5.6	17
Feb	6.0	0.4	3.2	-1.4	10.5	4	-3.8	22	2.3	24	5.8	4	13	0	63.1	139	21	13.8	7	67.3	79	8.8	24
Mar	8.4	0.8	4.6	-1.5	15.1	14	-4.1	28	2.8	19	8.3	5	11	0	52.4	127	18	13.2	18	125.9	105	9.2	22
Apr	11.3	2.4	6.9	-1.4	18.9	30	-1.7	21	7.3	26	7.0	19	5	0	14.5	28	11	2.5	3	175.3	114	12.4	29
May	12.8	3.7	8.3	-2.7	19.5	21	-3.3	3	7.3	12	9.9	21	3	0	55.5	125	12	20.1	11	190.4	102	14.7	23
June	17.8	7.4	12.6	-1.0	24.3	12	-0.7	19	12.8	24	11.2	8	1	0	9.4	15	4	4.7	25	186.6	113	14.9	19
July	18.3	9.5	13.9	-1.8	21.4	29	5.7	14	15.4	17	13.4	3	0	0	90.1	145	21	13.3	12	157.8	90	11.6	29
Aug	19.0	10.0	14.5	-1.1	23.3	25	4.4	30	15.0	8	12.8	9	0	0	76.4	115	18	11.1	15	132.8	79	12.6	23
Sep	16.4	7.1	11.8	-1.5	21.4	16	0.4	4	11.5	30	13.3	16	0	0	39.0	69	9	16.3	1	152.4	115	10.6	16
Oct	11.7	4.1	7.9	-2.1	15.7	15	-2.6	30	7.3	28	9.4	11	4	0	32.4	51	13	6.8	27	58.6	59	7.2	5
Nov	5.6	-0.3	2.7	-4.0	10.8	8	-6.0	15	0.3	15	7.7	8	16	0	40.1	55	15	15.8	12	70.2	103	6.0	9
Dec	5.7	1.2	3.5	-0.8	9.6	31	-3.5	9	1.7	12	6.7	31	9	0	133.7	218	24	19.5	3	34.8	61	5.9	7
1915	11.6	3.9	7.8	-1.7	24.3	6	-6.0	11	0.3	11	13.4	7	73	0	659.5	97	187	20.1	5	1392.6	95	14.9	6

DURHAM 1916

Month	Mean max	Mean min	Mean temp	Anom	Highest max	Date	Lowest min	Date	Lowest max	Date	Highest min	Date	Air frost	Days ≥ 25°C	Total pptn	Anom	Rain days	Wettest day	Date	Total sunshine	Anom	Sunniest day	Date
Jan	9.8	3.7	6.7	+2.7	14.9	6	-1.0	12	5.8	8	6.8	10	4	0	16.1	30	11	3.5	1	55.9	91	6.3	24
Feb	5.6	-0.2	2.7	-1.9	10.0	6	-4.6	21	1.4	26	3.4	6	19	0	73.5	162	19	10.5	28	76.0	89	8.3	17
Mar	4.4	-0.3	2.0	-4.1	11.7	31	-6.6	24	0.0	23	6.2	31	18	0	90.5	219	27	11.1	15	65.8	55	9.9	29
Apr	11.6	2.7	7.1	-1.1	17.8	26	-2.2	2	5.8	19	9.9	25	5	0	39.3	77	12	18.6	18	152.7	99	11.5	11
May	14.5	5.3	9.9	-1.1	25.1	20	-0.2	4	5.5	6	11.5	22	2	1	61.4	138	17	14.8	4	149.9	80	11.5	27
June	14.6	6.7	10.7	-2.9	20.0	23	0.6	17	8.2	13	11.9	24	0	0	42.0	69	18	8.9	10	97.5	59	11.2	17
July	19.5	9.9	14.7	-1.1	25.8	29	4.6	18	14.3	12	14.9	21	0	3	59.2	95	16	12.9	20	159.9	91	12.7	29
Aug	20.0	10.6	15.3	-0.3	28.8	5	5.1	5	13.6	29	14.7	14	0	3	81.5	123	15	15.9	16	142.2	85	14.0	5
Sep	16.3	7.6	11.9	-1.3	23.1	7	0.5	21	10.1	30	12.8	7	0	0	36.5	64	9	21.9	3	134.9	102	9.4	2
Oct	12.5	4.8	8.6	-1.4	18.4	6	-4.7	17	7.1	17	11.8	6	6	0	62.4	98	23	9.9	4	68.4	69	8.8	9
Nov	9.5	3.8	6.6	+0.0	15.6	13	-2.8	28	4.3	16	10.7	12	4	0	65.8	90	18	20.3	18	53.7	79	6.5	22
Dec	4.7	-0.9	1.9	-2.3	11.6	31	-8.8	20	-0.3	19	6.9	29	18	0	62.8	103	17	10.6	9	43.3	76	5.5	15
1916	11.9	4.5	8.2	-1.3	28.8	8	-8.8	12	-0.3	12	14.9	7	76	7	691.0	102	202	21.9	9	1200.2	81	14.0	8

DURHAM 1917

Month	Mean max	Mean min	Mean temp	Anom	Highest max	Date	Lowest min	Date	Lowest max	Date	Highest min	Date	Air frost	Days ≥ 25 °C	Total pptn	Anom	Rain days	Wettest day	Date	Total sunshine	Anom	Sunniest day	Date
Jan	3.3	0.0	1.7	−2.4	12.8	3	−3.8	10	0.0	29	9.1	1	23	0	80.4	151	24	15.7	7	19.5	32	6.5	4
Feb	3.8	−2.0	0.9	−3.6	10.5	26	−10.6	6	−0.3	11	3.8	24	17	0	21.9	48	12	7.5	20	46.3	54	7.5	26
Mar	5.9	−1.5	2.2	−4.0	12.3	24	−11.6	9	−0.4	8	5.6	17	27	0	52.5	127	17	14.6	10	106.5	88	9.4	23
Apr	8.5	0.3	4.4	−3.9	16.3	24	−11.1	2	−1.1	1	7.2	27	16	0	46.2	90	18	9.2	5	98.6	64	10.0	17
May	15.2	5.5	10.3	−0.7	23.2	27	−2.2	7	6.6	8	12.7	26	1	0	76.6	172	11	23.7	14	184.6	99	14.3	28
June	18.7	7.8	13.2	−0.4	24.7	13	3.5	6	14.0	3	13.2	17	0	0	34.5	56	14	9.9	20	220.3	133	15.2	30
July	20.1	9.9	15.0	−0.7	26.4	22	1.7	11	13.5	4	14.9	24	0	4	26.1	42	10	11.9	27	208.5	119	15.1	1
Aug	19.1	11.1	15.1	−0.4	24.4	6	5.0	31	15.0	9	14.4	8	0	0	157.0	237	24	19.5	8	108.6	65	9.8	19
Sep	17.3	8.9	13.1	−0.2	21.1	7	4.4	10	14.1	23	12.8	16	0	0	24.6	43	9	13.2	1	132.3	100	10.6	4
Oct	11.4	2.7	7.0	−3.0	19.7	1	−3.0	29	5.3	27	11.6	3	5	0	50.3	79	17	9.2	24	131.3	133	10.2	6
Nov	10.4	4.6	7.5	+0.8	13.8	29	−2.2	26	6.1	9	11.2	29	2	0	22.7	31	11	6.3	24	53.6	79	7.9	7
Dec	4.8	−0.3	2.2	−2.0	10.6	7	−6.9	4	0.0	4	6.4	13	18	0	67.4	110	15	15.5	16	53.3	94	5.8	3
1917	11.5	3.9	7.7	−1.7	26.4	7	−11.6	3	−1.1	4	14.9	7	109	4	660.2	97	182	23.7	5	1363.4	93	15.2	6

DURHAM 1918

Month	Mean max	Mean min	Mean temp	Anom	Highest max	Date	Lowest min	Date	Lowest max	Date	Highest min	Date	Air frost	Days ≥ 25 °C	Total pptn	Anom	Rain days	Wettest day	Date	Total sunshine	Anom	Sunniest day	Date
Jan	5.8	−0.3	2.8	−1.3	15.6	25	−12.3	14	−2.9	14	7.8	25	17	0	50.7	95	14	14.2	18	51.1	83	6.5	13
Feb	8.1	2.3	5.2	+0.6	13.8	22	−3.3	19	2.2	1	6.8	23	6	0	31.4	69	16	6.8	20	59.3	70	9.1	25
Mar	8.3	1.4	4.8	−1.3	18.0	23	−2.9	14	2.1	8	5.7	22	5	0	31.5	76	16	6.5	28	98.8	82	10.8	24
Apr	8.9	2.4	5.6	−2.6	17.7	26	−1.1	3	3.4	20	4.2	17	4	1	36.6	71	16	7.8	6	105.2	68	12.7	27
May	16.8	6.3	11.6	+0.6	25.2	22	−0.9	9	6.8	2	13.2	21	1	3	35.2	79	12	8.7	5	191.0	102	14.1	29
June	18.0	7.0	12.5	−1.1	26.4	2	3.3	16	13.9	15	11.7	7	0	3	8.1	13	8	4.2	6	200.6	121	13.0	1
July	19.7	9.6	14.7	−1.1	26.1	4	3.9	9	15.6	10	13.3	7	0	1	86.1	139	20	13.4	19	195.8	112	12.7	19
Aug	19.1	10.8	14.9	−0.6	23.9	20	4.4	24	12.2	31	15.6	21	0	0	35.8	54	14	7.3	27	163.6	98	13.3	20
Sep	14.2	6.4	10.3	−2.9	21.7	7	1.7	30	8.3	14	11.1	17	0	0	116.8	206	25	28.4	15	140.1	106	10.0	9
Oct	11.8	5.3	8.6	−1.5	16.1	10	0.6	23	8.3	12	11.7	3	0	0	49.6	78	18	10.3	19	77.6	79	9.8	4
Nov	8.0	0.9	4.4	−2.2	13.3	2	−3.9	12	1.7	22	8.3	2	14	0	29.4	40	14	7.6	1	60.3	89	7.4	6
Dec	7.9	2.3	5.1	+0.8	13.9	3	−3.9	3	2.8	21	9.4	2	8	0	62.4	102	23	14.0	15	48.0	84	5.1	17
1918	12.2	4.5	8.4	−1.1	26.4	6	−12.3	1	−2.9	1	15.6	8	55	5	573.6	84	196	28.4	9	1391.4	94	14.1	5

DURHAM 1919

Month	Mean max	Mean min	Mean temp	Anom	Highest max	Date	Lowest min	Date	Lowest max	Date	Highest min	Date	Air frost	Days ≥ 25 °C	Total pptn	Anom	Rain days	Wettest day	Date	Total sunshine	Anom	Sunniest day	Date
Jan	3.9	−0.5	1.7	−2.4	8.3	15	−5.0	19	0.6	19	3.3	9	14	0	69.4	130	25	11.4	4	44.8	73	6.4	26
Feb	3.9	−1.3	1.3	−3.2	7.8	24	−10.0	10	−1.7	8	3.9	22	16	0	40.5	89	16	6.0	20	54.4	64	7.1	28
Mar	5.8	−1.0	2.4	−3.8	11.1	2	−6.7	4	1.7	19	2.8	11	17	0	73.1	177	22	22.8	19	95.9	80	8.3	17
Apr	11.1	2.5	6.8	−1.5	18.9	18	−3.9	27	4.4	27	10.0	18	6	0	39.1	76	19	7.9	27	128.4	83	11.9	20
May	16.8	5.6	11.2	+0.2	23.9	28	−1.1	20	6.1	8	10.6	25	1	0	20.9	47	12	3.7	6	210.9	113	13.4	22
June	18.3	8.0	13.1	−0.5	27.2	7	3.9	3	11.1	30	12.8	6	0	4	22.5	37	16	6.4	23	195.0	118	14.2	9
July	17.4	8.4	12.9	−2.9	26.1	30	4.4	8	11.1	4	13.3	19	0	2	48.5	78	12	18.4	2	153.7	88	12.3	9
Aug	19.9	9.9	14.9	−0.6	27.2	10	3.9	28	9.4	28	14.4	18	0	4	59.1	89	11	22.4	28	175.1	104	12.1	6
Sep	17.1	7.2	12.2	−1.1	26.7	11	−0.6	28	12.2	1	14.4	5	1	1	19.7	35	13	5.6	11	133.9	101	10.5	10
Oct	11.8	2.6	7.2	−2.9	21.1	6	−1.7	15	3.3	28	6.1	2	13	0	119.7	188	18	22.3	27	111.1	113	8.9	26
Nov	5.3	−0.5	2.4	−4.2	13.3	23	−8.9	16	−1.1	12	10.0	23	13	0	91.6	125	27	7.4	8	48.7	72	7.2	18
Dec	6.3	0.9	3.6	−0.7	11.1	3	−4.4	26	1.7	10	4.4	5	5	0	57.6	94	20	13.0	15	39.6	70	6.0	7
1919	11.5	3.5	7.5	−2.0	27.2	6	−10.0	2	−1.7	2	14.4	8	74	11	661.7	97	211	22.8	3	1391.5	94	14.2	6

DURHAM 1920

Month	Mean max	Mean min	Mean temp	Anom	Highest max	Date	Lowest min	Date	Lowest max	Date	Highest min	Date	Air frost	Days ≥ 25 °C	Total pptn	Anom	Rain days	Wettest day	Date	Total sunshine	Anom	Sunniest day	Date
Jan	6.8	1.1	3.9	−0.2	12.2	17	−2.8	1	0.6	6	7.8	17	8	0	65.1	122	17	20.9	10	70.8	115	6.9	21
Feb	8.5	1.4	4.9	+0.4	13.9	18	−3.9	5	3.3	5	7.8	29	6	0	21.2	47	9	10.8	19	92.2	108	8.7	21
Mar	9.7	2.0	5.8	−0.3	16.1	21	−3.9	8	2.8	15	6.7	1	7	0	69.0	167	19	15.2	26	116.6	97	9.8	8
Apr	9.7	3.2	6.5	−1.8	13.9	15	−1.7	18	4.4	10	7.2	24	2	0	67.5	132	23	8.9	1	102.9	67	11.2	29
May	14.7	6.0	10.4	−0.6	22.8	23	−0.6	1	8.3	1	11.1	27	1	0	84.0	189	16	29.0	29	192.8	103	14.7	20
June	17.7	7.6	12.7	−0.9	23.3	17	−0.6	8	10.0	10	12.8	28	1	0	21.7	35	11	10.6	27	217.6	131	15.0	15
July	17.0	9.6	13.3	−2.5	21.1	20	6.7	27	10.0	4	12.8	2	0	0	108.2	174	23	23.3	3	105.9	60	14.2	19
Aug	17.5	8.0	12.8	−2.8	23.9	13	2.8	30	11.7	19	12.2	14	0	0	32.3	49	10	12.6	4	125.1	75	11.7	13
Sep	16.7	7.3	12.0	−1.3	20.6	13	1.7	23	11.1	20	11.7	6	0	0	26.7	47	16	5.2	30	121.1	91	11.0	10
Oct	13.0	5.5	9.3	−0.7	18.9	7	0.0	31	8.9	20	12.2	6	0	0	56.0	88	14	17.1	4	79.6	81	8.6	7
Nov	9.2	3.8	6.5	−0.1	14.4	9	−5.0	23	0.6	22	10.6	9	4	0	21.1	29	10	6.2	27	47.2	69	6.9	5
Dec	6.2	0.7	3.4	−0.8	11.1	31	−3.3	18	1.7	13	7.2	25	14	0	80.7	132	21	13.3	29	36.3	64	6.1	9
1920	12.2	4.7	8.5	−1.0	23.9	8	−5.0	11	0.6	1	12.8	6	43	0	653.5	96	189	29.0	5	1308.1	89	15.0	6

DURHAM 1921

Month	Mean max	Mean min	Mean temp	Anom	Highest max	Date	Lowest min	Date	Lowest max	Date	Highest min	Date	Air frost	Days ≥ 25 °C	Total pptn	Anom	Rain days	Wettest day	Date	Total sunshine	Anom	Sunniest day	Date
Jan	8.6	2.3	5.5	+1.4	14.4	4	-3.9	15	2.8	12	7.8	29	3	0	77.4	145	17	32.8	12	36.3	59	5.7	22
Feb	7.4	0.8	4.1	-0.5	13.3	16	-5.0	22	2.2	7	6.1	16	11	0	3.7	8	5	1.3	6	64.9	76	8.1	26
Mar	10.1	2.1	6.1	-0.1	13.9	21	-2.8	7	5.0	7	7.2	24	6	0	18.4	45	19	3.3	25	115.1	96	9.5	27
Apr	12.4	2.4	7.4	-0.9	21.1	1	-4.4	8	6.1	15	8.3	1	6	0	31.4	61	14	8.1	24	184.7	120	14.0	30
May	15.8	4.6	10.2	-0.8	23.9	25	-1.7	5	9.4	3	10.0	13	1	0	45.8	103	12	12.5	28	247.6	132	15.0	21
June	19.0	7.3	13.1	-0.5	28.3	25	1.7	28	13.3	4	12.8	25	0	3	19.6	32	8	5.2	26	214.5	130	15.1	3
July	21.2	10.4	15.8	+0.0	30.6	10	1.7	5	13.3	29	16.7	9	0	6	37.4	60	14	8.5	28	171.4	98	14.0	10
Aug	17.8	9.7	13.7	-1.8	22.8	1	3.9	30	11.1	29	15.6	1	0	0	104.6	158	19	19.3	29	113.6	68	12.5	30
Sep	17.7	8.1	12.9	-0.3	23.9	7	2.2	25	11.7	13	13.3	23	1	0	16.7	29	6	10.2	13	148.4	112	10.8	10
Oct	16.2	7.7	12.0	+1.9	22.2	6	-0.6	24	8.3	23	13.3	6	0	0	33.0	52	9	15.4	22	125.0	127	9.0	12
Nov	6.8	0.5	3.7	-3.0	15.0	4	-6.7	11	2.8	8	5.6	4	11	0	63.3	86	13	31.0	5	65.1	96	8.0	7
Dec	9.3	2.6	6.0	+1.7	14.4	17	-5.6	4	3.3	4	8.9	8	5	0	39.0	64	20	7.6	26	54.1	95	6.0	13
1921	13.5	4.9	9.2	-0.3	30.6	7	-6.7	11	2.2	2	16.7	7	44	9	490.3	72	156	32.8	1	1540.7	105	15.1	6

DURHAM 1922

Month	Mean max	Mean min	Mean temp	Anom	Highest max	Date	Lowest min	Date	Lowest max	Date	Highest min	Date	Air frost	Days ≥ 25 °C	Total pptn	Anom	Rain days	Wettest day	Date	Total sunshine	Anom	Sunniest day	Date
Jan	4.2	-1.2	1.5	-2.6	13.4	9	-10.8	18	-0.4	14	4.2	10	19	0	80.7	151	20	13.1	25	46.5	75	6.3	10
Feb	5.8	0.0	2.9	-1.7	13.4	25	-8.1	6	-1.6	6	8.1	25	14	0	48.2	106	19	10.1	2	76.0	89	8.7	20
Mar	7.1	0.7	3.9	-2.2	11.8	3	-3.6	31	2.9	21	4.2	6	8	0	39.5	96	22	6.1	27	79.1	66	8.8	31
Apr	8.8	0.2	4.5	-3.8	14.0	15	-4.7	7	4.6	1	4.7	15	15	0	49.5	97	15	9.8	14	156.8	102	11.6	20
May	16.3	6.6	11.5	+0.5	27.9	31	-0.3	2	8.4	1	13.1	21	1	1	13.5	30	6	7.1	22	197.0	105	14.4	29
June	18.1	7.6	12.9	-0.7	25.7	1	3.6	6	14.0	14	11.9	1	0	1	37.6	61	13	9.6	12	192.3	116	14.0	1
July	17.1	8.4	12.7	-3.0	20.7	18	4.2	12	11.8	8	11.4	13	0	0	115.2	186	23	30.8	5	120.4	69	12.6	7
Aug	16.6	8.7	12.6	-2.9	21.2	18	4.2	12	12.3	9	12.5	20	0	0	59.4	90	15	26.4	8	125.6	75	12.8	6
Sep	14.2	7.1	10.6	-2.6	17.9	5	1.4	23	10.1	21	11.4	6	0	0	95.8	169	18	24.1	21	64.6	49	10.7	29
Oct	11.1	4.3	7.7	-2.3	16.8	1	-0.3	26	6.2	29	10.3	4	1	0	38.0	60	21	4.9	3	92.5	94	8.6	6
Nov	9.2	2.4	5.8	-0.8	14.6	14	-3.1	12	5.1	4	6.9	17	8	0	18.9	26	11	4.5	3	96.9	142	8.3	4
Dec	7.5	2.1	4.8	+0.5	11.8	13	-3.1	9	2.9	31	10.3	13	7	0	86.5	141	18	27.1	20	40.2	71	5.3	26
1922	11.3	3.9	7.6	-1.8	27.9	5	-10.8	5	-1.6	2	13.1	5	73	2	682.8	100	201	30.8	7	1287.9	87	14.4	5

DURHAM 1923

Month	Mean max	Mean min	Mean temp	Anom	Highest max	Date	Lowest min	Date	Lowest max	Date	Highest min	Date	Air frost	Days ≥ 25 °C	Total pptn	Anom	Rain days	Wettest day	Date	Total sunshine	Anom	Sunniest day	Date
Jan	8.7	2.8	5.7	+1.7	14.0	30	-2.5	13	4.6	5	7.5	30	5	0	25.0	47	13	4.3	1	68.4	111	6.2	6
Feb	5.7	1.4	3.6	-1.0	13.4	4	-3.1	21	0.1	18	7.5	1	11	0	89.1	197	23	14.3	26	32.6	38	7.1	4
Mar	8.6	2.1	5.4	-0.8	18.4	27	-1.4	5	4.0	9	6.9	27	5	0	39.8	96	15	9.2	28	94.7	79	10.0	26
Apr	9.0	2.5	5.8	-2.5	14.0	4	-1.4	15	4.6	9	5.8	12	3	0	50.1	98	17	19.8	10	118.7	77	10.8	29
May	11.7	3.7	7.7	-3.3	20.1	3	-1.4	24	7.3	11	8.6	1	1	0	63.9	144	20	9.5	25	148.6	79	13.4	7
June	15.6	7.3	11.5	-2.1	23.4	23	2.5	15	9.6	4	13.6	23	0	0	24.4	40	11	7.2	3	129.2	78	13.2	23
July	20.7	11.6	16.2	+0.4	29.6	7	8.1	9	15.7	14	16.4	22	0	3	39.2	63	13	9.7	30	159.3	91	14.4	26
Aug	18.5	9.8	14.1	-1.4	24.0	9	3.6	29	12.9	23	15.3	9	0	0	76.0	115	18	22.6	29	169.6	101	13.0	4
Sep	15.7	7.6	11.6	-1.6	20.1	30	1.4	3	11.8	21	14.7	30	0	0	34.9	61	13	8.1	17	155.4	117	11.5	9
Oct	12.0	4.7	8.4	-1.7	16.8	1	-1.9	14	8.4	12	9.7	30	2	0	47.6	75	20	6.4	24	106.5	108	8.8	9
Nov	5.9	0.1	3.0	-3.6	13.4	13	-4.7	27	1.8	26	7.5	12	13	0	83.2	113	19	14.8	15	67.5	99	7.6	7
Dec	4.8	-0.2	2.3	-1.9	10.7	16	-5.3	5	0.7	4	6.9	17	17	0	58.1	95	18	10.5	25	54.8	96	6.3	23
1923	11.4	4.5	7.9	-1.5	29.6	7	-5.3	12	0.1	2	16.4	7	57	3	631.3	93	200	22.6	8	1305.3	89	14.4	7

DURHAM 1924

Month	Mean max	Mean min	Mean temp	Anom	Highest max	Date	Lowest min	Date	Lowest max	Date	Highest min	Date	Air frost	Days ≥ 25 °C	Total pptn	Anom	Rain days	Wettest day	Date	Total sunshine	Anom	Sunniest day	Date
Jan	5.8	0.4	3.1	-1.0	10.8	1	-5.8	10	-0.3	7	5.3	2	13	0	40.6	76	14	7.9	22	49.6	80	6.4	27
Feb	5.9	0.3	3.1	-1.5	10.8	4	-4.7	29	1.3	13	6.4	4	16	0	25.6	57	13	4.2	25	57.0	67	7.3	1
Mar	7.1	-2.3	2.4	-3.8	13.0	15	-7.5	3	1.3	2	5.3	23	27	0	24.6	60	8	14.9	23	123.3	102	9.7	20
Apr	10.1	1.8	5.9	-2.3	16.9	20	-2.5	10	5.2	1	8.6	25	8	0	44.8	88	16	9.0	28	132.5	86	13.0	15
May	13.3	6.2	9.7	-1.3	19.1	18	0.8	5	5.8	8	12.5	22	0	0	154.0	346	26	70.2	31	134.2	72	13.2	16
June	16.8	7.9	12.4	-1.2	21.9	24	0.3	3	9.7	1	13.6	26	0	0	50.1	82	16	14.3	1	151.3	91	13.9	14
July	18.6	10.3	14.5	-1.3	26.9	12	7.5	1	12.4	28	14.2	11	0	1	98.9	159	22	17.5	20	164.3	94	12.3	15
Aug	17.4	9.4	13.4	-2.2	22.4	11	5.3	9	13.6	17	13.6	1	0	0	59.0	89	16	15.3	17	123.5	74	10.9	15
Sep	15.8	8.7	12.2	-1.0	19.1	8	3.1	10	13.0	22	13.1	6	0	0	72.9	128	22	15.9	19	111.6	84	10.0	3
Oct	12.9	5.9	9.4	-0.6	18.6	14	-0.8	24	9.1	22	10.3	10	2	0	70.0	110	17	18.2	5	95.3	97	8.9	12
Nov	9.2	3.9	6.5	-0.1	13.0	23	-1.4	17	5.2	17	10.3	23	4	0	31.5	43	15	8.3	1	53.8	79	7.3	3
Dec	8.8	3.8	6.3	+2.0	12.4	18	0.3	10	4.7	28	10.3	18	0	0	55.5	91	16	11.0	2	37.2	65	5.2	6
1924	11.8	4.7	8.2	-1.2	26.9	7	-7.5	3	-0.3	1	14.2	7	70	1	727.5	107	201	70.2	5	1233.6	84	13.9	6

DURHAM 1925

Month	Mean max	Mean min	Mean temp	Anom	Highest max	Date	Lowest min	Date	Lowest max	Date	Highest min	Date	Air frost	Days ≥ 25 °C	Total pptn	Anom	Rain days	Wettest day	Date	Total sunshine	Anom	Sunniest day	Date
Jan	7.2	1.5	4.4	+0.3	11.6	17	−3.1	12	2.1	12	6.4	18	5	0	42.3	79	12	14.5	3	43.0	70	6.3	9
Feb	7.1	0.9	4.0	−0.5	12.1	8	−3.1	22	2.7	21	4.7	5	9	0	58.1	128	21	11.7	11	75.6	89	7.1	1
Mar	8.1	1.6	4.9	−1.3	12.1	19	−3.6	21	3.2	21	7.5	6	9	0	41.7	101	15	11.2	20	97.4	81	9.3	9
Apr	10.4	2.3	6.4	−1.9	15.4	12	−2.5	3	5.4	30	5.8	6	5	0	80.3	157	21	12.5	18	139.3	90	12.6	23
May	14.6	6.4	10.5	−0.5	19.3	16	−0.8	2	9.3	1	9.7	17	1	0	58.9	132	21	13.6	24	142.8	76	13.2	21
June	18.6	8.2	13.4	−0.2	30.4	10	3.6	19	11.0	24	11.9	10	0	2	1.7	3	6	0.4	1	194.1	117	15.2	1
July	21.5	11.2	16.3	+0.6	28.2	14	5.3	27	14.9	3	15.3	23	0	4	52.6	85	13	11.5	27	163.0	93	13.2	14
Aug	18.7	10.1	14.4	−1.1	22.7	15	4.7	26	14.9	22	14.7	13	0	0	66.9	101	15	31.2	21	137.4	82	13.7	16
Sep	14.9	6.4	10.7	−2.6	17.7	29	2.5	10	11.6	19	13.6	29	0	0	59.8	105	17	14.3	19	107.5	81	9.0	23
Oct	13.9	5.1	9.5	−0.5	21.6	5	−0.8	16	7.7	13	11.4	6	3	0	63.9	100	19	15.6	24	90.6	92	9.7	1
Nov	6.2	−0.5	2.9	−3.8	13.2	2	−6.4	15	0.4	28	9.2	2	20	0	57.8	79	20	17.1	7	73.8	108	7.0	11
Dec	4.8	−1.0	1.9	−2.3	12.7	29	−7.5	6	−1.8	5	5.8	29	18	0	79.4	130	17	14.6	19	61.3	108	5.6	11
1925	12.2	4.4	8.3	−1.2	30.4	6	−7.5	12	−1.8	12	15.3	7	70	6	663.4	98	197	31.2	8	1325.8	90	15.2	6

DURHAM 1926

Month	Mean max	Mean min	Mean temp	Anom	Highest max	Date	Lowest min	Date	Lowest max	Date	Highest min	Date	Air frost	Days ≥ 25 °C	Total pptn	Anom	Rain days	Wettest day	Date	Total sunshine	Anom	Sunniest day	Date
Jan	6.1	1.0	3.5	−0.5	11.0	25	−4.2	13	−0.1	17	5.8	27	12	0	53.1	99	18	7.9	28	38.5	62	6.0	30
Feb	8.2	2.7	5.5	+0.9	15.4	24	−2.5	13	1.0	10	6.9	23	6	0	40.2	89	19	8.2	6	35.5	42	8.0	28
Mar	9.4	2.0	5.7	−0.5	14.3	2	−3.6	16	3.2	4	9.2	8	5	0	19.7	48	12	4.5	19	107.1	89	9.6	10
Apr	13.1	3.6	8.3	+0.1	19.9	5	−3.1	12	7.7	29	8.1	7	2	0	23.0	45	11	7.1	25	103.1	67	11.4	12
May	13.1	4.3	8.7	−2.3	22.7	26	−3.1	19	7.7	7	11.4	26	3	0	42.6	96	19	16.2	26	167.4	90	13.0	16
June	17.6	8.3	12.9	−0.7	23.8	20	3.1	13	12.7	14	13.1	20	0	0	77.2	126	14	24.9	15	151.7	92	12.2	29
July	20.7	11.8	16.3	+0.5	28.8	14	5.3	26	14.3	6	17.5	12	0	3	80.5	130	12	27.2	18	161.0	92	13.9	11
Aug	20.4	10.7	15.6	+0.0	24.3	2	3.6	1	14.9	15	14.2	14	0	0	59.5	90	16	16.9	15	184.9	110	13.5	1
Sep	17.6	8.8	13.2	−0.1	24.9	19	0.3	26	13.8	25	14.7	17	0	0	56.4	99	11	15.3	8	144.2	109	10.0	22
Oct	11.0	3.1	7.1	−2.9	23.8	4	−5.3	25	3.8	27	14.7	4	9	0	70.3	110	17	11.6	24	121.7	123	9.4	18
Nov	8.2	1.7	4.9	−1.7	12.1	5	−5.3	1	4.3	2	6.4	11	7	0	89.7	122	23	10.1	1	50.6	74	6.1	24
Dec	7.3	1.4	4.3	+0.1	12.7	11	−3.6	13	2.7	21	7.5	10	8	0	17.7	29	8	5.6	13	64.2	113	6.1	29
1926	12.7	4.9	8.8	−0.6	28.8	7	−5.3	10	−0.1	1	17.5	7	52	3	629.9	93	180	27.2	7	1329.9	90	13.9	7

DURHAM 1927

Month	Mean max	Mean min	Mean temp	Anom	Highest max	Date	Lowest min	Date	Lowest max	Date	Highest min	Date	Air frost	Days ≥ 25 °C	Total pptn	Anom	Rain days	Wettest day	Date	Total sunshine	Anom	Sunniest day	Date
Jan	7.1	1.1	4.1	+0.0	12.7	9	-5.3	22	1.6	21	7.5	9	11	0	26.6	50	11	6.2	26	38.9	63	5.1	20
Feb	7.6	0.4	4.0	-0.6	11.6	3	-6.4	11	1.6	12	5.3	27	13	0	16.8	37	9	4.8	7	55.2	65	7.1	4
Mar	9.9	2.7	6.3	+0.1	16.6	21	-2.5	29	6.0	7	8.6	19	2	0	36.4	88	19	7.3	25	92.1	76	8.7	3
Apr	11.1	2.6	6.9	-1.4	16.0	17	-4.2	9	6.6	9	9.7	18	8	0	32.7	64	14	6.0	8	144.0	93	11.4	28
May	13.0	4.2	8.6	-2.4	21.6	7	-4.2	1	7.7	10	9.2	4	2	0	39.5	89	14	12.5	4	120.8	65	13.7	8
June	15.8	6.3	11.0	-2.6	22.7	16	-0.8	15	10.4	1	11.4	17	1	0	77.0	126	17	20.1	25	185.5	112	14.1	14
July	18.5	10.5	14.5	-1.3	23.2	10	4.7	19	12.1	1	14.2	4	0	0	96.5	156	16	28.0	21	134.9	77	15.2	10
Aug	19.0	10.2	14.6	-0.9	23.2	4	3.6	30	14.3	18	14.7	7	0	0	158.3	239	20	35.6	8	132.1	79	11.5	12
Sep	15.4	7.8	11.6	-1.6	22.7	2	2.5	28	11.6	26	14.7	6	0	0	92.4	163	18	26.7	20	103.2	78	10.6	2
Oct	13.6	5.5	9.5	-0.5	18.8	7	-1.9	21	8.8	12	11.4	26	2	0	36.2	57	14	6.8	21	101.2	103	10.0	3
Nov	8.5	2.4	5.4	-1.2	19.3	2	-3.1	9	3.2	8	13.1	2	10	0	53.4	73	21	12.0	7	65.6	96	6.9	4
Dec	3.4	-0.5	1.5	-2.8	8.8	6	-7.5	17	-1.8	19	3.6	8	12	0	70.7	116	17	12.9	21	16.5	29	5.4	28
1927	11.9	4.4	8.2	-1.3	23.2	7	-7.5	12	-1.8	12	14.7	8	61	0	736.5	108	189	35.6	8	1190.0	81	15.2	7

DURHAM 1928

Month	Mean max	Mean min	Mean temp	Anom	Highest max	Date	Lowest min	Date	Lowest max	Date	Highest min	Date	Air frost	Days ≥ 25 °C	Total pptn	Anom	Rain days	Wettest day	Date	Total sunshine	Anom	Sunniest day	Date
Jan	7.4	0.9	4.2	+0.1	12.7	21	-3.6	1	1.0	1	3.6	8	6	0	60.9	114	23	14.6	18	57.5	93	6.3	27
Feb	8.5	1.4	4.9	+0.4	12.7	15	-5.3	28	4.9	24	7.5	8	11	0	28.2	62	16	5.8	10	82.2	96	7.7	17
Mar	7.7	1.5	4.6	-1.6	13.2	20	-6.4	13	-0.1	11	5.8	5	9	0	56.3	136	26	13.3	29	48.2	40	9.5	26
Apr	11.4	2.7	7.0	-1.2	20.4	26	-1.9	20	4.3	14	8.6	10	7	0	18.5	36	16	4.1	20	126.3	82	10.6	27
May	12.9	4.8	8.8	-2.2	20.4	28	-1.9	11	8.2	16	11.9	28	2	0	40.6	91	15	7.6	13	119.7	64	13.0	5
June	15.5	6.7	11.1	-2.5	21.6	3	1.9	15	9.3	14	11.4	22	0	0	117.4	192	18	48.0	13	169.7	102	14.6	1
July	19.5	10.1	14.8	-1.0	24.9	15	4.7	8	15.4	1	14.2	14	0	0	38.9	63	8	27.6	26	232.9	133	14.9	15
Aug	19.1	10.2	14.7	-0.9	23.2	10	6.9	2	14.9	30	14.2	24	0	0	64.5	97	21	20.8	20	159.4	95	11.2	10
Sep	16.1	7.3	11.7	-1.6	23.2	5	1.9	21	9.3	28	15.3	4	0	0	17.2	30	13	4.6	10	132.7	100	11.0	19
Oct	12.5	5.3	8.9	-1.1	17.7	8	-1.9	14	8.2	11	11.4	8	2	0	62.5	98	16	19.1	11	96.1	98	8.2	28
Nov	10.1	4.0	7.1	+0.4	15.4	12	-1.4	9	4.9	28	8.6	12	2	0	56.9	77	20	10.8	16	60.5	89	7.9	4
Dec	5.7	-0.4	2.7	-1.6	12.7	26	-3.6	15	0.4	15	3.6	5	18	0	36.3	59	18	6.6	11	46.9	82	6.0	8
1928	12.2	4.5	8.4	-1.1	24.9	7	-6.4	7	-0.1	3	15.3	9	57	0	598.2	88	210	48.0	6	1332.1	90	14.9	7

DURHAM 1929

Month	Mean max	Mean min	Mean temp	Anom	Highest max	Date	Lowest min	Date	Lowest max	Date	Highest min	Date	Air frost	Days ≥ 25 °C	Total pptn	Anom	Rain days	Wettest day	Date	Total sunshine	Anom	Sunniest day	Date
Jan	3.9	−0.6	1.6	−2.5	11.0	30	−5.3	18	−0.1	16	6.9	31	19	0	39.8	75	19	9.2	9	21.4	35	5.4	16
Feb	2.5	−3.3	−0.4	−5.0	9.3	4	−11.9	14	−3.4	12	4.2	1	22	0	24.7	55	13	6.2	1	32.4	38	7.2	28
Mar	12.1	0.7	6.4	+0.2	21.6	27	−9.2	1	2.7	19	6.4	21	11	0	4.2	10	6	1.7	21	162.2	135	10.9	27
Apr	9.5	1.2	5.3	−2.9	14.9	6	−5.3	22	4.9	12	9.7	18	7	0	24.8	48	14	9.1	26	141.5	92	11.7	16
May	15.2	5.0	10.1	−0.9	22.7	23	−4.2	1	9.9	3	10.8	23	1	0	22.7	51	11	4.4	6	194.9	104	15.0	25
June	16.9	7.7	12.3	−1.3	22.1	11	2.5	5	11.0	4	12.5	19	0	0	40.5	66	14	9.5	12	172.6	104	14.0	26
July	20.3	9.7	15.0	−0.8	27.7	20	1.4	1	12.7	7	15.8	11	0	5	58.5	94	13	15.6	4	163.1	93	15.5	13
Aug	18.6	10.0	14.3	−1.2	22.7	31	5.8	13	14.9	19	13.6	30	0	0	71.6	108	22	22.5	6	127.9	76	13.1	25
Sep	19.3	8.6	14.0	+0.7	25.4	7	3.6	11	12.1	30	14.7	5	0	1	32.4	57	8	16.8	28	157.4	119	11.6	8
Oct	12.8	4.6	8.7	−1.3	16.6	12	−2.5	26	7.1	25	12.5	14	5	0	49.1	77	15	11.7	23	115.2	117	9.9	3
Nov	9.2	2.6	5.9	−0.8	13.8	11	−5.3	16	4.9	15	9.2	20	5	0	65.3	89	19	12.6	9	59.3	87	6.9	3
Dec	7.5	2.1	4.8	+0.5	12.7	13	−2.5	21	1.0	21	6.9	26	8	0	78.9	129	21	13.3	24	52.8	93	6.0	31
1929	12.3	4.0	8.2	−1.3	27.7	7	−11.9	2	−3.4	2	15.8	7	78	6	512.5	75	175	22.5	8	1400.7	95	15.5	7

DURHAM 1930

Month	Mean max	Mean min	Mean temp	Anom	Highest max	Date	Lowest min	Date	Lowest max	Date	Highest min	Date	Air frost	Days ≥ 25 °C	Total pptn	Anom	Rain days	Wettest day	Date	Total sunshine	Anom	Sunniest day	Date
Jan	7.5	1.2	4.4	+0.3	13.2	19	−3.6	26	2.7	16	7.5	19	10	0	89.7	168	20	16.5	13	55.6	90	6.3	15
Feb	4.5	−0.3	2.1	−2.5	9.3	28	−5.3	21	2.1	7	4.2	27	14	0	42.3	93	16	11.0	1	34.4	40	5.7	16
Mar	8.0	0.6	4.3	−1.9	14.9	28	−6.9	20	1.0	19	6.4	28	12	0	62.4	151	16	16.0	15	96.5	80	10.9	22
Apr	10.3	4.0	7.1	−1.1	16.0	25	0.3	23	4.9	4	8.1	24	0	0	54.9	107	16	9.0	2	78.7	51	12.3	22
May	13.4	4.8	9.1	−1.9	21.0	27	−1.9	10	7.7	9	9.2	20	2	0	38.6	87	18	11.6	7	159.6	85	14.2	27
June	19.1	8.7	13.9	+0.3	26.0	6	1.9	8	12.1	1	13.1	19	0	1	29.2	48	10	9.5	16	191.7	116	14.8	8
July	18.3	10.6	14.5	−1.3	23.8	1	6.9	11	11.0	22	14.2	1	0	0	183.9	297	19	44.7	22	130.3	74	12.5	5
Aug	19.6	10.5	15.0	−0.5	29.3	27	5.8	7	14.9	31	16.4	28	0	4	113.2	171	20	18.0	29	172.5	103	14.3	14
Sep	15.5	9.1	12.3	−1.0	21.0	23	2.5	16	11.0	15	14.7	23	0	0	77.7	137	20	15.2	26	83.3	63	10.0	24
Oct	13.0	6.0	9.5	−0.5	16.6	15	0.3	26	8.8	27	12.5	15	0	0	50.0	79	17	13.6	3	111.9	114	10.1	9
Nov	8.7	1.8	5.3	−1.4	15.4	9	−5.8	5	1.6	18	8.6	9	9	0	80.0	109	15	13.7	18	96.3	141	8.1	4
Dec	6.4	0.3	3.4	−0.9	11.6	18	−5.8	10	2.1	10	7.5	19	14	0	63.8	104	15	10.9	26	28.1	49	4.4	14
1930	12.0	4.8	8.4	−1.0	29.3	8	−6.9	3	1.0	3	16.4	8	61	5	885.7	130	202	44.7	7	1238.9	84	14.8	6

DURHAM 1931

Month	Mean max	Mean min	Mean temp	Anom	Highest max	Date	Lowest min	Date	Lowest max	Date	Highest min	Date	Air frost	Days ≥ 25°C	Total pptn	Anom	Rain days	Wettest day	Date	Total sunshine	Anom	Sunniest day	Date
Jan	5.7	−0.3	2.7	−1.4	12.1	16	−4.7	4	1.0	7	4.7	16	16	0	50.3	94	19	14.4	29	74.0	120	6.8	26
Feb	6.0	0.0	3.0	−1.5	13.2	25	−3.6	7	1.0	5	6.9	25	15	0	30.5	67	16	6.0	16	67.4	79	8.7	21
Mar	6.8	−1.1	2.9	−3.3	17.7	20	−7.5	2	−0.7	9	5.3	20	22	0	37.7	91	12	10.4	22	114.9	95	10.7	31
Apr	10.5	3.1	6.8	−1.5	16.0	10	−1.9	1	4.9	2	8.1	9	2	0	78.6	154	18	15.0	2	112.3	73	12.1	5
May	14.5	5.7	10.1	−0.9	20.4	30	−2.5	21	9.9	22	10.3	11	1	0	70.6	159	16	19.6	30	165.3	88	13.1	21
June	16.7	8.6	12.6	−1.0	22.7	28	1.4	25	9.3	3	13.6	27	0	0	92.0	150	20	20.9	2	132.6	80	15.5	29
July	18.6	10.8	14.7	−1.1	23.8	9	5.3	7	15.4	10	14.2	4	0	0	101.6	164	20	15.2	8	96.7	55	11.8	7
Aug	16.9	9.1	13.0	−2.6	23.2	6	4.7	23	13.2	8	13.1	1	0	0	56.1	85	16	9.9	19	126.8	76	11.9	26
Sep	15.1	6.6	10.8	−2.4	21.6	18	−0.3	11	11.0	4	11.9	16	1	0	84.4	149	15	27.0	2	70.1	53	10.3	18
Oct	12.7	4.0	8.4	−1.6	19.9	4	−4.2	26	6.0	30	14.2	5	7	0	27.0	42	12	9.7	29	134.9	137	9.1	15
Nov	9.6	3.5	6.6	−0.0	15.4	3	−2.5	17	4.9	22	8.6	4	3	0	53.3	73	17	11.0	26	37.1	54	5.6	21
Dec	7.9	1.1	4.5	+0.2	12.7	24	−4.7	1	0.4	30	6.9	24	12	0	25.0	41	8	7.1	2	55.8	98	5.6	13
1931	11.8	4.3	8.0	−1.4	23.8	7	−7.5	3	−0.7	3	14.2	7	79	0	707.1	104	189	27.0	9	1187.9	81	15.5	6

DURHAM 1932

Month	Mean max	Mean min	Mean temp	Anom	Highest max	Date	Lowest min	Date	Lowest max	Date	Highest min	Date	Air frost	Days ≥ 25°C	Total pptn	Anom	Rain days	Wettest day	Date	Total sunshine	Anom	Sunniest day	Date
Jan	9.0	1.8	5.4	+1.3	14.3	3	−6.9	28	1.0	28	11.4	3	12	0	37.7	71	12	7.1	16	78.5	127	6.2	8
Feb	7.1	1.2	4.2	−0.4	11.6	3	−4.2	18	1.0	10	5.8	1	8	0	26.4	58	14	6.6	11	67.6	79	8.2	27
Mar	8.0	0.9	4.4	−1.8	12.7	28	−5.8	12	3.8	2	4.7	20	7	0	56.7	137	17	14.3	27	80.8	67	10.4	18
Apr	10.1	1.8	5.9	−2.3	16.0	29	−4.2	3	5.4	2	6.9	29	6	0	32.5	63	21	6.5	21	150.0	97	12.3	27
May	12.3	5.6	9.0	−2.0	19.9	16	−1.9	8	7.1	1	10.8	20	1	0	104.2	234	24	21.3	21	86.6	46	11.6	14
June	16.9	8.0	12.4	−1.2	23.2	17	3.6	5	11.0	3	14.7	27	0	0	29.7	49	7	15.0	27	171.7	104	15.0	17
July	18.9	11.2	15.1	−0.7	27.1	10	6.9	26	13.8	14	15.3	10	0	1	110.6	178	18	31.6	11	121.4	69	13.0	9
Aug	19.7	11.1	15.4	−0.1	26.0	11	1.9	24	14.9	21	16.4	12	0	1	65.6	99	11	21.3	29	130.0	78	12.6	9
Sep	16.2	7.9	12.1	−1.2	24.3	15	−0.8	29	10.4	27	15.3	2	2	0	55.4	98	17	12.1	13	156.8	118	11.2	3
Oct	11.5	4.4	7.9	−2.1	14.9	7	−0.8	29	3.2	29	10.3	7	1	0	101.3	159	23	17.2	10	85.9	87	6.7	13
Nov	8.9	3.5	6.2	−0.5	14.9	2	−3.1	11	5.4	18	11.4	3	5	0	18.8	26	16	3.4	15	44.8	66	7.1	5
Dec	8.0	3.3	5.7	+1.4	13.8	17	−1.9	31	3.8	8	11.9	18	3	0	23.4	38	15	3.8	10	43.8	77	5.5	25
1932	12.2	5.1	8.6	−0.8	27.1	7	−6.9	7	1.0	1	16.4	8	45	2	662.3	97	195	31.6	7	1217.9	83	15.0	6

DURHAM 1933

Month	Mean max	Mean min	Mean temp	Anom	Highest max	Date	Lowest min	Date	Lowest max	Date	Highest min	Date	Air frost	Days ≥ 25 °C	Total pptn	Anom	Rain days	Wettest day	Date	Total sunshine	Anom	Sunniest day	Date
Jan	5.0	−0.9	2.0	−2.0	12.7	3	−9.2	23	−0.7	22	7.5	8	16	0	42.5	80	16	12.4	16	40.5	66	6.0	9
Feb	6.5	0.9	3.7	−0.8	13.2	4	−3.6	18	1.0	23	9.2	5	11	0	85.3	188	16	17.3	26	63.3	74	7.3	17
Mar	11.4	2.1	6.7	+0.6	18.8	28	−1.9	25	5.4	2	6.4	9	7	0	27.2	66	16	7.0	3	162.5	135	10.7	24
Apr	12.6	4.6	8.6	+0.4	19.3	8	−1.9	14	7.1	18	9.2	10	2	0	26.4	52	10	5.7	24	104.6	68	10.4	14
May	14.2	6.7	10.5	−0.5	21.0	20	1.4	14	7.7	2	11.4	21	0	0	59.3	133	14	8.0	6	100.9	54	12.5	22
June	19.5	9.2	14.3	+0.7	26.6	4	4.7	12	13.8	11	13.6	25	0	4	36.4	60	13	9.4	12	225.0	136	14.5	6
July	22.3	12.1	17.2	+1.4	29.9	3	8.6	21	16.0	13	15.3	25	0	4	44.7	72	13	18.3	19	192.2	110	15.3	3
Aug	21.9	11.7	16.8	+1.2	27.7	27	5.8	31	16.6	11	17.5	3	0	7	24.1	36	11	11.3	28	199.1	119	13.0	1
Sep	18.1	9.6	13.9	−0.6	24.5	5	4.7	13	13.4	30	14.2	4	0	0	54.2	95	12	22.4	25	131.8	99	10.9	2
Oct	12.6	6.3	9.5	−1.1	17.3	6	0.8	28	6.2	26	11.4	23	0	0	79.0	124	19	17.9	22	78.7	80	7.7	17
Nov	8.0	3.1	5.6		15.1	6	−1.9	13	3.4	26	8.6	7	3	0	90.5	123	21	14.9	2	44.6	65	6.5	2
Dec	4.2	0.2	2.2	−2.0	9.5	22	−4.2	21	−0.5	21	3.1	2	15	0	50.0	82	19	10.0	8	22.5	40	5.9	31
1933	13.0	5.5	9.3	−0.2	29.9	7	−9.2	1	−0.7	1	17.5	8	54	15	619.6	91	179	22.4	9	1365.7	93	15.3	7

DURHAM 1934

Month	Mean max	Mean min	Mean temp	Anom	Highest max	Date	Lowest min	Date	Lowest max	Date	Highest min	Date	Air frost	Days ≥ 25 °C	Total pptn	Anom	Rain days	Wettest day	Date	Total sunshine	Anom	Sunniest day	Date
Jan	7.1	1.1	4.1	+0.0	11.1	6	−3.3	25	1.1	25	5.6	7	6	0	30.1	56	16	8.9	13	73.4	119	6.2	19
Feb	8.4	1.6	5.0	+0.5	12.2	15	−3.9	26	3.9	26	6.1	7	8	0	10.9	24	4	8.7	27	101.6	119	8.0	28
Mar	8.1	0.7	4.4	−1.8	15.6	24	−3.3	1	4.4	10	3.9	20	7	0	64.6	156	15	18.3	12	96.8	80	10.1	27
Apr	10.2	3.1	6.7	−1.6	18.3	15	−5.6	7	5.0	5	7.8	15	3	0	94.0	184	16	18.7	26	107.3	70	11.7	21
May	15.0	5.4	10.2	−0.8	22.8	10	1.1	17	8.3	16	11.1	11	0	0	32.9	74	13	8.6	6	184.2	99	12.6	28
June	17.8	8.5	13.2	−0.5	25.0	16	3.9	1	11.7	5	13.9	17	0	1	53.3	87	13	14.1	22	161.6	98	12.9	11
July	21.7	11.4	16.6	+0.8	28.9	7	7.8	3	17.2	13	13.9	8	0	6	55.5	89	10	22.6	13	224.0	128	14.0	3
Aug	18.9	9.6	14.2	−1.3	22.2	5	3.3	31	15.0	29	13.3	2	0	0	61.8	93	18	22.6	6	173.8	104	12.5	4
Sep	18.1	8.3	13.2	−0.0	22.8	28	3.3	1	12.8	24	13.9	3	0	0	43.0	76	14	9.2	29	150.5	114	10.7	12
Oct	12.9	6.3	9.6	−0.4	19.4	7	−1.1	31	2.8	31	12.8	11	1	0	79.4	125	20	16.2	30	95.2	97	9.8	9
Nov	8.5	3.3	5.9	−0.7	13.3	27	−4.4	3	4.4	1	7.8	23	5	0	60.6	83	15	20.4	9	45.2	66	8.4	2
Dec	9.0	4.5	6.8	+2.5	12.2	8	−0.6	21	6.1	6	8.3	2	1	0	104.0	170	23	22.9	4	17.8	31	4.6	21
1934	13.0	5.3	9.2	−0.3	28.9	7	−5.6	4	1.1	1	13.9	6	31	7	690.1	101	177	22.9	12	1431.4	97	14.0	7

DURHAM 1935

Month	Mean max	Mean min	Mean temp	Anom	Highest max	Date	Lowest min	Date	Lowest max	Date	Highest min	Date	Air frost	Days ≥ 25 °C	Total pptn	Anom	Rain days	Wettest day	Date	Total sunshine	Anom	Sunniest day	Date
Jan	7.1	1.1	4.1	+0.0	12.2	2	-7.2	9	-1.1	9	7.2	2	10	0	57.2	107	16	15.0	26	40.0	65	5.7	12
Feb	8.0	1.9	5.0	+0.4	12.8	15	-5.6	24	2.8	27	7.8	19	6	0	81.3	179	15	18.8	24	44.3	52	9.0	26
Mar	9.5	2.9	6.2	+0.0	14.4	21	-2.2	12	2.8	9	7.2	25	4	0	15.2	37	11	2.9	16	100.2	83	10.3	27
Apr	10.2	3.2	6.7	-1.6	15.0	21	-2.8	7	6.1	7	7.8	10	4	0	81.3	159	19	31.7	21	106.0	69	11.3	6
May	12.5	4.1	8.3	-2.7	20.6	5	-2.2	19	8.3	14	8.9	3	3	0	20.4	46	5	7.4	13	170.9	91	12.5	19
June	18.2	9.1	13.6	+0.0	27.2	23	1.1	1	8.9	2	15.6	23	0	4	71.1	116	22	18.5	4	149.4	90	14.7	24
July	21.4	10.6	16.0	+0.2	28.3	13	4.4	31	18.3	19	15.0	27	0	2	7.7	12	6	2.6	19	229.5	131	14.6	8
Aug	21.4	10.7	16.1	+0.5	27.2	21	3.9	28	15.0	30	15.0	21	0	7	50.6	76	11	24.0	26	165.8	99	12.2	8
Sep	16.1	7.9	12.0	-1.2	19.4	1	1.1	8	10.6	24	12.8	13	0	0	135.1	238	15	55.8	21	139.0	105	10.9	19
Oct	12.4	4.7	8.5	-1.5	16.7	16	-2.2	23	8.3	21	10.0	14	3	0	91.9	144	20	17.4	4	74.7	76	8.3	20
Nov	8.6	3.0	5.8	-0.9	15.0	3	-1.1	14	2.2	13	6.1	21	2	0	103.2	141	25	28.5	17	33.9	50	6.4	6
Dec	4.6	0.4	2.5	-1.8	7.8	30	-7.8	24	1.1	23	3.9	10	14	0	45.2	74	17	9.2	24	54.1	95	6.3	3
1935	12.5	5.0	8.7	-0.7	28.3	7	-7.8	12	-1.1	1	15.6	6	46	13	760.2	112	182	55.8	9	1307.8	89	14.7	6

DURHAM 1936

Month	Mean max	Mean min	Mean temp	Anom	Highest max	Date	Lowest min	Date	Lowest max	Date	Highest min	Date	Air frost	Days ≥ 25 °C	Total pptn	Anom	Rain days	Wettest day	Date	Total sunshine	Anom	Sunniest day	Date
Jan	5.1	0.6	2.9	-1.2	11.7	9	-5.6	19	-0.6	16	5.6	9	9	0	86.6	162	21	15.6	19	59.5	96	6.4	18
Feb	4.6	-1.5	1.5	-3.0	10.6	16	-9.4	13	1.1	5	2.8	1	15	0	87.7	194	13	27.0	29	66.5	78	8.4	20
Mar	9.0	2.9	6.0	-0.2	16.1	22	-2.8	4	2.8	1	6.7	30	3	0	59.4	144	17	14.0	26	54.2	45	7.2	3
Apr	9.8	1.5	5.6	-2.6	15.6	25	-3.9	21	5.6	14	8.9	25	6	0	37.0	72	12	8.1	14	168.7	109	11.2	30
May	13.9	5.7	9.8	-1.2	19.4	15	0.6	1	8.3	24	9.4	26	0	0	42.3	95	14	21.0	24	152.7	82	14.0	11
June	17.8	7.7	12.8	-0.8	23.3	19	1.7	11	8.3	3	12.8	22	0	0	88.1	144	12	38.2	3	181.4	110	14.3	21
July	18.9	11.0	14.9	-0.8	22.2	6	7.8	22	15.6	11	14.4	7	0	0	77.0	124	20	11.3	23	127.2	73	11.5	8
Aug	20.4	10.9	15.7	+0.1	26.1	29	7.2	6	12.8	6	15.6	24	0	1	52.4	79	10	12.6	6	194.1	116	13.3	7
Sep	16.8	10.4	13.6	+0.4	21.7	13	4.4	29	12.2	26	15.0	3	0	0	68.6	121	16	18.5	24	75.5	57	8.8	5
Oct	12.9	4.9	8.9	-1.1	18.3	22	-0.6	4	8.9	27	11.1	22	1	0	34.5	54	15	7.4	26	2.6	94	9.0	18
Nov	8.6	1.7	5.2	-1.5	12.8	4	-5.6	24	1.7	24	5.6	3	6	0	62.0	84	16	15.0	11	62.0	91	7.2	3
Dec	7.5	1.7	4.6	+0.4	12.8	17	-3.9	7	0.6	7	8.9	20	9	0	45.0	74	13	18.4	13	55.4	97	5.9	4
1936	12.1	4.8	8.5	-1.0	26.1	8	-9.4	2	-0.6	1	15.6	8	49	1	740.6	109	179	38.2	6	1289.8	88	14.3	6

DURHAM 1937

Month	Mean max	Mean min	Mean temp	Anom	Highest max	Date	Lowest min	Date	Lowest max	Date	Highest min	Date	Air frost	Days ≥ 25 °C	Total pptn	Anom	Rain days	Wettest day	Date	Total sunshine	Anom	Sunniest day	Date
Jan	6.7	1.6	4.2	+0.1	12.2	22	−5.0	15	0.6	15	7.8	3	10	0	69.6	130	20	12.0	31	40.3	65	6.2	19
Feb	6.8	1.1	4.0	−0.6	12.2	14	−3.3	24	2.2	13	7.2	3	6	0	77.2	170	18	19.6	27	85.0	100	8.0	20
Mar	5.4	−0.4	2.5	−3.7	11.7	18	−5.0	16	1.1	8	4.4	20	17	0	78.9	191	26	14.7	11	63.7	53	9.1	26
Apr	10.8	4.4	7.6	−0.7	15.6	22	−2.2	27	5.0	2	8.9	10	2	0	56.4	110	18	13.9	16	71.3	46	11.1	21
May	15.4	6.6	11.0	−0.0	23.3	29	2.8	16	7.2	10	12.8	29	0	0	70.5	158	15	22.9	20	158.8	85	12.7	31
June	17.5	9.1	13.3	−0.3	23.3	10	5.0	30	12.8	3	13.9	5	0	0	43.2	71	15	6.7	14	133.4	81	15.6	26
July	19.7	11.4	15.5	−0.2	25.6	14	6.7	31	14.4	10	15.6	2	0	1	79.8	129	16	43.0	3	118.8	68	12.3	31
Aug	20.2	11.0	15.6	+0.1	26.1	3	5.0	27	15.0	21	14.4	3	0	2	38.0	57	11	18.3	14	137.3	82	13.0	7
Sep	16.5	7.7	12.1	−1.1	20.6	7	1.1	19	10.6	13	15.6	6	0	0	39.8	70	9	25.2	17	118.2	89	9.3	2
Oct	13.0	5.9	9.5	−0.6	16.7	1	0.0	19	7.8	20	10.6	7	0	0	42.2	66	13	15.9	23	61.9	63	9.4	4
Nov	8.1	3.0	5.6	−1.1	12.8	2	−2.2	29	3.3	20	9.4	2	6	0	34.1	46	13	12.5	19	46.9	69	6.5	13
Dec	4.4	−0.4	2.0	−2.3	12.2	24	−7.2	13	−0.6	20	5.0	2	13	0	161.4	264	23	18.9	11	30.2	53	5.8	25
1937	12.0	5.1	8.6	−0.9	26.1	8	−7.2	12	−0.6	12	15.6	7	54	3	791.1	116	197	43.0	7	1065.8	72	15.6	6

DURHAM 1938

Month	Mean max	Mean min	Mean temp	Anom	Highest max	Date	Lowest min	Date	Lowest max	Date	Highest min	Date	Air frost	Days ≥ 25 °C	Total pptn	Anom	Rain days	Wettest day	Date	Total sunshine	Anom	Sunniest day	Date
Jan	7.8	2.5	5.1	+1.0	12.8	23	−3.3	11	2.2	10	8.9	23	2	0	50.4	94	20	8.5	14	60.3	98	6.3	30
Feb	7.4	2.2	4.8	+0.2	12.2	26	−2.8	7	2.8	13	8.3	28	5	0	28.5	63	12	11.3	26	47.0	55	6.6	2
Mar	13.1	5.6	9.3	+3.2	16.1	13	−0.6	26	8.3	26	10.6	30	1	0	7.6	18	7	2.3	21	100.4	83	9.6	3
Apr	11.6	2.8	7.2	−1.0	17.2	5	−2.8	11	5.6	17	8.9	1	7	0	2.2	4	4	1.2	29	116.8	76	10.7	11
May	14.0	4.8	9.4	−1.6	18.9	6	−2.8	6	8.9	17	11.7	15	4	0	87.0	196	18	23.6	17	148.6	79	14.0	4
June	17.6	8.8	13.2	−0.4	23.3	18	4.4	16	9.4	2	12.8	24	0	0	68.6	112	14	28.3	1	188.7	114	12.4	9
July	18.6	10.0	14.3	−1.5	22.8	25	3.3	5	13.9	16	16.1	30	0	0	58.6	94	18	16.6	16	127.2	73	11.9	18
Aug	18.9	10.2	14.6	−1.0	24.4	11	1.7	23	14.4	28	15.6	8	0	0	40.3	61	14	9.8	24	139.3	83	13.5	1
Sep	17.2	8.7	12.9	−0.3	22.8	12	2.2	15	13.3	16	15.6	13	0	0	61.0	107	19	10.8	30	87.2	66	11.3	12
Oct	13.0	6.2	9.6	−0.4	17.8	13	0.6	28	9.4	26	11.7	13	0	0	79.9	125	19	16.5	3	102.2	104	8.5	14
Nov	10.9	5.0	8.0	+1.3	17.2	5	−0.6	27	5.6	20	13.3	13	1	0	59.5	81	18	8.0	18	79.4	117	6.6	15
Dec	6.2	1.4	3.8	−0.5	10.6	12	−4.4	22	−1.1	20	7.2	13	7	0	105.3	172	25	13.5	31	41.1	72	6.2	3
1938	13.0	5.7	9.3	−0.1	24.4	8	−4.4	12	−1.1	12	16.1	7	27	0	648.9	95	188	28.3	6	1238.2	84	14.0	5

DURHAM 1939

Month	Mean max	Mean min	Mean temp	Anom	Highest max	Date	Lowest min	Date	Lowest max	Date	Highest min	Date	Air frost	Days ≥ 25°C	Total pptn	Anom	Rain days	Wettest day	Date	Total sunshine	Anom	Sunniest day	Date
Jan	5.4	0.6	3.0	-1.1	11.7	8	-7.2	6	-0.6	6	8.3	8	15	0	121.8	228	23	15.7	1	42.1	68	6.2	5
Feb	8.2	2.3	5.2	+0.7	12.8	10	-7.2	3	0.6	2	8.9	11	3	0	18.5	41	12	4.9	28	65.4	77	7.9	26
Mar	8.7	2.0	5.3	-0.8	13.3	14	-1.7	10	3.3	27	6.1	14	3	0	72.3	175	20	20.6	27	112.4	93	10.2	22
Apr	11.9	3.8	7.9	-0.4	21.7	12	-0.6	6	5.6	4	8.9	13	1	0	28.4	55	14	7.7	24	166.6	108	12.4	19
May	15.2	5.6	10.4	-0.6	23.3	23	-1.1	3	8.3	17	10.6	27	1	0	33.8	76	10	15.7	16	181.5	97	14.5	23
June	18.3	7.9	13.1	-0.5	30.0	6	1.7	3	12.2	12	12.2	29	0	3	63.0	103	18	11.0	18	190.2	115	15.3	3
July	19.0	10.4	14.7	-1.1	23.3	4	3.3	24	15.6	8	15.6	29	0	0	79.0	127	21	23.5	9	125.0	71	13.5	24
Aug	19.8	11.5	15.6	+0.1	23.9	15	7.8	13	15.6	3	14.4	21	0	0	109.0	165	15	26.1	27	118.5	71	12.5	17
Sep	17.2	9.4	13.3	+0.0	22.8	4	1.7	28	10.6	29	14.4	2	0	0	35.0	62	15	20.5	2	96.0	72	9.7	5
Oct	11.8	4.1	7.9	-2.1	15.0	12	-1.7	1	6.1	26	8.9	13	3	0	142.6	224	20	36.0	12	100.0	101	8.9	20
Nov	10.3	4.5	7.4	+0.7	14.4	14	0.0	17	6.7	28	9.4	14	0	0	63.4	86	17	17.3	18	61.5	90	7.0	24
Dec	5.8	1.3	3.6	-0.7	13.3	1	-3.3	31	1.1	7	6.1	1	9	0	32.2	53	12	8.2	8	44.0	77	6.4	31
1939	12.6	5.3	9.0	-0.5	30.0	6	-7.2	1	-0.6	1	15.6	7	35	3	799.0	117	197	36.0	10	1303.2	88	15.3	6

DURHAM 1940

Month	Mean max	Mean min	Mean temp	Anom	Highest max	Date	Lowest min	Date	Lowest max	Date	Highest min	Date	Air frost	Days ≥ 25°C	Total pptn	Anom	Rain days	Wettest day	Date	Total sunshine	Anom	Sunniest day	Date
Jan	1.6	-4.6	-1.5	-5.6	7.2	8	-16.1	21	-2.8	17	1.7	8	28	0	40.9	77	15	8.1	31	56.6	92	6.7	17
Feb	3.4	-0.8	1.3	-3.3	13.9	27	-10.0	16	-1.7	16	3.9	23	15	0	50.8	112	18	9.8	28	30.6	36	5.9	23
Mar	9.0	1.5	5.2	-0.9	14.4	18	-4.4	7	3.9	29	6.7	22	11	0	46.8	113	15	18.3	12	113.2	94	9.8	30
Apr	10.7	2.6	6.7	-1.6	17.8	25	-3.9	18	5.6	22	8.9	27	5	0	39.6	77	16	11.1	2	106.4	69	12.1	15
May	16.5	5.6	11.0	+0.0	22.2	26	1.7	5	8.3	1	11.1	26	0	0	39.7	89	10	13.8	24	214.7	115	15.0	18
June	22.3	9.1	15.7	+2.1	30.0	8	5.0	28	15.0	22	13.3	12	0	8	33.2	54	6	16.0	8	297.0	179	15.8	5
July	18.6	9.0	13.8	-2.0	23.3	30	4.4	5	13.9	14	13.9	30	0	0	133.4	215	22	23.8	17	138.6	79	12.9	31
Aug	18.6	9.5	14.1	-1.5	24.4	4	3.9	24	15.0	23	13.3	31	0	0	21.2	32	7	8.8	21	158.5	95	12.9	17
Sep	16.1	7.1	11.6	-1.7	25.6	4	3.3	9	11.1	29	11.1	1	0	1	33.0	58	10	17.0	16	176.0	133	11.3	9
Oct	11.5	5.8	8.6	-1.4	15.0	6	-1.1	29	6.7	29	10.0	16	1	0	77.9	122	16	14.0	6	66.7	68	8.3	11
Nov	8.9	3.3	6.1	-0.5	13.3	2	-1.7	30	5.6	28	9.4	26	3	0	88.1	120	15	20.0	17	61.7	91	6.4	7
Dec	5.9	1.4	3.7	-0.6	12.2	16	-3.9	13	1.1	13	5.6	16	5	0	53.3	87	17	17.8	30	39.9	70	6.0	11
1940	11.9	4.1	8.0	-1.4	30.0	6	-16.1	1	-2.8	1	13.9	7	68	9	657.9	97	167	23.8	7	1459.9	99	15.8	6

DURHAM 1941

Month	Mean max	Mean min	Mean temp	Anom	Highest max	Date	Lowest min	Date	Lowest max	Date	Highest min	Date	Air frost	Days ≥ 25 °C	Total pptn	Anom	Rain days	Wettest day	Date	Total sunshine	Anom	Sunniest day	Date
Jan	1.8	−2.7	−0.4	−4.5	6.7	13	−13.3	5	−4.4	5	2.2	12	18	0	85.2	160	21	21.7	21	28.5	46	6.8	17
Feb	4.5	−1.6	1.4	−3.1	11.1	8	−10.0	26	0.0	20	3.9	8	15	0	152.5	337	18	50.0	19	70.3	82	8.0	10
Mar	7.1	−0.1	3.5	−2.7	12.2	14	−3.9	30	3.3	9	3.3	27	15	0	63.8	154	20	17.6	25	81.6	68	9.8	13
Apr	8.9	2.1	5.5	−2.8	13.9	12	−2.2	10	3.9	2	8.3	12	4	0	27.0	53	14	8.9	2	101.1	66	13.3	25
May	12.8	3.9	8.3	−2.7	16.1	11	−2.8	11	7.2	14	9.4	23	3	0	30.1	68	13	6.9	28	131.0	70	12.7	3
June	18.6	9.0	13.8	+0.2	29.4	22	2.2	11	9.4	2	17.2	30	0	2	21.7	35	10	13.1	22	86.3	112	14.1	24
July	22.0	11.5	16.7	+1.0	27.2	25	7.2	10	15.6	31	15.0	1	0	6	36.3	59	11	14.4	25	154.8	88	12.7	10
Aug	18.2	9.8	14.0	−1.6	24.4	3	5.6	24	15.6	1	12.2	3	0	0	58.2	88	16	11.5	9	154.0	92	11.7	26
Sep	18.4	9.9	14.2	+0.9	25.0	4	4.4	16	12.2	30	13.9	27	0	1	12.8	23	6	4.0	28	88.4	67	12.1	3
Oct	13.3	5.7	9.5	−0.5	18.9	2	−1.7	12	5.0	29	12.2	6	1	0	90.8	143	21	29.0	9	98.7	100	8.7	20
Nov	8.8	2.5	5.6	−1.0	13.9	24	−5.0	9	5.0	2	6.7	11	4	0	88.3	120	19	18.5	11	45.5	67	8.0	7
Dec	8.8	2.8	5.8	+1.6	12.2	13	−3.9	28	3.3	28	9.4	24	6	0	14.5	24	8	5.5	10	54.0	95	5.5	7
1941	11.9	4.4	8.2	−1.3	29.4	6	−13.3	1	−4.4	1	17.2	6	66	9	681.2	100	177	50.0	2	1194.2	81	14.1	6

DURHAM 1942

Month	Mean max	Mean min	Mean temp	Anom	Highest max	Date	Lowest min	Date	Lowest max	Date	Highest min	Date	Air frost	Days ≥ 25 °C	Total pptn	Anom	Rain days	Wettest day	Date	Total sunshine	Anom	Sunniest day	Date
Jan	3.0	−1.7	0.7	−3.4	10.6	3	−9.4	21	−3.9	22	8.3	3	21	0	49.3	92	18	10.5	4	12.5	20	2.4	8
Feb	2.8	−3.3	−0.2	−4.8	6.1	10	−10.0	8	−0.6	1	2.2	12	26	0	42.2	93	15	7.7	2	39.3	46	8.0	13
Mar	7.5	−0.6	3.5	−2.7	17.8	24	−5.6	8	−1.7	6	7.2	17	19	0	27.3	66	15	6.8	1	77.8	65	11.1	24
Apr	12.3	2.5	7.4	−0.8	21.1	16	−1.7	15	6.1	24	5.6	11	3	0	20.5	40	10	4.3	9	203.3	132	13.3	16
May	14.8	4.2	9.5	−1.5	20.6	17	−2.8	4	7.2	11	10.0	18	4	0	39.2	88	11	11.5	20	211.0	113	13.3	3
June	18.6	7.9	13.2	−0.4	28.9	6	1.1	19	9.4	13	13.3	6	0	4	28.5	47	8	12.1	13	172.7	104	15.1	3
July	19.3	9.3	14.3	−1.5	23.9	31	6.1	28	12.8	18	13.9	22	0	0	76.1	123	17	15.0	16	176.5	101	14.2	2
Aug	19.6	11.1	15.3	−0.2	28.9	28	7.8	5	13.9	4	15.6	26	0	2	51.2	77	18	8.6	30	95.6	57	11.3	16
Sep	17.0	8.1	12.5	−0.7	22.2	1	−0.6	27	7.8	25	13.3	4	1	0	67.3	119	14	16.0	25	151.6	114	11.6	9
Oct	14.1	6.0	10.0	+0.0	19.4	19	−2.2	28	7.8	30	12.8	18	1	0	43.4	68	14	15.1	25	106.5	108	9.2	10
Nov	8.4	1.7	5.1	−1.6	12.8	14	−3.3	19	3.9	22	5.6	25	6	0	16.8	23	14	5.5	5	49.8	73	7.1	13
Dec	8.3	1.5	4.9	+0.7	12.8	10	−7.2	3	1.1	30	8.3	9	9	0	52.8	86	22	11.4	11	28.1	49	6.7	2
1942	12.1	3.9	8.0	−1.4	28.9	6	−10.0	2	−3.9	1	15.6	8	90	6	514.6	76	176	16.0	9	1324.7	90	15.1	6

DURHAM 1943

Month	Mean max	Mean min	Mean temp	Anom	Highest max	Date	Lowest min	Date	Lowest max	Date	Highest min	Date	Air frost	Days ≥ 25 °C	Total pptn	Anom	Rain days	Wettest day	Date	Total sunshine	Anom	Sunniest day	Date
Jan	5.8	-0.5	2.7	-1.4	12.2	31	-6.7	8	-0.6	5	6.7	28	15	0	80.1	150	23	10.7	30	48.5	79	6.2	29
Feb	9.7	1.0	5.4	+0.8	12.8	14	-2.8	7	4.4	7	6.7	14	9	0	19.1	42	11	6.9	8	115.4	135	9.3	27
Mar	10.7	1.1	5.9	-0.3	13.9	1	-5.0	14	5.6	19	6.1	30	10	0	16.2	39	4	12.2	25	131.8	109	10.1	28
Apr	13.8	4.5	9.1	+0.9	17.8	3	-1.7	29	10.0	6	8.9	13	2	0	24.5	48	12	4.6	25	156.4	101	10.8	19
May	16.1	4.9	10.5	-0.5	22.2	17	-1.1	4	5.6	10	9.4	14	2	0	58.3	131	16	12.3	7	193.8	104	13.4	29
June	18.3	8.3	13.3	-0.3	22.8	24	3.9	18	12.8	6	12.8	11	0	0	29.9	49	14	7.6	4	170.6	103	15.3	13
July	19.7	9.0	14.4	-1.4	30.6	31	4.4	24	14.4	21	13.9	27	0	2	39.3	63	11	9.6	6	182.2	104	14.4	12
Aug	18.8	9.5	14.2	-1.4	25.6	1	5.0	26	14.4	28	14.4	1	0	1	98.4	149	21	25.2	5	120.5	72	9.9	30
Sep	16.6	7.3	12.0	-1.3	21.1	14	0.6	27	11.7	27	12.8	13	0	0	87.2	154	17	20.4	16	132.1	100	10.7	20
Oct	13.6	5.1	9.3	-0.7	17.2	1	-1.1	14	10.0	29	11.7	1	1	0	42.3	66	17	7.7	22	87.8	89	8.8	7
Nov	8.4	2.3	5.4	-1.3	12.8	1	-2.8	26	2.2	27	8.9	1	7	0	44.5	61	16	15.0	17	73.4	108	7.0	25
Dec	7.0	0.4	3.7	-0.6	12.8	27	-7.8	14	2.2	17	6.1	27	11	0	29.9	49	13	9.4	7	54.3	95	5.9	19
1943	13.2	4.4	8.8	-0.6	30.6	7	-7.8	12	-0.6	1	14.4	8	57	3	569.7	84	175	25.2	8	1466.8	100	15.3	6

DURHAM 1944

Month	Mean max	Mean min	Mean temp	Anom	Highest max	Date	Lowest min	Date	Lowest max	Date	Highest min	Date	Air frost	Days ≥ 25 °C	Total pptn	Anom	Rain days	Wettest day	Date	Total sunshine	Anom	Sunniest day	Date
Jan	8.7	1.6	5.2	+1.1	15.0	30	-6.1	11	0.6	11	7.8	2	10	0	22.5	42	12	7.5	24	51.4	83	6.1	15
Feb	6.4	0.2	3.3	-1.3	14.4	2	-8.3	29	1.7	27	7.8	2	11	0	55.1	122	18	16.7	26	49.9	59	7.1	28
Mar	9.4	0.5	5.0	-1.2	17.2	26	-4.4	4	2.8	4	5.0	18	11	0	9.7	23	11	2.2	20	116.1	96	10.9	26
Apr	13.7	4.2	8.9	+0.7	20.6	30	-5.6	1	6.1	2	8.3	23	1	0	54.4	106	11	22.4	3	125.9	82	11.1	26
May	15.0	4.9	9.9	-1.1	26.7	29	-2.8	7	7.8	15	12.2	27	1	1	43.9	99	14	8.5	19	145.1	78	12.4	11
June	16.9	7.2	12.1	-1.5	23.9	24	3.9	20	12.8	18	11.7	4	0	0	54.4	89	18	9.8	26	156.6	95	14.3	17
July	19.2	10.4	14.8	-1.0	23.9	16	7.8	12	14.4	18	14.4	3	0	0	86.4	139	18	21.1	27	75.3	43	11.1	17
Aug	20.4	11.1	15.7	+0.2	25.0	16	5.6	30	13.9	22	15.0	18	0	2	36.5	55	16	8.6	27	144.6	86	13.2	12
Sep	16.0	6.9	11.5	-1.8	21.1	17	1.1	10	12.2	2	12.2	15	0	0	131.4	232	18	20.1	2	129.7	98	10.0	27
Oct	11.8	5.1	8.5	-1.5	16.7	6	0.0	6	8.3	31	9.4	8	0	0	82.5	130	24	9.1	20	78.6	80	8.4	6
Nov	8.5	2.3	5.4	-1.3	13.9	23	-1.7	26	3.9	27	7.2	23	2	0	126.2	172	20	22.2	14	64.8	95	6.7	9
Dec	6.0	1.1	3.6	-0.7	11.7	17	-3.3	28	2.2	8	7.2	1	11	0	34.1	56	19	6.2	16	44.5	78	5.9	4
1944	12.7	4.6	8.6	-0.8	26.7	5	-8.3	2	0.6	1	15.0	8	47	3	737.1	108	199	22.4	4	1182.5	80	14.3	6

DURHAM 1945

Month	Mean max	Mean min	Mean temp	Anom	Highest max	Date	Lowest min	Date	Lowest max	Date	Highest min	Date	Air frost	Days ≥ 25 °C	Total pptn	Anom	Rain days	Wettest day	Date	Total sunshine	Anom	Sunniest day	Date
Jan	3.5	-2.5	0.5	-3.6	10.6	3	-12.8	24	-2.2	23	5.0	2	20	0	92.8	174	18	17.0	8	57.2	93	6.4	22
Feb	9.8	3.3	6.5	+2.0	15.0	26	-1.7	11	1.7	11	10.6	26	3	0	36.9	81	16	6.8	1	83.2	98	7.9	20
Mar	12.5	2.8	7.7	+1.5	20.0	24	-2.2	15	6.1	2	7.8	19	8	0	11.6	28	8	2.9	4	119.1	99	10.1	26
Apr	13.8	4.0	8.9	+0.6	21.1	16	-1.7	9	6.7	29	11.1	15	3	0	32.8	64	13	12.3	1	181.7	118	12.7	18
May	15.6	6.1	10.9	-0.1	23.9	12	-1.7	5	8.9	1	10.6	12	2	0	78.2	176	18	13.0	20	173.2	93	13.1	14
June	17.9	9.1	13.5	-0.1	21.7	21	3.9	13	13.9	5	13.3	21	0	0	53.6	88	23	9.9	22	184.0	111	11.8	19
July	21.3	11.4	16.3	+0.6	26.1	5	3.9	28	16.7	2	14.4	15	0	3	40.9	66	16	8.4	9	199.5	114	13.0	23
Aug	19.6	10.6	15.1	-0.4	28.3	2	7.2	3	13.9	6	15.0	28	0	3	85.5	129	13	32.6	29	135.8	81	13.2	2
Sep	18.0	9.1	13.5	+0.3	23.3	12	3.3	6	12.8	25	17.2	12	0	0	48.2	85	16	14.3	18	140.1	106	11.7	6
Oct	15.3	7.0	11.1	+1.1	20.6	15	0.6	14	11.1	26	12.2	21	0	0	93.8	147	15	37.4	21	105.1	107	8.5	14
Nov	10.0	4.7	7.3	+0.7	15.0	1	-2.2	16	6.1	16	10.0	3	1	0	22.3	30	14	3.4	2	46.4	68	6.8	8
Dec	6.7	1.8	4.3	+0.0	11.7	16	-3.3	31	2.2	9	7.8	16	6	0	34.3	56	16	7.2	23	42.3	74	5.8	5
1945	13.7	5.6	9.6	+0.2	28.3	8	-12.8	1	-2.2	1	17.2	9	43	6	630.9	93	186	37.4	10	1467.6	100	13.2	8

DURHAM 1946

Month	Mean max	Mean min	Mean temp	Anom	Highest max	Date	Lowest min	Date	Lowest max	Date	Highest min	Date	Air frost	Days ≥ 25 °C	Total pptn	Anom	Rain days	Wettest day	Date	Total sunshine	Anom	Sunniest day	Date
Jan	5.1	-0.4	2.3	-1.8	10.6	6	-8.3	16	0.6	3	6.7	11	15	0	33.3	62	20	9.4	29	42.5	69	5.8	14
Feb	8.4	2.2	5.3	+0.7	12.8	16	-6.1	26	3.3	21	7.2	18	8	0	8.1	18	9	2.0	5	100.2	117	9.4	27
Mar	8.8	1.4	5.1	-1.0	18.9	27	-3.9	2	2.2	3	9.4	19	12	0	35.9	87	12	6.2	2	102.9	85	9.2	27
Apr	13.9	4.6	9.2	+1.0	22.8	3	-1.1	10	7.2	25	11.1	4	1	0	46.0	90	13	11.6	4	150.6	98	12.3	15
May	13.4	4.8	9.1	-1.9	19.4	30	0.0	12	9.4	1	11.1	30	0	0	55.4	124	15	14.7	26	156.3	84	11.7	9
June	17.1	7.9	12.5	-1.1	27.2	22	2.8	12	12.8	10	12.8	23	0	2	43.7	71	16	10.0	9	184.4	111	13.5	30
July	20.3	10.8	15.5	-0.2	27.2	12	7.8	17	13.3	17	15.6	23	0	5	75.8	122	11	29.8	26	171.1	98	13.8	11
Aug	17.9	9.4	13.6	-1.9	23.3	5	2.8	16	13.9	10	13.9	5	0	0	91.9	139	17	21.9	12	124.1	74	12.0	16
Sep	16.3	9.1	12.7	-0.6	22.2	28	5.0	28	11.1	18	12.2	4	0	0	100.4	177	15	22.4	18	101.9	77	9.9	15
Oct	11.6	5.9	8.8	-1.3	17.2	3	-2.2	30	7.8	24	10.6	3	3	0	41.1	65	10	16.3	2	59.3	60	8.6	11
Nov	9.9	4.7	7.3	+0.7	18.9	4	0.6	23	5.6	14	11.7	4	0	0	121.4	165	26	16.7	20	34.9	51	7.0	13
Dec	5.7	0.8	3.2	-1.0	8.9	22	-2.8	21	2.2	10	4.4	1	11	0	69.6	114	20	11.4	11	53.2	93	5.4	2
1946	12.4	5.1	8.7	-0.7	27.2	6	-8.3	1	0.6	1	15.6	7	50	7	722.6	106	184	29.8	7	1281.4	87	13.8	7

DURHAM 1947

Month	Mean max	Mean min	Mean temp	Anom	Highest max	Date	Lowest min	Date	Lowest max	Date	Highest min	Date	Air frost	Days ≥ 25 °C	Total pptn	Anom	Rain days	Wettest day	Date	Total sunshine	Anom	Sunniest day	Date
Jan	4.7	0.1	2.4	-1.7	11.7	15	-8.3	30	-2.2	30	10.0	15	14	0	48.0	90	17	9.6	8	40.0	65	5.9	17
Feb	0.6	-2.9	-1.1	-5.7	4.4	25	-10.0	23	-2.8	7	0.0	2	26	0	80.5	178	21	11.4	25	42.8	50	9.1	24
Mar	4.7	-1.1	1.8	-4.3	13.3	28	-15.0	4	0.0	7	4.4	22	16	0	97.9	237	22	14.7	16	69.6	58	9.4	3
Apr	11.1	4.0	7.6	-0.7	16.1	16	-2.8	3	6.1	2	8.9	15	2	0	87.7	171	11	26.7	8	149.6	97	11.8	11
May	16.6	6.6	11.6	+0.6	27.2	29	2.8	2	5.0	4	11.7	26	0	1	49.6	111	16	18.2	30	170.1	91	13.3	28
June	18.5	10.1	14.3	+0.7	26.1	1	6.1	16	8.9	14	15.0	27	0	2	47.8	78	15	21.1	14	144.5	87	13.7	25
July	20.2	12.0	16.1	+0.3	26.1	31	8.3	7	15.0	6	16.1	15	0	2	56.2	91	18	21.0	26	173.8	99	13.7	5
Aug	23.0	11.4	17.2	+1.6	26.7	16	7.2	9	17.8	4	15.6	18	0	7	7.7	12	2	7.4	2	234.5	140	12.0	14
Sep	19.1	8.9	14.0	+0.7	23.9	3	2.2	30	13.9	30	15.0	11	0	0	42.7	75	10	14.8	20	146.2	110	9.5	24
Oct	14.6	6.5	10.5	+0.5	22.2	4	0.6	21	8.9	29	12.8	12	0	0	14.0	22	9	5.8	24	89.4	91	9.9	4
Nov	9.0	2.7	5.8	-0.8	16.1	21	-7.2	19	1.7	18	13.3	21	8	0	52.7	72	16	13.9	11	120.0	176	7.8	5
Dec	7.4	2.2	4.8	+0.5	13.3	12	-8.9	2	-1.7	2	7.2	12	7	0	38.3	63	16	7.7	5	46.9	82	5.9	29
1947	12.5	5.0	8.7	-0.7	27.2	5	-15.0	3	-2.8	2	16.1	7	73	12	623.1	92	173	26.7	4	1427.4	97	13.7	6

DURHAM 1948

Month	Mean max	Mean min	Mean temp	Anom	Highest max	Date	Lowest min	Date	Lowest max	Date	Highest min	Date	Air frost	Days ≥ 25 °C	Total pptn	Anom	Rain days	Wettest day	Date	Total sunshine	Anom	Sunniest day	Date
Jan	6.2	1.6	3.9	-0.2	12.2	1	-1.7	16	2.2	9	8.3	2	4	0	188.0	352	26	29.1	4	5.5	57	4.9	18
Feb	6.9	1.7	4.3	-0.3	12.8	8	-3.9	28	0.0	21	7.8	2	5	0	29.1	64	19	4.8	19	66.6	78	8.8	26
Mar	11.9	3.1	7.5	+1.3	18.3	9	-1.7	6	3.9	2	9.4	9	3	0	20.2	49	5	10.7	17	138.8	115	10.3	28
Apr	12.9	3.8	8.3	+0.1	19.4	26	0.6	5	5.6	29	8.3	27	0	0	35.1	69	11	9.9	29	168.8	109	12.6	25
May	14.9	4.6	9.7	-1.3	22.8	18	-1.7	1	8.9	4	9.4	10	4	0	29.2	66	12	7.7	22	202.2	108	14.8	18
June	16.2	8.2	12.2	-1.4	23.9	14	5.6	11	10.0	2	13.3	26	0	0	83.7	137	19	26.9	2	170.9	103	14.9	5
July	19.2	10.5	14.8	-0.9	27.8	31	5.6	24	12.2	9	16.1	26	0	5	42.8	69	12	16.4	31	150.6	86	12.7	28
Aug	17.4	9.9	13.7	-1.9	22.2	2	5.6	21	11.7	12	15.6	1	0	0	96.8	146	18	48.2	11	110.5	66	11.3	23
Sep	16.6	9.0	12.8	-0.5	20.6	27	0.0	23	9.4	23	15.0	27	0	0	98.8	174	12	55.6	3	110.8	84	10.6	6
Oct	13.5	5.6	9.5	-0.5	20.0	2	-3.9	28	7.2	26	14.4	2	3	0	39.5	62	15	6.5	1	91.6	93	9.6	2
Nov	9.4	3.4	6.4	-0.3	14.4	3	-5.0	24	2.8	24	11.7	14	7	0	34.1	46	14	6.1	11	61.3	90	7.7	5
Dec	7.2	1.6	4.4	+0.1	13.3	2	-5.6	20	-0.6	27	9.4	3	9	0	58.0	95	15	16.6	11	49.0	86	6.1	4
1948	12.7	5.2	9.0	-0.5	27.8	7	-5.6	12	-0.6	12	16.1	7	35	5	755.3	111	178	55.6	9	1356.6	92	14.9	6

DURHAM 1949

Month	Mean max	Mean min	Mean temp	Anom	Highest max	Date	Lowest min	Date	Lowest max	Date	Highest min	Date	Air frost	Days ≥ 25 °C	Total pptn	Anom	Rain days	Wettest day	Date	Total sunshine	Anom	Sunniest day	Date
Jan	7.9	1.1	4.5	+0.4	12.8	7	-4.4	3	2.2	3	7.2	16	11	0	14.5	27	13	4.7	15	65.1	105	7.2	24
Feb	8.6	1.7	5.1	+0.5	12.8	17	-6.1	5	3.9	6	8.9	14	6	0	12.6	28	10	3.8	11	108.4	127	7.3	11
Mar	8.3	1.3	4.8	-1.4	17.2	22	-2.2	6	1.7	8	8.9	22	11	0	28.3	68	17	6.8	3	115.1	96	10.7	23
Apr	14.0	5.6	9.8	+1.5	23.9	15	0.6	8	6.1	1	11.1	12	0	0	38.7	76	11	16.1	5	158.7	103	12.5	17
May	15.9	5.1	10.5	-0.5	23.3	13	0.6	10	10.0	3	9.4	23	0	0	15.5	35	12	2.3	25	208.9	112	14.3	13
June	19.5	7.9	13.7	+0.1	27.8	27	2.2	17	12.8	16	15.0	28	0	3	33.0	54	7	12.1	28	250.9	151	15.2	23
July	21.2	11.2	16.2	+0.4	30.0	11	7.2	9	12.2	15	16.1	27	0	8	50.8	82	11	24.5	13	198.4	113	14.4	11
Aug	20.5	11.3	15.9	+0.4	26.7	19	6.7	12	16.1	11	15.6	30	0	3	44.5	67	10	12.1	7	157.8	94	12.3	4
Sep	19.4	11.1	15.3	+2.0	27.2	5	6.7	2	13.9	20	17.2	5	0	1	20.0	35	11	5.7	22	106.5	80	10.0	3
Oct	14.4	7.3	10.9	+0.9	20.6	3	-2.8	28	6.1	28	13.3	3	2	0	61.9	97	15	27.2	25	92.2	94	8.7	27
Nov	8.9	3.3	6.1	-0.5	12.8	4	-2.8	20	3.9	19	7.8	30	2	0	100.2	136	22	19.6	17	55.7	82	7.7	8
Dec	7.9	2.3	5.1	+0.8	12.2	3	-2.2	12	2.8	11	8.3	27	4	0	52.0	85	17	21.6	9	59.1	104	6.4	15
1949	13.9	5.8	9.8	+0.4	30.0	7	-6.1	3	1.7	3	17.2	9	36	15	472.0	69	156	27.2	10	1576.8	107	15.2	6

DURHAM 1950

Month	Mean max	Mean min	Mean temp	Anom	Highest max	Date	Lowest min	Date	Lowest max	Date	Highest min	Date	Air frost	Days ≥ 25 °C	Total pptn	Anom	Rain days	Wettest day	Date	Total sunshine	Anom	Sunniest day	Date
Jan	6.4	1.8	4.1	+0.0	12.8	11	-7.8	26	0.0	27	8.9	11	11	0	41.6	78	10	13.7	3	22.8	37	4.0	26
Feb	7.0	0.8	3.9	-0.7	13.3	17	-8.3	1	2.8	25	9.4	16	9	0	76.8	170	18	16.2	12	95.9	112	9.0	26
Mar	11.1	2.8	6.9	+0.8	17.2	6	-2.8	1	5.0	13	7.2	31	2	0	20.8	50	12	5.8	18	148.4	123	10.3	25
Apr	10.6	2.6	6.6	-1.7	15.6	7	-3.9	25	6.1	24	7.2	21	6	0	33.2	65	16	8.8	26	132.4	86	12.5	25
May	12.9	5.9	9.4	-1.6	20.6	31	2.2	13	8.9	20	9.4	1	0	0	37.9	85	16	8.5	6	115.3	62	13.8	11
June	20.6	9.8	15.2	+1.6	28.9	7	3.9	15	13.3	14	15.0	7	0	4	19.6	32	10	5.9	21	234.1	141	14.7	11
July	19.8	10.8	15.3	-0.4	24.4	9	5.6	13	15.6	4	15.6	20	0	0	100.4	162	16	15.5	28	167.8	96	12.5	24
Aug	18.9	9.8	14.3	-1.2	23.9	6	6.1	13	13.3	31	14.4	11	0	0	122.3	185	20	16.6	30	148.8	89	11.9	16
Sep	15.5	8.3	11.9	-1.3	20.0	4	2.8	30	11.1	26	12.2	13	0	0	106.8	188	20	26.5	6	114.1	86	10.2	9
Oct	12.8	5.9	9.4	-0.6	18.9	5	-1.7	27	6.7	30	12.2	4	2	0	22.9	36	12	4.4	30	101.1	103	6.9	31
Nov	8.3	2.5	5.4	-1.2	12.2	8	-2.8	27	1.7	27	6.1	22	4	0	140.8	192	18	39.4	22	66.1	97	7.0	9
Dec	3.3	-1.2	1.0	-3.2	10.0	1	-6.1	30	-1.1	14	3.9	9	18	0	80.6	132	21	23.9	4	43.9	77	5.8	16
1950	12.3	5.0	8.6	-0.8	28.9	6	-8.3	2	-1.1	12	15.6	7	52	4	803.7	118	189	39.4	11	1390.7	94	14.7	6

DURHAM 1951

Month	Mean max	Mean min	Mean temp	Anom	Highest max	Date	Lowest min	Date	Lowest max	Date	Highest min	Date	Air frost	Days ≥ 25 °C	Total pptn	Anom	Rain days	Wettest day	Date	Total sunshine	Anom	Sunniest day	Date
Jan	5.5	−0.1	2.7	*−1.4*	11.7	20	−9.4	2	1.7	2	8.3	20	14	0	29.2	*55*	16	13.5	6	62.6	*101*	6.7	15
Feb	6.0	0.2	3.1	*−1.5*	10.0	28	−2.8	26	3.3	11	3.3	2	9	0	61.9	*137*	16	9.1	11	69.0	*81*	7.5	9
Mar	7.1	−0.3	3.4	*−2.8*	13.9	22	−4.4	27	1.7	11	4.4	23	19	0	85.5	*207*	18	23.1	17	88.5	*73*	9.0	4
Apr	10.3	1.7	6.0	*−2.3*	21.1	24	−3.3	22	6.7	1	6.1	24	4	0	28.6	*56*	15	6.9	29	161.3	*105*	12.2	22
May	11.8	4.6	8.2	*−2.8*	20.0	12	0.6	8	6.1	6	8.9	24	0	0	99.5	*224*	17	29.1	26	123.5	*66*	12.0	31
June	16.8	7.4	12.1	*−1.5*	23.9	30	0.6	1	11.7	24	13.3	14	0	0	23.0	*38*	11	6.7	25	193.4	*117*	15.3	5
July	20.3	10.7	15.5	*−0.3*	24.4	1	6.1	5	13.3	13	14.4	27	0	0	43.2	*70*	16	8.6	23	159.9	*91*	13.6	29
Aug	18.1	10.0	14.0	*−1.5*	22.2	1	6.7	28	12.2	12	13.9	3	0	0	105.5	*159*	18	24.8	11	131.3	*78*	12.4	1
Sep	17.5	8.9	13.2	*−0.0*	23.3	11	3.3	29	12.2	9	16.1	5	0	0	26.8	*47*	11	6.7	5	143.3	*108*	12.6	1
Oct	13.3	5.8	9.6	*−0.4*	20.6	16	−1.1	23	7.2	22	12.2	2	2	0	44.5	*70*	5	19.7	29	97.6	*99*	8.6	17
Nov	10.2	5.0	7.6	*+1.0*	13.9	1	−0.6	25	6.7	25	8.9	11	2	0	140.4	*191*	21	40.2	5	64.7	*95*	7.0	1
Dec	7.3	1.3	4.3	*+0.1*	11.7	4	−5.0	11	0.6	13	9.4	4	10	0	56.8	*93*	13	15.4	24	55.6	*98*	5.8	1
1951	12.0	4.6	8.3	*−1.1*	24.4	7	−9.4	1	0.6	12	16.1	9	60	0	744.9	*110*	177	40.2	11	1350.7	*92*	15.3	6

DURHAM 1952

Month	Mean max	Mean min	Mean temp	Anom	Highest max	Date	Lowest min	Date	Lowest max	Date	Highest min	Date	Air frost	Days ≥ 25 °C	Total pptn	Anom	Rain days	Wettest day	Date	Total sunshine	Anom	Sunniest day	Date
Jan	4.2	−0.7	1.8	*−2.3*	11.1	6	−6.7	28	−1.1	27	6.1	7	15	0	42.3	*79*	18	7.9	30	72.9	*118*	7.1	26
Feb	6.9	0.5	3.7	*−0.9*	12.2	23	−4.4	13	2.8	13	5.6	21	13	0	10.7	*24*	9	2.9	16	84.9	*100*	7.6	26
Mar	8.8	2.2	5.5	*−0.6*	12.8	8	−3.3	16	2.2	28	8.3	8	6	0	30.6	*74*	13	7.8	6	83.8	*70*	9.8	22
Apr	13.3	3.6	8.5	*+0.2*	21.1	18	−1.1	1	6.7	2	8.3	19	3	0	43.4	*85*	12	19.4	21	154.6	*100*	9.9	18
May	17.4	7.3	12.4	*+1.4*	25.0	16	1.7	8	8.3	3	11.7	27	0	1	27.0	*61*	8	15.3	31	209.9	*112*	15.3	29
June	18.9	8.7	13.8	*+0.2*	27.8	30	3.9	7	12.8	21	13.9	29	0	1	29.6	*48*	12	15.4	21	188.7	*114*	15.3	30
July	20.9	11.1	16.0	*+0.2*	27.8	21	5.6	4	13.9	16	15.6	25	0	5	16.0	*26*	11	6.1	27	205.3	*117*	14.5	4
Aug	19.5	10.7	15.1	*−0.5*	25.0	23	5.6	18	13.9	20	15.0	12	0	1	55.7	*84*	13	19.6	7	183.6	*109*	13.0	14
Sep	14.2	6.8	10.5	*−2.7*	18.9	23	1.1	7	8.3	29	12.8	23	0	0	107.3	*189*	23	21.8	28	95.6	*72*	7.2	2
Oct	11.4	5.2	8.3	*−1.7*	14.4	28	−1.1	15	8.3	19	8.9	3	1	0	73.4	*115*	20	17.3	1	113.6	*115*	8.4	6
Nov	6.6	1.1	3.8	*−2.8*	12.8	1	−5.6	25	−0.6	29	7.8	4	8	0	78.0	*106*	19	14.5	21	77.9	*114*	7.9	7
Dec	5.1	0.2	2.6	*−1.6*	9.4	9	−5.0	15	0.6	5	4.4	10	12	0	56.9	*93*	22	11.6	27	57.7	*101*	6.0	15
1952	12.3	4.7	8.5	*−0.9*	27.8	6	−6.7	1	−1.1	1	15.6	7	58	8	570.9	*84*	180	21.8	9	1528.5	*104*	15.3	5

DURHAM 1953

Month	Mean max	Mean min	Mean temp	Anom	Highest max	Date	Lowest min	Date	Lowest max	Date	Highest min	Date	Air frost	Days ≥ 25 °C	Total pptn	Anom	Rain days	Wettest day	Date	Total sunshine	Anom	Sunniest day	Date
Jan	6.8	1.5	4.2	+0.1	12.2	27	-6.1	5	1.1	5	8.9	28	7	0	13.4	25	10	3.3	2	49.1	80	6.5	18
Feb	7.3	1.6	4.5	-0.1	13.3	22	-4.4	8	-0.6	8	7.2	23	11	0	51.4	113	12	16.0	11	63.6	75	7.8	22
Mar	11.9	0.6	6.3	+0.1	21.7	24	-6.1	5	4.4	19	6.1	12	10	0	1.3	3	1	1.3	30	155.5	129	9.3	25
Apr	10.3	1.8	6.0	-2.2	18.3	23	-3.3	10	6.1	6	5.6	17	6	0	51.0	100	13	14.3	27	149.6	97	11.8	30
May	16.4	7.5	12.0	+0.9	25.0	25	0.0	9	9.4	12	13.3	25	0	1	42.2	95	10	21.8	25	187.9	100	13.1	2
June	16.7	8.6	12.6	-1.0	25.6	9	2.8	6	7.2	2	13.3	24	0	1	60.7	99	12	20.2	2	107.5	65	11.5	22
July	18.6	10.4	14.5	-1.3	23.3	5	6.7	31	15.0	2	12.8	5	0	0	58.3	94	21	9.9	12	142.6	81	11.4	4
Aug	19.9	10.6	15.2	-0.3	30.0	12	6.7	4	14.4	29	14.4	9	0	2	129.1	195	14	46.4	13	166.7	99	12.6	10
Sep	17.6	9.1	13.3	+0.1	25.6	8	3.3	13	14.4	24	14.4	1	0	1	30.8	54	12	6.6	1	131.8	99	10.8	8
Oct	12.7	4.5	8.6	-1.4	18.3	2	-1.1	29	7.8	30	15.0	1	1	0	41.3	65	10	18.2	13	88.5	90	7.9	11
Nov	10.3	5.3	7.8	+1.2	13.3	16	0.0	22	6.1	23	9.4	12	0	0	25.0	34	12	7.3	8	46.1	68	6.6	3
Dec	8.5	3.3	5.9	+1.6	13.3	4	-3.3	16	3.3	16	7.2	7	4	0	17.4	28	18	3.3	26	33.9	60	5.2	31
1953	13.1	5.4	9.2	-0.2	30.0	8	-6.1	1	-0.6	2	15.0	10	39	5	521.9	77	145	46.4	8	1322.8	90	13.1	5

DURHAM 1954

Month	Mean max	Mean min	Mean temp	Anom	Highest max	Date	Lowest min	Date	Lowest max	Date	Highest min	Date	Air frost	Days ≥ 25 °C	Total pptn	Anom	Rain days	Wettest day	Date	Total sunshine	Anom	Sunniest day	Date
Jan	5.1	0.2	2.7	-1.4	12.2	19	-5.0	8	-0.6	31	5.0	19	14	0	56.7	106	17	23.3	20	31.9	52	5.5	17
Feb	3.9	-1.3	1.3	-3.3	10.0	22	-10.6	6	-0.6	1	5.0	22	15	0	48.0	106	21	9.5	10	41.7	49	6.8	8
Mar	7.7	1.5	4.6	-1.6	13.9	11	-6.1	1	1.7	1	6.7	22	4	0	33.6	81	17	7.3	4	71.4	59	8.6	5
Apr	11.0	2.7	6.8	-1.4	15.6	11	-2.8	20	6.7	26	6.7	14	4	0	10.1	20	8	3.7	30	130.9	85	11.4	19
May	14.2	6.4	10.3	-0.7	22.8	27	1.7	8	8.3	3	13.3	28	0	0	120.0	270	15	35.2	28	129.7	69	9.8	15
June	15.7	8.9	12.3	-1.3	21.1	22	3.9	13	10.6	1	12.8	16	0	0	40.3	66	14	10.3	7	118.3	71	13.9	26
July	17.7	9.7	13.7	-2.1	22.2	24	5.0	6	13.3	4	15.0	24	0	0	47.0	76	15	10.7	4	142.0	81	11.8	25
Aug	16.8	10.0	13.4	-2.2	21.7	4	6.1	12	11.7	16	12.8	31	0	0	124.7	188	22	16.4	19	101.5	61	12.2	4
Sep	16.0	6.9	11.4	-1.8	23.3	1	-1.7	27	11.1	27	15.0	2	1	0	49.2	87	21	7.9	10	160.6	121	10.9	17
Oct	14.1	7.5	10.8	+0.8	18.9	2	0.0	25	7.8	26	14.4	18	0	0	106.4	167	21	32.0	23	74.2	75	8.1	6
Nov	9.1	3.1	6.1	-0.5	15.0	11	-1.7	17	6.7	6	7.8	4	3	0	90.7	124	22	21.3	5	75.9	111	7.5	14
Dec	8.6	2.9	5.8	+1.5	14.4	2	-5.0	8	2.8	7	8.9	19	7	0	54.3	89	15	30.6	8	50.8	89	5.2	15
1954	11.7	4.9	8.3	-1.2	23.3	9	-10.6	2	-0.6	1	15.0	7	48	0	781.0	115	208	35.2	5	1128.9	77	13.9	6

DURHAM 1955

Month	Mean max	Mean min	Mean temp	Anom	Highest max	Date	Lowest min	Date	Lowest max	Date	Highest min	Date	Air frost	Days ≥ 25 °C	Total pptn	Anom	Rain days	Wettest day	Date	Total sunshine	Anom	Sunniest day	Date
Jan	3.8	−0.7	1.6	−2.5	10.6	30	−12.2	14	−1.7	13	6.7	30	13	0	54.0	101	13	15.6	10	40.6	66	5.9	30
Feb	2.9	−2.2	0.4	−4.2	8.3	8	−12.2	20	−1.1	19	2.8	2	18	0	60.7	134	18	10.6	22	69.5	81	8.0	27
Mar	6.4	−0.7	2.8	−3.3	12.8	25	−5.6	20	2.2	1	3.9	25	16	0	42.4	103	16	14.3	23	129.2	107	9.4	17
Apr	13.3	4.1	8.7	+0.5	16.7	11	−1.1	2	8.3	17	9.4	12	2	0	14.1	28	9	2.9	23	155.8	101	11.8	13
May	12.5	4.1	8.3	−2.7	20.0	31	0.0	16	6.1	17	8.9	9	0	0	41.1	92	17	11.2	17	96.1	105	14.8	30
June	15.8	7.7	11.7	−1.9	21.1	21	1.7	10	9.4	12	12.8	29	0	0	63.8	104	13	23.2	7	175.0	106	15.2	18
July	22.0	10.7	16.4	+0.6	26.7	22	7.2	1	16.1	4	15.0	31	0	7	24.7	40	3	10.8	2	254.5	145	14.7	12
Aug	21.2	11.6	16.4	+0.8	25.6	23	5.6	9	15.6	28	16.7	19	0	5	15.6	24	9	6.5	21	166.9	100	13.1	26
Sep	18.0	8.9	13.4	+0.2	23.3	6	5.6	14	12.8	15	13.3	2	0	0	18.6	33	12	4.7	1	156.6	118	11.3	6
Oct	12.8	4.5	8.6	−1.4	18.3	8	−3.3	31	5.6	31	14.4	14	5	0	35.5	56	12	9.6	25	127.3	129	8.8	20
Nov	9.3	4.4	6.9	+0.2	15.0	4	−2.8	17	6.1	30	9.4	7	4	0	44.6	61	14	11.4	10	43.0	63	6.9	16
Dec	6.9	1.3	4.1	−0.1	12.8	6	−7.2	20	1.1	18	10.0	6	10	0	69.1	113	17	12.9	9	56.5	99	5.8	7
1955	12.1	4.5	8.3	−1.2	26.7	7	−12.2	1	−1.7	1	16.7	8	68	12	484.2	71	153	23.2	6	1571.0	107	15.2	6

DURHAM 1956

Month	Mean max	Mean min	Mean temp	Anom	Highest max	Date	Lowest min	Date	Lowest max	Date	Highest min	Date	Air frost	Days ≥ 25 °C	Total pptn	Anom	Rain days	Wettest day	Date	Total sunshine	Anom	Sunniest day	Date
Jan	5.3	−0.3	2.5	−1.6	12.8	3	−5.6	10	1.1	10	4.4	1	15	0	82.5	154	19	10.5	29	67.5	109	6.4	18
Feb	3.3	−3.1	0.1	−4.5	10.6	28	−10.6	25	−3.3	2	6.1	6	22	0	54.1	119	22	8.9	12	48.7	57	6.1	27
Mar	8.2	1.6	4.9	−1.3	15.0	26	−3.9	14	2.8	19	5.0	3	8	0	12.7	31	12	3.3	20	111.6	93	9.1	23
Apr	10.2	0.6	5.4	−2.9	16.7	2	−3.9	15	5.6	5	7.8	9	12	0	18.9	37	8	9.6	10	132.4	86	12.0	19
May	16.1	5.8	10.9	−0.1	19.4	22	1.7	20	10.0	18	10.0	5	0	0	13.7	31	8	5.4	17	214.6	115	14.7	25
June	16.0	7.9	12.0	−1.6	22.8	25	4.4	15	10.6	9	13.3	23	0	0	81.2	133	18	15.5	7	140.3	85	14.2	25
July	18.3	10.9	14.6	−1.2	21.7	22	5.0	12	12.8	30	14.4	25	0	0	47.2	76	13	11.5	13	152.5	87	13.5	10
Aug	16.1	8.6	12.3	−3.2	20.0	8	3.9	27	12.8	20	11.7	18	0	0	175.7	265	23	26.3	27	115.9	69	10.4	9
Sep	16.5	9.3	12.9	−0.3	23.3	13	2.8	1	12.2	2	13.3	22	0	0	48.4	85	11	13.3	2	89.1	67	9.0	17
Oct	12.9	5.8	9.3	−0.7	17.8	13	0.6	27	7.2	29	10.0	16	0	0	33.6	53	10	15.6	19	121.8	124	8.8	18
Nov	8.6	2.9	5.7	−0.9	12.2	5	−5.6	22	3.3	23	7.2	25	4	0	16.3	22	12	5.5	13	62.5	92	7.5	6
Dec	7.5	2.3	4.9	+0.7	13.3	7	−5.6	21	0.6	27	10.0	3	11	0	44.0	72	18	7.3	28	26.4	46	4.5	29
1956	11.6	4.4	8.0	−1.5	23.3	9	−10.6	2	−3.3	2	14.4	7	72	0	628.3	92	174	26.3	8	1283.3	87	14.7	5

DURHAM 1957

Month	Mean max	Mean min	Mean temp	Anom	Highest max	Date	Lowest min	Date	Lowest max	Date	Highest min	Date	Air frost	Days ≥ 25 °C	Total pptn	Anom	Rain days	Wettest day	Date	Total sunshine	Anom	Sunniest day	Date
Jan	7.8	2.2	5.0	+0.9	13.3	4	−5.6	19	2.8	14	9.4	5	6	0	43.5	81	22	6.3	31	51.3	83	6.0	6
Feb	7.1	0.2	3.6	−0.9	11.1	5	−5.6	19	2.8	23	6.7	8	13	0	57.2	126	16	9.0	23	91.7	108	8.8	20
Mar	11.2	4.4	7.8	+1.6	17.8	12	−1.7	25	4.4	5	9.4	16	3	0	25.9	63	13	4.1	6	87.7	73	10.8	21
Apr	11.9	3.0	7.4	−0.8	16.1	16	−1.7	29	6.7	8	8.9	4	5	0	6.9	13	8	2.8	21	122.6	79	11.3	28
May	13.9	4.8	9.4	−1.7	20.0	31	−1.1	7	7.2	9	8.3	13	2	0	39.5	89	12	8.7	14	179.0	96	14.8	31
June	19.3	7.7	13.5	−0.1	26.7	28	2.8	25	13.3	25	16.1	29	0	2	32.8	54	10	7.2	8	253.5	153	15.8	15
July	18.6	11.9	15.3	−0.5	22.8	1	7.8	3	14.4	10	14.4	7	0	0	76.7	124	19	15.3	12	97.7	56	11.8	2
Aug	17.7	10.8	14.3	−1.3	22.8	2	6.1	29	13.3	15	13.9	9	0	0	77.1	116	18	24.0	10	110.5	66	11.2	3
Sep	14.9	7.4	11.2	−2.1	18.3	1	4.4	15	10.6	30	11.7	1	0	0	85.7	151	18	24.1	22	118.4	89	9.2	5
Oct	13.6	5.6	9.6	−0.4	17.2	8	−1.7	21	9.4	17	9.4	14	2	0	38.0	60	12	14.5	29	84.3	86	8.3	19
Nov	9.0	4.0	6.5	−0.1	13.9	28	−3.3	7	6.1	15	9.4	28	2	0	32.5	44	10	11.6	4	61.1	90	7.9	1
Dec	6.8	1.2	4.0	−0.2	13.3	20	−3.9	3	1.1	15	7.2	20	9	0	34.4	56	11	12.2	10	43.2	76	5.9	3
1957	12.6	5.3	9.0	−0.5	26.7	6	−5.6	1	1.1	12	16.1	6	42	2	550.2	81	169	24.1	9	1301.0	88	15.8	6

DURHAM 1958

Month	Mean max	Mean min	Mean temp	Anom	Highest max	Date	Lowest min	Date	Lowest max	Date	Highest min	Date	Air frost	Days ≥ 25 °C	Total pptn	Anom	Rain days	Wettest day	Date	Total sunshine	Anom	Sunniest day	Date
Jan	5.0	−0.8	2.1	−2.0	10.6	16	−11.1	21	−2.2	21	7.2	28	14	0	34.9	65	14	6.3	24	74.9	121	7.0	30
Feb	6.2	0.4	3.3	−1.3	12.2	20	−6.1	7	0.0	7	6.1	15	14	0	94.1	208	20	17.0	7	66.6	78	8.1	16
Mar	5.4	−0.7	2.4	−3.8	13.9	30	−8.9	12	−0.6	9	6.1	4	18	0	55.3	134	18	9.9	27	92.6	77	7.8	16
Apr	10.8	2.5	6.7	−1.6	19.4	29	−5.0	12	2.2	3	8.9	19	9	0	30.3	59	12	10.4	4	114.5	74	11.1	10
May	13.6	4.9	9.2	−1.8	21.7	1	−0.6	13	6.7	12	11.1	8	1	0	91.8	206	18	24.4	14	136.3	73	11.9	20
June	15.8	8.2	12.0	−1.6	22.2	14	3.9	24	8.3	10	11.7	16	0	0	71.9	118	18	12.7	2	103.5	62	10.9	15
July	19.1	10.2	14.6	−1.1	27.2	8	4.4	25	13.9	1	13.3	3	0	1	51.2	83	19	12.9	15	154.3	88	14.6	8
Aug	19.1	10.6	14.9	−0.7	23.9	28	5.6	18	14.4	8	13.9	10	0	0	62.6	95	20	12.2	22	118.9	71	9.1	1
Sep	17.6	9.9	13.8	+0.5	22.8	5	2.8	27	13.9	27	13.9	2	0	0	43.8	77	14	9.7	29	109.0	82	8.3	3
Oct	13.7	5.8	9.8	−0.2	17.8	8	1.1	12	10.0	30	11.1	21	0	0	34.2	54	12	8.1	3	121.4	123	9.2	16
Nov	8.7	2.3	5.5	−1.1	13.9	2	−3.9	26	3.9	27	8.3	16	7	0	12.2	17	11	5.1	1	54.4	80	7.9	8
Dec	6.3	1.6	4.0	−0.3	12.2	28	−3.3	10	2.2	24	6.7	20	6	0	82.5	135	18	12.2	17	34.0	60	4.9	31
1958	11.8	4.6	8.2	−1.3	27.2	7	−11.1	1	−2.2	1	13.9	8	69	1	664.8	98	194	24.4	5	1180.4	80	14.6	7

DURHAM 1959

Month	Mean max	Mean min	Mean temp	Anom	Highest max	Date	Lowest min	Date	Lowest max	Date	Highest min	Date	Air frost	Days ≥ 25°C	Total pptn	Anom	Rain days	Wettest day	Date	Total sunshine	Anom	Sunniest day	Date
Jan	3.6	−2.3	0.7	−3.4	11.1	29	−6.7	17	−1.7	10	4.4	20	26	0	50.7	95	17	14.7	21	94.0	152	6.9	16
Feb	6.9	0.8	3.9	−0.7	13.3	27	−3.9	11	−0.6	7	6.7	26	13	0	15.1	33	3	11.6	21	35.9	42	8.0	19
Mar	9.8	3.1	6.5	+0.3	14.4	23	−1.1	4	4.4	17	5.6	3	1	0	14.5	35	12	2.8	3	69.3	58	8.8	15
Apr	13.0	3.7	8.3	+0.1	19.4	14	−1.1	4	8.9	7	9.4	13	4	0	38.2	75	13	8.3	25	129.7	84	10.2	12
May	16.4	6.1	11.2	+0.2	21.1	11	1.1	4	9.4	20	10.0	9	0	0	13.3	30	6	5.8	19	202.3	108	13.0	17
June	19.0	9.1	14.0	+0.4	26.7	14	3.9	14	12.2	29	12.8	23	0	1	61.1	100	13	16.6	27	184.7	112	15.7	14
July	21.6	10.9	16.3	+0.5	27.8	4	4.4	15	15.6	30	15.6	18	0	4	31.1	50	8	18.3	26	190.3	109	14.7	14
Aug	21.7	11.0	16.4	+0.8	28.9	20	4.4	17	17.2	28	17.8	26	0	7	6.4	10	4	5.6	13	185.7	111	13.3	17
Sep	19.4	7.8	13.6	+0.3	27.2	8	2.8	27	13.3	21	13.3	13	0	5	17.7	31	4	11.5	21	162.3	122	11.1	5
Oct	15.6	7.5	11.6	+1.5	23.3	3	1.1	7	7.2	27	12.2	5	0	0	25.0	39	14	7.0	26	116.2	118	8.4	6
Nov	9.5	3.2	6.3	−0.3	15.0	23	−6.1	12	4.4	27	8.9	23	4	0	89.6	122	18	13.7	15	45.5	67	7.4	6
Dec	7.5	2.2	4.9	+0.6	11.1	29	−0.6	2	4.4	4	5.6	31	1	0	77.1	126	23	12.7	4	25.8	45	5.5	28
1959	13.7	5.3	9.5	+0.0	28.9	8	−6.7	1	−1.7	1	17.8	8	49	17	439.8	65	135	18.3	7	1441.7	98	15.7	6

DURHAM 1960

Month	Mean max	Mean min	Mean temp	Anom	Highest max	Date	Lowest min	Date	Lowest max	Date	Highest min	Date	Air frost	Days ≥ 25°C	Total pptn	Anom	Rain days	Wettest day	Date	Total sunshine	Anom	Sunniest day	Date
Jan	6.0	0.4	3.2	−0.9	12.2	22	−3.3	26	0.6	13	9.4	23	14	0	107.5	201	23	18.0	28	44.4	72	6.9	25
Feb	5.9	−0.2	2.8	−1.7	14.4	29	−6.7	18	1.1	17	8.3	29	15	0	57.1	126	16	11.4	24	78.0	91	8.7	15
Mar	7.2	2.6	4.9	−1.3	11.7	1	−2.2	9	1.1	8	8.3	1	1	0	44.3	107	16	7.9	30	53.5	44	9.3	29
Apr	13.2	3.7	8.5	+0.2	18.9	22	−2.8	17	7.2	1	8.9	6	2	0	27.3	53	16	7.6	25	137.8	89	12.8	19
May	16.1	6.2	11.2	+0.1	21.1	26	−1.7	2	9.4	1	12.8	27	1	0	49.8	112	7	21.3	13	188.0	101	14.8	28
June	20.1	9.6	14.9	+1.3	26.1	22	5.6	13	15.6	13	15.6	18	0	2	5.3	9	7	2.0	15	237.6	143	15.1	20
July	18.4	10.0	14.2	−1.6	20.0	1	5.0	1	15.6	2	12.8	24	0	0	84.4	136	26	16.5	10	130.1	74	11.5	31
Aug	18.6	9.5	14.0	−1.5	21.7	7	5.6	9	14.4	13	13.3	5	0	0	59.3	90	17	14.2	24	155.2	93	13.9	15
Sep	16.2	8.6	12.4	−0.9	22.2	11	4.4	26	11.1	27	13.9	17	0	0	56.3	99	16	9.1	16	99.5	75	9.1	21
Oct	12.1	7.1	9.6	−0.4	17.8	4	1.7	17	9.4	30	11.1	3	0	0	172.4	271	24	21.3	8	40.9	42	5.5	13
Nov	8.7	2.6	5.7	−1.0	13.3	30	−5.0	19	3.9	19	7.2	2	2	0	71.0	97	21	10.9	25	53.6	79	6.8	18
Dec	5.9	0.8	3.4	−0.9	12.2	17	−4.4	7	1.7	13	7.2	1	9	0	48.6	79	21	10.7	9	53.5	94	6.2	21
1960	12.4	5.1	8.7	−0.7	26.1	6	−6.7	2	0.6	1	15.6	6	44	2	783.3	115	210	21.3	5	1272.1	86	15.1	6

DURHAM 1961

Month	Mean max	Mean min	Mean temp	Anom	Highest max	Date	Lowest min	Date	Lowest max	Date	Highest min	Date	Air frost	Days ≥ 25 °C	Total pptn	Anom	Rain days	Wettest day	Date	Total sunshine	Anom	Sunniest day	Date
Jan	5.5	0.1	2.8	−1.3	11.1	29	−5.0	15	1.7	17	5.6	13	10	0	103.4	194	24	29.7	2	48.6	79	7.0	30
Feb	9.2	2.7	5.9	+1.4	13.3	15	−1.1	5	4.4	21	6.1	11	3	0	28.4	63	11	6.4	6	81.1	95	8.1	7
Mar	12.3	3.8	8.1	+1.9	17.2	16	−0.6	20	5.6	30	10.0	13	1	0	23.4	57	4	16.0	29	133.7	111	9.8	23
Apr	11.9	4.8	8.4	+0.1	17.8	17	−2.2	4	5.6	4	9.4	13	2	0	59.7	117	22	11.7	1	81.6	53	9.3	24
May	13.8	5.8	9.8	−1.2	21.7	11	0.6	31	8.9	14	10.0	5	0	0	48.4	109	12	23.9	4	153.8	82	12.9	8
June	18.0	8.9	13.5	−0.2	26.1	30	1.7	14	11.7	11	14.4	24	0	1	47.6	78	10	27.9	10	147.3	89	14.2	29
July	18.0	9.9	13.9	−1.8	22.8	6	4.4	6	13.9	21	15.6	1	0	0	93.4	151	11	43.9	14	106.5	61	11.8	6
Aug	19.2	9.6	14.4	−1.1	26.1	29	3.3	12	15.0	8	13.3	29	0	1	81.9	124	12	37.1	10	164.5	98	11.8	31
Sep	18.4	9.6	14.0	+0.7	24.4	1	5.0	9	13.9	6	13.3	16	0	0	40.6	72	14	16.5	2	116.2	88	11.1	1
Oct	13.3	6.2	9.7	−0.3	17.2	6	1.7	30	8.3	19	11.1	7	0	0	76.5	120	19	10.9	19	92.9	94	8.7	12
Nov	8.3	2.5	5.4	−1.3	12.2	1	−2.8	10	6.1	15	7.2	1	6	0	26.5	36	18	3.8	12	65.9	97	7.7	3
Dec	4.0	−1.7	1.1	−3.1	12.2	12	−8.9	25	−0.6	27	5.6	11	20	0	80.0	131	17	26.4	29	29.2	51	4.9	7
1961	12.7	5.2	8.9	−0.5	26.1	6	−8.9	12	−0.6	12	15.6	7	42	2	709.8	104	174	43.9	7	1221.3	83	14.2	6

DURHAM 1962

Month	Mean max	Mean min	Mean temp	Anom	Highest max	Date	Lowest min	Date	Lowest max	Date	Highest min	Date	Air frost	Days ≥ 25 °C	Total pptn	Anom	Rain days	Wettest day	Date	Total sunshine	Anom	Sunniest day	Date
Jan	6.9	0.6	3.8	−0.3	11.1	25	−7.2	1	0.0	1	5.0	26	9	0	48.7	91	17	6.9	30	78.5	127	7.5	27
Feb	7.5	1.1	4.3	−0.3	13.9	20	−3.3	22	1.7	26	5.6	7	7	0	20.6	45	14	3.8	25	71.5	84	7.5	14
Mar	5.7	−1.8	2.0	−4.2	10.0	19	−9.4	7	1.7	3	3.3	27	18	0	27.4	66	20	4.1	4	103.2	86	10.2	14
Apr	11.3	2.7	7.0	−1.2	19.4	25	−0.6	1	6.1	4	6.1	26	1	0	52.9	103	14	19.6	21	161.6	105	12.4	25
May	13.8	5.5	9.6	−1.4	19.4	6	−0.6	1	8.9	26	10.0	7	1	0	59.0	133	16	11.9	18	177.8	95	13.8	23
June	17.5	7.5	12.5	−1.1	25.0	8	0.6	1	13.9	25	11.7	14	0	1	20.9	34	9	8.1	15	204.3	123	15.6	5
July	17.2	9.1	13.2	−2.6	21.7	9	4.4	17	12.8	14	12.8	20	0	0	58.1	94	15	33.8	10	116.0	66	10.7	20
Aug	17.3	9.3	13.3	−2.3	20.0	20	3.3	14	14.4	7	13.3	20	0	0	64.0	97	17	11.9	26	170.2	102	12.1	5
Sep	15.7	7.9	11.8	−1.5	21.7	2	3.3	13	10.6	20	11.7	3	0	0	70.2	124	19	15.5	11	92.2	70	9.4	17
Oct	13.7	6.2	10.0	−0.0	18.3	17	−1.7	27	6.1	26	10.0	2	2	0	13.2	21	8	4.6	25	99.1	101	8.2	17
Nov	7.9	1.4	4.6	−2.0	13.9	5	−5.6	21	1.1	19	8.9	6	9	0	72.8	99	19	28.2	17	66.0	97	7.4	15
Dec	4.7	−2.2	1.3	−3.0	12.8	15	−9.4	25	−2.8	3	5.6	15	23	0	67.6	110	17	11.2	8	69.0	121	7.2	16
1962	11.6	4.0	7.8	−1.7	25.0	6	−9.4	3	−2.8	12	13.3	8	70	1	575.4	85	185	33.8	7	1409.4	96	15.6	6

DURHAM 1963

Month	Mean max	Mean min	Mean temp	Anom	Highest max	Date	Lowest min	Date	Lowest max	Date	Highest min	Date	Air frost	Days ≥ 25 °C	Total pptn	Anom	Rain days	Wettest day	Date	Total sunshine	Anom	Sunniest day	Date
Jan	2.1	-3.2	-0.6	-4.6	5.0	26	-13.3	23	-2.8	22	2.2	27	26	0	74.4	139	17	15.0	3	56.2	91	5.6	22
Feb	1.5	-3.9	-1.2	-5.7	5.0	8	-10.0	25	-0.6	24	-0.6	11	28	0	71.8	159	18	14.0	7	59.9	70	8.7	25
Mar	8.2	1.3	4.7	-1.5	12.2	19	-11.7	2	2.2	2	7.2	18	8	0	42.7	103	17	7.9	29	108.8	90	9.4	26
Apr	11.4	4.3	7.9	-0.4	18.3	28	1.1	1	5.0	5	10.6	28	0	0	41.7	81	15	8.9	17	125.8	82	11.8	12
May	14.1	5.7	9.9	-1.1	20.6	31	1.1	4	9.4	2	10.0	8	0	0	24.6	55	14	7.1	10	154.9	83	14.1	28
June	17.5	9.6	13.6	-0.1	23.3	12	3.9	20	12.2	28	12.8	18	0	0	77.7	127	15	15.7	28	147.0	89	14.5	2
July	18.6	9.8	14.2	-1.6	25.0	30	5.0	14	13.9	2	14.4	24	0	1	33.3	54	15	8.4	4	150.1	86	14.8	20
Aug	17.1	9.2	13.2	-2.4	23.9	2	3.3	29	12.8	30	13.9	3	0	0	93.7	142	20	32.0	4	119.7	71	9.7	1
Sep	16.7	7.3	12.0	-1.3	22.8	17	0.6	20	12.8	2	11.7	24	0	0	45.2	80	10	10.7	25	183.2	138	11.5	16
Oct	13.5	6.3	9.9	-0.1	18.3	8	1.1	25	8.9	18	10.0	8	0	0	33.4	52	12	12.2	31	107.7	109	9.0	13
Nov	9.2	4.0	6.6	-0.0	12.8	5	-3.3	17	4.4	27	9.4	6	3	0	96.8	132	26	15.7	10	42.5	62	6.9	16
Dec	5.7	0.2	2.9	-1.3	10.6	30	-7.2	25	-0.6	20	5.6	3	10	0	37.3	61	14	7.9	14	33.3	58	4.8	26
1963	11.3	4.2	7.8	-1.7	25.0	7	-13.3	1	-2.8	1	14.4	7	75	1	672.6	99	193	32.0	8	1289.1	87	14.8	7

DURHAM 1964

Month	Mean max	Mean min	Mean temp	Anom	Highest max	Date	Lowest min	Date	Lowest max	Date	Highest min	Date	Air frost	Days ≥ 25 °C	Total pptn	Anom	Rain days	Wettest day	Date	Total sunshine	Anom	Sunniest day	Date
Jan	6.3	0.0	3.2	-0.9	12.2	31	-4.4	19	1.1	18	3.3	1	16	0	10.8	20	9	3.8	13	64.2	104	6.4	24
Feb	6.7	0.6	3.6	-0.9	12.8	27	-8.9	21	1.1	19	7.8	3	12	0	25.0	55	9	11.2	17	57.7	68	7.7	5
Mar	5.2	0.8	3.0	-3.2	8.3	23	-3.3	11	1.7	15	3.9	10	10	0	92.9	225	23	27.7	14	33.1	27	7.6	16
Apr	12.1	4.7	8.4	+0.2	17.8	27	-0.6	5	7.2	1	10.6	28	1	0	51.8	101	17	12.2	18	121.4	79	12.3	13
May	16.7	7.4	12.1	+1.1	22.2	17	3.9	1	11.7	31	11.7	12	0	0	24.6	55	10	9.1	31	212.4	114	13.2	26
June	16.7	8.8	12.7	-0.9	22.2	26	3.3	20	8.9	1	13.3	27	0	0	66.6	109	13	12.4	17	142.0	86	13.5	28
July	19.2	10.1	14.7	-1.1	23.9	17	3.3	6	15.0	8	15.0	25	0	0	41.5	67	10	27.4	18	162.2	93	12.3	16
Aug	19.0	9.7	14.4	-1.2	25.0	4	2.2	21	12.8	19	13.9	3	0	1	44.0	66	14	10.2	17	153.4	91	11.8	26
Sep	18.0	8.2	13.1	-0.2	22.8	4	1.1	21	14.4	17	14.4	10	0	0	39.5	70	12	20.8	16	146.0	110	8.8	23
Oct	12.5	4.5	8.5	-1.5	16.7	5	-1.1	10	6.7	23	10.6	20	1	0	36.9	58	12	19.6	10	94.6	96	7.7	12
Nov	9.9	2.9	6.4	-0.2	15.0	24	-5.6	10	1.7	30	10.0	25	8	0	28.7	39	15	6.1	29	66.5	98	6.8	1
Dec	5.5	-1.3	2.1	-2.1	12.2	8	-9.4	26	-1.1	25	4.4	7	19	0	54.9	90	21	12.7	12	44.5	78	6.2	3
1964	12.3	4.7	8.5	-0.9	25.0	8	-9.4	12	-1.1	12	15.0	7	67	1	517.2	76	165	27.7	3	1298.0	88	13.5	6

DURHAM 1965

Month	Mean max	Mean min	Mean temp	Anom	Highest max	Date	Lowest min	Date	Lowest max	Date	Highest min	Date	Air frost	Days ≥ 25°C	Total pptn	Anom	Rain days	Wettest day	Date	Total sunshine	Anom	Sunniest day	Date
Jan	5.5	-0.5	2.5	-1.6	10.6	7	-8.3	30	0.6	21	5.6	7	14	0	90.7	170	22	19.1	16	64.3	104	6.8	19
Feb	6.1	0.7	3.4	-1.2	10.0	12	-4.4	3	3.3	1	3.9	7	8	0	26.8	59	14	9.9	28	37.8	44	6.8	13
Mar	7.7	-0.4	3.7	-2.5	21.7	29	-12.2	3	1.1	1	5.6	28	16	0	51.2	124	16	17.5	23	103.4	86	11.3	29
Apr	11.8	2.6	7.2	-1.0	18.3	2	-2.2	22	6.7	20	6.1	4	4	0	60.3	118	17	8.9	14	169.7	110	11.2	30
May	14.5	5.7	10.1	-0.9	25.0	13	-0.6	19	8.9	17	10.0	14	1	1	60.7	136	20	12.2	17	141.0	75	11.4	9
June	18.3	8.5	13.4	-0.2	21.7	13	1.7	2	15.0	1	12.2	30	0	0	35.4	58	12	13.0	21	155.6	94	13.9	9
July	16.8	8.3	12.6	-3.2	21.1	22	3.9	31	12.2	13	12.2	20	0	0	93.9	151	17	37.3	24	86.9	50	12.7	16
Aug	18.2	8.7	13.5	-2.1	22.2	14	4.4	9	12.2	8	12.8	12	0	0	55.0	83	16	17.0	4	147.5	88	12.0	5
Sep	15.3	8.4	11.9	-1.4	20.0	22	2.8	8	11.1	28	12.8	22	0	0	95.6	168	22	15.7	28	72.4	55	8.3	21
Oct	13.6	6.3	10.0	-0.1	20.0	5	-0.6	19	8.9	22	12.2	28	1	0	20.4	32	8	9.4	31	75.6	77	9.1	15
Nov	6.2	0.6	3.4	-3.2	12.2	5	-5.6	16	0.6	25	6.7	10	16	0	186.1	253	23	24.6	26	56.7	83	6.4	2
Dec	5.4	-0.4	2.5	-1.7	12.2	17	-7.8	2	0.6	27	5.0	16	10	0	38.5	63	19	6.6	1	54.2	95	6.0	10
1965	11.6	4.0	7.8	-1.6	25.0	5	-12.2	3	0.6	1	12.8	8	70	1	814.6	120	206	37.3	7	1165.1	79	13.9	6

DURHAM 1966

Month	Mean max	Mean min	Mean temp	Anom	Highest max	Date	Lowest min	Date	Lowest max	Date	Highest min	Date	Air frost	Days ≥ 25°C	Total pptn	Anom	Rain days	Wettest day	Date	Total sunshine	Anom	Sunniest day	Date
Jan	3.9	-0.8	1.5	-2.5	11.1	28	-7.2	19	-1.1	18	5.6	30	19	0	50.6	95	19	6.6	14	24.4	40	6.1	3
Feb	6.0	0.6	3.3	-1.2	12.2	5	-6.1	18	-0.6	9	6.1	26	12	0	98.6	218	25	16.8	7	35.2	41	8.1	28
Mar	10.2	2.2	6.2	+0.1	13.9	15	-2.8	19	6.7	25	7.2	7	6	0	9.6	23	7	4.1	26	116.1	96	9.2	11
Apr	7.9	2.3	5.1	-3.1	21.1	30	-1.7	2	2.8	12	7.8	27	6	0	79.9	156	22	18.3	1	81.0	53	10.5	28
May	15.1	5.3	10.2	-0.8	23.3	2	0.0	29	8.9	5	8.9	14	0	0	57.0	128	13	22.4	11	180.7	97	14.0	15
June	18.5	10.0	14.3	+0.7	22.2	16	5.0	30	14.4	12	13.3	14	0	0	62.5	102	18	19.8	17	112.2	68	11.4	29
July	18.7	9.5	14.1	-1.6	25.6	22	5.6	18	13.9	26	12.2	9	0	1	67.5	109	21	15.5	31	165.9	95	12.9	18
Aug	17.3	9.7	13.5	-2.0	24.4	19	5.0	24	13.3	4	13.9	18	0	0	112.6	170	19	18.0	13	113.6	68	12.6	15
Sep	17.1	9.3	13.2	-0.1	21.1	19	3.3	16	12.8	22	14.4	11	0	0	27.1	48	12	7.6	4	108.4	82	10.3	21
Oct	12.6	5.7	9.2	-0.9	17.8	14	1.1	24	8.3	25	11.1	14	0	0	100.9	158	20	27.7	3	64.5	65	8.2	11
Nov	8.4	2.2	5.3	-1.3	11.7	13	-3.9	11	4.4	28	7.8	13	4	0	49.1	67	18	6.9	6	58.4	86	6.9	9
Dec	6.7	0.8	3.8	-0.5	11.1	7	-3.9	14	2.8	14	6.1	18	8	0	49.6	81	18	9.9	12	63.5	112	6.5	25
1966	11.9	4.8	8.3	-1.1	25.6	7	-7.2	1	-1.1	1	14.4	9	55	1	765.0	112	212	27.7	10	1123.9	76	14.0	5

DURHAM 1967

Month	Mean max	Mean min	Mean temp	Anom	Highest max	Date	Lowest min	Date	Lowest max	Date	Highest min	Date	Air frost	Days ≥ 25 °C	Total pptn	Anom	Rain days	Wettest day	Date	Total sunshine	Anom	Sunniest day	Date
Jan	6.3	0.2	3.3	−0.8	10.6	29	−8.3	8	1.1	6	5.6	30	12	0	32.3	60	15	7.6	22	77.6	126	6.3	26
Feb	7.8	2.2	5.0	+0.4	12.8	2	−1.7	13	3.9	13	9.4	3	4	0	53.0	117	13	23.9	22	80.8	95	7.3	23
Mar	9.6	3.2	6.4	+0.2	14.4	21	−2.2	31	6.1	9	7.2	6	2	0	23.5	57	11	8.9	9	154.1	128	10.7	23
Apr	11.5	3.2	7.3	−0.9	22.2	28	−5.0	1	6.1	6	7.2	30	3	0	37.9	74	16	8.1	6	113.6	74	12.7	17
May	12.6	4.8	8.7	−2.3	18.3	30	−3.9	3	7.2	13	7.8	6	2	0	97.9	220	27	9.9	29	119.7	64	14.2	30
June	17.9	8.0	13.0	−0.6	23.3	16	3.3	11	14.4	1	12.2	4	0	0	34.6	57	11	19.3	24	166.9	101	13.9	13
July	19.3	10.5	14.9	−0.9	23.3	11	6.1	22	12.8	14	15.6	31	0	0	69.3	112	12	24.6	14	160.2	91	12.4	8
Aug	19.0	10.2	14.6	−0.9	25.0	21	5.0	2	14.4	18	14.4	8	0	1	115.3	174	16	24.6	9	144.2	86	12.0	20
Sep	16.6	8.5	12.5	−0.8	20.0	26	5.0	14	13.3	15	12.8	26	0	0	52.8	93	17	12.4	22	102.6	77	10.7	7
Oct	12.8	6.1	9.4	−0.6	17.2	9	1.1	18	7.2	27	10.6	8	0	0	87.0	137	18	39.1	16	107.5	109	8.4	29
Nov	8.5	0.7	4.6	−2.0	14.4	11	−3.3	18	4.4	24	6.7	14	9	0	81.3	111	14	36.3	5	71.4	105	6.8	17
Dec	7.0	0.2	3.6	−0.6	14.4	1	−5.6	20	−0.6	8	8.3	23	13	0	34.3	56	14	10.2	9	72.8	128	6.2	7
1967	12.4	4.8	8.6	−0.8	25.0	8	−8.3	1	−0.6	12	15.6	7	45	1	719.2	106	184	39.1	10	1371.4	93	14.2	5

DURHAM 1968

Month	Mean max	Mean min	Mean temp	Anom	Highest max	Date	Lowest min	Date	Lowest max	Date	Highest min	Date	Air frost	Days ≥ 25 °C	Total pptn	Anom	Rain days	Wettest day	Date	Total sunshine	Anom	Sunniest day	Date
Jan	6.4	0.2	3.3	−0.8	12.8	19	−6.1	10	0.6	1	8.3	31	15	0	25.2	47	16	4.8	5	43.0	70	6.6	28
Feb	4.6	−2.0	1.3	−3.3	7.8	19	−6.1	24	0.6	4	1.7	1	23	0	54.2	120	14	13.0	5	69.3	81	7.2	28
Mar	9.7	2.1	5.9	−0.3	17.2	28	−2.8	1	1.7	1	7.2	24	8	0	30.1	73	12	7.6	20	117.8	98	11.4	21
Apr	12.2	2.2	7.2	−1.1	18.9	26	−5.6	4	4.4	2	8.3	20	11	0	47.7	93	12	20.6	16	181.0	117	13.1	26
May	12.7	3.6	8.2	−2.8	20.6	28	−2.2	9	6.1	8	7.8	15	3	0	49.3	111	15	13.7	25	120.9	65	11.1	13
June	18.6	8.4	13.5	−0.1	26.1	30	3.9	9	13.9	7	11.7	30	0	1	50.6	83	12	14.2	22	198.2	120	14.4	13
July	16.8	9.6	13.2	−2.6	20.0	19	5.0	19	12.8	15	15.6	1	0	0	102.4	165	15	25.9	1	81.9	47	8.9	19
Aug	17.8	10.7	14.3	−1.3	25.0	23	5.6	15	13.3	25	15.0	23	0	1	42.2	64	18	12.2	13	103.0	61	8.9	12
Sep	17.0	9.2	13.1	−0.2	21.1	6	5.0	4	13.3	22	12.8	11	0	0	113.6	200	16	28.2	12	114.6	86	10.4	3
Oct	14.5	8.6	11.6	+1.6	17.8	3	2.2	18	10.6	17	14.4	4	0	0	65.0	102	17	26.2	31	66.4	67	9.1	13
Nov	7.8	3.9	5.9	−0.8	12.2	25	−1.7	5	3.9	3	7.2	26	1	0	65.6	89	17	22.1	1	30.1	44	7.6	4
Dec	4.6	0.4	2.5	−1.7	8.9	22	−5.6	14	0.6	28	5.6	2	13	0	97.3	159	17	28.2	18	33.2	58	5.2	21
1968	11.9	4.7	8.3	−1.1	26.1	6	−6.1	1	0.6	1	15.6	7	74	2	743.2	109	181	28.2	9	1159.4	79	14.4	6

DURHAM 1969

Month	Mean max	Mean min	Mean temp	Anom	Highest max	Date	Lowest min	Date	Lowest max	Date	Highest min	Date	Air frost	Days ≥ 25 °C	Total pptn	Anom	Rain days	Wettest day	Date	Total sunshine	Anom	Sunniest day	Date
Jan	7.1	1.0	4.1	-0.0	13.3	22	-6.1	6	3.3	5	7.8	25	11	0	47.9	90	15	12.2	12	43.5	70	5.8	31
Feb	3.0	-2.6	0.2	-4.3	8.9	4	-11.7	18	-1.7	7	2.2	6	20	0	88.5	195	17	13.5	21	72.6	85	8.5	8
Mar	4.9	-0.5	2.2	-4.0	11.7	30	-5.6	5	1.1	16	5.6	30	16	0	85.9	208	19	16.3	13	63.5	53	9.7	6
Apr	10.7	2.1	6.4	-1.8	19.4	8	-1.7	5	5.6	20	8.9	10	6	0	45.5	89	16	13.0	21	154.8	100	11.7	4
May	13.1	5.9	9.5	-1.5	18.3	29	0.6	1	6.1	2	13.3	13	0	0	102.0	229	26	21.8	2	103.4	55	11.2	21
June	18.1	8.1	13.1	-0.5	23.3	12	1.7	6	11.7	4	11.7	18	0	0	78.0	128	13	23.9	23	202.5	122	14.4	8
July	20.4	11.4	15.9	+0.1	27.2	16	6.7	3	14.4	7	16.1	21	0	4	44.3	71	11	17.3	28	176.8	101	14.0	14
Aug	19.9	11.7	15.8	+0.2	26.7	11	6.7	24	15.6	22	16.1	10	0	1	35.8	54	12	8.4	21	140.3	84	12.2	5
Sep	17.1	9.9	13.5	+0.2	22.8	3	3.9	29	11.7	30	13.9	24	0	0	58.4	103	12	17.3	17	98.9	75	10.1	22
Oct	15.7	8.7	12.2	+2.2	23.9	10	2.8	1	11.1	21	13.9	9	0	0	9.6	15	5	2.5	23	94.3	96	8.9	9
Nov	6.9	1.0	3.9	-2.7	15.0	2	-4.4	28	-1.1	28	11.1	2	12	0	116.9	159	22	22.9	11	76.6	112	6.6	13
Dec	5.0	0.3	2.7	-1.6	10.0	7	-6.7	19	0.6	18	4.4	3	13	0	45.7	75	16	9.4	20	28.7	50	6.1	26
1969	11.8	4.7	8.3	-1.2	27.2	7	-11.7	2	-1.7	2	16.1	7	78	5	758.5	112	184	23.9	6	1255.9	85	14.4	6

DURHAM 1970

Month	Mean max	Mean min	Mean temp	Anom	Highest max	Date	Lowest min	Date	Lowest max	Date	Highest min	Date	Air frost	Days ≥ 25 °C	Total pptn	Anom	Rain days	Wettest day	Date	Total sunshine	Anom	Sunniest day	Date
Jan	5.0	0.0	2.5	-1.6	8.9	25	-10.6	7	-0.6	6	5.0	15	13	0	90.1	169	21	12.7	15	39.2	63	6.3	5
Feb	5.4	-1.2	2.1	-2.4	9.4	21	-6.7	15	1.1	13	3.3	23	14	0	47.7	105	15	8.6	19	114.1	134	8.3	11
Mar	7.5	-0.3	3.6	-2.6	12.8	17	-7.8	9	2.2	4	6.7	17	13	0	32.4	78	18	8.1	21	148.7	123	10.1	25
Apr	9.7	2.0	5.9	-2.4	15.6	17	-2.2	2	5.0	1	9.4	23	10	0	49.4	96	19	12.2	5	140.2	91	11.3	19
May	16.0	7.2	11.6	+0.6	21.7	5	4.4	1	9.4	8	10.6	19	0	2	22.7	51	11	6.4	10	181.8	97	12.1	25
June	20.3	9.4	14.9	+1.3	25.6	10	6.1	15	15.0	4	12.8	24	0	2	18.6	30	7	6.4	27	245.7	148	14.4	3
July	18.4	10.8	14.6	-1.2	25.6	8	5.6	22	13.3	25	15.0	6	0	2	46.7	75	13	11.7	18	151.2	86	12.5	10
Aug	20.1	11.3	15.7	+0.2	24.4	3	7.8	11	14.4	20	15.6	2	0	0	63.5	96	10	33.3	20	172.8	103	11.5	27
Sep	17.4	10.0	13.7	+0.4	23.3	24	2.8	15	13.3	21	13.3	8	0	0	34.3	60	16	7.1	8	123.2	93	9.6	1
Oct	13.9	6.2	10.1	+0.1	19.4	13	1.1	9	7.8	20	11.1	12	0	0	22.8	36	16	8.1	28	116.3	118	9.9	8
Nov	9.2	3.0	6.1	-0.6	15.0	1	-2.2	16	4.4	14	8.3	1	5	0	62.6	85	19	16.3	17	67.2	99	7.3	9
Dec	6.6	2.1	4.3	+0.1	11.7	18	-2.8	31	1.1	31	8.3	19	5	0	72.1	118	19	12.2	28	53.2	93	5.9	8
1970	12.5	5.1	8.8	-0.7	25.6	6	-10.6	6	-0.6	1	15.6	8	60	4	562.9	83	184	33.3	8	1553.6	105	14.4	6

DURHAM 1971

Month	Mean max	Mean min	Mean temp	Anom	Highest max	Date	Lowest min	Date	Lowest max	Date	Highest min	Date	Air frost	Days ≥ 25 °C	Total pptn	Anom	Rain days	Wettest day	Date	Total sunshine	Anom	Sunniest day	Date
Jan	6.3	1.1	3.7	−0.4	13.3	10	−5.6	4	−1.1	5	7.8	10	11	0	45.8	86	17	13.3	20	26.9	44	5.5	19
Feb	7.8	1.3	4.6	−0.0	13.1	3	−5.6	16	3.0	27	5.6	25	7	0	16.7	37	10	6.4	17	83.4	98	7.5	14
Mar	8.2	1.9	5.0	−1.2	12.7	24	−3.1	5	2.3	5	6.9	28	9	0	62.0	150	23	11.9	19	91.0	76	9.3	27
Apr	10.7	3.5	7.1	−1.2	16.9	21	−1.5	17	6.2	4	7.8	23	3	0	48.8	95	10	32.2	23	120.2	78	12.1	25
May	15.5	5.9	10.7	−0.3	19.7	11	−1.2	2	10.7	1	12.1	10	1	0	47.7	107	12	14.1	6	258.8	138	13.0	27
June	14.8	7.6	11.2	−2.4	21.1	30	1.6	10	9.6	8	12.7	25	0	0	59.0	96	17	12.7	19	139.0	84	11.5	26
July	20.7	11.1	15.9	+0.1	26.1	1	4.3	18	13.7	22	15.1	3	0	3	32.3	52	9	11.5	23	203.4	116	13.9	13
Aug	18.2	10.8	14.5	−1.0	21.6	20	7.2	16	11.7	14	14.3	9	0	0	127.7	193	21	50.3	13	123.4	74	11.6	2
Sep	18.2	9.3	13.8	+0.5	22.5	6	4.3	12	12.4	26	14.4	1	0	0	12.8	23	8	4.8	28	140.4	106	10.9	5
Oct	14.4	6.5	10.4	+0.4	20.6	2	−1.6	15	8.6	15	14.0	8	5	0	20.9	33	8	11.7	17	98.9	100	9.1	1
Nov	8.4	2.6	5.5	−1.1	15.9	1	−4.5	20	1.5	19	12.2	2	11	0	47.7	65	18	10.8	20	65.4	96	7.3	13
Dec	9.1	3.9	6.5	+2.2	14.2	20	−2.5	1	3.5	28	8.5	17	4	0	19.7	32	11	6.2	28	27.6	48	4.0	10
1971	12.7	5.5	9.1	−0.4	26.1	7	−5.6	1	−1.1	1	15.1	7	51	3	541.1	80	164	50.3	8	1378.4	94	13.9	7

DURHAM 1972

Month	Mean max	Mean min	Mean temp	Anom	Highest max	Date	Lowest min	Date	Lowest max	Date	Highest min	Date	Air frost	Days ≥ 25 °C	Total pptn	Anom	Rain days	Wettest day	Date	Total sunshine	Anom	Sunniest day	Date
Jan	5.1	1.2	3.1	−1.0	9.6	23	−8.5	31	0.0	31	6.5	23	10	0	116.0	217	24	23.9	26	33.1	54	6.3	21
Feb	5.7	1.1	3.4	−1.2	8.9	14	−13.1	1	1.3	1	4.7	17	6	0	56.0	124	20	12.7	15	33.4	39	8.2	14
Mar	8.9	1.9	5.4	−0.8	15.5	23	−1.5	14	3.2	13	6.7	23	4	0	51.9	126	17	8.6	26	99.8	83	9.3	21
Apr	11.7	4.4	8.1	−0.2	16.0	25	−1.1	21	7.1	24	8.5	2	1	0	30.6	60	16	6.1	5	108.3	70	12.1	20
May	13.9	5.7	9.8	−1.2	17.8	23	2.2	11	9.1	4	10.0	22	0	0	57.3	129	21	10.1	11	141.1	75	12.6	17
June	15.4	7.2	11.3	−2.3	18.7	16	3.9	8	11.3	10	12.6	25	0	0	59.6	97	23	7.1	10	126.6	76	13.1	28
July	19.0	9.8	14.4	−1.4	26.7	20	5.8	9	15.5	3	13.4	23	0	1	50.9	82	12	15.5	22	140.9	80	14.3	17
Aug	18.8	10.1	14.5	−1.1	23.7	24	5.8	19	14.5	25	13.6	6	0	0	14.6	22	6	5.8	7	169.1	101	12.2	21
Sep	15.5	6.9	11.2	−2.1	22.4	1	2.0	8	11.0	24	11.9	23	0	0	23.8	42	6	13.2	8	111.3	84	11.2	5
Oct	13.5	5.9	9.7	−0.3	18.4	8	−0.5	13	8.0	20	12.2	10	1	0	11.9	19	8	6.1	9	84.9	86	8.5	4
Nov	8.5	2.4	5.5	−1.2	17.0	6	−4.3	14	3.2	17	10.5	3	7	0	52.7	72	17	18.5	12	66.4	98	7.2	10
Dec	7.0	1.7	4.3	+0.1	13.0	14	−3.4	9	3.5	20	7.0	15	8	0	26.0	42	13	4.4	6	24.7	43	5.8	24
1972	11.9	4.9	8.4	−1.1	26.7	7	−13.1	2	0.0	1	13.6	8	37	1	551.3	81	183	23.9	1	1139.6	77	14.3	7

DURHAM 1973

Month	Mean max	Mean min	Mean temp	Anom	Highest max	Date	Lowest min	Date	Lowest max	Date	Highest min	Date	Air frost	Days ≥ 25 °C	Total pptn	Anom	Rain days	Wettest day	Date	Total sunshine	Anom	Sunniest day	Date
Jan	6.3	1.7	4.0	-0.1	11.0	2	-5.5	18	2.1	7	7.6	2	10	0	27.3	51	14	7.4	20	37.9	61	6.3	16
Feb	7.7	0.8	4.3	-0.3	14.0	20	-8.9	15	2.9	13	6.9	20	11	0	12.1	27	5	7.5	15	103.6	121	8.4	20
Mar	10.7	2.3	6.5	+0.3	17.2	23	-4.8	11	6.0	14	7.4	18	5	0	9.1	22	7	3.0	5	114.8	95	9.7	21
Apr	9.7	2.3	6.0	-2.3	16.8	15	-3.4	9	5.1	2	7.1	15	6	0	70.5	138	16	21.8	1	121.5	79	10.3	5
May	14.1	6.2	10.1	-0.9	21.3	27	0.9	16	8.9	20	12.0	27	0	0	56.8	128	14	13.1	4	163.2	87	14.1	16
June	18.8	9.9	14.4	+0.7	23.5	8	3.4	1	14.5	3	15.5	19	0	0	35.2	58	10	11.2	2	169.1	102	13.6	13
July	19.2	10.6	14.9	-0.9	24.4	5	4.9	23	13.5	17	14.4	30	0	0	78.3	126	13	33.7	16	125.1	71	12.6	31
Aug	19.0	10.1	14.5	-1.0	26.9	16	6.6	18	13.7	19	14.5	17	0	2	66.2	100	13	25.2	5	159.3	95	12.5	14
Sep	16.7	8.9	12.8	-0.5	23.1	8	2.8	11	10.4	30	15.4	7	0	0	48.8	86	13	16.3	22	119.3	90	10.3	6
Oct	11.8	5.2	8.5	-1.5	20.6	2	-0.4	12	6.9	17	11.1	3	1	0	41.0	64	16	6.1	14	71.7	73	9.7	17
Nov	7.9	1.7	4.8	-1.9	13.5	8	-3.6	28	0.9	26	9.1	9	10	0	21.6	29	7	10.4	9	89.3	131	7.6	5
Dec	7.0	0.5	3.7	-0.5	11.7	4	-9.6	1	2.1	1	7.9	4	12	0	36.5	60	13	17.6	7	65.4	115	6.4	31
1973	12.4	5.0	8.7	-0.7	26.9	8	-9.6	12	0.9	11	15.5	6	55	2	503.4	74	141	33.7	7	1340.2	91	14.1	5

DURHAM 1974

Month	Mean max	Mean min	Mean temp	Anom	Highest max	Date	Lowest min	Date	Lowest max	Date	Highest min	Date	Air frost	Days ≥ 25 °C	Total pptn	Anom	Rain days	Wettest day	Date	Total sunshine	Anom	Sunniest day	Date
Jan	7.6	1.8	4.7	+0.6	11.8	18	-3.6	1	0.1	1	7.1	19	5	0	74.7	140	18	10.9	8	49.1	80	6.4	20
Feb	7.5	2.1	4.8	+0.2	12.9	21	-1.5	4	2.6	4	5.7	16	7	0	45.3	100	18	6.9	14	57.2	67	7.6	7
Mar	7.7	1.6	4.7	-1.5	13.2	31	-1.4	30	3.7	9	4.4	24	7	0	52.3	127	15	13.1	3	98.5	82	9.9	30
Apr	9.6	2.6	6.1	-2.2	16.8	10	-0.7	22	4.8	12	4.8	24	3	0	16.5	32	4	7.1	28	96.0	62	11.6	14
May	15.1	5.7	10.4	-0.6	22.8	26	-0.5	7	10.3	4	9.7	27	1	0	19.7	44	9	7.0	23	188.9	101	13.4	15
June	16.7	7.9	12.3	-1.3	23.0	20	-0.5	28	12.3	5	12.6	2	1	0	34.4	56	9	12.1	16	151.2	91	14.3	21
July	18.2	9.8	14.0	-1.8	20.5	7	6.4	30	15.2	5	14.1	22	0	0	65.4	105	20	22.8	2	137.0	78	10.9	20
Aug	18.5	9.8	14.1	-1.4	21.6	6	4.9	4	15.0	30	14.3	25	0	0	48.4	73	13	13.3	10	160.2	96	12.2	27
Sep	15.0	7.4	11.2	-2.1	20.7	13	0.7	29	10.1	25	12.9	13	0	0	65.4	115	17	14.8	2	114.9	87	9.5	18
Oct	10.0	4.4	7.2	-2.8	12.8	26	0.4	31	4.0	30	7.1	8	0	0	76.1	120	20	10.1	3	63.6	65	8.1	27
Nov	8.5	3.0	5.7	-0.9	12.9	9	-2.6	18	4.5	19	7.5	9	4	0	55.1	75	19	11.7	21	66.8	98	6.1	25
Dec	9.6	3.6	6.6	+2.3	12.9	28	-2.5	13	2.7	11	10.2	8	4	0	35.7	58	18	10.6	25	61.5	108	5.1	9
1974	12.0	5.0	8.5	-1.0	23.0	6	-3.6	1	0.1	1	14.3	8	32	0	589.0	87	180	22.8	7	1244.9	84	14.3	6

DURHAM 1975

Month	Mean max	Mean min	Mean temp	Anom	Highest max	Date	Lowest min	Date	Lowest max	Date	Highest min	Date	Air frost	Days ≥ 25 °C	Total pptn	Anom	Rain days	Wettest day	Date	Total sunshine	Anom	Sunniest day	Date
Jan	8.8	2.4	5.6	+1.5	11.9	4	−2.1	19	4.6	18	8.2	1	6	0	43.9	82	17	6.6	30	61.5	100	5.8	30
Feb	6.3	0.8	3.6	−1.0	12.3	17	−3.2	28	1.9	9	4.1	13	12	0	21.6	48	8	9.4	13	39.9	47	6.8	2
Mar	7.6	0.7	4.1	−2.0	10.7	23	−2.7	1	4.3	27	3.9	12	13	0	45.0	109	19	8.5	10	94.9	79	8.4	20
Apr	11.6	3.7	7.6	−0.6	20.2	26	−2.4	10	4.5	8	9.5	22	5	0	59.5	116	21	12.1	18	117.7	76	11.7	30
May	12.2	3.7	8.0	−3.0	20.0	19	−0.6	4	7.6	14	7.1	10	1	0	49.6	111	11	13.1	7	156.0	83	13.8	23
June	18.6	7.8	13.2	−0.4	25.1	7	1.4	4	8.9	2	13.7	20	0	3	24.8	41	7	8.3	2	218.8	132	13.9	9
July	20.1	11.7	15.9	+0.1	24.5	28	5.9	5	15.1	8	15.3	14	0	16	61.5	99	17	19.2	14	144.3	82	13.7	6
Aug	22.8	12.3	17.6	+2.0	28.7	9	7.7	23	15.0	17	16.0	14	0	16	84.5	128	9	22.0	15	223.7	133	12.7	4
Sep	16.1	7.8	12.0	−1.3	22.0	2	2.5	16	11.3	27	12.4	23	0	0	78.6	138	17	35.8	27	122.5	92	11.5	1
Oct	12.5	5.1	8.8	−1.2	16.5	25	−0.7	13	8.2	29	12.0	5	3	0	20.5	32	8	7.7	2	88.5	90	8.8	7
Nov	8.7	2.0	5.3	−1.3	14.1	19	−3.7	14	5.0	14	7.2	5	7	0	34.2	47	16	6.9	1	73.5	108	5.9	21
Dec	8.4	2.3	5.4	+1.1	13.7	5	−3.8	13	3.1	17	6.5	5	3	0	22.5	37	7	6.3	1	63.4	111	6.3	27
1975	12.8	5.0	8.9	−0.5	28.7	8	−3.8	12	1.9	2	16.0	8	50	19	546.2	80	157	35.8	9	1404.7	95	13.9	6

DURHAM 1976

Month	Mean max	Mean min	Mean temp	Anom	Highest max	Date	Lowest min	Date	Lowest max	Date	Highest min	Date	Air frost	Days ≥ 25 °C	Total pptn	Anom	Rain days	Wettest day	Date	Total sunshine	Anom	Sunniest day	Date
Jan	7.4	2.1	4.7	+0.7	12.4	11	−3.1	25	0.6	28	9.1	13	10	0	46.2	87	13	9.5	2	53.0	86	6.2	23
Feb	6.5	1.7	4.1	−0.5	12.2	23	−1.8	14	2.2	16	7.0	26	5	0	17.5	39	16	3.9	6	40.6	48	6.7	10
Mar	7.1	1.1	4.1	−2.1	14.0	25	−2.7	24	2.9	23	6.8	28	8	0	26.0	63	14	4.2	21	100.6	84	10.6	26
Apr	10.7	3.5	7.1	−1.2	16.2	16	−2.2	8	4.2	2	8.2	18	4	0	31.4	61	10	11.3	1	105.9	69	10.6	20
May	14.5	6.5	10.5	−0.6	19.6	23	2.5	9	8.5	5	9.8	22	0	0	78.9	177	22	15.4	25	119.0	64	12.8	8
June	20.3	10.4	15.4	+1.7	29.3	29	4.3	3	13.7	18	15.4	9	0	7	24.9	41	8	17.4	1	212.6	128	14.4	25
July	22.1	10.7	16.4	+0.6	28.2	2	6.0	31	15.9	31	16.3	19	0	6	30.0	48	6	14.2	12	236.7	135	14.4	2
Aug	21.9	9.9	15.9	+0.3	27.4	25	5.9	24	17.2	1	13.9	14	0	2	18.4	28	3	16.2	28	220.5	132	12.2	10
Sep	15.4	8.6	12.0	−1.3	21.6	5	2.5	13	9.4	11	12.7	21	0	0	193.2	340	18	87.8	11	59.4	45	10.1	5
Oct	12.8	6.7	9.8	−0.3	18.0	11	2.0	13	9.4	30	11.3	1	1	0	142.6	224	27	29.3	14	70.2	71	8.8	6
Nov	8.0	1.5	4.8	−1.9	10.5	24	−3.5	14	2.5	30	7.1	25	11	0	22.8	31	11	6.8	15	61.2	90	7.2	3
Dec	4.0	−0.5	1.8	−2.5	7.5	8	−5.0	29	1.0	28	1.9	21	16	0	50.9	83	24	11.2	19	47.9	84	6.0	9
1976	12.6	5.2	8.9	−0.6	29.3	6	−5.0	12	0.6	1	16.3	7	54	15	682.8	100	172	87.8	9	1327.6	90	14.4	6

DURHAM 1977

Month	Mean max	Mean min	Mean temp	Anom	Highest max	Date	Lowest min	Date	Lowest max	Date	Highest min	Date	Air frost	Days ≥ 25 °C	Total pptn	Anom	Rain days	Wettest day	Date	Total sunshine	Anom	Sunniest day	Date
Jan	4.5	−0.8	1.9	−2.2	9.6	5	−9.8	13	0.9	12	3.3	26	17	0	101.0	189	20	24.0	14	52.4	85	7.0	29
Feb	5.8	0.9	3.4	−1.2	9.1	5	−3.5	1	2.5	24	3.5	11	6	0	63.6	140	19	11.5	9	49.4	58	8.6	28
Mar	9.2	2.7	5.9	−0.2	14.0	9	−2.0	29	4.5	28	6.8	3	5	0	54.8	133	19	8.1	26	73.6	61	7.9	4
Apr	10.3	2.6	6.5	−1.8	15.3	25	−2.9	18	5.2	7	9.0	22	5	0	33.0	64	20	6.5	6	158.2	103	9.6	2
May	13.6	4.7	9.2	−1.9	23.3	28	−1.3	1	9.0	13	9.4	23	1	0	35.3	79	9	13.0	2	197.7	106	14.3	27
June	15.9	7.2	11.5	−2.1	24.9	3	1.9	1	9.2	10	10.7	25	0	0	63.4	104	14	16.3	10	158.7	96	14.2	1
July	18.8	10.3	14.5	−1.2	25.8	7	5.3	15	13.2	12	15.3	23	0	2	9.4	15	8	3.4	26	135.8	78	14.3	5
Aug	18.6	10.1	14.3	−1.2	25.5	1	4.5	28	13.8	18	14.5	1	0	1	37.8	57	13	10.2	19	141.0	84	12.3	9
Sep	16.1	8.5	12.3	−1.0	21.5	6	2.7	13	9.7	22	14.0	28	0	0	16.3	29	11	5.8	25	86.0	65	10.3	3
Oct	13.5	7.7	10.6	+0.6	15.4	21	3.0	11	10.6	16	12.2	22	0	0	41.5	65	13	10.6	3	70.0	71	8.0	25
Nov	8.4	2.9	5.7	−1.0	15.0	10	−2.3	29	3.7	27	11.2	11	4	0	74.2	101	20	15.2	11	76.8	113	7.0	3
Dec	7.4	2.8	5.1	+0.8	14.2	23	−1.5	19	3.5	3	7.8	14	4	0	63.9	104	13	18.4	7	32.2	57	5.8	30
1977	11.9	5.0	8.4	−1.0	25.8	7	−9.8	1	0.9	1	15.3	7	42	3	594.2	87	179	24.0	1	1231.8	84	14.3	5

DURHAM 1978

Month	Mean max	Mean min	Mean temp	Anom	Highest max	Date	Lowest min	Date	Lowest max	Date	Highest min	Date	Air frost	Days ≥ 25 °C	Total pptn	Anom	Rain days	Wettest day	Date	Total sunshine	Anom	Sunniest day	Date
Jan	4.8	−0.9	1.9	−2.1	8.8	6	−6.3	18	1.1	19	3.2	6	17	0	80.8	151	20	15.6	10	60.1	97	6.5	18
Feb	3.7	−2.0	0.9	−3.7	11.0	28	−9.6	16	−0.6	12	5.4	26	19	0	71.3	157	24	14.3	1	49.6	58	6.5	20
Mar	9.9	2.9	6.4	+0.2	13.5	8	−1.7	5	5.2	17	7.1	11	3	0	29.5	71	15	7.3	31	108.4	90	10.7	30
Apr	9.4	2.4	5.9	−2.3	14.4	23	−3.8	11	4.5	10	5.7	2	5	0	34.4	67	13	12.3	26	97.7	63	9.6	19
May	15.4	6.0	10.7	−0.3	23.8	29	2.0	25	6.5	1	10.5	28	0	0	53.1	119	10	20.4	5	178.5	95	13.2	29
June	17.0	8.5	12.7	−0.9	25.2	1	2.3	14	11.3	13	14.3	2	0	1	32.8	54	12	5.2	21	150.4	91	12.7	18
July	17.8	9.8	13.8	−2.0	23.2	29	6.1	10	13.0	4	14.5	23	0	0	61.4	99	15	15.6	30	139.2	79	11.0	11
Aug	17.7	10.6	14.2	−1.4	24.2	19	6.9	30	13.0	30	14.4	22	0	0	92.7	140	16	35.4	5	105.2	63	12.5	19
Sep	16.6	10.0	13.3	+0.0	20.6	10	6.3	20	12.5	5	15.1	10	0	0	51.4	91	12	14.0	5	110.1	83	9.6	25
Oct	14.5	8.3	11.4	+1.4	21.5	11	1.3	18	9.5	17	14.5	6	0	0	20.3	32	12	6.4	15	89.8	91	9.0	7
Nov	10.7	4.7	7.7	+1.1	15.8	18	−6.3	30	1.8	30	12.5	5	5	0	24.7	34	9	10.5	23	64.7	95	6.9	26
Dec	4.7	0.1	2.4	−1.8	12.1	10	−8.0	1	−3.2	31	8.4	12	14	0	195.5	319	27	34.3	27	9.7	17	3.8	18
1978	11.8	5.0	8.4	−1.0	25.2	6	−9.6	2	−3.2	12	15.1	9	63	1	747.9	110	185	35.4	8	1163.4	79	13.2	5

DURHAM 1979

Month	Mean max	Mean min	Mean temp	Anom	Highest max	Date	Lowest min	Date	Lowest max	Date	Highest min	Date	Air frost	Days ≥ 25 °C	Total pptn	Anom	Rain days	Wettest day	Date	Total sunshine	Anom	Sunniest day	Date
Jan	2.4	−3.5	−0.6	−4.6	9.4	7	−10.9	5	−2.1	4	1.8	8	23	0	49.6	93	13	12.0	20	60.6	98	7.1	25
Feb	2.8	−1.1	0.9	−3.7	6.9	24	−5.7	15	−2.6	14	4.1	27	18	0	30.6	68	11	9.5	13	49.2	58	8.2	23
Mar	6.4	0.9	3.7	−2.5	11.9	25	−3.1	24	−0.3	17	7.1	3	13	0	166.0	402	20	50.6	28	90.6	75	10.2	23
Apr	10.0	3.4	6.7	−1.6	15.3	14	−0.1	2	4.9	7	6.7	14	1	0	44.7	87	20	6.6	9	91.7	59	10.4	30
May	12.8	4.6	8.7	−2.3	20.8	13	−1.3	5	6.0	1	10.2	15	4	0	96.5	217	18	17.4	20	150.3	80	12.5	14
June	17.8	9.1	13.4	−0.2	24.9	18	4.5	1	13.9	8	13.6	21	0	0	49.3	81	11	11.7	24	166.2	100	13.4	19
July	19.5	11.0	15.3	−0.5	23.5	27	7.3	1	15.5	20	15.1	28	0	0	25.3	41	11	9.4	30	160.4	92	12.7	12
Aug	18.0	9.8	13.9	−1.6	23.8	12	4.8	28	14.5	3	15.7	13	0	0	67.3	102	20	13.3	13	122.4	73	11.2	21
Sep	16.6	8.4	12.5	−0.8	22.8	1	1.3	29	13.1	21	13.5	26	0	0	39.2	69	10	9.1	26	152.2	115	10.5	5
Oct	13.5	7.5	10.5	+0.5	20.5	9	0.5	6	8.1	28	12.5	9	0	0	54.5	86	16	10.9	3	84.3	86	9.1	20
Nov	9.4	2.8	6.1	−0.6	15.6	3	−5.2	13	3.1	13	11.2	29	7	0	69.1	94	17	21.8	15	80.4	118	7.8	6
Dec	6.9	2.4	4.6	+0.4	13.2	4	−1.3	25	2.2	22	7.2	8	8	0	91.0	149	25	12.9	26	46.5	82	6.2	31
1979	11.4	4.6	8.0	−1.5	24.9	6	−10.9	1	−2.6	2	15.7	8	74	0	783.1	115	192	50.6	3	1254.8	85	13.4	6

DURHAM 1980

Month	Mean max	Mean min	Mean temp	Anom	Highest max	Date	Lowest min	Date	Lowest max	Date	Highest min	Date	Air frost	Days ≥ 25 °C	Total pptn	Anom	Rain days	Wettest day	Date	Total sunshine	Anom	Sunniest day	Date
Jan	4.0	−0.1	2.0	−2.1	8.1	30	−3.3	3	0.6	1	5.6	6	18	0	66.2	124	22	13.5	21	42.8	69	5.9	1
Feb	6.8	2.4	4.6	+0.0	11.8	17	−4.3	1	1.4	2	6.5	14	5	0	55.4	122	18	12.5	4	35.6	42	6.3	10
Mar	6.9	0.7	3.8	−2.4	12.0	1	−5.0	23	1.3	19	4.5	12	9	0	95.6	231	25	17.1	7	92.8	77	9.5	13
Apr	12.3	4.1	8.2	−0.0	20.0	13	−2.7	26	7.0	30	7.8	18	3	0	16.5	32	8	6.9	29	156.1	101	11.3	19
May	14.6	4.6	9.6	−1.4	23.0	19	−0.9	9	7.6	6	10.2	20	1	0	37.4	84	9	16.7	29	198.8	106	13.1	18
June	16.6	9.0	12.8	−0.8	26.0	4	3.2	23	11.5	12	17.8	5	0	2	191.4	313	22	33.4	11	140.0	85	12.2	6
July	18.0	10.0	14.0	−1.8	23.8	25	4.3	11	12.3	1	14.4	30	0	0	55.7	90	12	15.5	30	132.3	76	12.0	14
Aug	18.9	10.9	14.9	−0.7	23.8	14	6.5	25	14.5	30	14.8	15	0	0	92.7	140	14	22.6	14	132.4	79	12.0	21
Sep	17.5	10.0	13.8	+0.5	21.0	1	4.3	25	14.2	29	13.0	2	0	0	27.0	48	18	6.1	18	124.3	94	8.8	3
Oct	11.5	4.7	8.1	−1.9	16.9	1	−2.1	20	7.8	31	10.4	28	2	0	67.2	106	19	13.7	16	98.1	100	10.2	2
Nov	8.8	3.8	6.3	−0.3	14.1	20	−2.5	3	2.5	29	10.2	22	6	0	61.8	84	20	17.7	25	50.2	74	7.4	12
Dec	8.1	2.7	5.4	+1.2	12.2	10	−3.2	1	1.4	19	9.1	13	7	0	41.8	68	20	9.4	14	53.9	95	5.6	23
1980	12.0	5.2	8.6	−0.8	26.0	6	−5.0	3	0.6	1	17.8	6	51	2	808.7	119	207	33.4	6	1257.3	85	13.1	5

DURHAM 1981

Month	Mean max	Mean min	Mean temp	Anom	Highest max	Date	Lowest min	Date	Lowest max	Date	Highest min	Date	Air frost	Days ≥ 25°C	Total pptn	Anom	Rain days	Wettest day	Date	Total sunshine	Anom	Sunniest day	Date
Jan	7.1	0.1	3.6	-0.5	11.0	2	-4.8	16	1.5	6	4.9	9	20	0	28.3	53	15	7.6	16	63.1	102	6.9	27
Feb	5.5	0.3	2.9	-1.6	12.6	6	-4.2	11	1.1	22	6.4	7	13	0	31.7	70	11	12.3	27	59.0	69	7.7	10
Mar	9.6	3.1	6.4	+0.2	16.0	11	-1.5	4	3.0	3	11.1	8	5	0	113.5	275	22	31.7	21	65.6	54	5.9	22
Apr	10.8	3.1	6.9	-1.3	18.1	10	-2.1	23	5.2	24	10.0	30	3	0	49.7	97	14	19.4	24	137.7	89	11.7	13
May	15.1	6.3	10.7	-0.3	20.0	31	-1.0	5	8.1	3	10.2	13	2	0	47.7	107	17	13.1	27	153.1	82	12.5	15
June	16.5	8.8	12.6	-1.0	24.3	22	4.1	27	13.1	18	13.9	14	0	0	25.6	42	15	5.0	2	175.0	106	12.3	21
July	19.2	10.6	14.9	-0.9	24.1	30	6.5	25	13.2	23	13.6	9	0	0	36.4	59	12	9.4	22	148.0	84	12.3	30
Aug	20.3	10.8	15.6	+0.0	25.1	11	4.4	17	15.1	20	15.1	14	0	1	33.5	51	7	13.7	8	165.0	98	11.7	2
Sep	18.3	9.5	13.9	+0.6	23.5	10	6.5	26	15.0	23	13.5	8	0	0	117.2	206	15	26.6	25	154.5	117	9.8	3
Oct	10.9	3.3	7.1	-2.9	14.1	9	-3.6	15	6.9	15	11.5	1	5	0	95.0	149	16	38.1	1	146.1	148	9.9	10
Nov	10.5	3.7	7.1	+0.5	15.1	3	-2.3	7	5.7	7	10.1	23	3	0	56.6	77	14	10.9	23	69.2	102	6.5	24
Dec	2.9	-3.4	-0.2	-4.5	11.2	3	-12.5	18	-1.5	17	5.0	4	21	0	58.1	95	20	13.5	13	57.5	101	5.8	2
1981	12.2	4.7	8.4	-1.0	25.1	8	-12.5	12	-1.5	12	15.1	8	72	1	693.3	102	178	38.1	10	1393.8	95	12.5	5

DURHAM 1982

Month	Mean max	Mean min	Mean temp	Anom	Highest max	Date	Lowest min	Date	Lowest max	Date	Highest min	Date	Air frost	Days ≥ 25°C	Total pptn	Anom	Rain days	Wettest day	Date	Total sunshine	Anom	Sunniest day	Date
Jan	4.8	-2.2	1.3	-2.8	11.5	30	-16.1	11	-4.5	7	8.0	30	14	0	64.6	121	18	19.1	3	58.6	95	5.4	22
Feb	7.4	1.5	4.4	-0.1	13.1	28	-5.9	23	2.2	20	7.1	10	7	0	11.4	25	9	4.2	28	63.0	74	8.1	14
Mar	9.3	1.6	5.5	-0.7	16.9	26	-1.5	19	5.9	15	4.5	3	7	0	29.6	72	15	5.1	2	154.3	128	10.5	25
Apr	13.1	3.6	8.4	+0.1	20.2	27	-1.5	18	6.1	8	7.9	28	3	0	11.8	23	6	5.0	5	168.1	109	11.8	23
May	15.5	5.5	10.5	-0.5	22.8	31	-0.6	6	9.2	2	11.9	25	3	0	26.5	60	8	9.7	21	223.0	119	13.8	29
June	17.0	9.8	13.4	-0.2	27.0	5	5.0	15	10.5	22	14.7	6	0	4	113.7	186	21	31.0	4	107.4	65	12.3	5
July	19.2	11.2	15.2	-0.6	26.5	9	7.7	3	14.9	15	14.8	9	0	1	25.5	41	5	13.3	14	153.8	88	12.7	20
Aug	19.3	11.2	15.3	-0.3	25.3	3	5.6	28	14.8	20	16.3	6	0	2	59.5	90	16	16.3	30	162.2	97	11.0	14
Sep	18.0	8.7	13.4	+0.1	23.9	17	2.5	14	14.0	22	13.3	10	0	0	34.3	60	12	5.1	5	162.3	122	11.5	2
Oct	12.2	6.3	9.3	-0.7	16.6	3	-0.4	23	9.0	22	10.6	2	1	0	89.8	141	21	19.6	5	51.6	52	8.1	15
Nov	9.3	4.4	6.9	+0.2	16.0	1	-0.9	27	3.2	30	9.5	1	2	0	68.8	94	21	10.8	11	61.7	91	5.8	16
Dec	6.3	0.2	3.2	-1.0	12.6	15	-4.4	12	1.3	22	8.1	26	15	0	49.6	81	19	12.3	7	52.3	92	5.8	18
1982	12.6	5.2	8.9	-0.6	27.0	6	-16.1	1	-4.5	1	16.3	8	52	7	585.1	86	171	31.0	6	1418.3	96	13.8	5

DURHAM 1983

Month	Mean max	Mean min	Mean temp	Anom	Highest max	Date	Lowest min	Date	Lowest max	Date	Highest min	Date	Air frost	Days ≥ 25 °C	Total pptn	Anom	Rain days	Wettest day	Date	Total sunshine	Anom	Sunniest day	Date
Jan	8.5	2.9	5.7	+1.6	12.2	5	-3.0	23	2.6	30	7.4	12	5	0	30.3	57	14	8.3	12	82.6	134	6.9	27
Feb	4.2	-1.2	1.5	-3.1	7.9	28	-5.6	4	2.0	8	3.2	27	17	0	52.3	115	18	18.9	6	70.3	82	8.2	22
Mar	9.0	2.9	6.0	-0.2	13.6	17	-1.0	22	4.5	23	9.1	18	2	0	58.6	142	19	12.3	18	103.2	86	9.2	30
Apr	9.3	2.0	5.6	-2.6	13.2	15	-2.7	7	6.1	19	5.6	29	8	0	102.6	200	20	25.6	20	109.0	71	11.4	15
May	12.1	5.5	8.8	-2.2	17.0	31	2.0	23	6.8	2	8.1	7	0	0	102.4	230	23	29.3	27	104.4	56	12.6	23
June	17.0	8.7	12.8	-0.8	24.0	19	4.5	6	12.0	1	12.6	29	0	6	38.4	63	10	19.7	1	165.5	100	13.5	6
July	22.8	12.5	17.7	+1.9	29.1	14	9.3	1	15.0	19	16.3	15	0	21	20.2	33	7	6.3	17	195.6	112	11.9	21
Aug	21.1	11.4	16.3	+0.7	27.0	14	5.2	29	15.6	2	16.4	15	0	14	41.9	63	10	16.9	15	176.0	105	12.2	14
Sep	16.1	9.1	12.6	-0.7	21.5	1	1.7	25	11.0	11	12.9	28	0	6	61.4	108	16	25.7	10	119.1	90	9.9	6
Oct	13.2	6.1	9.7	-0.3	19.1	4	-2.1	22	8.1	28	14.5	4	3	0	36.2	57	13	18.7	15	124.1	126	8.8	19
Nov	9.2	4.2	6.7	+0.1	17.1	1	-4.2	30	4.0	21	9.1	8	6	0	33.0	45	11	9.5	2	53.3	78	6.9	22
Dec	8.1	2.2	5.1	+0.9	12.8	3	-5.0	11	1.7	10	10.1	28	7	0	98.1	160	20	28.3	8	48.5	85	5.9	26
1983	12.6	5.5	9.0	-0.4	29.1	7	-5.6	2	1.7	12	16.4	8	48	9	675.4	99	181	29.3	5	1351.6	92	13.5	6

DURHAM 1984

Month	Mean max	Mean min	Mean temp	Anom	Highest max	Date	Lowest min	Date	Lowest max	Date	Highest min	Date	Air frost	Days ≥ 25 °C	Total pptn	Anom	Rain days	Wettest day	Date	Total sunshine	Anom	Sunniest day	Date
Jan	4.8	-0.5	2.1	-2.0	11.1	12	-7.9	26	0.1	22	7.0	11	20	0	83.7	157	21	24.5	23	58.8	95	7.1	31
Feb	5.6	0.5	3.0	-1.5	11.0	29	-4.7	1	1.2	16	4.8	12	12	0	27.7	61	16	6.1	20	53.9	63	8.0	9
Mar	6.9	1.9	4.4	-1.8	12.7	6	-2.1	22	3.2	20	6.3	5	4	0	52.2	126	22	12.0	24	51.5	43	8.8	27
Apr	12.4	2.5	7.4	-0.8	19.6	25	-3.2	3	6.2	6	9.5	20	7	0	12.9	25	11	5.3	6	207.4	134	12.6	24
May	13.9	4.4	9.2	-1.9	20.4	24	-2.2	13	8.8	27	8.3	24	3	0	30.7	69	12	7.3	21	184.9	99	13.4	14
June	17.7	9.2	13.5	-0.1	25.0	19	5.2	10	12.1	5	13.9	17	0	1	36.8	60	9	10.5	4	159.5	96	12.7	25
July	21.2	10.3	15.8	+0.0	26.2	7	4.5	4	14.3	2	15.1	18	0	3	21.4	35	8	8.7	27	202.8	116	12.4	25
Aug	20.9	10.8	15.9	+0.3	26.8	19	5.7	26	13.7	3	15.5	24	0	2	55.7	84	8	22.5	3	165.1	98	11.7	19
Sep	16.0	9.3	12.6	-0.7	22.1	2	2.8	7	12.0	23	15.5	3	0	0	96.3	170	18	26.7	3	98.8	75	8.5	9
Oct	14.1	6.8	10.5	+0.5	19.7	14	1.3	27	10.0	20	13.3	14	0	0	39.8	63	13	9.4	24	128.4	130	9.0	4
Nov	9.4	5.2	7.3	+0.7	16.0	1	0.2	5	5.6	18	9.6	2	2	0	134.3	183	24	26.6	2	57.1	84	7.6	10
Dec	7.1	1.6	4.3	+0.1	12.5	23	-4.0	27	1.0	27	6.4	20	8	0	23.9	39	18	3.5	1	41.8	73	6.4	31
1984	12.5	5.2	8.8	-0.6	26.8	8	-7.9	1	0.1	1	15.5	8	54	6	615.4	90	180	26.7	9	1410.0	96	13.4	5

DURHAM 1985

Month	Mean max	Mean min	Mean temp	Anom	Highest max	Date	Lowest min	Date	Lowest max	Date	Highest min	Date	Air frost	Days ≥ 25 °C	Total pptn	Anom	Rain days	Wettest day	Date	Total sunshine	Anom	Sunniest day	Date
Jan	3.3	−1.4	1.0	−3.1	11.5	31	−6.9	27	0.0	26	4.2	31	22	0	82.0	154	20	12.6	1	37.0	60	7.0	23
Feb	5.5	−1.3	2.1	−2.5	10.7	26	−7.7	15	−0.8	11	5.4	2	19	0	5.7	13	6	2.0	13	78.0	91	8.7	15
Mar	7.7	0.6	4.2	−2.0	12.5	7	−4.1	18	2.4	1	6.6	31	12	0	62.1	150	16	9.5	29	128.6	107	9.7	25
Apr	11.0	4.0	7.5	−0.8	16.1	18	−2.5	24	5.7	9	8.9	19	4	0	74.8	146	19	13.2	7	114.3	74	12.5	24
May	13.5	5.9	9.7	−1.3	21.5	26	1.0	26	8.4	4	11.6	26	0	0	63.9	144	17	17.3	14	125.5	67	13.1	28
June	15.9	6.9	11.4	−2.2	19.1	30	2.1	7	11.7	7	11.6	23	0	0	34.6	57	17	6.4	12	153.7	93	13.7	1
July	19.0	10.7	14.9	−0.9	24.1	25	6.8	5	16.5	5	15.1	12	0	0	59.6	96	16	18.0	26	140.7	80	11.6	4
Aug	17.6	10.2	13.9	−1.6	20.7	27	6.5	15	14.6	11	13.0	28	0	0	81.5	123	21	24.3	4	152.1	91	12.2	12
Sep	17.5	9.5	13.5	+0.2	23.5	12	4.6	10	12.3	7	13.6	27	0	0	51.9	91	12	14.1	7	115.9	87	10.1	6
Oct	14.1	7.5	10.8	+0.8	24.1	1	−0.5	7	8.1	31	15.9	1	1	0	21.2	33	8	10.7	6	109.7	111	9.0	14
Nov	6.4	0.2	3.3	−3.3	12.9	8	−5.9	30	2.9	27	7.1	9	12	0	68.1	93	20	14.6	9	87.8	129	8.1	2
Dec	7.5	2.3	4.9	+0.6	14.5	1	−6.2	30	−1.8	27	9.5	16	12	0	40.5	66	18	7.1	17	41.9	74	6.4	27
1985	11.6	4.6	8.1	−1.4	24.1	7	−7.7	2	−1.8	12	15.9	10	82	0	645.9	95	190	24.3	8	1285.2	87	13.7	6

DURHAM 1986

Month	Mean max	Mean min	Mean temp	Anom	Highest max	Date	Lowest min	Date	Lowest max	Date	Highest min	Date	Air frost	Days ≥ 25 °C	Total pptn	Anom	Rain days	Wettest day	Date	Total sunshine	Anom	Sunniest day	Date
Jan	5.0	−0.3	2.3	−1.8	9.9	13	−6.4	5	0.1	4	4.9	13	14	0	63.1	118	22	11.1	1	63.8	103	7.5	25
Feb	1.8	−3.0	−0.6	−5.2	3.5	13	−11.8	11	−3.5	10	1.5	2	23	0	37.6	83	16	7.2	2	56.8	67	6.8	20
Mar	7.8	0.8	4.3	−1.9	11.8	22	−9.5	3	3.1	1	5.3	5	11	0	38.8	94	13	9.6	23	135.7	113	10.2	25
Apr	8.4	1.7	5.1	−3.2	13.5	26	−2.1	3	3.6	7	5.4	25	5	0	109.5	214	22	22.0	15	92.2	60	11.0	29
May	14.4	6.7	10.5	−0.5	19.9	1	3.7	16	10.8	5	11.4	26	0	0	65.5	147	18	20.4	6	192.6	103	11.8	1
June	18.1	8.9	13.5	−0.1	25.2	28	3.4	7	12.7	23	13.5	28	0	1	32.1	52	13	15.8	9	179.5	108	13.1	15
July	19.2	10.8	15.0	−0.8	26.6	16	7.3	11	15.7	22	16.6	15	0	2	24.2	39	9	6.2	23	133.4	76	11.5	20
Aug	15.9	8.7	12.3	−3.2	20.5	9	2.8	25	11.5	26	12.8	14	0	0	169.1	256	19	69.1	25	112.1	67	10.7	9
Sep	16.2	6.4	11.3	−2.0	23.5	30	1.0	11	13.0	12	13.2	29	0	0	28.0	49	5	22.1	2	192.7	145	11.1	13
Oct	13.7	6.3	10.0	−0.0	18.6	7	1.5	17	8.5	19	13.1	7	0	0	31.0	49	14	12.2	31	138.5	141	9.0	3
Nov	10.5	4.3	7.4	+0.7	14.2	7	−0.4	21	6.6	18	9.4	25	1	0	36.2	49	15	6.8	18	89.6	132	7.4	8
Dec	7.8	2.4	5.1	+0.8	13.8	4	−2.4	14	2.6	22	7.8	4	2	0	74.2	121	21	18.9	29	52.6	92	6.5	6
1986	11.6	4.5	8.0	−1.4	26.6	7	−11.8	2	−3.5	2	16.6	7	56	3	709.3	104	187	69.1	8	1439.5	98	13.1	6

DURHAM 1987

Month	Mean max	Mean min	Mean temp	Anom	Highest max	Date	Lowest min	Date	Lowest max	Date	Highest min	Date	Air frost	Days ≥ 25°C	Total pptn	Anom	Rain days	Wettest day	Date	Total sunshine	Anom	Sunniest day	Date
Jan	2.9	-1.5	0.7	-3.4	10.0	4	-7.1	13	-4.0	12	4.0	22	18	0	39.1	73	13	12.5	1	52.1	84	7.6	31
Feb	6.5	0.2	3.4	-1.2	13.1	28	-6.8	1	1.2	1	7.0	6	12	0	34.7	77	14	8.5	9	69.7	82	7.7	6
Mar	7.2	0.5	3.8	-2.3	13.2	30	-5.1	11	1.4	6	5.6	31	14	0	81.2	196	17	16.1	1	94.5	78	10.6	29
Apr	13.5	5.0	9.3	+1.0	20.7	27	0.5	2	4.8	1	9.7	19	0	0	59.4	116	11	15.0	7	158.2	103	13.0	28
May	13.3	4.9	9.1	-1.9	20.6	8	0.1	4	8.6	14	10.0	30	0	0	42.7	96	14	7.8	13	163.7	88	13.6	8
June	15.1	7.7	11.4	-2.2	21.3	28	1.4	16	11.2	8	15.5	29	0	2	92.4	151	25	13.4	2	89.4	54	9.9	30
July	19.0	11.1	15.0	-0.8	25.7	6	7.4	9	12.6	22	14.6	11	0	0	84.9	137	16	20.2	17	143.6	82	11.9	9
Aug	18.3	11.1	14.7	-0.8	24.5	20	4.5	10	12.5	26	16.5	21	0	0	53.7	81	15	11.4	21	117.2	70	11.5	21
Sep	16.7	8.5	12.6	-0.7	20.8	21	2.0	30	11.8	26	14.5	12	0	0	46.2	81	15	20.3	19	168.5	127	11.9	2
Oct	11.8	5.1	8.5	-1.6	14.5	5	-1.5	29	7.5	10	10.8	19	3	0	99.6	156	19	28.0	20	86.9	88	8.7	17
Nov	8.9	3.4	6.2	-0.5	14.3	2	-3.0	28	4.5	26	8.0	19	3	0	61.3	83	14	21.3	23	60.3	89	7.6	25
Dec	8.2	3.0	5.6	+1.4	15.1	20	-5.5	9	3.4	13	10.8	29	7	0	34.7	57	17	7.4	27	25.4	45	5.2	8
1987	11.8	4.9	8.3	-1.1	25.7	7	-7.1	1	-4.0	1	16.5	8	57	2	729.9	107	190	28.0	10	1229.5	83	13.6	5

DURHAM 1988

Month	Mean max	Mean min	Mean temp	Anom	Highest max	Date	Lowest min	Date	Lowest max	Date	Highest min	Date	Air frost	Days ≥ 25°C	Total pptn	Anom	Rain days	Wettest day	Date	Total sunshine	Anom	Sunniest day	Date
Jan	7.0	1.3	4.2	+0.1	12.9	9	-1.7	8	1.7	23	6.7	2	7	0	79.8	149	22	14.6	5	54.5	88	5.7	21
Feb	7.5	1.3	4.4	-0.1	12.6	19	-1.9	25	3.6	4	4.5	20	7	0	34.4	76	17	8.5	23	106.8	125	8.6	16
Mar	9.0	1.7	5.4	-0.8	12.7	20	-3.6	14	2.5	13	5.8	9	9	0	54.5	132	20	12.4	11	108.3	90	8.6	17
Apr	11.6	3.8	7.7	-0.5	17.9	19	-1.5	24	6.2	8	9.9	19	5	0	26.0	51	12	7.1	18	126.2	82	12.7	13
May	14.8	5.9	10.4	-0.7	19.5	15	1.6	20	10.1	17	8.7	9	0	0	58.0	130	18	7.7	11	174.3	93	14.0	15
June	18.4	9.4	13.9	+0.3	24.9	24	4.0	13	14.0	9	14.0	21	0	0	16.2	26	5	6.9	25	189.5	114	14.1	13
July	17.9	10.4	14.2	-1.6	22.5	23	6.8	3	14.2	28	13.8	1	0	0	115.1	186	22	20.1	28	163.3	93	12.3	13
Aug	19.3	10.2	14.8	-0.8	26.3	7	6.9	1	16.9	30	14.0	5	0	2	46.0	70	14	12.9	18	199.5	119	13.0	29
Sep	16.6	8.5	12.6	-0.7	23.7	7	1.6	30	12.6	28	13.4	8	0	0	44.3	78	14	11.5	1	145.6	110	11.5	3
Oct	13.0	6.7	9.8	-0.2	16.3	22	-0.5	29	9.1	29	10.6	19	3	0	75.8	119	18	20.4	18	98.8	100	8.4	28
Nov	9.0	2.0	5.5	-1.2	14.2	10	-4.0	3	1.5	21	7.7	10	6	0	64.7	88	13	15.6	29	87.8	129	7.6	4
Dec	10.1	4.4	7.3	+3.0	13.0	26	0.6	7	3.5	2	9.4	29	0	0	27.9	46	8	14.0	22	62.7	110	6.5	5
1988	12.9	5.5	9.2	-0.3	26.3	8	-4.0	11	1.5	11	14.0	6	37	2	642.7	94	183	20.4	10	1517.3	103	14.1	6

DURHAM 1989

Month	Mean max	Mean min	Mean temp	Anom	Highest max	Date	Lowest min	Date	Lowest max	Date	Highest min	Date	Air frost	Days ≥ 25 °C	Total pptn	Anom	Rain days	Wettest day	Date	Total sunshine	Anom	Sunniest day	Date
Jan	9.5	3.1	6.3	+2.2	13.4	27	−0.7	25	4.1	25	8.6	8	2	0	7.2	13	8	2.7	5	66.8	108	6.6	10
Feb	8.6	2.1	5.4	+0.8	12.2	18	−3.7	17	3.6	23	10.0	7	8	0	47.6	105	11	20.4	24	90.0	106	9.1	20
Mar	10.8	2.2	6.5	+0.3	16.5	27	−3.2	18	6.0	2	6.9	31	7	0	20.7	50	15	3.5	23	131.3	109	10.2	11
Apr	9.7	2.3	6.0	−2.3	14.5	30	−4.1	24	4.5	5	6.0	17	5	0	50.3	98	22	7.4	4	131.9	86	11.7	27
May	16.5	6.3	11.4	+0.4	22.0	7	1.3	12	10.5	12	12.0	18	0	0	7.7	17	8	2.8	12	246.0	132	14.7	26
June	19.0	8.0	13.5	−0.1	29.2	20	2.0	4	11.6	6	14.1	13	0	2	52.1	85	11	17.8	26	224.0	135	14.6	19
July	22.5	11.2	16.8	+1.1	27.5	22	6.7	6	15.3	8	15.6	29	0	8	10.1	16	5	3.5	8	237.2	135	14.5	3
Aug	20.1	10.7	15.4	−0.1	26.5	6	5.5	2	14.4	26	16.0	9	1		61.5	93	11	17.0	9	198.7	119	12.4	15
Sep	18.0	8.7	13.4	+0.1	23.3	6	1.8	26	13.6	8	13.0	22	0	0	11.7	21	9	7.1	12	126.7	96	10.5	23
Oct	14.7	6.9	10.8	+0.8	19.1	17	1.9	15	12.5	3	12.8	17	0	0	48.5	76	18	12.1	19	94.2	96	8.8	14
Nov	9.2	2.9	6.1	−0.6	13.9	14	−3.2	27	5.5	30	7.6	21	7	0	34.8	47	12	12.2	8	74.2	109	6.2	26
Dec	6.6	0.5	3.5	−0.7	11.6	24	−4.2	3	0.0	3	4.5	25	11	0	63.7	104	19	19.1	14	25.5	45	5.8	19
1989	13.8	5.4	9.6	+0.1	29.2	6	−4.2	12	0.0	12	16.0	8	40	11	415.9	61	149	20.4	2	1646.5	112	14.7	5

DURHAM 1990

Month	Mean max	Mean min	Mean temp	Anom	Highest max	Date	Lowest min	Date	Lowest max	Date	Highest min	Date	Air frost	Days ≥ 25 °C	Total pptn	Anom	Rain days	Wettest day	Date	Total sunshine	Anom	Sunniest day	Date
Jan	8.7	2.1	5.4	+1.3	14.0	15	−1.4	2	3.4	1	8.0	11	6	0	72.6	136	18	22.3	27	63.7	103	5.9	28
Feb	9.5	3.1	6.3	+1.8	13.9	5	−0.6	16	4.6	28	10.1	24	5	0	60.4	133	18	15.4	7	81.4	95	8.2	15
Mar	12.1	4.1	8.1	+2.0	20.0	18	−3.5	2	4.5	1	9.1	16	4	0	14.1	34	6	8.6	1	139.0	115	9.8	21
Apr	12.7	2.3	7.5	−0.8	21.5	30	−3.5	4	6.8	3	9.0	26	6	0	10.1	20	9	3.8	2	212.7	138	13.7	29
May	16.6	5.8	11.2	+0.2	24.9	4	0.8	26	10.8	18	10.6	30	0	0	42.1	95	12	12.4	9	195.9	105	13.4	3
June	17.3	8.1	12.7	−0.9	21.6	27	4.0	20	12.2	12	11.5	1	0	4	46.3	76	18	11.1	30	113.6	69	10.5	28
July	20.6	9.7	15.1	−0.6	28.0	20	3.1	24	14.8	1	15.6	30	0	5	30.9	50	6	11.7	4	223.3	127	14.0	24
Aug	22.3	11.7	17.0	+1.5	32.5	3	6.2	18	16.7	16	16.3	3	0	5	52.8	80	12	21.7	24	214.2	128	13.5	1
Sep	16.5	7.7	12.1	−1.2	20.5	3	2.8	27	12.5	20	14.2	3	0	0	43.2	76	12	16.1	29	124.8	94	11.1	4
Oct	13.8	7.8	10.8	+0.8	20.2	5	1.5	8	9.9	28	12.0	14	0	0	78.7	124	18	17.4	19	78.0	79	8.2	14
Nov	9.2	3.5	6.4	−0.2	14.9	16	−2.5	23	5.3	21	8.9	14	4	0	53.1	72	17	18.1	24	54.3	80	7.4	18
Dec	7.2	1.6	4.4	+0.1	11.9	22	−1.4	14	2.9	18	6.7	23	4	0	93.3	152	18	27.0	7	46.0	81	6.3	14
1990	13.9	5.6	9.8	+0.3	32.5	8	−3.5	3	2.9	12	16.3	8	29	9	597.6	88	164	27.0	12	1546.9	105	14.0	7

DURHAM 1991

Month	Mean max	Mean min	Mean temp	Anom	Highest max	Date	Lowest min	Date	Lowest max	Date	Highest min	Date	Air frost	Days ≥ 25 °C	Total pptn	Anom	Rain days	Wettest day	Date	Total sunshine	Anom	Sunniest day	Date
Jan	5.5	−0.9	2.3	−1.8	12.3	1	−5.5	14	0.8	27	3.0	21	20	0	47.1	88	15	13.0	8	84.7	137	6.7	15
Feb	4.9	−0.8	2.0	−2.5	12.5	24	−8.5	14	0.4	9	8.4	24	20	0	80.0	177	19	13.0	27	52.3	61	8.8	25
Mar	10.8	3.6	7.2	+1.0	17.6	13	−1.2	1	6.0	2	9.2	14	2	0	60.8	147	15	16.9	4	67.8	56	9.4	19
Apr	11.1	3.1	7.1	−1.2	16.5	11	−1.8	28	7.2	17	9.1	2	4	0	27.7	54	13	7.6	19	156.4	101	12.3	14
May	14.4	6.2	10.3	−0.7	19.7	20	0.4	9	10.1	4	12.0	21	0	0	19.3	43	8	5.9	3	124.3	66	11.3	9
June	15.7	6.8	11.3	−2.3	20.6	30	−0.6	8	10.2	2	12.7	25	1	0	53.9	88	20	8.3	2	152.2	92	10.5	5
July	21.0	11.4	16.2	+0.4	26.0	11	7.4	21	15.6	2	14.6	23	0	1	43.9	71	16	9.2	7	181.6	104	11.8	29
Aug	21.3	11.1	16.2	+0.7	25.5	22	8.1	3	17.2	6	16.5	10	0	3	21.2	32	10	9.6	6	191.9	114	12.5	20
Sep	18.1	7.5	12.8	−0.4	24.1	7	−0.5	12	10.2	29	13.9	14	1	0	24.3	43	11	5.3	28	155.8	118	12.1	7
Oct	12.6	5.8	9.2	−0.8	16.4	11	−0.5	21	8.9	19	10.4	13	1	0	50.0	79	16	13.1	31	78.3	79	9.5	3
Nov	9.1	1.6	5.4	−1.3	13.4	1	−3.9	17	4.1	13	6.8	1	8	0	60.0	82	16	19.6	18	62.2	91	7.2	5
Dec	7.7	−0.1	3.8	−0.4	15.1	22	−9.0	12	−0.4	11	7.6	23	14	0	44.6	73	10	9.2	21	65.4	115	6.2	8
1991	12.7	4.6	8.7	−0.8	26.0	7	−9.0	12	−0.4	12	16.5	8	71	4	532.8	78	169	19.6	11	1372.9	93	12.5	8

DURHAM 1992

Month	Mean max	Mean min	Mean temp	Anom	Highest max	Date	Lowest min	Date	Lowest max	Date	Highest min	Date	Air frost	Days ≥ 25 °C	Total pptn	Anom	Rain days	Wettest day	Date	Total sunshine	Anom	Sunniest day	Date
Jan	7.1	0.0	3.5	−0.6	12.6	1	−5.5	31	−0.6	23	7.3	7	15	0	22.7	43	7	8.4	8	65.4	106	6.7	28
Feb	9.1	1.7	5.4	+0.8	12.3	22	−5.4	1	4.5	18	7.7	6	8	0	34.0	75	12	12.3	3	102.0	120	9.0	28
Mar	10.4	3.0	6.7	+0.5	15.7	3	−1.0	15	6.6	13	7.5	20	3	0	62.7	152	18	17.1	30	98.1	81	8.9	3
Apr	11.9	4.1	8.0	−0.3	17.1	11	−0.2	2	4.5	2	7.5	18	2	0	95.0	186	20	43.0	1	124.2	81	11.8	27
May	16.6	6.6	11.6	+0.6	26.0	14	−1.2	11	10.4	22	11.0	14	1	1	19.0	43	9	7.4	9	220.2	118	14.0	17
June	19.7	9.5	14.6	+1.0	26.5	29	2.9	17	11.9	5	14.4	30	0	3	20.2	33	11	5.7	4	194.0	117	14.4	12
July	19.6	10.6	15.1	−0.7	23.5	16	6.1	15	12.8	3	15.1	17	0	0	54.1	87	14	26.2	3	168.7	96	12.0	10
Aug	18.8	9.8	14.3	−1.2	22.0	5	3.1	29	15.3	8	13.5	2	0	0	68.8	104	17	22.1	8	184.1	110	12.3	6
Sep	15.7	8.5	12.1	−1.2	20.5	28	3.9	21	11.3	24	11.5	19	0	0	97.7	172	23	45.8	21	101.6	77	11.6	1
Oct	10.5	3.2	6.8	−3.2	17.0	8	−2.6	30	6.4	25	8.3	6	7	0	61.0	96	20	13.1	2	79.4	81	9.7	8
Nov	9.8	2.4	6.1	−0.5	14.5	5	−4.4	14	4.9	14	10.5	7	6	0	60.9	83	18	8.3	9	78.3	115	6.8	12
Dec	5.9	0.0	2.9	−1.3	11.6	13	−6.7	31	0.2	29	8.1	15	16	0	44.7	73	11	11.3	1	27.2	48	4.0	25
1992	12.9	4.9	8.9	−0.5	26.5	6	−6.7	12	−0.6	1	15.1	7	58	4	640.8	94	180	45.8	9	1443.2	98	14.4	6

DURHAM 1993

Month	Mean max	Mean min	Mean temp	Anom	Highest max	Date	Lowest min	Date	Lowest max	Date	Highest min	Date	Air frost	Days ≥ 25 °C	Total pptn	Anom	Rain days	Wettest day	Date	Total sunshine	Anom	Sunniest day	Date
Jan	8.3	1.1	4.7	+0.6	13.1	17	-6.4	3	0.5	3	6.1	17	10	0	40.1	75	24	7.9	23	51.7	84	6.2	31
Feb	8.2	2.1	5.1	+0.6	16.8	7	-4.0	27	3.1	28	8.3	7	7	0	15.6	34	8	6.0	28	54.3	64	7.4	27
Mar	10.1	1.8	6.0	-0.2	16.5	13	-5.0	26	3.6	3	7.6	18	8	0	10.1	24	10	2.8	31	126.6	105	10.5	26
Apr	11.8	4.3	8.1	-0.2	20.2	30	-1.8	15	7.5	12	8.5	23	2	0	94.9	185	18	19.0	9	110.8	72	9.7	15
May	14.3	5.8	10.0	-1.0	19.4	17	0.4	15	8.1	14	10.3	30	0	0	92.7	208	13	41.0	13	133.0	71	13.5	7
June	18.1	8.9	13.5	-0.1	24.6	9	2.9	15	13.7	12	12.4	9	0	0	31.4	51	14	5.4	13	162.4	98	14.2	27
July	18.6	10.2	14.4	-1.3	22.1	3	6.1	12	14.8	14	14.6	4	0	0	42.0	68	13	9.8	15	162.9	93	11.2	6
Aug	18.6	9.1	13.8	-1.7	23.0	31	5.1	23	14.3	23	12.8	9	0	0	89.4	135	13	39.2	4	162.1	97	12.0	13
Sep	15.2	7.7	11.5	-1.8	25.1	1	1.8	28	10.9	27	12.0	13	0	1	117.6	207	14	34.7	14	76.2	57	9.9	17
Oct	11.0	4.2	7.6	-2.4	15.4	3	-3.3	18	7.2	16	10.0	7	5	0	73.6	116	19	22.9	6	88.2	90	9.7	16
Nov	5.9	0.7	3.3	-3.4	12.1	4	-12.0	24	0.4	24	6.9	6	12	0	64.6	88	13	25.5	13	34.5	51	4.8	12
Dec	6.3	1.0	3.7	-0.6	12.9	18	-3.4	21	1.8	26	9.3	3	15	0	77.7	127	20	21.7	12	67.3	118	5.8	7
1993	12.2	4.7	8.5	-1.0	25.1	9	12.0	11	0.4	11	14.6	7	59	1	749.7	110	179	41.0	5	1230.0	83	14.2	6

DURHAM 1994

Month	Mean max	Mean min	Mean temp	Anom	Highest max	Date	Lowest min	Date	Lowest max	Date	Highest min	Date	Air frost	Days ≥ 25 °C	Total pptn	Anom	Rain days	Wettest day	Date	Total sunshine	Anom	Sunniest day	Date
Jan	6.5	1.0	3.8	-0.3	10.0	24	-2.6	8	2.7	5	5.5	21	12	0	70.8	133	21	13.7	5	72.6	118	6.8	14
Feb	4.7	-1.3	1.7	-2.9	8.6	10	-8.5	22	0.5	15	4.0	1	15	0	53.4	118	17	15.7	25	55.5	65	6.8	19
Mar	10.0	3.0	6.5	+0.3	14.7	22	-0.9	21	6.1	18	8.9	8	3	0	31.7	77	15	11.9	31	142.6	118	10.4	19
Apr	11.0	3.4	7.2	-1.1	20.6	29	-2.1	11	5.8	8	12.5	28	4	0	37.8	74	14	4.3	20	169.4	110	11.2	10
May	12.9	5.0	8.9	-2.1	17.3	30	-1.1	28	8.2	15	7.7	12	1	0	45.9	103	8	26.7	21	157.8	84	13.3	12
June	17.7	9.0	13.4	-0.3	26.0	28	4.6	8	13.7	8	13.5	25	0	1	22.2	36	11	4.9	3	205.9	124	13.2	13
July	21.3	10.6	16.0	+0.2	26.5	12	5.0	19	17.1	2	15.0	26	0	2	44.2	71	12	12.6	24	183.4	105	12.4	23
Aug	18.4	10.0	14.2	-1.4	23.5	4	5.8	31	15.4	10	16.9	4	0	0	49.1	74	15	18.7	3	141.2	84	12.1	26
Sep	15.3	7.6	11.4	-1.8	21.0	2	3.3	28	11.5	19	11.6	29	0	0	69.2	122	16	13.0	24	116.6	88	12.7	2
Oct	12.6	5.0	8.8	-1.2	17.7	14	-0.2	4	8.0	3	10.9	7	1	0	58.7	92	12	18.2	2	98.9	100	7.3	14
Nov	11.5	5.2	8.4	+1.7	14.9	13	-1.5	29	5.9	29	9.6	26	2	0	61.9	84	15	14.7	4	58.8	86	7.8	1
Dec	8.1	2.4	5.2	+1.0	15.0	10	-2.4	22	1.4	31	11.0	12	7	0	66.7	109	18	10.9	30	60.5	106	6.2	23
1994	12.5	5.1	8.8	-0.7	26.5	7	-8.5	7	0.5	2	16.9	8	45	3	611.6	90	174	26.7	5	1463.2	99	13.3	5

DURHAM 1995

Month	Mean max	Mean min	Mean temp	Anom	Highest max	Date	Lowest min	Date	Lowest max	Date	Highest min	Date	Air frost	Days ≥ 25 °C	Total pptn	Anom	Rain days	Wettest day	Date	Total sunshine	Anom	Sunniest day	Date
Jan	6.3	0.3	3.3	−0.8	11.1	31	−5.2	3	0.6	1	5.3	14	14	0	75.2	141	20	14.8	25	59.4	96	6.5	11
Feb	8.3	2.7	5.5	+0.9	12.4	27	−1.4	24	2.5	9	7.6	4	4	0	62.1	137	17	13.1	22	93.7	110	10.2	26
Mar	8.1	0.9	4.5	−1.7	14.2	31	−3.5	4	2.5	28	6.2	23	14	0	39.5	96	18	6.2	1	159.8	133	10.2	12
Apr	12.3	3.6	7.9	−0.3	17.0	14	−1.6	29	7.1	18	10.6	7	8	0	27.5	54	9	10.3	22	173.4	112	12.4	12
May	15.3	5.8	10.5	−0.5	24.2	4	−1.6	12	8.5	16	10.8	27	2	0	31.2	70	13	7.4	14	186.8	100	11.6	4
June	17.2	8.3	12.8	−0.9	26.0	29	5.5	22	11.1	11	15.0	20	0	1	16.2	26	12	4.1	7	168.2	102	14.6	27
July	21.6	11.7	16.7	+0.9	27.5	29	5.6	3	15.6	2	17.3	20	0	7	24.5	40	12	6.5	13	209.9	120	14.1	9
Aug	23.1	11.3	17.2	+1.7	30.3	21	5.9	9	16.4	26	16.0	2	0	13	14.2	21	7	4.5	12	249.3	149	13.7	15
Sep	16.7	8.7	12.7	−0.6	21.0	10	0.1	28	11.1	27	12.8	1	0	0	86.2	152	18	20.0	7	122.5	92	11.0	29
Oct	15.6	8.0	11.8	+1.8	22.3	8	−0.6	29	11.2	29	12.9	9	2	0	25.2	40	10	6.2	3	122.9	125	9.2	8
Nov	10.0	4.0	7.0	+0.4	16.0	7	−2.3	5	2.7	17	9.8	24	5	0	116.1	158	16	43.2	15	48.4	71	6.6	18
Dec	4.5	−0.9	1.8	−2.5	9.1	3	−10.4	28	−1.3	28	5.6	3	15	0	87.6	143	19	23.4	3	30.1	53	5.7	27
1995	13.2	5.4	9.3	−0.1	30.3	8	−10.4	12	−1.3	12	17.3	7	64	21	605.5	89	171	43.2	11	1624.4	110	14.6	6

DURHAM 1996

Month	Mean max	Mean min	Mean temp	Anom	Highest max	Date	Lowest min	Date	Lowest max	Date	Highest min	Date	Air frost	Days ≥ 25 °C	Total pptn	Anom	Rain days	Wettest day	Date	Total sunshine	Anom	Sunniest day	Date
Jan	5.1	1.9	3.5	−0.6	10.2	13	−2.6	27	0.8	26	8.9	13	5	0	29.9	56	13	5.4	26	7.6	12	3.6	28
Feb	5.7	−0.7	2.5	−2.1	13.5	29	−4.3	28	−0.5	6	5.5	16	19	0	66.7	147	17	21.1	11	90.4	106	9.5	27
Mar	5.4	1.2	3.3	−2.9	10.0	6	−2.3	13	0.3	12	5.0	2	9	0	25.9	63	13	10.5	11	29.3	24	8.6	7
Apr	11.5	3.7	7.6	−0.7	17.2	9	−3.6	1	4.6	12	9.0	16	8	0	33.8	66	10	15.5	30	110.4	72	11.4	5
May	12.6	3.9	8.3	−2.8	20.5	30	−1.8	8	6.9	1	10.4	29	3	0	36.0	81	15	7.8	1	190.3	102	13.5	31
June	18.2	7.9	13.1	−0.5	23.7	16	3.1	23	13.0	20	12.8	10	0	0	8.6	14	8	2.7	25	217.0	131	14.0	15
July	20.1	10.2	15.2	−0.6	27.3	21	5.5	17	15.0	2	15.1	26	0	3	46.8	75	11	22.3	23	182.0	104	14.3	17
Aug	20.7	10.6	15.6	+0.1	29.0	18	5.0	4	17.0	30	16.6	19	0	2	71.8	108	11	21.1	23	179.6	107	12.7	12
Sep	16.4	8.8	12.6	−0.7	21.1	2	2.9	14	13.1	20	11.6	3	0	0	24.6	43	11	9.6	1	103.7	78	10.9	15
Oct	14.3	6.9	10.6	+0.6	19.0	13	2.4	17	9.5	29	11.6	7	0	0	36.3	57	18	6.8	28	107.4	109	8.0	29
Nov	7.9	1.8	4.8	−1.8	16.4	2	−3.1	10	3.2	10	12.4	3	13	0	77.3	105	20	15.0	19	87.6	129	7.4	18
Dec	5.1	0.4	2.7	−1.5	10.5	1	−4.2	6	1.4	31	4.0	16	13	0	62.5	102	18	18.0	18	39.5	69	6.1	2
1996	11.9	4.7	8.3	−1.1	29.0	8	−4.3	2	−0.5	2	16.6	8	70	5	520.2	76	165	22.3	7	1344.8	91	14.3	7

DURHAM 1997

Month	Mean max	Mean min	Mean temp	Anom	Highest max	Date	Lowest min	Date	Lowest max	Date	Highest min	Date	Air frost	Days ≥ 25 °C	Total pptn	Anom	Rain days	Wettest day	Date	Total sunshine	Anom	Sunniest day	Date
Jan	5.3	−0.2	2.5	−1.5	10.9	14	−3.4	21	0.9	7	8.0	13	19	0	13.8	26	14	3.1	5	44.7	72	7.4	29
Feb	9.0	2.2	5.6	+1.0	12.2	23	−1.1	3	5.7	1	5.4	10	3	0	71.6	158	17	14.1	17	96.8	113	8.4	15
Mar	11.8	3.9	7.9	+1.7	14.2	15	−3.9	4	8.1	2	9.6	16	4	0	12.4	30	9	4.5	27	146.7	122	11.8	31
Apr	12.7	3.8	8.2	−0.0	20.4	30	−1.3	4	8.0	25	9.1	29	5	0	5.8	11	8	1.9	5	135.5	88	12.1	15
May	14.6	4.9	9.8	−1.3	23.9	30	−0.9	9	6.5	5	10.5	31	4	0	54.5	122	16	14.0	5	195.9	105	14.3	15
June	16.0	9.0	12.5	−1.1	21.0	8	6.1	16	11.1	14	14.9	7	0	0	174.4	285	22	31.8	25	103.4	62	13.4	28
July	20.8	11.1	15.9	+0.2	25.6	9	7.2	21	13.5	1	14.4	7	0	1	67.0	108	13	13.4	26	209.4	120	13.8	1
Aug	22.0	12.3	17.2	+1.6	29.5	12	6.3	31	15.4	24	16.7	21	0	7	48.1	73	13	18.6	31	186.2	111	12.8	21
Sep	17.2	8.0	12.6	−0.7	21.4	10	2.0	23	13.4	26	13.0	3	0	0	11.3	20	9	5.0	11	171.2	129	11.3	15
Oct	13.1	5.4	9.3	−0.8	19.9	18	−3.8	29	8.0	28	12.7	5	3	0	31.7	50	15	6.0	9	120.2	122	9.2	4
Nov	10.2	5.0	7.6	+1.0	15.5	1	−1.2	14	6.4	22	10.0	17	2	0	68.2	93	19	12.7	28	33.0	48	7.1	12
Dec	7.9	2.7	5.3	+1.1	12.5	10	−3.2	2	2.5	2	6.7	9	4	0	82.8	135	18	19.0	19	26.6	47	7.0	1
1997	13.4	5.7	9.5	+0.1	29.5	8	−3.9	3	0.9	1	16.7	8	44	8	641.6	94	173	31.8	6	1469.6	100	14.3	5

DURHAM 1998

Month	Mean max	Mean min	Mean temp	Anom	Highest max	Date	Lowest min	Date	Lowest max	Date	Highest min	Date	Air frost	Days ≥ 25 °C	Total pptn	Anom	Rain days	Wettest day	Date	Total sunshine	Anom	Sunniest day	Date
Jan	6.9	1.4	4.2	+0.1	13.5	9	−2.5	27	2.1	19	9.0	10	6	0	92.0	172	20	14.4	18	44.2	72	7.2	16
Feb	10.9	4.3	7.6	+3.0	16.7	13	−2.3	3	5.7	28	10.1	12	5	0	6.1	13	6	2.3	28	83.2	98	8.8	22
Mar	11.0	3.1	7.1	+0.9	16.8	28	−4.9	1	6.1	1	8.0	30	7	0	59.3	143	14	15.1	3	84.3	70	8.5	5
Apr	10.6	3.4	7.0	−1.2	15.6	22	−3.0	14	5.7	10	8.5	23	5	0	151.0	295	25	25.2	2	118.0	77	11.6	25
May	16.3	7.3	11.8	+0.8	22.0	18	2.1	2	11.5	11	10.7	24	0	0	71.8	161	15	35.2	28	175.9	94	13.9	17
June	16.8	9.1	12.9	−0.7	25.3	20	1.9	4	9.5	2	16.2	21	0	1	97.8	160	18	21.2	2	127.0	77	13.3	21
July	18.5	10.4	14.4	−1.3	22.3	4	5.3	7	15.5	1	13.4	5	0	0	52.6	85	20	12.8	20	136.5	78	11.5	11
Aug	18.7	11.0	14.8	−0.7	25.1	11	4.7	27	13.5	23	14.8	7	0	0	44.9	68	12	10.6	25	158.6	95	11.7	18
Sep	16.7	10.3	13.5	+0.2	23.0	20	5.2	12	13.0	13	14.4	8	0	1	53.4	94	14	12.7	30	88.9	67	9.7	19
Oct	12.8	5.9	9.4	−0.6	17.7	13	−1.0	20	9.5	18	13.0	22	0	0	96.9	152	20	19.3	24	120.8	123	9.0	12
Nov	8.3	1.7	5.0	−1.7	14.0	8	−2.5	19	4.0	18	8.8	9	6	0	64.9	88	19	31.9	2	76.1	112	7.0	10
Dec	7.8	1.5	4.6	+0.4	14.5	14	−3.0	21	1.9	5	8.0	17	8	0	27.3	45	15	5.6	12	37.1	65	6.3	5
1998	12.9	5.8	9.4	−0.1	25.3	6	−4.9	3	1.9	12	16.2	6	38	2	818.0	120	198	35.2	5	1250.6	85	13.9	5

DURHAM 1999

Month	Mean max	Mean min	Mean temp	Anom	Highest max	Date	Lowest min	Date	Lowest max	Date	Highest min	Date	Air frost	Days ≥ 25 °C	Total pptn	Anom	Rain days	Wettest day	Date	Total sunshine	Anom	Sunniest day	Date
Jan	7.7	1.6	4.6	+0.5	12.2	5	-3.6	12	3.9	9	6.4	20	6	0	57.5	108	19	12.5	7	68.5	111	7.6	28
Feb	8.2	1.4	4.8	+0.2	12.6	4	-3.5	8	2.6	7	7.8	4	10	0	15.6	34	11	7.2	28	108.5	127	9.2	22
Mar	9.9	3.1	6.5	+0.3	16.1	31	-2.1	28	4.4	7	7.8	24	2	0	92.7	224	17	16.6	6	95.0	79	9.5	30
Apr	12.6	4.0	8.3	+0.0	18.1	1	-4.4	14	4.0	13	10.6	6	6	0	42.5	83	13	16.7	20	147.8	96	12.2	29
May	15.8	7.6	11.7	+0.7	20.7	27	2.6	2	12.7	1	12.2	29	0	0	50.2	113	12	11.8	7	149.5	80	14.0	18
June	16.7	8.3	12.5	-1.1	22.7	25	4.1	5	10.6	7	13.2	19	0	0	67.2	110	17	15.6	2	144.3	87	12.3	25
July	20.8	11.5	16.2	+0.4	26.0	8	4.1	27	16.1	28	16.1	9	0	2	12.1	20	8	4.4	19	201.2	115	14.1	26
Aug	19.0	10.8	14.9	-0.7	28.0	2	2.7	22	13.7	9	16.0	4	0	3	81.3	123	17	23.0	25	146.2	87	12.2	2
Sep	19.3	10.1	14.7	+1.4	25.4	5	0.2	18	13.7	27	16.3	2	0	2	53.4	94	16	14.1	19	168.5	127	11.3	3
Oct	13.5	7.1	10.3	+0.3	19.3	10	-0.7	6	9.9	19	14.2	10	1	0	37.8	59	16	14.6	21	84.3	86	8.3	5
Nov	10.4	4.4	7.4	+0.8	14.0	1	-2.1	21	5.0	17	10.2	1	2	0	45.0	61	18	15.0	5	56.0	82	6.5	2
Dec	6.3	0.7	3.5	-0.8	11.7	6	-7.4	20	0.3	19	6.3	1	13	0	65.8	108	19	12.1	2	77.4	136	5.9	19
1999	13.3	5.9	9.6	+0.2	28.0	8	-7.4	12	0.3	12	16.3	9	40	7	621.1	91	183	23.0	8	1447.2	98	14.1	7

DURHAM 2000

Month	Mean max	Mean min	Mean temp	Anom	Highest max	Date	Lowest min	Date	Lowest max	Date	Highest min	Date	Air frost	Days ≥ 25 °C	Total pptn	Anom	Rain days	Wettest day	Date	Total sunshine	Anom	Sunniest day	Date
Jan	7.9	1.9	4.9	+0.8	12.2	17	-2.6	26	4.3	4	5.8	29	6	0	34.2	64	15	13.5	12	86	139		
Feb	8.7	2.4	5.6	+1.0	12.9	4	-3.5	20	3.6	16	8.4	5	6	0	29.5	65	15	6.1	11	101	118		
Mar	10.8	3.6	7.2	+1.0	16.7	8	-2.7	26	5.8	4	10.3	8	4	0	21.0	51	11	7.6	23	125	104		
Apr	11.0	3.4	7.2	-1.0	17.7	30	-3.3	15	4.7	3	8.3	20	5	0	149.8	293	23	18.4	13	128	83		
May	15.2	6.4	10.8	-0.2	23.9	15	2.1	28	9.7	10	12.5	15	0	0	46.0	103	16	19.2	26	185	99		
June	17.2	9.4	13.3	-0.3	28.1	19	5.3	7	8.8	3	14.5	19	0	3	89.4	146	18	54.0	3	149	90		
July	17.8	10.4	14.1	-1.7	23.9	21	4.6	17	13.1	10	13.8	30	0	0	49.4	80	17	8.8	9	122	70		
Aug	20.2	10.9	15.6	+0.0	24.9	25	5.0	22	16.8	19	16.3	14	0	0	43.8	66	12	17.8	20	181	108		
Sep	17.3	9.7	13.5	+0.2	21.9	23	2.7	19	14.0	19	13.3	12	0	0	84.2	148	21	20.8	19	98	74		
Oct	13.0	6.0	9.5	-0.5	18.2	3	-0.3	31	6.7	30	10.2	1	1	0	119.2	187	24	24.8	29	101	102		
Nov	8.8	3.4	6.1	-0.5	13.5	28	0.7	20	4.8	23	8.5	29	0	0	148.4	202	22	37.0	6	65	95		
Dec	6.8	2.3	4.5	+0.3	13.1	5	-6.4	27	-0.3	28	9.5	12	11	0	72.4	118	24	16.8	7	58	102		
2000	12.9	5.8	9.4	-0.1	28.1	6	-6.4	12	-0.3	12	16.3	8	33	3	887.3	130	218	54.0	6	1399	95		

DURHAM 2001

Month	Mean max	Mean min	Mean temp	Anom	Highest max	Date	Lowest min	Date	Lowest max	Date	Highest min	Date	Air frost	Days ≥ 25 °C	Total pptn	Anom	Rain days	Wettest day	Date	Total sunshine	Anom	Sunniest day	Date
Jan	5.7	0.1	2.9	−1.2	8.3	23	−3.3	14	1.6	18	3.9	24	15	0	46.2	87	21	12.0	2	81	131		
Feb	7.4	−0.4	3.5	−1.1	11.9	20	−4.4	1	0.8	1	5.3	7	16	0	103.6	229	19	19.4	4	98	115		
Mar	7.5	0.1	3.8	−2.4	14.3	10	−10.8	3	1.4	3	6.7	11	11	0	43.4	105	19	10.4	27	112	93		
Apr	10.5	3.0	6.8	−1.5	14.8	25	−2.1	21	6.6	12	8.4	2	3	0	59.8	117	24	18.0	6	121	78		
May	16.9	6.4	11.6	+0.6	29.0	23	0.1	1	10.2	15	12.2	27	0	1	15.0	34	8	4.0	16	237	127		
June	16.8	8.9	12.8	−0.8	24.7	26	3.6	3	10.9	16	15.1	29	0	0	38.2	62	11	9.2	9	155	94		
July	20.1	12.0	16.0	+0.2	25.6	3	4.9	16	13.1	14	17.5	3	0	2	42.7	69	13	12.3	13	164	94		
Aug	20.3	11.7	16.0	+0.5	25.0	15	6.8	28	14.5	8	17.9	14	0	1	68.3	103	19	28.6	7	164	98		
Sep	16.0	9.7	12.9	−0.4	21.1	5	7.1	17	11.1	18	14.3	2	0	0	72.3	127	21	12.6	19	96	72		
Oct	15.6	9.9	12.7	+2.7	19.7	12	5.2	19	11.7	31	13.9	12	0	0	76.8	121	19	20.0	21	110	112		
Nov	11.0	4.2	7.6	+0.9	15.1	29	−0.9	10	4.2	8	10.4	25	3	0	35.8	49	16	7.6	7	66	97		
Dec	6.0	0.3	3.2	−1.1	10.1	1	−4.4	31	0.5	10	10.0	1	17	0	54.4	89	19	13.0	5	70	123		
2001	12.8	5.5	9.2	−0.3	29.0	5	−10.8	3	0.5	12	17.9	8	65	4	656.5	97	209	28.6	8	1474	100		

DURHAM 2002

Month	Mean max	Mean min	Mean temp	Anom	Highest max	Date	Lowest min	Date	Lowest max	Date	Highest min	Date	Air frost	Days ≥ 25 °C	Total pptn	Anom	Rain days	Wettest day	Date	Total sunshine	Anom	Sunniest day	Date
Jan	7.5	1.5	4.5	+0.4	13.1	21	−6.9	3	−0.6	2	7.3	21	12	0	37.0	69	19	8.0	25	48	78		
Feb	9.3	2.8	6.0	+1.5	13.1	1	−1.0	14	5.0	23	7.0	2	4	0	83.6	185	24	23.0	25	83	97		
Mar	11.1	2.8	6.9	+0.8	15.1	21	−2.0	14	4.6	9	7.3	4	4	0	37.4	90	16	8.6	15	124	103		
Apr	13.3	4.1	8.7	+0.5	19.8	23	−2.0	11	9.5	6	10.5	23	4	0	19.4	38	14	4.6	30	186	121		
May	15.1	7.6	11.3	+0.3	22.6	16	1.3	5	10.2	6	11.8	22	0	0	53.0	119	23	7.0	3	155	83		
June	17.5	9.9	13.7	+0.1	21.3	1	5.6	4	14.3	7	14.9	17	0	0	47.0	77	14	18.0	14	154	93		
July	19.3	10.6	14.9	−0.9	25.5	29	5.1	6	13.7	1	16.0	29	0	1	79.6	128	17	30.6	30	144	82		
Aug	20.5	12.7	16.6	+1.1	26.8	17	7.0	27	14.2	10	16.5	14	0	0	91.2	138	13	22.8	1	141	84		
Sep	17.4	9.5	13.5	+0.2	21.1	11	4.6	24	13.6	14	13.1	13	0	0	25.2	44	9	15.0	9	128	97		
Oct	12.4	5.4	8.9	−1.1	21.6	1	−2.5	20	6.8	29	11.4	2	3	0	96.0	151	21	21.8	20	91	92		
Nov	9.6	4.6	7.1	+0.5	14.1	5	−0.9	18	5.1	29	9.8	2	1	0	83.8	114	20	16.2	1	52	76		
Dec	6.4	3.4	4.9	+0.7	11.1	24	−4.1	19	1.7	20	7.4	24	4	0	92.2	151	22	16.2	29	32	56		
2002	13.3	6.3	9.8	+0.3	26.8	8	−6.9	1	−0.6	1	16.5	8	32	2	745.4	110	212	30.6	7	1338	91		

DURHAM 2003

Month	Mean max	Mean min	Mean temp	Anom	Highest max	Date	Lowest min	Date	Lowest max	Date	Highest min	Date	Air frost	Days ≥ 25 °C	Total pptn	Anom	Rain days	Wettest day	Date	Total sunshine	Anom	Sunniest day	Date
Jan	7.0	2.2	4.6	+0.5	14.4	26	−3.7	31	0.7	30	9.8	27	9	0	66.0	124	21	20.0	2	64	104		
Feb	6.8	−0.7	3.1	−1.5	11.5	25	−5.5	15	0.6	16	5.9	9	17	0	18.0	40	15	4.8	28	104	122		
Mar	12.1	2.0	7.0	+0.9	16.8	24	−2.6	19	7.1	13	8.4	10	8	0	17.6	43	9	8.6	10	172	143		
Apr	14.0	4.3	9.1	+0.9	24.1	16	−1.1	8	7.8	6	9.5	4	5	0	24.0	47	11	5.8	27	188	122		
May	15.8	7.7	11.7	+0.7	24.0	31	2.4	10	11.3	2	14.1	30	0	0	44.4	100	17	9.8	2	178	95		
June	19.6	10.9	15.2	+1.6	23.6	1	6.7	3	16.1	30	16.1	2	0	0	61.0	100	15	26.6	30	201	121		
July	20.4	12.8	16.6	+0.8	25.9	13	7.1	13	15.9	1	16.7	17	0	1	50.4	81	16	12.4	25	169	96		
Aug	21.0	12.2	16.6	+1.1	28.5	9	6.7	15	15.4	28	16.9	10	0	5	15.9	24	7	7.7	28	201	120		
Sep	18.3	8.9	13.6	+0.3	24.0	17	3.2	24	12.0	22	16.0	18	0	0	41.5	73	13	11.6	19	160	121		
Oct	12.8	5.0	8.9	−1.1	17.9	1	−0.6	22	7.4	22	10.2	9	2	0	40.7	64	16	9.5	22	129	131		
Nov	10.5	4.4	7.5	+0.8	15.7	6	−3.7	24	3.4	22	12.5	19	4	0	35.0	48	16	8.4	30	70	103		
Dec	7.7	0.9	4.3	+0.1	12.8	5	−8.1	31	1.4	31	9.4	25	14	0	65.0	106	20	23.0	1	67	118		
2003	13.8	5.9	9.9	+0.4	28.5	8	−8.1	12	0.6	2	16.9	8	59	6	479.5	70	176	26.6	6	1703	116		

DURHAM 2004

Month	Mean max	Mean min	Mean temp	Anom	Highest max	Date	Lowest min	Date	Lowest max	Date	Highest min	Date	Air frost	Days ≥ 25 °C	Total pptn	Anom	Rain days	Wettest day	Date	Total sunshine	Anom	Sunniest day	Date
Jan	7.6	2.4	5.0	+0.9	12.8	19	−5.9	1	2.0	28	9.4	20	7	0	97.0	182	24	18.8	31	51	83		
Feb	8.8	3.0	5.9	+1.3	15.4	3	−2.8	29	3.0	26	11.1	5	8	0	29.0	64	15	11.4	28	95	111		
Mar	10.4	2.9	6.7	+0.5	15.9	16	−5.4	1	4.7	12	10.5	16	4	0	24.5	59	14	8.0	18	107	89		
Apr	13.2	5.9	9.5	+1.3	21.7	25	−0.7	9	8.8	28	9.5	26	1	0	54.4	106	20	25.1	18	113	73		
May	18.3	8.4	13.4	+2.4	24.2	15	0.4	26	12.5	25	14.6	19	0	0	21.8	49	12	7.7	20	199	106		
June	18.7	10.1	14.4	+0.7	24.0	13	3.3	21	13.0	19	15.8	15	0	0	67.6	111	14	17.8	17	170	103		
July	19.0	10.9	15.0	−0.8	24.6	29	8.2	6	13.3	8	15.9	29	0	0	62.8	101	19	13.6	4	154	88		
Aug	20.0	12.5	16.3	+0.7	25.3	6	5.9	22	16.1	20	16.8	9	0	2	156.8	237	21	52.2	9	152	91		
Sep	17.6	9.7	13.7	+0.4	25.8	5	4.4	25	14.1	22	15.9	4	0	1	19.8	35	17	3.2	14	154	116		
Oct	12.6	6.6	9.6	−0.4	16.5	7	2.5	15	9.5	19	9.3	30	0	0	120.2	189	26	21.8	16	92	93		
Nov	10.0	4.3	7.2	+0.5	13.7	7	−4.9	21	4.0	19	10.6	7	6	0	18.2	25	18	3.6	15	54	79		
Dec	8.3	1.7	5.0	+0.7	12.7	22	−2.6	20	2.7	20	9.0	23	10	0	19.8	32	12	11.6	30	69	121		
2004	13.7	6.5	10.1	+0.7	25.8	9	−5.9	1	2.0	1	16.8	8	36	3	691.9	102	212	52.2	8	1410	96		

DURHAM 2005

Month	Mean max	Mean min	Mean temp	Anom	Highest max	Date	Lowest min	Date	Lowest max	Date	Highest min	Date	Air frost	Days ≥ 25 °C	Total pptn	Anom	Rain days	Wettest day	Date	Total sunshine	Anom	Sunniest day	Date
Jan	8.4	2.7	5.5	+1.5	13.0	9	−2.2	22	3.5	24	7.9	4	6	0	35.6	67	23	6.6	7	62	100		
Feb	6.9	1.3	4.1	−0.4	13.2	3	−2.4	16	2.7	24	5.4	5	9	0	41.5	92	20	7.2	11	76	89		
Mar	9.7	3.7	6.7	+0.5	15.7	17	−2.7	6	4.3	1	9.8	17	6	0	44.8	108	17	20.0	27	68	56		
Apr	11.7	4.1	7.9	−0.4	16.0	29	−1.1	19	6.1	15	8.4	30	1	0	82.0	160	16	39.8	15	147	95		
May	14.9	6.2	10.5	−0.5	19.4	2	−0.8	11	9.6	16	13.2	26	1	0	20.4	46	16	7.0	3	206	110		
June	18.6	10.0	14.3	+0.7	26.7	19	1.4	7	10.1	12	16.8	19	0	2	36.2	59	14	22.4	19	172	104		
July	19.7	11.7	15.7	−0.1	28.2	12	7.9	8	13.3	30	14.7	17	0	4	72.6	117	13	26.8	28	157	90		
Aug	—	—	(15.4)	(−0.2)	22.1	10	7.6	2	16.1	12	13.6	11	0	0	34.7	52	15	11.3	11	176	105		
Sep	17.4	9.5	13.4	+0.2	22.2	12	2.9	24	12.3	15	15.7	8	0	0	63.8	112	15	15.2	9	151	114		
Oct	14.5	8.9	11.7	+1.7	18.7	10	1.9	14	10.0	23	13.4	28	0	0	74.6	117	18	13.0	21	58	59		
Nov	9.3	2.4	5.8	−0.8	15.6	2	−4.6	23	2.0	22	10.6	3	11	0	71.6	97	21	11.0	25	107	157		
Dec	7.3	1.2	4.3	+0.0	12.7	11	−9.0	29	−0.7	29	8.8	11	10	0	37.8	62	18	9.0	5	63	111		
2005	13.2	6.0	9.6	+0.1	28.2	7	−9.0	12	−0.7	12	16.8	6	44	6	615.6	91	206	39.8	4	1443	98		

DURHAM 2006

Month	Mean max	Mean min	Mean temp	Anom	Highest max	Date	Lowest min	Date	Lowest max	Date	Highest min	Date	Air frost	Days ≥ 25 °C	Total pptn	Anom	Rain days	Wettest day	Date	Total sunshine	Anom	Sunniest day	Date
Jan	6.3	1.3	3.8	−0.3	11.9	10	−4.4	30	1.3	30	6.4	19	9	0	20.6	39	18	5.8	18	50	81		
Feb	7.0	1.1	4.0	−0.5	12.4	4	−4.9	1	0.9	1	6.6	7	9	0	43.1	95	17	10.3	23	67	79		
Mar	6.7	0.8	3.8	−2.4	13.5	26	−5.2	3	0.5	12	10.5	27	14	0	79.2	192	22	10.0	14	95	79		
Apr	12.0	3.9	8.0	−0.3	16.6	25	−1.3	5	7.7	4	8.1	25	3	0	24.0	47	14	5.5	2	166	108		
May	15.1	6.3	10.7	−0.3	21.9	11	−0.4	14	8.8	22	12.0	5	1	0	84.4	190	20	20.0	21	183	98		
June	19.7	9.9	14.8	+1.2	26.2	11	5.9	27	13.4	25	14.8	17	0	1	13.0	21	7	5.2	24	190	115		
July	24.2	12.4	18.3	+2.5	30.5	17	5.3	14	16.6	8	16.3	6	0	17	10.0	16	6	3.4	31	273	156		
Aug	19.2	11.5	15.3	−0.2	26.3	6	7.6	8	13.2	13	17.9	6	0	1	57.4	87	20	12.8	2	133	79		
Sep	20.0	11.8	15.9	+2.6	26.7	21	4.1	8	16.9	25	15.0	1	0	0	35.0	62	16	7.8	2	157	118		
Oct	15.5	8.6	12.1	+2.0	17.9	13	3.7	24	12.0	27	11.8	15	0	0	53.2	84	25	14.8	25	90	91		
Nov	11.0	4.5	7.8	+1.1	13.9	8	−0.8	3	7.3	18	9.4	6	1	0	58.2	79	16	15.6	11	101	148		
Dec	7.6	2.2	4.9	+0.7	12.9	4	−4.7	21	2.0	26	9.2	14	9	0	79.3	130	22	24.2	11	67	118		
2006	13.7	6.2	9.9	+0.5	30.5	7	−5.2	3	0.5	3	17.9	8	46	20	557.4	82	203	24.2	12	1572	107		

DURHAM 2007

Month	Mean max	Mean min	Mean temp	Anom	Highest max	Date	Lowest min	Date	Lowest max	Date	Highest min	Date	Air frost	Days ≥ 25 °C	Total pptn	Anom	Rain days	Wettest day	Date	Total sunshine	Anom	Sunniest day	Date
Jan	9.5	3.1	6.3	+2.2	13.4	31	−1.3	23	3.0	23	8.3	4	5	0	65.1	122	20	10.4	17	81	131		
Feb	8.5	2.1	5.3	+0.7	13.7	1	−4.2	6	1.6	8	8.0	1	8	0	70.0	155	19	14.8	10	80	94		
Mar	10.1	3.3	6.7	+0.5	15.0	28	−2.6	21	4.5	20	8.8	12	4	0	20.8	50	15	9.2	4	141	117		
Apr	15.5	6.0	10.7	+2.5	23.3	15	0.2	4	8.8	3	12.4	25	0	0	12.6	25	7	5.8	24	195	126		
May	14.8	6.9	10.9	−0.1	18.6	31	2.4	1	10.0	4	11.8	24	0	0	50.0	112	15	13.2	13	152	81		
June	17.4	9.9	13.6	+0.0	23.7	11	5.6	1	10.9	14	12.7	22	0	0	120.8	198	19	19.0	15	127	77		
July	19.0	10.9	14.9	−0.9	21.5	24	5.9	10	14.3	21	14.2	14	0	0	100.2	162	20	22.0	18	160	91		
Aug	19.2	10.6	14.9	−0.7	25.5	5	5.4	9	14.4	19	16.6	4	0	2	30.6	46	9	17.2	14	192	115		
Sep	17.4	8.8	13.1	−0.2	23.4	6	2.6	18	10.4	26	14.4	6	0	0	39.4	69	12	8.2	21	145	109		
Oct	14.2	6.1	10.1	+0.1	18.6	12	−2.3	24	8.4	22	13.6	12	3	0	13.0	20	9	8.2	8	117	119		
Nov	10.3	3.9	7.1	+0.5	17.0	2	−2.2	24	5.2	23	12.5	2	3	0	61.2	83	19	21.6	20	75	110		
Dec	6.5	1.1	3.8	−0.4	13.6	4	−3.4	17	−0.5	14	7.9	28	14	0	52.4	86	20	15.4	9	47	83		
2007	13.5	6.0	9.8	+0.3	25.5	8	−4.2	2	−0.5	12	16.6	8	37	2	636.1	94	184	22.0	7	1512	103		

DURHAM 2008

Month	Mean max	Mean min	Mean temp	Anom	Highest max	Date	Lowest min	Date	Lowest max	Date	Highest min	Date	Air frost	Days ≥ 25 °C	Total pptn	Anom	Rain days	Wettest day	Date	Total sunshine	Anom	Sunniest day	Date
Jan	7.9	2.8	5.3	+1.2	12.4	18	−2.1	12	2.2	3	8.7	26	7	0	113.4	212	21	21.0	21	52	84		
Feb	8.9	0.7	4.8	+0.2	14.7	12	−6.5	20	2.7	19	7.7	22	11	0	13.4	30	12	3.7	4	118	138		
Mar	9.1	1.8	5.5	−0.7	13.8	31	−3.6	5	4.9	26	5.9	2	7	0	38.4	93	23	11.0	29	138	115		
Apr	11.3	3.3	7.3	−1.0	17.6	3	−3.0	6	6.1	6	10.6	26	4	0	83.6	163	22	14.8	29	129	84		
May	15.9	7.1	11.5	+0.5	23.7	10	−0.9	20	10.7	16	11.6	30	1	0	21.2	48	11	9.6	29	200	107		
June	17.8	9.1	13.4	−0.2	24.2	8	4.8	21	13.7	1	12.7	29	0	0	77.0	126	16	13.4	21	169	102		
July	20.1	11.4	15.8	−0.0	26.1	27	6.9	21	13.9	11	15.5	15	0	2	134.0	216	16	27.6	6	173	99		
Aug	19.1	12.2	15.6	+0.1	22.1	7	7.7	21	14.5	14	17.0	30	0	0	95.4	144	23	15.0	16	106	63		
Sep	16.6	9.6	13.1	−0.2	20.7	20	4.1	5	12.7	16	13.5	11	0	0	97.7	172	19	40.0	5	102	77		
Oct	12.8	5.2	9.0	−1.0	18.0	10	−2.5	30	6.5	28	10.8	20	2	0	52.6	83	13	12.2	31	127	129		
Nov	9.0	3.7	6.4	−0.3	14.0	14	−5.0	29	1.5	30	9.7	15	6	0	34.4	47	14	9.0	9	70	103		
Dec	6.4	0.6	3.5	−0.7	12.3	21	−4.4	1	−0.3	31	7.8	22	12	0	57.0	93	15	13.2	13	61	107		
2008	12.9	5.6	9.3	−0.2	26.1	7	−6.5	2	−0.3	12	17.0	8	50	2	818.1	120	205	40.0	9	1445	98		

DURHAM 2009

Month	Mean max	Mean min	Mean temp	Anom	Highest max	Date	Lowest min	Date	Lowest max	Date	Highest min	Date	Air frost	Days ≥ 25 °C	Total pptn	Anom	Rain days	Wettest day	Date	Total sunshine	Anom	Sunniest day	Date
Jan	6.0	0.4	3.2	−0.9	10.8	11	−5.8	6	2.4	1	6.6	12	11	0	39.4	74	22	6.6	24	59	96		
Feb	7.6	2.0	4.8	+0.2	13.7	17	−4.0	8	1.0	4	7.7	24	12	0	35.6	79	14	21.0	2	63	74		
Mar	11.0	3.0	7.0	+0.9	17.2	18	−5.3	5	5.4	28	8.1	23	4	0	21.2	51	10	7.8	27	166	138		
Apr	13.7	4.3	9.0	+0.8	18.6	20	−0.7	12	7.9	15	8.7	24	1	0	36.8	72	8	24.6	28	154	100		
May	16.0	6.8	11.4	+0.4	23.0	29	2.9	11	11.4	15	10.3	28	0	0	38.2	86	16	9.6	15	218	117		
June	18.0	8.9	13.5	−0.1	25.6	22	3.5	5	8.3	5	14.5	29	0	1	77.8	127	20	17.6	5	181	109		
July	19.7	11.4	15.6	−0.2	26.8	1	8.0	11	14.3	17	15.9	1	0	2	168.8	272	23	44.4	16	178	102		
Aug	20.1	12.1	16.1	+0.5	24.3	19	8.9	8	16.7	28	17.1	20	0	0	37.2	56	17	12.0	31	166	99		
Sep	17.9	9.5	13.7	+0.4	23.7	12	3.0	17	11.5	16	13.9	22	0	0	14.4	25	9	6.0	2	135	102		
Oct	14.0	7.1	10.5	+0.5	16.3	31	0.0	9	10.8	17	11.5	25	0	0	46.2	73	19	7.2	31	86	87		
Nov	10.4	4.7	7.5	+0.9	14.5	19	0.4	10	4.5	30	11.8	20	0	0	146.6	200	27	28.4	29	77	113		
Dec	5.1	−0.4	2.3	−1.9	10.6	5	−7.8	22	0.6	18	5.0	7	16	0	86.0	141	28	16.6	16	71	125		
2009	13.3	5.8	9.6	+0.1	26.8	7	−7.8	12	0.6	12	17.1	8	44	3	748.2	110	213	44.4	7	1554	105		

DURHAM 2010

Month	Mean max	Mean min	Mean temp	Anom	Highest max	Date	Lowest min	Date	Lowest max	Date	Highest min	Date	Air frost	Days ≥ 25 °C	Total pptn	Anom	Rain days	Wettest day	Date	Total sunshine	Anom	Sunniest day	Date
Jan	3.5	−1.1	1.2	−2.9	9.2	18	−8.9	8	−0.4	8	2.9	18	16	0	59.8	112	24	8.0	21	59	96		
Feb	4.6	−1.0	1.8	−2.8	7.8	14	−4.7	21	0.7	21	3.2	26	20	0	67.6	149	24	17.6	26	55	64		
Mar	9.7	2.0	5.9	−0.3	15.3	18	−4.6	8	4.6	6	7.7	18	10	0	68.2	165	17	13.2	29	127	105		
Apr	13.5	4.1	8.8	+0.5	19.8	28	−1.8	2	7.3	2	11.2	28	2	0	12.4	24	8	4.0	2	169	110		
May	15.0	5.2	10.1	−0.9	27.2	23	−2.5	12	8.9	10	12.0	22	3	2	24.0	54	14	6.2	5	200	107		
June	19.4	9.5	14.4	+0.8	24.6	22	3.1	15	11.9	9	14.0	24	0	0	56.0	92	13	16.6	9	208	126		
July	20.3	12.0	16.2	+0.4	24.0	10	5.9	23	16.2	22	15.5	27	0	0	62.0	100	12	31.8	20	144	82		
Aug	19.1	10.1	14.6	−1.0	23.3	20	4.2	31	13.9	13	13.8	7	0	0	44.8	68	15	14.6	13	159	95		
Sep	17.8	9.8	13.8	+0.6	21.4	3	4.9	25	11.9	24	14.3	11	0	0	75.8	134	18	16.8	6	132	100		
Oct	13.0	6.4	9.7	−0.3	16.9	7	−0.7	25	8.4	23	12.0	9	1	0	62.2	98	21	14.6	3	104	106		
Nov	7.6	2.0	4.8	−1.8	16.1	4	−8.7	28	0.2	27	8.7	5	9	0	153.3	209	28	19.4	2	79	116		
Dec	2.7	−3.4	−0.3	−4.6	8.4	11	−10.4	21	−1.5	17	5.2	11	21	0	40.7	66	23	11.6	27	64	112		
2010	12.2	4.6	8.4	−1.0	27.2	5	−10.4	12	−1.5	12	15.5	7	82	2	726.8	107	217	31.8	7	1500	102		

DURHAM 2011

Month	Mean max	Mean min	Mean temp	Anom	Highest max	Date	Lowest min	Date	Lowest max	Date	Highest min	Date	Air frost	Days ≥ 25 °C	Total pptn	Anom	Rain days	Wettest day	Date	Total sunshine	Anom	Sunniest day	Date
Jan	6.3	0.5	3.4	−0.7	12.4	13	−6.5	7	2.3	2	9.3	16	13	0	36.0	67	22	8.2	26	63	102		
Feb	8.4	2.8	5.6	+1.0	12.8	25	−1.1	11	2.5	21	8.1	25	2	0	58.2	128	22	8.4	11	56	66		
Mar	10.5	2.1	6.3	+0.1	18.7	23	−4.2	3	5.9	7	6.8	22	8	0	24.8	60	13	16.0	11	135	112		
Apr	16.5	5.9	11.2	+2.9	21.8	23	−0.2	28	10.7	13	11.1	6	1	0	7.2	14	4	4.4	1	212	137		
May	16.0	7.4	11.7	+0.7	20.0	7	−0.8	4	12.7	3	12.2	7	1	0	40.0	90	15	17.2	7	199	106		
June	18.6	8.8	13.7	+0.1	26.7	27	2.5	10	12.7	4	14.5	27	0	2	48.4	79	18	9.2	20	186	112		
July	19.4	9.7	14.6	−1.2	23.6	4	5.4	14	15.7	20	13.3	6	0	0	65.6	106	14	12.4	17	164	94		
Aug	18.6	10.6	14.6	−1.0	23.2	1	7.5	9	13.8	26	14.9	2	0	0	125.0	189	24	37.2	6	133	79		
Sep	18.1	10.2	14.1	+0.9	24.8	30	4.4	15	14.3	16	13.3	3	0	0	28.4	50	13	7.0	16	149	112		
Oct	14.6	8.2	11.4	+1.4	25.3	1	1.5	20	10.2	12	15.2	3	0	1	53.8	84	17	12.0	11	105	107		
Nov	11.4	6.0	8.7	+2.1	15.6	3	−2.5	7	6.9	20	10.8	4	2	0	27.2	37	19	4.6	4	61	90		
Dec	7.5	2.3	4.9	+0.7	12.9	22	−1.9	10	2.2	16	10.5	26	6	0	51.6	84	22	8.0	30	57	100		
2011	13.8	6.2	10.0	+0.6	26.7	6	−6.5	1	2.2	12	15.2	10	33	3	566.2	83	203	37.2	8	1520	103		

DURHAM 2012

Month	Mean max	Mean min	Mean temp	Anom	Highest max	Date	Lowest min	Date	Lowest max	Date	Highest min	Date	Air frost	Days ≥ 25 °C	Total pptn	Anom	Rain days	Wettest day	Date	Total sunshine	Anom	Sunniest day	Date
Jan	7.7	1.5	4.6	+0.5	11.7	2	−5.0	16	0.6	15	8.0	12	8	0	30.4	57	16	6.4	2	82	133		
Feb	8.1	1.8	4.9	+0.4	17.4	28	−6.6	8	0.3	10	9.2	23	11	0	10.2	23	14	4.6	9	77	90		
Mar	13.7	3.6	8.7	+2.5	21.8	28	−2.5	6	4.3	4	8.9	10	3	0	15.0	36	9	6.2	4	180	149		
Apr	10.8	2.9	6.9	−1.4	15.3	1	−2.7	5	7.7	3	6.4	2	4	0	134.4	263	24	26.2	3	104	67		
May	15.2	5.9	10.5	−0.5	25.2	28	−2.0	6	8.5	3	12.9	31	1	1	66.2	149	17	17.8	9	184	98		
June	16.1	8.7	12.4	−1.2	21.5	27	3.5	4	9.3	2	14.4	28	0	0	136.6	223	24	26.6	28	102	62		
July	18.3	10.9	14.6	−1.2	22.8	24	5.8	31	13.6	10	17.4	24	0	0	98.0	158	18	18.2	5	136	78		
Aug	19.9	11.3	15.6	+0.1	23.5	14	1.9	31	13.8	30	16.1	18	0	0	103.4	156	18	23.8	15	149	89		
Sep	16.8	8.7	12.7	−0.5	22.7	2	1.6	22	10.9	24	15.1	4	0	0	116.2	205	13	57.4	24	154	116		
Oct	12.3	4.4	8.4	−1.7	16.1	21	−0.2	27	6.9	26	10.1	24	2	0	83.6	131	19	27.4	11	107	109		
Nov	9.2	2.9	6.1	−0.6	13.4	12	−3.1	30	3.7	4	9.4	14	6	0	124.8	170	18	48.0	26	74	109		
Dec	6.6	1.1	3.8	−0.4	12.1	28	−4.7	14	1.3	13	5.0	23	10	0	99.2	162	25	21.4	20	63	111		
2012	12.9	5.3	9.1	−0.3	25.2	5	−6.6	2	0.3	2	17.4	7	45	1	1018.0	150	215	57.4	9	1412	96		

DURHAM 2013

Month	Mean max	Mean min	Mean temp	Anom	Highest max	Date	Lowest min	Date	Lowest max	Date	Highest min	Date	Air frost	Days ≥ 25 °C	Total pptn	Anom	Rain days	Wettest day	Date	Total sunshine	Anom	Sunniest day	Date
Jan	5.5	0.8	3.1	-0.9	12.1	2	-7.6	16	-0.8	16	8.8	4	13	0	80.9	151	23	10.0	26	49	79		
Feb	6.0	0.2	3.1	-1.5	10.4	3	-3.7	28	1.6	13	4.9	4	13	0	27.2	60	12	8.0	23	81	95		
Mar	4.9	-0.2	2.4	-3.8	11.9	2	-3.9	31	0.7	23	4.3	3	15	0	59.4	144	21	20.0	17	71	59		
Apr	10.9	3.3	7.1	-1.2	16.9	14	-2.2	3	4.4	8	10.1	15	3	0	22.8	45	14	4.6	26	164	106		
May	15.0	5.8	10.4	-0.6	22.8	31	2.0	23	10.6	23	10.6	20	3	0	100.8	227	19	29.2	17	171	91		
June	18.1	8.9	13.5	-0.1	21.7	19	4.7	3	15.9	10	12.9	12	0	0	23.0	38	13	7.4	23	167	101		
July	23.2	12.8	18.0	+2.2	28.4	9	8.1	2	16.8	2	16.6	24	0	6	52.8	85	10	38.6	27	227	130		
Aug	20.6	11.9	16.3	+0.7	25.1	1	6.5	7	16.9	5	15.8	2	0	1	71.2	108	12	21.6	5	173	103		
Sep	17.1	8.8	13.0	-0.3	24.0	5	3.4	8	12.4	6	14.1	3	0	0	84.6	149	15	45.6	6	133	100		
Oct	14.4	8.8	11.6	+1.5	19.4	7	4.1	30	10.7	14	13.5	7	0	0	100.2	157	22	20.2	13	81	82		
Nov	9.2	2.2	5.7	-0.9	13.2	11	-1.3	23	6.0	19	6.0	1	5	0	54.0	74	17	21.0	20	93	137		
Dec	8.8	3.1	6.0	+1.7	12.4	12	-1.1	26	5.1	24	9.7	9	3	0	65.4	107	20	13.8	23	63	111		
2013	12.8	5.5	9.2	-0.3	28.4	7	-7.6	1	-0.8	1	16.6	7	52	7	742.3	109	198	45.6	9	1473	100		

DURHAM 2014

Month	Mean max	Mean min	Mean temp	Anom	Highest max	Date	Lowest min	Date	Lowest max	Date	Highest min	Date	Air frost	Days ≥ 25 °C	Total pptn	Anom	Rain days	Wettest day	Date	Total sunshine	Anom	Sunniest day	Date
Jan	7.1	2.3	4.7	+0.6	10.1	25	-1.5	12	4.1	30	6.9	8	6	0	91.2	171	26	9.4	5	56	91		
Feb	8.3	2.8	5.5	+1.0	11.5	23	-2.3	28	5.5	11	6.7	23	2	0	61.2	135	23	13.0	14	88	103		
Mar	10.9	3.2	7.0	+0.9	16.2	13	-3.8	24	6.6	28	9.2	7	6	0	33.0	80	17	5.2	28	135	112		
Apr	13.4	5.7	9.6	+1.3	19.3	29	-1.7	19	6.6	2	10.4	6	2	0	56.0	109	16	11.4	25	126	82		
May	15.6	7.7	11.7	+0.6	21.9	18	-1.4	3	9.1	1	10.5	28	1	0	78.0	175	19	13.4	28	155	83		
June	19.0	10.3	14.6	+1.0	22.5	18	5.9	25	14.0	4	14.1	18	0	0	47.2	77	13	19.0	7	155	94		
July	21.8	11.5	16.7	+0.9	27.4	26	6.1	6	18.3	6	15.8	20	0	2	54.4	88	11	21.6	4	214	122		
Aug	18.6	10.2	14.4	-1.1	22.5	6	3.9	25	15.4	23	14.8	6	0	0	79.4	120	21	17.6	14	179	107		
Sep	18.2	9.9	14.0	+0.8	22.6	5	5.4	22	13.7	20	13.7	18	0	0	15.8	28	9	7.0	5	113	85		
Oct	14.9	7.8	11.4	+1.4	19.8	3	1.8	29	10.4	6	13.7	19	0	0	57.2	90	18	13.0	6	95	96		
Nov	10.5	4.9	7.7	+1.0	14.5	1	-0.8	6	6.5	25	11.5	1	2	0	63.6	87	27	15.0	16	50	73		
Dec	8.1	1.9	5.0	+0.7	13.9	18	-2.8	4	3.3	26	9.8	18	11	0	27.0	44	19	5.4	10	84	148		
2014	13.9	6.5	10.2	+0.7	27.4	7	-3.8	3	3.3	12	15.8	7	30	2	664.0	98	219	21.6	7	1450	98		

DURHAM 2015

Month	Mean max	Mean min	Mean temp	Anom	Highest max	Date	Lowest min	Date	Lowest max	Date	Highest min	Date	Air frost	Days ≥ 25 °C	Total pptn	Anom	Rain days	Wettest day	Date	Total sunshine	Anom	Sunniest day	Date
Jan	6.7	1.6	4.1	+0.0	13.5	1	−3.2	30	0.0	18	6.1	10	9	0	50.4	94	23	12.8	14	85	138		
Feb	7.7	1.0	4.4	−0.2	12.1	8	−2.8	10	2.8	2	7.1	19	10	0	15.4	34	14	2.4	5	99	116		
Mar	10.0	2.2	6.1	−0.1	13.9	20	−1.7	9	5.8	14	8.9	8	6	0	39.8	96	18	17.6	12	128	106		
Apr	13.7	3.4	8.5	+0.3	19.7	23	−1.7	27	8.3	3	9.4	25	2	0	21.0	41	12	8.2	12	219	142		
May	14.4	6.0	10.2	−0.8	18.5	22	−1.2	1	9.0	2	9.5	11	1	0	70.6	159	22	13.2	8	181	97		
June	17.8	8.0	12.9	−0.7	27.0	30	1.3	15	12.7	14	13.4	17	0	1	28.4	46	11	6.8	8	193	117		
July	19.1	10.5	14.8	−1.0	31.0	1	5.0	26	13.9	27	16.0	2	0	1	86.2	139	26	14.8	26	167	95		
Aug	19.4	11.1	15.3	−0.3	22.7	21	6.0	16	14.6	14	16.7	21	0	0	81.0	122	20	15.8	22	173	103		
Sep	16.7	7.6	12.1	−1.2	20.4	19	3.5	16	12.4	8	11.1	21	0	0	38.0	67	15	9.6	14	156	118		
Oct	14.2	6.2	10.2	+0.2	20.9	2	1.4	2	10.9	3	12.3	7	0	0	69.8	110	16	15.8	6	90	91		
Nov	11.4	5.7	8.5	+1.9	18.1	1	−2.2	22	3.7	21	13.7	11	1	0	104.4	142	23	18.6	14	51	75		
Dec	10.9	5.0	7.9	+3.7	15.9	19	−1.2	13	2.9	13	11.3	17	2	0	120.2	196	25	33.6	25	41	72		
2015	13.5	5.7	9.6	+0.1	31.0	7	−3.2	1	0.0	1	16.7	8	31	2	725.2	107	225	33.6	12	1583	107		

DURHAM 2016

Month	Mean max	Mean min	Mean temp	Anom	Highest max	Date	Lowest min	Date	Lowest max	Date	Highest min	Date	Air frost	Days ≥ 25 °C	Total pptn	Anom	Rain days	Wettest day	Date	Total sunshine	Anom	Sunniest day	Date
Jan	7.2	2.2	4.7	+0.6	14.2	24	−2.6	17	1.4	14	10.6	25	10	0	116.6	218	26	28.6	4	41	66		
Feb	7.9	0.8	4.3	−0.2	12.3	1	−3.2	24	4.4	13	7.2	5	13	0	30.2	67	17	4.8	6	107	125		
Mar	9.7	2.2	6.0	−0.2	13.7	21	−1.8	8	5.1	9	6.4	23	5	0	52.4	127	15	14.2	9	114	95		
Apr	10.9	2.9	6.9	−1.3	17.9	21	−1.3	28	6.4	27	6.4	12	5	0	83.2	163	22	15.4	12	160	104		
May	15.6	6.7	11.1	+0.1	22.4	8	0.0	19	10.1	13	10.8	18	2	0	39.0	88	11	17.2	25	188	101		
June	17.8	9.5	13.7	+0.0	25.9	7	5.1	3	10.6	2	14.6	20	0	1	49.8	81	15	8.6	29	128	77		
July	20.2	12.0	16.1	+0.3	29.2	19	5.7	2	15.8	28	18.9	20	0	2	34.6	56	14	7.0	4	176	100		
Aug	20.2	12.0	16.1	+0.6	24.5	23	6.7	10	16.5	28	15.9	7	0	0	69.8	105	15	16.2	27	195	116		
Sep	19.2	11.3	15.3	+2.0	27.3	13	6.0	18	14.1	26	17.2	7	0	2	35.6	63	14	13.4	3	140	106		
Oct	13.9	6.8	10.4	+0.3	18.3	4	1.9	3	9.1	23	10.5	28	0	0	52.2	82	20	7.6	13	101	102		
Nov	8.2	1.7	4.9	−1.7	15.3	14	−3.4	26	3.6	9	10.2	15	8	0	86.0	117	18	35.0	21	88	129		
Dec	9.0	3.1	6.0	+1.8	14.8	7	−3.4	5	2.9	5	9.5	8	5	0	46.4	76	17	9.8	31	58	102		
2016	13.3	5.9	9.6	+0.2	29.2	7	−3.4	7	1.4	1	18.9	7	43	5	695.8	102	204	35.0	11	1496	102		

DURHAM 2017

Month	Mean max	Mean min	Mean temp	Anom	Highest max	Date	Lowest min	Date	Lowest max	Date	Highest min	Date	Air frost	Days ≥ 25 °C	Total pptn	Anom	Rain days	Wettest day	Date	Total sunshine	Anom	Sunniest day	Date
Jan	6.5	0.7	3.6	−0.5	11.9	10	−3.3	27	−0.3	26	6.5	16	16	0	33.2	62	17	6.4	14	66	107		
Feb	8.3	3.1	5.7	+1.2	13.5	19	−1.0	28	2.8	9	8.5	20	1	0	59.7	132	17	9.8	22	65	76		
Mar	11.7	4.1	7.9	+1.8	17.0	25	−0.7	26	6.9	3	10.8	31	3	0	42.4	103	16	11.8	3	139	115		
Apr	13.6	4.3	9.0	+0.7	20.0	9	−1.2	18	8.9	16	8.3	12	3	0	29.0	57	9	7.8	25	173	112		
May	17.3	8.0	12.7	+1.7	26.1	25	1.8	10	10.6	7	14.0	26	0	3	19.8	44	9	6.4	27	195	104		
June	19.0	11.1	15.1	+1.5	27.8	18	7.5	26	12.4	28	14.7	17	0	3	103.4	169	18	28.6	6	167	101		
July	19.2	11.1	15.2	−0.6	24.1	17	8.1	8	14.0	5	14.9	20	0	0	71.2	115	21	13.2	23	167	95		
Aug	19.1	10.9	15.0	−0.6	22.0	17	6.1	21	15.7	8	15.5	23	0	0	49.0	74	17	18.6	8	176	105		
Sep	16.8	9.3	13.1	−0.2	19.7	4	3.5	19	14.1	18	13.7	5	0	0	89.4	158	21	28.0	12	116	88		
Oct	15.3	8.9	12.1	+2.1	19.5	13	1.1	30	10.6	29	15.1	14	0	0	29.0	46	16	11.6	19	84	85		
Nov	9.5	2.9	6.2	−0.4	14.3	22	−1.2	6	2.3	30	11.2	1	3	0	85.8	117	15	39.2	22	95	140		
Dec	7.4	1.4	4.4	+0.1	12.7	6	−6.9	12	0.3	10	9.5	25	14	0	25.2	41	15	6.2	29	71	125		
2017	13.7	6.3	10.0	+0.5	27.8	6	−6.9	12	−0.3	1	15.5	8	40	6	637.1	94	191	39.2	11	1514	103		

DURHAM 2018

Month	Mean max	Mean min	Mean temp	Anom	Highest max	Date	Lowest min	Date	Lowest max	Date	Highest min	Date	Air frost	Days ≥ 25 °C	Total pptn	Anom	Rain days	Wettest day	Date	Total sunshine	Anom	Sunniest day	Date
Jan	6.7	1.6	4.2	+0.1	13.9	28	−5.7	8	3.4	9	7.6	29	8	0	55.0	103	21	11.4	4	66	107		
Feb	5.5	−0.2	2.6	−1.9	10.3	20	−4.4	28	−0.8	28	3.7	20	18	0	39.0	86	16	8.4	10	87	102		
Mar	7.3	0.9	4.1	−2.0	12.6	25	−5.1	1	−0.1	1	6.0	22	12	0	88.4	214	19	16.2	4	85	71		
Apr	12.3	5.1	8.7	+0.4	23.7	19	−1.0	5	5.0	2	11.4	19	1	0	73.5	144	15	21.6	4	124	80		
May	17.3	7.4	12.3	+1.3	26.1	7	0.9	1	11.5	22	12.5	8	0	1	26.4	59	7	9.8	12	248	133		
June	19.9	9.7	14.8	+1.2	26.6	25	7.0	10	14.2	15	14.0	3	0	4	31.2	51	8	17.2	2	211	127		
July	22.9	12.9	17.9	+2.1	27.6	7	8.9	2	17.3	9	17.0	24	0	8	52.1	84	8	22.5	16	230	131		
Aug	20.2	12.1	16.1	+0.6	25.1	2	4.1	31	14.9	24	17.7	3	0	1	49.8	75	15	12.0	12	162	97		
Sep	17.6	9.1	13.4	+0.1	24.2	2	3.4	29	13.8	21	14.9	3	0	2	46.8	82	11	28.4	20	144	109		
Oct	14.2	5.9	10.1	+0.1	21.1	10	−0.7	29	7.1	30	13.5	13	2	0	53.4	84	16	15.6	12	126	128		
Nov	10.3	4.9	7.6	+1.0	13.6	28	−1.8	22	5.7	21	9.7	7	1	0	56.4	77	23	17.4	20	68	100		
Dec	8.6	2.5	5.5	+1.3	12.5	6	−1.6	4	4.4	14	7.7	7	5	0	43.8	72	18	10.6	15	58	102		
2018	13.6	6.0	9.8	+0.3	27.6	7	−5.7	1	−0.8	2	17.7	8	47	14	615.8	91	177	28.4	9	1609	109		

DURHAM 2019

Month	Mean max	Mean min	Mean temp	Anom	Highest max	Date	Lowest min	Date	Lowest max	Date	Highest min	Date	Air frost	Days ≥ 25 °C	Total pptn	Anom	Rain days	Wettest day	Date	Total sunshine	Anom	Sunniest day	Date
Jan	7.2	1.2	4.2	+0.1	12.2	13	−6.8	31	1.9	29	8.0	13	13	0	12.0	22	14	4.2	26	78	126		
Feb	10.6	2.2	6.4	+1.8	16.8	26	−6.6	3	2.6	1	9.5	21	6	0	30.6	68	12	8.8	8	128	150		
Mar	11.1	3.8	7.4	+1.3	16.6	21	−0.8	8	6.4	10	8.8	21	1	0	52.8	128	16	11.8	15	145	120		
Apr	12.9	3.5	8.2	−0.0	22.3	20	−2.0	11	7.4	3	7.8	21	4	0	30.2	59	8	7.2	1	185	120		
May	15.3	6.0	10.7	−0.4	21.0	14	−0.9	10	8.0	8	13.6	31	1	0	37.2	84	18	11.4	8	186	99		
June	17.5	9.4	13.5	−0.1	26.8	29	3.9	10	10.7	13	14.9	30	0	1	108.4	177	15	27.6	12	153	92		
July	21.6	12.9	17.2	+1.4	32.9	25	6.5	8	17.8	7	17.9	26	0	4	70.2	113	16	26.6	26	172	98		
Aug	20.4	12.5	16.5	+0.9	28.1	25	9.1	21	15.7	11	15.5	1	0	3	81.2	123	20	17.4	8	189	113		
Sep	17.6	9.1	13.3	+0.1	22.7	21	2.0	8	12.5	29	13.5	24	0	0	84.2	148	19	21.4	30	159	120		
Oct	12.0	5.9	8.9	−1.1	15.2	11	0.6	29	7.5	25	10.4	11	0	0	88.4	139	18	14.4	25	92	93		
Nov	8.1	3.2	5.7	−0.9	12.2	4	−3.8	30	3.6	19	8.1	26	7	0	106.6	145	25	17.0	4	48	70		
Dec	8.0	2.8	5.4	+1.2	12.4	5	−2.6	1	4.0	12	7.8	7	7	0	29.0	47	18	6.0	18	66	116		
2019	13.5	6.0	9.8	+0.3	32.9	7	−6.8	1	1.9	1	17.9	7	39	8	730.8	107	199	27.6	6	1601	109		

DURHAM 2020

Month	Mean max	Mean min	Mean temp	Anom	Highest max	Date	Lowest min	Date	Lowest max	Date	Highest min	Date	Air frost	Days ≥ 25 °C	Total pptn	Anom	Rain days	Wettest day	Date	Total sunshine	Anom	Sunniest day	Date
Jan	9.2	3.2	6.2	+2.1	14.0	22	−0.5	19	5.5	9	7.5	6	1	0	33.0	62	11	14.0	8	75	121		
Feb	8.4	2.3	5.4	+0.8	11.3	15	−1.3	6	5.0	13	7.4	1	3	0	87.2	193	22	22.0	15	94	110		
Mar	10.4	2.2	6.3	+0.1	16.6	25	−3.0	6	5.4	12	6.9	8	7	0	20.6	50	13	6.2	11	156	130		
Apr	14.4	4.4	9.4	+1.1	21.2	15	−2.3	14	7.8	13	9.8	6	1	0	3.8	7	5	1.8	29	218	141		
May	17.4	6.9	12.2	+1.1	24.9	20	−0.6	6	8.6	10	12.5	22	1	0	18.0	40	11	6.4	1	256	137		
June	18.2	10.1	14.2	+0.6	27.1	24	4.1	9	11.4	7	15.5	23	0	3	82.6	135	19	33.8	11	156	94		
July	19.3	11.0	15.1	−0.6	30.8	31	7.5	12	15.0	7	15.5	18	0	1	55.6	90	21	9.8	10	129	74		
Aug	19.5	12.1	15.8	+0.3	26.9	11	3.4	31	12.7	29	16.0	12	0	3	110.8	167	21	30.8	27	131	78		
Sep	17.2	8.8	13.0	−0.3	26.4	15	1.8	28	10.1	24	14.4	8	0	1	57.0	100	15	20.2	23	149	112		
Oct	13.1	6.2	9.6	−0.4	17.1	20	1.1	2	9.5	27	10.2	21	0	0	96.6	152	24	26.8	3	70	71		
Nov	11.3	5.1	8.2	+1.6	18.5	5	−2.0	28	6.1	28	11.8	18	2	0	29.0	39	17	4.8	15	64	94		
Dec	6.8	2.1	4.5	+0.2	12.5	18	−2.4	29	2.5	28	8.6	19	7	0	111.6	182	27	14.4	4	47	83		
2020	13.8	6.2	10.0	+0.5	30.8	7	−3.0	3	2.5	12	16.0	8	22	8	705.8	104	206	33.8	6	1545	105		

DURHAM 2021

Month	Mean max	Mean min	Mean temp	Anom	Highest max	Date	Lowest min	Date	Lowest max	Date	Highest min	Date	Air frost	Days ≥ 25°C	Total pptn	Anom	Rain days	Wettest day	Date	Total sunshine	Anom	Sunniest day	Date
Jan	5.1	−0.4	2.3	−1.8	10.0	19	−4.7	9	1.1	8	4.5	20	16	0	137.2	257	22	18.6	20	61	99		
Feb	7.5	1.1	4.3	−0.2	14.5	27	−9.4	12	−0.1	13	7.2	21	10	0	79.8	176	21	18.2	2	80	94		
Mar	11.0	3.9	7.5	+1.3	20.8	31	−1.4	1	3.2	2	10.4	31	4	0	24.0	58	15	7.6	10	113	94		
Apr	12.1	0.7	6.4	−1.9	17.4	23	−3.2	10	7.3	6	6.7	27	12	0	14.4	28	8	4.2	27	229	148		
May	14.3	5.0	9.6	−1.4	20.4	29	−1.0	6	9.1	4	8.7	27	2	0	86.8	195	19	16.0	3	158	84		
June	19.6	10.1	14.8	+1.2	24.3	24	4.1	22	12.6	25	14.9	10	0	5	28.8	47	5	16.2	24	190	115		
July	21.6	12.8	17.2	+1.4	28.5	17	8.5	1	16.0	6	16.0	18	0	0	69.0	111	14	22.8	4	178	102		
Aug	19.4	11.5	15.5	−0.1	22.5	24	6.4	3	15.5	26	15.8	21	0	0	44.8	68	20	10.8	21	122	73		
Sep	19.6	11.4	15.5	+2.2	28.0	8	5.5	30	13.6	28	15.2	10	0	2	44.4	78	12	12.8	14	134	101		
Oct	14.6	8.1	11.3	+1.3	20.9	7	0.5	16	10.8	21	14.7	8	0	0	88.0	138	18	26.8	5	93	94		
Nov	11.0	4.7	7.9	+1.2	15.9	9	−2.1	28	2.6	28	10.2	19	3	0	51.0	69	14	26.4	26	78	115		
Dec	7.4	2.2	4.8	+0.6	14.1	30	−0.9	19	2.3	18	11.0	31	4	0	66.2	108	22	16.8	4	38	67		
2021	13.6	5.9	9.8	+0.3	28.5	7	−9.4	7	−0.1	2	16.0	7	51	7	734.4	108	190	26.8	10	1474	100		

References

1. Manley, G., 1953: The mean temperature of central England, 1698–1952. *Quarterly Journal of the Royal Meteorological Society*, 79(342): pp. 558–567. doi: 10.1002/qj.49707934222
2. Manley, G., 1974: Central England temperatures: monthly means 1659 to 1973. *Quarterly Journal of the Royal Meteorological Society*, 100(425): pp. 389–405. doi: 10.1002/qj.49710042511
3. Parker, D.E., T.P. Legg, and C.K. Folland, 1992: A new daily Central England temperature series, 1772–1991. *International Journal of Climatology*, 12(4): pp. 317–342. doi: 10.1002/joc.3370120402
4. UNESCO. Durham Cathedral. Available from: https://whc.unesco.org/en/list/370/. Accessed 23 January 2020
5. Brickstock, R.J., 2007: *Durham Castle: fortress, palace, college*. Huddersfield: Jeremy Mills Publishing
6. Bott, M.H.P., 1967: Geophysical investigations of the northern Pennine basement rocks. *Proceedings of the Yorkshire Geological Society*, 36: pp. 139–168
7. Dunham, K.C., 1944: The genesis of the north Pennine ore deposits. *Quarterly Journal of the Geological Society*, 90: pp. 689–720
8. Whittow, J.B., 1992: *Geology and scenery in Britain*. London: Chapman and Hall
9. Kenworthy, J.M., T.P. Burt, and N.J. Cox, 2007: Durham University Observatory and its meteorological record. *Weather*, 62(10): pp. 265–269. doi: 10.1002/wea.86
10. Wheeler, D. and J. Mayes, 1997: *Regional climates of the British Isles*. London and New York: Routledge
11. Manley, G., 1936: The climate of the northern Pennines: the coldest part of England. *Quarterly Journal of the Royal Meteorological Society*, 62(263): pp. 103–115. doi: 10.1002/qj.94706226310
12. Burt, T.P. and B.P. Horton, 2003: The climate of Malham Tarn. *Field Studies*, 10: pp. 635–652
13. Burt, T.P. and E.J.S. Ferranti, 2012: Changing patterns of heavy rainfall in upland areas: a case study from northern England. *International Journal of Climatology*, 32(4): pp. 518–532. doi: 10.1002/joc.2287
14. Manley, G., 1939: On the occurrence of snow-cover in Great Britain. *Quarterly Journal of the Royal Meteorological Society*, 65(278): pp. 2–27. doi: 10.1002/qj.49706527803
15. Manley, G., 1942: Meteorological observations on Dun Fell, a mountain station in Northern England. *Quarterly Journal of the Royal Meteorological Society*, 68(295): pp. 151–166. doi: 10.1002/qj.49706829502
16. Manley, G., 1943: Further climatological averages for the northern Pennines, with a note on topographical effects. *Quarterly Journal of the Royal Meteorological Society*, 69(302): pp. 251–261. doi: 10.1002/qj.49706930203
17. Crisp, D.T., 1966: Input and output of minerals for an area of Pennine moorland: the importance of precipitation, drainage, peat erosion and animals. *Journal of Applied Ecology*, 3: pp. 327–348
18. Burt, T.P. and J. Holden, 2010: Changing temperature and rainfall gradients in the British uplands. *Climate Research*, 45: pp. 57–70

19. Manley, G., 1941: The Durham meteorological record, 1847–1940. *Quarterly Journal of the Royal Meteorological Society*, **67**(292): pp. 363–380. doi: 10.1002/qj.49706729209

20. Manley, G., 1952: *Climate and the British scene*. London and Glasgow: Collins

21. National River Flow Archive NRFA. NRFA website for Station 24001. Available from: https://nrfa.ceh.ac.uk/data/station/info/24001. Accessed 31 March 2021

22. Kenworthy, J.M., 1994: *The Durham University Observatory meteorological record: 150 years of Durham weather*, in *Observatories and climatological research*, Chapter 3, B.D. Giles and J.M. Kenworthy, Editors. Durham: University of Durham: Dept of Geography: Occasional Publication No. 29, pp. 12–22

23. Wolfendale, A.W., 1992: Durham and the new astronomies. *Quarterly Journal of the Royal Astronomical Society*, **33**: pp. 311–320

24. Durham University Library, 2016: Catalogue of the records of Durham University Observatory. Available at http://reed.dur.ac.uk/xtf/view?docId=ark/32150_s108612n525.xml

25. Cliver, E.W., 2006: The 1859 space weather event: then and now. *Advances in Space Research*, **38**: pp. 119–129

26. Clark, S., 2009: *The sun kings: the unexpected tragedy of Richard Carrington and the tale of how modern astronomy began*. Princeton: Princeton University Press

27. Plummer, J.J. and R.H. Scott, 1873: On some results of temperature observations at Durham. *Quarterly Journal of the Royal Meteorological Society*, **1**(8): pp. 241–246. doi: 10.1002/qj.4970010802

28. Kenworthy, J.M., 1985: *The Durham University record and Gordon Manley's work on a longer temperature series for north-east England*, in *The climatic scene*, M. Tooley and G.M. Sheail, Editors. London: George Allen & Unwin

29. Durham Observatory, 1850: Manuscript observation registers for Durham Observatory, volume 1 (M1, 1843–1847) and Volume 2 (M2, 1848–1850). Scanned copies available online through Durham University Library at
    M1- https://iiif.durham.ac.uk/index.html?manifest=t2m44558d54g
    M2- https://iiif.durham.ac.uk/index.html?manifest=t2m1544bp33r

30. Fewster, E., 1991: *A handlist of records of Durham University Observatory*. Palace Green: Durham University Library

31. Endfield, G.H., L. Veale, and A. Hall, 2015: Gordon Valentine Manley and his contribution to the study of climate change: a review of his life and work. *Wiley Interdisciplinary Reviews: Climate Change*, **6**(3): pp. 287–299. doi: 10.1002/wcc.334

32. Lamb, H.H., et al., 1981: The life and work of Professor Gordon Manley (1902–1980). *Weather*, **36**(8): pp. 220–231. doi: 10.1002/j.1477-8696.1981.tb05407.x

33. Ratcliffe, R.A.S., 1993: Pen portraits of presidents–Professor Gordon Manley, MA, DSc. *Weather*, **48**(8): pp. 267–268. doi: 10.1002/j.1477-8696.1993.tb05906.x

34. Tooley, M.J. and G.M. Sheail, eds., 1985. *The climatic scene*. London: George Allen & Unwin

35. Baxter, E.F., 1956: Durham University Observatory *Weather*, **11**(7): pp. 218–222. doi: 10.1002/j.1477-8696.1956.tb00345.x

36. Manley, G., 1980: The northern Pennines revisited: Moor House, 1932–78. *Meteorological Magazine*, **109**: pp. 281–292

37. Diver, A., 1969: Observing the winter weather at Great Dun Fell. *Weather*, **24**(2): pp. 75–77. doi: 10.1002/j.1477-8696.1969.tb03147.x

38. Veale, L. and G. Endfield, 2014: The helm wind of Cross Fell. *Weather*, **69**(1): pp. 3–7. doi: 10.1002/wea.2165

39. Manley, G., 1945: The helm wind of Cross Fell, 1937–1939. *Quarterly Journal of the Royal Meteorological Society*, 71(309–310): pp. 197–219. doi: 10.1002/qj.49707130901

40. Parker, D.E., 2010: Uncertainties in early Central England temperatures. *International Journal of Climatology*, 30(8): pp. 1105–1113. doi: 10.1002/joc.1967

41. Lamb, H.H., 1980: Obituary—Gordon Manley. *Quarterly Journal of the Royal Meteorological Society*, 106(449): pp. 656–657. doi: 10.1002/qj.49710644926

42. Giles, B.D. and J.M. Kenworthy, eds., 1994. *Observatories and climatological research.* Durham: University of Durham, Dept of Geography

43. Kenworthy, J.M. and J.M. Walker, 1997: *Colonial observatories and observations: meteorology and geophysics.* Occasional Publication No. 31. Durham: University of Durham, Dept of Geography

44. Kenworthy, J.M., N.J. Cox, and A.N. Joyce, 1997: *Computerisation and analysis of the Durham Observatory meteorological record: Final Report to the Leverhulme Trust.* Reference F/128/Q. Durham: University of Durham

45. Eglise, M., 2003: A monthly temperature series for Durham from 1784. PhD thesis, Durham: University of Durham, Dept of Geography

46. Manley, G., 1946: Temperature trend in Lancashire, 1753–1945. *Quarterly Journal of the Royal Meteorological Society*, 72(311): pp. 1–31. doi: 10.1002/qj.49707231102

47. International Organization for Standardization (ISO), 1975: *Standard atmosphere, ISO 2533:1975.* Geneva: International Organization for Standardization (ISO)

48. Smith, D., 2019: *The James Losh diaries, 1802–1833: life and weather in early nineteenth century Newcastle-upon-Tyne.* Newcastle upon Tyne: Cambridge Scholars Publishing

49. Burt, S. and T. Burt, 2019: *Oxford weather and climate since 1767.* Oxford: Oxford University Press

50. Lamb, H.H., 1950: Types and spells of weather around the year in the British Isles: annual trends, seasonal structure of the year, singularities. *Quarterly Journal of the Royal Meteorological Society*, 76(330): pp. 393–429. doi: doi:10.1002/qj.49707633005

51. Perry, A.H., 1976: *Synoptic climatology,* in *The climate of the British Isles,* T.J. Chandler and S. Gregory, Editors. London: Longman, pp. 8–38

52. Shellard, H.C., 1976: *Wind,* in *The climate of the British Isles,* T.J. Chandler and S. Gregory, Editors. London: Longman, pp. 39–73

53. Jones, P.D., C. Harpham, and K.R. Briffa, 2013: Lamb weather types derived from reanalysis products. *International Journal of Climatology*, 33(5): pp. 1129–1139. doi: 10.1002/joc.3498

54. Burt, T.P. and N.J.K. Howden, 2013: North Atlantic Oscillation amplifies orographic precipitation and river flow in upland Britain. *Water Resources Research*, 49: pp. 3504–3515

55. Burt, S., 2007b: The Highest of the Highs … Extremes of barometric pressure in the British Isles, Part 2—the most intense anticyclones. *Weather*, 62(2): pp. 31–41. doi: 10.1002/wea.35

56. Burt, S., 2007a: The Lowest of the Lows … Extremes of barometric pressure in the British Isles, Part 1—the deepest depressions. *Weather*, 62(1): pp. 4–14. doi: 10.1002/wea.20

57. Baxter, P.J., 2005: The East Coast big flood, 31 January–1 February 1953: a summary of the human disaster. *Philosophical Transactions: Mathematical, Physical and Engineering Sciences*, 363(1831): pp. 1293–1312

58. Slivinski, L.C., et al., 2021: An evaluation of the performance of the twentieth century reanalysis version 3. *Journal of Climate*, 34(4): pp. 1417–1438. doi: 10.1175/jcli-d-20-0505.1

59. Burt, S., 1982: Heavy rainfall and snowstorms, 23–26 April 1981. *Weather*, 37(4): pp. 108–115. doi: 10.1002/j.1477-8696.1982.tb03572.x

60. Eden, P. 2009: Coronation weather. Available from: https://www.weatheronline.co.uk/reports/philip-eden/Coronation-Weather.htm. Accessed 23 January 2021

61. Burt, S., 1992: The exceptional hot spell of early August 1990 in the United Kingdom. *International Journal of Climatology*, **12**(6): pp. 547–567. doi: 10.1002/joc.3370120603

62. Met Office, 1989: *Monthly and annual totals of rainfall 1986 for the United Kingdom.* Bracknell: Met Office

63. Kendon, M. and M. McCarthy, 2021: The United Kingdom's wettest day on record—so far—3 October 2020. *Weather*, **76**(9): pp. 316–319. doi: 10.1002/wea.3910:

64. Kendon, M., 2012: When was the warmest day of 2011? *Weather*, **67**(2): pp. 44–45. doi: 10.1002/wea.1889

65. Booth, B.J., 1970: The *Royal Charter. Weather*, **25**(12): pp. 550–553. doi: 10.1002/j.1477-8696.1970.tb04108.x

66. Skidmore, I., 1979: *Anglesey & Lleyn shipwrecks*. Swansea: Christopher Davies (Publishers) Ltd

67. Fitzroy, R., 1860: Notice of 'The *Royal Charter* Storm' in October 1859. *Proceedings of the Royal Society of London*, **10**: pp. 561–567. doi: 10.1098/rspl.1859.0111

68. Walker, M., 2012: *History of the Meteorological Office*. Cambridge: Cambridge University Press

69. Lamb, H.H., 1991: *Historic storms of the North Sea, British Isles and Northwest Europe.* Cambridge: Cambridge University Press

70. Symons, G.J., 1881: On the gale of October 13th–14th, 1881, over the British Isles. *Quarterly Journal of the Royal Meteorological Society*, **8**(41): pp. 1–17. doi: 10.1002/qj.4970084102

71. Eden, P., 2008: *Great British weather disasters*. London and New York: Continuum

72. Burt, S. and M. Kendon, 2016: December 2015—an exceptionally mild month in the United Kingdom. *Weather*, **71**(12): pp. 314–320. doi: 10.1002/wea.2800

73. Harwood, W.A., 1947: Extremes of low temperature [December 1879]. *Meteorological Magazine*, **76**: pp. 44–45, 114

74. Burt, P.J.A., 2004: The great storm and the fall of the first Tay Rail Bridge. *Weather*, **59**(12): pp. 347–350. doi: 10.1256/wea.199.04

75. Burt, S., 1997: The Altnaharra minimum temperature of −27.2°C on 30 December 1995. *Weather*, **52**(5): pp. 134–144. doi: 10.1002/j.1477-8696.1997.tb06294.x

76. Burt, T.P., P.D. Jones, and N.J.K. Howden, 2015: An analysis of rainfall across the British Isles in the 1870s. *International Journal of Climatology*, **35**: pp. 2934–2947. doi: 10.1002/joc.4184

77. Burt, S., 1980: Snowfall in Britain during winter 1978/79. *Weather*, **35**(10): pp. 288–301. doi: 10.1002/j.1477-8696.1980.tb04708.x

78. Burt, S., 1980: Rainfall in the United Kingdom during 1978. *Journal of Meteorology, UK*, **5**: pp. 37–61

79. World Meteorological Organization (WMO) Commission for Climatology, 2018: Guidelines on the definition and monitoring of extreme weather and climate events. Available from: http://www.wmo.int/pages/prog/wcp/ccl/documents/

80. Burt, T.P. and B.P. Horton, 2007: Inter-decadal variability in daily rainfall at Durham (UK) since the 1850s. *International Journal of Climatology*, **27**(7): pp. 945–956. doi: 10.1002/joc.1443

81. Scott, R.H., 1885: On the measurement of sunshine. *Quarterly Journal of the Royal Meteorological Society*, **11**(55): pp. 205–216. doi: 10.1002/qj.4970115503

82. Kendon, M. and J. Prior, 2011: Two remarkable British summers—'perfect' 1911 and 'calamitous' 1912. *Weather*, **66**(7): pp. 179–184. doi: 10.1002/wea.818

83. Oman, L., 2005: Climatic response to high-latitude volcanic eruptions. *Journal of Geophysical Research*, **110**(D13103): pp. 1–13. doi: 10.1029/2004jd005487

84. Kalnay, E., et al., 1996: The NCEP/NCAR 40-Year Reanalysis Project. *Bulletin of the American Meteorological Society*, 77(3): pp. 437–472. doi: 10.1175/1520-0477(1996)077<0437:Tnyrp>2.0.Co;2

85. Wheeler, D.A., 1984: The July 1983 'heatwave' in north-east England. *Weather*, 39(6): pp. 178–181. doi: 10.1002/j.1477-8696.1984.tb06758.x

86. Vautard, R., et al., 2020: Human contribution to the record-breaking June and July 2019 heat waves in Western Europe. *Environmental Research Letters*, 15(094077). doi: 10.1088/1748-9326/aba3d4

87. Burt, S., 2004: The August 2003 heatwave in the United Kingdom: Part 1—Maximum temperatures and historical precedents. *Weather*, 59(8): pp. 199–208. doi: 10.1256/wea.10.04A

88. Cinderey, M., 2005: The North Yorkshire–Teesside storm of 10 August 2003. *Weather*, 60(3): pp. 60–65. doi: 10.1256/wea.117.04

89. IPCC Intergovernmental Panel on Climate Change, 2018: *Summary for Policymakers*, in *Global Warming of 1.5 °C. An IPCC Special Report on the impacts of global warming of 1.5 °C above pre-industrial levels and related global greenhouse gas emission pathways, in the context of strengthening the global response to the threat of climate change, sustainable development, and efforts to eradicate poverty*. Geneva: World Meteorological Organization

90. Hawkins, E., et al., 2020: Observed emergence of the climate change signal: from the familiar to the unknown. *Geophysical Research Letters* [online], 47(6). doi: 10.1029/2019GL086259

91. Frame, D., et al., 2017: Population-based emergence of unfamiliar climates. *Nature Climate Change*, 7: pp. 407–411. https://doi.org/10.1038/nclimate3297

92. Wuebbles, D.J., et al., eds., 2017. *Climate science special report: fourth national climate assessment, volume I*. Washington, DC: U.S. Global Change Research Program. doi: 10.7930/J0J964J6

93. NASA Earth Observatory. *NASA Earth Observatory glossary*. Available from: https://earthobservatory.nasa.gov/glossary/l/n. Accessed on 7 July 2020

94. Burt, T.P., 2009: Homogenising the rainfall record at Durham for the 1870s. *Hydrological Sciences Journal*, 54: pp. 199–209

95. Noone, S., et al., 2016: Homogenization and analysis of an expanded long-term monthly rainfall network for the Island of Ireland (1850–2010). *International Journal of Climatology*, 36(8): pp. 2837–2853. doi: 10.1002/joc.4522

96. Barker, L.J., et al., 2019: Historic hydrological droughts 1891–2015: systematic characterisation for a diverse set of catchments across the UK. *Hydrology and Earth System Sciences*, 2019: pp. 1–31. doi: 10.5194/hess-2019-202

97. Ellis, W., 1877: Results derived from the sunshine records obtained at the Royal Observatory, Greenwich, by means of Campbell's self-registering sun dial, during the year ending April 30th, 1877. *Quarterly Journal of the Royal Meteorological Society*, 3(24): pp. 460–467. doi: 10.1002/qj.4970032406

98. Stokes, G.G., 1880: Description of the card supporter for sunshine recorders adopted at the Meteorological Office. *Quarterly Journal of the Royal Meteorological Society*, 7: pp. 83–94

99. Burt, T.P., 1994: Long-term study of the natural environment—perceptive science or mindless monitoring? *Progress in Physical Geography*, 18(4): pp. 475–496. doi: 10.1177/030913339401800401

100. Manley, G., 1958: The great winter of 1740. *Weather*, 13(1): pp. 11–17. doi: 10.1002/j.1477-8696.1958.tb05086.x

101. Kington, J., 1988: *The weather journals of a Rutland squire: Thomas Barker of Lyndon Hall*. Oakham: Rutland Record Society

102. Kington, J., 2010: *Climate and weather*. New Naturalist Series. London: Collins

103. Reynolds, D.J., 1991: The weather in south Staffordshire 1739–1754 from the diary of Dr Richard Wilkes: Part 1: 1739–1740. *Journal of Meteorology, UK*, **16**: pp. 299–305

104. Rennison, R.W., 2001: The great inundation of 1771 and the rebuilding of the North-East's bridges. *Archaeological Aeliana*, **5**: pp. 269–291

105. Archer, D.R., F. Leesch, and K. Harwood, 2007: Assessment of severity of the extreme River Tyne flood in January 2005 using gauged and historical information. *Hydrological Sciences Journal*, **52**(5): pp. 992–1003

106. Thordarson, T. and S. Self, 2003: Atmospheric and environmental effects of the 1783–1784 Laki eruption: a review and reassessment. *Journal of Geophysical Research: Atmospheres*, **108**(D1): pp. AAC 7-1–AAC 7-29. doi: 10.1029/2001JD002042

107. Ward, P.L., 2009: Sulfur dioxide initiates global climate change in four ways. *Thin solid films*, **517**: pp. 3188–3203 [VEI numbers are listed in Appendix A: supplementary data]

108. Witze, A. and J. Kanipe, 2014: *Island on fire: The exatrrordinary story of Laki, the forgotten volcano that turned eighteenth-century Europe dark*. London: Profile Books

109. Briffa, K.R., et al., 1998: Influence of volcanic eruptions on Northern Hemisphere summer temperature over the past 600 years. *Nature*, **393**: pp. 450–455. doi: 10.1038/30943

110. Dawson, A., 2009: *So foul and fair a day: A history of Scotland's weather and climate*. Edinburgh: Birlinn Ltd

111. Raible, C.C., et al., 2016: Tambora 1815 as a test case for high impact volcanic eruptions: Earth system effects. *Wiley Interdisciplinary Reviews: Climate Change*, 7(4): pp. 569–589. doi: 10.1002/wcc.407

112. Klingaman, W.K. and N.P. Klingaman, 2013: *The year without summer: 1816 and the volcano that darkened the world and changed history*. New York: St Martin's Press

113. Wood, G.D., 2014: *Tambora: The eruption that changed the world*. Princeton and Oxford: Princeton University Press

114. Oppenheimer, C., 2003: Climatic, environmental and human consequences of the largest known historic eruption: Tambora volcano (Indonesia) 1815. *Progress in Physical Geography: Earth and Environment*, 27(2): pp. 230–259. doi: 10.1191/0309133303pp379ra

115. Oppenheimer, C., 2011: *Eruptions that shook the world*. Cambridge: Cambridge University Press

116. Post, J.D., 1977: *The last great subsistence crisis in the Western world*. Baltimore: Johns Hopkins

117. Veale, L. and G.H. Endfield, 2016: Situating 1816, the 'year without summer', in the UK. *The Geographical Journal*, **182**(4): pp. 318–330. doi: doi:10.1111/geoj.12191

118. Guevara-Murua, A., et al., 2014: Observations of a stratospheric aerosol veil from a tropical volcanic eruption in December 1808: is this the unknown 1809 eruption? *Climate of the Past*, **10**(5): pp. 1707–1722. doi: 10.5194/cp-10-1707-2014

119. Pike, W.S., 1988: The polar low that led to widespread snow and the extremely cold Christmas of 1860 in Britain. *Journal of Meteorology, UK*, **13**(134): pp. 373–383

120. Slivinski, L.C., et al., 2019: Towards a more reliable historical reanalysis: Improvements for version 3 of the Twentieth Century Reanalysis system. *Quarterly Journal of the Royal Meteorological Society*, **145**(724): pp. 2876–2908. doi: 10.1002/qj.3598

121. Hollis, D., et al., 2019: HadUK-Grid—A new UK dataset of gridded climate observations. *Geoscience Data Journal*, **6**: pp. 151–159. doi: 10.1002/gdj3.78

122. Wheeler, D., 1983: The work of Thomas Backhouse: Victorian meteorologist. *Weather*, **38**(8): pp. 240–246. doi: 10.1002/j.1477-8696.1983.tb03708.x

123. Harding, C., 1887: The storm and low barometer of December 8th and 9th, 1886. *Quarterly Journal of the Royal Meteorological Society*, **13**(63): pp. 201–215. doi: 10.1002/qj.4970136303

124. Marsh, T.J., B.J. Greenfield, and J.A. Hannaford, 2005: The 1894 Thames flood—a reappraisal. *Proceedings of the Institution of Civil Engineers*, **158**(WM3): pp. 103–110

125. Bayard, F.C. and W. Marriott, 1895: The frost of January and February 1895 over the British Isles. *Quarterly Journal of the Royal Meteorological Society*, **21**(95): pp. 141–160. doi: 10.1002/qj.4970219502

126. Pike, W.S., 1995: Rivalry on ice: skating and the memorable 1894/95 winter. *Weather*, **50**(2): pp. 48–54. doi: 10.1002/j.1477-8696.1995.tb06076.x

127. Lempfert, R.G.K., 1907: The high barometer of January 1907. *Meteorological Magazine*, **42**: pp. 1–3

128. Moyes, W.A., 1996: *Hatfield 1846–1996: a history of Hatfield College in the University of Durham*. Durham: Hatfield College Trust

129. Wheeler, D.A., 1991: The great north-eastern snowstorm of February 1941. *Weather*, **46**(10): pp. 311–314. doi: 10.1002/j.1477-8696.1991.tb07066.x

130. Catchpole, A.J.W., 1963: The Houghall frost hollow. *Meteorological Magazine*, **92**: pp. 121–129

131. Simpson, I.R. and P.D. Jones, 2012: Updated precipitation series for the UK derived from Met Office gridded data. *International Journal of Climatology*, **32**(15): pp. 2271–2282. doi: 10.1002/joc.3397

132. Grindley, J., 1980: *Rainfall*, in *Atlas of drought in Britain, 1975–1976*, J.C. Doornkamp, K.J. Gregory, and A.S. Burn, Editors. London: Institute of British Geographers, pp. 27–28

133. Miles, M.K., 1977: Atmospheric circulation during the severe drought of 1975/76. *Meteorological Magazine*, **106**: pp. 154–164

134. Perry, A.H., 1980: *Dominant pressure patterns*, in *Atlas of drought in Britain, 1975–1976*, J.C. Doornkamp, K.J. Gregory, and A.S. Burn, Editors. London: Institute of British Geographers, pp. 13–14

135. Roach, W.T. and J.L. Brownscombe, 1984: Possible causes of the extreme cold during winter 1981–82. *Weather*, **39**(12): pp. 362–372. doi: 10.1002/j.1477-8696.1984.tb06746.x

136. Lawrence, M.B., 1987: Annual summary of the 1986 Atlantic hurricane season. *Monthly Weather Review*, **115**(9): pp. 2158–2160

137. Met Office, 2013: *UK Monthly Climate Summary—Annual 2012*. Available from: https://digital.nmla.metoffice.gov.uk/IO_95a0c079-6e1f-4454-a4a3-468b94522ff2/. Accessed 14 May 2021

138. Burt, T.P., 2014: Mass movement: a case study of the Wear valley. *Geography Review* **27**(4): pp. 38–41

139. Kendon, M., et al., 2021: State of the UK climate 2020. *International Journal of Climatology*, **41**. doi: 10.1002/joc.7285

140. McCarthy, M., L. Armstrong, and N. Armstrong, 2021: A new heatwave definition for the UK. *Weather*, **74**: pp. 382–387. doi: 10.1002/wea.3629

141. Field, M., 2010: Pen portraits of presidents—William Ellis. *Weather*, **65**(9): pp. 250–252. doi: 10.1002/wea.524

142. Met Office, 2019: *Met Office MIDAS Open: UK Land Surface Stations Data (1853–current)*. Chilton: Centre for Environmental Data Analysis (CEDA). Available from: http://catalogue.ceda.ac.uk/uuid/dbd451271eb04662beade68da43546e1

143. Symons, G.J., 1897: Meteorological instruments in 1837, and in 1897. *Quarterly Journal of the Royal Meteorological Society*, **23**(103): pp. 205–220. doi: 10.1002/qj.49702310302

144. Austin, J.F. and A. McConnell, 1980: James Six F.R.S.: two hundred years of the Six's self-registering thermometer. *Notes and Records of the Royal Society of London*, **35**(1): pp. 49–65

145. Glaisher, J., 1868: Description of thermometer stand by James Glaisher. *Symons's Meteorological Magazine*, **3**: pp. 155–157

146. Laing, J., 1977: Maximum summer temperatures recorded in Glaisher stands and Stevenson screens. *Meteorological Magazine*, **106**: pp. 220–228

147. Margary, I.D., 1924: Glaisher stand versus Stevenson screen: a comparison of forty years' observations of maximum and minimum temperature as recorded in both screens at Camden Square, London. *Quarterly Journal of the Royal Meteorological Society*, **50**(211): pp. 209–226. doi: 10.1002/qj.49705021109

148. Thorne, P.W., et al., 2016: Reassessing changes in diurnal temperature range: a new data set and characterization of data biases. *Journal of Geophysical Research: Atmospheres*, **121**(10): pp. 5115–5137. doi: 10.1002/2015jd024583

149. Met Office, 2019: Scanned archive of rainfall Ten Year Sheets, 1677–1960. Available from: https://digital.nmla.metoffice.gov.uk/SO_d383374a-91c3-4a7b-ba96-41b81cfb9d67/. Accessed 12 July 2021

150. Hawkins, E., et al., 2022: Rainfall rescue: millions of historical monthly rainfall observations taken in the UK and Ireland rescued by citizen scientists. *Geoscience Data Journal*, doi: 10.1002/gdj3.157

151. Burt, S., 2010: British rainfall 1860–1993. *Weather*, **65**(5): pp. 121–128. doi: 10.1002/wea.603

152. Harris, R., 1985: *Variations in the Durham rainfall and temperature record, 1847–1981*, in *The climatic scene*, M. Tooley and G.M. Sheail, Editors. London: George Allen & Unwin, pp. 39–59

153. Burt, S., 2021: A twice-daily barometric pressure record from Durham Observatory in north-east England, 1843–1960. *Geoscience Data Journal* [online]. https://doi.org/10.1002/gdj3.135

154. Compo, G.P., et al. 2019: The International Surface Pressure Databank version 4. Research Data Archive at the National Center for Atmospheric Research. Boulder: Computational and Information Systems Laboratory: Available from: https://doi.org/10.5065/9EYR-TY90

155. Meliconi, I., 2004: Browning, John (1830/31–1925), scientific instrument maker, in *Oxford Dictionary of National Biography*. Oxford: Oxford University Press. Available from: https://www.oxforddnb.com/view/10.1093/ref:odnb/9780198614128.001.0001/odnb-9780198614128-e–66149

156. Banfield, E., 1991: *Barometer makers and retailers 1660–1900*. Trowbridge: Baros Books

157. Cram, T.A., et al., 2015: The International Surface Pressure Databank version 2. *Geoscience Data Journal*, **2**(1): pp. 31–46. doi: 10.1002/gdj3.25

158. Hawkins, E., et al., 2019: Hourly weather observations from the Scottish Highlands (1883–1904) rescued by volunteer citizen scientists. *Geoscience Data Journal*, **6**: pp. 160–173. doi: 10.1002/gdj3.79

159. Meteorological Office, 1980: *Measurement of atmospheric pressure*, in *Handbook of Meteorological Instruments: Volume 1*. London: Her Majesty's Stationery Office

160. World Meteorological Organization (WMO), 2018: WMO No.8 - *Guide to Meteorological Instruments and Methods of Observation (CIMO guide)*. 2018 edition - Volume I: Measurement of Meteorological Variables. Geneva: World Meteorological Organization. Available from: https://library.wmo.int/index.php?lvl=notice_display&id=12407. English: also available in French, Spanish, and Russian. Accessed 21 June 2021

161. Meteorological Office, 1956: *Handbook of Meteorological Instruments: Part 1, Instruments for surface observations*. Fifth impression 1969. London: Her Majesty's Stationery Office

162. Bolton, D., 1980: The computation of equivalent potential temperature. *Monthly Weather Review*, **108**: pp. 1046–1053

163. Lewis, R.P.W., 1982: The *Daily Weather Report* and associated publications: 1860–1980. *Meteorological Magazine*, **111**: pp. 103–121

164. Craig, P.M. and E. Hawkins, 2020: Digitizing observations from the Met Office Daily Weather Reports for 1900–1910 using citizen scientist volunteers. *Geoscience Data Journal*, 7: pp. 116–134. 10.1002/gdj3.93: doi: 10.1002/gdj3.93

165. World Meteorological Organization (WMO), 2019: *Manual on Codes*. International Codes, Volume I.1: Annex II to the WMO Technical Regulations, Part A–Alphanumeric Codes. Geneva: World Meteorological Organization. Available from: https://library.wmo.int/doc_num.php?explnum_id=10235. Present weather ww code is 4677 and is given in full on page A-356

166. Bennett, J.A., 1990: *Church, state and astronomy in Ireland: 200 years of Armagh Observatory*. Armagh: The Armagh Observatory

167. Whipple, G.M. and W.H. Dines, 1888: Report of the wind force committee on experiments with anemometers conducted at Hersham. *Quarterly Journal of the Royal Meteorological Society*, 14(68): pp. 253–263. doi: 10.1002/qj.4970146801

168. Wilson-Barker, C.T.D., 1899: Comparison of estimated wind force with that given by instruments. *Quarterly Journal of the Royal Meteorological Society*, 25(109): pp. 13–19. doi: 10.1002/qj.49702510902

169. Pike, W.S., 1990: The Royal Meteorological Society's wind force committee. *Weather*, 45(3): p. 90. https://doi.org/10.1002/j.1477-8696.1990.tb05061.x

170. Pike, W.S., 1989: One hundred years of the Dines pressure-tube anemometer. *Meteorological Magazine*, 118: pp. 209–214

171. Davis, N.E., 1968: An optimum summer weather index. *Weather*, 23: pp. 305–317

Index